Constructed Wetlands for Industrial Wastewater Treatment

Challenges in Water Management Series

Editor:
Justin Taberham
Independent Consultant and Environmental Advisor, London, UK

Titles in the series:

Urban Water Security
Robert C. Brears
2016
ISBN: 9781119131724

Water Resources: A New Water Architecture
Alexander Lane, Michael Norton and Sandra Ryan
2017
ISBN: 9781118793909

Industrial Water Resource Management
Pradip K. Sengupta
2017
ISBN: 9781119272502

Handbook of Knowledge Management for Sustainable Water Systems
Meir Russ
2018
ISBN: 9781119271635

Smart Water Technologies and Techniques
David A. Lloyd Owen
2018
ISBN: 9781119078647

Constructed Wetlands for Industrial Wastewater Treatment

Edited by

Alexandros I. Stefanakis

Bauer Resources, Schrobenhausen, Germany;
Bauer Nimr LLC, Muscat, Oman; and
German University of Technology in Oman, Muscat, Oman

Registered Office(s)
John Wiley & Sons, Inc., 111 River Street, Hoboken, NJ 07030, USA
John Wiley & Sons Ltd, The Atrium, Southern Gate, Chichester, West Sussex, PO19 8SQ, UK

Editorial Office
The Atrium, Southern Gate, Chichester, West Sussex, PO19 8SQ, UK

For details of our global editorial offices, customer services, and more information about Wiley products visit us at www.wiley.com.

Wiley also publishes its books in a variety of electronic formats and by print-on-demand. Some content that appears in standard print versions of this book may not be available in other formats.

Library of Congress Cataloging-in-Publication Data:

Names: Stefanakis, Alexandros I., 1982- editor.
Title: Constructed wetlands for industrial wastewater treatment / edited by Dr. Alexandros I. Stefanakis.
Description: Hoboken, NJ : John Wiley & Sons, 2018. | Series: Challenges in water management series | Includes
    bibliographical references and index. | Identifiers: LCCN 2018008634 (print) | LCCN 2018013538 (ebook) |
    ISBN 9781119268413 (pdf) | ISBN 9781119268413 (epub) | ISBN 9781119268345 (cloth)
Subjects: LCSH: Constructed wetlands. | Sewage–Purification.
Classification: LCC TD756.5 (ebook) | LCC TD756.5 .C655 2018 (print) | DDC 628/.742–dc23
LC record available at https://lccn.loc.gov/2018008634

Cover Design: Wiley
Cover Image: Treatment Wetland Bed at Nimr Water Treatment Plant, Oman.
Courtesy of BAUER Group, www.bauer.de/bre

Set in 10/12pt WarnockPro by SPi Global, Chennai, India

Printed in Singapore by C.O.S. Printers Pte Ltd

10 9 8 7 6 5 4 3 2 1

Contents

Series Foreword – Challenges in Water Management

The World Bank in 2014 noted:

Water is one of the most basic human needs. With impacts on agriculture, education, energy, health, gender equity, and livelihood, water management underlies the most basic development challenges. Water is under unprecedented pressures as growing populations and economies demand more of it. Practically every development challenge of the twenty-first century – food security, managing rapid urbanization, energy security, environmental protection, adapting to climate change – requires urgent attention to water resources management.

Yet already, groundwater is being depleted faster than it is being replenished and worsening water quality degrades the environment and adds to costs. The pressures on water resources are expected to worsen because of climate change. There is ample evidence that climate change will increase hydrologic variability, resulting in extreme weather events such as droughts floods, and major storms. It will continue to have a profound impact on economies, health, lives, and livelihoods. The poorest people will suffer most.

It is clear there are numerous challenges in water management in the twenty-first century. In the twentieth century, most elements of water management had their own distinct set of organisations, skill sets, preferred approaches and professionals. The overlying issue of industrial pollution of water resources was managed from a "point source" perspective.

However, it has become accepted that water management has to be seen from a holistic viewpoint and managed in an integrated manner. Our current key challenges include:

- The impact of climate change on water management, its many facets and challenges – extreme weather, developing resilience, storm-water management, future development and risks to infrastructure.
- Implementing river basin/watershed/catchment management in a way that is effective and deliverable.
- Water management and food and energy security.
- The policy, legislation and regulatory framework that is required to rise to these challenges.
- Social aspects of water management – equitable use and allocation of water resources, the potential for "water wars", stakeholder engagement, valuing water and the ecosystems that depend upon it.

This series highlights cutting-edge material in the global water management sector from a practitioner as well as an academic viewpoint. The issues covered in this series are of critical interest to advanced level undergraduates and Masters Students as well as industry, investors and the media.

Justin Taberham, CEnv
Series Editor
www.justintaberham.com

List of Contributors

Rouzbeh Abbassi
Australian Maritime College (AMC)
University of Tasmania
Launceston
Australia

Emmanuel Aboagye-Nimo
School of Environment and Technology
University of Brighton
Brighton
UK

A. Akcil
Mineral-Metal Recovery and Recycling
(MMR&R) Research Group
Mineral Processing Division
Department of Mining Engineering
Suleyman Demiral University
Isparta
Turkey

Christos S. Akratos
Department of Civil Engineering
Democritus University of Thrace
Xanthi
Greece

António Albuquerque
Universidade da Beira Interior
Materiais Fibrosos e Tecnologias Ambientais
(FibEnTech-UBI)
Rua Marquês d'Ávila e Bolama
Covilhã
Portugal

J.A. Álvarez
AIMEN
Spain

C.A. Arias
Department of Bioscience – Aquatic Biology
Aarhus University
Aarhus
Denmark

M. Aulinas
Life MinAqua PM
Grup Fundació Ramon Noguera
Girona
Spain

J.M. Bayona
Intituto de Diagnóstico Ambiental y Estudios
del Agua
IDAEA-CSIC
Spain

Eloy Bécares
Departamento de Biodiversidad y Gestión
Ambiental
Facultad de Ciencias Biológicas y Ambientales
Universidad de León
León
Spain

Johan Blom
Tauw bv
Handelskade
The Netherlands

Floris Boogaard
Hanze University of Applied Sciences
(Hanze UAS)
Zernikeplein
Groningen
The Netherlands;
Department of Water Management
Faculty of Civil Engineering and Geosciences
Delft University of Technology
Delft
The Netherlands

R. Bresciani
Iridra Srl
Florence
Italy

Roman Breuer
Bauer Resources GmbH
BAUER-Strasse 1
Schrobenhausen
Germany

H. Brix
Department of Bioscience – Aquatic Biology
Aarhus University
Aarhus
Denmark

E. Bruun
Orbicon A/S
Ringstedvej 20
Roskilde
Denmark

Ricardo Sidrach-Cardona
Instituto de Medio Ambiente
Universidad de León
León
Spain

P. Carvalho
Department of Bioscience – Aquatic Biology
Aarhus University
Aarhus
Denmark

A.J.P. Carvalho
Chemistry Department
School of Sciences and Technology
University of Évora
Évora
Portugal;
Évora Chemistry Centre
University of Évora
Évora
Portugal

Ashutosh Kumar Choudhary
Department of Applied Science and Humanities
HSET
Swami Rama Himalayan University
Dehradun
India

Saeed Dehestani
Environmental Health Research Center
Kurdistan University of Medical Sciences
Sanandaj
Kurdistan
Iran

D. De la Varga
Sedaqua (Spin-off from University of A Coruña)
Spain;
University of A Coruña
Spain

A. Dordio
Chemistry Department
School of Sciences and Technology
University of Évora
Évora
Portugal;
MARE – Marine and Environmental Research Centre
University of Évora
Évora
Portugal

Mark Wellington Fitch
Missouri University of Science and Technology
Civil Engineering
Rolla
USA

Susan Flash
PurEnergy Operating Services
LLC
Onondaga Blvd.
Syracuse
USA

M. Folch
Soil Science Laboratory
Faculty of Pharmacy
University of Barcelona
Barcelona
Spain

Arlindo C. Gomes
Universidade da Beira Interior
Materiais Fibrosos e Tecnologias Ambientais
(FibEnTech-UBI)
Rua Marquês d'Ávila e Bolama
Covilhã
Portugal

Justus Harding
Bauer Resources GmbH
Schrobenhausen
Germany

John Hanlon
PurEnergy Operating Services
LLC
Onondaga Blvd.
Syracuse
USA

Marco Hartl
Environmental Engineering and Microbiology
Research Group
Department of Civil and Environmental
Engineering
Universitat Politécnica de Catalunya -
BarcelonaTech
Barcelona
Spain

Joseph Hogan
Ecol-Eau
Montreal
Canada

Vasiliki Ioannidou
School of Engineering & the Built Environment
Birmingham City University
Birmingham
UK

K.B.S.N. Jinadasa
Department of Civil Engineering
University of Peradeniya
Sri Lanka

Christopher H. Keller
Wetland Solutions
Inc.
Gainesville
USA

R.M. Kilian
Kilian Water Ltd.
Denmark

U. Kappelmeyer
Helmholtz Centre for Environmental
Research-UFZ
Department of Environmental Biotechnology
Leipzig
Germany

Naresh Kumar
Department of Geological and Environmental
Sciences
Stanford University
Stanford
USA

Satish Kumar
Department of Paper Technology
Indian Institute of Technology Roorkee
Saharanpur Campus
Saharanpur
India

F. Masi
Iridra Srl
Florence
Italy

T.A.O.K. Meetiyagoda
R&D Unit
CETEC Pvt Ltd
Kandy
Sri Lanka

V. Matamoros
Intituto de Diagnóstico Ambiental y Estudios
del Agua
IDAEA-CSIC
Spain

Barada Kanta Mishra
Department of Environmental and
Sustainability
CSIR-Institute Minerals and Materials
Technology
Bhubaneswar
India

Ashirbad Mohanty
Environmental Biotechnology
Division Helmholtz-Zentrum für
Umweltforschung -UFZ
Leipzig
Germany

I. Moodley
Industrial Mine Water Research Unit
(IMWaRU)
Centre in Water Research Development
(CIWaRD)
School of Chemical and Metallurgical
Engineering (CHMT)
University of the Witwatersrand
Johannesburg
South Africa

E. Navarro
Universidad Tecnológica de Izúcar de
Matamoros
México

Wun Jern Ng
Nanyang Environment and Water Research
Institute and School of Civil
and Environmental Engineering
Nanyang Technological University
Singpore

Ioannis E. Nikolaou
Business Economic and Environmental
Technology Lab
Department of Environmental Engineering
Democritus University of Thrace
Xanthi
Greece

S. Nielsen
Orbicon A/S
Ringstedvej 20
Roskilde
Denmark

A. Pascual
AIMEN
Spain;
University of A Coruña
Spain

R. Pastor
Cátedra UNESCO de Sostenibilidad-UPC
Spain

Anna Pedescoll
Departamento de Biodiversidad y Gestión
Ambiental
Facultad de Ciencias Biológicas y Ambientales
Universidad de León
León
Spain

Stephane Prigent
Bauer Resources GmbH
BAUER-Strasse 1
Schrobenhausen
Germany

A. Rizzo
Iridra Srl
Florence
Italy

M. Salgot
Soil Science Laboratory
Faculty of Pharmacy
University of Barcelona, Barcelona
Spain

Olga Sánchez
Departament de Genètica i Microbiologia
Edifici C
Universitat Autònoma de Barcelona
Cerdanyola del Vallès
Barcelona
Spain

C. Sheridan
Industrial Mine Water Research Unit
(IMWaRU)
Centre in Water Research Development
(CIWaRD)
School of Chemical and Metallurgical
Engineering (CHMT)
University of the Witwatersrand
Johannesburg
South Africa

Adam Sochacki
Environmental Biotechnology Department
Faculty of Power and Environmental
Engineering
Silesian University of Technology
Poland;
Centre for Biotechnology
Silesian University of Technology
Poland;
Department of Applied Ecology
Faculty of Environmental Sciences
Czech University of Life Sciences Prague
Kamýcká
Czech Republic

M. Soto
Sedaqua (Spin-off from University of A Coruña)
Spain;
University of A Coruña
Spain

Rogério Simões
Universidade da Beira Interior
Materiais Fibrosos e Tecnologias Ambientais
(FibEnTech-UBI)
Rua Marquês d'Ávila e Bolama
Covilhã
Portugal

Pratiksha Srivastava
Australian Maritime College (AMC)
University of Tasmania
Launceston
Australia

Alexandros I. Stefanakis
Bauer Resources GmbH
BAUER-Strasse 1
Schrobenhausen
Germany;
Bauer Nimr LLC
Muscat
Oman;

Department of Engineering
German University of Technology in Oman
Athaibah
Oman

Athanasia G. Tekerlekopoulou
Department of Environmental & Natural
Resources Management
University of Patras
Agrinio
Greece

Fulya Aydın Temel
Department of Environmental Engineering
Faculty of Engineering
Giresun University
Giresun
Turkey

Yalçın Tepe
Department of Biology
Faculty of Science and Arts
Giresun University
Giresun
Turkey

Martin Thullner
Department of Environmental Microbiology
UFZ – Helmholtz Centre for Environmental
Research
Leipzig
Germany

A. Torrens
Soil Science Laboratory
Faculty of Pharmacy
University of Barcelona
Barcelona
Spain

Dion Van Oirschot
RietLand bvba
Van Aertselaerstraat 70
Minderhout
Belgium

María Hijosa-Valsero
Departamento de Biodiversidad y Gestión
Ambiental
Facultad de Ciencias Biológicas y Ambientales
Universidad de León
León
Spain;
Present address: Instituto Tecnológico Agrario
de Castilla y León (ITACyL)
Centro de Biocombustibles y Bioproductos
Polígono Agroindustrial
León
Spain

Joost van den Bulk
Tauw bv
Handelskade
The Netherlands

Dimitrios V. Vayenas
Department of Chemical Engineering
University of Patras
Patras
Greece;
Institute of Chemical Engineering and High
Temperature Chemical Processes
(FORTH/ICE-HT)
Platani
Patras
Greece

Gladys Vidal, Catalina Plaza de Los Reyes and Oliver Sáez
Engineering and Biotechnology Environmental Group (GIBA-UDEC)
Environmental Science Faculty and Center EULA–Chile
University of Concepción
Concepción-Chile

Asheesh K. Yadav
Department of Environmental and Sustainability
CSIR-Institute Minerals and Materials Technology
Bhubaneswar
India

Preface

Constructed Wetlands are already considered today an established treatment technology worldwide. There are thousands facilities across all continents and numerous new ones are designed and constructed every year. Although they are not that old as conventional treatment technologies, Constructed Wetlands are now gaining the attention they deserve among professionals, institutions and stakeholders. Intensified research of the last 25 years provided reliable and effective designs, initially for the treatment of domestic and municipal wastewater.

However, as the performance of full-scale facilities was closely observed, it was gradually realized that these ecological treatment systems possess an even higher treatment capacity. Thus, the application of stronger wastewaters, i.e., industrial wastewaters with higher pollutant loads, started slowly but increasingly to attract the interest of engineers and professionals. Therefore, industrial wastewaters of different origin gradually became the core of new developments in Wetland Technology. It can now be said that, although research on domestic/municipal wastewater is still ongoing (for example, to optimize the removal of nutrients), the current challenges lie in the effective treatment of industrial wastewaters.

Industrial wastewaters have a far more complex composition than domestic or municipal ones, with a larger variety of pollutants of varying nature and properties, depending on the industrial process they originate from. As it is easily understood, this makes it more difficult to develop an appropriate, effective design. As a result, research, trials and full-scale industrial applications of Constructed Wetlands are today in the forefront of new developments in the field of Ecological Engineering.

This is the reason that resulted in the need for this book. With the increasing number of industrial applications of Constructed Wetlands, either for research purposes or full-scale facilities, and considering the numerous industrial wastewater sources, it is difficult to follow all new developments. Relative information on this subject is scattered among articles in scientific journals and few chapters in edited books. This book, with the simple but targeted title, aims to cover exactly that need. The idea is to present in a single reference, summarized information and the state-of-the-art knowledge on the use of Constructed Wetlands in the industrial sector through case studies, research outcomes and review chapters.

The result is more than satisfying: 26 chapters including the Introductory chapter, with 73 authors from 22 different countries provide a comprehensive coverage of the wide range of applications of these sustainable systems for the treatment of industrial wastewaters. This book is the first one in the international literature that presents such integrated information solely dedicated to the industrial applications of the Constructed Wetland technology.

I hope that the readers, either professionals, environmental engineers and consultants, academics, researchers, scientists, government agency employees, students or industrial stakeholders/entities, will find the information provided here valuable for their work and tasks and useful for the fulfillment of their goals. It is the intention of all contributors that this book will further boost the dissemination of Wetland Technology, proving its remarkable treatment potential to the international audience.

Muscat, 16 July 2017

Alexandros I. Stefanakis

Acknowledgements

This work was produced over a period of less than two years. During that time, a large number of individuals contributed significant material. The number of chapter proposals received was almost double than the final number of chapters. I tried to give the opportunity to each proposed contributor to be included in the book; hence, I encouraged interactions and jointly written chapters, as you can see in the Table of Contents. I apologize mightily if not all proposals were included in this first edition. I extend heartfelt gratitude and acknowledgement to the 73 authors of the chapters, since without them this book would not become a reality. Finally, special gratitude should be given to the staff of Wiley with whom I worked together these two years to bring this book project into life.

Introduction to Constructed Wetland Technology

Alexandros I. Stefanakis

Bauer Resources GmbH, BAUER-Strasse 1, Schrobenhausen, Germany
Bauer Nimr LLC, Muscat, Oman
Department of Engineering, German University of Technology in Oman, Athaibah, Oman

1 From Natural to Constructed Wetlands

Constructed Wetlands technology is not a "new" development, if we consider the exact definition of this word. Wetlands have been used – one way or another – by humans for hundreds (or even thousands) of years. Until recently, people didn't realize this utilization of wetland systems or their benefits and role not just for humans but for the whole planetary ecosystem. The first contact of humans and natural wetlands took place thousands of years ago: the first civilizations were established close to wetland environments, since these provided them with various economic and vital sources. Early civilizations (e.g., the Mesopotamian and Egyptian) were developed near marshes and rivers. Wetlands provided raw materials for simple up to more complicated constructions and discoveries. For example, dried reeds were used to build houses, papyrus was used to make paper or even construct ships, etc. Therefore, the exploitation of wetlands by humans was already a reality. In other words, wetlands represent a critical life source for humans and wildlife, with a great contribution to our quality of life.

Wetlands are considered today as natural systems with great ecological significance, which provide habitats for numerous species and support their life. This is why natural wetlands are often called as "the Earth's kidneys" or "the biological supermarket" [1, 2], since they are among the most productive natural environments on Earth. They also act as filters, retaining the pollutants from the water that flows through on its way to lakes, streams, and oceans. Natural wetlands fulfill a series of multiple functions with important value for humanity; for example, carbon dioxide fixation by wetland plants is beneficial for the global climate, while supporting food chains adds a value (e.g., fishing) to human activity. The values of wetlands can be distinguished into three main types: ecological, socio-cultural and economic, which together define the Total Value of Wetlands (Figure 1). In general, natural wetlands offer services for [3–5]:

- Enrichment of groundwater aquifers.
- Control/amendment of flood incidents (protective buffers).
- Retention of sediments and other substances.
- Absorption of carbon dioxide.

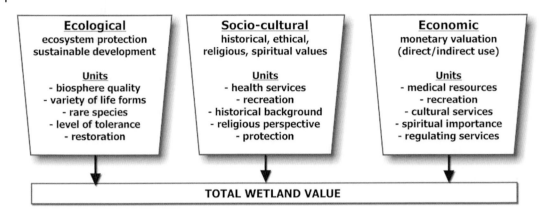

Ecological	Socio-cultural	Economic
ecosystem protection sustainable development	historical, ethical, religious, spiritual values	monetary valuation (direct/indirect use)
Units	**Units**	**Units**
- biosphere quality	- health services	- medical resources
- variety of life forms	- recreation	- recreation
- rare species	- historical background	- cultural services
- level of tolerance	- religious perspective	- spiritual importance
- restoration	- protection	- regulating services

TOTAL WETLAND VALUE

Figure 1 Individual type of values and respective criteria and value units for the determination of the Total Value of Wetlands [4, 5].

- Heat storage and release.
- Absorption of solar radiation and respective support to food chains.

Wetlands represent a considerable section of national and local economies, contributing with the provision of resources for recreational activities, pollution control and flood protection. Despite these benefits, the value of natural wetlands has only recently been recognized. Today it is understood that natural wetlands have the ability to receive and control flood incidents and alleviate their possible negative impacts. They have been used for the discharge of wastewater for centuries, for example, during the Minoan time in the Greek island of Crete (2000 BC), where advanced sewerage collection systems in Knossos and Zakros Palaces were collecting and discharging wastewater to natural wetland sites [6]. However, this use of wetlands as disposal sites resulted in their degradation in many areas around the world.

The water purification capacity of wetlands was gradually recognized and today it is known that wetlands are able to eliminate and transform various pollutants (organics, nutrients, trace elements, etc.) through physical, biological, and chemical processes. The numerous ecological and economic benefits of wetlands and their appreciation encouraged the study of wetland capacities for various technological applications. This observation of the natural wetlands resulted in the investigation of human-made wetland ecosystems and their purification functions. Several studies revealed the potential of wetland ecosystems for pollution reduction. However, it is clear today that the use of natural wetlands for treatment purposes is not a preferred or even a legal solution.

The basic concept of building constructed wetland (CWs) systems is to replicate the various naturally occurring wetland processes in a beneficial way and under controlled conditions. Of special interest here are water storage, flood protection and water quality improvement. CWs are man-made structures and they are built in a way to resemble and operate similarly to a natural wetland. A common definition of these systems is that constructed wetlands are, "*Man-made complexes of saturated substrate, emergent and submerged vegetation, animal life and water that simulate natural wetlands for human use and benefits*" [7].

Generally, it can be said that wetland technology is a relatively recent development (i.e., of the last 40–50 years) compared to conventional treatment methods which have been in use for more than

80 years. Despite this relatively long period, it is only during the last 20–30 years that a tremendous increase of interest in CWs has occurred. This could also be related to the rapid increase of interest on environmental issues. Thus, research on CWs and observation of the first full-scale systems assisted in improving the fundamental knowledge and basic understanding of the processes taking place within these bioreactors and in optimizing their efficiency.

2   The Need for Sustainable Solutions

Constructed Wetlands are a very interesting development in the field of ecological engineering during the 20th century, which could be attributed to two main facts. First, for almost a century, wastewater treatment in the developed world has been implemented via conventional centralized facilities. These heavy installations are energy-consuming with a maximum lifetime of 25–30 years, which means that new investment is required for new facilities to replace the old ones. On the other hand, low-income regions usually cannot afford the construction of such large centralized facilities and the technical expertise to run them. Thus, in both developed and developing regions alternative treatment techniques are required, apparently for different reasons, which should combine acceptable performance, cost-efficiency and, of course, the – recently added – sustainability parameter.

Hence, natural treatment systems, such as constructed wetlands, can be a solution that satisfies these parameters. The current status of the technology proved their high levels of performance. They don't depend on energy-consuming and expensive processes (since renewable sources are mostly used) and they don't demand synthetic raw materials, but mainly natural ones, e.g., plants and aggregate materials, providing from this point of view an ecological treatment. Also, the idea of using plants to purify wastewater is always viewed as innovative and attractive for society, which can therefore increase their social acceptability.

Conventional and natural treatment systems serve the same goal, i.e., wastewater treatment, but under a different approach. Natural systems provide decentralized treatment services, which may be seen as an alternative approach to treat wastewater at or near the source. This approach further satisfies the concept of sustainable development, meaning that the same function (i.e., wastewater treatment) is achieved in a more economic, environmentally friendly, and energy-efficient way.

3   Constructed Wetlands or Conventional Systems – Pros and Cons

Conventional technologies provide effective wastewater treatment, but they come with some undesired impacts. Usually, they require extensive sewer collection networks to bring the wastewater into a centralized facility. This relates to some obvious environmental and economic concerns. Conventional treatment plants usually have an unattractive appearance, including large mechanical parts (ventilators, dosing schemes, pumps, etc.), extensive use of concrete and steel and odor and noise generation. As a result, they require large amounts of energy, which means respectively high greenhouse gas emissions. In addition, the investment costs and, especially, the operation and maintenance costs are usually high. Moreover, the daily production of surplus sludge demands further handling and management, which adds to the total operational costs.

On the other hand, Constructed Wetlands can be characterized as a cost-efficient treatment technology with a usually lower or similar investment cost, and significantly reduced costs for operation and maintenance than conventional treatment methods. Global experience from many countries has shown that operational costs of CW facilities can be up to 90% lower compared to conventional/mechanical plants. Moreover, CWs operate without the need for the addition of chemical substances, which is not typically the case in conventional treatment plants. Locally available resources are used for the construction of a CW facility (i.e., high in-country value), which is relatively easy and simple to build in the absence of big and complex infrastructure. Energy is required in very low amounts, e.g., for the lighting of the facility, and, possibly, for the operation of few pumps. The use of pumps can even be avoided if the natural ground slope is exploited to use the gravity flow along the system. Additionally, there is no need for specialized staff to run the facility and only periodic inspection is required. Furthermore, they have a prolonged useful lifetime, which extends to 30 years or even more.

Wastewater treatment in Constructed Wetlands does not practically generate any by-product. Sludge can be accumulated and dewatered within the system. In conventional treatment plants, large amounts of excess sludge that need to be managed and stabilized are produced on a daily basis. Although the sludge volume represents only a small portion (1–3%) of the total treated wastewater volume, its management and handling can reach up to 50% of the total facility costs [5]. However, it should be noted that periodic sludge removal (e.g., 1–2 times/year) should take place in a CW facility if there is a pretreatment stage (e.g., sedimentation or Imhoff tank), but this cost is usually lower than the sludge production and handling cost in conventional systems. The only product that could potentially be viewed as by-product in CW facilities is the plant biomass. Usually, the produced plant biomass is harvested and collected annually or every few years. However, this biomass can further be exploited as biofuel for energy or for compost production.

At this point, it should be mentioned that the technology of Constructed Wetlands should not be viewed as directly competing with conventional technologies. Given that the main limitation of CWs is the larger area demands, they cannot completely replace conventional treatment facilities. Therefore, they can be viewed as complementary to conventional facilities, providing the benefit of decreasing the required number and capacity of centralized conventional plants by implementing several onsite treatment plants based on Wetland Technology.

However, Constructed Wetlands are appropriate for a series of installations of different scales. Due to their flexible design, they can easily be built on most sites. They become very competitive systems for wastewater generated from single households or residential complexes. In rural, remote, insular and mountainous areas, where usually no sewer system exists and it is difficult to build a centralized plant, CWs can be an extremely attractive alternative from both an economic and environmental point of view. They can effectively serve villages and small or medium cities/settlements up to few thousands of population. And, of course, for onsite application in numerous industrial facilities, CWs can be an ideal solution. Table 1 summarizes the main characteristics of conventional treatment plants and Constructed Wetlands.

Despite the long list of advantages, CWs have also limitations – as any technology. The major one is that CWs require a larger land area compared to a conventional treatment facility. Although continuous research managed to reduce the total footprint of CW facilities and optimize their design (e.g., the use vertical flow systems and lately of artificially aerated systems), area demands are still higher (e.g., 3–10 times) for a CWs solution than for conventional systems [5]. Furthermore, false design could also result in odor issues and the occurrence of a water surface in subsurface systems. However, it should be noted that if properly designed and constructed, CWs generally do not create odor issues.

Table 1 Main characteristics and comparison of conventional treatment methods and Constructed Wetlands [5, 8–11]

	Conventional treatmentmethods	Constructed Wetlands
Investment	Moderate – High	Moderate
Facility	Many/large mechanical parts and equipment	No mechanical parts (maybe pumps)
Operational costs	High	Low
Application scale	Small to large	Small to medium
Performance	High quality effluent	High quality effluent
Raw materials	Use of non-renewable materials during construction (concrete, steel, polymer membrane material etc.) and operation (electricity, chemicals)	Almost exclusive use of renewable sources (solar, wind) – "ecological" character
Corrosion resistance	Low	High (medium)
Odour nuisance	Medium to high	Low
Energy input	High	Low
Use of chemicals	Required	Not required
Greenhouse gas emissions	High	Low
Lifetime	5–10 years (MBR), 20–25 years (activated sludge)	25–30 years or more
Monitoring	Need for frequent monitoring	Need for periodical monitoring
Staff during operation	Demand for specialized personnel	No specialized personnel needed
Maintenance	High needs/costs – regular membrane failures, reinvestment after 10 years	Low, e.g., reed harvesting every 2–5 years
Response to flow variations	Higher/shock inflow rates negatively affect the performance	Robust to short-term high flow variations
Robustness to toxic substances, e.g. oil	Toxic pollutants may lead to system breakdown	Robust to some toxic constituent, e.g. heavy metals
Recovery period	Prolonged time to re-gain full treatment performance	Robust with no downtime
By-products	Large daily volumes of sludge, which need handling on a daily basis	No sludge by-product
Appearance	Unattractive	Aesthetically accepted
Social responsibility	Low	High
Biodiversity enhancement	No	Yes
In-country value	Only 20% of materials/equipment sourced within the country	More than 80% of materials/equipment sourced within the country

4 Classification of Constructed Wetlands

Based on their functions, Constructed Wetlands can be classified into three main areas [5, 12–14]:

- **Habitat creation:** systems designed to provide an upgraded wildlife habitat and to enhance the existing ecological benefits, e.g., attracting birds and creating green spaces, while addressing water/wastewater treatment. Four different CW types exist here: ponds, marshes, swamps and ephemeral wetlands.
- **Flood control:** systems that function as runoff receivers during flood incidents and increase the stormwatwer storage capacity, especially in urban areas.
- **Wastewater treatment:** systems designed and operated to receive and treat wastewater of different origin.

The most widely used classification of CWs is based on the water flow direction and the type of vegetation used (Figure 2). Also, based on the flow path across the CW system, there are two general types:

- Free Water Surface Constructed Wetlands (FWS CWs), also called Surface Flow Constructed Wetlands (SFCWs) and
- Subsurface Flow Constructed Wetlands (SSF CWs).

A further classification of SSF CWs can be made according to the flow path direction between horizontal (HSF CWs) or vertical (VFCWs) flow systems. The type of vegetation used in CWs is also one of the main characteristics of CWs systems (Figure 2). Therefore, CWs can be also classified according to the wetland plant species used as:

- Emergent macrophyte wetlands
- Submerged macrophyte wetlands and
- Floating treatment wetlands (FTWs).

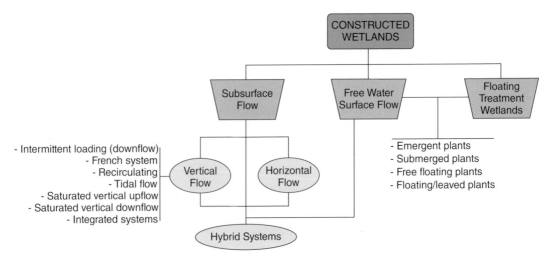

Figure 2 Classification of Constructed Wetlands [5, 13].

Free Water Surface Constructed Wetlands

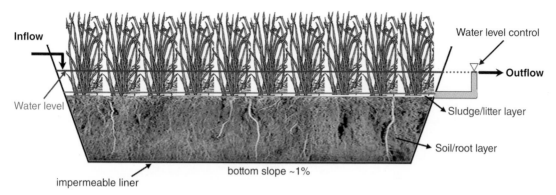

Figure 3 Schematic representation of a typical cross-section of a Free Water Surface Constructed Wetland.

Among these types, CWs with rooted emergent macrophytes are the most widely used. A brief description of the different constructed wetland designs is now presented.

4.1 Free Water Surface Constructed Wetlands (FWS CWs)

This type of CW is a shallow basin or channel containing a soil layer of 30–40 cm thickness, in which the macrophytes are planted. Common plants species used are common reeds (*Phragmites australis*), cattails (*Typha* spp.), bulrush (*Scirpus* spp.) and herbs (*Juncus* spp.) [5, 15]. The bottom of the basin (as in all CW systems) is covered by a geo-textile/geo-membrane or clay material to prevent wastewater leakage to the groundwater. A water column of 10–50 cm in depth exists above the soil layer, which means that water is exposed to the atmosphere and the solar radiation. Water level can be adjusted at the outlet of the system (Figure 3). The water flows horizontally through the plant stems and rhizomes and comes into contact with the top layer of the soil, the different plant parts and the associated biofilm, which enables pollutant removal through various physical, biological and chemical processes. FWS CWs could attract mosquitos if the water remains almost stagnant inside the system due to false design or improper construction.

The performance of this wetland type is good for suspended solids (SS) and biochemical oxygen demand (BOD) removal and satisfactory removal of nitrogen (N) and pathogens, but phosphorus (P) removal is usually limited [5, 13, 14, 16]. FWS CWs have been applied for the treatment of primary and secondary municipal effluents, but mainly for polishing treated effluents, stormwatter and highway runoff, as well as for agricultural effluents [5, 15]. This type is also used for produced water treatment, i.e., water containing petroleum hydrocarbons [17, 18]. FWS CWs have higher area demands compared to the other wetland types, but they resemble natural wetlands to the most.

4.2 Horizontal Subsurface Flow Constructed Wetlands (HSF CWs)

HSF CWs are more widely used in Europe than in the USA [19]. They are basins containing gravel material planted with common reeds (*Phragmites australis*) or other wetland plant species such as *Typha* (e.g., *latifolia, angustifolia*) and *Scirpus* (e.g., *lacustris, californicus*) [20, 21]. The substrate

Horizontal Subsurface flow Constructed Wetlands

Figure 4 Schematic representation of a typical cross-section of a Horizontal Subsurface Flow Constructed Wetland.

used is rock and gravel of different origin and composition. In this CW type, there is no water surface exposed to the atmosphere and the water level is kept 5–10 cm below the gravel surface (Figure 4). The water flows horizontally through the pores of the substrate media and comes into contact with the media grains, the plant roots and the attached biofilm [5]. Thus, respective health risks due to possible human contact with the wastewater and mosquito issues are limited in this CW type [14]. The substrate layer thickness varies from 30 to 100 cm [14, 22]. The bottom of the bed is usually covered with an impermeable geo-membrane and has a slight slope (1–3%). As for FWS CWs, the uniform distribution of the wastewater across the wetland width at the inflow point is a key parameter for the proper function of the system, while step-feeding of the wastewater at different points along the wetland length could enhance the performance [23]. HSF CWs have the advantage of lower area demands compared to FWS CWs, although capital costs might be higher [5, 14].

This CW type has been proved to be very effective in the treatment of municipal wastewater, removing SS and organic matter (BOD) at high rates, although nutrient removal (nitrogen, phosphorus) is usually lower [5, 13, 14, 22]. Various modifications of the system design have been proposed in order to improve the performance, such as effluent recirculation [19], wastewater step-feeding [23], water level raising [19] and effluent treatment with gravity filters containing special substrate [19, 24]. HSF CWs are applied for the treatment of a wide range of industrial wastewater, e.g., mine drainage, dairy, swine, olive mills, landfill leachate, cork effluent, contaminated groundwater, hydrocarbons, etc. [25–29].

4.3   Vertical Flow Constructed Wetlands (VFCWs)

At the early years of wetland technology, FWS and HSF CWs were the dominant types, mainly due to higher costs for VFCWs construction and operation. However, the interest in VFCWs was gradually increased with time; especially when the higher oxygen transfer capacity of this system was realized compared to the other types. Today, there are various available modifications of VF systems applied or under investigation (for much more detail, see [5]). VFCWs are mainly used in Europe, especially in Denmark, Austria, Germany, France, and the UK [5, 14]. The most common setup is a basin containing several layers of gravel and sand with increasing gradation from top to the

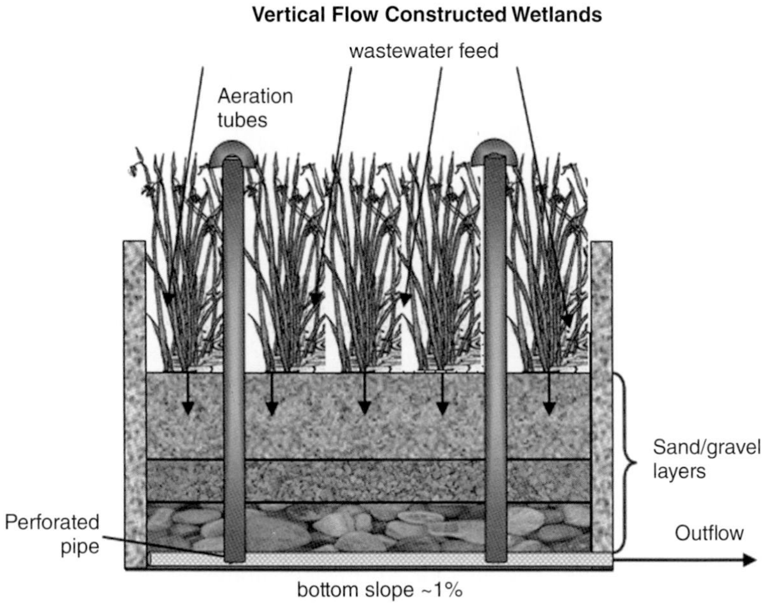

Vertical Flow Constructed Wetlands

Figure 5 Schematic representation of a typical cross-section of a Vertical Flow Constructed Wetland [5, 30, 31].

bottom [5, 30] (Figure 5). The total thickness of the substrate varies from 30 to 180 cm [5]. Usually, the top layer of the bed is a sand layer. Plants are established in the gravel surface or in the sand layer (if any). Common reeds (*Phragmites australis*) and cattails (*Typha latifolia*) are the two most widely used plant species [5, 21]. The bottom of the bed is covered by a geo-membrane/geo-textile material and has a slight slope of 1–2%. VFCWs also contain perforated vertical aeration tubes, which are connected at the bottom of the bed with the drainage collection pipeline system. These aeration tubes allow for the better aeration of the deeper parts of the bed [30].

The most commonly used mode is that of intermittent loading; wastewater is applied uniformly across the entire surface of the bed in batches and drains vertically by gravity [5, 24, 30]. VFCWs have smaller surface area demands compared to HSF and FWS CWs. Due to the better aeration capability, they are very effective in organic matter (BOD) and ammonia nitrogen removal [5, 13, 14]. Phosphorus removal remains limited and alternative modifications have been proposed for the performance improvement, e.g., an additional stage with gravity filters containing bauxite, zeolite or other reactive material for effluent treatment [24, 31]. Their overall effectiveness has enabled the use of this CW type for the treatment of wastewater with different origins, e.g., domestic, municipal, industrial, agro-industrial and landfill leachate [5, 27, 29].

4.4 Floating Treatment Wetlands (FTWs)

This type is a relatively new version of wetland technology, which combines both a traditional CW system and a pond [32]. These systems consist of a floating element (usually made of a plastic material) on which the plants are established (Figure 6). Thus, these systems optically look like floating islands. As in the other CW types, the plants develop a deep and dense root system within the underlying

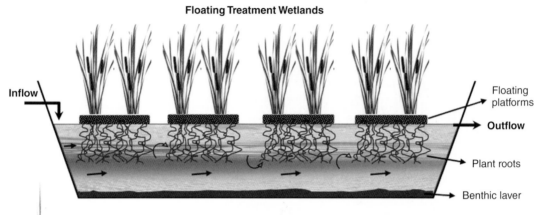

Figure 6 Schematic representation of a typical cross-section of a Floating Treatment Wetland [5].

water column [5, 33]. Since the combined system of the porous plastic mat and the plants floats on the water surface, the system is not affected by water level fluctuations. FTWs have been used for water purification in rivers, channels, lakes etc., and also for stormwater, domestic and municipal wastewater treatment [5, 32].

4.5   Sludge Treatment Wetlands (STWs)

These systems represent a special application of CW technology for the dewatering and management of excess sludge produced in conventional wastewater treatment plants. STWs appear as alternative systems to mechanical dewatering methods, such as centrifuges or filter presses. They practically are vertical flow constructed wetlands (Figure 3) modified for wastewater sludge dewatering and drying [3, 5–37]. STWs can be rectangular or trapezoidal excavated basins filled with gravel and planted with reeds. Usually the basin contains 1–2 gravel layers and a sand layer on top, although there have been systems designed without a top sand layer [34, 36, 38]. The total thickness of the porous media layers varies between 30–70 cm [5].

Sludge is applied on top of the bed in feeding cycles: a typical feeding scheme is 2–10 days of sludge feeding followed by 1–3 weeks of resting or even longer [5, 34, 35, 38]. The beds can be constructed with concrete or can be excavated basins covered with a liner, typically with a material of low-permeability (i.e., HDPE geo-membrane covered in both sides with geo-textile or clay) [5]. The bottom layer also contains a network of perforated plastic pipes for the collection of the drained water that flows vertically through the bed body. This bottom pipe network is connected with the atmosphere with vertical perforated plastic aeration pipes, which are extended above the top substrate layer.

On top of the bed organic solids are gradually accumulating. With the alternating feeding/resting periods, the accumulated solids are dewatered and, in the long-term, they are converted to a stabilized material. After few years of operation (depending on the loading rate and the climatic conditions), the residual sludge layer is removed and reused as valuable biosolids, e.g., in agriculture as fertilizer. STWs achieve high rates of sludge dewatering (volume reduction up to 96%; [5, 34, 37], through evapotranspiration and draining processes [39].

4.6   Aerated Constructed Wetlands

Given that the level of oxygen inside the system affects many pollutant aerobic removal processes and is often the limiting factor of the performance, many modifications have been tested and applied to increase oxygen concentration. One modification that increasingly attracts interest over the last years uses artificial means for oxygen supply in subsurface CW beds. The concept is based on the use of an air pump (usually a small blower) to provide compressed air [40]. This alternative has been tested in HSF CWs, where oxygen availability is lower [41–43]. The same modification has also been applied in saturated VFCW systems [40, 44, 45]. Although wastewater aeration is a common practice for other treatment methods (e.g., in activated sludge systems), aeration of gravel beds appeared only during the last 10–15 years. Effective aeration in SSFCWs is implemented with uniform distribution of small air quantities across the bottom of the bed. The main advantage of bottom additional aeration in VFCWs is that the combined action of the vertical downward drainage of the wastewater with the upflow movement of air bubbles results in a very good air/water mixture in the bed.

Improved results are reported in intensified saturated VFCWs receiving primarily treated wastewater with low effluent pollutant concentrations (<5 mg/L BOD_5 and <5 mg/L NH_4^+-N) compared to non-aerated systems [46]. The increased oxygen availability makes these systems appropriate not only for domestic/municipal wastewater treatment, but also for stronger wastewater – especially industrial effluents. Generally, artificial aeration appears as an attractive alternative when an additional oxygen amount is desirable. The main advantage is that, due to the enhanced removal processes via the artificial aeration, the area demand becomes significantly lower, which has a positive impact on the respective construction costs. Renewable energy sources, e.g., wind or solar power, could be a potential sustainable solution to cover the small needs for energy for the artificial aeration equipment [47, 48].

5   Design Considerations of Constructed Wetlands

CWs construction is generally considered easy; however, in reality the proper design of a CW facility is not as easy as it may seem to the non-expert eyes. Currently, there are still not unanimously accepted guidelines, or a widely applied methodology. System design tends to differ not only from country to country, but also among designers and experts/engineers. Personal experience is a key parameter, especially in the more demanding and complicated industrial wastewater applications. However, there are some general rules and some basic design considerations used in the design process, especially for simple applications, e.g., domestic wastewater, including meteorological, topographical and operational parameters such as [5]:

- Climatic conditions of the area where the system will be installed.
- Topographical information in order to choose the most appropriate installation site and ensure gravity flow (if possible).
- Geological structure of the area to ensure the stability of the bunds.
- Availability of the required land.
- Legal permits that are required.
- Any ecologically sensitive areas in the vicinity or wildlife habitats.
- Current and future wastewater flow and volumes.

- Any legal limits that apply in the area for the effluent quality or the desired treatment performance.
- Appropriate or desired treated effluent reuse application.
- A nearby water body to receive the treated effluent.
- Total costs.

As it is understood, most of these general factors are common prior the implementation of any treatment technology. There are three main design parameters for constructed wetland systems, which are usually taken into account: unit area demand (m²/pe); organic and hydraulic loading; and oxygen transfer capacity.

The unit area demand expresses the surface area demand (m²) per person equivalent (pe). This parameter is generally accepted as a good indication of the land area demands. Apparently, this parameter is affected by climatic conditions. However, even for the same region different values are commonly proposed based on the experience of different individuals/experts. Generally, VFCWs have lower area demands (1–3 m²/pe) than HSF CWS (5–10 m²/pe) [5]. Organic (e.g., g BOD$_5$ or COD/m²/yr) and hydraulic (m³/m²/d or m/d) loads are also widely used expressions and can be very helpful in the estimation of the optimum load to avoid operational problems. The hydraulic loading rate (HLR or q) is also widely used and is calculated by the ratio of the inflow rate (m³/day) to the surface area (m²) of the system. Finally, oxygen transfer capacity (OTC) provides important information on the oxidation potential of the system, especially for organic matter decomposition and nitrification.

Another important parameter, mostly applied in horizontal CWs, is the hydraulic retention time (HRT), which is given as the ratio of volume (m³) to flow rate (m³/d). Its value is crucial since it defines the time of direct contact between wastewater and the wetland parts (porous media grains, plant roots, biofilm) and, thus, the extent of the various removal/transformation processes. The selection of an appropriate HRT is directly related to the system surface area (and so to the unit area demand) and performance. Currently, there are many published studies and results from pilot and full-scale installations. For example, typical HRT in HSF CWs varies between 5–10 days, depending on the climatic conditions, the level of treatment etc. Aerated HSF CWs can be designed with much lower HRT (e.g., < 1 day). Proper dimensioning of horizontal beds is also very important to ensure the hydraulic efficiency of the system. HSF and FWS CWs usually have a rectangular plan view with a varying width to length ratio of 1:3–5. Longer length than width is preferred in order to ensure plug flow hydraulics.

Generally, designing and sizing a wetland bed varies from simple "rule of thumb" to more complex models. The plug-flow first order reaction equation (known as the kC* model) was widely used in the past. A simple equation for the calculation of the required surface area for BOD$_5$ removal is the following:

$$A = \frac{Q[\ln(C_o/C_e)]}{K_T dn} \tag{1}$$

Where:

A = the surface area of the bed (m²);
d = the saturated depth of the bed (m);
n = the substrate porosity (decimal fraction);
K_T = the first-order areal rate constant (m/d);
Q = the average daily flow rate (m³/d);

C_o = the mean influent concentration (mg/L); and
C_e = the required effluent concentration (mg/L).

Other advanced design approaches are lately frequently used such as the *PkC** model [14]. The required wetland area can be calculated using the following equation:

$$A = \frac{PQ_i}{k_A}\left[\left(\frac{C_i - C^*}{C_o - C^*}\right)^{1/P} - 1\right] = \frac{PQ_i}{k_V h}\left[\left(\frac{C_i - C^*}{C_o - C^*}\right)^{1/P} - 1\right] \tag{2}$$

Where:

A = the surface area of the bed (m^2);
P = apparent number of tanks-in-series (TIS) (–);
Qi = influent flow rate (m^3/d);
k_A = modified first-order areal rate coefficient (1/d);
C_i = inlet concentration (mg/L);
C_o = outlet concentration (mg/L);
C^* = background concentration (mg/L);
P = apparent number of tanks-in-series (TIS) (–);
k_V = modified first-order volumetric rate coefficient (1/d);
h = wetland water depth (m).

This approach can give more accurate designs, since it considers the rate coefficients and temperature correction. However, it includes several variables, which are not always known to allow for the design using this equation.

Usually, BOD$_5$ is used as the main target parameter, but other pollutants such as suspended solids (SS), total or ammonia nitrogen and total phosphorus have also been used. The parameters used for the evaluation of CWs performance are the same pollutant indicators as for every wastewater treatment technology, such as organic matter (BOD$_5$ and COD), nitrogen compounds (total nitrogen, ammonia nitrogen, nitrate, nitrite), phosphorus (total phosphorus, ortho-phosphate), coliform bacteria (*E. coli*, faecal coliforms) and heavy metals. Physical characteristics (e.g., dissolved oxygen, electrical conductivity, pH etc) are also used to describe CW operating conditions.

In Sludge Treatment Wetlands, the design is based on the quality of the raw sludge and the local climatic conditions, as well as on the annual surplus sludge production of the wastewater treatment plant [5]. As for other CW types, the design and sizing of STWs follows mostly empirical observations and personal experience of the designer. Again, there are no commonly accepted guidelines and the design varies from country to country. The basic design parameter is the sludge loading rate (SLR; kg dm/m^2/yr), which expresses the annual dry mass (dm/yr) or dry solids (ds/yr) that will be applied to the bed per surface unit (m^2). It is directly related to the climate of the installation area. For example, proposed SLR for Denmark is 60 kg dm/m^2/yr for activated sludge and 50 kg dm/m^2/yr for sludge with higher fat content [35]. Higher SLRs have been proposed for the Mediterranean region: up to 90 kg dm/m^2/yr in Greece [34], 45 kg dm/m^2/yr in Italy [49] and 55–110 kg dm/m^2/yr in Spain [50, 51]. Stefanakis et al. [5] presented an overview of SLRs applied in various countries around the world. The operational life time of STWs can last up to 30 years and more and it is divided in 2–3 phases of 8–12 years, based on the applied SLR. After the completion of each phase, the bed is left to rest for a period (a few months up to 1 year), then the accumulated sludge residual is removed and a new feeding cycle begins [5, 34, 35].

Two important design features of CW facilities are the plant species and the substrate media. The most common emergent plant species used are common reeds (*Phragmites australis*) and cattails (*Typha latifolia*) and also *Scirpus* spp., due to their worldwide occurrence. However, other locally available species may also be used; for example in tropical regions bamboo has been tested. Generally, the selected species should be well adapted to the local climatic conditions, tolerant against the various pollutants and able to uptake certain constituents such as nitrogen. Indigenous species are always preferred for use in CWs and not exotic ones, to avoid ecological risks such as invasion of the new species and/or diseases [5].

The selection of an appropriate substrate medium is also important for the system performance. The grain size should be carefully selected, especially for subsurface systems, since clogging problems due to low porosity and high hydraulic loads might occur and deteriorate the system efficiency. An ideal substrate would also have the capacity of removing some constituents from wastewater by various processes (e.g., ion exchange, adsorption, precipitation). The substrate layer is where the plants are established; thus, it supports plant growth and enhances the bed stability, provides filtration effects and together with the plants supports the various transformation/removal processes [5, 13]. Media used in CW systems include natural materials (e.g., minerals, rocks and soils), synthetic materials (e.g., synthetic zeolites, activated carbon) and industrial by-products (e.g., slags, blast furnace) [5].

Proper design and construction of a CW system allows for an effective and reliable performance. Problems such as bed clogging, water runoff from the surface or limited plant development may be caused by inadequate design or construction [5]. Typical problems such as pump/valve failure may occur as in conventional plants [52]. Slightly increased maintenance time and more advanced skills may be required only in the case of more complex modifications, such as artificially aerated beds.

6 Constructed Wetlands as a Sustainable Solution for the Industrial Sector

Generally, the term "green" for such a technology includes specific factors such as effective treatment, robustness, no by-products, recycled/reuse of materials, minimum energy consumption, use of renewable energy sources, no use of chemicals and minimum environmental nuisance [53]. The sustainable character of a treatment system is defined by its economic viability, technical feasibility, environmental protection contribution and social acceptance [54]. Additionally, it should close the flow cycle of materials, i.e., provide the option for safe reuse of the treated effluents. The approach of sustainable sanitation systems integrates aspects such as public health and hygiene, protection of the environment and natural resources, technological and operational parameters, financial parameters and socio-cultural aspects [53, 54].

As described above, it can be said that the concept of Constructed Wetlands itself places them in the sustainability field. Although nowadays the term "sustainability" is used so often that its content is confusing or misinterpreted, its use here refers to the integration of environmental aspects in the treatment process. Especially for the industry, the claim for the sustainable character of wetland systems relates to the advantage of promoting both economic growth, as well as protection of ecosystems and public health. This is the real benefit that wetland technology can bring to the industry.

Low energy consumption and use of natural materials (gravel, soil, sand and plants) are two crucial factors for the sustainable character of the system. One of the largest wetland projects is the Everglades restoration, USA, where CWs were used to remove nutrients from agricultural runoff entering the Everglades [55]. This project showed in the most emphatic way the sustainable dynamic of CWs. Similar projects of sustainable design using wetland technology have been developed for the removal of non-point source pesticide pollution in river catchments [56] and diffuse pollution at the catchment scale [57]; the protection of coastal zones from human activities and pollution in China [58]; and the treatment of produced water from oilfields in desert environments [17, 18], among others.

Effective treatment and high removal efficiencies significantly decrease the pollution load discharged to final receivers (surface and ground water bodies), thus limiting the risk for ecosystem and aquatic life degradation. The construction of CW systems usually involves locally available natural materials (gravel/sand, plants), with minimum use of synthetic or non-renewable materials. This means that the use of raw materials does not include significant energy-consuming and pollution generating processes. The minimum energy consumption in CW facilities preserves natural resources and minimizes pollution generation, especially when the main energy source is non-renewable (e.g., fossil fuels). These environmental benefits can be translated to low levels of greenhouse gas emissions [5, 52].

Quantification of greenhouse gas emissions during construction and operation phases is important for the estimation of the ecological impact and the global warming potential. Several studies report that CWs have a slightly lower environmental impact during the construction phase compared to other conventional methods (e.g., activated sludge, trickling filters), but a much lower impact for the operation phase [11]. Life cycle analysis studies on CW systems and comparison with conventional treatment methods have shown that the global warming potential of CWs is lower in terms of CO_2 emissions [5, 59, 60]. It is also interesting that in CW systems the major portion of the environmental impact occurs during the construction phase [59, 60], while in conventional treatment systems the environmental impact of the operational phase is higher than that of the construction phase [59, 61, 62]. Among the various CW types, FWS systems produce the lowest CO_2 emissions, VFCWs the lowest CH_4 emissions, while N_2O emissions are reported to be comparable in all CW types [63]. Hybrid CW systems (i.e., combination of different CW types) achieve higher removal efficiencies and minimum greenhouse gas emissions at the same time.

The ecological character of CWs is also enhanced by the fact that they promote biodiversity; they provide a habitat for various wetland organisms; water savings; and multiple hydrological functions [52]. For example, it is reported that a large FWS CW system built in a desert environment in the Middle East, i.e., in a former arid and dry area, for the treatment of water containing petroleum hydrocarbons from oilfields, provides a habitat for more than 120 different migratory bird species [17].

These findings become even more obvious in sludge dewatering applications using Sludge Treatment Wetlands [5]. The comparison with conventional mechanical methods, such as centrifuges (considering all aspects of sludge management, i.e., transport, investment, construction, raw materials, energy consumption) reveals that STWs have the lowest environmental impact [51]. As for CWs for wastewater treatment, the major impact occurs during the construction phase. If the CW basin is simply earth-excavated or built with recycled concrete, the impact of the construction phase can be reduced. Centrifuges have the highest environmental impact, since their operation requires high energy input. It is reported that the overall environmental impact of Sludge Treatment Wetlands is 500 times lower than centrifuges and 2,000 times lower than sludge transport to a centralized dewatering facility [5, 51]. Moreover, studies have shown that the final digested sludge product after treatment in STWs (biosolids) is a well stabilized and non-phytotoxic material, which can be reused, e.g., as fertilizer in agriculture [51, 64, 65].

The economic aspect is also important for the characterization of a treatment method as sustainable. The major costs during the construction phase of CWs include earthworks (excavation and fill of the basin), the filter media and the bottom liner [5, 14]. Most of the items are usually locally available, which can decrease the transportation costs, especially for short distances. The liner and mechanical equipment (e.g., pumps) may represent a higher portion of the costs if they are imported. Labor costs also vary from country to country. Generally, construction costs of CW facilities are comparable or slightly lower to those of conventional treatment plants [5, 52]. For small-scale applications (e.g., up to 1,000 pe) CWs offer an economic advantage in terms of investment, but as the population served increases, the investment costs become comparable mainly due to the higher land requirements. However, the main economic benefit of CWs, even for higher flow rates, is the significantly reduced costs for operation and maintenance due to significantly lower energy consumption and equipment used [5, 52, 53, 59].

Finally, the social aspect of CWs – the last component of sustainability – as treatment systems is steadily increasing. The green, aesthetic appearance of CW facilities compared to the conventional treatment plants (Figure 7) makes them more acceptable by society. Many industries and municipal-private companies select more and more often CW technology for the treatment of wastewater produced in their premises, as a mean to enhance their green profile and incorporate the CW system into their corporate social responsibility plan.

7 Scope of this Book

Wetland technology is attracting more and more the interest of stakeholders and institutions as an effective wastewater treatment technology. After almost 20–25 years of intensified research efforts

Figure 7 Typical view of a Constructed Wetland implemented in the industrial sector (courtesy of Bauer Resources GmbH).

and an exponentially increasing number of successful full-scale applications across the world, it is already considered an established technology. It can be said that for the majority of applications (i.e., mainly domestic and municipal wastewater) current efforts focus on performance and design optimization, for example, how to maintain high efficiency in nutrient removal (especially phosphorus) in the long run and to reduce the area demands without jeopardizing the performance.

As the treatment capacity of constructed wetland systems was recognized, it was reasonable to investigate the feasibility of these systems in other applications with higher pollutant loads than domestic/municipal wastewaters. Thus, the interest is gradually shifting to applications in the industrial sector. Over the previous few years there has been an apparent increase of studies testing different wetland systems in various industrial applications. It can be stated that the new challenges for wetland technology occur now in the industrial sector. And the results so far indicate that there is a respectively high potential for wetland systems to be further expanded in various industrial sectors.

Based on this, this book is the first in the published literature to summarize and present various applications and case studies of Constructed Wetlands in the industry. In the 25 chapters of the book you will find several different industrial applications and an interesting discussion on the various aspects of wetland technology. Since there is a large variety of wetland systems applied in industries, the book is divided into different sections. Each section covers a specific industrial sector, i.e., petrochemical industry, food industry, agro-industry, mine drainage and leachate, wood and leather industry, pharmaceuticals industry etc. and includes a number of chapters, each one presenting a different application. Some individual applications on specific industrial areas are presented under a separate section, as they can be seen as novel applications. Finally, a last section includes two chapters, which for the first time discusses the construction and HSE aspects of Constructed Wetlands facilities, and the integration of wetland technology within a Corporate Social Responsibility context with the industry's strategic management.

References

1 Barbier EB, Acreman M, Knowler D. Economic valuation of wetlands – A guide for policy makers and planners. Gland, Switzerland: Ramsar Convention Bureau; 1997.
2 USEPA. Constructed Treatment Wetlands. EPA 843-F-03-013, U.S. D.C., USA: Environmental Protection Agency: Office of Water, Washington; 2004.
3 MEA (Millennium Ecosystem Assessment). Ecosystems and Human Well-being: Wetlands and Water Synthesis. Washington DC: MEA, World Resources Institute; 2005. Available at: www .millenniumassessment.org.
4 De Groot R, Stuip M, Finlayson M, Davidson N. Valuing Wetlands: Guidance for valuing the benefits derived from wetland ecosystem services. Ramsar Technical Report No. 3, CBD Technical Series No. 27. Gland, Switzerland: Ramsar Convention Secretariat; 2006.
5 Stefanakis AI, Akratos CS, Tsihrintzis VA. Vertical Flow Constructed Wetlands: Eco-engineering Systems for Wastewater and Sludge Treatment. Oxford, UK: Elsevier Science; 2014.
6 Angelakis AN, Koutsoyiannis D, Tchobanoglous G. Urban wastewater and stormwater technologies in ancient Greece. Water Res. 2005; 39:210–220.
7 Hammer DA, Bastian RK. Wetlands ecosystems: natural water purifiers? In: Hammer DA, editor. Constructed wetlands for wastewater treatment. USA: Lewis Publisher; 1989.

8 Fenu A, de Wilde W, Gaertner M, Weemaes M, de Gueldre G, De Steene B. Elaborating the membrane life concept in a full scale hollow-fibers MBR. J Membrane Sci. 2012; 421–422: 349–354.

9 Ozgun H, Dereli RK, Ersahin ME, Kinaci C, Spanjers H, van Lier JB. A review of anaerobic membrane bioreactors for municipal wastewater treatment: Integration options, limitations and expectations. Sep Puri Technol. 2013; 118:89–104.

10 Ayala DF, Ferre V, Judd SJ. Membrane life estimation in full-scale immersed membrane bioreactors. J. Membrane Sci. 2011; 378(1–2): 95–100.

11 Georges K, Thornton A, Sadler R. Transforming Wastewater Treatment to Reduce Carbon Emissions: Resource efficiency programme, Evidence Directorate. Report SC070010/R2. Environmental Agency, Rio House, Waterside Drive, Aztec West, Almondsbury, Bristol, UK; 2009.

12 Knight RL. Wildlife habitat and public use benefits of treatment wetlands. Water Sci Technol. 1997; 35(5):35–43.

13 Vymazal J. Removal of nutrients in various types of constructed wetlands. Sci Total Environ. 2007; 380(1–3):48–65.

14 Kadlec RH, Wallace SD. Treatment Wetlands, 2nd edn. Boca Raton, FL: CRC Press; 2009.

15 Vymazal J. Emergent plants used in free water surface constructed wetlands: A review. Ecol Eng. 2013; 61(B):582–592.

16 Kotti IP, Gikas GD, Tsihrintzis VA. Effect of operational and design parameters on removal efficiency of pilot-scale FWS constructed wetlands and comparison with HSF systems. Ecol Eng. 2010; 36:862–875.

17 Stefanakis AI, Al-Hadrami A, Prigent S. Treatment of produced water from oilfield in a large Constructed Wetland: 6 years of operation under desert conditions. In: Proceedings, 7th International Symposium for Wetland Pollutant Dynamics and Control (WETPOL), Montana, USA, August 21–25; 2017.

18 Breuer R, Headley TR, Thaker YI. The first year's operation of the Nimr Water Treatment Plant in Oman-Sustainable produced water management using wetlands. Soc Petrol Eng. 2012; 156427.

19 Stefanakis AI, Akratos CS, Gikas GD, Tsihrintzis VA. Effluent quality improvement of two pilot-scale, horizontal subsurface flow constructed wetlands using natural zeolite (clinoptilolite). Microp Mesop Mat. 2009; 124(1–3):131–143.

20 Vymazal J. Constructed Wetlands for wastewater treatment: five decades of experience. Environ Sci Technol. 2011; 45:61–69.

21 Vymazal J. Vegetation development in subsurface flow constructed wetlands in the Czech Republic. Ecol Eng. 2013; 61(B):575–581.

22 Akratos CS, Tsihrintzis VA. Effect of temperature, HRT, vegetation and porous media on removal efficiency of pilot-scale horizontal subsurface flow constructed wetlands. Ecol Eng. 2007; 29:173–191.

23 Stefanakis AI, Akratos CS, Tsihrintzis VA. Effect of wastewater step-feeding on removal efficiency of pilot-scale horizontal subsurface flow constructed wetlands. Ecol Eng. 2011; 37(3):431–443.

24 Brix H, Arias AC. The use of vertical flow constructed wetlands for on-site treatment of domestic wastewater: New Danish guidelines. Ecol Eng. 2005; 25:491–500.

25 Vymazal J. The use constructed wetlands with horizontal sub-surface flow for various types of wastewater. Ecol Eng. 2009; 35:1–17.

26 Tatoulis T, Stefanakis AI, Frontistis Z, Akratos CS, Tekerlekopoulou AG, Mantzavinos D, Vayenas DV. Treatment of table olive washing water using trickling filters, constructed wetlands and electrooxidation. Env Sci Pollut Res. 2016; 1–8: doi10.1007/s11356-016-7058-6.

27 Wu S, Wallace S, Brix H, Kuschk P, Kirui WK, Masi F, Dong R. Treatment of industrial effluents in constructed wetlands: Challenges, operational strategies and overall performance. Environ Pollut. 2015; 201:107–120.

28 Tanner CC, Clayton JS, Upsdell MP. Effect of loading rate and planting on treatment of dairy farm wastewaters in constructed wetlands – I. removal of oxygen demand, suspended solids and faecal coliforms. Water Res. 1995; 29:17–26.

29 Sultana MY, Akratos CS, Pavlou S, Vayenas DV. Constructed wetlands in the treatment of agro-industrial wastewater: A review. Hemijska Industrija 2015; 69:127–142.

30 Stefanakis AI, Tsihrintzis VA. Effects of loading, resting period, temperature, porous media, vegetation and aeration on performance of pilot-scale vertical flow constructed wetlands. Chem Eng J. 2012; 181–182:416–430.

31 Stefanakis AI, Tsihrintzis VA. Use of zeolite and bauxite as filter media treating the effluent of Vertical Flow Constructed Wetlands. Microp Mesop Mater. 2012; 155:106–116.

32 Van de Moortel AMK, Meers E, De Pauw N, Tack FMG. Effects of vegetation, season and temperature on the removal of pollutants in experimental floating treatment wetlands. Water Air Soil Pollut. 2010; 212(1–4):281–297.

33 Tanner CC, Headley TR. Components of floating emergent macrophyte treatment wetlands influencing removal of stormwater pollutants. Ecol Eng. 2011; 37:474–486.

34 Stefanakis AI, Tsihrintzis VA. Effect of various design and operation parameters on performance of pilot-scale sludge drying reed beds. Ecol Eng. 2012; 38:65–78.

35 Nielsen S, Bruun EW. Sludge quality after 10–20 years of treatment in reed bed systems. Env Sci Pollut Res. 2015; 22(17): 12885–12891.

36 Brix H. Sludge Dewatering and Mineralization in Sludge Treatment Reed Beds. Water. 2017; 9(3): 160.

37 Masciandaro G, Peruzzi E, Nielsen S. Sewage sludge and waterworks sludge stabilization in sludge treatment reed bed systems. Water Sci Technol. 2017; 75(10).

38 Stefanakis AI, Akratos CS, Melidis P, Tsihrintzis VA. Surplus activated sludge dewatering in pilot-scale Sludge Drying Reed Beds. J Hazard Mater. 2009; 172(2–3):1122–1130.

39 Stefanakis AI, Tsihrintzis VA. Dewatering mechanisms in pilot-scale sludge drying reed beds: Effect of design and operational parameters. Chem Eng J. 2011; 172:430–443.

40 Wu S, Kuschk P, Brix H, Vymazal J, Dong R, Wu H. Development of constructed wetlands in performance intensifications for wastewater treatment: A nitrogen and organic matter targeted review. Water Res. 2014; 57:40–55.

41 Chazarenc F, Gagnon V, Comeau Y, Brisson J. Effect of plant and artificial aeration on solids accumulation and biological activities in constructed wetlands. Ecol Eng. 2009; 35:1005–1010.

42 Zhang L-Y, Zhang L, Liu Y-D, Shen Y-W, Liu H, Xiong Y. Effect of limited artificial aeration on constructed wetland treatment of domestic wastewater. Desalination 2010; 250:915–920.

43 Fan J, Zhang B, Zhang J, Ngo HH, Guo W, Liu F, Guo Y. Intermittent aeration strategy to enhance organics and nitrogen removal in subsurface flow constructed wetlands. Bioresour Technol. 2013; 141:117–122.

44 Fan J, Wang W, Zhang B, Guo Y, Ngo HH, Guo W, Zhang J, Wu H. Nitrogen removal in intermittently aerated vertical flow constructed wetlands: Impact of influent COD/N ratios. Bioresour Technol. 2013; 143:461–466.

45 Dong H, Qiang Z, Li T, Jin H, Chen W. Effect of artificial aeration on the performance of vertical-flow constructed wetland treating heavily polluted river water. J Environ Sci. 2012; 24(4): 596–601.

46 Nivala J, Wallace S, Headley T, Kassa K, Brix H. Oxygen transfer and consumption in subsurface flow treatment wetlands. Ecol Eng. 2012; 61(B):544–554.

47 Nivala J, Headley T, Wallace S, Bernhard K, Brix H, van Afferden M, Müller RA. Comparative analysis of constructed wetlands: The design and construction of the ecotechnology research facility in Langenreichenbach, Germany. Ecol Eng. 2013; 61B:527–543.

48 Tao M, He F, Xu D, Li M, Wu Z. How artificial aeration improved sewage treatment of an integrated vertical-flow constructed wetland. Pol J Environ Stud. 2010; 19(1):183–191.

49 Bianchi V, Peruzzi E, Masciandaro G, Ceccanti B, Mora Ravelo S, Iannelli R. Efficiency assessment of a reed bed pilot plant (*Phragmites australis*) for sludge stabilisation in Tuscany (Italy). Ecol Eng. 2011; 37(5):779–785.

50 Uggetti E, Llorens E, Pedescol A, Ferrer I, Castellnou R, García J. Sludge dewatering and stabilization in drying reed beds: characterization of three full-scale systems in Catalonia, Spain. Bioresour Technol. 2009; 100(17):3882–3890.

51 Uggetti E, Ferrer I, Molist J, García J. Technical, economic and environmental assessment of sludge treatment wetlands. Water Res. 2011; 45:573–582.

52 Langergraber G. Are constructed treatment wetlands sustainable sanitation solutions? Water Sci Technol. 2013; 67(10):2133–2140.

53 Brix H. How green are aquaculture, constructed wetlands and conventional wastewater treatment systems? Water Sci Technol. 1999; 40(3):45–50.

54 SunSanA. Towards more sustainable sanitation solutions – SuSanA Vision Document. Sustainable Sanitation Alliance (SuSanA); 2008. Available at www.susana.org/docs_ccbk/susana_download/2-267-en-susana-statement-version-1-2-february-2008.pdf.

55 Guardo M, Fink L, Fontaine TD, Newman S, Chimney M, Baerzotti R, Goforth G. Large-scale Constructed Wetlands for nutrient removal from stormwater runoff: an Everglades restoration project. Environ Manage. 1995; 19(6):879–889.

56 Schulz R, Peall SKC. Effectiveness of a Constructed Wetland for retention of nonpoint-source pesticide pollution in the Lourens river catchment, South Africa. Environ Sci Technol. 2001; 35:422–426.

57 Harrington R, O'Donovan G, McGrath G. Integrated Constructed Wetlands (ICW) working at the landscape-scale: the Anne Valley project, Ireland. Ecol Inform. 2013; 14:104–107.

58 Zuo P, Wan SW, Qin P, Du J, Wang H. A comparison of the sustainability of original and constructed wetlands in Yancheng Biosphere Reserve, China: implications from emergy evaluation. Environ Sci Policy. 2004; 7:329–343.

59 Dixon A, Simon M, Burkitt T. Assessing the environmental impact of two options for small-scale wastewater treatment: comparing a reedbed and an aerated biological filter using a life cycle approach. Ecol Eng. 2003; 20(4):297–308.

60 Machado AP, Urbano L, Brito AG, Janknecht P, Salas JJ, Nogueira R. Life cycle assessment of wastewater treatment options for small and decentralized communities. Water Sci Technol. 2007; 56(3):15–22.

61 Ortiz M, Raluy RG, Serra L, Uche J. Life cycle assessment of water treatment technologies: wastewater and water-reuse in a small town. Desalination 2007; 204:121–131.

62 Renou S, Thomas JS, Aoustin E, Pons MN. Influence of impact assessment methods in wastewater treatment LCA. J Clean Prod. 2008; 16:1098–1105.

63 Mander Ü, Dotro G, Ebie Y, Towprayoon S, Chiemchaisri C, Nogueira SF, Jamsranjav B, Kasak K, Truu J, Tournebize J, Mitsch WJ. Greenhouse gas emission in constructed wetlands for wastewater treatment: a review. Ecol Eng. 2014; 66:19–35.

64 Stefanakis AI, Komilis DP, Tsihrintzis VA. Stability and maturity of thickened wastewater sludge treated in pilot-scale Sludge Treatment Wetlands. Water Res. 2011; 45:6441–6452.

65 Stefanakis AI, Tsihrintzis VA. Heavy metal fate in pilot-scale Sludge Drying Reed Beds under various design and operation conditions. J Hazard Mater. 2012; 213–214:393–405.

Part I

Petrochemical and Chemical Industry

1

Integrated Produced Water Management in a Desert Oilfield Using Wetland Technology and Innovative Reuse Practices

Alexandros I. Stefanakis[1,2,3], Stephane Prigent[1] and Roman Breuer[1]

[1] *Bauer Resources GmbH, BAUER-Strasse 1, Schrobenhausen, Germany*
[2] *Bauer Nimr LLC, Muscat, Oman*
[3] *Department of Engineering, German University of Technology in Oman, Athaibah, Oman*

1.1   Introduction

The water that is produced during the exploration and production of oil and gas represents one of the largest industrial waste streams worldwide [1]. This water occurs not only during the crude oil recovery, but also during other forms of fossil energy recovery including shale gas, oil sands and coal bed methane [2]. Produced water may include water from the reservoir, natural formation water and water injected into the formation, along with any chemical substances used during the production and treatment processes. Even after the majority of the oil and gas has been extracted, produced water is typically contaminated with residual hydrocarbons. Stricter environmental policies drive the vision of environmentally friendly treatment of produced water. The North American region is the largest market for produced water treatment; more than 20 billion bbl (barrels of oil) of produced water are annually produced in the USA alone [3]. The worldwide production of produced water associated with hydrocarbon recovery exceeds 77 billion bbl per annum [4]. The Middle East market remains a key growth area due to increasing awareness and fresh water shortage.

In the oil and gas industry, produced water management represents a major challenge, since if it is not properly managed it can have an adverse impact to the environment [5, 6]. Produced water quality varies among different oil fields and wells, depending on the produced hydrocarbon type and the geological formations [2]; however, specific compounds are usually present in this water, such as oil and grease, salts (i.e., total dissolved solids, salinity) and other organic and inorganic compounds (e.g., emulsion breakers, chemical additives, solvents, heavy metals etc.). The levels of these constituents in produced water usually define the required management options to be selected. Due to the high salt content and petroleum hydrocarbons, produced water cannot be released to the environment, since it can affect the soil salinity and plant productivity [7]. As the levels of salinity increase, so does the treatment cost. Therefore, desalination of produced water is a widely applied method to improve its quality and make it appropriate for reuse options [8].

Large volumes of produced water are generated as an associated co-product of oil production in many countries; the management of which often imposes a limitation on oil production. In many cases, a portion of this water is re-injected into reservoirs to maintain pressure for the oil wells. The remaining volume is typically disposed of into shallow aquifers or via Deep Well Disposal (DWD), which are environmentally undesirable and operationally energy intensive. The most common practice for produced water management is disposal either into the ocean (mainly for offshore production activities) or underground deep injection (for onshore activities) [9, 10].

However, produced water is gradually viewed as a useful by-product, while its environmental impact lead to more stringent discharge standards [11]. Various produced water handling methods have been developed such as mechanical/chemical technologies including membrane filtration technology [2, 12–14], thermal technologies [11, 15, 16], aerated filters [11, 17], flotation [11, 18, 19] and electroagulation and electrodialysis [2, 11, 15, 20, 21], among others. However, the main disadvantage of these methods is the high operational and maintenance costs due to high-energy consumption and frequent mechanical failure, which respectively affects their performance [22]. Mechanical wastewater treatment methods, such as activated sludge and sequential batch reactor, also require high-energy consumption technologies with high operating costs. Experience of implementing conventional technologies throughout remote areas in developing countries (where many oil resources are found) shows inadequate treatment due to high maintenance cost, lack of local expertise and poor governance [23].

Constructed Wetlands (CWs) are natural wastewater treatment systems and have the advantage of significantly decreasing the capital and operation costs compared to mechanical systems. Especially over the last two decades, the green technology of CWs attracted attention as a natural process for produced water treatment. CWs are well-accepted as a reliable wastewater treatment technology in Europe and North America to treat a wide range of wastewater, such as domestic and municipal wastewaters [24–27]. The realization of their high treatment capacity enabled the use of this technology for the treatment of various industrial wastewaters [28].

The petrochemical industry is one of the fields where wetland technology is rapidly developing, as this sector is looking into new approaches and technologies to improve water efficiency. Existing knowledge and experience indicates that CWs can provide an effective, cost-efficient and ecological solution for the treatment of water contaminated with petroleum hydrocarbons, additives and phenols [29], as well as for produced water [30, 31]. Wetland systems are particularly appropriate as a remediation technology for oil production fields in remote areas, where land availability is usually high. Currently there are only a few wetland systems in various facilities such as refineries, oil and gas fields and pumping stations [29, 33] and a few others in the USA [34], in Sudan [35] and in China [36].

As water scarcity is becoming more and more topical around the world, water reuse opportunities have become of great interest [37, 38]. Wastewater, either municipal or industrial, is increasingly viewed as an additional source that can be added to the water balance and provide a new source of good quality water. Industrial wastewater reuse represents a high technical challenge, considering the varying nature of the different pollutants found in industrial effluents. Among the various types, produced water from oil fields is probably the most challenging due to the toxic and inorganic pollutants present in this water and the fact that most of these water sources occur in remote and/or desert environments. Water reuse for agricultural irrigation is often viewed as a positive means towards water recycling due to the potentially large volumes of water that can be exploited in this way. Recycled water can have the advantage of being a constant and reliable water source, while it reduces the amount of fresh water extracted from the environment [39, 40]. The major concern, however, has to do with the potential impact of the quality of the recycled water, both on the irrigated crop as well as on the end users of the crops [40]. Water reuse applications are usually viewed as an environmentally friendly practice, serving the sustainable water management approach and as economically beneficial [41, 42].

Produced water reuse is a challenging task due to the variety of pollutants present in this water, e.g., high levels of salinity (i.e., total dissolved solids – TDS), oil and grease, dispersed oil, heavy metals, dissolved organic compounds and chemicals that may have been used in the production. However, irrigation with treated produced water could be an attractive and sustainable option for produced water management. This option could represent an innovative approach, especially if it is applied in regions with limited water resources that would offer added value to this water source. Treated wastewater availability for reuse purposes is highly dependent of the water balance of the system. In arid climates where rainfall is low or non-existent, ET is the most important factor in water balance and leads to a decrease of the outflow [43] and an increase of the pollutant levels and HRT [24, 44]. At the same time, the reuse of treated produced water for irrigation of beneficial plants could close the loop of resources use, and eliminate waste production. Under this frame, there is a need to identify plant species that can be irrigated with treated produced water containing a relatively high salt content.

To the best of our knowledge, the number of scientific publications in the international literature is limited for this industrial application of wetland technology. The good treatment capacity of the existing systems worldwide for produced water applications have enabled some research efforts [45–47]. One of the largest wetland systems worldwide treating produced water from an oilfield exists in Oman [30–32, 48]. Therefore, the aim of this chapter is to present this unique case study of one of the largest Constructed Wetland facilities in the world, located in Oman, designed and built for the treatment of produced water from an oilfield and also to highlight how different approaches to integrate reuse practices of the treated water will make this facility a global example of circular economy.

1.2 Constructed Wetland for Produced Water Treatment

1.2.1 Location and Description

Oil production in Oman is associated with large volumes of water (oil production water, OPW), where the ratio of water to oil can be as high as 1:10 after separation. Only 40% of the production water is utilized for the maintenance of the reservoir pressure by injection while the remainder was disposed of into shallow aquifers in the past, and now into deep aquifers. The Nimr oilfield itself is located in the southern part of Oman about 700 km away from its capital Muscat and produces an oil water ratio of 1/10. Deep wells have been used in the Nimr oilfield, where the water is disposed by booster pumps into aquifers. However, over the last three decades, both methods of disposal progressively became unacceptable for various environmental reasons. This activity demands high-energy consumption in an area with limited power supply [31]. One of the major concerns was the possibility of contaminating the exploitable groundwater resources with toxic organic and inorganic contaminants. Disposal into shallow aquifers was phased out in 2005 due to environmental issues, leaving Petroleum Development Oman (PDO) with a deep water disposal option only [31, 49]. This stipulated the prime need to re-assess disposal practices and evaluate potential methods of treatment and utilization.

PDO decided to proceed with a large-scale application of wetland technology for management of produced water in its Nimr oil field. In 2008, BAUER was awarded a Design, Build-Own, Operate and Transfer (DBOOT) contract to develop the Nimr Water Treatment Plant (NWTP) and a 32 km^2

plot was made available for the purpose of setting up the facilities as well as a pipeline to connect the oil production facilities with the water treatment plant, which was commissioned in November 2010 and will operate for at least 20 years (Figure 1.1).

The Nimr treatment facility is a hybrid system, incorporating elements of natural systems (green infrastructure) with traditional treatment technologies (grey infrastructure). The reed beds form a wetland built in a previously arid desert. This wetland is now a habitat for migratory birds, fish and other wildlife, providing a series of ecosystem services, which in combination create a system that offers the most resilience.

The NWTP is currently treating 115,000 m³/day of produced water from a nearby oilfield using 360 ha of Surface Flow Constructed Wetland (SFCW) and 500 ha of downstream evaporation ponds (EPs). The size of this system makes it is one of the world's largest constructed wetlands. Produced water is sent through a pipeline to the Turn-Over-Point (TOP) of the plant, where separation and recovery of the majority of oil from the produced water takes place, using a series of passive hydro-cyclone oil separators. Then, the produced water is distributed in the SFCW via a long buffer pond. The 360 ha of SFCWs are divided into nine parallel tracks, each consisting of four (90 ha terraces with 10 ha wetland cells) wetland terraces in series, operating with gravity flow without any pumps (Figure 1.2). The treated water flows into a series of EPs, which are used for disposing the majority of the treated produced water, where evaporation results in salt formation, which can be processed into industrial grade salt as an end product.

A special mineral sealing layer has been developed using locally available soil material to reduce the environmental and cost impact of High Density Polyethylene (HDPE) liner. The mineral sealing layer was used in all wetland cells, while the HDPE liner is only used in the inlet buffer pond. Figure 1.3 shows the installation of the mineral sealing layer and the HDPE liner in the buffer pond and the installation of the mineral sealing layer in the wetland cells.

The Wetland system was initially planted with common reeds (*Phragmites australis*) as the only plant species. Currently, four more local plant species have already been introduced into the system, i.e., *Typha domingensis, Schoenoplectus littoralis, Juncus rigidus* and *Cyperus* spp., to enhance the biomass production and the resilience of the ecosystem, which makes this Wetland system a polyculture. These plant species are widely used in SFCWs worldwide [50]. Samples of native wetland plant species were collected throughout Oman (from wadis and coastal lagoons), propagated in the onsite nursery and then planted in the SFCW. More than 2 million plant seedlings have been planted to date. It should be noted that the SFCW is now well integrated in the local environment, accepted by the wildlife and provides a comfortable stop-over for more than 120 migratory bird species between Asia and Africa (Figure 1.4).

1.2.2 Weather Station

Weather data are recorded through a Davis Vantage Pro 2 weather station located onsite (Figure 1.5). Each weather station contains a rain collector (self-emptying tipping-bucket), temperature sensor (platinum wire thermostat), humidity sensor (film capacitor element), an anemometer (vane anemometer with wind cups) and a solar radiation sensor. An integrated sensor collects the outside weather conditions by a Vantage Pro 2 console in 30-minute intervals. The data was recorded by a USB Data Logger and thereafter readout by the DAVIS software Weather Link. The data are downloaded off the weather station at the beginning of each month. Monthly average from 2013 to 2016 air temperature, humidity, wind speed, solar radiation and rainfall are then calculated

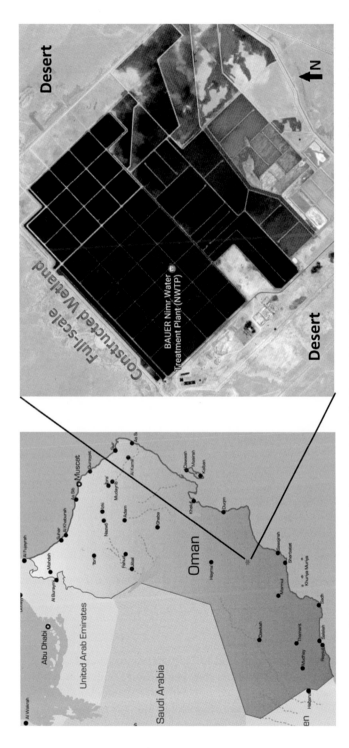

Figure 1.1 Location of the Constructed Wetland treatment plant in the desert of the south eastern Arabian peninsula in Oman (left) and aerial picture of the Constructed Wetland system and the surrounding desert (right).

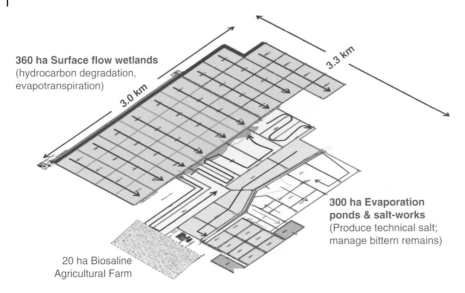

360 ha Surface flow wetlands
(hydrocarbon degradation,
evapotranspiration)

3.0 km

3.3 km

**300 ha Evaporation
ponds & salt-works**
(Produce technical salt;
manage bittern remains)

20 ha Biosaline
Agricultural Farm

Figure 1.2 General schematic overview of the NWTP, showing each stage and all current activities in the system.

Figure 1.3 Installation of the HDPE liner in the buffer pond (left) and of the mineral sealing layer in the wetland cells (right).

1.2.3 Chemical Analyses

The Constructed Wetland facility is monitored on a quarterly basis and samples are collected at various points along the system (TOP, inlet terrace 1, inlet terrace 2, inlet terrace 3, inlet terrace 4, outlet terrace 4). Grab samples are collected in glass/polyethylene bottles of 400 mL volume to determine Oil in Water (OiW) content, physicochemical (water temperature, pH, electrical conductivity; EC, dissolved oxygen; DO, and oxidation reduction potential; ORP) and all other chemical

Figure 1.4 Aerial view of the inlet buffer pond and the SFCW cells. [4]

Figure 1.5 Weather station close to the SFCW cells. [5]

Table 1.1 Laboratory analyses and respective analytical method for samples taken in the NWTP during regular monitoring campaigns.

Parameter	Unit	Analytical Method
Water temperature (T)	°C	PHC10101 probe, HACH
pH	–	PHC10101 probe, HACH
Electrical Conductivity (EC)	mS/cm	CDC40101 probe, HACH
Oxidation Reduction Potential (ORP)	mV	LDO10101 probe, HACH
Dissolved Oxygen (DO)	%	LDO10101 probe, HACH
Oil in Water (OiW)	mg/L	Hexane solvent extraction, HACH
Chemical Oxygen Demand (COD), Boron (B), Total Nitrogen (TN), Nitrite (NO2-N), Nitrate (NO3-N), Total Phosphorus (TP), Chloride (Cl), Suspended Solids (SS), Total Sulfur, Sulphide (S), Calcium (Ca), Magnesium (Mg), Sodium (Na), Potassium (K), Zinc (Zn), Bromide (Br), Barium (Ba), Lithium (Li), Iron (Fe), Lead (Pb), Manganese (Mn).	mg/L	External Laboratory

parameters (chemical oxygen demand, COD; boron, B; total nitrogen, TN; nitrite, NO2-N; nitrate, NO3-N; total phosphorus, TP; total dissolved solids, TDS; and heavy metals). Portable instruments (HACH HQ30d, Germany) are used for the onsite measurement of physicochemical parameters and OiW is measured onsite using the hexane solvent extraction method (HACH). All other chemical parameters are measured in an external laboratory in Europe according to European standards. Table 1.1 summarizes the main parameters and respective analytical methods for water analyses.

1.3 Results and Discussion

1.3.1 Weather Data

Table 1.2 summarizes typical mean monthly weather data for the area of the NWTP from 2013 to 2017. As it is obvious, average air temperature increases from January to June; after that, it decreases from July to December. Lower humidity values are usually observed during warmer months (April till October). It is also noticeable that practically no rainfall takes place in the area. These values are indicative for the local desert climate.

1.3.2 Water Quality

Since its start-up and until mid-2017, the NWTP has effectively received and treated more than 150 million m³ of produced water. At the same time, the recovered oil volume reached approximately 660,000 barrels, proving the high performance of the various treatment infrastructures.

Table 1.3 presents the typical characteristics of the inflow produced water quality at the TOP. The inlet produced water is brackish with influent TDS concentration close to 7,000 mg/L [30–32]. The

Table 1.2 Mean monthly weather data (air temperature, relative humidity, wind velocity, solar radiation and total rainfall) collected at the project location between 2013 and 2017.

Month	Air temperature (°C)	Relative humidity (%)	Wind velocity (m/s)	Solar-radiation (W/m^2)	Total rainfall (mm)
January	19.8	59.5	2.5	186.5	0.1
February	21.4	60.7	2.7	247.0	0.5
March	25.1	52.8	3.2	243.0	0.3
April	28.3	60.6	4.7	271.1	0.7
May	31.6	54.5	2.8	266.1	0.3
June	32.8	40.9	3.1	235.5	0.0
July	29.6	60.7	4.1	223.2	1.7
August	29.8	61.9	4.0	217.9	1.1
September	28.7	64.0	4.2	220.1	0.3
October	28.1	61.9	3.0	225.6	0.7
November	24.7	66.9	2.6	208.0	0.0
December	21.2	57.5	1.8	204.5	0.1

main pollutant of concern in produced water is Oil in Water (OiW). OiW concentration at the TOP is on the average close to 350 mg/L (in some cases it exceeds even 500 mg/L), while the produced water is low in nutrients concentration, i.e., total nitrogen and phosphorus concentrations are lower than 2.5 mg/L. More than 85% of the oil is recovered at the front-end of the system using passive hydrocyclones and skimmers. The residual oil hydrocarbons concentration (on average 30 ppm) is routed with gravity from the inlet buffer pond to the SFCW cells and is biologically degraded within the wetlands, producing an effluent with oil-in-water below the limit value (<0.5 mg OiW/L) [32]. Figure 1.6 shows the gradual removal of OiW along the wetland length. As it is obvious, the wetland systems provide excellent polishing of the pre-treated produced water, resulting in complete removal of OiW in the final outflow. It has been found that the rhizosphere in the wetland system is rich in hydrocarbon-degrading bacteria [48], resulting in high rates of oil removal. The reed stems act as a physical filter for trapping floating oil, which is subsequently biodegraded by microorganisms growing on the surface of the reed stems, roots and the soil surface.

Besides the removal of residual hydrocarbons from the produced water, the constructed wetland also allows for the volume reduction via the high evapotranspiration rate of the wetland plants. The current design of the NWTP is at a zero-discharge system with the intention of producing industrial salt as an end-product. The goal is to exploit the high evapotranspiration (ET) rate of the wetland plants in the wetland cells to significantly reduce the produced water volume, before its discharge into the unplanted evaporation ponds. As a result of the high ET rate in the wetland (which results on average in 40% of water loss), the TDS concentration increases along the wetland length and reaches a value of up to 12,000 mg/L at the wetland outflow [31, 32]. Despite this high level of salinity, the lack of nutrients and the general climatic conditions in the area, especially at the downstream cells of the wetland system, the wetland plant species used in the system are so far capable of surviving.

Table 1.3 Inflow produced water quality at the TOP (Turn-Over-Point).

TOP-Water Analysis (2011–2016)

Parameter	Unit	Average	St dev
Total Dissolved Solids	mg/L	6810	648
Electrical Conductivity	μs/cm	13,073	1,045
pH	–	7.55	0.07
Temperature	(°C)	23.65	0.49
Chloride as Cl	mg/L	3991.0	493.3
Suspended Solids	mg/L	18.9	21.2
Oil in Water	mg/L	280	150
BOD	mg/L	15.7	14.7
COD	mg/L	121.6	93.0
Total Nitrogen as N	mg/L	2.46	1.66
Ammonia Nitrogen as N	mg/L	1.30	0.93
Nitrite as N	mg/L	0.03	0.03
Nitrate as N	mg/L	0.08	0.07
Total Phosphorus as TP	mg/L	0.03	0.03
Boron as B (Dissolved)	mg/L	4.5	1.2
Total Sulphur as SO_4 (Dissolved)	mg/L	488	773
Sulphide as S	mg/L	9.3	15.5
Calcium as Ca (Dissolved)	mg/L	96.4	31.3
Magnesium as Mg (Dissolved)	mg/L	41.1	43.0
Sodium as Na (Dissolved)	mg/L	2580	651
Potassium as K (Dissolved)	mg/L	39.7	10.9
Zinc as Zn (Total)	mg/L	1.38	6.77
Bromide as Br	mg/L	13.0	7.6
Barium as Ba (Dissolved)	mg/L	22.7	111.7
Lithium as Li (Dissolved)	mg/L	0.16	0.10
Iron as Fe (Dissolved)	mg/L	0.24	0.22
Lead as Pb (Total)	mg/L	0.00	0.00
Manganese as Mn (Total)	mg/L	0.18	0.20

BOD_5 concentration is always very low at the inflow (around <50 mg/L) and is completely removed in the outflow. COD is also removed in the system, which could be attributed to the potential release of organic matter in the wetland system (e.g., algae, plant litter, bacteria). However, it should be noted that the COD/BOD ratio in the wetland system is around 8–10, indicating that the majority of COD in the produced water is likely not readily biodegradable. Moreover, produced water is naturally low in nutrients concentration (2.5 mg/L of total nitrogen and 1.3 mg/L of ammoniacal nitrogen), which is also completely removed in the bed. Phosphorus is practically absent in the produced water (<0.5 mg/L).

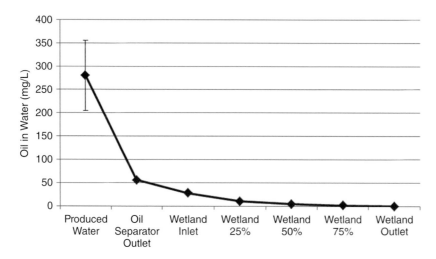

Figure 1.6 Mean OiW concentration along the different treatment stages of the NWTP.

1.3.3 Environmental Performance

Due to the operation of the NWTP, PDO has been able to shut down five of the twelve high pressure pumps that are used to dispose produced water from the oil field into deep-lying aquifers. Considering also that the NWTP is a gravity-based system with close to zero energy demand for the water treatment processes, it is estimated that the energy saved could add up to the equivalent of around 23 billion ft^3 of gas over a 10-year period.

Moreover, the Constructed Wetland is a biological system and, thus, acts as a sink for CO_2 by fixing carbon in the plant structures through photosynthesis. The estimated capacity of an annual fixture is 15,000 t CO_2 [51]. At the same time, it is known that a wetland system can produce a yield of 4–44 t/ha of dry biomass per year [52]. Considering a conservative productivity of approximately 10 t/ha for the Nimr area, the dry biomass produced at the NWTP could be estimated at 360 t/year. This translates to 1,521,000 kWh/annum heat energy content related to the reed plants [30].

Energy within the NWTP is consumed during operation for the instrumentation (flow metering), office and accommodation facilities (including water supply, air conditioning, kitchen, etc.), as well as the Oil/Water Separator and the Reverse Osmosis system [30, 31]. The gravity flow along the wetland minimizes the power requirements down to zero. Table 1.4 presents a comparison of the power requirements for the deep disposal wells, for a mechanical wastewater treatment plant and for the Constructed Wetland system for the total operational period of 20 years. Compared to the deep disposal wells, the Wetland facility uses only 1/50 of the energy consumed including all related infrastructure facilities, which is directly related to significant reduction of carbon emissions. A reduction of carbon emissions of more than 1.5 million tons CO_2, or 99% compared to the other options, can be achieved using the Constructed Wetland technology [31]. It should be mentioned that the NWTP contributes by approximately 4.26% to the Oman's overall Intended Nationally Determined Contributions to reduce emissions by 2%.

The selection of the mineral sealing layer to line the bottom of the wetland cells instead of the HDPE liner is also considered a more environmentally friendly solution. Although the transport and installation of the HDPE liner has a lower impact than the sealing layer installation, the overall balance

Table 1.4 Energy consumed and respective carbon emissions by the Constructed Wetland system and other produced water management options, i.e., deep wells and mechanical wastewater treatment methods.

Energy consumption	[kWh/m³]	CO₂ emissions in 20-year operation
Deep Wells	up to 4.0	3,200,000 MWh 1,700,000 t CO_2
Mechanical Wastewater Treatment Plant	0.8–1.0	700,000 MWh 390,000 t CO_2
Constructed Wetland	<0.1	4,000 MWh 2,150 t CO_2

showed that lining the cell bottom using local material accounts for only 21% of the impact caused by the use of the HDPE liner, due to the higher environmental footprint of the raw materials and the production process of the HDPE [30]. Therefore, this also contributes significantly to the reduction of the environmental footprint of the whole project. Specifically, the installation of the mineral sealing layer translates to a carbon footprint of 1 million tons, while the installation of a HDPE sealing layer would have resulted in a footprint of 5 million tons.

It is also worth mentioning that the large Constructed Wetland system and the series of evaporation ponds provide a valuable habitat for migratory and resident birds and other wildlife. The routine monitoring campaigns and incidental observations so far resulted in the identification of more than 120 different bird species in and around the wetland cells and ponds. Given that the site is located in the middle of the East Asia/East Africa flyway, such a large water body in the middle of the desert apparently represents an attractive island refuge, especially for those birds migrating between Asia and Africa.

1.4 Treated Effluent Reuse for Saline Irrigation

A large-scale experimental project, co-funded by PDO, is ongoing at the NWTP (Figure 1.4) covering 22 ha, to investigate the establishment and growth of 13 different plant species irrigated with oilfield produced water treated in the adjacent Constructed Wetland system under desert field conditions (Figure 1.7). Two main irrigation areas of 11 ha each, named flood irrigation and overhead irrigation, respectively, have been constructed in this area. Each irrigation area consists of three blocks of about 3.7 ha each, receiving three different water qualities with the use of a pumping system [53, 54].

As shown in Figure 1.6, the oil content decreases as the water moves through the wetland terraces. Thus, the OiW is significantly decreased at the downstream cells of the wetland after 50% of the wetland length. However, other parameters such as boron and salinity increase along the length of the SFCW and may have a potential impact on plant growth. Thus, it was decided to apply three different water qualities, pumping produced water from three respective points along the wetland length (Figure 1.2) with low/minimum hydrocarbon concentration: water after the second terrace,

Figure 1.7 Aerial view of the agriculture research project area at NWTP.

after the third terrace and the final effluent [53]. Two general methods of irrigation are used: flood and overhead. Sprinkler, drip and bubbler systems are three different techniques used for the overhead irrigation.

Perennial and annual plant species have been selected according to the climate and water characteristics, as well as the potential commercial value of the end products. Different methods of planting were implemented according to the agronomic recommendations for each type of plant species. Planting methods used were seed sowing and transplanting. The criteria to select plant species (Table 1.5) were:

- Tolerance to brackish water (up to 12,000 ppm TDS).
- Compatible with the hot and dry desert climate.
- Tolerance to water logging due to flood irrigation.
- Plants with valuable commercial end-product, e.g., biofuel, timber or carbon credits.
- Non-invasive and with limited risk for the local biodiversity.

The project irrigation area is mostly covered with a bed rock, thus it was decided to backfill the flood irrigation area with approximately 65 cm thickness of local. The soil addition was necessary to provide substrate for plant root development. A 35 cm top soil layer of screened red soil (0-20 mm) was added above the construction material to enable the planting of the different plant species. Ridges were then created by cutting furrows with construction equipment. Local compost was placed at each marked plant station for the perennial trees at a rate of 1.5 kg per plant station. The same rate was also applied to the annual plant stations on the top of each soil pile. After the application, the compost was mixed with the soil at each plant station. Local compost was applied before planting in order to improve the soil structure and the water holding capacity of the soil as well as to release nutrients to the plants for optimum plant growth.

Table 1.5 Perennial, annual plant species and grasses investigated in the saline agriculture research project at NWTP [53, 54].

Plant species	Common name	End product	Growth form
Acacia nilotica	Acacia (Qarat)	Wood/honey wax	Perennial tree
Acacia ampliceps	Acacia	Wood/honey wax	Perennial tree
Casurina equisetifolia	Casurina	Wood/windbreak	Perennial tree
Conocarpus lancifolius	Kuwaiti tree	Wood/windbreak	Perennial tree
Eucalyptus camaldulensis	Red river gum	Wood/windbreak	Perennial tree
Prosopis cineraria	Ghaf	Wood	Perennial tree
Distichlis spicata	Distichlis grass	Forage – landscaping	Grass
Paspalum vaginatum	Salt grass	Forage – landscaping	Grass
Cotton spp.	Cotton	Textile	Annual
Brassica napus	Canola	Oil – Biofuel	Annual
Cyamopsis tetragonoloba	Guar	Guar gum	Annual
Ricinus communis	Castor	Oil – Biofuel	Annual
Salicornia bigelovii	Dwarf Saltwort	Oil - Biofuel	Annual

Table 1.6 Monitoring schedule for the different tasks during the project period [53].

Monitoring task	Frequency
Water flow rate	Daily
Water quality	Bi-weekly
Soil quality	Every year
Plant parameters	
Perennial plants	Every three months
Annual plants	Every four weeks
Leaf sampling (all plants)	Every six months
Dry weight (annual grasses)	Every three months

The project started in early 2016 and will run for three continuous years. During the operational period, regular monitoring of the water balance, water quality, soil quality and plant growth parameters and yield takes place, according to the schedule shown in Table 1.6. The project is currently in the second year of operation.

The first monitoring results indicate that flood irrigation method favors plant establishment. Perennial tree species present the highest survival rates, higher compared to those under bubbler irrigation method. Regarding all perennial tree species, the outcome is positive and the survival rates ranged between 60% and 90%, with the exception of *Acacia ampliceps*, which did not manage to survive [54].

Annual grass species established well under flood and bubbler irrigation method (establishment rates up to 99%). *Distichlis spicata* showed a lower establishment rate only under bubbler irrigation (14%). Cotton plants and *Ricinus communis* also present good establishment rates under both irrigation methods (flood and bubbler). However, the rest of the annual species, i.e., *Brassica napus*, *Cyamopsis tetragonoloba* and *Salicornia bigelovii* present a negative overall outcome under both flood and overhead irrigation methods, with low survival rates or no survival at all [54].

In general, so far 10 out of the 13 plants species have been established well in the field. Future tasks and experiments will determine the limiting factor of the germination/plant establishment (i.e., water salinity, water quality, climatic conditions). The successful implementation of this research project will provide significant information on the plant species that can survive under these specific environmental conditions (water quality, climate) and also provide a beneficial product, in order to minimize the waste production at the NWTP and close the materials cycle. Ultimately, the plant species that combine site condition tolerance with commercial value will be selected for irrigation with the total outflow volume of the Constructed Wetland system.

1.5 Conclusions

A large-scale Surface Flow Constructed Wetland (SFCW) has been designed, constructed and is currently operating by Bauer in Nimr, Oman for produced water treatment under desert climatic conditions. This facility has shown on a technical and commercial basis that the Constructed Wetland technology is a highly competitive solution for industrial applications. The highly energy-efficient and extremely reliable Natural Treatment system in Nimr provides a free-of-oil treated effluent, while it converted a previously arid desert into a massive reed-vegetated area of 360 ha where migratory birds, fish and other wildlife find habitat. Since the start-up of the Nimr wetland facility, five of the twelve deep well disposal sites have shut down, saving billions of cubic feet of gas. The technical and environmental performance of this unique facility has made it a landmark for the oil and gas sector and for the wider environmental community. Among the many international awards given to the project, has been the prestigious Global Water Award, which was presented in 2011 by former United Nations Secretary-General Kofi Annan. Ongoing research activities for the reuse of the treated effluent for saline irrigation of beneficial plants in the desert will further highlight the potential of wetland technology to be further integrated with sustainable activities, making the Nimr Water Treatment Plant a unique showcase of circular economy principles applied in practice.

References

1 Aditya R. Produced water treatment market by application (onshore and offshore), by treatment types (physical, chemical, membrane and others), and by geography. Global Trends & Forecast to 2019; 2016.

2 Jain P, Sharma M, Dureja P, Sarma PM, Lal B. Bioelectrochemical approaches for removal of sulfate, hydrocarbon and salinity from produced water. Chemosphere. 2017; 166:96–108.

3 Clark CE, Veil JA. Produced water volumes and management practices in the United States. U.S. Department of Energy, Office of Fossil Energy, National Energy Technology Laboratory [Internet], 2009. Available at www.osti.gov/bridge.

4 Duraisamy RT, Beni AH, Henni A. State of the art treatment of produced water. In: Elshorbagy W, Chowdhury RK, Editors. Water Treatment. Croatia: InTech; 2013. pp. 199–222.

5 Hladik ML, Focazio MJ, Engle M. Discharges of produced waters from oil and gas extraction via wastewater treatment plants are sources of disinfection by-products to receiving streams. Sci Total Environ. 2014; 466–467:1085–1093.

6 Torres L, Yadav OP, Khan E. A review on risk assessment techniques for hydraulic fracturing water and produced water management implemented in onshore unconventional oil and gas production. Sci Total Environ. 2016; 539:478–493.

7 Kim KH, Jahan SA, Kabir E, Brown RJ. A review of airborne polycyclic aromatic hydrocarbons (PAHs) and their human health effects. Environ Int. 2013; 60:71–80.

8 Schaffer DL, Chavez ALH, Ben-Sasson M, Castrillion S, Yip NY, Elimelech M. Desalination and Reuse of High-Salinity Shale Gas Produced Water: Drivers, Technologies, and Future Directions. Environ Sci Technol. 2013; 47:9569–9583.

9 API. Overview of exploration and production waste volumes and waste management practices in the United States. Washington, DC: ICF Consulting for the American Petroleum Institute; 2000.

10 Arthur J, Langhus B, Patel C. Technical summary of oil and gas produced water treatment technologies. Tulsa, Oklahoma, USA: ALL; 2005. pp. 1–53.

11 Igunnu ET, Chen GZ3. Produced water treatment technologies. Int J Low Carbon Tech. 2013; 9:157–177.

12 Xu P, Drewes JE, Heil D. Beneficial use of co-produced water through membrane treatment: technical-economic assessment. Desalination. 2008; 225:139–155.

13 Weschenfelder SE, Borges CP, Campos JC. Oilfield produced water treatment by ceramic membranes: Bench and pilot scale evaluation. J Membrane Sci. 2015; 495:242–251.

14 Munirasu S, Haija MA, Banat F. Use of membrane technology for oil field and refinery produced water treatment – A review. Process Saf Environ. 2016; 100:183–202.

15 Fakhru'l-Razi A, Alireza P, Luqman CA, Dayang RAB, Sayed SM, Zurina ZA. Review of technologies for oil and gas produced water treatment. J Hazard Mater. 2009; 170:530–551.

16 Estrada JM, Bhamidimarri R. A review of the issues and treatment options for wastewater from shale gas extraction by hydraulic fracturing. Fuel. 2016; 182:292–303.

17 Su D, Wang J, Liu K, Zhou D. Kinetic Performance of Oil-field Produced Water Treatment by Biological Aerated Filter. Chin J Eng. 2007; 15(4):591–594.

18 Da Silva SS, Chiavone-Filho O, de Barros Neto EL, Foletto EL. Oil removal from produced water by conjugation of flotation and photo-Fenton processes. J Environ Manage. 2015; 147:257–263.

19 Saththasivam J, Loganathan K, Sarp S. An overview of oil–water separation using gas flotation systems. Chemosphere. 2016; 144:671–680.

20 Mousa IE. Total petroleum hydrocarbon degradation by hybrid electrobiochemical reactor in oilfield produced water. Mar Pollut Bull. 2016; 109(1):356–360.

21 An C, Huang G, Yao Y, Zhao S. Emerging usage of electrocoagulation technology for oil removal from wastewater: A review. Sci Total Environ. 2017; 579:537–556.

22 Vlasopoulos N, Memon FA, Butler D, Murphy R. Life cycle assessment of wastewater treatment technologies treating petroleum process waters. Sci Total Environ. 2006; 367(1):58–70.

23 Mustafa A. Constructed Wetland for wastewater treatment and reuse: a case study of developing country. Int J Environ Sci Devel. 2013; 4(1): 20–24.

24 Kadlec RH, Wallace SD. Treatment Wetlands. 2nd edn. Boca Raton, FL: CRC Press; 2009.

25 Stefanakis AI, Akratos CS, Tsihrintzis VA. Effect of wastewater step-feeding on removal efficiency of pilot-scale horizontal subsurface flow constructed wetlands. Ecol Eng. 2011; 37:431–443.

26 Stefanakis AI, Akratos CS, Tsihrintzis VA. Vertical Flow Constructed Wetlands: Eco-engineering Systems for Wastewater and Sludge Treatment. Oxford, UK: Elsevier Science; 2014.

27 Wu S, Kuschk P, Brix H, Vymazal J, Dong R. Development of constructed wetlands in performance intensifications for wastewater treatment: A nitrogen and organic matter targeted review. Water Res. 2014; 57:40–55.

28 Wu S, Wallace S, Brix H, Kuschk P, Kirui WK, Masi F, Dong R. Treatment of industrial effluents in constructed wetlands: Challenges, operational strategies and overall performance. Environ Pollut. 2015; 201:107–120.

29 Stefanakis AI Seeger E, Dorer C, Sinke A, Thullner M. Performance of pilot-scale horizontal subsurface flow constructed wetlands treating groundwater contaminated with phenols and petroleum derivatives. Ecol Eng. 2016; 95:514–526.

30 Breuer R, Grissemann E. Produced water treatment using wetlands - reducing the environmental impact of oilfield operations. Soc Petrol Eng J. 2011; 140124.

31 Breuer R, Headley TR, Thaker YI. The first year's operation of the Nimr Water Treatment Plant in Oman – Sustainable produced water management using wetlands. Soc Petrol Eng 2012; 156427.

32 Stefanakis AI, Al-Hadrami A, Prigent S. Treatment of produced water from oilfield in a large Constructed Wetland: 6 years of operation under desert conditions. In: Proceedings, 7th International Symposium for Wetland Pollutant Dynamics and Control (WETPOL), Montana, USA, August 21–25; 2017.

33 Knight RL, Robert H, Kadlec H, Ohlendorf M. The Use of treatment wetlands for petroleum industry effluents. Environ Sci Technol. 1999; 33(7):973–980.

34 Wallace S, Schmidt M, Larson E. Long term hydrocarbon removal using treatment wetlands. In: SPE Annual Technical Conference and Exhibition, Denver, Colorado, USA, 30 October–2 November; 2011.

35 Saad ASG, Khadam MA, Agab MA. Biological method for treatment of petroleum water oil content in Sudan. University of Khartoum; 2009. Available at http://research.uofk.edu/multisites/ UofK_research/images/stories/research/PDF/BESBC/biological%20treatment.pdf.

36 Ji GD, Sun TH, Ni JR. Surface flow constructed wetland for heavy oil-produced water treatment. Bioresource Technol. 2007; 98:436–441.

37 Borin M, Bonaiti G, Giardini L. Controlled Drainage and Wetlands to Reduce Agricultural Pollution. J Environ Qual. 2001; 30:1330–1340.

38 Leto C, Tuttolomondo T, La Bella S, Leone R, Licata M. Effects of plant species in a horizontal subsurface flow constructed wetland – phytoremediation of treated urban wastewater with *Cyperus alternifolius* L. and *Typha latifolia* L. in the West of Sicily (Italy). Ecol Eng. 2013; 61A:282–291.

39 Toze S. Reuse of effluent water – benefits and risks. Agr Water Manage. 2005; 80:147–159.

40 Stefanakis AI. Ecological impact of water reuse. In: Eslamian S, ed. Handbook of Urban Water Reuse. Boca Raton, FL, USA: CRC Press, Taylor & Francis Group; 2015. pp. 219–228.

41 Asano T, Burton FL, Leverenz HL, Tsuchihashi R, Tchobanoglous G. Water Reuse: Issues, Technologies, and Applications. New York: McGraw-Hill, Metcalf & Eddy Inc., AECO; 2007.

42 Atherton JG. Health and environmental aspects of recycled water. In: Doelle HW, Rokem JS, Berovic M, eds. Biotechnology X, Encyclopedia of Life Support Systems (EOLSS), Developed under the Auspices of the UNESCO, Eolss Publishers, Oxford, UK; 2011. p. 138.

43 Grismer ME, Tausendschoen M, Shepherd HL. Hydraulic characteristics of a subsurface flow constructed wetland for winery effluent treatment. Water Environ Res. 2001; 73(4): 466-477(12).

44 Chazarenc F, Naylor S, Comeau Y, Merlin G, Brisson J. Modeling the effect of plants and peat on evapotranspiration in constructed wetlands. Int J Chem Eng. 2010; 412734.

45 Castle J, Wasser Z, Rodgers J, Spacil M, Alley B, Horner J, Pardue M. Pilot-scale Constructed Wetlands systems for treating energy-produced waters. In: Water/Energy Sustainability Symposium, Pittsburgh, Pa, September 29; 2010.

46 Alley BL, Willis B, Rodgers Jr J, Castle JW. Water depths and treatment performance of pilot-scale free water surface constructed wetland treatment systems for simulated fresh oilfield produced water. Ecol. Eng. 2013; 61(A):190–199.

47 Pardue MJ, Castle JW, Rodgers Jr JH, Huddleston III, GM. Treatment of oil and grease in produced water by a pilot-scale constructed wetland system using biogeochemical processes. Chemosphere 2014; 103:67–73.

48 Abed RMM, Al-Kharusi S, Prigent S, Headley T. Diversity, distribution and hydrocarbon biodegradation capabilities of microbial communities in oil-contaminated cyanobacterial mats from a constructed wetland. PLoS ONE 2014; 9(12):e114570.

49 Al-Masfry R, van den Hoek P, Verbeek P, Schaapveld M, Baaljens T, van Eijden J, Al-Lamki M, Beizen E. Technology reaches water disposal. July 30, 2007. Available at www.epmag.com/EP-Magazine/archive/Technology-reaches-water-disposal_530.

50 Vymazal J. Emergent plants used in free water surface constructed wetlands: A review. Ecol Eng. 2013; 61(B):582–592.

51 Breuer R, Al Asmi SR. Nimr Water Treatment Project – Up scaling A Reed Bed Trail To Industrial Scale Produced Water Treatment. Soc Petrol Eng. 2009; 126265.

52 Barz M Wichtmann W. Utilisation of Common Reed as an Energy Source. 15th European Biomass Conference and Exhibition, Berlin; 2007.

53 Prigent S, Al-Hadrami A, Headley T, Al-Harrasi W, Stefanakis AI. The reuse of Wetland-treated oilfield produced water for saline irrigation. International Conference of the International Desalination Association (IDA) on Water Reuse and Recycling, Nice, France, September 25–27; 2016.

54 Stefanakis AI, Al-Hadrami A, Prigent S. Reuse of oilfield produced water treated in a Constructed Wetland for saline irrigation under desert climate. In: 7th International Symposium on Wetland Pollutant Dynamics and Control (WETPOL), Montana, USA, August 21–25; 2017.

2

Constructed Wetlands Treating Water Contaminated with Organic Hydrocarbons

Martin Thullner[1], Alexandros I. Stefanakis[2,3,5] and Saeed Dehestani[4]

[1] *Department of Environmental Microbiology, UFZ – Helmholtz Centre for Environmental Research, Leipzig, Germany*
[2] *Bauer Resources GmbH, BAUER-Strasse 1, Schrobenhausen, Germany*
[3] *Department of Engineering, German University of Technology in Oman, Athaibah, Oman*
[4] *Environmental Health Research Center, Kurdistan University of Medical Sciences, Sanandaj, Kurdistan, Iran*
[5] *Bauer Nimr LLC, Muscat, Oman*

2.1 Introduction

Hydrocarbons are one of the most commonly found water contaminants in modern societies due to their extensive use and widespread occurrence. This group of contaminants contains simple organic compounds (comprising only carbon and hydrogen), but there is a large variety of compounds with different chemical and physical properties. They can be classified into three main categories; aromatics, aliphatic and alicyclic. Total petroleum hydrocarbons refer to compounds derived from petroleum sources and processing, such as diesel, petrol, kerosene and lubricating oils. Lighter hydrocarbon compounds (i.e., with less than 16 carbon atoms) generally include substances with higher solubility and volatility, e.g., benzene. Many of these substances are well (e.g., benzene, toluene, ethylbenzene and xylenes) or even highly soluble (e.g., MTBE and alcohols). They can also be toxic and represent an environmental hazard [1].

Hydrocarbon contamination is a common problem that affects groundwater or surface water quality in many regions around the world. These compounds can be found in water as free floating, dissolved, emulsified or even adsorbed to suspended solids. Large hydrocarbon molecule structures tend to be free, while the smaller ones can emulsify with water. Hydrocarbon contamination occurs in areas where industrial activities are located, such as the chemical–petrochemical industry, oil production and refineries, electricity generation plants, manufacturing industry, plastics and steel production and water cooling plants, where environmental release of hydrocarbon products takes place. The worldwide increasing demand for oil and gas increases the pressure on the natural environment. Accidental spills, improper treatment or even illegal disposal introduce these organic compounds to the environment. Other introduction routes of hydrocarbons into the natural environment can be stormwater runoff, spills from roads, fueling deposits (i.e., tank farms, airports etc.) and transportation, among others [2, 3]. Hydrocarbon contamination also occurs near petrol stations, where fossil fuels are used. The extent and migration of hydrocarbon contamination also depends on the interactions with other water pollutants.

Microorganisms that are naturally present in water and soils are capable of degrading these substances. Various reactions and transformation processes contribute to hydrocarbon decomposition. However, the natural ability to degrade these pollutants cannot deal with the increased hydrocarbon loads in water caused by human activities. Besides the risks to human health, hydrocarbons can affect the respiration and reproduction of fishes, destroy algae and plankton and generally adversely affect aquatic ecosystems.

Among the various hydrocarbon compounds, fuel hydrocarbons such as BTEX compounds (benzene, toluene, ethylbenzene, and xylenes), MTBE (methyl-tert-butyl-ether) and phenolic compounds are commonly found in water [4, 5]. MTBE and BTEX compounds are soluble and mobile in groundwater, while both are considered as toxic compounds with negative implications for humans (for example, BTEX is a known human carcinogen). Therefore, both pollutants have regulated concentration limits in drinking water (200 and 5 μg/L, respectively [6, 7]). Phenols are also found in many waters such as effluents from oil refineries and the petrochemical industry, as well in other wastewaters, e.g., tanneries [8], olive mills [9], and pulp and paper mills [10], while the usage of pesticides and disinfectants also contributes to phenol contamination [11]. The toxic activity of phenols also resulted in its regulation in water (0.001 μg/L; [12])

Due to the importance and related risks of these hydrocarbon compounds, various physical and chemical techniques and technologies have been developed to remediate hydrocarbon contaminated water, e.g., membrane separation, adsorption onto porous media (e.g., activated carbon, zeolites), advanced oxidation processes (e.g., H_2O_2/O_3, H_2O_2/UV, Fenton process), chemical oxidation, *in situ* air stripping and vapor extraction [3, 13–15].

Although most of these techniques and methods can be effective in the removal of these compounds, they may require significant expertise and complex mechanical equipment, they have high investment, operation and maintenance costs, they require high external energy input and also they may be accompanied by operational safety risks, which makes them financially or technically infeasible, especially in small-scale facilities [15]. Therefore, the use of alternative and, especially, sustainable technologies becomes more attractive.

Constructed Wetlands (CWs) are considered an effective eco-tech treatment method. CWs can provide high levels of performance, reduced construction costs and significantly reduced operation and maintenance costs compared to conventional mechanical technologies [1, 5]. They are characterized as ecological treatment systems, i.e., environmentally friendly, with multiple environmental, economic, and social advantages. CWs have been effectively applied for the treatment of domestic and municipal wastewater. But their good treatment capacity shifted the interest towards their use in various industrial wastewater treatment projects. Hence, waters from the petroleum industry contaminated with hydrocarbons appear as another promising application of wetland technology. The aim of this chapter is to summarize the available information and status of technology regarding the use of CWs for the treatment of light hydrocarbons with focus on MTBE, benzene-BTEX and phenolic compounds.

2.1.1 Benzene Removal in Constructed Wetlands

Benzene (a BTEX compound; C_6H_6) is a constituent derived from gasoline production and one of the most frequently detected organic contaminants in groundwater, which poses a risk to human health [16, 17]. It is soluble in water (approx. 1,780 mg/L in 20°C) and, thus, can be extremely mobile in groundwater formations [18]. Benzene is the most toxic and water soluble BTEX compound that can be biologically degraded, particularly in the presence of oxygen [19]; it is biodegraded under oxic [20] and even hypoxic conditions [21]. Under anoxic conditions, benzene is highly recalcitrant [22, 23].

Various physical, chemical and biological processes can facilitate the transformation and removal of benzene from waters. Volatilization, sorption, and dilution practically relocate the contaminant between different phases. Plants are also capable of assimilating benzene in their biomass through their transpiration and transferring and releasing it to the atmosphere. Among all the removal processes, microbial degradation is considered the most important, which allows for the complete degradation of this contaminant.

The biodegradation of petroleum hydrocarbons in the environment is strongly affected by physical and chemical parameters, such as temperature, oxygen availability, salinity, nutrients, pH, pressure, as well as the chemical composition, physical state, and concentration of the contaminant. Biological parameters, such as the composition and adaptability of the microbial population, also play an important role in biodegradation [24].

Constructed Wetlands systems have been proven to be a very promising sustainable technological approach for remediation of water contaminated with benzene. As a nature-based remediation technology, CWs have the capacity to remove organic contaminants, such as benzene, through a combination of physical, chemical and – mainly – biological processes, as a result of the interactions between the plants and the microorganisms developed in their rhizosphere [25–28]. CWs can provide the appropriate conditions and environment for the development of the microbial community, capable of directly biodegrading organics or catalyzing chemical reactions through biological removal processes occurring within the root zone [29]. Bacteria capable of degrading volatile organics such as benzene, toluene, ethylbenzene and p-xylene have been identified in the rhizosphere [30, 31]. Among them, the most abundant are the *Pseudomonas* species, accounting for up to 87% of the gasoline-degrading microorganisms in contaminated aquifers [32]. Moreover, few compound-specific isotope analysis studies implemented in order to determine the biodegradation pathway of benzene in CWs, have shown that the largest fraction of benzene is degraded aerobically [17, 33].

Additionally, plants used in CW systems also have the capacity to uptake compounds such as benzene (phytotransformation) and concentrate them in their biomass (phytoextraction), which also contributes to the removal of benzene from the contaminated water. Plant transpiration releases the previously assimilated compounds to the atmosphere. Plant uptake mainly depends on the compound lipophilicity. The relatively low octanol-water partition coefficient $\log K_{OW}$ value for benzene (2.13) lies within the range (0.5–3) for possible plant uptake [34–36].

Volatilization of the highly volatile benzene can also be important for the removal of this contaminant in CWs. Benzene can be released directly from the soil/substrate surface and through phytovolatilization via the stems and leaves of the plants. Phytovolatilization is a relevant process for substances quickly translocated ($\log K_{OW} < 3.5$), and with a high vapor pressure ($V_p > 1.01$ kPa) [37] or a high Henry's Law constant (dimensionless $K_{AW} \gg 10^{-5}$) [38]. The values (25°C) for benzene ($K_{AW} = 2.22 \times 10^{-1}$; $V_p = 12.66$ kPa) [34] indicate that phytovolatilization could be a mechanism for benzene removal. Experimental investigations have shown that phytovolatilization is a significant emission path for MTBE and benzene together with the direct volatilization via the soil surface in a constructed wetland [39]. It is reported that for a retention time higher than 1 day, volatilization becomes the dominant removal mechanism in vertical flow (VF) CWs [40]. However, Chen et al. [41] estimated the contribution of volatilization in the removal of benzene to be only 1.1% in horizontal subsurface flow (HSF) CWs, while low volatilization rates (<5%) are also reported by Seeger et al. [42]. Similar results (volatilization rate < 1%) are also reported for VFCWs [43].

Table 2.1 presents selected studies on benzene removal using different wetland designs and setups. Almost all CW types have been tested, with the majority being subsurface systems, either with

Table 2.1 Basic information of studies on Constructed Wetland systems treating waters containing benzene and the observed removal efficiencies.

Wetland type and dimensions (L W × D; m)	Plant species	Substrate	Flow rate (m³/day)	HRT (days)	C_{in} (mg/L)	Removal (%)	Reference
HSF (6 × 1 × 0.5)	Phragmites australis	Sand (25%) Gravel 67% lignite (10%)	0.113	6	0.026	93	[33]
VF (H = 0.75, d = 0.1)	Phragmites australis	Stones, gravel, sand		1	1.3	85–95	[40]
HSF (5.9 × 1.1 × 1.2)	Phragmites australis	Gravel	0.528	10	10.2	72–82	[1]
HSF (5 × 1.1 × 0.6)	Phragmites australis	Fine gravel, charcoal, ferric oxides	0.144		20	81–43 (summer-winter)	[28]
Plant root mat (water depth 0.15 m)	Phragmites australis	–	0.166		20	99–18 (summer-winter)	
HSF pilot (7 × 1.7 × 1.1) full scale HSF+FWS (1.9 ha, d = 0.9)	Salix, Phragmites, Schoenoplectus, Juncus, Cornus Typha angustfolia	Gravel, sand	5.4 and 6,000	1	0.17 (benzene) 0.47 (BTEX)	100 (below detection limit)	[44]
HSF (4.8 × 7.2 × 0.6)	Phragmites australis, Typha latifolia	Clay soil, stones, gravel	1	1.5	0.6	57	[45]
HSF (5 × 1.1 × 0.6)	Phragmites australis	Gravel	0.144		up to 24	1,900 mg/d	[46]
HSF (5 × 1.1 × 0.6)	Phragmites australis	Gravel	0.144	6	13	100 (summer)	[41]
Plant root mat (water depth 0.15)	Phragmites australis	–	0.144			100 (summer)	
HSF (7 × 1.7 × 1.1)	Phragmites, Scirpus, Juncus, Cornus	Washed soil, sand, pea, gravel	5.5		0.395	61–81	[47]
VF (0.75, d = 0.1)	Phragmites australis	Stones, gravel, sand		3.5	1	73–89	[48]
SF (1 × 0.6 × 0.8)	Phragmites karka	Gravel, soil	0.144	8	66–45	48	[49]
2 stages VF (2.3 × 1.75 × 1.75)	Salix alba	Granular media, clay, zeolite	up to 1.9		13.9	100	[43]

horizontal or vertical flow. The overall performance proves that wetlands are capable of removing benzene (or BTEX) from water.

A wide range of benzene concentrations was found in the literature for contaminated water treated in CWs. In general, reported efficiency was of a high level, and in some cases almost complete removal of benzene is reported. Given that benzene is mainly removed through aerobic biodegradation, it seems that the VF mode could be a preferable setup, since this system provides better aeration conditions due to the water feeding regime, i.e., water is applied across the bed surface and drains vertically with gravity [5].

Among the various designs of natural treatment systems tested for benzene removal are also aerated ponds and plant root mats [28]. These systems have also been found very effective in benzene removal, even higher than HSF CWs, with reported removal rates reaching almost 100%; i.e., reduction of an influent concentration of 20 mg/L to an effluent concentration of 1 µg/L [23]. It is reported that in plant root mats only 1% of benzene loss could be attributed to volatilization, while benzene loss was removed through aerobic mineralization, i.e., through aerobic microbial degradation.

The removal of benzene seems to be regulated by seasonal variations. Various studies report that higher removal rates are observed with high temperatures. Temperature values higher than 15°C and, especially, during summer period were found to enhance the system performance in HSF CWs [1, 28, 41]. The same was also found in VFCW systems [40, 50]. It is reported that the seasonal benzene removal efficiency in VFCWs was negatively correlated with seasonal variations of the dissolved oxygen in the effluent, whilst positively correlated to the seasonal variations of pH and redox values [50]. Since oxygen is the limiting factor for the removal [28], higher temperatures (i.e., spring and summer months) promote the plant growth, hence enhance the development of the microbial community and the respective biodegradation rate.

The presence of plants seems to play a role in the removal process. Better removal rates in planted than in unplanted systems are generally reported in the literature [1, 28, 41, 51–53], mainly due to the presence of the root zone. For example, 33% higher monochlorobenzene removal rate was obtained in the planted HSF filter (*Phragmites australis*) than in the unplanted control bed [26, 54]. Ranieri et al. [45] report, on average, 5% higher benzene removal rates in the HSF *Phragmites* system than in the *Typha* system, and 23% higher than the unplanted one. Similar conclusions regarding the role of plants are also reported for surface flow (SF) CWs [49].

The positive impact of plants is attributed to the improved oxygen supply (due to root oxygen release into the rhizosphere), which enhances the biodegradation, and the direct plant uptake. Moreover, a strong decrease of the concentration of benzene with increasing height of the plant is also reported [46]. However, in VFCWs the role of plants seems to be of less importance. VFCWs planted with *Phragmites australis* had no significant impact on benzene removal compared to the unplanted beds [40, 50]. In these systems, the vertical drainage with gravity and the relatively short contact time probably affects the extent of the removal processes, something that has been observed in similar systems designed to treat municipal wastewater [55]. Generally, the role of plants in pollutant removal is considered as indirect, i.e., to function as a carbon supplier for microbe metabolism, offer attachment sites for microbes on their extended root system and transfer oxygen through their roots.

To conclude, CW systems with various designs have been effectively applied for the removal of benzene from various contaminated water sources. The dominant removal mechanism is biodegradation. VFCWs appear to be the optimum design due to the aerobic conditions created in this system, although high removal rates have been achieved in HSF CWs too. The available studies provide good indications of the removal capacity, but further investigations are required to formulate solid design standards, as the efficiency depends on the influent load and the seasonal variations.

2.2 MTBE Removal in Constructed Wetlands

Methyl tert-butyl ether (MTBE; $C_5H_{12}O$) is a widely used fuel oxygenate, which was introduced as a replacement of lead additives. MTBE is highly water soluble (solubility of 48 g/L at 25°C; [56]) and volatile (dimensionless Henry's law coefficient of 0.026 at 25°C; [57]), which enables its spreading within the water cycle and the atmosphere. The global production of MTBE reached 21 megatons in 1999 [58], decreased thereafter to 12 megatons in 2011 due to the ban of MTBE in the USA [59], but it is about to rise again due to an increased demand in the Asia-Pacific region.

Main industrial releases of MTBE to the environment are consequently from refineries. Although most of these releases are to the atmosphere, there are also significant amounts discharged into surface water or injected into underground formations and aquifers [13, 60]. In addition, storage tank releases, pipe line leakages and spills at industrial sites, fueling facilities and during transportation are relevant point sources leading to contamination of surface and subsurface water bodies [13], while diffuse sources from urban runoff and atmospheric deposition [57] are less relevant.

MTBE has a relatively high persistence in the environment [3, 13, 56], which combined with its high solubility promotes its spreading within the aqueous environment. As a consequence, since the 1990s MTBE is a frequently detected groundwater contaminant in Europe and North America [18, 60–63]. MTBE is mainly a problem for taste and odor of the water, but also more serious human health concerns exist [13]. It is classified as health threat by the US Environmental Protection Agency (EPA) and drinking water concentration limit is 200 µg/L in the US or in Germany [43]. It is thus necessary to remove MTBE from industrial wastewater to avoid further contamination and to treat MTBE contaminated water appropriately before its potential use as drinking water resource.

Treatment options for water bodies containing MTBE include a variety of abiotic techniques (see [3, 13] for a detailed review of such techniques); adsorption is a commonly used approach to remove MTBE from treated water using mainly carbon but also other sorbents like resins, silica, or diatomite [64–67]. While such sorption approaches do not produce any undesired by-products, they do not lead to a destruction of the contaminant but transfer it only to a different compartment. The same applies to air stripping technologies [64, 66], the efficiency of which is also limited by MTBE being less volatile than other common volatile organic contaminants. More recently, membrane technologies have been suggested to overcome this limitation [59]. Alternatively, advanced oxidation techniques provide the opportunity for a complete mineralization of MTBE. This however, requires expensive reactants, specific reaction conditions and may nevertheless lead to the formation of undesired by-products. Although cost efficiencies of the different abiotic techniques vary with the type and amount of treated water [64, 66], they impose a major limitation to the application of these techniques, especially when large water bodies have to be treated.

Biological techniques provide alternative options for the removal of MTBE from contaminated water. The most important biological MTBE removal mechanism is microbial degradation. The ability of microorganisms to degrade MTBE has been extensively investigated in laboratory studies (see reviews provided, for example, by [3, 56, 68–70]). These studies confirm the potential for microbial degradation of MTBE and its metabolites at aerobic conditions. However, site specific factors may cause strong variations of the in situ occurrence and dynamics of such aerobic MTBE degradation [71, 72] and only few microorganisms are able to use MTBE as its sole carbon source [73–75]. Also in anaerobic conditions microbial MTBE degradation is thermodynamically feasible and degradation has been reported for laboratory studies [56]; degradation rates are however much slower and an accumulation of tert-butyl alcohol (TBA) or other intermediates may occur. The occurrence of such MTBE degradation processes in situ has been investigated with respect to the natural attenuation of groundwater contaminations. Although compared to the available laboratory studies, the number of such field studies is smaller; available studies show that in situ biodegradation (i.e. microbial degradation) is possible under (micro) aerobic [76] and under anaerobic conditions [77]. However, anaerobic conditions are often not sufficiently supporting for high degradation rates, and additional oxygen supply is needed to speed-up the degradation process [78–80]. Another biological treatment option is the phytoremediation of MTBE by plants. A number of studies show that plants such as poplars [81, 82], alfalfa [83], weeping willows [84] or conifers [85] are able to remove MTBE from groundwater via their root system. However, in these studies and in [86] no evidence was found that MTBE is degraded within the plants, but the MTBE is transpired to the atmosphere, leading to potentially undesired VOC emissions there. In turn, plant toxicity effects are apparently not an issue [84].

The fact that aerobic biodegradation of MTBE is a feasible removal technique suggests that constructed wetlands promoting aerobic conditions are an efficient treatment option for MTBE contaminated water, as long as the wetland plants do not promote too many atmospheric emissions. However, only a small number of studies are available, which address the fate of MTBE in constructed wetlands.

In the attempt to remove petroleum hydrocarbons from refinery wastewater (here contaminated groundwater recovered from a former refinery), Ferro et al. [87] and Bedessem et al. [88] tested a pilot scale CW system consisting of four parallel, subsurface flow treatment units operated in an upward flow mode with a mean hydraulic residence time (HRT) of 1 day. The wetland units (surface area approx. 10 m^2, depth approx. 1 m) were filled with sand, planted either with a mixture of willows, reed bulrush, rush and dogwood or with roots and shoots contained in sods from a mature wetland, allowing for a forced subsurface aeration. The applied water contained (after pre-treatment) typically 1–1.5 mg/L MTBE with average total concentration of other hydrocarbons (including BTEX) of 45 mg/L. Results of the study showed that average MTBE removal ranged depending on the operation mode between 15 and 30% (which was lower than for the other monitored hydrocarbon species). The higher removal percentages were found for the aerated systems and the additional layer of mature wetland sod also had a positive influence on removal performance.

MTBE removal of 40% was also reported for wetlands of size 1 ha, used for the treatment of urban wastewater [89]. However, inflow concentrations of MTBE were very low (0.1 µg/L) and close to the detection limit.

More recently the ability of CWs to remove MTBE from contaminated water was analyzed in detail in a pilot scale facility located at the site of a former refinery where groundwater is heavily contaminated with benzene, MTBE and ammonium. Tested systems included among others aerated ponds [23], a series of HSF CWs and VFCWs, which were treating the contaminated groundwater from that site. Two different types of VFCWs were tested [43]: an unplanted system using expanded clay as main matrix material and a planted system (willows; *Salix alba*) using zeolites as main matrix material. Both systems had a surface area of 4 m^2 and a depth of 1.75 m. Water was applied in hourly pulses of varying magnitude injected 15 or 25 cm underneath the top of the matrix filling, which resulted in a transient unsaturated downward flow. The hydraulic loading rate (HLR), thus, also varied between 60 and 480 L/m^2/d, leading to HRTs of a few hours for the more dynamic unplanted, expanded clay-filled system, while the planted, zeolite-filled system exhibited HRTs of up to two days [27, 90, 91]. Both systems were fed with contaminated groundwater containing an average concentration of 3 mg/L MTBE (as well as 14 mg/L benzene and 51 mg/L ammonium). Results of the study show that MTBE removal in the unplanted system decreased from 97% for the lowest HLR to 75% for the highest HLR. In contrast, the unplanted system allowed for at least 93% removal for all tested loading rates [43]. For benzene, these systems exhibited even higher removal rates (see above).

A detailed analysis of the flow, transport and reactive processes in the two systems in combination with numerical model simulations showed that nearly all of the observed MTBE removal rates can be attributed to biodegradation and that volatile emissions are negligible [27, 90]. The latter required the uppermost layers of the unsaturated system to be an efficient biofilter for vapor phase organic compounds, a hypothesis which was confirmed by laboratory experiments [92]. Based on numerical model results, it was also shown that filter material and depth of the contaminated water injection are crucial factors for the efficiency of such treatment systems, as they have to balance opposing effects: high oxygen supply and high HRT to promote biodegradation and sufficient separation from the open atmosphere and fast drainage to avoid atmospheric emissions [93].

HSF CWs tested within this pilot scale facility had a surface area of approx. 5 m^2 and a depth of approx. 0.5 m filled with fine gravel and planted with common reed (*Phragmites australis*) [28]. The HLR was approx. 30 L/m^2/d and the HRT 6 days. In addition, the performance of a plant root mat with the same surface area was tested. Treated water contained an average concentration of 4 mg/L MTBE (as well as 20 mg/L benzene and 45 mg/L ammonium; note that differences between average concentration for different treatment systems at the same site are caused by differences of the experimental time periods). Removal performance of the wetland systems showed high seasonal variations with highest values obtained during the summer months. For MTBE, the gravel filled wetlands led to a removal of 17% during summer, which was much lower than values observed for benzene and ammonium and only slightly higher than the MTBE removal observed for an unplanted control system (8%) and could not be improved by filter material additives (charcoal, iron(III)) [28]. In turn, the plant root mat allowed for a MTBE removal of 82% during this period of time.

An analysis of the individual removal processes showed that although volatilization fluxes were promoted by the wetland plants [39] these fluxes did not contribute significantly to the observed removal in the gravel filled wetlands, which was mainly attributed to biodegradation and accumulation in the plants [42]. For the plant root mat, volatilization was responsible for a major part of the observed removal due to the direct contact of aqueous and gaseous phase. Subsequent studies at the same site [41, 94] generally confirmed these results and showed that apparently due to an adaptation of the microbial community also the planted gravel filter allowed for higher MTBE removal (up to 33% [94] and up to 93% [41]) after operation periods of 3–4 years. Using the same contaminated groundwater for a laboratory-scale CW experiment, it was shown that MTBE removal (and the removal of benzene and ammonium) could be enhanced by combining the CW with microbial electrochemical technology [95].

In conclusion the available studies show that MTBE removal in wetlands is possible. Biodegradation is the most relevant removal processes but degradation rates are lower than for other common petroleum hydrocarbons. VFCW systems appear more efficient for MTBE removal, but up to now the limited number of available field studies does not allow for a full assessment of the fate of MTBE in different types of wetland.

2.3 Phenol Removal in Constructed Wetlands

Phenol, also known as carbolic acid, monohydroxybenzene or phenylalcohol, is an aromatic organic compound with molecular type C_6H_5OH. Phenol is a white crystalline solid with a limited solubility at room temperature (water solubility at 15°C is 82 g/L, $logK_{ow}$ 1.46). However, it is soluble in most organic solvents [96]. It is one of the derivatives of benzene, in which a phenyl group ($-C_6H_5$) is bonded to a hydroxyl group (-OH). In turn, phenol has different derivatives; its methyl (cresols) and dimethyl derivates (xylenols). Chlorinated phenols are considered as priority pollutants and have been classified as hazardous pollutants because of their harmful potential to human health [97]. Since phenol does not adhere strongly to the soil and because of its relatively high solubility in water and low vapor pressure, after releasing into the soils it is likely to move to groundwater.

The major part of phenol has a natural source; however, it is originated from anthropogenic sources too, such as production and use of phenol and its products, particularly phenolic resins and caprolactam, exhaust gases and residential wood burning. Furthermore, phenols and phenolic compounds are discharged into the environment from industrial effluents including coal tar, gasoline, plastic, oil refinery, coal transformation, paper, rubber proofing, disinfectant, medicinal preparations such as mouthwashes, pharmaceutical and steel industries and domestic wastewater, agricultural run-off and chemical spills [100, 101]. Phenol is also released from synthetic fuel manufacturing, paper pulp mills, wood treatment facilities and olive mills [102–105]. In general, 40–80% of the total COD in industrial wastewaters is attributed to phenolic compounds [106]. Consumption of phenol contaminated water is harmful and causes a serious problem for humans, plants, animals and microorganisms. Its antibiosis effect and phytotoxicity has been reported [98, 99] and a concentration limit value of 1 µg/L for phenol in drinking water has been regulated [96].

Phenol is removed from water through various physical, chemical and biological methods. The most important biological methods include activated sludge, trickling filters, oxidation ponds and lagoons. Chemical methods include chemical oxidation (oxidation using air, chlorine, chlorine dioxide, ozone, and hydrogen peroxide), coagulation, flocculation, photo catalytic and electrolytic oxidation, and advanced oxidation processes. Physical methods include adsorption, reverse osmosis, solvent extraction and ion exchange. Conventional methods are mostly physicochemical processes, but they have various disadvantages such as high costs for chemicals and energy and emission of different hazardous by-products associated with oxidation processes. In adsorption processes, phenol is adsorbed into the adsorbent and is not eliminated from the aqueous media. Moreover, these methods are not cost-efficient, since they use a lot of energy [107–109].

Biological methods using microorganisms are considered the most efficient and attractive technology for treatment of wastewaters containing phenol and phenolic compounds. Phenol can be biodegraded under both aerobic (faster) and anaerobic conditions (slower) with microorganisms utilizing phenolic compounds as an energy and as sole carbon source. A number of phenol-degrading microorganisms (bacteria, fungi, yeast and algae) have been isolated in different environments including aerobic degraders such as *Pseudomonas aeruginosa* [111], *Alcaligenes eutrophus* [112], *Arthrobacter* [113] *Pseudomonas fluorescens* [111] *Acinetobacter* sp. [114] *Pseudomonas putida* [115], and anaerobic degraders such as *Desulfobacterium phenolicum* sp. [116].

Constructed Wetlands have been used as an alternative, sustainable technology for treatment of phenol contaminated waters. The first studies on phenol removal in subsurface flow (SSF) systems were carried out in microcosms planted with bulrushes, and degradation rates of 5–20 $g/m^3/d$ were reported [117]. Abira et al. [10] investigated the performance of a SSF CW for phenol removal from pulp and paper mill wastewater under different HRTs. The results showed decreasing removal efficiency with increasing HRT. This could be attributed to oxygen and nutrient deficiencies resulted from longer retention time. Herouvim et al. [9] studied the phenol removal from olive mill wastewater in pilot-scale VFCWs. The influent concentrations of phenol and COD were 2,841 and 14,120 mg/L, respectively. The phenol surface load in VFCWs varied from 17 to 997 g/m^2/d. The obtained results indicated that phenol and COD removal rates follow the same trend. Moreover, it was shown that temperature and surface loading have the same effect on the removal of COD and phenols. They concluded that COD and phenol removal is mainly attributed to microbial biofilm activity.

Several physicochemical and biological processes are involved in phenol removal in CWs. The main mechanisms include biodegradation, plant uptake, sorption, and volatilization [5, 118–122]. The extent of a removal process is affected by temperature variations, organic matter and nutrient content, available electron acceptors, and oxygen availability [5]. Volatilization, phytovolatilization, plant uptake, phytoaccumulation, sorption and sedimentation are non-destructive processes, whereas phytodegradation and microbial degradation are destructive processes [25]. Removal mechanisms of organic contaminants are predominantly affected by physico-chemical properties of organic contaminants. Since the vapor pressure (Pv) and Henry coefficient (H) of phenol at 25°C is 0.2–0.5 [hPa] and 0.03–0.3 [Pa × m^3 × mol/L], respectively, direct volatilization and phytovolatilization are expected to be moderate for phenol [5, 25, 122]. It has been reported that since in SSF CWs the water is kept below the surface of the bed, direct contaminant volatilization is expected to be limited in comparison with free water surface (FWS) CWs, where water is in a direct exposure to the atmosphere [25].

On the other hand, the octanol-water partition coefficient, log K_{ow}, governs the uptake of phenol. Organics that are moderately hydrophobic compound with log K_{ow} ranging from 0.5 to 3 are most likely to be taken up by plants [123]. Therefore, phenol with log K_{ow} ranging 1.5–2 [25] is expected to be taken up by plants. Moreover, plants' resistance to phenol is a key factor for the plant uptake process in CWs. Related studies pointed out that *Phragmites australis* can tolerate phenol toxicity and low concentration of phenol increases the plant growth [124]. Commonly used plants in CWs for treatment of wastewater containing phenolic compounds include *L. multiflorum* [121], *Juncus effuses* [125], *Typha latifolia* [126], *Phragmites mauritianus*, *Cyperus immensus*, *Cyperus papyrus* and *Typha domingensis* [10]. However, it was reported that there was no significant difference in phenol removal efficiency between planted and unplanted units [10].

Polprasert et al. [122] investigated the efficiency and suitable operating conditions of a FWS CW system planted with *Typha* and located in tropics. The influent phenol concentrations varied from 25–700 mg/L. They argued that volatilization contributes 26–37% to phenol removal at a HRT of 5–7 days. The results also indicated that when the influent phenol concentration was 400 mg/L, the system could effectively remove it. Moreover, it has been found that if phenol removal via volatilization were excluded, the extent of phenol removal by the combined biodegradation, plant uptake and adsorption processes was 76–78%. They also concluded that phenol accumulated at the *Typha* roots, followed by the stems and leaves.

Phenol adsorption to the clay soils such as montmorillonite, kaolinite and illite has also been reported (CCME, 1999). The rate of phenol biodegradation decreases due to adsorption. However, sorption to clay surface is reversible (CCME, 1999). The organic carbon partition coefficient (K_{oc}) is a valuable indicator for predicting sorption behavior of organic contaminants (Imfeld et al., 2009). Since relative to other mono-substituted benzene derivatives, K_{ow} and K_{oc} of phenol are low, therefore, sorption to organic matter is expected to be low. Furthermore, the sorption capacity in acidic soil is also decreased [127]. The extent of phenol sorption is affected by the characteristics of the filter media, interaction with metals, pH, and particulate and dissolved organic matter [5]. Additionally, the applied media also affect the performance of CWs for phenol removal [126]. In a related study, where gravel and rice husk-based media were tested for phenol removal in a HSF CW planted with cattails (*Typha latifolia*), it is reported that a husk-based wetland had a better performance than a gravel-based one. This was attributed to more rhizomes created in the rice husk-based unit, which resulted in more micro-aerobic zones for degradation of phenol [126]. Finally, operational and design parameters affect the removal of phenolic compounds. Avila et al. [128] found that reduced conditions diminish the removal efficiency of bisphenol A. They pointed out that batch operation resulted in higher removal compared with a permanently saturated mode. It has been shown that the final effluent of batch operated reactors had a higher redox potential than the effluent of control and anaerobic reactors.

In a wetland ecosystem, biodegradation takes place under both aerobic and anaerobic conditions [129]. Activity of phenol oxidase, which is active under aerobic conditions, is related to phenolic compounds and is highly sensitive to the oxygen availability. However, many phenolic compounds also inhibit the enzymatic activity [129]. It has been found that since phenol oxidase requires oxygen as an electron acceptor, there is a proportional relationship between enzyme activity and concentration of oxygen in riparian wetland [130]. As the water drains, oxygen is introduced into the soil and in this way the aeration is enhanced and the activity of phenol oxidase is increased [129]. It has also been found that detrital layers provide better conditions for phenol oxidase activity and decline with soil depth [129]. In a study where the addition of lime to increase the pH was examined in CWs treating coffee processing wastewater cultivated with ryegrass (*Lolium multiflorum Lam*), results showed that artificial aeration improved the phenol removal efficiency. However, it was found that the role of plants was more important than the addition of oxygen, because of their role in the development of the microbial community around the root zone [121].

Microorganisms in SSF systems are planktonic and sessile. Kurzbaum et al. [118] investigated the efficiency of phenol biodegradation by planktonic *Pseudomonas pseudoalcaligenes* isolated from CWs vs. root and gravel biofilm. They chose *Zea mays* for the experiments. Initial concentration of phenol varied from 10 to 525 mg/L. Complete phenol degradation was reached up to 80 mg/L. However, bacterial inhibition was observed at higher phenol concentration due to the inhibitory effect of the intermediate compound catechol on *P. pseudoalcaligenes*. The obtained results indicated that performance of biofilm and planktonic cells of *P. pseudoalcaligenes* in phenol removal are different from each other. Low-molecular-weight phenol is believed to be removed rapidly by planktonic bacteria, whereas diffusion and mass transfer limit its availability [118]. However it is mentioned that biofilm microorganisms play a more important role in CW systems, due to enhancement of bacterial biomass [118]. They also concluded that the efficiency of roots and gravel biofilm was almost equal.

Kurzbaum et al. [119] investigated the relative contribution of plant roots, microbial activity and porous media for phenol removal in a SSF CW planted with *Phragmites australis*. They found that root biofilm significantly contributed to phenol removal in CW system, whereas the contribution of abiotic mechanisms was found to be negligible [119]. One study by Toyama et al. [131] found that organic chemical degradation by aquatic-bacterial associations was accelerated. The study was aimed to investigate the contribution of *Spirodela polyrrhiza* and selective bacteria in its rhizosphere to degradation of three aromatic compounds including phenol, aniline and 2,4-dichlorophenol. The result clarified that in the case of phenol, indigenous bacteria in the rhizosphere zone significantly contribute to phenol degradation. In aniline degradation, bacteria both in the rhizosphere and bulk water played an important role. However, 2,4-dichlorophenol was removed mainly via plant uptake and degradation. They concluded that removal mechanisms are affected by substrates.

It was also found that microbial density and activity depends on the macrophyte species. A higher value was reported for *Phalaris* compared to *Typha* and *Phragmites*. This can be attributed to differences in root oxygen release with different plant species [132]. Kurzbaum et al. [133] presented a comparison of plant root biofilm, gravel attached biofilm and planktonic microbial population in phenol removal within CWs. They found that the highest phenol biodegradation took place for the gravel-attached biofilm followed by root-attached biofilm and planktonic population. The authors concluded that since the gravel bed created a higher surface area, a higher number of specific degrading bacteria presented at the gravel-attached biofilm revealing higher performance in phenol removal. Similar phenol removal kinetics was reported for a root-attached population alone and the planktonic bacterial population alone. Since intermediates resulted from phenol degradation such as catechol and hydroquinone may hamper the phenol biodegradation in the CW, application of other technologies such as photocatalytic methods are also suggested as a pretreatment before introducing industrial wastewater into the CWs. Herrera et al. [134] combined TiO_2-photocatalysis with CWs. They concluded that this pretreatment can be applied as an efficient method to remove phenolic intermediates. Obtained results indicated that after 6–8 h, the concentration of phenol can reach appropriate values (below 50 ppm) for the wetland influent.

The fate of phenolic compounds (phenol and m-cresol) was also tested in HSF CW systems treating contaminated groundwater [1]. Phenols were completely removed, the pilot beds receiving a load of 314.5 and 45.5 $mg/m^2/d$ for phenol and m-cresol, respectively. The planted beds performed better than the unplanted one, which indicated the positive role of the plants, i.e., promoting microbial community development that degrades the phenolic compounds.

Based on the physico-chemical properties of phenol, the presence of a different variety of indigenous microorganism in CWs, where different environmental conditions are expected, and regarding the different metabolic pathways for phenol, wetland technology can be applied for the treatment of industrial wastewater containing phenol. Various processes are involved in phenol removal from wetland systems due to the complexity of these engineered systems. Although the mechanisms of phenol removal in CWs are more or less known, further studies are necessary to understand better the predominant and minor mechanisms. This, in turn, will result in better design, operation and higher efficiency of phenol removal in CWs.

2.4 Combined Treatment of Different Compounds

It is very common that some or even all of the previously investigated hydrocarbon compounds (i.e., benzene – BTEX, MTBE, phenols) are simultaneously present in the contaminated water source. MTBE in groundwater is often found together with other gasoline contaminants, usually BTEX [135]. In this case, the water composition becomes more complex and appropriate treatment design should be carefully selected, since there are many interactions/interferences that may appear between the various contaminants.

Deeb et al. [136] found that the presence of ethylbenzene or xylenes (BTEX compounds) in mixtures with MTBE completely inhibited MTBE degradation, while benzene and toluene partially inhibited MTBE degradation. MTBE degradation did not increase to higher rates until benzene and toluene were almost entirely degraded. This study suggested that BTEX and MTBE degradation takes places via two independent and inducible pathways [136]. Wang and Deshusses [137] also reported that a single BTEX compound or BTEX mixtures inhibited MTBE degradation in a biotrickling filter, but not completely.

Raynal and Pruden [69] reported that the microbial community composition regulates the simultaneous removal of MTBE and BTEX. They found complete inhibition of MTBE degradation by BTEX in batch reactors inoculated with distinct enrichment cultures of both MTBE and BTEX. On the other hand, in the semi-batch reactor, the MTBE biodegradation rate was almost three times higher in the presence of BTEX as in the batch reactor, but slower than MTBE biodegradation in the absence of BTEX. Thus, the authors suggested that MTBE bioremediation in the presence of BTEX is feasible and depends on the culture composition and the reactor configuration. According to Sedran et al. [138], the presence of BTEX did not have a significant effect on MTBE degradation in batch conditions, but slightly affected TBA (Tertiary butyl alcohol; a MTBE intermediate) degradation. Also, under continuous flow conditions, all compounds degraded simultaneously. Wang and Deshusses [137] also suggested that simultaneous biodegradation of MTBE, BTEX and TBA is feasible if an appropriate bacterial mixture is used.

Similar results have also been found in CW systems treating water contaminated with both MTBE and BTEX/benzene. The vast majority of information regarding this issue comes from the experimental facility in Leuna, Germany (SAFIRA-project; remediation research in regionally contaminated aquifers). This area is one of the oldest and largest industrial facilities in Germany with chemical, petrochemical and manufacturing industries. Many studies have been published from this experimental facility, while most of the studies conclude that the presence of BTEX compounds inhibits to a varying degree the biodegradation of MTBE. Low MTBE removal (17%) compared to benzene (81%) was found in a gravel HSF CW system [28], which could be a result of hindered microbial attack due to the compound structure [56], of overall low growth yields of MTBE utilizing microorganisms [78], and inhibition of MTBE degradation by BTEX compounds [136]. It is assumed that after benzene reduction to values below 3.54 mg/L, MTBE degradation was no longer inhibited by benzene, and thus, very low outflow MTBE concentrations could be achieved [28]. Smaller differences are reported for a HSF CW system; 24–100% and 16–93% for benzene and MTBE, respectively [41]. Higher mass removal was also found in a plant root mat for benzene (98%; 544 mg/m^2/d) than MTBE (78%; 54 mg/m^2/d) [94]. Reiche et al. [39] reported benzene and MTBE volatilization fluxes below 10% of the total mass removal in the planted HSF CWs with gravel matrix at the Leuna site, whereas for an aerated trench system the mass loss due to emission amounted to 1% for benzene and 53% for MTBE [23]. Moreover, complete removal of both compounds was achieved in a 2-stage VFCW system in Leuna [43]. The combination of a first-stage (unplanted) vertical roughing filter with a second-stage vertical polishing planted filter resulted in MTBE and benzene effluent concentrations of 5 ± 10 and 0.6 ± 0.2 µg/L, respectively, for a respective inflow of $2,970 \pm 816$ and $13,966 \pm 1,998$ µg/L.

Other studies on petroleum-contaminated groundwater treatment in CWs [87, 88] and aerated trench systems [23] showed similar results and reported much smaller MTBE removal rates compared to benzene removal. On the other hand, Keefe et al. [89] found the removal of benzene and MTBE in CWs were more or less similar (between 30–40% for both contaminants), but this could be attributed to the low influent concentrations, which were close to the detection limits.

Finally, Stefanakis et al. [1] investigated the fate of MTBE, benzene and phenolic compounds (phenol and m-cresol) simultaneously present in the contaminated groundwater in HSF CW systems. Results showed a complete removal of the two phenolic compounds (influent concentrations of 15 and 2 mg/L or 314.5 and 45.5 mg/m^2/d for phenol and m-cresol, respectively) without any alteration in the MTBE and benzene removal rates (20.2 and 334.6 mg/m^2/d, respectively). This is the first study showing that CWs can be used to effectively remove different hydrocarbons simultaneously. The planted bed presented higher contaminant removal rates, which confirmed the positive role of plants. Moreover, the major portion of the removal took place in the first part of wetland length, which indicates that the HSF CW system could potentially receive higher influent loads.

References

1 Stefanakis AI, Seeger E, Dorer C, Sinke A, Thullner, M. Performance of pilot-scale horizontal subsurface flow constructed wetlands treating groundwater contaminated with phenols and petroleum derivatives. Ecol Eng. 2016; 95:514–526.

2 Langwaldt JH, Puhakka JA. On-site biological remediation of contaminated groundwater: a review. Environ Pollut. 2000; 107:187–197.

3 Levchuk I, Bhatnagar A, Sillanpää M. Overview of technologies for removal of methyl tert-butyl ether (MTBE) from water. Sci Total Environ. 2014; 476–477:415–433.

4 Wu Y, Lerner DN, Banwart SA, Thornton SF, Pickup RW. Persistence of fermentative process to phenolic toxicity in groundwater. J Environ Qual. 2006; 35:2021–2025.

5 Stefanakis AI, Akratos CS, Tsihrintzis VA. Vertical flow constructed wetlands: Eco-engineering systems for wastewater and sludge treatment, 1st edn. Amsterdam, The Netherlands: Elsevier; 2014.

6 USEPA. List of Drinking Water Contaminants and MCLs; 2009. Available from www.epa.gov/ground-water-and-drinking-water/table-regulated-drinking-water-contaminants.

7 USEPA. Integrated Risk Information System (IRIS); 2015. Available from www.epa.gov/iris.

8 Costa CR, Botta CMR, Espindola ELG, Olivi P. Electrochemical treatment of tannery wastewater using DSA® electrodes. J Hazard Mater. 2008; 153(1–2):616–627.

9 Herouvim E, Akratos CS, Tekerlekopoulou A, Vayenas DV. Treatment of olive mill wastewater in pilot-scale vertical flow constructed wetland. Ecol. Eng. 2011; 37:931–939.

10 Abira MA, Van Bruggen JJA, Denny P. Potential of a tropical subsurface constructed wetland to remove phenol from pre-treated pulp and papermill wastewater. Water Sci Technol. 2005; 51(9):173–176.

11 Stottmeister U, Kuschk P, Wiessner A. Full-scale bioremediation and long-term monitoring of a phenolic wastewater disposal lake. Pure Appl Chem. 2010; 82:161–173.

12 WHO. Guidelines for Drinking-Water Quality. 4th edn. Switzerland: WHO; 2011.

13 Deeb RA, Chu K-H, Shih T, Linder S, Suffet I, Kavanaugh MC, et al. MTBE and other oxygenates: environmental sources, analysis, occurrence, and treatment. Environ Eng Sci. 2003; 20(5):433–447.

14 Hodaifa G, Ochando-Pulido JM, Rodriguez-Vives S, Nartinez-Ferez A. Optimization of continuous reactor at pilot scale for olive-oil mill wastewater treatment by Fenton-like process. Chem Eng J. 2013; 220:117–124.

15 Stefanakis AI, Thullner M. Fate of phenolic compounds in Constructed Wetlands treating contaminated water. In: Ansari AA et al., eds. Phytoremediation. Switzerland: Springer International Publishing; 2016, pp. 311–325.

16 Galbraith D, Gross SA, Paustenbach D. Benzene and human health: a historical review and appraisal of associations with various diseases. Crit Rev Toxicol. 2010; 40(S2):1–46.

17 Rakoczy J, Remy B, Vogy C, Richnow HH. A bench-scale Constructed Wetland as a model to characterize benzene biodegradation processes in freshwater wetlands. Environ Sci Technol. 2011; 45:10036–10044.

18 Squillace PJ, Zogorski JS, Wilber WG, Price CV. Preliminary assessment of the occurrence and possible sources of MTBE in groundwater in the United States, 1993-1994. Environ Sci Technol. 1996; 30(5):1721–1730.

19 Alexander M, Biodegradation and Bioremediation, 2nd edn. New York: Academic Press; 1999.

20 Agteren MHV, Keuning S, Janssen DB. Handbook on biodegradation and biological treatment of hazardous organic compounds. The Netherlands: Kluwer Academic Publishers; 1998.

21 Yerushalmi L, Lascourreges JF, Guiot SR. Kinetics of benzene biotransformation under microaerophilic and oxygen-limited conditions. Biotechnol Bioeng. 2202; 79(3):347–355.

22 Foght J. Anaerobic biodegradation of aromatic hydrocarbons: pathways and prospects. J Mol Microbiol Biotechnol. 2008; 15(2-3):93–120.

23 Jechalke S, Vogt C, Reich N, Franchini AG, Borsdorf H, Neu TR, Richnow HH. Aerated treatment pond technology with biofilm promoting mats for the bioremediation of benzene, MTBE and ammonium contaminated groundwater. Water Res. 2010; 44:1785–1796.

24 Zhou E, Crawford RL. Effects of oxygen, nitrogen, and temperature on gasoline biodegradation in soil. Biodegradation 1995; 6:127–140.

25 Imfeld G, Braeckevelt M, Kuschk P, Richnow HH. Monitoring and assessing processes of organic chemicals removal in constructed wetlands. Chemosphere 2009; 74:349–62.

26 Braeckevelt M, Reiche N, Trapp S, Wiessner A, Paschke H, Kuschk P, Kaestner M. Chlorobenzene removal efficiencies and removal processes in a pilot-scale constructed wetland treating contaminated groundwater. Ecol Eng. 2011; 37:903–913.

27 De Biase C, Reger D, Schmidt A, Jechalke S, Reiche N, Martinez-Lavanchy PM, Rosell M, Van Afferden M, Maier U, Oswald SE, Thullner M. Treatment of volatile organic contaminants in a vertical flow filter: Relevance of different removal processes. Ecol Eng. 2011; 37:1292–1303.

28 Seeger E, Kuschk P, Fazekas H, Grathwohl P, Kaestner M. Bioremediation of benzene-, MTBE- and ammonia-contaminated groundwater with pilot-scale constructed wetlands. Environ Pollut. 2011; 159:3769–3776.

29 Stottmeister U, Wießner A, Kuschk P, Kappelmeyer U, Kästner M, Bederski O, Müller RA, Moormann H. Effects of plants and microorganisms in Constructed wetlands for wastewater treatment. Biotechnol. Adv. 2003; 22:93–117.

30 Sugai SF, Lindstrom JE and Braddock JF, Environmental influences on the microbial degradation of Exxon Valdez oil on the shorelines of Prince William Sound, Alaska. Env Sci Technol. 1997; 31:1564–1572.

31 Pardue JH, Kassenga G, Shin WS. Design approaches for chlorinated VOC treatment wetland. In: Means JL, Hinchee RE, eds. Ohio: United States; Battelle Press: Columbus, 2000, pp. 301–308.

32 Ridgeway HF, Safarik J, Phipps D, Carl P and Clark D, Identification and catabolic activity of well-derived gasoline degrading bacteria and a contaminated aquifer. Appl Environ Microbiol 1990; 56:3565–3575.

33 Braeckevelt M, Rokadia H, Imfeld G, Stelzer N, Paschke H, Kuschk P, Kaestner M, Richnow HH, Weber S. Assessment of in situ biodegradation of monochlorobenzene in contaminated groundwater treated in a constructed wetland. Environ Pollut. 2007; 148:428–437.

34 Briggs GG, Bromilow RH, Evans AA. Relationships between lipophilicity and root uptake and translocation of non-ionized chemicals by barley. Pestic Sci. 1982; 13:495–504.

35 Schirmer M, Butler BJ, Barker JF, Church CD, Schirmer K. Evaluation of biodegradation and dispersion as natural attenuation processes of MTBE and benzene at the Borden field site. Phys Chem Earth PT B: Hydrology, Oceans and Atmosphere. 1999; 24(6): 557–560.

36 Pilon-Smits E. Phytoremediation. Annu Rev Plant Biol. 2005; 56:15–39.

37 Burken JG, Schnoor JL. Distribution and volatilization of organic compounds following uptake by hybrid poplar trees. Int J Phytoremediat. 1999; 1:139–151.

38 Trapp S, Karlson U. Aspects of phytoremediation of organic pollutants. J Soils Sediments. 2001; 1:37–43.

39 Reiche N, Lorenz W, Borsdorf H. Development and application of dynamic air chambers for measurement of volatilization fluxes of benzene and MTBE from constructed wetlands planted with common reed. Chemosphere. 2010; 79(2):162–168.

40 Eke PE, Scholz M. Benzene removal with vertical-flow constructed treatment wetlands. J Chem Technol Biotechnol. 2008; 83:55–63.

41 Chen Z, Kuschk P, Reiche N, Borsdorf H, Kaestner M, Koeser H. Comparative evaluation of pilot scale horizontal subsurface-flow constructed wetlands and plant root mats for treating groundwater contaminated with benzene and MTBE. J Hazard Mater. 2012; 209–210:510–515.

42 Seeger E, Reiche N, Kuschk P, Borsdorf H, Kaestner M. Performance evaluation using a three compartment mass balance for the removal of volatile organic compounds in pilot scale Constructed Wetlands. Environ Sci Technol. 2011; 45:8467–8474.

43 Van Afferden M, Rahman KZ, Mosig P, De Biase C, Thullner M, Oswald SE, Mueller RA. Remediation of groundwater contaminated with MTBE and benzene: The potential of vertical-flow soil filter systems. Water Res. 2011; 45:5063–5074.

44 Wallace S, Kadlec R. BTEX degradation in a cold-climate wetland system. Water Sci Technol. 2005; 51(9):165–171.

45 Ranieri E, Gikas P, Tchobanoglous G. BTEX removal in pilot-scale horizontal subsurface flow constructed wetlands. Desalin Water Treat. 2013; 51(13–15):3032–3039.

46 Mothes F, Reiche N, Fiedler P, Moeder M, Borsdorf H. Capability of headspace based sample preparation methods for the determination of methyl tert-butyl ether and benzene in reed (*Phragmites australis*) from constructed wetlands. Chemosphere 2010; 80:396–403.

47 Bedessem ME, Ferro AM, Hiegel T. Pilot-scale Constructed Wetlands for petroleum-contaminated groundwater. Water Environ Res. 2007; 79(6):581-586.

48 Tang X, Scholz M, Emeka P, Huang S. Nutrient removal as a function of benzene supply within vertical-flow constructed wetlands. Environ Technol. 2010; 31(6):681–691.

49 Ballesteros Jr F, Vuong TH, Secondes MF, Tuan PD. Removal efficiencies of constructed wetland and efficacy of plant on treating benzene. Sustain Environ Res. 2016; 26:93–96.

50 Tang X, Eke P, Scholz M, Huang S. Processes impacting on benzene removal in vertical-flow constructed wetlands. Bioresour Technol. 2009; 100:227–234.

51 Haberl R, Grego S, Langergraber G, Kadlec RH, Cicalini AR, Dias SM, Novais JM, Aubert S, Gerth A, Thomas H, Hebner A. Constructed wetlands for treatment of organic pollutants. J. Soils Sediments 2003; 3(2):109–124.

52 Gerhardt KE, Huang XD, Glick BR, Greenberg BM. Phytoremediation and rhizoremediation of organic soil contaminants: potential and challenges. Plant Sci. 2009; 176:20–30.

53 Ranieri E, Gorgoglione A, Montanaro C, Iacovelli A, Gikas P. Removal capacity of BTEX and metals of constructed wetlands under the influence of hydraulic conductivity. Desalin Water Treat. 2014; Doi: 10.1080/19443994.2014.951963.

54 Braeckevelt M, Mischel G, Wiessner A, Rueckert M, Reiche N, Vogt C, Schultz A, Paschke H, Kuschk P, Kaestner M. Treatment of chlorobenzene-contaminated groundwater in a pilot-scale constructed wetland. Ecol. Eng. 2008; 33(1):45–53.

55 Stefanakis AI, Tsihrintzis VA. Effects of loading, resting period, temperature, porous media, vegetation and aeration on performance of pilot-scale Vertical Flow Constructed Wetlands. Chem Eng. 2012; 181-182:416–430.

56 Schmidt TC, Schirmer M, Weiss H, Haderlein SB. Microbial degradation of methyl tert-butyl ether and tert-butyl alcohol in the subsurface. J Contam Hydrol. 2004; 70(3–4):173–203.

57 Baehr AL, Stackelberg PE, Baker RJ. Evaluation of the atmosphere as a source of volatile organic compounds in shallow groundwater. Water Resour Res. 1999; 35(1):127–36.

58 Krayer von Kraus M, Harremoes P. MTBE in petrol as a substitute for lead. In: Hearremoes P, ed. Late Lessons from Early Warnings: The Precautionary Principle 1896–2000 Environmental Issue Report. 22. Copenhagen, Denmark: Office for Official Publications of the European Communities; 2001.

59 Kujawa J, Cerneaux S, Kujawski W. Removal of hazardous volatile organic compounds from water by vacuum pervaporation with hydrophobic ceramic membranes. J Membr Sci. 2015; 474:11–19.

60 Johnson R, Pankow J, Bender D, Price C, Zogorski J. MTBE – To what extent will past releases contaminate community water supply wells? Environ Sci Technol. 2000; 34(9):210A–217A.

61 Schmidt TC, Morgenroth E, Schirmer M, Effenberger M, Haderlein SB. Use and occurrence of fuel oxygenates in Europe. In: Diaz AF, Dorgos DL, eds. Oxygenates in Gasoline: Environmental Aspects. Washington, DC, USA: ACS; 2002, pp. 58–79.

62 Morgenstern P, Versteegh AFM, de Korte GAL, Hoogerbrugge R, Mooibroek D, Bannink A, et al. Survey of the occurrence of residues of methyl tertiary butyl ether (MTBE) in Dutch drinking water sources and drinking water. J Environ Monitor. 2003; 5(6):885.

63 Klinger J, Stieler C, Sacher F, Brauch H-J. MTBE (methyl tertiary-butyl ether) in groundwaters: Monitoring results from Germany. J Environ Monitor. 2002; 4(2):276–279.

64 Wilhelm MJ, Adams D, Curtis JG, Middlebrooks EJ. Carbon adsorption and air-Stripping removal of MTBE from river water. J Env Eng-Asce. 2002; 128(9):913–823.

65 Davis SW, Powers SE. Alternate sorbents for removing MTBE from gasoline-contaminated ground water. J Env Eng-Asce. 2000; 126(4):354–360.

66 Sutherland J, Adams C, Kekobad J. Treatment of MTBE by air stripping, carbon adsorption, and advanced oxidation: technical and economic comparison for five groundwaters. Water Res. 2004; 38(1):193–205.

67 Aivalioti M, Vamvasakis I, Gidarakos E. BTEX and MTBE adsorption onto raw and thermally modified diatomite. J Hazard Mater. 2010; 178(1–3):136–143.

68 Deeb RA, Scow KM, Alvarez-Cohen L. Aerobic MTBE biodegradation: an examination of past studies, current challenges and future research directions. Biodegradation. 2000; 11(2-3):171–186.

69 Raynal M, Pruden A. Aerobic MTBE biodegradation in the presence of BTEX by two consortia under batch and semi-batch conditions. Biodegradation. 2008; 19(2):269–282.

70 Stocking AJ, Deeb RA, Flores AE, Stringfellow W, Talley J, Brownell R, et al. Bioremediation of MTBE: a review from a practical perspective. Biodegradation. 2000; 11:187–201.

71 Moreels D, Bastiaens L, Ollevier F, Merckx R, Diels L, Springael D. Evaluation of the intrinsic methyl tert-butyl ether (MTBE) biodegradation potential of hydrocarbon contaminated subsurface soils in batch microcosm systems. FEMS Microbiol Ecol. 2004; 49(1):121–128.

72 Schirmer M, Butler BJ, Church CD, Barker J, Nadarajah N. Laboratory evidence of MTBE biodegradation in Borden aquifer material. J Contam Hydrol. 2003; 60:229–249.

73 Müller RH, Rohwerder T, Harms H. Carbon conversion efficiency and limits of productive bacterial degradation of methyl tert-butyl ether and related compounds. Appl Environ Microbiol. 2007; 73(6):1783–1791.

74 Rohwerder T, Muller RH, Weichler MT, Schuster J, Hubschmann T, Muller S, et al. Cultivation of *Aquincola tertiaricarbonis* L108 on the fuel oxygenate intermediate tert-butyl alcohol induces aerobic anoxygenic photosynthesis at extremely low feeding rates. Microbiology. 2013; 159(Pt 10):2180–2190.

75 Ferreira NL, Malandain C, Fayolle-Guichard F. Enzymes and genes involved in the aerobic biodegradation of methyl tert-butyl ether (MTBE). Appl Microbiol Biotechnol. 2006; 72(2):252–262.

76 Martienssen M, Fabritius H, Kukla S, Balcke GU, Hasselwander E, Schirmer M. Determination of naturally occurring MTBE biodegradation by analysing metabolites and biodegradation by-products. J Contam Hydrol. 2006; 87(1–2):37–53.

77 Wilson JT, Adair C, Kaiser P, Kolhatkar R. Anaerobic Biodegradation of MTBE at a gasoline spill site. Ground Water Monit Remed. 2005; 25(3):103–115.

78 Salanitro JP, Johnson PC, Spinnler GE, Maner PM, Wisniewski HL, Bruce C. Field-scale demonstration of enhanced MTBE bioremediation through aquifer bioaugmentation and oxygenation. Environ Sci Technol. 2000; 34(19):4152–4162.

79 Wilson RD, Mackay DM, Scow KM. In Situ MTBE biodegradation supported by diffusive oxygen release. Environ Sci Technol. 2002; 36(2):190–199.

80 Smith AE, Hristova K, Wood I, Mackay DM, Lory E, Lorenzana D, et al. Comparison of biostimulation versus bioaugmentation with bacterial strain PM1 for treatment of groundwater contaminated with Methyl Tertiary Butyl Ether (MTBE). Environ Health Persp. 2005; 113(3):317–322.

81 Hong MS, Fermayan WF, Dortch IJ, Chiang CY, McMillan SK, Schnoor JL. Phytoremediation of MTBE from a groundwater plume. Environ Sci Technol. 2001;35:1231–1239.

82 Rubin E, Ramaswami A. The potential for phytoremediation of MTBE. Water Res. 2001; 35(5):1348–1353.

83 Zhang Q, Davis LC, Erickson LE. Transport of methyl tert-butyl ether through alfalfa plants. Environ Sci Technol. 2001; 35(4):725–731.

84 Yu XZ, Gu JD. Uptake, metabolism, and toxicity of methyl tert-butyl ether (MTBE) in weeping willows. J Hazard Mater. 2006; 137(3):1417–1423.

85 Arnold CW, Parfitt DG, Kaltreider M. Field note phytovolatilization of oxygenated gasoline-impacted groundwater at an underground storage tank site via conifers. Int J Phytoremediation. 2007; 9(1):53–69.

86 Trapp S, Yu X, Mosbaek H. Persistence of methyl tertiary butyl ether (MTBE) against metabolism by Danish vegetation. Environ Sci Pollut Res. 2006; 10(6):357–360.

87 Ferro AM, Kadlec RH, Deschamp J. Constructed Wetland system to treat wastewater at the BP Amoco former Casper refinery: pilot scale project. The 9th International Petroleum Environmental Conference; Albuquerque, NM, USA; 2002.

88 Bedessem ME, Ferro AM, Hiegel T. Pilot-Scale Constructed Wetlands for petroleum-contaminated groundwater. Water Environ Res. 2007; 79(6):581–586.

89 Keefe SH, Barber LB, Runkel RL, Ryan JN, McKnight DM, Wass RD. Conservative and reactive solute transport in constructed wetlands. Water Resour Res. 2004; 40(1): DOI: 10.1029/2003WR002130.

90 De Biase C, Carminati A, Oswald SE, Thullner M. Numerical modeling analysis of VOC removal processes in different aerobic vertical flow systems for groundwater remediation. J Contam Hydrol. 2013;154:53–69.

91 Reger D. Analysis of contaminant degradation in vertical soil filter systems. Freiberg, Germany: Technical University Bergakademie Freiberg; 2009.

92 Khan AM, Wick LY, Harms H, Thullner M. Biodegradation of vapor-phase toluene in unsaturated porous media: column experiments. Environ Pollut. 2016; 211:325–331.

93 De Biase C, Maier U, Baeder-Bederski O, Bayer P, Oswald SE, Thullner M. Removal of volatile organic compounds in Vertical Flow Filters: predictions from reactive transport modeling. Ground Water Monit Remed. 2012; 32(2):106–121.

94 Seeger EM, Maier U, Grathwohl P, Kuschk P, Kaestner M. Performance evaluation of different horizontal subsurface flow wetland types by characterization of flow behavior, mass removal and depth-dependent contaminant load. Water Res. 2013; 47(2):769–780.

95 Wei M, Rakoczy J, Vogt C, Harnisch F, Schumann R, Richnow HH. Enhancement and monitoring of pollutant removal in a constructed wetland by microbial electrochemical technology. Bioresour Technol. 2015; 196:490–499.

96 WHO (World Health Organization). Phenol. Geneva, Switzerland; 1994.

97 Van der Oost R, Beyer J, PE V. Fish bioaccumulation and biomarkers in environmental risk assessment: a review. Environ Toxicol Pharmacol. 2003; 13:57–149.

98 Capasso R, Cristinzo G, Evidnete A, Scognamiglio F. Isolation, spectroscopy and selective phytotoxic effects of polyphenols from vegetable waste waters. Phytochem. 1992; 31:4125–4128.

99 Gonzalez DM, Moreno E, Sarmiento JQ, Ramos-Cormenzana A. Studies on antibacterial activity of waste waters from olive mills (Alpechin): inhibitory activity of phenolic and fatty acids. Chemosphere. 1990; 20:423–432.

100 Lin SU, Juang RS. Adsorption of phenol and its derivatives from water using synthetic resins and low-cost natural adsorbents: A review. J Environ Manage. 2009; 90:1336–1349.

101 Girelli AM, Mattei E, Messina A. Phenols removal by immobilized tyrosinase reactor in on-line high performance liquid chromatography. Anal Chim Acta. 2006; 580:271–277.

102 Parkhurst BR, Bradshaw AS, Forte JL, Wright GP. An evaluation of the acute toxicity to aquatic biota of a coal conversion effluent and its major components. Bull Environ Contam Toxicol. 1979; 23:349–356.

103 Keith LH. Identification of organic compounds in unbleached treated Kraft paper mill wastewaters. Environ Sci Technol. 1976; 10:555–564.

104 Goerlitz DF, Troutman DE, Gody EM, Franks BJ. Migration of wood-preserving chemical in contaminated groundwater in a sand aquifer at Pensacola, Florida. Environ Sci Technol. 1985; 19:955–961.

105 Kougias PG, Kotsopoulos TA, Martzopoulos GG. Effect of feedstock composition and organic loading rate during the mesophilic co-digestion of olive mill wastewater and swine manure. Renew Energ. 2014; 69:202–207.

106 Veeresh GS, Kumar P, Mehrotra I. Treatment of phenol and cresols in upflow anaerobic sludge blanket (UASB) process: a review. Water Res. 2005; 39:154–170.

107 Chandanalakshmi MVV, Sridevi V. A review on biodegradation of phenol from industrial effluents. J Ind Pollut Contr. 2009; 25:13–27.

108 Si L, Ruixue K, Lin S, Sifan L, Shuangchun Y. Study on treatment methods of Phenol in industrial wastewater. Int J Sci Eng Res. 2013; 4(5):230–232.

109 Mishra A, James H Clark, eds. Green materials for sustainable water remediation and treatment. Cambridge, UK: RSC Publishing, 2013.

110 Sridevi V, Chandana L, Manasa M, Sravani M. Metabolic pathways for the biodegradation of phenol. Int J Eng Sci Adv Technol 2012; 2(3):695–705.

111 Oboirien BO, Amigun B, Ojumu TV, Ogunkunle OA, Adetunji OA, Betiku E, et al. Substrate inhibition kinetics of phenol degradation by *Pseudomonas aeruginosa* and *Pseudomonas* fluorescence. Biotechnol. 2005; 4(1):56–61.

112 Leonard D, Lindley ND. Carbon and energy flux constraints in continuous cultures of *Alcaligenes eutrophus* grown on phenol. Microbiol. 1998; 144:241–248.

113 Baradarajan A, Vijayaraghavan S, Srinivasaraghavan T, Musti S, Kar S, Swaminathan T. Biodegradation of phenol by arthrobacter and modeling of the kinetics. Bioprocess Eng. 1995; 12:227–229.

114 Abd-El-Haleem D, Beshay U, Abdelhamid AO, Moawad H, Zaki H. Effects of mixed nitrogen sources on biodegradation of phenol by immobilized *Acinetobacter* sp. strain W-17. Afr J Biotechnol. 2003; 2(1):8–12.

115 Reardon KF, Mosteller DC, Rogers JDB. Biodegradation kinetics of benzene, toluene, and phenol as single and mixed substrates for *Pseudomonas putida* F1. Biotechnol Bioeng. 2000; 69(4):385–400.

116 Bak F, Widdel F. Anaerobic degradation of phenol and phenol derivatives by *Desulfobacterium phenolicum* sp. nov. Arch Microbiol. 1986; 146:177–180.

117 Kadlec RH, Wallace SD. Treatment Wetlands, 2nd edn. Boca Raton, USA: CRC Press, Taylor & Francis Group; 2009.

118 Kurzbaum E, Kirzhner F, Sela S, Zimmels Y, Armon R. Efficiency of phenol biodegradation by planktonic *Pseudomonas pseudoalcaligenes* (a constructed wetland isolate) vs. root and gravel biofilm. Water Res. 2010; 44:5021–5031.

119 Kurzbaum E, Zimmels Y, Kirzhner F, Armon R. Removal of phenol in a constructed wetland system and the relative contribution of plant roots, microbial activity and porous bed. Water Sci Technol. 2010; 62(1327–1334).

120 Poerschmann J, Schultze-Nobre L. Sorption determination of phenols and polycyclic aromatic hydrocarbons in a multiphase constructed wetland system by solid phase microextraction. Sci Total Environ. 2014; 482-483:234–240.

121 Rossmann M, de Matos AT, Abreu EC, Borges AC. Performance of constructed wetlands in the treatment of aerated coffee processing wastewater: Removal of nutrients and phenolic compounds. Ecol Eng. 2012; 49:264–269.

122 Polprasert C, Dan NP, Thayalakumaran N. Application of constructed wetlands to treat some toxic wastewaters under tropical conditions. 1996; 34(11):165–171.

123 Alkorta I, Garbisu C. Phytoremediation of organic contaminants in soils. Bioresour Technol. 2001; 79:273–276.

124 Hübner TM, Tischer S, Tanneberg H, Kuschk P. Influence of phenol and phenanthrene on the growth of *Phalaris arundinacea* and *Phragmites australis*. 2000; 2(4):331–342.

125 Schultze-Nobre L, Wiessner A, Wang D, Bartsch C, Kappelmeyer U, Paschke H, et al. Removal of dimethylphenols from an artificial wastewater in a laboratory-scale wetland system planted with *Juncus effusus*. Ecol Eng. 2015; 80:151–155.

126 Tee HC, Seng CE, Noor AM, Lim PE. Performance comparison of constructed wetlands with gravel- and rice husk-based media for phenol and nitrogen removal. Sci Total Environ. 2009; 407:3563–3571.

127 CCME. Canadian soil quality guidelines for the protection of environmental and human health: Phenol. In: Canadian environmental quality guidelines, Canadian Council of Ministers of the Environment, Winnipeg; 1999.

128 Avila C, Reyes C, Bayona JM, Garcia J. Emerging organic contaminant removal depending on primary treatment and operational strategy in horizontal subsurface flow constructed wetlands: Influence of redox. Water Res. 2013; 47:315–325.

129 Reddy KR, Delaune RD. Biogeochemistry of wetlands: Science and applications. Boca Raton, FL, USA: CRC Press, Taylor & Francis Group; 2009.

130 Pind A, Freeman C, Lock MA. Enzymic degradation of phenolic materials in peatlands-measurement of phenol oxidase activity. Plant Soil. 1994; 159:227–231.

131 Toyama T, Yu N, Kumada H, Sei K, Ike M, Fujita M. Accelerated aromatic compounds degradation in aquatic environment by use of interaction between *Spirodela polyrrhiza* and bacteria in its rhizosphere. J Biosci Bioeng. 2006; 101(4):346–353.

132 Gagnon V, Chazarenc F, Comeau Y, Brisson J. Influence of macrophyte species on microbial density and activity in constructed wetlands. Water Sci Technol. 2007; 56(3):249–254.

133 Kurzbaum E, Kirzhner F, Armon R. Performance comparison of plant root biofilm, gravel attached biofilm and planktonic microbial populations, in phenol removal within a constructed wetland wastewater treatment system. Water SA. 2016; 42(1):166–170.

134 Herrera Melián JA, Araña J, Ortega JA, Martín Muñoz F, Tello Rendón E, Pérez Peña J. Comparative study of phenolics degradation between biological and photocatalytic systems. J Sol Energy Eng. 2008; 130(4):1003/1–7.

135 Wang X, Deshusses MA. Biotreatment of groundwater contaminated with MTBE: interaction of common environmental co-contaminants. Biodegradation 2007; 18:37–50.

136 Deeb RA, Hu HY, Hanson JR, Scow KM, Alvarez-Cohen L Substrate interactions in BTEX and MTBE mixtures by an MTBE-degrading isolate. Environ Sci Technol. 2001; 35:312–317.

137 Wang X, Deshusses MA. Biotreatment of groundwater contaminated with MTBE: interaction of common environmental co-contaminants. Biodegradation 2007; 18:37–50.

138 Sedran MA, Pruden A, Wilson GJ, Suidan MT, Venosa AD. Effect of BTEX on degradation of MTBE and TBA by mixed bacterial consortium. J Environ Eng. 2002; 128(9):830–835.

Part II

Food and Beverage Industry

3

Aerated Constructed Wetlands for Treatment of Municipal and Food Industry Wastewater

A. Pascual[1,3], D. De la Varga[2,3], M. Soto[2,3], D. Van Oirschot[4], R.M. Kilian[5], J.A. Álvarez[1], P. Carvalho[6], H. Brix[6] and C.A. Arias[6]

[1] AIMEN, Spain
[2] Sedaqua (Spin-off from University of A Coruña), Spain
[3] University of A Coruña, Spain
[4] Rietland bvba, Van Aertselaerstraat 70, Minderhout, Belgium
[5] Kilian Water Ltd., Denmark
[6] Department of Bioscience – Aquatic Biology, Aarhus University, Aarhus, Denmark

3.1 Introduction

Constructed Wetlands (CWs) are engineered wastewater treatment systems that have been designed and constructed to mimic processes that occur in natural wetlands. Vegetation, soils, and their associated microbial assemblages are combined to effectively treat wastewater [1].

CWs are shallow basins, generally from 0.3 to 1.0 m. Wastewater can circulate freely, like natural ponds, and this kind of CW is called a free water surface (FWS) system, with aquatic vegetation rooted in the bottom, or floating plants. Another type of CW are planted beds filled with sand or gravel, and they are called subsurface flow systems (SSF). Depending on the flow direction, they are horizontal flow (HF) or vertical flow (VF) systems.

HF are permanently flooded, water flows horizontally and is not exposed to the atmosphere level as it is maintained under the surface (about 1–5 cm). On the other hand, VF wetlands are intermittently pulse-loaded, on top, and wastewater percolates through the unsaturated substrate. Aeration pipes connecting the atmosphere to a manifold of perforated drainage pipes are installed to provide a pathway for air to be drawn into the substrate from the bottom of the bed. Thus, air enters the bed from either the top or the bottom and maintains aerobic conditions in the bed. This approach provides a significant improvement of subsurface oxygen availability compared to HF designs.

Engineered treatment wetlands are other options of CWs systems that might include "reciprocating", also known as "tidal flow" or "fill-and-drain" wetlands. As the wetland bed is drained, air is drawn into the bed [2], oxygenating the exposed biofilms on the wetland substratum. This improves the treatment performance compared to systems with a static water lever [3, 4]. Mechanical aeration of SSF wetlands using air distribution pipes installed at the bottom of the wetland bed has also been utilized as a means to increase oxygen transfer in wetland treatment systems. They are called (artificially) aerated wetlands or Forced Bed Aeration Wetlands (FBA®).

Constructed Wetlands for Industrial Wastewater Treatment, First Edition. Edited by Alexandros I. Stefanakis.
© 2018 John Wiley & Sons Ltd. Published 2018 by John Wiley & Sons Ltd.

CWs have been used for wastewater treatment for more than fifty years to treat different types of polluted waters around the world. CWs became a widely accepted technology to deal with both point and non-point sources of water pollution as they offer a technical, low-energy, and low-operational-requirements alternative to conventional treatment systems, besides being able to meet discharge standards. Used initially to treat municipal wastewaters, the application of CWs has been expanded to the treatment of industrial effluents, agricultural wastewaters, livestock farm effluents, landfill leachate and stormwater runoff, among others [5–7].

The processes involved in pollutant removal include sedimentation, sorption, precipitation, evapotranspiration, volatilization, photodegradation, diffusion, plant uptake, and microbial degradation processes such as nitrification, denitrification, sulphate reduction, carbon metabolization, among others [8].

CW systems treat industrial effluents from petrochemical, dairy, meat processing, abattoir, and pulp and paper factory production. Brewery, winery, tannery and olive mills wastewaters have been recently added to CW applications. CWs can be applied to several and different kinds of industrial wastewaters, including acid mine wastewater with low organic matter content and landfill leachate. Vymazal [9] reported the use of CWs for the treatment of industrial wastewaters with influent concentrations up to 10,000–24,000 mg of chemical oxygen demand (COD)/L and up to 496 mg NH_4^+/L. However, there are no general rules for selecting the most suitable type of CW for a certain industrial wastewater or even urban wastewater. Every single case must be studied according to several conditions: type of wastewater, land availability, amount of flow and pollutant load, outlet discharge limits, etc. [8].

Industrial wastewaters differ substantially in composition from municipal sewage, as well as among themselves. Industrial wastewaters can present very high concentrations of organics, total suspended solids (TSS), ammonia and other pollutants; therefore the use of CWs almost always requires some kind of pretreatment. The BOD/COD ratio is a parameter which tentatively indicates the biological degradability. If this ratio is greater than 0.5, the wastewater is easily biodegradable, such as wastewaters from dairies, breweries, the food industry, abattoirs or starch and yeast production. The BOD/COD ratio for these wastewaters usually ranges between 0.6 and 0.7 but could be as high as 0.8. On the other hand, wastewaters with a low BOD/COD ratio and, thus, low biodegradability are represented, for example, by pulp and paper wastewaters. Tentative comparison of the industrial wastewater strength with municipal sewage could be done on the basis of population equivalent (PE: 60 g BOD_5 per person per day).

3.2 Aerated Constructed Wetlands

Oxygen availability to support aerobic processes is the main limitation in HF CWs, especially when nitrification (and subsequent total nitrogen removal) is a treatment objective [10]. As a result, to increase the oxygen availability, CWs have evolved into more effective treatment systems by installing an aeration system capable of transferring sufficient oxygen to perform aerobic processes. Design variants now span from completely passive systems (HF), to moderately engineered systems (unsaturated VF systems with pulse loading) up to highly engineered or intensified systems, with increased pumping, water level fluctuation, or forced aeration [11].

As a result, most of the treatment wetland design and operational modifications developed in the last decade aim at improving subsurface oxygen availability. The simplest (most passive) modification is the construction of shallow HSF flow beds, highlighted by Garcia et al. [12]. Their findings suggest that by limiting the depth of the HF bed, all of the wastewater is forced through the root zone. Their results show improved treatment performance for COD, BOD_5, and NH_4-N in shallow beds (27 cm water depth) compared to deeper ones (50 cm water depth). However, recent studies suggest that this positive effect of shallow beds is limited to low surface loading rates [13, 14].

Recirculation of treated effluent has also been shown to improve removal of ammonium nitrogen and organic matter [15–19]. Operational adaptations to improve subsurface oxygen availability include water level fluctuations such as batch loading [20–22], "fill-and-drain", "reciprocating", or "tidal flow" [23–28]. A step forward is the use of active aeration (e.g., a network of air distribution pipes installed at the bottom of the bed connected to a blower pump to supply atmospheric air) which has also been applied to HF to constructed wetland beds [29–32] and saturated VF systems [33, 34], often showing a more than ten-fold increase of removal rates compared to passive systems. Most of the reports on intensified treatment wetland designs come from private engineering companies which hold patents. However, the potential use of intensified treatment wetland is widely recognized, and design guidance and parameters have yet to be determined [35].

As indicated, VF CWs have predominant aerobic conditions, while HF CWs mainly presented anaerobic conditions. Combining both types of CW in hybrid systems could achieve complete nitrogen removal, so in more recent years, interest in the study of multi-step and hybrid systems has increased [9, 36, 37]. The most commonly used hybrid system is the two-step VF-HF CW, which has been used for treatment of both sewage and industrial wastewaters [9, 38, 39]. In general, all types of hybrid CWs are comparable with single VF CWs in terms of NH_3-N removal rates whilst they are more efficient in TN removal than single HF or VF CWs [9]. However, even in hybrid VF+HF systems, the TN removal remains low [9, 40, 41]. The effectiveness of alternating aerobic and anaerobic conditions in VF-HF hybrid systems was evaluated by Gaboutloeloe et al. [40], who reported that the most limiting factor of these systems was nitrate accumulation, mainly caused by the depletion of carbon during the aerobic phase. Tanner et al. [41] pointed out that the endogenous organic carbon supply from plant biomass decay and root-zone exudation has often been found to be insufficient to achieve full denitrification in VF+HF hybrid systems. In order to solve this handicap and improve TN removal, several authors studied the effect of step-feeding in tidal and saturated VF CWs [42–45]. Tanner et al. [41] proved the use of carbonaceous bioreactors, which incorporate a slow-release source of organic C (e.g., wood chips) aiming to increase denitrification. Recirculation has been employed in various configurations [46–50] in order to increase simultaneous nitrification and denitrification processes in either a single CW unit or in the two-step HF+VF system. Artificial aeration in hydraulic saturated units, attaining to only part of the system or timed, has a high potential as an alternative to enhance TN removal [32, 51–53].

3.2.1 Oxygen Transfer at the Water–Biofilm Interface

Early HF wetland designs were based on the Root Zone Method (RZM) [54]. Plant-mediated oxygen transfer was thought to be a key mechanism in RZM designs, but actual oxygen transfer rates generally did not meet these design expectations [55] and the systems often clogged. This led to the development of VF wetlands in the late 1980s. However, if VF wetlands are hydraulically or organically overloaded, ponding of wastewater occurs. This effectively cuts off air circulation and promotes clogging, which dramatically reduces oxygen transfer [56].

Mechanisms for oxygen transfer in treatment wetlands include atmospheric diffusion, plant-mediated oxygen transfer, and oxygen transfer at the water–biofilm interface [10]. Research in the recent years identified several design and operation factors, which improve oxygen transfer at the water–biofilm interface, such as artificial aeration [51, 52, 57, 58] and fill-and-drain operations [3, 59]. Since the rate of air circulation (and thus oxygen transfer) is related to the frequency of water level fluctuation in filling and draining systems, internal recycling to rapidly fill and drain multiple wetland compartments is often employed.

From a pilot-scale research facility in Langenreichenbach, Germany, Nivala et al. [10] estimated oxygen consumption rates (OCR) for the main CW designs. Measured OCRs ($g/m^2/d$) were in the range of 0.5–13 for HF CW, 8–59 for VF CWs and 11–88 for intensified CW systems. Similar or even higher OCRs were previously reported in literature. However, as pointed out by Nivala et al. [10], those rates may not necessarily be sustainable over the long term operation of the system. Intensifying oxygen input in the CWs through the use of artificial aeration combine the advantage of maintaining low energy consumption and clogging prevention [31]. Artificial aeration strategies can vary extensively from partial to total aeration in relation to time and space, and from low to high intensity. Although some small and large field applications of aerated CW have been reported, properly described experiences of artificial aerated CWs are mainly limited to a few laboratory and pilot scale systems [31, 32, 51, 52, 57].

Aerated HCW varied from 0.3 m to 1 m in depth [32, 51, 52]. Probably, an efficient aeration process requires a higher depth in order to reach a high oxygen transfer rate (OTR) enhanced by sufficient long contact time between the supplied air and WW. Automated aeration devices were used in some cases, setting dissolved oxygen (DO) concentration set point to activate air pumps in the range of 0.2–0.6 mg/L [51]. In other cases, continuous aeration was provided, over the overall wetland bed surface [10, 52] or only near the inlet zone [32]. However, intermittent aeration or a spatial segregation of aerated and non-aerated zones has been considered possible in order to reach simultaneous nitrification and denitrification [32, 43, 53, 60, 61]. In efficient aerated HF CW, nitrification occurs when the aeration system is turned on, while denitrification requires anoxic conditions, which could be obtained by ceasing aeration. Aeration intensities were reported for VF CW by Pan et al. [62] and Maltais-Laundry et al. [32] ranging from 0.12 to 0.76 $m^3/m^2/h$. These authors found that oxygen utilization efficiency decreased when the aeration intensity increased.

3.2.2 Benefits of Artificial Aeration in Constructed Wetlands

Important factors affecting the treatment performance include the flow type, substrate characteristics, plant species, hydraulic loading rate (HLR) and temperature. HLR, related to the space available for the water to flow through the CW, is a principal parameter for the design and operation of CW. Sakadevan and Bavor [63] reported that the removal of pollutants in a CW was improved by decreasing HLR when the applied hydraulic retention time (HRT) ranged from 4 to 15 days. A lower HLR implies more contact time and more treatment stability, however, it occupies a larger land area [5].

Physical processes such us sedimentation and decantation, important in particulate organic matter removal, are mostly unaffected by winter conditions. However, biological processes are temperature dependent, and winter removal performances of HF CWs for nitrogen and soluble organic matter, both highly driven by biological activity, may be reduced [59, 64].

Besides lower winter temperature, low oxygen availability, which is already a common limiting factor in HF CWs during the growing season, may be even more so in winter. Oxygen solubility is higher in colder water [65], but gas exchange in HF CWs may be reduced by the additional insulation layer and the fact that plants are dormant. Low oxygen content results in low aerobic organic matter decomposition [28, 66–68]. This leads to fermentation processes [69, 70] that can represent, in certain overload cases, the main way of organic matter decomposition [71]. Moreover, the nitrification step represents the main limiting factor for N removal in HF CWs because of low oxygen availability [72]. In addition, to the TN concentration, the form of N is also often a crucial factor affecting the receiving water body. For instance, besides being toxic to aquatic biota, the associated nitrogenous biochemical oxygen demand of NH_4^+-N can depress DO levels.

Although in CWs, oxygen availability may be enhanced by the presence of macrophytes through diffusion of oxygen via the aerenchym to the rhizomes [66], the exact contribution of plants remains in debate [52, 73–76]. It was reported that the contribution of plants to pollutant removal was usually less than 10% [77], although it has been found to be important for nutrient removal in low loaded systems [77, 78]. Caldheiros et al. [80] also found that there was no significant difference in pollutant removal between the planted and unplanted wetlands during a 17-month operation period. The primary role of plants is to hold the wetland components in place, preventing erosion and landscape integration. Therefore, artificial aeration appears necessary when the CW is operated under a high HLR.

3.2.3 Dissolved Oxygen Profile along CWs

DO plays an important role in the activity of microbes in wetlands. To achieve the simultaneous removal of organic matter (COD) and nutrients (N, P), the aerobic and anoxic regions in wetlands need optimization depending on wastewater characteristics and operational manipulation.

Dong et al. [58] compared different aeration strategies in three VF CWs: non-aeration (NA), continuous-aeration (CA) and intermittent-aeration (IA), to treat heavily polluted river water. The VF CWs were continuously fed from a feed tank using a metering pump. Although the VF CWs have higher oxygen mass transfer efficiency than the HF ones, the DO concentrations (averaged over three tested HLRs) in the 5–40 cm region above the reactor bottom were below 1 mg/L in NA, which could inhibit the nitrification process.

For the CA, DO concentrations ranged from 1.3 to 2.2 mg/L (in the 5–20 cm profile) and from 3.8 to 4.4 mg/L (in the 40–60 cm profile). However, for the IA, DO concentrations varied from 0.8 to 1.1 mg/L and from 2.5 to 2.8 mg/L in the mentioned DO profiles.

It was reported that no obvious nitrification was observed when the DO concentration was lower than 0.5 mg/L [81]. According to the DO values, artificial aeration significantly improved the oxygen availability in the VF CWs. Although all DO concentrations in IA and CA appeared to exceed that required for anoxic condition (i.e., 0.2–0.5 mg/L), anoxic regions could still exist in the aerated VF CWs due to the spatial stratification of biofilms in both IA and CA operation modes, and particularly in the IA operation mode, which would facilitate denitrification.

3.2.4 TSS Removal

Several authors found that supplemental aeration of CWs had a positive effect on TSS reduction [82, 83]. Ouellet-Plamondon et al. [31] concluded that artificial aeration may have reduced matter accumulation by increasing degradation kinetics and prevented clogging.

3.2.5 COD Removal

COD removal is related to HLR. The increase of HLR makes the removal efficiencies of COD decrease. Increasing HLR would reduce the contact time between wastewater and microbes, enhance the detachment of microbes off substrate surfaces, and decrease the oxygen availability [84]. Although organic matter can be degraded both aerobically and anaerobically by heterotrophic bacteria in the wetlands depending on local DO concentrations, aerobic degradation is usually more important [85]. Dong et al. [58] found that COD removal efficiency was positively correlated with the aeration condition: CA > IA > NA, continuous-aeration, intermittent-aeration and non-aeration conditions, respectively (see Section 3.3.3).

As reported by Ouellet-Plamondon et al. [31], during summer, there was a slight improvement in COD removal in planted mesocosms compared to unplanted ($p < 0.01$), but no effect of artificial aeration, regardless of the presence of plants. In winter, the expected reduction in COD removal in non-aerated mesocosms was totally compensated with a significant improvement in aerated meso-cosms, both for planted and unplanted units. The added oxygen in winter probably counterbalanced the reduction of removal kinetics due to temperature and plants dormancy [31].

When oxidation decreases, the amount of residual inert organic matter accumulated increases and aggregates in filtration matrix changing the hydraulic conditions by reducing HRT [3, 84] and biological properties [87]. Increasing oxygen availability with artificial aeration could enhance mineralization and reduce hydraulic clogging [88]. Sulphate reduction, a typical diagnostic of poor oxygen conditions in CWs [89], could also be inhibited by artificial aeration. Thus, for organic matter removal, their results suggest that artificial aeration in HF CWs could be beneficial in winter, when plants are dormant.

3.2.6 Nitrogen Removal

Nitrogen removal in CWs occurs through adsorption, assimilation into biomass, ammonia volatilization and coupled nitrification/denitrification, of which the nitrification/denitrification process is the most important [90, 91].

Since nitrification and denitrification are two operationally separate processes (either temporally or spatially), which respectively require aerobic and anoxic conditions, the rate of nitrification significantly impacts the removal of TN. The removal efficiency of TN significantly dropped with an increase of HLR. Artificial aeration significantly improved the oxygen availability and thus enhanced the removal of NH_4-N in the VF CWs. Intermittent aeration was optimal for TN removal, which facilitated denitrification due to both spatial and temporal formations of anoxic zones in the VF CWs. Although continuous aeration achieved the highest nitrification rate, the denitrification process was notably suppressed due to an excessive oxygen supply that artificial aeration significantly enhanced NH_4-N removal in VF CWs [53, 58].

Besides all this, average TKN removal in winter was lower than in summer, most likely because of the lower winter temperature, which was well under optimal temperature for nitrifying activity [92]. In winter, artificial aeration improved TKN removal for all mesocosms ($p < 0.01$), with a more pronounced improvement for unplanted units [31].

3.3 HIGHWET Project

The HIGHWET project was addressed to improve the capacity and effectiveness of CWs as high-rate and sustainable wastewater treatment system. HIGHWET aimed to perform and validate new approaches based on the combination of the hydrolytic up-flow sludge bed (HUSB) anaerobic digester and CWs with forced aeration for decreasing the required surface of conventional HF CWs and improving the final effluent quality. For this purpose, two demonstration plants were designed and constructed in Spain and Denmark. The first configuration (A Coruña, NW of Spain) consisted of a HUSB and HF CWs for raw municipal wastewater treatment [53], while second configuration (at KT Food, nearby Aarhus, Denmark) consisted of a combination of a HUSB and hybrid (FV-HF) CWs for treatment of high load organic industrial wastewater. The effect of effluent recirculation, aeration regime and different phosphorus adsorbent materials was planned to be checked in both plants. The authors report in this chapter the results obtained in the KT Food HIGHWET plant.

3.3.1   KT Food Pilot Plant

KT Food is a food producing company located at Randersvej 147, in the town of Purhus (56° 33' 38.58" N 9° 51' 26.08" E) in Denmark. Additional to the production of food, the site also generates water from a small dairy farm and domestic activities. To meet the discharge standards demanded by the environmental authorities, a wastewater treatment system was built. The plant was constructed in August 2014 as a research plant funded by the European Union under the FP7 grant agreement N° 605445. After technical, environmental considerations and to meet the discharge demands, it was decided to design and construct the treatment plant using a combination of an anaerobic digester as primary treatment, followed by two parallel treatment trains (aerated line and non-aerated line) of constructed wetlands, several wells to allow controlled recirculation of treated waters and additional wells to host reactive media to remove P before discharge The conceptual design of the system installed is shown in Figure 3.1.

The wastewater is collected from the house and the two industrial plants, homogenized in a well where fat and grease is removed. Thereafter, wastewater is pumped to the treatment plant where flow is measured using an ultrasound digital flow meter. Once the research project is finished, all the wastewater produced at the site is being treated by the plant.

After the first well, water is transported to a second well where pollutant concentration can be increased by adding a prepared feedstock solution in order to reach the desired concentrations for the research project. The primary treatment consisted of a Hydrolytic Upflow Sludge Blanket (HUSB) digester. After the HUSB, water flows to a pumping well that is fitted with two pumps, where direction and flow volume can be selected to any of the two treatment trains. Both treatment trains consist of a VF CW followed by HF beds and phosphorous removal wells. The surface of the VF beds is 16 m², while both of the HF beds are 3 m². In the eastern train, the VF bed is fitted with forced aeration. while the bed of the western train is passively aerated. Both of the HF beds are fitted with forced aeration. The aeration systems installed to supply air to the beds use individual compressors that provide atmospheric air to increase the oxygen availability, and improve the aerobic processes of pollutant degradation. Additionally, the aeration time and cycles to each one of the beds can be controlled with automatic timers according to the operation planned.

3.3.2   Research Operational Plan of KT Food Treatment Plant

During the research phase, the WWTP operated with different loading schemes where pollutant loadings, aeration cycles and recirculation were modified to obtain the largest amount of data possible and to determine the treatment capacity. Sampling strategy was to take grab samples from nine different points along the treatment train (Figure 3.1). Five sampling campaigns were performed

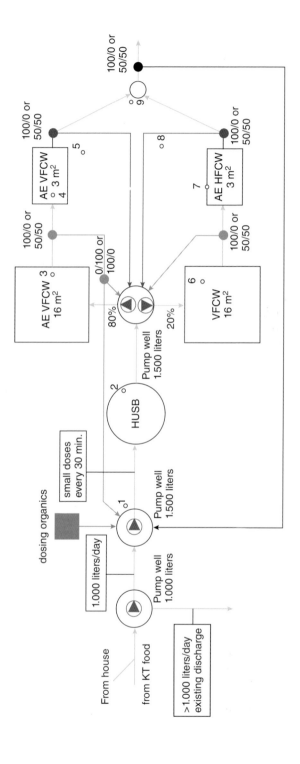

Figure 3.1 Conceptual design of the system installed at KT Food. Empty circle and numbers indicate the sampling points along the treatment trains 1) inlet; 2) after the HUSB; 3) after the aerated VF; 4) after the aerated HF bed; 5) after the P removal filtered filled with Tobermorite; 6) after the non-aerated bed; 7) after the aerated bed; 8) after the P removal filter filled with Polonite; 9) final effluent. Treatment trains: aerated line (1–2–3–4–5) and non-aerated line (1–2–6–7–8).

Table 3.1 Planned exploitation parameters for each of the sampling campaigns.

Operation parameter[a]	Campaign				
	1	**2**	**3**	**4**	**5**
Month	January	March	May	July	September
AE VF CW – AE HF CW (L/d)[b]	800	800	800	800	1440
VF CW – AE HF CW (L/d)[c]	200	200	200	200	360
Recirculation	No	No	No	No	80%
VF CW aeration time (h on/h off)	24/0	24/0	4/4	6/2	24/0
HF CW aeration (h on/h off)	24/0	24/0	24/0	24/0	24/0

[a] Planned influent concentration was 5000 mg COD/L, 500 mg TN/L and 30 mg TP/L along the whole study.
[b] AE VF CW – AE HF CW = Aerated Vertical Flow Constructed Wetland – Aerated Horizontal Flow Constructed Wetland.
[c] VF CW = Vertical Flow Constructed Wetland.

for different aeration schemes and effluent recirculation (Table 3.1). Analyses were carried out as described in Standard Methods [91].

Any change of operational parameters implies the need of acclimation time so that the processes can become stable and performance is optimized. Therefore a period of three to four weeks acclimation time was allowed between the measuring campaigns.

After the first samples were collected during plant start-up (data not shown), it was evident that the wastewater produced at the site did not reach the aimed high concentration to achieve the organic or nutrient overloading stated in the exploitation plan. Therefore, it was necessary to install a system that could supply a prepared solution to reach the planned pollutant and hydraulic loadings. The system was built using a 1 m³ tank, and a time controlled dosing pump that fed the solution to the well located before the HUSB. The loading solution was prepared using a blending of fresh pig manure, molasses, starch, urea and fertilizer. The volumes of each component were calculated to reach the planned loading and were monitored regularly to maintain a constant loading. The flow was controlled and always was close to the desired overall flow of 1,000 L/d.

After the initial adaptation period, plant equipment including aeration pumps and a dosing system functioned without any problems, in spite of the low temperatures and the snow that covered the system, which is to be expected for the winter period in Denmark (Figure 3.2e). As it can be seen in Figure 3.2, no plant development was present during Campaigns 1 and 2, but plant development started in April, before Campaign 3 carried out in May.

According to the exploitation plan, the last campaign included increasing the flow by recirculating 80% of the treated water that went through the aerated VF bed. That means that during the campaign, the overall influent flow to the beds was of 1,800 L/d. The flow to the forced aerated bed was 1,440 L/d and to the passively aerated bed 360 L/d, while the hydraulic loading rates were 9 cm/d and 2.25 cm/d, respectively.

Figure 3.2 Satellite image showing the location of the treatment plant (a); a general view of the plant once it was established (b); details of HF bed (c); detail of P removal unit (d); state of the system during Campaign 1 at winter 2015 (e); state of the non-aerated VF bed during Campaign 2 (f); the two VF beds during Campaign 3 (aerated VF bed: g; passively aerated VF bed: h).

3.3.2.1 Campaign 1

Campaign 1 was carried out during winter 2015. Averages and the standard deviation of the evaluated parameters are presented in the following tables (Tables 3.2–3.4). Even though environmental temperature was below or close to 0°C, the wastewater temperature was always above freezing in the beds. Temperature was uniform along the components of the treatment. Electrical Conductivity (EC) decreased along the treatment. pH was in the range of 6.7–10.6 in the different beds, being the highest after the Polonite well. As expected, DO was low at the inlet and after the HUSB. As water went through the system, the DO increased to reach oxygen saturation.

Table 3.3 presents the results of the average concentrations of TSS, COD and BOD_5 during the campaign. TSS concentration varied between 96 mg/L at the inlet to 3.9 at the effluent. The overall

Table 3.2 In-situ monitored parameters during Campaign 1.

Sampling place	Temperature (°C)		EC (µS/cm)		pH		DO (mg/L)	
	Aver	Stdv	Aver	Stdv	Aver	Stdv	Aver	Stdv
1: Inlet	7.9	2.0	1,775	285	6.9	0.9	3.4	3.3
2: After HUSB	7.8	1.0	1,325	445	6.7	1.3	0.4	0.2
3: After aerated VF bed	7.4	1.8	817	208	8.0	0.2	10.2	1.8
4: After HF bed	6.8	1.7	697	203	8.3	0.3	12.1	0.8
5: After Tobermorite	6.7	1.5	721	182	9.3	0.1	9.0	1.9
6: After non-aerated VF bed	6.8	1.9	733	233	8.1	0.2	8.6	2.5
7: After HF bed	6.3	1.9	566	180	8.2	0.2	12.4	1.0
8: After Polonite	6.2	1.9	504	156	10.6	0.5	10.3	0.8
9: Effluent	6.5	2.1	597	188	9.8	0.5	9.0	2.0

Table 3.3 Average TSS, COD and BOD_5 in the system during Campaign 1.

Sampling place	TSS (mg/L)		COD (mg/L)		BOD_5 (mg/L)	
	Aver	Stdv	Aver	Stdv	Aver	Stdv
1: Inlet	96	49.17	5538	381	2567	603
2: After HUSB	40	12.02	2541	459	1317	580
3: After aerated VF Bed	12	7.3	50	36	4.0	1,0
4: After HF bed	4.8	1.3	70	21	5.0	6.1
5: After Tobermorite	5,8	1.6	39	21	18.0	14.7
6: After non-aerated VF bed	8	2.2	119	33	12.7	9.6
7: After HF bed	5	1.4	52	29	3.3	3.2
8: After Polonite	8	3.2	63	35	3.3	4,0
9: Effluent	3.9	4.6	51	16	4.7	2.5

Table 3.4 Average nutrient concentrations and performance along the system during Campaign 1.

Sampling place	NH$_4$-N (mg/L)		NO$_3$-N (mg/L)		TN (mg/L)		TP-P (mg/L)	
	Aver	Stdv	Aver	Stdv	Aver	Stdv	Aver	Stdv
1: Inlet	116	69	49	14	489	71	32	17.6
2: After HUSB	87	82	12	15	232	7	22	8.3
3: After aerated VF Bed	0	0	8.8	7.5	54	2	3	0.7
4: After HF bed	0	0	8.5	6.5	58	3	2	0.4
5: After Tobermorite	1	2	8.5	7.1	50	2	2	0.4
6: After non-aerated VF bed	1	1	6.7	5.3	121	2	1	0.2
7: After HF bed	0	0	8.1	4.9	83	5	0.8	0.4
8: After Polonite	0	1	4.2	2.3	66	2	0.4	0.2
9: Effluent	0	1	2.8	2.8	40	0	1	0.3

removal of TSS was 96% while the reduction in the HUSB was around 60%. There is further reduction along the system and the final concentration is sufficient to meet any discharge standard. A high reduction of COD occurred in the HUSB where 50% of the COD was removed. Further COD removal happened in the aerated bed reaching 98% removal. The removal between the HUSB and the non-aerated bed was also high, reaching 95%. After the two VF beds, there were low removal but it can be explained by the low COD concentrations after the VF beds. Average BOD$_5$ concentration during the campaign at the influent was around 2,600 mg/L and around 5 mg/L at the effluent, with an overall removal of 99%. Between the influent and the HUSB, the removal of BOD$_5$ reached 49%. After the HUSB, the removal of BOD$_5$ reached 99%, both in the aerated VF bed and in the non-aerated bed.

Conversion of nitrogen compounds and total phosphorus is given in Table 3.4. Nitrification in the system was effective and the overall nitrification was close to 100%. The nitrification process occurred mainly while the water was in the VF beds. Simultaneous to nitrification, denitrification was also taking place along the treatment and the overall denitrification rate was 94%. Denitrification occurred in the HUSB where 75% of the NO$_3$-N was removed. P removal in the system occurred in all the structures, reaching up to 97%. The two reactive materials tested showed that they can produce effluents with concentrations below 1 mg/L.

3.3.2.2 Campaign 2

The results obtained for Campaign 2 are presented in Tables 3.5–3.7. Table 3.5 shows the average temperature along the structures which are affected by the external temperature, ranging from around 12°C to 9°C. EC was higher at the influent and decreased along the treatment. pH was around 8 but increased above 11 after the Polonite tank. DO concentration was low at the influent but increased along the treatment to reach nearly DO saturation concentrations after the first VF beds in both treatment trains.

In spite of TSS concentration in the influent was higher than the previous campaign, TSS concentration after the HUSB was already 51% lower (Table 3.6). Further removal occurred along the treatment reaching an overall TSS removal of 97%. COD was around 6,000 mg/L with an overall removal of 99% at the effluent. The HUSB removed 71% and the VF beds were able to remove the rest of the COD.

Table 3.5 In-situ measured parameters during Campaign 2.

Sampling place	Temperature (°C)		EC (µS/cm)		pH		DO (mg/L)	
	Aver	Stdv	Aver	Stdv	Aver	Stdv	Aver	Stdv
1: Inlet	11.8	0.9	1,773	549	7.8	0.5	1.6	2.1
2: After HUSB	10.2	0.9	1,503	212	8.1	0.5	0.8	0.5
3: After aerated VF Bed	10.4	0.2	1,214	266	8.6	0.1	12.0	0.2
4: After HF bed	9.6	0.6	1,222	266	8.6	0.1	9.7	0.4
5: After Tobermorite	9.3	0.5	982	203	9.3	0.1	9.7	0.4
6: After non-aerated VF bed	9.8	0.5	1,490	413	8.2	0.1	9.3	0.6
7: After HF bed	9.2	1.0	962	228	8.6	0.2	12.1	0.3
8: After Polonite	8.8	0.6	709	87	11.2	0.1	10.7	0.4
9: Effluent	9.4	0.5	829	320	9.8	0.5	10.0	1.0

Table 3.6 Average TSS, COD and BOD_5 in the system during Campaign 2.

Sampling place	TSS (mg/L)		COD (mg/L)		BOD_5 (mg/L)	
	Aver	Stdv	Aver	Stdv	Aver	Stdv
1: Inlet	235	56	6108	1547	3500	1061
2: After HUSB	115	20	1748	161	1625	530
3: After aerated VF Bed	19	8,0	57	8	8,5	3,5
4: After HF bed	6,7	3,1	49	5	1,5	0,7
5: After Tobermorite	8,4	0,4	45	5	1,0	0,0
6: After non-aerated VF bed	11	7,5	67	20	3,0	1,4
7: After HF bed	6	1,8	37	1	2,5	0,7
8: After Polonite	10	3,2	28	2	1,0	0,0
9: Effluent	8,2	1,0	38	3	1,0	0,0

Similarly, BOD_5 at the influent was on average 3,500 mg/L with a removal of 54% in the HUSB. The VF beds removed on average more than 99% of the remaining BOD_5 leaving very little BOD to be removed in the following structures.

Regarding nutrient removal (Table 3.7), different behavior took place compared to the previous campaign. Dynamics of the removal along the beds were different than in previous campaigns, reaching an overall removal of 94%. NH_4-N average concentration in the inlet was around 100 mg/L, with only 10% removal in the HUSB. In the aerated VF the removal was close to 60%. The NH_4-N removal in the non-aerated bed was less effective and only 46% was removed. The rest of the system continued to remove NH_4-N to reach a final concentration of 6 mg/L. Nitrate in the system was removed in all the structures, especially in the HUSB, where all the NO_3-N was removed. Along the

Table 3.7 Average nutrient concentrations (mg/L) and performance along the system during Campaign 2.

Sampling place	NH_4-N		NO_3-N		TN		TP-P	
	Aver	Stdv	Aver	Stdv	Aver	Stdv	Aver	Stdv
1: Inlet	101	46	26	5.3	152	61	26	12.9
2: After HUSB	89	4	1.1	0.2	89	34	16	3.3
3: After aerated VF Bed	34	9	23	8.9	58	11	5	0.3
4: After HF bed	25	4	24	3.8	61	9	4	0.3
5: After Tobermorite	9	6	21	1	38	3	2	0.2
6: After non-aerated VF bed	48	32	33	13.4	89	13	3	0.9
7: After HF bed	7	6	24	5.4	30	6	1	0.1
8: After Polonite	0	0	16	1.6	19	1	0.4	0.1
9: Effluent	6	5	18	0.8	28	2	2	0.7

bed there was an increase of NO_3-N as a result of nitrification in the structures. Both of the tested media presented good P removal capacity, reaching 92% through the overall system.

3.3.2.3 Campaign 3

After Campaign 2, aeration time for the aerated VF bed was set to intermittent aeration (4 hours on, 4 hours off). The results obtained Campaign 3 are presented in Tables 3.8–3.10. The third campaign took place in May when temperatures began to increase and weather was milder. The plants in all the beds began to grow due to the noticeable effect of the season being more effective in the aerated beds because of the higher water flow that allowed better nutrient availability. Even though weather was milder, it was not reflected in the water temperature. This can be explained by the low temperatures

Table 3.8 In-situ monitored parameters during Campaign 3.

Sampling points	Temperature (°C)		pH		EC (µS/cm)		DO (mg/L)	
	Aver	Stdv	Aver	Stdv	Aver	Stdv	Aver	Stdv
1: Inlet	12.1	0.6	7.6	0.6	1883	746	0.7	0.4
2: After HUSB	10.3	0.5	7.7	0.3	1539	289	0.7	0.4
3: After aerated VF Bed	10.5	0.2	8.0	0.1	1394	140	6.3	0.9
4: After HF bed	9.6	0.7	8.7	0.0	1352	64	11.9	0.1
5: After Tobermorite	9.6	0.4	9.3	0.0	1126	65	9.5	0.2
6: After non-aerated VF bed	9.9	0.6	8.2	0.1	1671	244	9.1	0.6
7: After HF bed	9.1	1.2	8.7	0.0	1063	130	12.0	0.3
8: After Polonite	8.4	0.5	11.3	0.0	763	50	10.6	0.3
9: Effluent	9.3	0.6	9.6	0.3	981	124	9.6	0.8

Table 3.9 Average TSS, COD and BOD₅ concentrations in the system during Campaign 3.

Sampling point	TSS (mg/L)		COD (mg/L)		BOD₅ (mg/L)	
	Aver	Stdv	Aver	Stdv	Aver	Stdv
1: Inlet	117	14	5,268	917	4,167	4,404
2: After HUSB	25.2	6.4	2,055	1516	1,383	693
3: After aerated VF bed	22.6	26.8	211	37.5	44	25
4: After HF bed	33.6	21.7	156	13.5	6	2
5: After Tobermorite	33.8	31.7	156	6.8	6	2
6: After non-aerated VF bed	86.5	23.0	174	60	3	2
7: After HF bed	101.7	39.3	82	6.1	2	0
8: After Polonite	102.7	36.9	61	2.9	2	2
9: Effluent	47.2	31.0	130	24.0	3	3

Table 3.10 Average nutrient concentrations and performance along the system during Campaign 3.

Sampling Point	NH_4-N (mg/L)		NO_3-N (mg/L)		TN (mg/L)		TP (mg/L)	
	Aver	Stdv	Aver	Stdv	Aver	Stdv	Aver	Stdv
1: Inlet	250	162.2	113.5	16.5	382	404	25.1	18.5
2: After HUSB	197	67.4	3.6	4.6	201	419	18.2	12.2
3: After aerated VF Bed	5.2	2.3	1.2	0.1	11	12	5.2	0.4
4: After HF bed	0.12	0.021	5.0	0.6	9	4	4.0	0.3
5: After Tobermorite	0.14	0.019	4.5	0.6	11	19	2.2	0.1
6: After non-aerated VF bed	0.10	0.010	34	7.6	42	6	2.9	0.2
7: After HF bed	BDL	BDL	36	5.7	40	12	1.2	0.1
8: After Polonite	0.7	0.0	40	6.5	45	21	0.5	0.1
9: Effluent	0.3	0.0	26	6.6	27	11	1.8	0.1

reached at night. DO concentrations were as expected, with anaerobic conditions in the inlet and at the outlet of the HUSB. Except for the aerated VF bed, DO concentrations were close to saturation. The lower DO in the aerated VF bed effluent can be explained by the fact that during this campaign, aeration was carried out in cycles, with 4 hours of aeration and 4 hours with no aeration. Even though DO was lower, it was still above 60% saturation.

TSS in the raw water was within the expected limits of raw wastewater (Table 3.9). At the beginning of this campaign, a problem with the HUSB occurred, due to the presence of grease in the water surface. It was rapidly skimmed and removed from the reactor before the sampling campaign. The final effluent was about 34 mg/L. The aerated VF bed produced effluent COD concentration higher than the passively aerated bed. However, it should be considered that the aerated bed was loaded with four times the loading compared to the passively aerated bed. The targeted COD at the inlet was as planned and removal in the HUSB was effective. After the HUSB and through the two treatment trains, removal of COD was high with similar concentrations in the effluent of both VF beds. During this campaign, BOD concentration was higher than the previous campaign. This can be explained by possible changes in the food processing, as the loading solution was prepared as usual. The HUSB removed more than half of the BOD concentration. Along the system BOD was removed to low concentrations. The highest concentration of 45 mg/L was observed in the aerated VF bed.

Nitrogen species concentrations were below the targeted 500 mg/L TN (Table 3.10). This can be explained due to uncertainty about the actual concentrations of the pig manure used to prepare the solution. It can vary depending on the storage, weather conditions and the washing practices in the farm. Removal of NH_4-N was low through the HUSB. After the HUSB, removal of ammonia was effective in both trains. Results presented in Table 3.10 show that NO_3-N was denitrified in the HUSB and also in the wetlands during the shut-off of aeration periods. The passively aerated bed did not show the same effective denitrification and the effluent had a NO_3-N concentration of 40 mg/L. The same dynamics were followed by the TN.

3.3.2.4 Campaign 4

As indicated in Table 3.1, after Campaign 3, the aeration time for the aerated VF bed was set to 6 hours on, 2 hours off. Working to these conditions, the results obtained for Campaign 4 are presented in Tables 3.11–3.13. During the fourth campaign, ambient temperature increased so water temperature along the system was affected increasing to around 17°C (Table 3.11). pH behaved similarly to the previous campaigns with around 7 along the treatment and changing when water went through the P removal media. DO was low at the inlet and after the HUSB. After the aerated VF bed, DO was low but it increased as water went through the different structures of the treatment plant.

Raw water TSS was lower if compared to the previous campaign and the HUSB removed more than half of the influent TSS (Table 3.12). This suggests that skimming the grease and fat had a positive effect. The aerated bed further removed additional TSS so the concentration in the effluent was below the discharge limits. COD was around the targeted concentration and about half was removed by the

Table 3.11 *In-situ* monitored parameters during Campaign 4.

Sampling points	Temperature (°C)		pH		EC (µS/cm)		DO (mg/L)	
	Aver	Stdv	Aver	Stdv	µS/cm	Stdv	mg/L	Stdv
1: Inlet	17.0	0.3	6.8	0.7	1,690	1,196	2.2	1.9
2: After HUSB	17.3	0.4	6.2	0.5	1,059	416	1.4	1.0
3: After aerated VF Bed	17.4	0.3	8.3	0.1	1,043	82	1.5	0.4
4: After HF bed	17.3	0.4	9.2	0.0	1,039	78	8.0	0.1
5: After Tobermorite	16.8	0.2	9.7	0.0	1,070	68	5.5	1.0
6: After non-aerated VF bed	16.8	0.3	8.5	0.1	641	93	6.6	0.2
7: After HF bed	16.5	0.4	9.2	0.0	792	122	8.5	0.1
8: After Polonite	16.8	0.4	11.4	0.6	914	146	8.1	0.4
9: Effluent	17.0	0.8	10.4	0.3	991	108	7.7	0.6

Table 3.12 Average TSS, COD and BOD$_5$ in the system during Campaign 4.

Sampling point	TSS (mg/L) Aver	Stdv	COD (mg/L) Aver	Stdv	BOD$_5$ (mg/L) Aver	Stdv
1: Inlet	117	14	4,771	1,658	2,250	0
2: After HUSB	25.2	6.4	2,516	743	1,375	106
3: After aerated VF bed	22.6	26.8	99	66.8	26	34
4: After HF bed	33.6	21.7	55	21.4	1	1
5: After Tobermorite	33.8	31.7	52	7.8	2	2
6: After non-aerated VF bed	8.5	2.3	63	26	5	1
7: After HF bed	10.7	3.3	42	10.0	6	6
8: After Polonite	10.7	3.9	37	21.2	0	0
9: Effluent	4.2	3.0	53	4.1	2	1

HUSB. The aerated bed removed around 90%. No considerable further removal was archived in this train. Through the other treatment train, the passively aerated bed performed well and was able to remove COD down to 50 mg/L. BOD followed the same pattern as COD in spite of the fact that the BOD/COD ratio was lower than in previous campaigns. The aerated bed produced an effluent, with 26 mg/L being further removed along the following structures. The treatment train with the passively aerated bed performed well reaching BOD concentrations down to 10 mg/L after the bed.

Nitrogen species in the inlet were close to the targeted concentration of 500 mg TN/L (Table 3.13). Through the HUSB there was considerable denitrification and the NO$_3$-N present was denitrified effectively. The HUSB did not remove NH$_4$-N. Nitrification seemed to be effective in the aerated bed, with inlet NH$_4$-N concentrations around 239 mg/L and outlet concentrations of 5 mg/L. NO$_3$-N was also as low as 1 mg/L suggesting that the intermittent aeration can enhance the N removal. On the

Table 3.13 Nitrogen and phosphorus concentrations in the different points during Campaign 4.

Sampling point	NH$_4$-N (mg/L) Aver	Stdv	NO$_3$-N (mg/L) Aver	Stdv	TN (mg/L) Aver	Stdv	TP (mg/L) Aver	Stdv
1: Inlet	250	32.4	113	16.5	382	104	33.4	13.2
2: After HUSB	239	13.5	3.6	4.6	243	79	21.3	3.3
3: After aerated VF bed	5.2	2.3	1.2	0.1	7	2	5.7	1.7
4: After HF bed	0.01	0.021	5.0	0.6	6	4	5.1	0.6
5: After Tobermorite	0.01	0.019	4.5	0.6	5	2	3.9	0.6
6: After non-aerated VF bed	0.01	0.010	34	7.6	45	6	2.6	0.4
7: After HF bed	0.000	0.000	36	5.7	48	12	2.3	0.4
8: After Polonite	0.0	0.0	40	6.5	46	18	1.1	0.0
9: Effluent	0.0	0.0	26	6.6	34	21	2.3	0.7

other treatment train, wastewater was nitrified effectively, but denitrification did not occur at the same rate with intermittent aeration. No further considerable denitrification was registered in the treatment train.

3.3.2.5 Campaign 5

The fifth campaign included recirculation of the effluent of the aerated bed back to the pumping well to increase the hydraulic loading on the beds (Tables 3.14–3.16). The calculated hydraulic loading increased corresponds to around 80% more water to each one of the beds (Table 3.1). Initially and when the flow was increased, both beds presented an increase in TSS and the release of biofilm from the media was evident. A decrease in TSS concentration along time and no further biofilm was present in the effluent when the campaign started. Temperature was similar to the previous campaign

Table 3.14 In-situ monitored parameters during Campaign 5.

Sampling point	Temperature (°C)		pH		EC (µS/cm)		DO (mg/L)	
	Aver	Stdv	Aver	Stdv	Aver	Stdv	Aver	Stdv
1: Inlet	16.0	0.4	6.6	1.0	2,208	1,333	1.0	0.6
2: After HUSB	15.7	0.3	6.5	0.9	1,130	228	1.7	0.3
3: After aerated VF bed	16.2	0.4	8.9	0.2	1,057	94	3.7	3.2
4: After HF bed	15.8	0.3	9.1	0.1	1,058	68	7.0	0.5
5 After Tobermorite	15.3	0.5	10.0	0.0	1,102	30	5.3	0.4
6 After non-aerated VF bed	15.1	0.4	8.3	0.1	924	118	6.2	0.6
7 After HF bed	14.9	0.6	9.1	0.4	1,163	67	8.4	0.1
8 After Polonite	14.7	0.8	11.1	0.3	1,220	44	7.8	0.4
9 Effluent	15.4	0.4	10.1	0.2	1,143	51	6.8	0.3

Table 3.15 Average TSS, COD and BOD$_5$ in the system during Campaign 5.

Sampling point	TSS mg/L		COD mg/L		BOD$_5$ mg/L	
	Aver	Stdv	Aver	Stdv	Aver	Stdv
1: Inlet	217	145	5,268	917	4,267	404
2: After HUSB	96.0	9.2	2,055	1,516	1,167	419
3: After aerated VF bed	23.3	7.6	211	37.5	22	12
4: After HF bed	7.6	4.4	156	13.5	3	1
5: After Tobermorite	9.6	5.0	156	6.8	4	1
6: After non-aerated VF bed	12.0	4.7	174	60	9	4
7: After HF bed	6.4	6.0	82	6.1	2	1
8: After Polonite	8.9	6.7	61	2.9	1	1
9: Effluent	6.9	5.6	130	24.0	3	1

Table 3.16 Average nutrient concentrations and performance along the system during Campaign 5.

Sampling point	NH$_4$-N (mg/L)		NO$_3$-N (mg/L)		TN (mg/L)		TP (mg/L)	
	Aver	Stdv	Aver	Stdv	Aver	Stdv	Aver	Stdv
1: Inlet	367	171.4	125	15.0	493	157	40	25
2: After HUSB	316	198.0	3.6	2.4	320	196	15	2.2
3: After aerated VF bed	5.9	2.1	3.2	1.2	9	1	6.3	1.2
4: After HF bed	0.00	0.00	12.9	0.5	13	0	6.2	0.5
5: After Tobermorite	3.8	1.2	26.4	9.4	30	11	5.5	0.2
6: After non-aerated VF bed	0.1	0.1	53	3.0	53	3	5.7	1.3
7: After HF bed	BDL	BDL	119	21.6	119	22	2.6	0.4
8: After Polonite	0.1	0.1	133	7.0	133	7	1.0	0.3
9: Effluent	2.9	2.1	60	17.1	63	16	1.3	0.9

because ambient temperature was mild. pH was close to neutral except when water went through the P removal material, which increased pH and in the case of Polonite up to 11. DO was low for raw wastewater and through the HUSB and relatively low after effluent of the aerated bed when measured concentration was below 4 mg/L. During this campaign 100% saturation was never achieved through the other structures.

TSS influent concentration was around 200 mg/L and the HUSB removed around 2/3 of the TSS. After the aerated VF Bed, the TSS concentration was already down to 20 mg/L. While water went through the other structures, concentration continued to drop and the final effluent was more than enough below the discharge requirements. COD influent reached the targeted concentration and the HUSB removed half of the concentration. Through the aerated bed, an additional 90% was removed and no considerable further removal occurred in the treatment train. The treatment train fitted with the non-aerated bed showed similar performance. BOD influent concentration was relatively high if compared to previous concentration, but the HUSB was able to remove around 60% of the load. The two treatment trains had no difficulty dealing with the BOD and final effluent reached concentrations close to the detection limit.

The inlet TN concentration target was reached. The HUSB did not nitrify but nitrification took place in the aerated bed. Further nitrification happened through the treatment and the wastewater was nitrified at the end of the process. Denitrification occurred in the HUSB and also in the aerated bed. After the aerated bed, no denitrification was evident. The passive aerated bed denitrified a fraction, but no further denitrification happened in this treatment train.

3.3.3 Comparison of Results

Applied SLR (g/m^2/d) in the aerated line of KT Food HIGHWET project were 2.5 ± 1.8, 92 ± 14, 58 ± 7, 9.1 ± 3.5, 7.8 ± 4.2 and 0.8 ± 0.1 for TSS, COD, DBO$_5$, TN, NH$_4^+$-N and TP, respectively, whilst SLRs were four times lower in the non-aerated line (i.e., 0.6 ± 0.4, 23 ± 4, 15 ± 2, 2.3 ± 0.9 and 2.0 ± 1.0 for TSS, COD, DBO$_5$, TN and NH$_4^+$-N, respectively). Thus, the non-aerated line operated at conservative design loading rates and reached satisfactory contaminant removal, usually from 90 to 99% of TSS, COD, BOD_5 and ammonia. Similar or even higher percentage removal rates were obtained in the aerated line, operated at four times higher loading rates.

TN removal reached $43 \pm 7\%$ in the HUSB digester, due to the denitrification of influent nitrate. Overall, TN removal was $85 \pm 7\%$ in the non-aerated line and $91 \pm 9\%$ in the aerated line. The aerated VF CW unit reached $80 \pm 27\%$ TN removal ($91 \pm 10\%$ excluding Campaign 2), whilst the non-aerated VF CW unit reached $58 \pm 36\%$ TN removal ($73 \pm 17\%$ excluding Campaign 2). TN removal was not found in the small size aerated HF units.

TP removal reached $39 \pm 14\%$ in the HUSB digester, whilst overall TP removal was $98 \pm 1\%$ in the non-aerated line and $90 \pm 3\%$ in the aerated line. Both the aerated and non-aerated VF CW units noticeable contributed to TP removal, reaching $72 \pm 10\%$ and $82 \pm 13\%$, respectively. Additional TP removal took place in the aerated HF units, reaching $18 \pm 12\%$ and $42 \pm 25\%$ for HF1 and HF2, respectively. According to Vymazal [82], these TP removal rates obtained under average loading rates of 0.8 g $TP/m^2/d$ may be considered very satisfactory. Finally, the P removal units with Polonite as phosphorus adsorbant material reached $56 \pm 5\%$ TP removal whilst TP removal in the unit with Tobermorite decreased from about 50% at Campaigns 2 and 3 to 11% at Campaign 5.

Therefore, the aerated line was successful in treating a four times higher loading rate and with similar or higher treatment efficiency than the non-aerated VF CW unit for organic matter and nitrogen removal and only slightly lower for phosphorus removal. The HUSB efficiently contributed to TSS, COD, BOD and nitrate removal. These high percentage removal rates were obtained at organic SLR in the range of referred studies for different kind of industrial wastewaters while ammonia and TN loading rates were higher in the aerated line of the KT Food HIGHWET plant.

Studies on industrial wastewater treatment on artificially aerated CWs are scarce. Results for coffee processing wastewater [94], dairy parlor wastewater [83], aquaculture effluent [95] and dye containing wastewater [96, 97] are summarized below.

Rossmann et al. [94] treated coffee processing wastewater (CPW) with aerated and non-aerated influent previously to pilot-scale HF CWs. The applied organic load during the experiment was 89 g $COD/m^2/d$, and the HRT was 12 d. Removal efficiencies of COD, BOD and TSS ranged from 87.9 to 91.5, from 84.4 to 87.7 and from 73.7 to 84.8%, respectively. Aeration of CPW in the storage tank for 2.5 days did not affect the removal efficiencies of organic matter in the CWs, which agrees with previous findings of Zhang et al. [51], due to low redox values and anoxic conditions in spite of the aeration. In this study [94], phosphorus removal (54.3–72.1%) was statistically different among treatments, with better performance for the aerated planted system, and worse for the non-aerated unplanted.

The feedlot runoff and dairy parlor wastewater in Burlington (Vermont, USA) was treated in four HF CWs (non-aerated unplanted CW1, aerated planted CW2, non-aerated planted CW3, and aerated planted CW4) of 225 m^2 each in an experiment carried out by Tunçsiper et al. [83]. HRT in CWs ranged from 3 to 16 days. Over the four years of monitoring, the CWs operated with surface loading rate of 210 g $BOD_5/m^2/d$ and 70 g $TSS/m^2/d$ in average. Average BOD_5 removals were 83%, 78%, 84% and 86% for CW1, CW2, CW3 and CW4, respectively. The authors of this study concluded that supplemental aeration of CWs had a positive effect on BOD_5 reduction.

Aquaculture effluent under high HLR was assessed in Jingzhou city (China) by Zhang et al. [95]. Two parallel, identical hybrid wetland systems (CW 1+2), each with down, up and HF chambers were constructed in the field. The HLR was approximately 8.0 m/day, giving a theoretical HRT of 0.96 h. For the wetland with diffused-air enhancement, there was a significant decrease in COD and NH_4^+-N concentrations after filtration. Further, the aeration significantly increased the levels of DO,

ORP, nitrite, and TN, while significantly decreasing the levels of EC, COD, NH_4^+-N, and TP concentrations in the outflow compared to the non-aerated treatment. High organic loading rates of 132 and 146 g $COD/m^2/d$ were applied for the non-aerated (stage 1) and aerated (stage 2) conditions. Concentration of COD in the effluent of aerated wetland was significantly lower than in the non-aerated wetland. TN removal was higher in the non-aerated wetland in which sedimentation of organic N was determined to be the main process of TN removal. On the other hand, TN removal was dominated by ammonium removal in the aerated stage. In the non-aerated wetland, NH_4^+-N outlet concentrations were generally higher than in the inlet, observing an opposite trend in the aerated wetland. The authors concluded that denitrification process was contained with short HRT (0.96 h) even though carbon source seemed to be enough for denitrification. Concerning P removal, higher percentage reductions were observed in the aerated wetland.

Ong et al. [97] studied the mineralization of diazo dye (Reactive Black 5, RB5) in wastewater using recirculated up-flow CW reactor in Malaysia. The HRT was 2 days. COD removal in the aerated reactor (92%) was higher than that in the non-aerated reactor (83%) whilst RB5 removal efficiency presented the opposite trend (81 and 89%, respectively).

Ong et al. [96] conducted other experiment to study the removal of azo dye Acid Orange 7 (AO7) in three parallel lab-scale CWs of 0.3 m height and 0.18 m of diameter. The CWs were planted with *Phragmites australis*, and there were aerated (A), non-aerated (B) and non-aerated unplanted (C). With an HRT of 2 days, COD removal in the aerated and non-aerated CWs was 95% and 62%, respectively (both higher than COD removal in control unplanted CW). The three CWs removed more than 94% of AO7, being slightly higher in the aerated one. The ammonia removal was significantly higher in the aerated CW (86%) than in the non-aerated CW (14%) which, additionally, performed better than the control CW (4%).

3.4 Conclusions

This study reports the effect of effluent recirculation, aeration regime and different phosphorus adsorbent materials in a system that combines a HUSB, hybrid (FV-HF) CWs and two different phosphorus adsorbent materials for treatment of industrial wastewater. Applied SLR ($g/m^2/d$) in the aerated line were 2.5 ± 1.8, 92 ± 14, 58 ± 7, 9.1 ± 3.5, 7.8 ± 4.2 and 0.8 ± 0.1 for TSS, COD, DBO_5, TN, NH_4^+-N and TP, respectively, whilst SLRs were four times lower in the non-aerated line (i.e. 0.6 ± 0.4, 23 ± 4, 15 ± 2, 2.3 ± 0.9 and 2.0 ± 1.0 for TSS, COD, DBO_5, TN and NH_4^+-N, respectively). The non-aerated line reached satisfactory contaminant removal, usually from 90 to 99% of TSS, COD, BOD_5 and ammonia. Similar or even higher percentage removal rates were obtained in the aerated line. TN removal reached $43 \pm 7\%$ in the HUSB digester, due to the denitrification of influent nitrate. Overall, TN removal was $85 \pm 7\%$ in the non-aerated line and $91 \pm 9\%$ in the aerated line. The aerated VF CW unit provided $80 \pm 27\%$ TN removal, whilst the non-aerated VF CW unit reached $58 \pm 36\%$ TN removal. Overall TP removal was $98 \pm 1\%$ in the non-aerated line and $90 \pm 3\%$ in the aerated line. Both the aerated and non-aerated VF CW units noticeable contributed to TP removal, reaching $72 \pm 10\%$ and $82 \pm 13\%$, respectively. Additional TP removal was obtained in Polonite unit ($56 \pm 5\%$) at 0.2 g $TP/m^2/d$ during the whole study whilst TP removal in Tobermorite unit at 0.87 g $TP/m^2/d$ decreased from about 50% to 11% after 6 month of treatment. These results showed that the aerated VF CW was successful in treating a four times higher loading rate and with similar or higher treatment efficiency than the non-aerated VF CW.

Acknowledgements

This work was supported by the EU's seventh framework programme for research, technological development and demonstration under grant agreement n° 605445.

References

1 Vymazal J. Constructed wetlands for treatment of industrial wastewater: A review. Ecol Eng. 2015; 73:724–751.

2 Green M, Friedler E, Ruskol Y, Safrai I. Investigation of alternative method for nitrification in constructed wetlands Water Sci Technol. 1997; 35(5):63–70.

3 Tanner C, Sukias J, Upsdell M. Substratum Phosphorus Accumulation during Maturation of Gravel-bed constructed wetlands. Water Sci. Technol. 1999; 40:147.

4 Leibowitz SG. Isolated wetlands and their functions: an ecological perspective, Wetlands BioOne 2003; 23(3):517–531.

5 Kadlec RH, Knight RL, Vymazal J, Brix H, Cooper P, Haberl R. Constructed Wetlands for Pollution Control: Processes, Performance, Design and Operation. IWA Specialist Group on Use of Macrophytes in Water Pollution Control, Scientific and Technical Report 8; 2000.

6 Vymazal J. The use constructed wetlands with horizontal sub–surface flow for various types of wastewater. Ecol Eng. 2009; 35(1):1–17.

7 Vymazal J. Constructed wetlands for wastewater treatment: five decades of experience. Environ Sci Technol. 2011; 45(1):61–69.

8 De la Varga D, Soto M, Arias CA, Oirschot DV, Kilian R, Pascual A, Álvarez JA. Constructed Wetlands for Industrial Wastewater Treatment and Removal of Nutrients. In: Technologies for the Treatment and Recovery of Nutrients from Industrial Wastewater, 2016; pp. 202–230.

9 Vymazal J. The use of hybrid constructed wetlands for wastewater treatment with special attention to nitrogen removal: A review of a recent development. Water Res. 2013; 47:4795–4811.

10 Nivala J, Headley T, Wallace S, Bernhard K, Brix H, van Afferden M, Müller R. Comparative analysis of constructed wetlands: The design and construction of the ecotechnology research facility in Langenreichenbach, Germany. Ecol Eng. 2012; 61(B): 527–543.

11 Fonder N, Headley T. Systematic classification, nomenclature and reporting for constructed treatment wetlands. In: Vymazal J, editor. Water and Nutrient Management in Natural and Constructed Wetlands. Springer Science + Business Media B.V.: Dordrecht; 2010, pp. 191–219.

12 García J, Aguirre P, Barragán J, Mujeriego R. Effect of key design parameters on the efficiency of horizontal subsurface flow constructed wetlands Ecol Eng. 2003; 25(4):405–418.

13 De la Varga D, Ruiz I, Soto M. Winery wastewater treatment in subsurface constructed wetlands with different bed depths. Water Air Soil Pollut. 2013; 224:1485–1497.

14 De la Varga D, Díaz MA, Ruiz I, Soto M. Avoiding clogging in constructed wetlands by using anaerobic digesters as pre-treatment. Ecol Eng. 2013; 52:262–269.

15 Arias CA, Brix H. Initial experience from a compact vertical flow constructed wetland treating single household wastewater. In: Vymazal J, ed. Natural and Constructed Wetlands: Nutrients, Metals and Management. Backhuys Publishers, Leiden, The Netherlands; 2005, pp. 52–64.

16 García-Pérez A, Grant B, Harrison M. Water quality effluent from a recirculating, vertical-flow constructed wetland. Small Flows Quart. 2006; 7(4):34–38.

17 Gross A, Sklarz MY, Yakirevich A, Soares MI. Small scale recirculating vertical flow constructed wetland (RVFCW) for the treatment and reuse of wastewater Water Sci Technol. 2008; 58(2):487–494.

18 Konnerup D, Trang NTD, Brix H. Treatment of fishpond water by recirculating horizontal and vertical flow constructed wetlands in the tropics. Aquaculture 2011; 313(1–4):57–64.

19 Torrijos V, Gonzalo OG, Trueba-Santiso A, Ruiz I, Soto M. Effect of bypass and effluent recirculation on nitrogen removal in hybrid constructed wetlands for domestic and industrial wastewater treatment. Water Res. 2016; 103:92–100.

20 Stein OR, Hook PB, Beiderman JA, Allen WC, Borden DJ. Does batch operation enhance oxidation in subsurface flow constructed wetlands. Water Sci Technol. 2003; 48(5):149–156.

21 Corzo A, Pedescoll A, Álvarez E, García J. Solids accumulation and drainable porosity in experimental subsurface flow constructed wetlands with different primary treatments and operating strategies. In: Billore SK, Dass P, Vymazal P, editors. Proceedings, 11th International Conference on Wetland Systems for Water Pollution Control, Vikram University and IWA, Indore, India; 1–7 November 2008, pp. 290–295.

22 Põldvere E, Karabelnik K, Noorvee A, Maddison M, Nurk K, Zaytsev I, Mander Ü. Improving wastewater effluent filtration by changing flow regimes – Investigations in two cold climate pilot scale systems. Ecol Eng. 2009; 35(2):193–203.

23 Behrends LL. Patent: Reciprocating subsurface flow constructed wetlands for improving wastewater treatment. United States: US 5,863,433; 1999.

24 McBride GB, Tanner CC. Modelling biofilm nitrogen transformations in constructed wetland mesocosms with fluctuating water levels Ecol Eng. 2000; 14(1–2):93–106.

25 Austin DC. Patent: Tidal vertical flow wastewater treatment system and method. United States: US 6,896,805 B2; 2005.

26 Sun G, Gray KR, Biddlestone AJ, Cooper DJ. Treatment of agricultural wastewater in a combined tidal flow: downflow reed bed system. Water Sci Technol. 1999; 40(3):139–146.

27 Ronen T, Wallace SD. TAYA – Intensive wetland technology facilitates the treatment of high loads of organic pollutants and ammonia. In: Masi F, Nivala J, editors. Proceedings, 12th International Conference on Wetland Systems for Water Pollution Control, IRIDRA S.r.l. and IWA, Venice, Italy; 4–8 October 2010, pp. 872–878.

28 Wu S, Zhang D, Austin D, Dong R, Pang C. Evaluation of a lab-scale tidal flow constructed wetland performance: oxygen transfer capacity, organic matter and ammonium removal. Ecol Eng. 2011; 37(11):1789–1795.

29 Wallace SD. Patent: System for removing pollutants from water. United States: US 6,200,469 B1; 2001.

30 Higgins JP. The use of engineered wetlands to treat recalcitrant wastewaters., Southampton, United Kingdom; WIT Press. 2003: pp. 137–160.

31 Ouellet-Plamondon C, Chazarenc F, Comeau Y, Brisson J. Artificial aeration to increase pollutant removal efficiency of constructed wetlands in cold climate. Ecol Eng. 2006; 27:258–264.

32 Maltais-Landry G, Maranger R, Brisson J, Chazarenc F. Nitrogen transformations and retention in planted and artificially aerated constructed wetlands. Water Res. 2009; 43:535–545.

33 Murphy C, Cooper D. An investigation into contaminant removal in an aerated saturated vertical flow constructed wetland treating septic tank effluent. In: Vymazal J, ed. Joint Meeting of Society of Wetland Scientists, WETPOL, and Wetlands Biogeochemistry, Czech University of Life Sciences, Prague, Czech Republic; 3–8 July 2011: p. 224.

34 Wallace SD, Liner MO. Design and performance of the wetland treatment system at Buffalo Niagara International Airport. IWA Specialist Group on Use of Macrophytes in Water Pollution Control. No. 38:36–42; 2011.

35 Kadlec RH, Wallace SD. Treatment Wetlands. 2nd ed. Boca Raton, USA: CRC Press, Taylor & Francis Group; 2009.

36 Ávila C, Garfí M, García J. Three-stage hybrid constructed wetland system for wastewater treatment and reuse in warm climate regions. Ecol Eng. 2013; 61:43–49.

37 Vymazal J, Kröpfelová L. Wastewater Treatment in Constructed Wetlands with Horizontal Subsurface Flow. Dordrecht: Springer; 2008.

38 Stefanakis AI, Akratos CS, Tsihrintzis VA. Vertical flow constructed wetlands: Eco-engineering systems for wastewater and sludge treatment, 1st edn. Amsterdam, The Netherlands: Elsevier; 2014.

39 Kim SH, Cho JS, Park JH, Heo JS, Ok YS, Delaune RD, Seo DC. Long-term performance of vertical-flow and horizontal-flow constructed wetlands as affected by season, N load, and operating-stage for treating nitrogen from domestic sewage. Environ Sci Pollut. Res. 2016; 23:1108–1119.

40 Gaboutloeloe G, Chen S, Barber M, Stöckle C. Combinations of horizontal and vertical flow constructed wetlands to improve nitrogen removal. Water Air Soil Pollut. 2009; 9:279–282.

41 Tanner CC, Sukias JPS, Headley TR, Yates CR, Stott R. Constructed wetlands and denitrifying bioreactors for on-site and decentralised wastewater treatment: Comparison of five alternative configurations. Ecol Eng. 2012; 42:112–123.

42 Yang Y, Zhao YQ, Wang SP, Guo XC, Ren YX, Wang L, Wang XC. A promising approach of reject water treatment using a tidal flow constructed wetland system employing alum sludge as main substrate. Water Sci Technol. 2011; 63(10):2367–2373.

43 Hu YS, Zhao YQ, Zhao XH, Kumar JLG. Comprehensive analysis of step–feeding strategy to enhance biological nitrogen removal in alum sludge-based tidal flow constructed wetlands. Bioresour Technol. 2012; 111:27–35.

44 Fan J, Liang S, Zhang B, Zhang J. Enhanced organics and nitrogen removal in batch-operated vertical flow constructed wetlands by combination of intermittent aeration and step feeding strategy. Environ Sci Pollut Res. 2013; 20:2448–2455.

45 Wang Z, Liu C, Liao J, Liu L, Liu Y, Huang X. Nitrogen removal and N$_2$O emission in subsurface vertical flow constructed wetland treating swine wastewater: Effect of shunt ratio. Ecol Eng. 2014; 73:446–453.

46 Brix H, Arias CA, Johansen NH. Experiments in a two–stage constructed wetland system: nitrification capacity and effects of recycling on nitrogen removal. In: Vymazal J, ed. Wetlands – Nutrients, Metals and Mass Cycling. Leiden, The Netherlands: Backhuys Publishers; 2003, pp. 237–258.

47 Brix H, Arias CA. The use of vertical flow constructed wetlands for on-site treatment of domestic wastewater: New Danish guidelines. Ecol Eng. 2005; 25:491–500.

48 Ayaz SÇ, Aktaş Ö, Fındık N, Akça L, Kınacı C. Effect of recirculation on nitrogen removal in a hybrid constructed wetland system. Ecol Eng. 2012; 40:1–5.

49 Foladori P, Ruaben J, Ortigara ARC. Recirculation or artificial aeration in vertical flow constructed wetlands: A comparative study for treating high load wastewater. Bioresour Technol. 2013; 149:398–405.

50 Vázquez MA, De la Varga D, Plana R, Soto M. Vertical flow constructed wetland treating high strength wastewater from swine slurry composting. Ecol Eng. 2013; 50:37–43.

51 Zhang L, Zhang L, Liu Y, Shen Y, Liu H, Xiong Y. Effect of limited artificial aeration on constructed wetland treatment of domestic wastewater. Desalination. 2010; 250(3):915–920.

52 Nivala J, Wallace S, Headley T, Kassa K, Brix H, van Afferden M, Müller R. Oxygen transfer and consumption in subsurface flow treatment wetlands. Ecol Eng. 2013; 61:544–554.

53 Pascual A, De la Varga D, Arias CA, Van Oirschot D, Kilian R, Álvarez JA, Soto M. Hydrolytic anaerobic reactor and aerated constructed wetland systems for municipal wastewater treatment – HIGHWET project. Environ Technol. 2017; 38(2):209–219.

54 Kickuth R. Abwasserreinigung in Mosaikmatritzen aus aeroben und anaeroben Teilbezirken. In: Moser, F. (ed.) Grundlagen der Abwasserreinigung. MÚnchen, Wien: Verlag Oldenburg; 1981, pp. 630–665.

55 Brix H. Gas exchange through the soil–atmosphere interphase and through dead culms of Phragmites australis in a constructed reed bed receiving domestic sewage. Water Res. 1990; 24:259–266.

56 Platzer C, Mauch K. Soil Clogging in Vertical-flow Reed Beds – Mechanismns, Parameters, Consequences and …Solutions? Water Sci. Technol. 1997; 35(5):175–181.

57 Chazarenc F, Gagnon V, Comeau Y, Brisson J. Effect of plant and artificial aeration on solids accumulation and biological activities in constructed wetlands. Ecol Eng. 2009; 35(6):1005–1010.

58 Dong H, Qiang Z, Li T, Jin H, Chen W. Effect of artificial aeration on the performance of vertical-flow constructed wetland treating heavily polluted river water. J Environ Sci. 2012; 24(4):596–601.

59 Pedescoll A, Corzo A, Álvarez E, Puigagut J, García J. Contaminant removal efficiency depending on primary treatment and operational strategy in horizontal subsurface flow treatment wetlands. Ecol Eng. 2011; 37:372–380.

60 Tao W, Wang J. Effects of vegetation, limestone and aeration on nitritation, anammox and denitrification in wetland treatment systems. Ecol Eng. 2009; 35:836–842.

61 Li HB, Li YH, Sun TH, Wang X. The use of a subsurface infiltration system in treating campus sewage under variable loading rates. Ecol Eng. 2012; 38:105–109.

62 Pan JZ, Li WC, Ke F, Wang L, Li XJ. Oxygen transfer efficiency of four kinds substrates applied in artificial aeration vertical-flow wetland. In Chinese. Env Sci. 2009; 30(2):402–406.

63 Sakadevan K, Bavor HJ. Nutrient removal mechanisms in constructed wetlands and sustainable water management. Water Sci Technol. 1999; 40:121–128.

64 Kadlec RH, Reddy KR. Temperature effects in treatment wetlands. Water Environ Res. 2001; 73:543–557.

65 Stein OR, Hook PB. Temperature, plants and oxygen: how does season affect constructed wetland performance? J. Environ Sci Health A. 2005; 40(6–7):1331–1342.

66 Brix H. Constructed wetlands for municipal wastewater treatment in Europe. In: Mitsch WJ, ed. Global wetlands: old world and new. Amsterdam: Elsevier Science; 1994: 325–333.

67 Reed SC, Brown D. Subsurface flow wetlands – a performance evaluation. Water Environ Res. 1995; 67244–67248.

68 IWA (International Water Association). Constructed Wetlands For Pollution Control: Processes, Performance, Design and Operation. London: Scientific and Technical Report No 8. IWA Publishing; 2000, pp. 156.

69 Mander Ü, Kuusemets V, Lõhmus K, Mauring T, Teiter S, Augustin J. Nitrous oxide, dinitrogen and methane emission in a subsurface flow constructed wetland. Water Sci Technol. 2003; 48(5):135–142.

70 Johansson AE, Gustavsson AM, Oquist MG, Svensson BH. Methane emissions from a constructed wetland treating wastewater – seasonal and spatial distribution and dependence on edaphic factors. Water Res. 2004; 38:3960–3970.

71 Baptista JDC, Donelly T, Rayne D, Davenport RJ. Microbial mechanisms of carbon removal in subsurface flow wetlands. Water Sci Tech. 2003; 48(5):127–134.

72 Kuschk P, Wiessner A, Kappelmeyer U, Weissbrodt E. Annual cycle of nitrogen removal by a pilot-scale subsurface horizontal flow in a constructed wetland under moderate climate. Water Res. 2003; 37(1):4236–4242.

73 Bedford BL, Bouldin DR, Beliveau BD. Net oxygen and carbon dioxide balances in solutions bathing roots of wetland plants. J Ecol. 1991; 79:943–959.

74 Sorrell BK, Armstrong W. On the difficulties of measuring oxygen release by root systems of wetland plants. J Ecol. 1994; 82:177–183.

75 Armstrong W, Cousins D, Armstrong J, Turner DW, Beckett PM. Oxygen distribution in wetland plant roots and permeability barriers to gas–exchange with the rhizosphere: a microelectrode and modeling study with *Phragmites australis*. Ann Bot. 2000; 86:687–703.

76 Wießner A, Kuschk P, Kaestner M, Stottmeister U. Abilities of halophyte species to release oxygen into rhizospheres with varying redox conditions in laboratory-scale hydroponic systems. Int. J. Phytoremediation 2002; 4:1–15.

77 Olson RK. Created and Natural Wetlands for Controlling Non-Point Source Pollution. Man-Tech Environmental Technologies, Inc., USEPA, Corvallis, OR, USA; 1993.

78 Tanner C Plants as ecosystem engineers in subsurface-flow treatment wetlands. Water Sci Technol. 2001; 44(11–12):9–17.

79 Carballeira T, Ruiz I, Soto M. Effect of plants and surface loading rate on the treatment efficiency of shallow subsurface constructed wetlands. Ecol Eng. 2016; 90:203–214.

80 Calheiros CSC, Rangel AOSS, Castro PML. Constructed wetland systems vegetated with different plants applied to the treatment of tannery wastewater. Water Res. 2007; 41(8):1790–1798.

81 Vymazal J. Removal of nutrients in various types of constructed wetlands. Sci Total Environ. 2007; 380:48–65.

82 Redmond E, Just CL, Parkin GF. Nitrogen removal from wastewater by an aerated subsurface-flow constructed wetland in cold climates. Water Environ Res. 2014; 86(4):305–313.

83 Tunçsiper B, Drizo A, Twohig E. Constructed wetlands as a potential management practice for cold climate dairy effluent treatment. USA CATENA (135); 2015, pp. 184–192.

84 Toet S, Van Logtestijn RSP, Kampf R, Schreijer M, Verhoeven JTA. The effect of hydraulic retention time on the removal of pollutants from sewage treatment plant effluent in a surface-flow wetland system. Wetlands. 2005; 25(2):375–391.

85 Vymazal J. Nitrogen removal in constructed wetlands with horizontal sub-surface flow – can we determine the key process? In: Vymazal J, ed. Nutrient cycling and retention in natural and constructed wetlands. Backhuys: Leiden; 199, pp. 1–17.

86 He Q, Mankin KR. Seasonal variation in hydraulic performance of rock plant filters. Environ Technol. 2001; 22:991–999.

87 Machate T, Noll H, Behrens H, Kettrup A. Degradation of phenanthrene and hydraulic characteristics in a constructed wetland. Water Res. 1997; 31(3):554–560.

88 Carballeira T, Ruiz I, Soto M. Aerobic and anaerobic biodegradability of accumulated solids in horizontal subsurface flow constructed wetlands. Int Biodet Biodeg. 2017; 119:396–404.

89 Faulkner SP, Richardson CJ. Physical and chemical characteristics of freshwater wetland soils. In: Hammer DA, ed. Constructed Wetlands for Wastewater Treatment: Municipal, Industrial and Agricultural; 1989.

90 Nurk K, Truu J, Truu M, Mander Ü. Microbial characteristics and nitrogen transformation in planted soil filter for domestic wastewater treatment. J Environ Sci Health 2005; 40:1201–1214.

91 Saeed T, Sun G. A review on nitrogen and organics removal mechanisms in subsurface flow constructed wetlands: Dependency on environmental parameters, operating conditions and supporting media. J Environ Manage. 2012; 112:429–448.

92 Spieles DJ, Mitsch WJ. Macroinvertebrate community structure in high–and low–nutrient constructed wetlands. Wetlands BioOne 2000; 20(4):716–729.

93 APHA. Standard Methods for the Examination of Water and Wastewater. 21st edn, Washington: American Public Health Association/American Water Works Association/Water Environment Federation; 2005.

94 Rossman M, Teixeira A, Carneiro E, Fonseca F, Carraro A. Effect of influent aeration on removal of organic matter from coffee processing wastewater in constructed wetlands. J Environ Manage 2013; 128:912–919.

95 Zhang S, Li G, Chang J, Li X, Tao L. Aerated enhanced treatment of aquaculture effluent by three-stage, subsurface flow constructed wetlands under a high loading rate. Pol J Environ Stud 2015; 23(5):1821–1830.

96 Ong SA, Ho LN, Wong YS, Chen SF, Dugil DL, Samad S. Semi-batch operated constructed wetlands planted with *Phragmites australis* for treatment of dyeing wastewater. J Eng Sci Technol. 2011; 6:619:627.

97 Ong SA, Ho LN, Wong YS, Chen SF, Viswanathan M, Bahari R. Mineralization of diazo dye (Reactive Black 5) in wastewater using recirculated up-flow constructed wetland reactor. Desalination Water Treat. 2012; 46:312–320.

4

Treatment of Wineries and Breweries Effluents using Constructed Wetlands

F. Masi, A. Rizzo and R. Bresciani

Iridra Srl, Florence, Italy

4.1   Introduction

Wine and beer are two of the most popular and consumed beverages worldwide, with around 250 million and 1.34 billion hectolitres of wine and beer produced yearly, respectively [1, 2]. Both wineries and breweries generate relevant fluxes of wastewater, which need to be managed and treated in order to avoid dangerous environmental impacts on receiving water bodies [3].

Although wineries and breweries generate a wastewater with similar characteristics from a chemical point of view, the widespread use of constructed wetlands (CWs – or treatment wetlands, TWs) to treat these effluents had opposite trends. CWs to treat winery wastewater are nowadays widely applied with many successful applications in different countries (such as France, Italy, Germany, Spain, USA, South Africa), from the simplest single-stage horizontal subsurface flow systems for small size wineries to complex multistage schemes for medium and large sized ones. On the other hand, very few applications of CWs to treat brewery wastewater are reported nowadays, highlighting the lack of development of CW technologies as a potential solution for this type of wastewater.

This chapter aims to provide insightful information for CW system designers for both winery and brewery wastewater. Therefore, an analysis of winery and brewery wastewater production and characterization is provided, with particular attention to the analogies and differences of the two wastewater types. Subsequently, the CW application for the two cases are briefly described, exposing the most advanced solutions proposed for wineries and reporting all the CW experiences for breweries available to date. Finally, an interesting cost-benefit comparison among CW solutions for wineries is proposed, in order to guide the designers to choose the proper solution for different winery contexts (winery size, available area, energy consumption, carbon footprint). Moreover, future perspectives of CW system application for brewery wastewater treatment are provided, particularly in terms of the possible reasons behind the scarce development of CW in this field; treatability of brewery wastewater in comparison to winery wastewater; and lessons to be learnt from CWs for wineries useful for future design of CWs for breweries.

Constructed Wetlands for Industrial Wastewater Treatment, First Edition. Edited by Alexandros I. Stefanakis.
© 2018 John Wiley & Sons Ltd. Published 2018 by John Wiley & Sons Ltd.

4.2 Wastewater Production and Characterization

4.2.1 Wineries

Wineries produce wastewater: (i) during the grape processing (vintage and racking) period, in which high wastewater volumes are generated; and (ii) outside the grape processing period due to bottling and container cleaning activities, which generate an almost continuous lower volume of wastewater. On average, 1 L of wine produces about 1.6–2.0 L of wastewater and 5–10 g of chemical oxygen demand (COD) [1]. Due to seasonal-dependent activities and different industrial process chains, wineries produce a wastewater characterized by high fluctuation in terms of qualities and quantities, as confirmed by the winery wastewater characterization encountered in literature and summarized in Table 4.1. Organics, solids, and nitrogen at very high peaks (especially compared with the common targets aimed at for discharge treated wastewater) reported in Table 4.1 are due to the grape processing phase during vintage, which is typically two months long.

The organic load driven by winery wastewater consists of highly soluble sugars, 25 different alcohols, acids and recalcitrant high molecular weight compounds (e.g., polyphenols), 26 tannins, and lignins [4]. A total of 90% of the organic matter present is due to ethanol and sugars (fructose and glucose) [1]. Different varieties of grapes and of the vinification conditions influence the phenolic composition of wines and also of the produced wastewater. In general, the major phenolic compounds present in grape wastes, and therefore identified in wastewater, are anthocyanins, catechins, glycosides of flavonoids and phenolic acids [5].

4.2.2 Breweries

Breweries generate wastewater from wort boiling, and stabilization and clarification industrial processes. Moreover, brewery processes can also generate wastewater from weak wort and discharged residual beer [6]. Despite differences due to different adopted technologies, one litre of beer generates on average 3–10 litres of wastewater [6]. The characterization of brewery wastewater reported in the literature is summarized in Table 4.2.

The high organic load is due to organics generated from dissolved carbohydrates, and the alcohol from wasted beer, while the high suspended solid loads can be attributed to spent maize, malt, and yeasts [7]. Nitrogen and phosphorous in brewery wastewater are linked to the raw material and the amount of yeast released in the effluent. In particular, nitrogen arises from malt and adjuncts, plus

Table 4.1 Winery wastewater characterization from the literature (adapted from Masi et al. [1]).

		Min	Max	Mean	Common treatment targets
COD	mg/L	340	49,103	14,570	120–160
BOD$_5$	mg/L	181	22,418	7,071	20–40
TSS	mg/L	190	18,000	1,695	35–80
pH	–	3.5	7.9	4.9	6–8
N (NH$_4^+$ + NO$_3^-$)	mg/L	2.88	364	26	15–35

BOD, biochemical oxygen demand; COD, chemical oxygen demand; TSS, total suspended solids.

Table 4.2 Brewery wastewater characterization from literature
(adapted from Vymazal [7]).

		Min	Max	Common treatment targets
COD	mg/L	750	80,000	120–160
BOD$_5$	mg/L	500	64,000	20–40
TSS	mg/L	100	3,000	35–80
NH$_4{}^+$	mg/L	1	8	15–35
TKN	mg/L	67	216	15–35
TP	mg/L	17	216	2

BOD, biochemical oxygen demand; COD, chemical oxygen demand; TKN,
total Kjehldahl nitrogen; TP, total phosphorus; TSS, total suspended solids.

a possible contribution of nitric acid used for cleaning on-site. Phosphorous can also originate from cleaning agents [6]. Usually, heavy metal concentrations are quite low [6, 7], while the pH could be very low, in between 3 and 4 [7].

4.3 Applications and Configurations

4.3.1 Wineries

CWs started to be used to treat winery wastewater in the early 1990s in the USA, and a few years later the first applications of this technology were also adopted in France, Italy, Germany, and Spain [1]. A crucial role in spreading the CW application for winery wastewater treatment was the dissemination of the results, which demonstrated the suitability of technologies in terms of both removal efficiencies and additional advantages (e.g., low cost, low maintenance, energy savings) [8–17]. Different schemes have been adopted in the function of different wineries production cycles. Successful applications of CWs for wineries wastewater with average COD removal efficiencies of 60–99% have been reported in the literature for a wide variability of: (i) adopted pretreatments (e.g., coarse sand filters, facultative ponds, septic tanks, simple equalization, anaerobic digesters); (ii) CW area, from 15 m^2 of pilot scale to 3,000 m^2 of multistage solutions; (iii) hydraulic retention times (HRTs), from 3 to 14 days; (iv) organic loading rates (OLRs) during peak season, from 34.5 to 630 g$_{COD}$/m^2/d [1].

Generally, the complexity of the CW treatment scheme is function of the size of the wineries, which can be expressed in terms of quantity of produced wine per year (hL/year). For instance, a census of about 100 CW systems located in Northern and Central Italian wineries done by Masi et al. [1] in 2015 revealed that very often the simplest scheme is adopted when the productivity is lower than 2,000 hL of wine per year: septic or Imhoff tanks (with also an equalization role) plus a single-stage CW (usually horizontal subsurface flow, with some cases of vertical subsurface flow). On the other hand, bigger wineries produce very high organic and suspended solid loads concentrated in a few months, and single-stage solutions require a huge amount of area to avoid the risk of clogging. According to the available literature, three different approaches can be adopted to limit the CW requested area and the risk of CW clogging in case of big wineries: (i) a passive approach, with preliminary composting of the wastewater as pretreatment, provided by nature-based solutions (French Reed Bed – FRB, [9]); (ii) a passive approach, with preliminary composting of the wastewater as pretreatment, provided by technological solutions (anaerobic digester – AD; [15–17]) (iii) an active solution, with aerobic technological reactors and subsequent sludge composting on CWs. Since the multistage CW systems represent the state of the art in terms of available CW technology solutions for winery wastewater, their real scale applications are analysed in detail.

4.3.1.1 Multistage CW with Nature-Based Composting as Pretreatment for Wastewater: An Italian Case Study

The Italian case study proposed here describes the multistage CW treatment plant for Casa Vinicola Luigi Cecchi & Sons Winery (Castellina in Chianti – Siena), designed by Iridra Srl. This winery only performs bottling and aging of wine in cellar activities, since the wine is produced elsewhere. Therefore, the wastewater to be treated is generated only by washing of bottles, tanks, silos, ground floors, etc. Consequently, the up to 100 m³/d of wastewater generated is characterized by a constantly high organic matter content (3,800 mg$_{COD}$/L on average), low nutrient loads, light acidity, and daily fluctuations linked to the high variability of the activities performed during the day.

The CW was originally designed and realized in 2001 with a simpler scheme to treat up to 35 m³/d: Imhoff tank, single horizontal subsurface flow (HF) stage, final free water surface (FWS) stage. However, starting from 2006, winery productivity increased up to 70 m³/d, putting stress on the CW system and causing severe clogging after 2–3 years of overload. As a consequence, the treatment plant was upgraded in 2009 leading to the current configuration pictured in Figure 4.1: an equalization tank; French Reed Bed (FRB) system for raw wastewater as first stage; four parallel HF beds (the old refurbished bed, HF1, and the three new ones in a single hydraulically separated bed, HF2) as second stage; the existing FWS as third stage; an optional sand filter as fourth stage for gardening reuse purposes. The results of the first three years after the upgrading in 2009 were reported by Masi et al. [18] in 2014, showing a COD removal efficiency up to 96% during the peak loads (range 96–99% during all seasons) and an overall surface removal efficiency up to 232 g$_{COD}$/m²/d.

The greater novelty of this system is the adoption of a composting preliminary phase developed with the FRB nature-based solution. FRB systems are vertical subsurface flow CWs filled with a coarse material (gravel) and planted with *Phragmites australis*. The FRB is fed in batches with raw wastewater, which is filtered on the top of the FRB surface generating the so-called deposit layer [19]. The deposit layer retains the majority of the TSS contained in the winery wastewater, which mineralizes under aerobic conditions during the resting period of the bed; in other words, the top of the

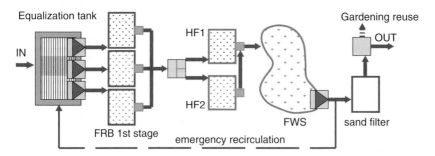

Figure 4.1 Schematization of the multistage CW treatment plant of the Italian case study, in which nature-based composting as pre-treatment with FRB is adopted.

FRB surface works as an aerobic composting reactor. In order to provide sufficient resting periods, the three beds are alternatively fed in batches. The movement of the reeds due to wind, which creates fractures within the deposit layer, and the bottom drainage system maintain aerobic conditions and avoid odor issues. After 15–20 years the deposit layer needs to be removed (estimation done on domestic wastewater treatment, while data on winery wastewater are still needed due to the recent application); after that the FRB can start again for a new phase. The removed material of the deposit layer can be used as soil amendment in agriculture after proper quality tests. The FRB remains in aerobic conditions, and the wastewater percolating within the FRB is treated with classical aerobic biofilm processes such as hydrolysis, aerobic organic matter degradation, and nitrification.

The Italian case of Castellina in Chianti highlights the possibilities of also using FRB for wineries wastewater, since the previous experiences with FRBs were only for domestic wastewater. Indeed, the efficiency of the first FRB stage was very high, with average COD removal efficiencies of 75% under very high OLRs (56–205 $g_{COD}/m^2/d$). Overall, the FRB approach avoided the typical CW primary treatment, with the old Imhoff tank dismissed. This led to reduced costs in terms of sludge disposal. Finally, the adoption of FRB as first stage eliminated the clogging problem in the second stage HF bed.

4.3.1.2 Multistage CW with Technological Composting as Pretreatment for Wastewater: A Spanish Case Study

The Spanish case study is situated in Galicia (northwest Spain), and regards the combination of an anaerobic digester (AD) plus CWs to treat the wastewater from a winery with a production capacity of 3,150 hl of white wine per year. The registered average raw wastewater characteristics are 520 mg_{TSS}/L, 2,107 mg_{COD}/L, and 1,199 mg_{BOD5}/L. The details of the treatment plant are (Figure 4.2): (i) a 6 m^3 hydrolytic upflow sludge blanket (HUSB) AD as pretreatment; (ii) a 50 m^2 vertical subsurface flow (VF) CW bed as first stage, planted with *Phragmites australis*; (iii) three HF CWs in parallel (100 m^2 each) as second stage, planted with *Juncus effusus*, and with one bed shallower (HF1 – depth of 0.35 m) than the other two (HF2 and HF3 – depth of 0.65 m). The CW treatment plant commenced operation in 2008, and was closely monitored for more than 2 years, with results described by Serrano et al. [15] in 2011 and de la Varga et al. [16, 17] in 2013, and reviewed by Masi [1] in 2015. The overall average removal efficiencies of the system are 86.8% for TSS, 73.3% for COD, 74.2% for BOD$_5$, 52.4% for TKN, 55.4% for NH_4^+-N, and 17.4% for phosphates. However, the higher organic loading rate during vintage seasons led to an overall decrease of removal efficiencies near 50%.

The peculiarity of this case study is the adoption of a technological component for reducing the organic load by anaerobic digestion as pretreatment, in a more efficient way if comparing with common primary treatment tools such as septic or Imhoff tanks. For a proper AD functioning, the

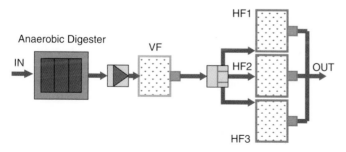

Figure 4.2 Schematization of the multistage CW treatment plant of the Spanish case study.

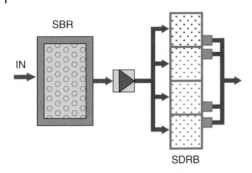

Figure 4.3 Schematization of the multistage CW treatment plant of the Spanish case study, in which a technological composting as pretreatment with FRB is adopted.

raw wastewater is accumulated in a 20 m^3 storage tank and then pumped with an average flow rate of 6.8 m^3/d to the bottom of the AD. The AD has shown average removal efficiencies of 76.4% for TSS, 26.3% for COD, and 21.3% for BOD$_5$. As a consequence, the TSS concentration influent to the CW system remained below 200 mg/l, and no clogging was observed during the two years of monitoring at the high surface loading rate of the overall system. In terms of environmental impact and a resources recovery oriented approach, this configuration can offer the advantage of providing a combustible gas as a product from the AD (mainly for its methane content) that can be used for producing energy.

4.3.1.3 Multistage CW with Technological Aerobic Reactor and Subsequent Composting on CW: A French Case Study

The French case study describes the Cantemerle winery, situated in Gironde. The Cantemerle winery produces 2,700 hL of wine per year, and is equipped with a treatment plant designed to treat 6.5 m^3/d of wastewater, 58.5 kg$_{BOD5}$/d, and 117 kg$_{COD}$/d. The treatment plant consists of (Figure 4.3): (i) a buffer tank of 10 m^3, with an HRT of 1.5–2 days; (ii) a screening unit, with sieve of 1 mm; (iii) a sequencing batch reactor (SBR) with a volume of 50 m^3; (iv) four sludge drying reed bed filters (SDRB) with a surface of 14 m^2 each, used to manage the sludge produced by the SBR. The removal efficiencies monitored during one and a half years after commissioning (2005–2006) are very high as reported by Masi et al. [1], with average removal efficiency for COD equal to 99%.

In this solution, the wastewater is mainly treated by the SBR technological reactor, while the SDRB both composts the sludge and polishes the SBR effluent. SDRBs are a particular application of CWs, born to dewater the sludge produced by centralized activated sludge wastewater treatment plants [20]. The SDRB technology is mostly comparable to a VF CW and exploits the same principle of the FRB; indeed, SDRB consists of a drainage layer made with porous media of different sizes (sand, gravel, small stones), in which the sludge is filtered and remains at the top of the SDRB. To this aim, the SDRB embankments are higher compared to classical VF CWs (0.6 m over the gravel top layer for the Cantemerle case study), in order to have enough storage volume to accumulate the sludge over the years. A bottom drainage system collects the leachate and ensures aeration of the bed during the resting periods, maintaining the best conditions for SDRB functioning. *Phragmites australis* are planted in the beds, catalyzing and helping the processes of dewatering (through transpiration) and mineralization of the sludge. Moreover, the macrophytes also prevent clogging in the root zone and help to maintain aerobic conditions within the deposited sludge on top thanks to the wind-driven movements, limiting odor issues and maintaining the hydraulic conductivity. The dewatered sludge at the top of the SDRB is removed when the freeboard results are almost filled, and can be reused as fertilizer, according to limits imposed by laws and the specific content.

Although the solution is more expensive in terms of required energy, the combination of SBR and SDRB effectively fits the scope of winery wastewater treatment. In particular, the SDRB applied for dewatering the sludge produced by winery wastewater treatment was very effective, since the huge organic and suspended solid loads are produced in a few months (typically the 2 months of the vintage season), leaving enough resting period (usually 10 months) to properly manage the sludge and SDRB's clogging avoidance.

4.3.2 Breweries

Although CWs are reported as a possible option to treat brewery wastewater by Simate et al. [21], only two applications are available in the literature as recently reviewed by Vymazal [7], in 2014.

The South African Miller brewery plant at the Ibhayi Brewery in Port Elisabeth (South Africa) was testing a pilot plant for treating brewery wastewater, composed of an anaerobic digester and an integrated algal pond system [22, 23]. This treatment plant didn't meet the South African discharge limits; therefore an additional HF CW has been added for final polishing. The HF CW occupies an area of 56 m^2 (divided into four channels), and was planted with *Typha compensis* and *Phragmites australis* in sequential blocks. The preliminary results were encouraging, with overall effluent concentrations below the discharge limits (75 mg/L for COD, 3 mg/L for ammonia, 15 mg/L for nitrate, and 10 mg/L for phosphate).

A FWS CW is also reported to be in operation for The Coors Brewery in Golden (Colorado, USA) by Kadlec and Wallace [24], but without any specifications about purposes and treatment plant characteristics.

4.4 Discussion and Conclusions

4.4.1 Advantages and Disadvantages of Different Multistage CW Treatment Plants

The choice among different multistage CW solutions is usually done on the basis of three concepts: area available, energy required and operational costs, and sustainability. In order to provide useful insights for future designers, these three parameters are analyzed for the three multistage systems previously described. For each solution, the area, energy, and carbon footprints are estimated and correlated to the yearly amount of COD treated. The data used to make the comparison are summarized in Table 4.3. For both AD+VF+HF and SBR+SDRB systems, the duration of the vintage period is assumed to be 60 days long. For SBR+SDRB, only the effect of the vintage season is considered. The SBR+SDRB energy consumption is calculated from the person equivalent (PE) treated, estimated by dividing the yearly carbon load treated per PE daily organic load (130 g COD/PE/d; [25]) and per 365 day of a year. The yearly SBR+SDRB energy consumption per PE is assumed equal to 67.5 kWh/PE/year [25]. The AD+VF+HF energy consumption is calculated assuming a small pumping system with a power of 1 kW, suitable for pumping the average wastewater flow. The AD+VF+HF pump is assumed to work for the same time as the FRB+HF+FWS treatment plant. The energy price is taken equal to 0.128 €/kWh, i.e., the value reported for industries in the Euro area by the Eurostat for 2014. On the basis of European Environmental Agency data from 1992 to 2007, the average carbon footprint strongly varies in the different geographical areas (from 320 g$_{CO2}$/kWh of Russian Federation to 786 g$_{CO2}$/kWh of China), due to different national energetic strategies; in this case, the European Union value of 389 g$_{CO2}$/kWh is assumed.

Table 4.3 Data used for area, energy, and carbon footprint from [1, 3, 16, 17].

Parameter	Unit	FRB+HF+FWS	AD+VF+HF	SBR+SDRB
Area	m^2	3010	352	81
Area VF			50	
Area HF			300	
Pump power	kW	7[a]	1[b]	
Influent COD average concentration	mg_{COD}/L	3800	2107	
Surface loading rates in vintage (only VF)	$g_{COD}/m^2/d$		593	
Surface loading rates outside vintage (all CWs)	$g_{COD}/m^2/d$		30.4	
Influent mass load per day during vintage	kg_{COD}/d			117
Influent winery wastewater quantity during vintage	$[m^3/d]$			6.5
Influent average winery wastewater quantity	$[m^3/d]$	90	6.8	
Day of wastewater production in vintage	[d]		60[b]	60[b]
Day of wastewater production outside vintage	[d]	252[a]	192[b]	0[b]
Time of pump functioning per day	[min/d]	100[a]	100[b]	

[a] Personal information
[b] Assumption.

Table 4.4 Results from the area, energy, and carbon footprint analysis of proposed multistage CW solutions.

Parameter	Unit	FRB+HF+FWS	AD+VF+HF	SBR+SDRB
Average COD removal	[%]	96-99	73 (average) 50 (peak)	99
COD mass treated per year	$[t_{COD}/y]$	86.2	3.8	7.0
AE equivalent	[PE]	1816	81	148
Energy consumption	[kWh/y]	2940	320	11836
Total cost per energy consumed	[€/y]	376	41	1515
Area footprint	$[m^2/t_{COD}]$	35	92	12
Energy footprint	$[€/t_{COD}]$	4	11	182
Carbon footprint	$[kg_{CO2}/t_{COD}]$	13	33	554

The results of the analysis and the removal efficiencies of the systems are summarized in Table 4.4. As expected, the minimization of area footprint is achieved by the SBR+SDRB solution; however, FRB+HF+FWS does not appear as the most extensive solution, since the area footprint of AD+VF+HF is higher. On the other hand, the smaller energy footprint provided by FRB+HF+FWS is also confirmed. The energy footprint of AD+VF+HF is of the same order of magnitude, but higher than the FRB+HF+FWS solution. As expected, the energy footprint of SBR+SDRB is one order of magnitude higher compared to other CW multistage systems. As a consequence of the energy footprint, the carbon footprint follows the same trend, with FRB+HF+FWS resulting as the best solution in terms of environmental impact reduction. These results confirm that the more nature-based solutions are preferable when sufficiently cheap lands are available, obtaining by their adoption a relevant saving in operational costs and a reduction of the environmental impacts linked to the wastewater treatment. At the same time, the more technological solution (SBR+SDRB) remains highly attractive when there are land availability limitations, both for geomorphological issues and economic reasons. Finally, AD+VF+HF still needs to be further optimized to become a widely spread solution for winery wastewater treatment, especially in terms of removal efficiencies.

4.4.2 Future Perspectives of CW for Brewery Wastewater Treatment

The main cause of the lower number of applications of CWs for breweries, in comparison to the wineries, can be reasonably assumed to be the location of the facilities, very often inside urban settlements, where the land availability can be a limiting factor. As the brewery sector is one of the most water-demanding industries on a worldwide scale, development of treatment schemes are aimed reaching a similar quality of effluents to that of the water input to the productive cycle. All the different technologies that are based on biological processes are almost ineffective regarding salt content, and this can limit their application, as high salinity is not appropriate for closed-loop reuse. Therefore combinations of technologies are recommended, where the biological treatment (including CWs) can still play an important role for the reduction of organic matter, solids and nutrients before treating effluent in a quaternary treatment for high quality water production and internal reuse. This approach could save about 70% of the water consumption for any litre of produced beer.

The extension of the successful experiences of CW systems applied to winery effluents to the breweries is not immediate, and must take into consideration the very different operational phases along a normal year, where the wineries have to deal with a short period of a few months with high loading peaks and a further long period, where the system can in some ways regenerate, while the breweries present a very constant release of wastewater with the same chemical composition. All the CW systems analyzed for the wineries are in any case showing that solids and organic content can be efficiently removed at various levels of removal that depend on the complexity and the extension of the treatment scheme. If combined with further treatment, such as membrane bioreactors, electrochemical methods, nanofiltration, reverse osmosis (for citing some), CWs can be very effective in optimizing the treatment approach and reduce both the investment and operational costs of the whole sequence.

References

1 Masi F, Rochereau J, Troesch S, Ruiz I, Soto M. Wineries wastewater treatment by constructed wetlands: a review. Water Sci Technol. 2015; 71(8):1113–1127.

2 Source FA. 2002 World beer production. BIOS International. 2003; 8(2):47–50.

3 Serrano L, De la Varga D, Ruiz I, Soto M. Winery wastewater treatment in a hybrid constructed wetland. Ecol Eng. 2011; 37(5):744–753.

4 Arienzo M, Christen EW, Quayle WC. Phytotoxicity testing of winery wastewater for constructed wetland treatment. J Hazard Mater. 2009; 169(1):94–9.

5 Lafka TI, Sinanoglou V, Lazos ES. On the extraction and antioxidant activity of phenolic compounds from winery wastes. Food Chem. 2007; 104(3):1206–1214.

6 Olajire AA. The brewing industry and environmental challenges. J Clean Prod. 2012.

7 Vymazal J. Constructed wetlands for treatment of industrial wastewaters: a review. Ecol Eng. 2014; 73:724–751.

8 Shepherd HL, Grismer ME, Tchobanoglous G. Treatment of high-strength winery wastewater using a subsurface-flow constructed wetland. Water Env Res. 2001; 1:394–403.

9 Masi F, Conte G, Martinuzzi N, Pucci B. Winery high organic content wastewaters treated by constructed wetlands in Mediterranean climate. In: Proceedings, 8th International Conference on Wetland Systems for Water Pollution Control 2002; Sep 16, pp. 274–282.

10 Müller DH, Dobelmann JK, Hahn H, Pollatz T, Romanski K, Coppik L. The application of constructed wetlands to effluent purification in wineries. In: Proceedings, 8th International Conference on Wetland Systems for Water Pollution Control 2002; Sep 16, pp. 599–605.

11 Rochard J, Ferrier VM, Kaiser A, Salomon N. The application of constructed wetlands in the viticultural sector: experimentation on a winery effluent treatment device. In: Proceedings, 8th International Conference on Wetland Systems for Water Pollution Control 2002; Sep 16, pp. 494–503.

12 Grismer ME, Carr MA, Shepherd HL. Evaluation of constructed wetland treatment performance for winery wastewater. Water Env Res. 2003; 1:412–1421.

13 Mulidzi AR. Winery wastewater treatment by constructed wetlands and the use of treated wastewater for cash crop production. Water Sci Technol. 2007; 56(2).

14 Vymazal J. The use constructed wetlands with horizontal sub-surface flow for various types of wastewater. Ecol Eng. 2009; 35(1):1–7.

15 Serrano L, De la Varga D, Ruiz I, Soto M. Winery wastewater treatment in a hybrid constructed wetland. Ecol Eng. 2011; 37(5):744–753.

16 De la Varga D, Ruiz I, Soto M. Winery wastewater treatment in subsurface constructed wetlands with different bed depths. Water Air Soil Pollut. 2013; 224(4):1–3.

17 De la Varga D, Díaz MA, Ruiz I, Soto M. Avoiding clogging in constructed wetlands by using anaerobic digesters as pretreatment. Ecol Eng. 2013; 52:262–269.

18 Masi F, Bresciani R, Bracali M. A new concept of multistage treatment wetland for winery wastewater treatment: long-term evaluation of performances. In: The Role of Natural and Constructed Wetlands in Nutrient Cycling and Retention on the Landscape; Springer International Publishing. 2015, pp. 189–201.

19 Molle P. French vertical flow constructed wetlands: a need of a better understanding of the role of the deposit layer. Water Sci Technol. 2014; 69(1).

20 Nielsen S. Sludge drying reed beds. Water Sci Technol. 2003; 48(5):101–109.

21 Simate GS, Cluett J, Iyuke SE, Musapatika ET, Ndlovu S, Walubita LF, Alvarez AE. The treatment of brewery wastewater for reuse: State of the art. Desalination. 2011; 273(2):235–247.

22 Crous L, Britz P. The use of constructed wetland technology in the treatment and beneficiation of brewery effluent for aquaculture. In: 13th IWA International Conference on Wetland Systems for Water Pollution Control, Venice, Italy; 2010, Oct 4, pp. 1254–1259.

23 Jaiyeola AT, Bwapwa JK. Treatment technology for brewery wastewater in a water-scarce country: A review. S Afr J Sci. 2016; 112(3-4):1–8.

24 Kadlec RH, Wallace S. Treatment wetlands. Boca Raton: CRC Press; 2009.

25 Masotti L. Depurazione delle acque. Tecniche ed impianti per il trattamento delle acque di rifiuto (Wastewater treatment. Technologies and plants for wastewater treatment). Calderini Editore – Il Sole. 2011;24.

5

Treatment of Effluents from Fish and Shrimp Aquaculture in Constructed Wetlands

Yalçın Tepe[1] and Fulya Aydın Temel[2]

[1] *Department of Biology, Faculty of Sciences and Arts, Giresun University, Giresun, Turkey*
[2] *Department of Environmental Engineering, Faculty of Engineering, Giresun University, Giresun, Turkey*

5.1   Introduction

Farming of fish and shrimp, which culture over 600 aquatic species worldwide including finfish (e.g., sea bream, salmon, trout, carp), crustaceans (e.g., lobster, crayfish, shrimp, crabs, prawn), and mollusks (e.g., clams, mussels, oysters), is an evolving sector of agriculture that creates important economic opportunities in many rural communities plagued by unemployment (Figure 5.1) [1]. Seafood are excellent nutritional foods with high level of protein, vitamins, fatty acids and minerals for current and future generations. Aquaculture must keep growing to meet future seafood demands, which is expected to increase by 70% within the next 30 years. Aquaculture is able to bring significant economic development in both rural and urban areas by improving family income, providing employment opportunities from culture to marketing and reducing problems of food supply and food safety [2]. On the other hand, aquatic resources are limited and the growing aquaculture sector must reconsider the serious concerns of resource allocation, environmental impact, and sustainability. Effluent water may cause negative effects on the environment with its high organic matter, nutrients, and suspended solids.

Constructed Wetlands have been used for wastewater treatment in a wide range of purposes, and are becoming available for use in aquaculture. However, special consideration needs to be given to the nutrients, organic matter, and suspended solids, together with the potential volume of the discharge in order to make Constructed Wetlands feasible for aquaculture operations. Wetland systems may eliminate important quantities of organic matter, suspended solids, phosphorus, nitrogen, trace elements, and microorganisms detected in effluents. Considering these reductions in environmental impact, wetland construction seems promising for the future of aquaculture. Besides these benefits, Constructed Wetland systems have also reasonable investment costs, low energy use and maintenance needs, and benefits for the enlarged wildlife habitat.

5.1.1   Concerns in Aquaculture

As natural fish stocks are being fished to their sustainable limit and beyond, aquaculture is estimated to keep growing, already affording about 30% of fishery production [3]. The development of the

Constructed Wetlands for Industrial Wastewater Treatment, First Edition. Edited by Alexandros I. Stefanakis.
© 2018 John Wiley & Sons Ltd. Published 2018 by John Wiley & Sons Ltd.

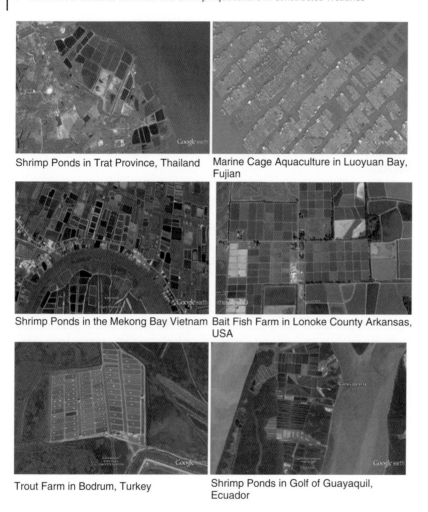

Shrimp Ponds in Trat Province, Thailand

Marine Cage Aquaculture in Luoyuan Bay, Fujian

Shrimp Ponds in the Mekong Bay Vietnam

Bait Fish Farm in Lonoke County Arkansas, USA

Trout Farm in Bodrum, Turkey

Shrimp Ponds in Golf of Guayaquil, Ecuador

Figure 5.1 Different aquaculture farms through the world (raceways, cages, ponds). (Image source: Google Earth).

aquaculture industry needs to be compatible with the environment and ecologically sustainable. Some of the potential environmental concerns regarding aquaculture are listed below [4–6]:

(a) Habitat loss, destruction of wetlands, mangroves and lagoons by aquaculture projects; the destruction and loss of marshes and mangroves for shrimp pond construction and agricultural fields for freshwater fish culture are the main concerns.
(b) Habitat modification; modification of agricultural land for aquaculture ponds.
(c) Surface water deterioration and pollution due to aquaculture pond effluents.
(d) Misuse of drugs; antibiotics for disease control of fish and shrimp.
(e) Wasteful use of fishmeal for aquaculture production.
(f) Salination of streams, aquifers, and soils by pond discharge.
(g) Misuse of ground and surface water for aquaculture production processes.

(h) Escapes of cultured fish and interactions with wild populations; disease infection.
(i) Adverse effects of non-native escaped fish on biodiversity, effects of birds, otters and other predators, and drift of aquatic organisms in pumps and pipes.
(j) Conflicts with other resource users and disturbance of neighboring communities, such as hotels and touristic facilities.
(k) Excessive harvesting of wild seed/spawners and damage to catch.

Considering the above and several additional potential adverse effects, water deterioration and pollution by aquaculture effluents is apparently the most important concern, and in most nations, this issue has attracted the highest official attention [7, 8].

Raceways and ponds, where shrimp and fish production are mostly carried out, may produce effluents when they are harvested or after flooding [9]. Recirculated production systems are not technologically or cost-effectively achievable to manage in most fish and shrimp farms without water discharge and renewal.

Aquaculture effluents include significant amounts of organic matter, plankton, suspended solids, nutrients such as nitrogen, and phosphorus, and require oxygen demand that can further deteriorate the quality of receiving surface waters. Therefore, an applicable and affordable wastewater treatment process is vital to sustain aquaculture development in the world.

5.2 Overview of Aquaculture and Effluent Treatment

Aquaculture techniques, an ancient practice, were first seen among the indigenous Gunditjmara people in Victoria, Australia, who had cultured eels as early as 6,000 BC by converting almost a hundred km^2 of volcanic flood fields into a series of channel complexes, reservoirs and ponds. They had captured and preserved eels into handmade woven traps to consume later on, all year round [10]. The Japanese began to cultivate seaweed as early as 1670 in Tokyo Bay by using oyster shells and bamboo sticks to provide anchoring surfaces for spots [11].

Mechanically sophisticated and biologically complex, recirculating aquaculture systems (RAS) reuse water by passing it across various filter systems to eliminate organic matter out of the water and then returning it back into the culture tanks [12]. The initial scientific research on RAS carried out in Japan back in the 1950s by focusing on biofilter design for carp culture because, they needed to use locally-limited water resources more efficiently. European and American scientists endeavored to modify technology using biofilters, trickling filters, etc. which were initially developed for domestic wastewater treatment, then became involved in marine structures for fish and shrimp production.

Aquaponics is a relatively new fish production facility, which unites known tank aquaculture practices with hydroponics, i.e., cultivating plants in water, in a reciprocal environment [13]. Polyculture aquaculture systems and early examples of aquaponics existed in Thailand, South China and many Far Eastern countries, where rice was cultivated with fish such as common carp [14]. Aztecs grew crops on artificial agricultural islands as early as 1150–1350 BC in harmony with fish, using nutrient-rich mud and water from the canals which were the architects of aquaponics. This kind of primitive aquaponics, called chinampa, was constructed by staking out the shallow lake bottom and fencing in the rectangle with wattle. Fish waste settled to the bottom of the ponds and was then collected to fertilize plants. These floating or stationary gardens had excessive harvest yields with four or more harvests per year [15].

The term aquaponics is often attributed to the several projects of the researchers in the New Alchemy Institute at North Carolina State University. In 1969, the culmination of their efforts was the construction of a prototype Bioshelter, the "Ark" [13].

Excessive use of pelleted feed and expansion of the aquaculture area are the main negative environmental effects of modern intensive aquaculture [6]. Pellet feed is partially converted into fish biomass and the rest released into the water as suspended organic solids, carbon dioxide, ammonia, phosphates and other compounds [3, 16, 17], which can result in a significant contribution of organic and inorganic matter to the aquatic biota [18]. Formulation of more digestible diets with lower nutrient levels is definitely important to reduce this effluent load. However, treatment of effluents is an inevitable necessity to compensate for new administrative demands and the pressure coming from environmentalists [19].

5.2.1 Effluent Water Quality Considerations

The water quality parameters of most concern in fish and shrimp aquaculture are inorganic suspended solids (ISS), total suspended solids (TSS), organic matter (OM), biochemical oxygen demand (BOD), dissolved oxygen (DO), total phosphorus (TP), pH and nitrogenous compounds [20].

Governmental agencies issue and enforce standards for effluent water quality regulations and monitor the individual effluent outfalls. The main purpose of regulating these water quality standards is to prevent the negative impacts of effluents on receiving natural waters [21]. Standards specify limits for selected water quality variables and may contain the following critical water quality restrictions: pH, 6 to 9; DO, 5 mg/L or more; BOD_5, 30 mg/l or less; TSS, 50 mg/L or less.

At low pH, the amount of mucus on gill surfaces increases, since gill tissue is the primary target organ of acid stress. Pond pH changes during the day in response to photosynthesis. Intense photosynthesis may increase the pH up to 9 in the afternoon, while pH of poorly buffered waters may fall to 6 in the early morning.

BOD is a measure of the rate of oxygen consumption by the plankton and bacteria in a pond water. The BOD is commonly used in estimating the pollution strength of the effluent. Shrimp and fish ponds typically have BOD values of 5–10 mg/L, and aeration is needed when BOD exceeds 20 mg/L. As BOD is an indicator of OM accumulation in ponds, the greater BOD, the greater degree of enrichment of pond water with OM.

Toxic metabolites such as CO_2, ammonia, and H_2S may reach harmful concentrations as a result of metabolic activity by organisms in ponds.

Shrimp and most fish species can survive in water containing up to 60 mg/L of CO_2 if DO concentration is maintained at high levels. High levels of CO_2 are frequently encountered after plankton die-offs which lower DO.

The major source of ammonia in pond water is the direct excretion of ammonia by fish and shrimp. The combination of high total ammonia nitrogen concentration (TAN) and high pH can result in ammonia toxicity to fish and crustaceans in ponds. The effluent, which does not comply with the permit or standards, must be treated by some procedure that will result in higher quality effluent.

The treatment of fish and shrimp aquaculture effluents are crucial for a recirculating aquaculture system [22, 23]. There are several conventional treatment technologies, and physical, chemical, and biological methods which are applied in aquaculture systems.

Removal of solids takes place by physical processes (sedimentation and mechanical separation). Mechanical separation technologies used for aquaculture effluent are static and rotation microscreens. Pollutant removals achieved removal of 50–74% of solids, 49.3–63% of TP, and 10–42.7 of TN with the microscreens tested in several studies [24–29]. Sedimentation designed to generate non-turbulent flow conditions includes off-line settling basins, quiescent zones, sludge cones, and full flow settling basins. Solid (97%) and phosphorus removal (34%); 75–80% of solids and 99.9% of pathogenic viruses are achieved according to research [25, 30]. There are several disadvantages such as large surface area, high initial capital expenditure, clogging, energy, untreated dissolved pollution, only the removal of solids, large range of removal efficiencies, while physical methods have some advantages including simple/understandable design principles, simple/automatic operation, compact/smaller land area [31].

Biological methods are anaerobic processes such as anaerobic fixed bed and fluidized bed reactors, and upflow anaerobic sludge blanket (UASB) reactor, while aerobic processes include the activated sludge system, rotating biological contactor, trickling filter, aerated lagoon, and integrated bioprocess. Nitrification and denitrification of NO_3-N and NO_2-N into nitrogen gas, and the oxidation of organic matter are provided by these technologies [32–36], which assist in phosphorus removal [22]. The list of biological systems applied for aquaculture effluents is given in Table 5.1. The high COD/BOD removal can be achieved by biological treatment technologies (more than 60% and 80%

Table 5.1 Performance of biological treatment technologies for fish and shrimp farm effluents.

Process	Industry	Wastewater characteristics (mg/L)	Organic loading	Removal (%)	Ref
Activated sludge	Fish processing		0.5 kg $BOD_5/m^3/d$	BOD_5: 90–95	[90]
Rotating biological contactor	Fish cannery	pH: 6–7, BOD: 5,100, COD: 6,000–9,000, TSS: 2,000, TKN: 750	0.018–0.037 kg $COD/m^2/d$	COD: 85–98	[91]
Trickling filter	Squid processing	BOD: 2–3,000	0.08–0.4 kg $BOD_5/m^3/d$	BOD_5: 80–87	[92]
Aerated lagoon	Fish processing		0.5 kg $BOD_5/m^3/d$	BOD_5: 90–95	[90]
Anaerobic filter	Seafood processing		0.3–0.99 kg $COD/m^3/d$	COD: 78–84	[93]
Anaerobic filter	Seafood processing Tuna condensate	VA: 3,340	1.67 kg $COD/m^3/d$	COD: 60	[93]
Anaerobic digester	Tuna cooking	COD: 34,500, Cl⁻: 14,000, TS: 4,000	4.5 kg $COD/m^3/d$	COD: 80	[94]
Anaerobic digester	Mussel cooking	COD: 18,500, TS: 1,400, Cl⁻: 13,000	4.2 kg $COD/m^3/d$	COD: 75–85	[94]
Denitrification tank	Tilapia culture	Nitrate-N: 45.77, Ammonia: 0.17, Nitrite: 0.03, pH: 7.02, Alkalinity: 165		Nitrate: 85	[33]

(Continued)

Table 5.1 (Continued)

Process	Industry	Wastewater characteristics (mg/L)	Organic loading	Removal (%)	Ref
Upflow anaerobic filter	Saline wastewater	Cl⁻: 7500; TAN: 2,000; Free Ammonia: 3,000	5 kg COD/m³/d	Ammonia: >80	[95]
Biofloc Technology	African catfish, *C. gariepinus* culture	Protein: 32–34%		Ammonia: 98.7	[96]
Biofloc Technology	Tilapia ponds	C/N:20		TAN: 80.49–95, Nitrite: 78.79–83.33	[97]
Sequencing batch reactors-external compartment of BFT	Simulated fish pond water	TAN: 13–14		Ammonia: 98	[98]
Pretreatment +anaerobic digester + activated sludge bioreactor	Tuna processing	pH: 6.96, BOD: 3,300, COD: 5,553, TKN: 440, TSS: 1,575, fat: 1,450	1.2 kg COD/m³/d	COD: 85–95	[99]
Rotating biological contactor	Tilapia wastewater	TAN: 3.5		TAN: 40	[100]
Sequencing batch reactors	Shrimp wastewater	COD: 1,201, TS: 13.1, Ammonia: 101.7, Nitrate: 33.3, Nitrite: 260, Salinity: 2.6 ppt, pH: 7.8		COD: 97	[101]

for anaerobic and aerobic treatment, respectively) [32]. Although biological methods provide high efficiency, unstable performances, high energy demand, sludge production, frequent maintenance requirements, and inadequate removal of nitrogenous compounds constitute their negative points [22, 37, 38].

Removal of nitrogenous compounds including ammonia, nitrite and nitrate are also achieved by chemical treatment technologies alongside biological methods. The methods are sorption, ion exchange, reverse osmosis and electrodialysis [39]. Although ion exchange and sorption processes are effective for saline water treatment, they have higher operation costs including expensive adsorbent materials, the regeneration process of adsorbent, extra operational steps to remove anions, energy combustion, capital investment, maintenance requirements, and solid waste generation [40, 41]. Reverse osmosis has several advantages such as a low production cost, environmental friendly consequences, and high permeability efficiencies. It is used to treat organic compounds, ions, and proteins from aquaculture wastewater, sea water, and brackish water. The most important disadvantage is the energy cost during the operation [42]. Electrodialysis has some advantages such as no osmotic pressure, lower energy, much higher brine concentration, no regeneration compared to reverse osmosis and other technologies. However, there are a lot of disadvantages such

Table 5.2 List of chemical/electrochemical treatment systems for fish and shrimp farm effluents.

Process	Industry	Wastewater characteristics (mg/L)	Removal (%)	Ref
Ultra-low pressure asymmetric polyethersulfone(PES) membrane	Aquaculture wastewater	TSS: 260, TDS: 83.75, TP: 1.076, TAN:0.432	TAN: 75.42–85.30, TP: 83.85–96.49	[102]
Wind driven reverse osmosis system	Culture water of *Oreochromis niloticus* (tilapia)	TAN: 0.40–1.20	TAN: 92	[42]
Wind driven reverse osmosis system	Culture water of tilapia (both seawater and freshwater fish culture)	Ammonia: 0.40–0.71, Nitrite: 0.02–0.07, Nitrate: 0.30–0.67	Nitrogen: 90–97	[103]
Ion exchange membrane bioreactor	Oceanarium	Nitrate: 251–380, HCO_3^-: 140, Cl^-: 18,980, SO_4^{2-}: 2649, PO_4^{3-}: 5, Br^-: 65, $H_2BO_3^-$: 26, F^-:100, Na^+:10556, K^+: 380, Ca^{2+}: 400, Mg^{2+}:1272, Sr^{2+}:13, total salinity: 34,580	Nitrate: 41.61–89.36	[104]
Batch Electrolysis	Synthetic nitrate solutions	Nitrate: 100, NaCl: 0.5	Nitrate: 90.30	[105]
Batch Electrolysis	Synthetic aquaculture wastewater	Nitrite: 5, NaCl: 2,000, pH: 7	Nitrite: 99	[106]
Batch Electrolysis	Synthetic nitrate solutions	Nitrate: 100, NaCl: 500, pH: 7	Nitrate: 87.1	[107]
Batch, divided two electrode cell with cation and anion permeable membrane	Aquaculture wastewater	pH: 7.5–7.7, temperature: 24±1, Nitrate: 120–195.5, TOC: 9.2–11.2	Nitrate: 94.8, TOC: 97.3	[108]
Batch electrolysis	Pond water	TP: 0.296, TN: 4.4, COD: 46, BOD: 10, SS: 68	TP: 89.86, COD: 92.4, BOD: 78, SS: 97.1	[109]
Batch electrolysis	Seawater	pH: 6.9, COD: 54.8, TAN: 8, Nitrite: 80, Nitrate: 403.20, Cl: 26,167, Sulphate: 35,000	TAN: 100, COD: 88	[110]

as prohibitively expensive investment cost/energy combustion, inadequate removal capacities of microorganisms and organics [39, 43]. The list of chemical/electrochemical treatment systems is given in Table 5.2. New research methods are now being investigated to treat aquaculture effluent, because of the disadvantages of conventional treatment technologies.

5.3 Use of Constructed Wetlands for Treatment of Fish and Shrimp Aquaculture Effluents

Constructed Wetlands (CWs) are engineered systems with contained design, construction, and operation [44]. CWs are complex systems including biology, hydraulics and chemistry and have been applied for the treatment of aquaculture effluents during last two decades [45]. Due to the success achieved in the treatment of wastewaters such as domestic sewage, stormwater, agricultural, industrial, mine drainage, landfill leachate, polluted river water, and urban runoff, CWs have increased in popularity in recent years [22, 46]. CWs have become alternative treatment systems to traditional treatment technologies especially in rural settlements, industries, hotels, etc. because of their advantages such as low energy consumption/maintenance requirements, easy operation/maintenance, cost-effectiveness, landscape aesthetics, reuse and benefits of increased wildlife habitat [47–49].

The main factors of CWs are wetland vegetation, filter materials, hydrology, and microbial communities [50]. Hydrology is a key factor due to the effects of the vegetation type, microbial activity, and cycling of nutrients in filter material. Microorganisms play a role in nutrient transformation, while the type of vegetation affects the nutrient uptake capacity [51]. The most important role of macrophytes is to transfer oxygen via their root zones to the treatment bed [52]. Moreover, they contribute to treatment bed stabilization, uptake and store nutrients, increase porosity, hinder channelized flow, enhance aesthetics, and isolate the bed against freezing [46, 47].

Macrophytes are emergent plants, submergent plants, free floating plants and floating leaved plants [53]. The species such as *Typha* spp., *Phragmites* spp., *Scirpus* spp., *Juncus* spp., *Eleocharis* spp., *Iris* spp. are among emergent species that are used commonly. *Ceratophyllum demersum*, *Hydrilla verticillata*, *Myriophyllum verticillatum*, *Potamogeton crispus* and *Vallisneria natans* are given as examples of submergent plant species. The floating leaved plants are mainly *Marsilea quadrifolia*, *Nymphaea tetragona*, *Trapa bispinosa* and *Nymphoides peltata*. *Lemna minor*, *Hydrocharis dubia*, *Salvinia natans* and *Eichhornia crassipes* are free-floating plants [46, 53, 54]. The vegetation selection is made depending on various factors such as aims of project, climate conditions, wastewater characteristics, geographical distribution, availability of the plants in the region, agronomic management costs and maintenance [55]. CWs show a natural treatment based on biological activities between macrophytes and microorganisms (fungi, bacteria, algae) and their interactions in the filter media (gravel, sand, detritus, soil).

Physical (filtration, sedimentation, adsorption, and volatilization), chemical (precipitation, adsorption, degradation) and biological (plant metabolism, microbial interactions, natural die-off, microbial mediates reactions) treatment mechanisms can be employed together to remove various pollutions or to recover the water quality in CWs for wastewater treatment [51, 53, 56]. CWs remove contaminants as organic matter, inorganic matter, trace organics, pharmaceutical contaminants, pathogens, total phosphorus, ammonium, nitrite, nitrate, solids, fecal coliform bacteria (FC), and heavy metals [49, 57, 58].

The first studies were conducted on the application of wetland vegetation (*Phragmites australis*) by Seidel in 1952–1956 in Germany for the treatment of various types of wastewater (phenol, dairy and livestock) [50]. In the 1970s and 1980s, CWs were solely constructed for treatment of municipal wastewaters [59]. The first full-scale constructed wetland designed to treat the runoff from a dunghill was built in Prague in 1989. Although many of the problems experienced such as the lack of knowledge about design and operation, encouraging results were obtained [60]. Since the 1990s, CWs used to treat the numerous type of wastewaters have developed as an applicable and alternative treatment technology [59, 61]. There are now more than 1,000 CWs in China, more than 10,000 in America, more than 50,000 in Europe, and more than 100,000 around the world. Over million cubic meters of wastewater per day are treated by CWs worldwide [46, 62, 63].

CWs are classified typically into two types according to the hydrology, free water surface (FWS) and subsurface flow (SF) constructed wetlands [64]. SFCWs are subdivided into horizontal flow (HF) and vertical flow (VF) [65]. CWs can be combined as hybrid systems to exploit the special advantages of the various CWs types.

5.3.1 Free Water Surface Constructed Wetlands (FWS CWs)

FWS CWs mimic the natural wetlands [66]. The wastewater surface is in contact with the atmosphere. Oxygen is also supplied by diffusion via the photosynthetic activities and air–water interface. FWS CWs have aerobic, anoxic and anaerobic zones. FWS CW was constituted for the treatment of wastewater from the town and to preserve the water quality of Lake Balaton in Hungary in 1968. In 1975, this system was applied to treat industrial wastewater at Amoco Oil Company's Mandan Refinery in North Dakota [61].

These systems include a bed/channel, water/wastewater at a relatively shallow depth (20–40 cm), substrate such as geotechnical materials, 20–30 cm of rooting soil/clay, and vegetation that can be planted only as one species or a mixture of wetland vegetations (Figure 5.2) [59, 67]. The horizontal flow path through the system is intended, but wind affects the flow direction [59]. Pollution removal occurs by reactions in wastewater and the upper sediment zone [68, 69]. The main removal mechanisms are aggregation, sedimentation and surface adhesion [61]. Removal efficiencies can be achieved above 70% for TSS, COD, BOD, and pathogens. The nitrogen and phosphorous removal efficiencies are 40–50%, and 40–90%, respectively [69].

There are two type system with and without sand/soil layers [59]. Topography, climate conditions, wastewater compositions, flows, loads, and wildlife activity are design factors of FWS CWs. There are many design variables containing hydraulic retention time, total area, size, depth, plant species, shape, inlet/outlet location, and flow pattern. Their low investment/operating costs, and intelligible construction, operation and maintenance are the major advantages of FWS CWs, while the larger area requirement is the main limit of the system [70].

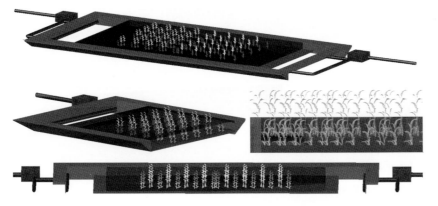

Figure 5.2 Schematic representation of Free Water Surface Constructed Wetland.

5.3.2 Subsurface Flow Constructed Wetlands (SFCWs)

SFCWs have filter media usually consisting of a gravel matrix different from FWS CWs [71]. The water level remains below the top of the filter material. The filter media also support the roots of the wetland plants [59]. The treatment performance may be faster for SFCWs than FWS CWs because the filter material ensures more available surface area for interactions. Therefore, SFCW design is smaller compared to FWS CWs for the same wastewater conditions [64].

SFCWs have typically a depth of 0.3–0.9 m due to root access of plants to provide oxygen transfer in the whole treatment bed [59, 72]. Removal of pollution is achieved by passing wastewater through filter media planted with wetland plants [49]. SFCWs are more effective for biologically driven processes due to microorganisms attached on the filter materials and plant roots [73]. The use of SFCWs (horizontal or vertical) is increasing all over the world for industrial wastewater treatment [49].

The wastewater flows horizontally and continuously through the treatment bed in Horizontal Subsurface Flow Constructed Wetlands (HSFCWs), promoted by the bottom slope. On the other hand, in Vertical Subsurface Flow Constructed Wetlands (VSFCWs) wastewater feeding is made intermittently and is distributed vertically across of the filter media [61, 65]. Treated wastewater is collected typically below the bed surface of 0.3–0.6 m at the outlet in HSFCWs, and at the bottom of the bed in VSFCWs (Figure 5.3) [64, 74].

Nitrogen removal is less effective due to insufficient oxygen in HSFCWs, but it has been successfully applied for secondary treatment to remove BOD and solids. On the other hand, VSFCWs are very effective for all pollutant parameters [49, 75]. VSFCWs became recently more popular because the intermittent application and vertical drainage enhance the aerobic conditions in the treatment bed [72]. VSFCWs have smaller land demands compared to HSFCWs, but VSFCWs usually need more maintenance and operation efforts such as pumps, timers, electric and mechanical devices [61].

The most important disadvantages of CWs encountered in construction is the large surface area requirement due to low hydraulic loading rates and clogging of the gravel media in operation [68, 76, 77]. The primary treatment includes settlers, Imhoff tanks or Hydrolytic Up Flow Sludge Blanket

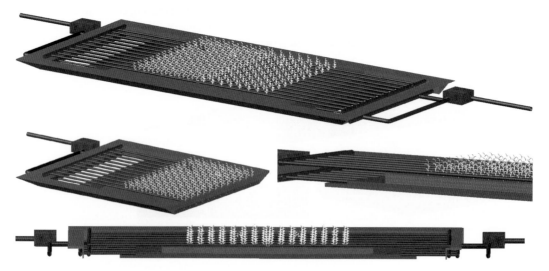

Figure 5.3 Presentation of Vertical Subsurface Flow Constructed Wetland.

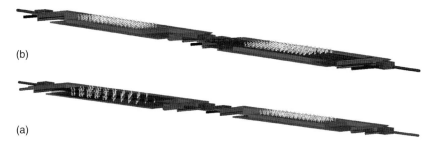

(b)

(a)

Figure 5.4 Presentations of Hybrit Systems: (a) FWSCWs-HSFCWs, (b) HSFCWs-VSFCWs.

(HUSB) reactors, which are applied to prevent clogging [77]. Clogging problems depend on hydraulics of SFCWs and can be delayed by improved design, operation and maintenance [76]. In colder climatic conditions, performance of biologically driven processes decreases because of the microorganisms and plant metabolic activity losses [73]. However, SFCWs have some exceptions related to public access, mosquitos, or wildlife compared to FWS CWs. In SFCWs, the deciduous plant leaves on the filter media provide isolation against freezing during cold climates [72].

5.3.3 Hybrid Systems (HS)

CWs can be combined as an HS to exploit the specific advantages of different types (Figure 5.4) [75]. The purification of many wastewaters is difficult in a single stage CW. In such cases, FWS-SF, VSF-HF, VSF-HSF-FWS or other combinations can be applied to achieve higher treatment effects, especially for nitrogen from various types of wastewaters [50, 61, 69]. The performance achieved is TSS removal of 94%, BOD of 84%, COD of 86%, NH_4-N of 80%, TP of 55%, NO_3-N of 64%, and TN of 67% in 11 hybrid systems [69]. Hence, there is an increasing interest in HS.

A summary of CWs applications for fish and shrimp farm effluents treatment is given in Table 5.3.

Schulz et al. [78] investigated HSFCWs for treating rainbow trout farm effluents generated by *Oncorhynchus mykiss*. Three HSFCWs were established, and filled with sands of 1–2 mm diam. The inlet/outlet sides were packed with coarse-grained gravel (16–32 mm diam) to provide horizontal flow and facilitate drainage. *Phragmites australis* was planted with the planting density being 20 plant/m^2 in wetland units (L:1.40 × W:1.00 × H:0.70). The studies were performed on three various hydraulic loading rates (1, 3, 5 L/s) and hydraulic retention times (1.5, 2.5, 7.5 h). Good removals of TSS (95.8–97.3%), COD (64.1–73.8%), TP (49–68.5%), TN (20.6–41.8%), NH_4-N (72.5–91.4%) were achieved. The best treatment performance was found at an HRT of 7.5 h [78].

Lymbery et al. [79] studied the treatment performance of HSFCWs planted with *Juncus kraussii* for rainbow trout (*Oncorhynchus mykiss*) aquaculture effluent. The sixteen HSFCWs (1 m^2) were set up in series. The wetland units were filled with basalt gravel (diam of 15 mm) over floor cover material after being leveled with slope of 5‰. Four treatment combinations were created as HSHN, HSLN, LSHN, LSLN by using NaCl and filtrate from culture tank. The maximum removal percentages of TN and TP, were 69% and 88.5% in LSLN, respectively [79].

Sindilariu et al. [80] investigated the treatment of aquaculture effluents in SFCWs. *Oncorhynchus mykiss*, *Salmo trutta*, *Salvelinus fontinalis* were generated in an aquaculture facility. The aquaculture effluent was transferred first to sedimentation basins (8 units) and then to VSFCWs. VFCWs were filled with local gravel (4–8 mm). The per cell treatment area was 35.8 m^2 (total of 215 m^2). Each cell

Table 5.3 A summary of CWs studies for fish and shrimp farm effluents treatment.

CW design	Industry	Wastewater characteristics (mg/L)	Removal (%)	Country	Ref
FWSCWs-SFCWS	Aquaculture wastewater	NH_4-N: 0.8–3.31, NO_2-N: 0.03–0.647, NO_3-N: 0.26–2.66, TIN: 0.45–4.48, PO_4-P : 2.39–10.45	NH_4-N: 86-98, NO_2-N: >99, NO_3-N: 82–99, TIN: 95–98, PO_4-P: 32–71	Taiwan	[22]
FWSCWs-SFCWS	Shrimp aquaculture	BOD_5: 5, SS: 36, Chl-a: 58, TAN: 0.21, NO_2-N: 0.05, NO_3-N: 0.41, PO_4-P: 8.45, turbidity: 7.9, pH:7.83, DO: 5.4	BOD_5: 24, SS: 71, TAN: 57, PO_4-P: 5.4, NO_3-N: 68, NO_2-N: 90, turbidity: 63, *Chl-a*: 88	Taiwan	[38]
FWSCWs-SFCWS	Shrimp aquaculture	BOD_5: 3.1-7.1, SS: 12.9–22.3, TAN: 0.23-0.29, NO_2-N: 0.10–0.26, NO_3-N: 5.88-39.9, PO_4-P: 1.06–3.70, turbidity: 1.7-3.8, pH:7.8-8.0, DO: 6.5–7.8	BOD_5: 37–54, SS: 55–66, TAN: 64–66, NO_2-N: 83–94, turbidity: 91–99	Taiwan	[86]
HSFCWs	Fishpond water	COD: 116–132, BOD_5: 16–27, TAN: 0.31–0.85, TN: 6.2–9.7, TP: 1.0–2.5	BOD: 50, COD: 50	Vietnam	[87]
HSFCWs	Rainbow trout farm effluent	TSS: 14.15, COD: 41.01, TP: 0.347, TN: 2.40, NH_4-N: 0.61	TSS: 95.8–97.3, COD: 64.1–73.8, TP: 49–68.5, TN: 20.6–41.8, NH_4-N:72.5–91.4	Germany	[78]
HSFCWs	Trout farm	BOD_5: 2.41, COD: 7, TOC: 2.71, TSS: 2.70, NO_3-N: 4.95, TN: 5.12, TAN: 140.63 µg/L, NO_2-N: 16.60 µg/L, TP: 58.35 µg/L, PO_4-P:31.31 µg/L, pH: 7.73, EC: 723 µS/cm	BOD_5: 36.9, COD: 24.3, TOC: 9.2, TSS: 34.4, TAN: 86.9, NO_2-N: 35.5, TP: 38.1	Germany	[80]
HSFCWs	Tilapia production	DO: 3.20, NH_4-N: 5.23, NO_2-N: 0.39, NO_3-N: 1.50, TP: 4.37, TCOD: 41, SCOD: 32, TSS:6.9	DO: 93.2, NH_4-N: 7.5, NO_2-N: 90.8, NO_3-N: 75.9, TP: 0, TCOD: 12.5, SCOD: 25.4, TSS: 90	USA	[83]
HSFCWs	Inland saline aquaculture	TN: 730–2966.7, TP: 160–1270	TN:69, TP: 88.5	Australia	[79]
HSFCWs	Trout farm	TN: 6.4, PO_4-P: 0.041, BOD_5: 6.90, COD: 14.20, TSS: 7.17, TAN: 0.75, NO_3-N: 4.84, TP: 0.25, NO_2-N: 0.011	TN: 10, PO_4-P: 209, BOD_5: 88.7, TSS: 90.1, COD: 67.2, TAN: 87.8, NO_3-N: 13, TP: 43.4, NO_2-N: 100	Germany	[82]
HSFCWs	Freshwater fish farm effluent	TSS: 187, COD: 373, BOD_5: 104, TKN: 12.41, NH_4^+: 1.39, NO_3^-: 0.99, TP: 2.69, o-PO_4: 1.78	TSS: 99.47, COD: 91.15, BOD_5: 99, TKN: 89.61, NH_4^+: 81, NO_3^-: 69.70, TP: 91.08, o-PO_4: 95.51	Canada	[85]

(Continued)

Table 5.3 (Continued)

CW design	Industry	Wastewater characteristics (mg/L)	Removal (%)	Country	Ref
VSFCWS	Fishpond water	COD: 116–132, BOD$_5$: 16–27, TAN: 0.31–0.85, TN: 6.2–9.7, TP: 1.0–2.5	BOD: 50, COD: 50	Vietnam	[87]
VSFCWs	Aquaculture ponds	TN: 2.84, NH$_4^+$: 0.52, NO$_3$-N: 0.41, TP: 0.35, PO$_4$-P: 0.05, BOD$_5$: 5.8, TSS: 21.5, *Chl-a*: 30.6	TN: 54.6, NH$_4^+$: 61.5, NO$_3$-N: 68, TP: 80.1, PO$_4$-P: 20, BOD$_5$: 70.5, TSS: 81.9, *Chl-a*: 91.9	China	[81]
VSFCWs	Channel catfish culture effluent	NH$_4$-N: 0.62, NO$_2$-N: 0.041, NO$_3$-N: 0.058, TN: 1.21, TP: 0.09, COD: 36.5, BOD$_5$: 4.3, TOC: 15.8, TSS: 12.4, *Chl-a*: 25.5 µg/L	NH$_4$-N: 34.2, NO$_2$-N: 48.7, NO$_3$-N: 3.1, TN: 48.2, TP: 16.7, COD: 25.6, BOD$_5$: 55.6, TOC: 19.5, TSS: 57.6, *Chl-a*: 81.6	China	[84]
VSFCWS-HSFCWs	Shrimp aquaculture	pH: 786–7.92, DO: 5.07–5.14, TN: 13.5–14.3, TAN: 1.34–1.39, NO$_2$-N: 0.13–0.15, NO$_3$-N: 2.62–3.42, TP: 0.08, COD: 10.5–11.0, TSS: 200–220	DO: 3.8, TN: 66.8, TAN: 70.8, NO$_2$-N:85.1, NO$_3$-N: 58.7, TP:23.8, COD: 26.7, TSS: 65.9	China	[37]
HSFCWs	Channel catfish pond effluent	TAN: 0.337, NO$_2$-N: 0.041, NO$_3$-N: 0.543, TKN: 1.61, TP: 0.162, BOD: 5.61, SS: 34.5, VSS: 12.5	TAN: 71.2, NO$_2$-N: 43.9, NO$_3$-N: 52.7, TKN: 45.3, TP: 68.5, BOD: 39.6, SS: 75.3, VSS: 68.8	USA	[111]
FWSCWs	Seafood wastewater	BOD: 332–389, SS: 54–124, ammonia: 31–58, TN: 95–124, TP: 58–63	BOD: 91–99, SS: 52–90, TN: 72-92, TP: 72–77	Bangkok	[112]
FWSCWs	Saline wastewater	BOD: 97.5–108.9, SS: 65.2–89.4, NH$_3$-N: 19.5–24.3, TP: 7.8–8.8	BOD:44.4–67.9, SS: 41.4–70.4, NH$_3$-N: 18.0–65.3, TP: 12.2–40.5	Thailand	[113]
HSFCWs	Shrimp aquaculture	DO:5.7, pH: 8.3, TDS: 2.7, TAN:1.1, TP: 0.31, TSS: 47, ISS: 36, BOD: 10.6, VSS: 49	BOD: 17, TAN: 67, VSS: 48, ISS: 76, TSS: 65, TP: 31	Texas	[114]
HSFCWs	Salmonid hatcheries wastewater	BOD: 38.53, TP: 1.61, TAN: 0.67, TSS: 256.60, SS: 1.39	BOD: 81.6, TP: 82.3, TAN: 75.1, TSS: 91.3, SS: 95.4	USA	[115]

was planted with *Phragmites communis* with a coverage of 35%, *Phalaris arundinacea* with 35% and different swamp and land plants with 30%. COD, BOD$_5$, TOC, TAN, TP, NO$_2$-N and TSS parameters were analyzed over two years and removals found were 24.3%, 36.9%, 9.2%, 86.9%, 38.1%, 35.5% and 34.4%, respectively [80].

Li et al. [81] set up a group of VSFCWs to treat the effluents from four aquaculture ponds. The established system had five ponds consisting of a control pond and a reservoir pond, and two parallel VSFCWs (L:10 × W:8 x H: 1.1 m) consisting of downflow cells and upflow cells. VSFCWs were filled with gravel of 1–16 mm and were planted with *Canna indicia* in the downflow CWs, *Acrorus calamus*, *Typha latifolia* and *Agrave sisalana* in upflow CWs. TN, NH_4^+, NO_3-N, TP, PO_4-P, BOD_5, TSS and *Chl-a* were analyzed in water samples. The mean removal rates were 54.6%, 61.5%, 68%, 80.1%, 20%, 70.5%, 81.9% and 91.9%, respectively. They reported that although they achieved a good treatment performance, many mechanisms were still unknown [81].

Sindilariu et al. [82] investigated the effects of three hydraulic loading rates on the treatment performance of SFCWs for the effluent of a trout farm generating *Oncorhynchus mykiss* for 6 months. The treatment system included a primary sedimentation tank of 9.6 m² and 6 SFCWs of 23.9 m² (total surface area of 143.4 m²). Two duplicate SFCWs were installed with 0.9, 1.8, 3.9 L/s, respectively. The maximum removal percentages were 10% for TN, 29% for PO_4-P, 88.7% for BOD_5, 67.2% for COD, 90.1% for TSS by 0.9 L/s; 87.8% for TAN, 13% for NO_3-N, 43.4% for TP by 1.8 L/s; 100% by 3.9 L/s for NO_2-N. They reported that, SFCWs were found as an effective technology for treatment of trout farm effluent [82].

Zachritz et al. [83] examined the treatment performance of SFCWs for tilapia production effluent. The tilapia production system had two culture tanks and a sedimentation tank. A SFCW (L:8.4 × W:6.4 × D: 0.9 m) was modified into the system. The wetland unit was filled with lava rock (diam of 380 mm), had a porosity of 54%, and was planted with a mixture of *Canna* lillies and bulrush. The water discharged from SFCWs was routed back in culture tanks. The system performance (clarifier and wetland) was 93.2% for DO, 4.5% for NH_4-N, 90.7% for NO_2-N, 66.4% for NO_3-N, 67.2% for TSS over a 36-month period [83].

Zhang et al. [84] examined the removal performance of VSFCWs consisting of downflow and upflow beds to treat the channel catfish culture effluent generated by *Ictalurus punctatus*. The RAS system had five culture ponds, and two VSFCWs. The wetlands had a total area of 320 m² and were divided into four equal sizes and planted with *Acorus calamus*, *Canna indica* L., *Typha latifolia* L. and sisal. The filter media was washed gravel of 0.8–6.4 cm diam. The removal performances were 34.2% for NH_4-N, 48.7% for NO_2-N, 3.1% for NO_3-N, 48.2% for TN, 16.7% for TP, 25.6% for COD, 55.6% for BOD_5, 19.5% for TOC, 57.6% for TSS and 81.6% for *Chl-a* [84].

Lin et al. [22] established a pilot scale constructed wetland including two stages for removal of inorganic nitrogen and phosphorous from aquaculture effluent. This system was a hybrid system consisting of FWS CWs and SFCWs. FWS and SFCWs were made in cast iron tanks of dimensions 5 × 1 × 0.8 m (L × W × H) and set up in series. They were covered with an impermeable plastic material. The treatment beds consist of two layers; 30 cm of local soil, 40 cm of water for FWSCWs, and 60 cm of gravel (diam: 10–20 mm), 40 cm of water for SFCWs over cover materials. Three plant species were used: *Phragmites australis* in SFCWs, *Ipomoe aquatica* and *Paspalum vaginatum* in the first and second half of FWSCWs, respectively. The study was performed under five different hydraulic loading rates for 8 months. Temperature, NH_4-N, NO_2-N, NO_3-N, TIN, and PO_4-P were monitored in the inlet of FWS CWs and the outlet of FWS and SFCWs. According to the results, hydraulic loading rates had little effect on the removal performance: 86–98% for NH_4-N, >99% for NO_2-N, 82–99% for NO_3-N, 95–98% for TIN and 32–71% for phosphate [22].

Lin et al. [38] set up a pilot scale constructed wetland system to treat shrimp aquaculture effluent. The hybrid system included FWS CW and SFCW and was integrated into a culture tank that was generating *Litopenaeus vannamei*, and was operated with a hydraulic loading rate of 0.3 m/day. Both FWSCW and SFCW had the following dimensions: W: 1 × L: 5 × H: 0.8 m. The effluent of the SFCW was recirculated in the culture tank. The parts of the whole system were covered with impermeable liners. The removal performances were 24% for BOD_5, 71% for SS, 88% for *Chl-a*, 57% for TAN, 90% for NO_2-N, 68% for NO_3-N, 5.4% for PO_4-P, 63% for turbidity [38].

Naylor et al. [85] investigated the treatment performance of freshwater fish farm effluent by SFCWs. They set up 20 plastic basins of a surface a 1 m². Three substrate types (i.e., slag, limestone, granite with or without peat) were used as filter media in various combinations. The wetland cells were planted with *Phragmites australis* and *Typha latifolia*. Four wetland cells were used as control units. The minimum and maximum removal efficiencies were 96.8% and 99.5% for TSS, 52.8% and 91.15% for COD, 68.3% and 99% for BOD, 40.13% and 89.6% for TKN, 41% and 81% for NH_4-N, 44.4% and 69.71% for NO_3-N, 85.9% and 91.1% for TP, 5.1% and 95.5% for PO_4-P [85].

Lin et al. [86] established a hybrid system consisting of FWS CWs and SFCWs. The aim of the study was to treat aquaculture wastewater generated by an intensive shrimp culture. The established system was a merchant-scale recirculating system. The system (L:15.2 × W:2.1 m) was divided lengthways into two equal parts using a brick concrete wall. The hybrid system formed a U shape with four cells: a settling cell, a FWS CW cell, a SFCW cell and a sump. the FWS CW and SFCW were planted with *Typha angustifolia* L. and *Phragmites australis*, respectively. The SFCW was filled with river gravel (diameter 10–20 mm) of 80 cm depth, and had a water level of 65 cm. The FWS CW was filled with local soil of 30 cm depth and had an average water depth of 40 cm. SS, *Chl-a*, turbidity, BOD_5, TAN, NO_2-N, NO_3-N, PO_4-P were analyzed in the samples for two seasons, called Phase 1 (warm season) and Phase 2 (cold season), respectively. The removal performances were 37% for BOD_5, 66% for SS, 66% for TAN, 94% for NO_2-N, and 99% for turbidity in Phase 1, as well as 54% for BOD_5, 55% for SS, 64% for TAN, 83% for NO_2-N, and 91% for turbidity in Phase 2 [86].

Konnerup et al. [87] studied the treatment of fishpond wastewater generated by the polyculture of Nile tilapia and common carp by using HSF and VSFCWs. Two HSFCWs and two VSFCWs were integrated into the fishpond. HSFCWs (W:0.85 × L:3.7 × D:0.3 m) were made in wood and metal plates, covered by a plastic liner, and filled with gravel (30–50 mm). VSFCWs (D: 0.7 m) were a circular plastic container (diameter of 1.25 m) and had a aeration pipe. The beds consisted of three layers: 0.1 m stone (diameter 50–100 mm), 0.4 m gravel (diameter 30–50 mm) and 0.2 m (diameter 10–20 mm) from bottom to top. All wetland units were planted with *Canna × generalis* cultivars. Three hydraulic loading rates were applied, with influent concentrations of COD 116–132 mg/L, BOD_5 16–27 mg/L, TAN 0.31–0.85 mg/L, TN 6.2–9.7 mg/L and TP 1.0–2.5 mg/L. The removal efficiencies of BOD and COD were found 50% in all type wetland units [87].

5.4 Conclusions

Fish and shrimp culture has increased very rapidly all over the world, while volumes of wastewater produced by aquaculture are also very large, and the wastewater pollution has become a very serious problem to the natural aquatic ecosystem [5, 88]. Aquaculture produces waste in the form of solids (uneaten feed, feces, etc.) and dissolved material, which is transported out of the rearing system with the husbandry water and could result in serious eutrophication and impact on the aquatic ecosystem [89]. Removal performances of CWs were reported according to the findings. The results indicate that CW is the best alternative, effective and viable wastewater treatment system, economically and environmentally, to solve aquaculture effluent treatment problems. When these systems are designed and constructed in accordance with the rules, the malfunction risk is low, and CWs may be utilized successfully in secondary and tertiary treatment.

References

1 Troell M, Naylor RL, Metian M, Beveridge M, Tyedmers PH, Folke C, et al. Does aquaculture add resilience to the global food system? Proc Natl Acad Sci. 2014;111(37):13257–13263.

2 Xie B, Qin J, Yang H, Wang X, Wang YH, Li TY. Organic aquaculture in China: A review from a global perspective. Aquaculture. 2013;414–415:243–253.

3 Boyd CE. Guidelines for aquaculture effluent management at the farm-level. Aquaculture. 2003;226(1–4):101–12.

4 Dierberg FE, Kiattisimkul W. Issues, impacts, and implications of shrimp aquaculture in Thailand. Environ Manage. 1996;20(5):649–666.

5 Naylor RL, Goldburg RJ, Primavera JH, Kautsky N, Beveridge MCM, Clay J, et al. Effect of aquaculture on world fish supplies. Nature. 2000;405:1017–1024.

6 Edwards P. Aquaculture environment interactions: Past, present and likely future trends. Aquaculture. 2015;447:2–14.

7 Boyd CE, Queiroz J, Lee J, Rowan M, Whitis GN, Gross A. Environmental assessment of channel catfish Ictalurus punctatus farming in Alabama. J World Aquac Soc. 2000;31(4):511–544.

8 Brooks BW, Riley TM, Taylor RD. Water quality of effluent-dominated ecosystems: ecotoxicological, hydrological, and management considerations. Hydrobiologia. 2006;556(1):365–379.

9 Boyd CE, Lim C, Queiroz J, Salie K, de Wet LAM. Best management practices for responsible aquaculture. USAID, 2008.

10 Turcios A, Papenbrock J. Sustainable Treatment of aquaculture effluents – what can we learn from the past for the future? Sustainability. 2014;6(2):836–856.

11 Neori A, Chopin T, Troell M, Buschmann AH, Kraemer GP, Halling C, et al. Integrated aquaculture: rationale, evolution and state of the art emphasizing seaweed biofiltration in modern mariculture. Aquaculture. 2004;231(1–4):361–391.

12 Masser MP, Rakocy J, Losordo TM. Recirculating Aquaculture Tank Production Systems Management of Recirculating Systems. South Reg Aquac Cent. 1999;452:1–12.

13 Martan E. Polyculture of Fishes in Aquaponics and Recirculating Aquaculture. Aquaponics. 2008;1(48):28–33.

14 Binh CT, Phillips MJ, Demaine H. Integrated shrimp-mangrove farming systems in the Mekong delta of Vietnam. Aquac Res. 1997;28(8):599–610.

15 Crossley PL. Sub-irrigation in wetland agriculture. Agric Human Values. 2004;21(2–3):191–205.

16 True B, Johnson W, Chen S. Reducing phosphorus discharge from flow-through aquaculture I: facility and effluent characterization. Aquac Eng. 2004;32(1):129–144.

17 Baccarin AE, Camargo AFM. Characterization and evaluation of the impact of feed management on the effluents of Nile tilapia (*Oreochromis niloticus*) culture. Brazilian Arch Biol Technol. 2005;48(1):81–90.

18 Stephens WW, Farris JL. A biomonitoring approach to aquaculture effluent characterization in channel catfish fingerling production. Aquaculture. 2004;241(1–4):319–330.

19 Henry-Silva GG, Camargo AFM. Efficiency of aquatic macrophytes to treat Nile tilapia pond effluents. Sci Agric. 2006;63(5):433–438.

20 Páez-Osuna F. The environmental impact of shrimp aquaculture: causes, effects, and mitigating alternatives. Environ Manage. 2001;28(1):131–140.

21 USEPA. Guidelines for Water Reuse. United States Environmental Protection Agency; 2012, pp. 1–252.

22 Lin Y, Jing S, Lee D, Wang T. Nutrient removal from aquaculture wastewater using a constructed wetlands system. Aquaculture. 2002;209(1–4):169–184.

23 Lawson TB. Fundamentals of Aquacultural Engineering. Boston/Dordrecht/London: Kluwer Academiz Publishers; 1995, pp. 1–639.

24 Bergheim A, Sanni S, Indrevik G, Hølland P. Sludge removal from salmonid tank effluent using rotating microsieves. Aquac Eng. 1993;12(2):97–109.

25 Bergheim A, Cripps SJ, Liltved H. A system for the treatment of sludge from land-based fish-farms. Aquat Living Resour. 1998;11(4):279–287.

26 Lekang OI, Bergheim A, Dalen H. An integrated wastewater treatment system for land-based fish-farming. Aquac Eng. 2000;22(3):199–211.

27 M'kinen T, Lindgren S, Eskelinen P. Sieving as an effluent treatment method for aquaculture. Aquac Eng. 1988;7(6):367–377.

28 Shpigel M, Gasith A, Kimmel E. A biomechanical filter for treating fish-pond effluents. Aquaculture. 1997;152(1–4):103–117.

29 Schulz C, Gelbrecht J, Rennert B. Treatment of rainbow trout farm effluents in constructed wetland with emergent plants and subsurface horizontal water flow. Aquaculture. 2003;217(1–4):207–221.

30 Fladung E. Humboldt-University Fischproduktionsanlagen (Forellenrinnenanlagen) in die Vorfluter. Humboldt-University; 1993.

31 Snow A, Anderson B, Wootton B. Flow-through land-based aquaculture wastewater and its treatment in subsurface flow constructed wetlands. Environ Rev. 2012;20(1):54–69.

32 Chowdhury P, Viraraghavan T, Srinivasan A. Biological treatment processes for fish processing wastewater – A review. Bioresour Technol. 2010;101(2):439–449.

33 Pungrasmi W, Playchoom C, Powtongsook S. Optimization and evaluation of a bottom substrate denitrification tank for nitrate removal from a recirculating aquaculture system. J Environ Sci. 2013;25(8):1557–1564.

34 Samocha TM, Fricker J, Ali AM, Shpigel M, Neori A. Growth and nutrient uptake of the macroalga Gracilaria tikvahiae cultured with the shrimp *Litopenaeus vannamei* in an Integrated Multi-Trophic Aquaculture (IMTA) system. Aquaculture. 2015;446:263–271.

35 Van Rijn J. The potential for integrated biological treatment systems in recirculating fish culture – A review. Aquaculture. 1996;139(3–4):181–201.

36 Van Rijn J. Waste treatment in recirculating aquaculture systems. Aquac Eng. 2013;53:49–56.

37 Shi Y, Zhang G, Liu J, Zhu Y, Xu J. Performance of a constructed wetland in treating brackish wastewater from commercial recirculating and super-intensive shrimp growout systems. Bioresour Technol. 2011;102(20):9416–24.

38 Lin Y-F, Jing S-R, Lee D-Y. The potential use of constructed wetlands in a recirculating aquaculture system for shrimp culture. Environ Pollut. 2003;123(1):107–13.

39 Mook WT, Chakrabarti MH, Aroua MK, Khan GMA, Ali BS, Islam MS, et al. Removal of total ammonia nitrogen (TAN), nitrate and total organic carbon (TOC) from aquaculture wastewater using electrochemical technology: A review. Desalination. 2012;285:1–13.

40 Shrimali M, Singh K. New methods of nitrate removal from water. Environ Pollut. 2001;112(3):351–359.

41 Webb JM, Quintã R, Papadimitriou S, Norman L, Rigby M, Thomas DN, et al. Halophyte filter beds for treatment of saline wastewater from aquaculture. Water Res. 2012;46(16):5102–5114.

42 Liu CCK, Xia W, Park JW. A wind-driven reverse osmosis system for aquaculture wastewater reuse and nutrient recovery. Desalination. 2007;202(1–3):24–30.

43 Strathmann H. Electrodialysis, a mature technology with a multitude of new applications. Desalination. 2010;264(3):268–288.

44 Vymazal J. Removal of nutrients in various types of constructed wetlands. Sci Total Environ. 2007;380(1–3):48–65.

45 Song Z, Zheng Z, Li J, Sun X, Han X, Wang W, et al. Seasonal and annual performance of a full-scale constructed wetland system for sewage treatment in China. Ecol Eng. 2006;26(3):272–282.

46 Vymazal J. Plants used in constructed wetlands with horizontal subsurface flow: A review. Hydrobiologia. 2011;674(1):133–156.

47 Coleman J, Hench K, Garbutt K, Sexstone A, Bissonnette G, Skousen J. Treatment of domestic wastewater by three plant species in constructed wetlands. Water Air Soil Pollut. 2001;128(3–4):283–295.

48 Ayaz SÇ, Akça L. Treatment of wastewater by natural systems. Environ Int. 2001;26(3):189–195.

49 Abou-Elela SI, Golinielli G, Abou-Taleb EM, Hellal MS. Municipal wastewater treatment in horizontal and vertical flows constructed wetlands. Ecol Eng. 2013;61(A):460–468.

50 Vymazal J. Horizontal sub-surface flow and hybrid constructed wetlands systems for wastewater treatment. Ecol Eng. 2005;25(5):478–490.

51 Kivaisi AK. The potential for constructed wetlands for wastewater treatment and reuse in developing countries: a review. Ecol Eng. 2001;16(4):545–560.

52 Ayaz SÇ, Akça L. Treatment of wastewater by natural systems. Environ Int. 2001;26(3):189–195.

53 Wu H, Zhang J, Ngo HH, Guo W, Hu Z, Liang S, et al. A review on the sustainability of constructed wetlands for wastewater treatment: Design and operation. Bioresour Technol. 2015;175:594–601.

54 Vymazal J. Emergent plants used in free water surface constructed wetlands: A review. Ecol Eng. 2013;61(B):582–592.

55 Leto C, Tuttolomondo T, La Bella S, Leone R, Licata M. Effects of plant species in a horizontal subsurface flow constructed wetland – phytoremediation of treated urban wastewater with *Cyperus alternifolius* L. and *Typha latifolia* L. in the West of Sicily (Italy). Ecol Eng. 2013;61(A):282–291.

56 Mashauri DA, Mulungu DMM, Abdulhussein BS. Constructed wetland at the university of Dar es Salaam. Water Res. 2000;34(4):1135–1144.

57 Kantawanichkul S, Duangjaisak W. Domestic wastewater treatment by a constructed wetland system planted with rice. Water Sci Technol. 2011;64(12):2376–2380.

58 Bilgin M, Şimşek İ, Tulun Ş. Treatment of domestic wastewater using a lab-scale activated sludge/vertical flow subsurface constructed wetlands by using Cyperus alternifolius. Ecol Eng. 2014;70:362–365.

59 Farooqi IH, Basheer F, Chaudhari RJ. Constructed Wetland System (CWS) for Wastewater Treatment. Proceedings of Taal 2007: The 12th World Lake Conference; 2008, pp. 1004–1009.

60 Vymazal J. The use of sub-surface constructed wetlands for wastewater treatment in the Czech Republic: 10 years experience. Ecol Eng. 2002;18(5):633–646.

61 Vymazal J. Constructed Wetlands for Wastewater Treatment : A Review. In: Sangupta M, Dalwani R, eds. Jaipur, Rajasthan, India: The 12th World Lake Conference (Taal 2007); 2008, pp. 965–980.

62 Yan Y, Xu J. Improving Winter Performance of Constructed Wetlands for Wastewater Treatment in Northern China: A Review. Wetlands. 2014;34(2):243–253.

63 Türker OC, Vymazal J, Türe C. Constructed wetlands for boron removal: A review. Ecol Eng. 2014;64:350–359.

64 USEPA. Subsurface Flow Constructed Wetlands for WasteWater Treatment: A Technology Assessment. United States Environmental Protection Agency; 1993, pp. 1–87.

65 Avila C, Matamoros V, Reyes-Contreras C, Piña B, Casado M, Mita L, et al. Attenuation of emerging organic contaminants in a hybrid constructed wetland system under different hydraulic loading rates and their associated toxicological effects in wastewater. Sci Total Environ. 2014;470–471:1272–1280.

66 Bhamidimarri R, Shilton A, Armstrong I, Jacobson P, Scarlett D. Constructed wetlands for wastewater treatment: the New Zealand experience. Wal Sci Tech. 1991;24(5):247–253.

67 Healy MG, Rodgers M, Mulqueen J. Treatment of dairy wastewater using constructed wetlands and intermittent sand filters. Bioresour Technol. 2007;98(12):2268–2281.

68 Jenssen PD, Maehlum T, Krogstad T. Potential use of constructed wetlands for wastewater treatment in northern environments. Water Sci Technol. 1993;28(10):149–157.

69 Zhang DQ, Jinadasa KBSN, Gersberg RM, Liu Y, Ng WJ, Tan SK. Application of constructed wetlands for wastewater treatment in developing countries – a review of recent developments (2000–2013). J Environ Manage. 2014;141:116–31.

70 USEPA. Manual Constructed Wetlands Treatment of Municipal Wastewaters. Cincinnati, Ohio: United States Environmental Protection Agency; 2000, pp. 1–166.

71 Sindilariu P-D, Brinker A, Reiter R. Factors influencing the efficiency of constructed wetlands used for the treatment of intensive trout farm effluent. Ecol Eng. 2009;35(5):711–722.

72 USEPA. Wastewater Technology Fact Sheet Wetlands : Subsurface Flow. Washington, DC: United States Environmental Protection Agency; 2000, pp. 1–9.

73 Regmi TP, Thompson AL, Sievers DM. Compatarive studies of vegatated and non-vegetated submerged flow wetlands treating primary lagoon effluent. Trans ASAE. 2003;46(1):17–27.

74 Hua SC. The use of constructed wetlands for wastewater treatment. Malaysia Office: Wetlands International; 2003, 24 pp.

75 Cui L-H, Ouyang Y, Chen Y, Zhu X-Z, Zhu W-L. Removal of total nitrogen by Cyperus alternifolius from wastewaters in simulated vertical-flow constructed wetlands. Ecol Eng. 2009;35(8):1271–1274.

76 Pedescoll A, Sidrach-Cardona R, Sánchez JC, Carretero J, Garfi M, Bécares E. Design configurations affecting flow pattern and solids accumulation in horizontal free water and subsurface flow constructed wetlands. Water Res. 2013;47(3):1448–1458.

77 Corbella C, Puigagut J. Effect of primary treatment and organic loading on methane emissions from horizontal subsurface flow constructed wetlands treating urban wastewater. Ecol Eng. 2015;80:79–84.

78 Schulz C, Gelbrecht J, Rennert B. Treatment of rainbow trout farm effluents in constructed wetland with emergent plants and subsurface horizontal water flow. Aquaculture. 2003;217(1–4):207–221.

79 Lymbery AJ, Doupé RG, Bennett T, Starcevich MR. Efficacy of a subsurface-flow wetland using the estuarine sedge *Juncus kraussii* to treat effluent from inland saline aquaculture. Aquac Eng. 2006;34(1):1–7.

80 Sindilariu PD, Schulz C, Reiter R. Treatment of flow-through trout aquaculture effluents in a constructed wetland. Aquaculture. 2007;270(1–4):92–104.

81 Li G, Wu Z, Cheng S, Liang W, He F, Fu G, et al. Application of constructed wetlands on wastewater treatment for aquaculture ponds. Wuhan Univ J Nat Sci. 2007;12(6):1131–1135.

82 Sindilariu PD, Wolter C, Reiter R. Constructed wetlands as a treatment method for effluents from intensive trout farms. Aquaculture. 2008;277(3–4):179–184.

83 Zachritz WH, Hanson AT, Sauceda JA, Fitzsimmons KM. Evaluation of submerged surface flow (SSF) constructed wetlands for recirculating tilapia production systems. Aquac Eng. 2008;39(1):16–23.

84 Zhang, Shi-yang; Zho , Qiao-hong; Xu, Dong; He, Feng; Cheng, Shui-ping; Liang, Wei; Du, Cheng; Wu Z. Vertical-flow constructed wetlands applied in a recirculating aquaculture system for channel catfish culture: effects on water quality and zooplankton. Polish J Environ Stud. 2010;19(5):1063–1070.

85 Naylor S, Brisson J, Labelle MA, Drizo A, Comeau Y. Treatment of freshwater fish farm effluent using constructed wetlands: The role of plants and substrate. Water Sci Technol. 2003;48(5):215–222.

86 Lin YF, Jing SR, Lee DY, Chang YF, Chen YM, Shih KC. Performance of a constructed wetland treating intensive shrimp aquaculture wastewater under high hydraulic loading rate. Environ Pollut. 2005;134:411–21.

87 Konnerup D, Trang NTD, Brix H. Treatment of fishpond water by recirculating horizontal and vertical flow constructed wetlands in the tropics. Aquaculture. 2011;313(1–4):57–64.

88 Hargreaves JA, Tucker CS, Thornton ER, Kingsbury SK. Characteristics and sedimentation of initial effluent discharged from excavated levee ponds for channel catfish. Aquac Eng. 2005;33(2):96–109.

89 Bonsdorff E, Blomqvist EM, Mattila J, Norkko A. Coastal eutrophication: Causes, consequences and perspectives in the Archipelago areas of the northern Baltic Sea. Estuar Coast Shelf Sci. 1997;44:63–72.

90 Carawan RE. Processing Plant Waste Management Guidelines – Aquatic Fishery Products. North Carolina State University, Department of Food Science, Ph.D.; 1991.

91 Najafpour GD, Zinatizadeh AAL, Lee LK. Performance of a three-stage aerobic RBC reactor in food canning wastewater treatment. Biochem Eng J. 2006;30(3):297–302.

92 Park E, Enander R, Barnett SM, Lee C. Pollution prevention and biochemical oxygen demand reduction in a squid processing facility. J Clean Prod. 2001;9(4):341–349.

93 Prasertsan P, Jung S, Buckle K a. Anaerobic filter treatment of fishery wastewater. World J Microbiol Biotechnol. 1994;10(1):11–13.

94 Mendez R, Lema JM, Soto M. Treatment of Seafood Processing Wastewaters in Mesophilic and Thermophilic Anaerobic Filters. Water Environ Res. 1995;67(1):33–45.

95 Guerrero L, Omil F, Méndez R, Lema JM. Treatment of saline wastewaters from fish meal factories in an anaerobic filter under extreme ammonia concentrations. Bioresour Technol. 1997;61(1):69–78.

96 Abu Bakar NS, Mohd Nasir N, Lananan F, Abdul Hamid SH, Lam SS, Jusoh A. Optimization of C/N ratios for nutrient removal in aquaculture system culturing African catfish (*Clarias gariepinus*) utilizing Bioflocs Technology. Int Biodeterior Biodegrad. 2015;102:100–106.

97 Crab R, Kochva M, Verstraete W, Avnimelech Y. Bio-flocs technology application in over-wintering of tilapia. Aquac Eng. 2009;40(3):105–112.

98 De Schryver P, Verstraete W. Nitrogen removal from aquaculture pond water by heterotrophic nitrogen assimilation in lab-scale sequencing batch reactors. Bioresour Technol. 2009;100(3):1162–1167.

99 Achour M, Khelifi O, Bouazizi I, Hamdi M. Design of an integrated bioprocess for the treatment of tuna processing liquid effluents. Process Biochem. 2000;35(9):1013–1017.

100 Brazil BL. Performance and operation of a rotating biological contactor in a tilapia recirculating aquaculture system. Aquac Eng. 2006;34(3):261–274.

101 Boopathy R, Bonvillain C, Fontenot Q, Kilgen M. Biological treatment of shrimp aquaculture wastewater using a sequencing batch reactor. Int Biodeterior Biodegradation. 2007;59:16–19.

102 Ali N, Mohammad AW, Jusoh A, Hasan MR, Ghazali N, Kamaruzaman K. Treatment of aquaculture wastewater using ultra-low pressure asymmetric polyethersulfone (PES) membrane. Desalination. 2005;185(1–3):317–326.

103 Qin G, Liu CCK, Richman NH, Moncur JET. Aquaculture wastewater treatment and reuse by wind-driven reverse osmosis membrane technology: A pilot study on Coconut Island, Hawaii. Aquac Eng. 2005;32(3–4):365–378.

104 Matos CT, Sequeira AM, Velizarov S, Crespo JG, Reis MAM. Nitrate removal in a closed marine system through the ion exchange membrane bioreactor. J Hazard Mater. 2009;166(1):428–434.

105 Li M, Feng C, Zhang Z, Lei X, Chen R, Yang Y, et al. Simultaneous reduction of nitrate and oxidation of by-products using electrochemical method. J Hazard Mater. 2009;171(1–3):724–30.

106 Lin SH, Wu CL. Electrochemical removal of nitrite and ammonia for aquaculture. Water Res. 1996;30(3):715–721.

107 Li M, Feng C, Zhang Z, Yang S, Sugiura N. Treatment of nitrate contaminated water using an electrochemical method. Bioresour Technol. 2010;101(16):6553–6557.

108 Virkutyte J, Jegatheesan V. Electro-Fenton, hydrogenotrophic and Fe2+ ions mediated TOC and nitrate removal from aquaculture system: Different experimental strategies. Bioresour Technol. 2009;100(7):2189–2197.

109 Feng C, Sugiura N, Shimada S, Maekawa T. Development of a high performance electrochemical wastewater treatment system. J Hazard Mater. 2003;103(1–2):65–78.

110 Díaz V, Ibáñez R, Gómez P, Urtiaga AM, Ortiz I. Kinetics of electro-oxidation of ammonia-N, nitrites and COD from a recirculating aquaculture saline water system using BDD anodes. Water Res. 2011;45(1):125–134.

111 Schwartz MF, Boyd CE. Constructed wetlands for treatment of channel catfish pond effluents. Progress Fish-Culturist. 1995;57(4):255–266.

112 Sohsalam P, Englande AJ, Sirianuntapiboon S. Seafood wastewater treatment in constructed wetland: Tropical case. Bioresour Technol. 2008;99(5):1218–1224.

113 Klomjek P, Nitisoravut S. Constructed treatment wetland: a study of eight plant species under saline conditions. Chemosphere. 2005;58(5):585–593.

114 Tilley DR, Badrinarayanan H, Rosati R, Son J. Constructed wetlands as recirculation filters in large-scale shrimp aquaculture. Aquac Eng. 2002;26(2):81–109.

115 Michael JH. Nutrients in salmon hatchery wastewater and its removal through the use of a wetland constructed to treat off-line settling pond effluent. Aquaculture. 2003;226(1–4):213–225.

6

Evaluation of Treatment Wetlands of Different Configuration for the Sugarcane-Mill Effluent under Tropical Conditions

E. Navarro[1], R. Pastor[2], V. Matamoros[3] and J.M. Bayona[3]*

[1]*Universidad Tecnológica de Izúcar de Matamoros, México*
[2]*Cátedra UNESCO de Sostenibilidad-UPC, Spain*
[3]*Intituto de Diagnóstico Ambiental y Estudios del Agua, IDAEA-CSIC, Spain*

6.1 Introduction

Sugarcane (*Saccharum officinarum* L.) production is a globally important crop, which represents the 85% of the total sugar production localized in tropical and subtropical countries, crystalline sugar and bio-ethanol being the main final products [1]. According to the Food and Agricultural Organization (FAO), sugarcane was cultivated on an area of 26.94 million hectares (ha) in 121 countries with a total cane production of 1.91 billion metric tons in 2013 (FAOSTAT 2015: http://faostat3.fao.org/browse/Q/QC/E). Since 1990s, the total global area under sugarcane production has been annually increased, with Brazil, India, China, Pakistan and Thailand being the top five producers. Figure 6.1 shows a steady increase in the sugarcane production worldwide mainly due to a gain of both production area and crop productivity.

Generally sugarcane mills include attached ethanol distilleries and integrate the production of sugar cane and bioethanol. Sugar mills and bioethanol production consume a large amount of fresh water. Typically, a sugar mill with annexed alcohol distilleries releases an average of 155 L of stillage and 250 kg bagasse per 1,000 kg of sugarcane to obtain 12 L of bioethanol and 95 kg of sugar [3].

Bagasse is the main by-product, followed by process wastewater. Usually it includes preliminary filtration and sedimentation of suspended solids, flow and load equalization for a biological treatment (anaerobic and aerobic) and nutrient removal in a few cases.

Molasses is another by-product of the sugar industry, which is currently being used to produce bioethanol. However, this process produces a large volume of spent wash water and strategies to minimize water consumption are of great interest. Molasses wastewater is a dark brown effluent conventionally treated by anaerobic digestion to generate methane followed by an aerobic trickling filter or activated sewage sludge. Besides organic matter, molasses contain high concentrations of nitrogen, phosphorus and potassium, which enable its use for direct field fertilization, at a rate of about 60 m^3/ha and substitution of traditional fertilizers, as well as a soil quality amendment [4, 5].

*Corresponding author

Constructed Wetlands for Industrial Wastewater Treatment, First Edition. Edited by Alexandros I. Stefanakis.
© 2018 John Wiley & Sons Ltd. Published 2018 by John Wiley & Sons Ltd.

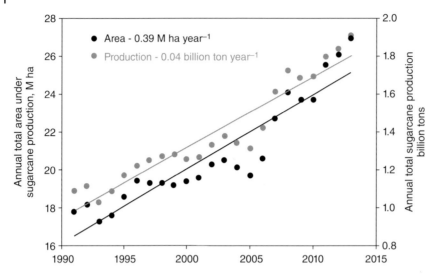

Figure 6.1 Temporal trend in the sugarcane production worldwide according with the production area and its productivity reprinted (permission pending) from Bhatnagar and Kesari [2].

The wastewater from the mill and process houses is combined with the wastewater from the boiler house to form a large amount of combined wastewater from the sugar mill, which has enormous potential for pollution. In view of the volume of generated wastewater and its strength, sugar industries are considered as one of the most polluting agri-food industries if the process effluents are not properly treated [2].

Typically, sugarcane wastewater is characterized by high chemical oxygen demand (COD), biological oxygen demand (BOD$_5$), and total soluble solids which impairs a dark color. Wastewater from the sugar industry generally contains carbohydrates, phenolic compounds, nutrients, oil and grease, chlorides, sulfates, and heavy metals [6]. Characteristic properties of sugarcane molasses stillage are reported in Table 6.1. These values are typical of high strength industrial wastewater and require a multi-step treatment process. Although the concentrations of heavy metals in sugarcane effluents are rather low, mercury concentration might exceed some of the environmental quality standards due to the high toxicity of this trace element [12].

Environmental impact associated with sugar mill effluents related to water and land pollution has been reported. Sugar mill effluents contain remarkably high concentrations of harmful substances including soluble salts and heavy metals such as Fe, Cu, Zn, Cu, Zn, Mn and Pb. The long-term use of this sugar mill effluent for irrigation must be discouraged as it results in the contamination of soils and crops. If the effluent infiltrates into subsoil and reaches the groundwater, it will disturb the groundwater quality. On the other hand, eutrophication of surface water may also create environmental conditions that favor the growth of toxin-producing cyanobacteria [13]. However, if the sugar industry effluent is properly treated, it can be reused for crop irrigation (Figure 6.2).

Due to the high cost of water, its limited availability and environmental regulations, the minimization of freshwater consumption and wastewater discharge into the environment is of great interest. El-Halwagi and Manousiouthakis [15] introduced the concept of mass exchange networks to deal with separation systems. Their work is focused on the design of optimal mass exchanger networks to transfer currents between a set of rich streams and set of lean streams. They developed the synthesis

Table 6.1 Physical–chemical characteristics of sugar cane molasses [7–11].

Parameter	Units	Sugarcane molasses[a]	Anaerobic treated molasses[b]	Sugar cane effluent
pH	–	3.8–4.5	nr	nr
TSS	mg/L	572–100,000	2490–4332	350
COD	mg/L	23,000–135,000	35,880–68,540	3,500–10,000
BOD$_5$	mg/L	10,000–89,800	483–803	4,000–7,000
TN	mg/L	210	1008–1752	nr
TKN	mg/L	1,600–2,975	nr	53
NH$_4$-N	mg/L	432–722	486–799	nr
NO$_3$-N	mg/L	720	8.6–17.7	nr
TP	mg/L	2630	53.1–97.7	4.8
PO$_4$-P	mg/L	68–1,100	nr	nr
SO$_4$-S	mg/L	6,250–8,220	nr	nr
Phenols	mg/L	380	nr	nr
Color density	absorbance at 475 nm	nr	56.9–94.3	nr
Electrical conductivity	µS/cm	3999	nr	nr

[a] Spentwash molasses current.
[b] Anaerobic treatment anaerobic pond HRT 60 days.
nr: not reported.

of mass exchange networks by analogy with the synthesis of heat exchanger networks. A number of studies use these concepts in the wastewater treatment design [16]. Wang and Smith [17] extended the concept of pinch to wastewater application using a graphical representation and techniques on superstructures of alternative designs.

Ideally, the zero-discharge approach (e.g., no effluent) would be theoretically feasible to achieve in the sugar cane industry, taking into account that as much as 70% of the raw material is water. Since the mean water intake by sugar mills is recorded at 60 m^3 per 100 t of cane [18], it can be inferred that most of the sugar mill should cope without additional water sources.

Five steps have been identified in approaching a close-to-ideal yet practical zero-liquid-discharge and they are listed below:

1) Minimize freshwater abstraction and wastewater generation.
2) Segregation and holding of the major wastewater streams to facilitate their reuse with minimal pretreatment.
3) Reuse and recycling of the wastewater streams without compromising the product quality or plant integrity.
4) Effluent treatment by physical–chemical or biological processes.
5) Disposal in accordance with environmental standards.

Currently in the sugar industry efforts are made to reduce both volumes of process water, as water disposed to the environment, which undoubtedly affects the characteristics of the wastewater treatment process. Therefore, in this chapter, we are focusing on the steps 3 and 4 of those described above,

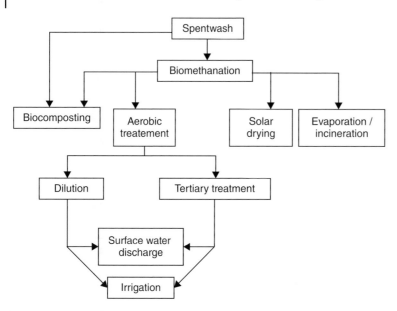

Figure 6.2 Water management of the spentwash effluents in the sugarcane industry. Reprinted (permission pending) from Satyawali and Balakrishnan, 2016 [14].

i.e., water consumption minimization and methodologies used to treat sugarcane and bioethanol effluent. Special focus will be devoted to the constructed wetlands (CWs) used as secondary or tertiary treatment. In this regard, the experience obtained in the application of CWs in different sugarcane production countries will be presented. In fact, the application of CWs to the agro-industrial wastewater treatment has been critically evaluated [19], but similar information is not available for sugarcane effluents. Moreover, limited experience exists regarding CWs to treat sugar beet industries in cold and temperate climates [20]. However, we will focus on sugarcane effluent treatment since the effluent characteristics and environmental conditions are completely different, since sugarcane production is limited to tropical climates.

6.2 Modeling Water Consumption Minimization

A mathematical programming model has been developed to represent the basic process characteristics and constraints, and it has been validated by application to production scenarios [21] which will not be covered in this chapter. Basically, the model represents mass balance in each production unit. The user must specify all units that generate wastewater and diverse wastewater treatment options. The model automatically identifies the following recycle and reuse options for each water stream: (i) reuse with regeneration; (ii) reuse without regeneration; (iii) recycle with regeneration; and (iv) recycling without regeneration.

The mathematical model consists of a linear programming problem. The objective function to be minimized contains the following terms, cost associated to freshwater supplies (Cfw_i), wastewater treatment cost (Cwt_j) and water discharge taxes (Cd_i).

The model can be formulated as follows:

Objective function

$$\min = \sum_{j} Cwt_j * AT_{ji} + \sum_{i} Cfw * FW_i + \sum_{ii'} Cre_{ii'} * RE_{ii'} + \sum_{i} Cd_i * D_i$$

Subject to the following constraints:

(a) The concentration of contaminants in the water that is supplied to the processes must be within specified limits. This constrain is expressed mathematically as:

$$FW_i + \sum_{j} AT_{ji} \ \varphi_{cj} + \sum_{i'} RE_{i'i} \ \gamma_{ci'} \le in_i \phi_{ci} \quad \forall c, i. \tag{6.1}$$

(b) The flow of water in and out of each process *i* is stated by the following two equations:

$$FW_i + \sum_{j} AT_{ji} + \sum_{i'} RE_{i'i} = in_i \quad \forall c, i \tag{6.2}$$

$$D_i + \sum_{i'} RE_{ii'} + \sum_{j} TV_{ij} = out_i \ \forall i \tag{6.3}$$

(c) It is assumed that all the water sent to treatment is transformed into treated water as:

$$\sum_{i} TV_{ji} = \sum_{i} AT_{ji} \ \forall j \tag{6.4}$$

(d) The water treatment produces a reduction in the concentration of contaminants as stated:

$$(1 - \alpha_{cj}) \sum_{i} TV_{ij} \gamma_{ci} = AT_j \varphi_{cj} \ \forall c , \forall j \tag{6.5}$$

6.2.1 First Approach to Linearity

This problem formulation involves nonlinearities associated to the two variables: $AT_{ji} \ \varphi_{cj}$. A first attempt to solve this problem was to consider φ_{cj} as constant. That is the concentration of contaminant *c* after the water treatment *j* is forced to be below or equal to a pre-defined value. This approach is only acceptable if the real concentrations of all contaminants are very close to their pre-defined values.

It can be argued that the LP problem obtained is so easy and fast to solve that solving this problem several times by means of a heuristic to define the new value of φ_{cj} may be worthwhile. This approach is also useful in *what-if* situations when a facility is operating with pre-defined values of φ_{cj} and wants to check the consequences of the definition of new values for φ_{cj}.

6.2.2 A MILP Approach to the Problem

A new approach consisting of a Mixed Integer Linear Programming (MILP) model was developed to avoid the estimation of rough values for φ_{cj}. Considering that every water treatment has a process tolerance that must be taken into account, it is proposed the definition of bands of operation for the concentration of contaminants. For each band *k* of the concentration of contaminants *c*, minimum (φmin_{ckj}) and maximum (φmax_{ckj}) values are defined preferably according to tolerance of the water treatment process *j*. As a consequence a binary variable (X_{ckji}) and an auxiliary variable (XA_{ckji}) are introduced in the model, where:

$$XA_{ckji} = X_{ckji} * AT_{ji} \quad \forall c, k, j, i. \tag{6.6}$$

Therefore equations (6.1) and (6.5) are changed in the proposed formulation forcing the concentration of contaminants to be within minimum and maximum values.

$$FW_i + \sum_k \sum_j XA_{ckji} \, \varphi max_{ckj} + \sum_{i'} RE_{i'i} \, \gamma_{ci'} \le in_i \phi_{ci} \quad \forall c, i \tag{6.7}$$

$$(1 - \alpha_{cj}) \sum_i TV_{ij} \gamma_{ci} \le \sum_k \sum_i XA_{ckji} \, \varphi max_{ckj} \quad \forall c, j \tag{6.8}$$

$$(1 - \alpha_{cj}) \sum_i TV_{ij} \gamma_{ci} \ge \sum_k \sum_i XA_{ckji} \, \varphi min_{ckj} \quad \forall c, j \tag{6.9}$$

The resulting linearization of equations (6.1) and (6.5) by expression 6 forces the introduction of new equations to enforce the linearity of the model. When the binary variable X_{ckji} is equal to 1 it is stated that there is a flow of treated water with concentration *k* of contaminant *c* from water treatment *j* to process *i*. Equation (6.10) was introduced in the model to guaranty that when X_{ckji} is equal to 1 we will have $AT_{ji} = XA_{ckji}$.

$$- in_i(1 - X_{ckji}) \le AT_{ji} - XA_{ckji} \le in_i(1 - X_{ckji}) \quad \forall c, k, j, i \tag{6.10}$$

When the binary variable X_{ckji} is equal to zero there is no flow of treated water with concentration *k* of contaminant *c* from water treatment *j* to process *i*. Therefore the auxiliary variable XA_{ckji} is also equal to zero as stated in equation (6.11). This equation also forces the amount of flow expressed in the auxiliary variable to be within desired minimum (L_j) and maximum values (in_i).

$$L_j X_{ckji} \le XA_{ckji} \le in_i X_{ckji} \quad \forall c, k, j, i \tag{6.11}$$

If there is a flow of water from water treatment *j* to process *i* then the concentration of each contaminant *c* will be within one of the defined bands *k*. Equation (6.12) expresses this statement mathematically.

$$\sum_k X_{ckji} \le 1 \quad \forall c, j, i \tag{6.12}$$

It is defined in equation (6.13) that if there is a flow of contaminant *c* from water treatment *j* to process *i*, then there will also be a flow of the other contaminants *c'* from water treatment *j* to process *i*.

$$\sum_k XA_{ckji} = \sum_k XA_{c'kji} \quad \forall c, \, c' \ne c, j, i \tag{6.13}$$

The model proposed identified significant results in the fresh water consumption. Moreover, the combined effect of lowering freshwater costs simultaneously reducing the effluent contaminants, leads to maximum reuse or recycled water at minimum cost. Furthermore, the automated wastewater treatment unit selection that has the most efficient method of contaminant removal enhances the reuse opportunities and contributes to additional savings [22].

In the following sections, different methodologies that have been used for sugarcane effluent treatment focused on CWs will be assessed.

6.3   Type of Effluent and Pretreatment

Sugarcane and bioethanol effluents are characterized by their high wastewater strength but the latter usually contains higher levels of organic matter and nutrients (Table 6.1). Consequently, a multistep approach is required to achieve the targeted water quality standards for discharge in surface waters. Constructed wetlands have been successfully used as secondary treatment in combination with other processes used to remove suspended particles and organic matter. High concentrations of suspended particles are conducive to clogging of the porous media. Some of the pretreatment methods are listed below.

6.3.1   Physical–Chemical Methods

Among the physical–chemical methods used for effluent treatment, adsorption, coagulation and flocculation, ozone oxidation, membrane, electrochemical oxidation and evaporation/combustion are well suited for this type of effluent. However, the high operation and maintenance cost of some of these processes, makes them of limited applicability for the sugarcane effluents involving high volumes of wastewater containing high COD and BOD loads. Another important aspect is the significance and effectiveness of these pretreatment processes combined with CWs being an aspect not fully addressed in many publications. For instance, coagulation is a successful method to reduce the suspended solids (SS) but it uses a high pH value (pH>8) which might be not compatible with the CW vegetation [19]. Electrochemical oxidation is useful to remove COD but it is more efficient to treat the CW effluent [23].

6.3.2   Intensive Biological Processes

Due to the high BOD_5 concentration found in sugarcane effluents, the anaerobic digestion with biogas recovery has been extensively employed [14]. These methodologies reduce the organic pollution load and bring the BOD_5 down to the 80–95% of the original value. Different treatment methodologies have been evaluated for anaerobic digestion of high strength industrial wastewater [24] and some of the results achieved for sugarcane effluents are discussed below (Table 6.2).

6.3.2.1   Suspended Bed Reactor
Upflow anaerobic sludge blanket (UASB) reactor has been used for the treatment of many industrial effluents including sugarcane molasses spent wash achieving up the 75% of the COD removal. Most of the UASB systems are operated under mesophilic conditions; however thermophilic ones lead to a higher methanogenic activity.

6.3.2.2   Fixed Bed Reactor
This involves the immobilization of microbial biomass on an inert support to limit the lost of biomass and to enhance the bacterial activity per reactor volume unit. It provides higher COD removal at lower hydraulic retention time (HRT) and better tolerance to toxic and organic shock loadings. Polyurethane, clay brick, granular activated carbon and polyvinyl chloride have been employed resulting in the 67–98% of COD reduction [34].

Table 6.2 Performance of various anaerobic reactors for molasses distillery wastewater.

Reactor configuration	COD loading (kg COD/m^3/d)	HRT (d)	COD (%)	BOD (%)	Reference
UASB	24	2.1	75	nr	(25)
Thermophilic UASB	Up to 86.4	nr	60		(26)
Thermophilic UASB	Up to 28	nr	39–67	>80	(27)
Downflow fixed film reactor	14.2–20.4	3.3–2.5	85–97	60–73	(28)
Two-phase thermophilic process					
Acidogenesis	4.6–20.0	2	nr	nr	(29)
Methanogenesis	nr	15.2	nr	nr	
Diphasic (upflow) fixed film reactor (clay brick granules support)	22	3	71.8	nr	
Diphasic (upflow) fixed film reactor (granular activated carbon support)	21.3	4	67.1	nr	(30)
Downflow filter	8	nr	55–85	nr	(31)
Two-stage bioreactor (anaerobic)	7	nr	71	86	(32)
Anaerobic contact filter (in series)	nr	4	73–98	nr	(29)
Upflow blanket filter	9–11	11–12	70	nr	(33)

nr: not reported.

6.3.2.3 Fluidized Bed Reactor

This contains an appropriate medium such as sand, gravel, or PVC for bacterial attachment and growth. These reactors can be operated in upflow or downflow modes by applying high fluid velocities. A COD reduction along the 75–95% range could be achieved by using ground perlite [14].

6.3.3 Extensive Biological Processes

Anaerobic lagoons in series are the simplest option to pretreat sugarcane effluents [6]. Twelve stabilization ponds of 4.05 ha have been built to treat the effluent of sugar cane where the first ponds are anaerobic. However, large surface area requirement, odor emission and chances of groundwater pollution restrict its widespread use. The post-anaerobic treatment stage effluent still yields to a high COD/BOD loading and is generally followed by an aerobic treatment such as solar drying. Other treatment processes that have been demonstrated for biomethanated distillery effluent are aquaculture, biocomposting, fungal, bacterial and algal treatment [14]. However, the CW technology has been successfully used and will be presented in the following sections.

6.4 Constructed Wetlands (CWs)

The success of CWs in treating process wastewaters containing high-strength organic matter depends on several factors related primarily to organic loading rate (OLR), HRTs, the tolerance of selected macrophytes to possible phytotoxic contaminants in the process wastewater, and plant biofilm activity. In this regard, it has been acknowledged that CWs in tropical or subtropical climates, work under optimum conditions all through the year due to the almost constant high temperatures.

CWs of different configurations and scale (mesocosmos, pilot and field) have been evaluated as a secondary or tertiary treatment of urban wastewaters in subtropical and arid climates without land space limitations such as Australia [35]. For instance, in Queensland, free-water surface constructed wetland (FWS-CW) with a diversity of macrophyte types offers a great potential for effluent polishing. The high macrophyte growth rate is conductive to high removal of nutrients, in particular nitrogen and fecal coliform producing effluents close to 100 colony forming units (cfu) per 100 mL acceptable for agricultural irrigation.

In high rainfall climates, it can affect the CW performance due to dilution of the influent occasioned by heavy rainfall infiltration. In the following section, different experiences in the implementation of CWs for sugarcane or molasses effluent treatment are presented. Note that in most of these studies, the removal efficiency is not corrected by evapotranspiration leading to underestimation of the system's efficiency. Evapotranspiration rates in tropical countries can be from 4.7 mm/day to 7–25 mm/day in hot arid climates [36, 37].

Another important issue regarding field CWs is the hydraulic assessment of its bed flow properties under the variable operational conditions in addition to contaminant removal measurements [38]. In this regard, tracer experiments are strongly recommended when pilot scale is used.

6.4.1 Case Studies

A summary of CWs performance to remove organic matter and nutrients is presented in Tables 6.3 and 6.4, respectively. In the following sections, different case studies are reported for the countries where the experiments were performed.

6.4.1.1 India

Trivedy and Nakate [40] performed a laboratory scale experiment using *T. latipholia* to treat diluted distillery effluent. A bed zone (0.3 m × 0.3 m × 1.5 m) filled with 75% sand and gravel and 25% soil was evaluated. The system yields a 76% of COD removal in a HRT of 7 days, which increased marginally to 78% in 10 days. On the other hand, the BOD removal was 22 and 47% on a HRT of 7 and 10 days respectively.

Billore et al. [41] have evaluated a four-celled horizontal subsurface flow (HSSF) CW for the treatment of a distillery effluent after anaerobic treatment. The post-anaerobic effluent had about 2.5 g/L of BOD and nearly 14 g/L of COD. A pretreatment chamber was used to capture the suspended solids. All the cells were filled with gravel and one planted with *T. latipholia* and the other with *P. karka*. The HRT was 14.4 days achieving the 64%, 85%, and 42% removal of COD, BOD_5, and total solids respectively.

Table 6.3 Evaluation of constructed wetland efficiency to remove suspended solids and organic matter from sugar cane effluents.

Influent pretreatment	Scale	Configuration	Time span evaluation (d)	Macrophyte	HRT (d)	HLR (mm/d)	TSS (%)	COD (%)	BOD$_5$ (%)	Ref.
Anaerobic	Full-scale	HSSF	730	Typha spp. Phragmites spp.	nr	500	nr	82–85	nr	[39]
nr	Mesocosms	nr	nr	Typha spp.	7/10	nr	nr	76/78	22/47	[40]
Anaerobic	Mesocosms	HSSF	nr	Typha spp. Phragmites spp.	14.4	nr	42	64	85	[41]
nr	Mesocosms	VFCW	nr	Phargmites spp. Typha spp.	7	nr	nr	77–98	89–99	[42]
Dilution	Mesocosms	HSSF	nr	Pontederia spp.	2.5, 5	nr	nr	80	82–87	[8]
Activated sludge	Mesocosms	VFCW HSSF	nr	Phragmites spp. Typha spp.	1.7/5	nr	nr	86/92	92/94	[43]
Anaerobic	Microcosms	FWS	42	Cyperus spp. Thalia spp.	7	nr	90–93	67	88–89	[9]
Stabilization ponds (9)	Pilot	FWS	365	Cyperus spp. Echinochloa spp.	nr	17/27	63/47	41/47	nr	[44]

nr: not reported.

Table 6.4 Constructed wetland to treat nutrients from sugar cane wastewater effluents.

Influent characteristics	Scale	Configuration	Time Span (d)	Plant	HRT (d)	HLR mm/d	NH$_4$-N (%)	NO$_3$-N (%)	TKN (%)	TP (%)	Ref.
Stabilization pond	Pilot	FWS	360	Cyperus spp. Echinochloa spp.	nr	17–27	35–44	nr	nr	44	[44]
Dilution	Mesocosms	HSSF	55	Pontederia spp.	2.5, 5	nr	nr	56–59	73–76	0	[8]
Anaerobic	Microcosms	SFCW	42	Cyperus spp. Thalia spp.	nr	nr	77–82	94–95	nr	70–76	[9]
nr	Mesocosms	VFCW	nr	Phargmites spp. Typha spp.	7	nr	100/89	nr	100/75	nr	[42]
Activated sludge	Mesocosms	VFCW HSSF	nr	Phragmites spp. Typha spp.	1.7/5	nr	37/56	nr	21/32	nr	[43]

nr: not reported.

Pratik and Nishith [42] evaluated a vertical flow constructed wetland (VFCW) planted with *P. australis* and *T. angustata* by using local soil to treat the sugar cane effluent. This study was performed at mesocosmos scale (0.8 cm length × 0.54 cm wide × 0.73 cm height). The HRT was 7 days and the best results were obtained with *P. australis* leading to a COD and BOD removal of 98.4% and 89.2% respectively.

6.4.1.2 Kenya

Bojcevska and Tonderski [45] and Tonderski et al. [44] evaluated the impact of loading rates and two macrophyte species (e.g. *Cyperus papyrus* and *Echinochloa pyramidalis*) in a FWS CW in western Kenya at 1269 m asl for a 12-month experiment. Four CWs were operated at HRT of 75 mm d^{-1} and four at 225 mm/day coming from a stabilization pond of a sugar factory. Seasonality is characterized by three periods, short rain, dry and long rain, and had a significant impact on the mass removal rate of TDP, NH$_4$-N and TSS. COD removal ranged from 41 to 47% depending on the HLR (Table 6.4).

6.4.1.3 Mexico

Olguin et al. [8] (reported a pilot study under greenhouse conditions in Xalapa, Veracruz (1580 m asl). A HSSF CW (3 m length × 0.3 m wide × 0.5 m deep) using volcanic porous substrate (particle 4 cm; porosity 50%) was evaluated. The influent was a diluted sugarcane molasses stillage and pH adjusted to 6.0 and the OLR was 47–95 g COD/m^2/d. The aquatic plant selected was pickerelweed (*Pontederia sagittata)* endemic in this region, and non-planted controls were also included. Removal rates for COD and BOD$_5$, were 80% and 82–87%, respectively depending on the HRT.

We have conducted an evaluation study of different CW technologies to treat the effluent of a sugarcane factory located at Atencingo (Puebla, Mexico) at 1280 m asl where wastewater is treated by an activated sewage with forced aeration process before its discharge to a small river (Nexapa). A schematic diagram of the mesocosmos used is presented in Figure 6.3.

In this study, a comprehensive evaluation at mesocosmos scale was performed of flow direction (VFCW and HSSF CWs) and shape (0.85 m length; 0.33 m wide; 0.5 m depth vs 0.62 m length; 0.33 m wide; 0.5 m depth) operated in the semi-batch mode. The following design variables were evaluated:

1) Flow direction.
    a. VF CW partially saturated (80%) with a resting period of 16 hours.
    b. HSSF CW intermittent fed at 20% volume with a resting period of 5 days.
2) HRT.
    a. Short: 1.7 days in the VF
    b. Long: 5 days
3) Type of macrophyte.
    a. *Phragmites australis*
    b. *Typha latifolia*
4) Type of substrate.
    a. Ashed crushed gravel
    b. Volcanic (*tezontle*). It is a highly porous material from volcanic origin widespread in this region.
5) Size substrate (diameter)
    a. 5 cm,
    b. 2.5 cm
    c. 1 cm

In this study, we kept the water level constant by refilling to compensate the losses due to evapotranspiration. The semi-batch operation mode was chosen since it provides an extended aeration of the gravel bed. It allows aerobic biodegradation of the organic matter and nitrification in typically anaerobic systems such as HFSS [46, 47]. The main results achieved were as follows:

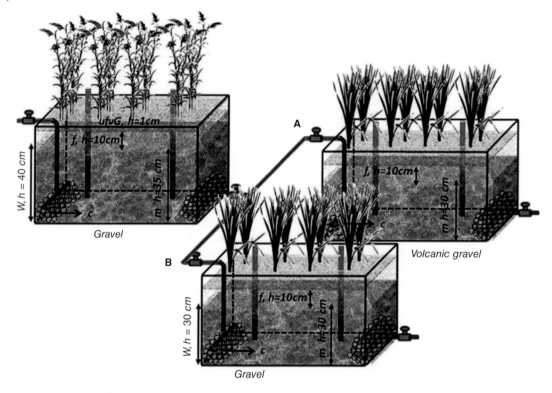

Figure 6.3 Outline of the mesocosms used to treat an effluent of Atencingo a sugarcane factory in Mexico. Mesocosm configurations: A – Horizontal subsurface flow; B – Shallow vertical flow. Notes: c – coarse, φ= 5 cm; m – medium, φ= 2.5 cm; f – fine, φ= 1 cm; ufvG –ultrafine volcanic gravel, φ = 2–3 mm; W – water.

- Despite the high variability of the organic load of the inflow to the CWs system, high removal efficiency of the BOD_5 and COD, 86–92% was in the range of 92–94% respectively.
- The best results, considering the lower HRT and smaller footprint, were obtained with the VFCW treatment, with a resting period between cell loadings with wastewater of 16 h.

6.4.1.4   South Africa
Schumann [39] evaluated the performance of HSSF CW for the direct treatment of sugarcane wastewater (Table 6.4). This study was conducted in a full-scale plant (2,000 m^2) for 18 months by using two plant species (bulrushes and *Typha* spp.) and no statistical differences were found between them. However, the flow rate was reduced due to clogging problems, but irrespective of plant species and flow rate, the COD removal efficiency ranged from 82 to 85%.

6.4.1.5   Thailand
Sohsalam and Sirianuntapiboon [9] evaluated a FWS with *Cyperus involucratus* and *Thalia dealbata* to treat anaerobic treated-molasses wastewater. The system performance was evaluated at different OLR. At the optimum conditions (612 kg BOD_5/ha/d) the TSS, BOD_5, and COD removal were 90–93%, 88–89% and 67% respectively.

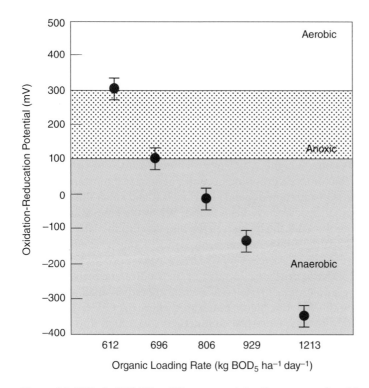

Figure 6.4 ORP of a FWS CW at different organic loading rates reprinted from Sohsalam and Sirianuntapiboon [9].

6.4.2 Effects of Design and Operation on the COD, BOD and Nutrient Removal

In this section the impact of the different operational modes in the CW performance is presented.

Hydraulic retention time. For a specific class of CWs, the BOD_5 removal will depend on the HRT. Usually typical HRTs are as follows: FWS > HSSF > VF CW. Usually the higher the HRT, the higher is the BOD removal for a specific class of CW. However, VF CWs outperformed the other CWs (FWS and HSSF) in term of mass of COD per surface area.

Effect of the organic loading rate (OLR). This chiefly depends on the hydraulics of the system. In fact, the FWS is very sensitive to the organic overloading. For instance, it was reported that higher than 61.2 g $BOD_5/m^2/d$, the removal efficiency and plant growth declined [9]. It could be attributed to the redox potential declining from aerobic to strictly anaerobic due to a decrease in the oxygen concentration in the water column [9] (Figure 6.4). The same authors found a statistically significant decline of one log for the viable biofilm bacteria in the rizosphere of the three macrophytes investigated (*Cyperus involucratus* and *Thalia dealbata*) when the OLR steadily increased from 612 to 1213 kg $BOD_5/m^2/d$.

A significant positive linear relationship between the mass removal rate of TSS, TP, NH_4-N, and the mass load was found in pilot scale FWSs fed with the anaerobic effluent of a stabilization pond [45]. In the same pilot plant, Odinga et al. [48] measured the foliar total nutrient concentrations (nitrogen and phosphorus). They found that even though the nutrients were removed from the wetland, they were not accumulated in the foliar system of the CW macrophytes. These results suggest that the nutrient removal might take place in the rizosphere.

HSSFs CW allow a higher OLR than FWS, but the VFCWs can be fed at higher OLR than SSFCWs. The VF CWs can be fed with OLRs from 63 to 102 g/m^2/d of COD whereas HSSF CW from 15 to 49 g/m^2/d [43].

The temporal changes in BOD$_5$-to-COD ratio is a useful parameter to evaluate the ability of the system to progressively degrade more recalcitrant organic compounds in the process wastewater due to acclimation of microbiota in the biofilm. This parameter steadily increases during the first month of plant operation [49].

Effect of vegetation. Usually planted CWs offer a higher efficiency than non-planted or control units up to 30% in terms of COD removal, but biomass cropping becomes necessary if nutrient removal is also required. However, the biomass production depends on the OLR and small differences in the removal rate among the macrophyte species have been found. When the FWS was fed with anaerobic treated molasses, the higher the OLR the lower was the biomass productivity [9].

Type of vegetation. This has been evaluated for a number of macrophytes including: (a) *P. australis*; (b) *T. latifolia*; (c) *Pontederia sagittata*; (d) *Cyperus papyrus*; (e) *Cyperus involucratus*; (f) *Echinochloa pyramidalis*; (g) *T. angustata* and (h) *Thalia dealbata*. In most of the studies one or two species are evaluated. However, in many case studies the differences observed are rather small and consequently the recommended species are the endemic plants of the area where the CW is developed, and also their tolerance of dissolved salts (e.g., electrical conductivity), and high nutrient loads should be considered. In this regard, *Phragmites* spp. and *Typha* spp. are the most widely used due to their high resilience to high COD, TKN and TP concentration levels.

Type of substrate. Although the gravel bed is the most widely used, other media have been evaluated. The homogeneity of the diameter is an important design parameter to avoid preferential pathways. Navarro et al. [43] found no significant differences of the measured parameters between gravel bed and porous volcanic medium (e.g., tezontle) cells in the VF and HSSF modes.

6.4.3 Other Water Quality Parameters

6.4.3.1 Turbidity
Turbidity is an indicator of the water quality and widely used in the legislation (e.g., California Title 22 on water reuse). Navarro et al. [43] compared the VFCW and the HSSF at different regimes of operation. The best results were obtained at HSSF mode at characteristic design flow rates (14.3 L/m^2/d) achieving values below to 10 NTU according with the Mexican guideline (2016 Mexican Federal Law of Rights concerning National Waters).

6.4.3.2 Pigments
Wastewater effluent from molasses and sugarcane mills possesses a dark intense color that needs to be removed before its disposal into the surface waters or its reuse in agriculture. Color intensity removal has been evaluated in different CW configuration. In the case of the FWS, three macrophytes were evaluated (e.g. *C. involucratus, T. angustifolia, T. dealbata*) and 4 OLR from 612 to 1213 kg BOD$_5$/ha/d. Although, the differences obtained with the different plant species were not significantly high, the highest color removal was obtained with *C. involucratus* at the lowest OLR [9].

6.4.3.3 Sulfate
High concentrations of sulfates have been detected in the molasses effluents (Table 6.1) compared to other industrial effluents [20], which positively correlates with the EC ($r^2 = 0.7998$) and chloride

($r^2 = 0.8339$). These results suggest a high contribution of these ions to the EC of wastewater. Sulfate removal form wastewater is of interest as a precursor of reduced sulfur species which are responsible for off-flavors, corrosion and freshwater toxicity at different trophic levels [50].

The SO_4^{2-} removal in a HSSF CW was in the range 68–69% [8]. In a comparative study between different CW configurations, we found higher removal efficiency in a VF CW with about a 50% of reduction of this parameter in the best operating conditions.

6.4.3.4   Nitrogen Removal

Optimum nutrient removal has been obtained when biomass is cropped. Accordingly, the following results are given where the removal of different nitrogen species are recorded under these operational conditions:

- The highest removal efficiencies of nitrogen species were obtained in a FWS CW operated at different OLR in the optimum operational conditions; the NH_4-N, NO_3-N removal was in the range of 77–82% and 94–95%, respectively [9].
- A slightly lower removal was obtained in an HSSF CW, where the TKN removal was 73–76% and NO3-N ranged from 56–59% [8].
- In a comparison between the macrophytes *Cyperus papyrus* and *Echinochloa pyramidalis* in a FWS CW, NH_4-N removal was higher when the CW was planted with the former macrophyte [45].

6.4.3.5   Phosphorus

There are conflicting data regarding P removal in sugarcane effluents. Whereas in an HSSF CW, P was not removed [8] other authors reported 79% TP elimination [41] in a CW of similar configuration. Removal in the range of 70–76% has been reported in a FWS CW [9].

6.5   Research Needs

Some mathematical models, which consider the cost of the fresh water, its treatment and wastewater discharge taxation, are available and could be used for environmentally and economically sustainable management of water in the sugarcane industry.

Since sugarcane is a seasonal crop, the CW needs to be operated with alternative sources of wastewater during the off-harvest season, and the period of acclimation when the influent wastewater is changed to sugarcane or molasses effluents needs to be optimized by steadily increasing the OLR.

Mid-term studies regarding substrate accumulation of refractory to biodegradation organic matter, and leading to a reduction of hydraulic conductivity with the time of operation conducting to clogging, are needed.

The fate of heavy metals, which have been already detected in industrial effluents and agrochemicals taken up by the crop during the CW treatment, is largely unknown.

Acknowledgements

The authors acknowledge the AECID 11-CAP2-1756 for funding and Prof. Dr. Lluis Puigjaner for research encouragement and technical assistance during the model development.

References

1 Hess T, Aldaya M, Fawell J, Franceschini H, Ober E, Schaub R, Schulze-Aurich J. Understanding the impact of crop and food production on the water environment – using sugar as a model. J. Sci. Food Agric. 2014; 94: 2–8.

2 Bhatnagar A, Kesari KK. Multidisciplinary approaches to handling wastes in sugar industries. Water Air Soil Pollut. 2016; 227(11):1–30.

3 Siddiqui WA, Waseem MA. A comparative study of sugar mill treated and untreated effluent – A case study. Orient J. Chem. 2012; 28(4): 1899–1904.

4 Körndorfer GH, Nolla A, Gama AJM. Manejo, aplicación y valor fertilizante de la vinaza para caña de azúcar y otros cultivos. Tecnicaña 2010; 24:23–28 (in Spanish).

5 Vázquez Brito YL, Castro Lizazo I, López O. Effect of vinasse application on some soil physical indicators Pardo sialítico carbonated cultivated tomato (*Solanum lycopersicum* L) cv. Amalia. Rev Ingen Agríc. 2014; 4:24–29 (in Spanish).

6 Rais M, Sheoran A. Treatment of sugarcane industry effluents: Science & technology issues. M. Rais Int. J. Engineer. Res. Appl. January 2015; 5(1):11–19.

7 Tewari PK, Batra VS, Balakrishnan. Water management initiatives in sugarcane molasses based distilleries in India. Resour. Conserv. Recycl 2007; 52:351–367.

8 Olguin EJ, Sánchez-Galván G., González-Portela RE, López-Vela M. Constructed wetland mesocosms for the treatment of diluted sugarcane molasses stillage from ethanol production using *Pontederia sagittata*. Water Res. 2008; 42:3659–3666.

9 Sohsalam P, Sirianuntapiboon S. Feasibility of using constructed wetland treatment for molasses wastewater treatment. Bioresour. Technol. 2008; 99:5610–5616.

10 Mancini IM, Boari G, Trulli E. Integrated biological treatment for high strength agro-industries wastewaters. Proceed. 4th Intl Conf. Wetland Systems for Water Pollution Control, Guangzhou, China, 1994, pp. 589–598.

11 Díaz M, Martínez C, Navarro A, Paumier V. Geoquímica del medio ambiente en la región de Izúcar de Matamoros e impacto ambiental originado por la actividad industrial. Parte I. Efluentes industriales. Trabajo Q29, Memorias de la 2da Jornada Científica y de Calidad del CEINPET (CD-ROM), Cuba, Diciembre 2003. © 2003 CEINPET, ©2003 IDICT, ISBN 959-234-039-0.

12 Qureshi MA, Mastoi GM. The physiochemistry of sugar mill effluent pollution of coastlines in Pakistan. Ecol. Eng. 2015; 75:137–144.

13 Poddar PK, Sahu O. Quality and management of wastewater in sugar industry. Appl Water Sci 2015; 1–8.

14 Satyawali Y, Balakrishnan M. Wastewater treatment in molasses-based alcohol distilleries for COD and color removal: A review. J. Environ. Manag. 2008; 86:481–497.

15 El-Halwagi M, Manousiouthakis V. Synthesis of mass exchange networks. AIChE J, 1989; 35 (8):1233–1244.

16 Rossiter AP. Waste Minimization Through Process Design, 1995, McGraw-Hill, New York.

17 Wang I, Smith R. Design of distributed effluent treatment. Chem. Engin. Sci., 1994; 49(18): 3127–3145.

18 Palazzo A. Mill-wide water management in the South African sugar industry. Proc. S. Afr. Sug. Technol. Ass. 2004; 78:419–426.

19 Sultana M-Y, Akratos ChS, Vayenas DV, Pavlou S. Constructed wetlands in the treatment of agro-industrial wastewater: A review. Hem. Ind. 2015, 69(2):127–142.

20 Vymazal J. Constructed wetlands for treatment of industrial wastewaters: A review. Ecol. Eng. 2014; 73:724–751.

21 Pastor R. Estrategias de optimization para el reuso del aga en procesos industrials. Aplicaciones a la industria alimentaria. PhD Dissertation, 2003. UPC, Barcelona (in Spanish).

22 Pastor R, Benqlilou Ch, Paz D, Cárdenas G., Espuña A, Puigjaner A. Design optimization of constructed wetlands for wastewater treatment. Resour. Conserv. Recycl. 2003; 37:193–204.

23 Grafias P, Xekoukoulotakis NP, Mantzavinos D, Diamadopoulos E. Pilot treatment of olive pomace leachate by vertical-flow constructed wetland and electrochemical oxidation: An efficient hybrid process. Water Res. 2010; 44:2773–2780.

24 Rajeshwari KV, Balakrishnan M, Kansal A, Lata K, Kishore VVN. State-of-the-art of anaerobic digestión technology for industrial wastewater treatment. Renew. Sustain. Ener. Rev. 2000; 4:135–156.

25 Sánchez Riera F, Córdoba P, Siñeriz F. Use of the UASB reactor for the anaerobic treatment of stillage from sugarcane molasses. Biotechnol. Bioengin. 1985; 27 (12):1710–1716.

26 Wiegant WM, Claassen JA, Lettinga G. Thermophilic anaerobic digestion of high strength wastewaters. Biotechnol. Bioenginen 1985; 27:1374–1381.

27 Harada, H, Uemura, S, Chen AC, Jayadevan J. Anaerobic treatment of a recalcitrant wastewater by a thermophilic UASB reactor. Bioresou Technol. 1996; 55 (3):215–221.

28 Bories A, Raynal J, Bazile F. Anaerobic digestion of high-strength distillery wastewater (cane molasses stillage) in a fixed-film reactor. Biol. Wastes 1988; 23 (4):251–267.

29 Seth R, Goyal .K, Handa BK. Fixed film biomethanation of distillery spentwash using low cost porous media. Resour. Conserv. Rec. 1995; 14 (2):79–89.

30 Goyal SK, Seth, R, Handa BK, Diphasic fixed-film biomethanation of distillery spentwash. Bioresour. Technol. 1996; 56 (2–3):239–244.

31 Athanasopoulos N, Anaerobic treatment of beet molasses alcoholic fermentation wastewater in a downflow filter. Resour. Conservat. 1987; 15 (1–2):147–150.

32 Vlissidis, A, Zouboulis AI. Thermophilic anaerobic digestion of alcohol distillery wastewaters. Bioresour Technol. 1993; 43 (2):131–140.

33 Bardiya, MC, Hashia R, Chandna S. Performance of hybrid reactor for anaerobic digestion of distillery effluent. J Indian Assoc. Environ. Manag. 1995; 22 (3):237–239.

34 Vijayaraghavan K, Ramanujam TK. Peformance of anerobic contact filter in series for treating distillery spentwash. Bioprocess. Biosystem Engn 2000; 22(2):109–114.

35 Greenaway M. The role of constructed wetlands in secondary effluent treatment and water reuse in subtropical and arid Australia. Ecol. Engen. 2005, 25:501–5019.

36 Abira MA, Ngirigacha HW, van Bruggen JJA. Preliminary investigation of the potential of four tropical emergent macrophytes for treatment of pre-treated pulp and papermill wastewater in Kenya. Water Sci. Technol. 2003; 48:223–231.

37 Bodin H. Wastewater treatment in constructed wetlands: Effects of vegetation, hydraulics and data analysis methods. May 2013. PhD Dissertation No 1509. Linköping University. Sweden.

38 Grisner ME, Carr MA, Shepherd HL. Evaluation of constructed Wetland Treatment Performance for Winery Wastewater. Water Environ. Res. 2003, 75(5):412–421.

39 Schumann GT. Artificial wetlands for the treatment of mill effluent. Proceed. South African Sugar Technologist Assoc. June 1991; 228–233.

40 Trivedy RK, Nakate SS. Treatment of a diluted distillery waste by constructed wetlands. Indian J Environ. Protect. 2000; 10:749–753.

41 Billore SK, Singh H, Ram HK, Sharma JK, Singh VP, Nelson RM, Dass P. Treatment of molasses based distillery effluent in a constructed wetland in central India. Water Sci. Technol. 2001; 44(11–12):441–448.

42 Pratik P, Nishith D. Phytoremediation of sugar industry effluent using *Typha angustata* and *Phragmites australis* through constructed wetland. J. Chem. Biol, Phys Sci. Section D Environ. Sci. Nov 2013–Jan 2014; 4(1): 846–851.

43 Navarro E et al. Optimization of constructed wetlands for Atencingo sugarcane effluent treatment. (In preparation).

44 Tonderski KS, Grönlund E, Billgren Ch, Raburu Ph. Management of sugar effluent in the Lake Victoria region. Ecohydrol. Hydrobiol. 2007; 7(3–4):345–351.

45 Bojcevska H, Tonderski K. Impact of loads, season, and plant species on the performance of a tropical constructed wetland polishing effluent from sugar factory stabilization ponds. Ecol. Eng. 2007; 29:66–76.

46 Stein OR, Hook PB, Biederman JA, Allen WC, Borden DJ. Does batch operation enhance oxidation in subsurface constructed wetlands?. Water Sci Technol. 2003; 48(5):149–156.

47 Ávila C, Reyes C, Bayona JM, García J. Emerging organic contaminant removal depending on primary treatment and operational strategy in horizontal subsurface flow constructed wetlands: influence of redox. Water Res. 2013; 47:315–325.

48 Odinga Ch, Otieno F, Adeyemo J. Investigating the effectiveness of aquatic plants (*Echinocloa* L and *Cyperus* L) in removing nutrients from wastewater: The case of Chemelil constructed wetland-Kenya. Intl. J. Phys. Sci. August 2011; 6(16): 3960–3970.

49 Grisner ME. Plants in constructed wetlands help to process agricultural processing wastewater. California Agriculture. April–June 2011; 65(2): DOI: 10.3733/ca.v065n02p73.

50 Huang Y, Ortiz L, Aguirre P, García J, Mujeriego R, Bayona JM. Effect of design parameters in horizontal flow constructed wetland on the behaviour of volatile fatty acids and volatile alkylsulfide. Chemosphere. 2005; 59:769–777.

7

Treatment of Effluents from Meat, Vegetable and Soft Drinks Processing using Constructed Wetlands

Marco Hartl[1], Joseph Hogan[2] and Vasiliki Ioannidou[3]

[1] *GEMMA – Environmental Engineering and Microbiology Research Group, Department of Civil and Environmental Engineering, Universitat Politècnica de Catalunya – BarcelonaTech, Barcelona, Spain*
[2] *Ecol-Eau, Montreal, Canada*
[3] *School of Engineering & the Built Environment, Birmingham City University, Birmingham, UK*

7.1 Treatment of Slaughterhouse and Meat Processing Wastewater

The worldwide consumption of meat is increasing as part of a global dietary transition, driven by rising incomes and urbanization. Besides the negative effects on human health, this process has also a severe impact on the environment and earth's water resources [2]. Fresh water is a limited resource and accounts only for a fraction of the earth's water reservoirs. Hence, it is of uttermost importance to protect this vital resource from excessive abstraction and pollution [1].

Slaughterhouse wastewater (SWW) is produced in large volumes during the slaughtering of animals and the cleaning of the slaughterhouse facilities. In addition, meat processing wastewater accounts for 24% of the freshwater consumption of beverage and food industries [3].

SWW has a very complex composition and contains high organic loads – due to blood, fat, manure, and undigested stomach contents – and, therefore, significant concentrations of total organic carbon (TOC), biological oxygen demand (BOD), chemical oxygen demand (COD), total suspended solids (TSS), total phosphorus (TP), total nitrogen (TN), sodium (Na), pathogenic and non-pathogenic microorganisms, while it fluctuates in temperature and pH, the latter caused by cleaning detergents and disinfectants (Table 7.1) [3]. Additionally, residues from veterinary pharmaceutical products can be present [4].

The selection of a suitable treatment of SWW is mainly dependent on wastewater characteristics, the best available technology (BAT) and the targeted effluent quality according to the prevailing regulations [3]. There are several types of individual or combined processes in order to treat SWW successfully. Constructed Wetlands (CWs) are used in a number of locations for treatment of this wastewater, alone or in combination with other technologies [5].

Scientific studies on CWs for SWW treatment have already been conducted as early as in the 1980s by research teams in New Zealand [6] and Australia [7].

Finlayson and Chick [7] tested the treatment performance of three emergent aquatic plant species (*Typha* spp., *Phragmites australis* and *Scirpus validus*) planted in an experimental trench system with horizontal flow, treating poultry SWW. All three trenches efficiently removed TSS (83–89%) and turbidity (58–67%). *Phragmites* and *Scirpus* trenches showed better oxygenation of the anaerobic inflow,

Table 7.1 General characteristics of slaughterhouse wastewater [3].

Parameter	Range	Mean
TOC (mg/L)	70–1,200	546
BOD_5 (mg/L)	150–4,635	1,209
COD (mg/L)	500–15,900	4,221
TN (mg/L)	50–841	427
TSS (mg/L)	270–6,400	1,164
pH	4.90–8.10	6.95
TP (mg/L)	25–200	50
Orto–PO_4 (mg/L)	20–100	25
Orto–P_2O_5 (mg/L)	10–80	20
K (mg/L)	0.01–100	90
Color (mg/L Pt scale)	175–400	290
Turbidity (FAU[a])	200–300	275

[a] FAU, formazine attenuation units.

whereas the *Scirpus* trench was significantly more effective than both other trenches in reducing TN (74%) and TP (79%), as well as potassium (56%), chloride (53%) and sodium (34%), during the one-month investigation. However, it has to be noted that the three trenches had different lengths (two of 20 m and one of 15 m) and were operated with different theoretical hydraulic retention times (tHRTs), with 2.7 ± 0.2 days in *Typha*, 3.6 ± 0.3 days in *Phragmites* and 3.0 ± 0.2 days in the *Scirpus* trench.

Van Ostroom and Cooper [6] started their experiments with three subsurface flow (SSF) systems (*Schoenoplectus validus*, *Glyceria maxima*, and unplanted) and three free water surface (FWS) trenches (mixtures of *Typha*, *Glyceria*, and *Schoenoplectus*). Due to earlier problems with nitrogen removal [8], FWS systems with floating mats (three times *Clyceria maxima* and one control out of nylon fabric) were tested in order to improve nitrogen removal from a nitrified meat processing effluent. The planted systems were finally able to remove 46–49% nitrogen with a hydraulic loading rate of 5.7 cm/d, which was twice as much as in the unplanted system. The main removal mechanisms were denitrification and accumulation in plant biomass, accounting for 87 and 13% of the removal, respectively. COD consumption by denitrification was higher than the available concentration in the inlet, indicating that between 24 and 48% of the COD was supplied by the plants [9].

Rivera et al. [10] implemented a two-stage system for treatment of SWW from a slaughterhouse north of Mexico City, Mexico. The system consisted of a settling tank and an anaerobic digester (AD) for primary treatment, followed by a HSSF (horizontal subsurface flow) CW system for secondary treatment. The CW had an HRT of 1.7 days and was planted with *Phragmites australis* and *Typha latifolia* in alternate strips. The SWW was successfully treated, resulting in removal rates of 88, 87, 89 and 73% for BOD_5, COD, SS and organic nitrogen, respectively. Removal rates were poor for orthophosphate, inorganic nitrogen species. The removal of pathogens was good with >99% total and fecal coliforms removed/inactivated. The CW was essential for the improvement of the SWW quality due to the high variability of the AD effluent. The final effluent could be used for irrigation of ornamental plants and recreational lands [10].

The first CW in Ecuador was built to treat the wastewater of a town's slaughterhouse. It consists of a primary settler, two HSSF beds in series; a first stage for "roughening" and a second for "polishing" of water. With an HRT of 2 days, inflow of 20 m^3/d and area of 1,200 m^2, the system's effluent reached a quality close to drinking water. The system is also a great example for stakeholder involvement since it is built and operated by the inhabitants themselves, using only local materials [11].

Another full-scale CW system improving SWW in Hidalgo, Mexico consists of a primary sedimentation tank, an anaerobic lagoon and a SSF CW in series. The CW had an HRT of 10.6 days and an organic loading rate (OLR) of 0.0818 COD/m^2/d and accounted for around 30% of the organic matter removal, resulting in a total removal efficiency of BOD$_5$, COD and TSS of 91, 89 and 85%, respectively. However, the system could not meet Mexican environmental regulations in terms of fecal coliforms, BOD$_5$ and TSS. In addition, organic nitrogen was removed by 80%, but ammonia-nitrogen was only removed by 9% in the wetland stage. Laboratory tests suggested a post-treatment of the effluent in a sand filter followed by disinfection with sodium hypochlorite (NaOCl) [12].

Another HSSF system was used for SWW treatment in a cold climate in Nematekas, Lithuania. After a primary treatment using physico-chemical methods and settling ponds, the wastewater had pollution characteristics similar to domestic wastewater. The following two HSSF CW beds in series had a HLR between 8.5 and 10.6 cm/d and reduced BOD$_5$, TN and TP by 41, 44, 84%, respectively [13].

Tanner et al. [14] used duplicates of five planted (*Schoenoplectus tabernaemontani*) CW mesocosms operated in series, in order to investigate the influence of different wastewater characteristics (e.g., COD:N ratio) on nitrification, denitrification and COD removal. One duplicate received meat processing wastewater. The systems had an overall HRT of 3.41 days and HLR of 2.3 cm/d. The wastewater showed a high level of TN with 4.5 g per m^2/d, and a mean TN removal of 50%. The theoretical additional oxygen demand (NBOD) for the meat processing wastewater was comparatively high with 70% of the total oxygen demand (e.g., the NBOD of dairy wastewater was only ca. 20 to 35%). TN removal amounted to 50%, whereas 79% of removal was attributed to denitrification and only 21% to plant uptake. Interestingly, nitrification and denitrification happened concurrently with COD removal, indicating that in SSF CWs these processes are likely to be much more closely coupled as shown in specialized laboratory studies [14].

Another study looking closer at the fate of nitrogen in HSSF CWs was performed by Kadlec et al. [15], using a stable isotope 15N as a pulse fed tracer. The above described cascade mesocosms [14] were used to treat primary meat processing water. However, this time two small field-scale CWs were also investigated (one unplanted and one planted with *Schoenoplectus tabernaemontani*). The field-scale systems had an nHRT of ca. 4 days and an HLR of 3.2 cm/d and were also fed with nitrogen-rich meat processing water (NH$_4$-N = 122, Total Kjeldahl nitrogen (TKN) = TN = 129, TP = 17, COD = 239 g/m^3). The 15N ammonium tracer impulse needed ca. 120 days to move through the system and only 1–16% of the tracer was recovered after 25 days. The main part of the tracer was found in plants and sediments with 6–48 and 28–37%, respectively. The interchange between sediments and water was identified as the main delay mechanism. In addition, nitrogen taken up in leaves and roots delayed the release, since the plants need to decompose in order to bring back nitrogen to the sediments on this pathway. These processes greatly increase the nitrogen residence time in comparison to the HRT, resulting in long response delays to changes in nitrogen loading and attenuation of short time differences in loading [15].

Soroko [16] investigated the treatment of pig slaughterhouse wastewater using a hybrid of two VF CWs and a HF CW in series all planted with common reed (*Phragmites australis*). This setup achieved removal rates of BOD_5, COD and N_{tot} of 99, 97 and 78%, respectively. However, the effluent N_{tot} concentration was still several times too high for Polish regulations. In a second experiment, the effluent after the second VF CW was recirculated to a sedimentation tank before the first VF. The recirculation was applied in order to increase nitrate and total nitrogen removal, as well as to increase the HLR (4–5 cm/d in the second VF CW) and thereby reduce the risk of clogging. Total nitrogen removal amounted to 86, 92 and 97% for recirculation rates of 100, 150 and 200%, respectively. However, denitrification in the last HF bed was low since practically no organic matter was left for denitrifiers, which was possibly caused by uneven flow or too coarse filling material [16].

Carreau et al. [17] evaluated the performance of a SF CW treating wastewater from a small slaughterhouse in the cold climate of Nova Scotia, Canada. Two cells (29.25 m^2 each) in series were planted with *Typha latifolia*. Average hydraulic loading was low with an HRT of 111 days and an HLR of 0.5 cm/d. However, monthly variability was high with HLRs reaching up to more than 4 cm/d on slaughter days. Mass removals were good for BOD_5, TSS, total dissolved solids (TDS), TP, TKN, and total ammonia nitrogen (TAN) with 95, 72, 81, 88, 97, 87 and 87%. With the dilution of the receiving water BOD_5 and TSS limits were met and average *E. coli* levels were below local limits. During the 2-year study, the system performed better during summer seasons, indicating temperature dependency [17].

A summary of all described CW case studies is shown in Tables 7.2 and 7.3.

Experiments with pilot scale CW systems treating slaughterhouse and meat processing water started in the 1980s in New Zealand and Australia. After several pilot scale studies, the first full scale implementations were set up in the 1990s and systems were built in different climates around the world, from rather arid climates in Australia or Mexico to cold and wet climates in Lithuania or Poland, as well as in a tropical rainforest in Ecuador. Seven out of the ten CW systems outlined here were operated with HSSF, two with a FWS and one in with a hybrid (VF and HF beds in series) flow regime.

A direct comparison between systems is difficult since wastewater characteristics, CW design (flow, scale, hydraulic and organic loading, etc.) and other influences like climate vary from case to case. In addition, local regulations differ, and even if treatment efficiencies for main parameters are high, the prevailing regulations might not be met, as seen in the examples of Gutiérrez-Sarabia et al. [12] for coliforms and Soroko [16] concerning nitrogen levels.

What can be seen in all examples is that CWs need a pretreatment of wastewater from slaughterhouses and meat processing plants, consisting at least of some kind of settling or sedimentation tank. However, the majority of different CW designs have led to satisfying results, from simple systems using only one kind of flow regime (mostly HSSF) to more complex setups including different CW beds in series, or recycling of the effluent [16]. A major advantage of CW systems is that they can be constructed using local materials and work force. In addition, maintenance needs are low compared to "grey" conventional treatment technologies, but proper management is nevertheless crucial for a successful long-term operation [11]. In general, CWs have the potential to treat wastewater from slaughterhouses and meat processing plants, and are part of the solution to the goal of a cleaner environment and water bodies, as well as the recycling of our world's limited water resources.

Table 7.2 General information and hydraulic and loading rates of the ten presented CW case studies.

CW design	Scale	Waste-water	Pretreatment	Location	Koppen climate class	Area (m²)	Inflow (m³/d)	HRT (d)	HLR (cm/d)	OLR (g COD/m²/d)	Reference
HSSF trenches	Pilot	Slaughter-house	Storage tank with mechanical aerator	Griffith, NSW, Australia	Semi-arid (BSk)	27–36[b]	30–50	2.7–3.6	83–185[b]	–	[7]
FWS with floating mats	Pilot	Meat processing	Anaerobic treatment and aerated tank	Hamilton, New Zealand	Oceanic (Cfb)	2.88[b]	0,164[b]	7	5.7	23.5	[9]
HSSF	Full	Slaughter-house	Settling tank anaerobic. digester	Pachuca, Mexico	Semi-arid (BSk)	600[b]	35	1.7	5,83[b]	47–124[b]	[10]
HSSF (2 beds in series)	Full	Slaughter-house	Settling tank	Shushufindi, Ecuador	Tropical rainforest (Af)	1,200	20	2	1.7	3.1–6.6[b]	[11]
HSSF	Full	Slaughter-house	Sedimentation tank, anaerobic. lagoon	Hidalgo, Mexico	Subtrop. highland climate (Cwb)	1,144	65	4.22	5.7	81.8	[12]
HSSF (2 beds in series)	Full	Meat processing	Physico-chemical and pond settling	Nematekas, Lithuania	Humid continental (Dfb)	940	80–100	–	8.5–10.6	–	[13]
HSSF (5 mesocosms in series)	Experimental	Meat processing	Primary treatment	Hamilton, New Zealand[a]	Oceanic (Cfb)	0.353	0.008	3.41	2.3	8.7	[14]
HSSF	Field	Meat processing	Primary treatment	Hamilton, New Zealand	Oceanic (Cfb)	18	–	4	3.2	–	[15]
Hybrid (2 VSSF + 1 HSSF bed)	Pilot	Slaughter-house	Storage and 2-chamber sediment. tank	Wroclaw, Poland	Humid continental (Cfb)	25	0.15	–	4–5	47.8	[16]
SF CW (two cells in series)	Full	Slaughter-house	Septic tank	Nova Scotia, Canada	Warm humid continental (Dfb)	58.5	0.38	111	0.5	–	[17]

[a] Beneath horticultural plastic.
[b] Calculated.

Table 7.3 Influent concentrations (mg/L) and treatment efficiency (%) of the ten presented CW case studies.

BOD$_5$ (mg/L)	COD (mg/L)	TSS (mg/l)	TN (mg/L)	TP (mg/L)	BOD$_5$ (%)	COD (%)	TSS (%)	TN (%)	TP (%)	Reference
–	–	184–229	–	–	–	–	83–89	14–56	37–61	[7]
38	405	321	197	–	32–41	66–69	–	46–49	–	[9]
112–1,560	806–2,120	–	–	–	89	87	89	–	–	[10]
185–288	187–396	–	–	–	98[a]	97[a]	–	–	–	[11]
585	1,440	421	–	29.4	74	77	55	–	0	[12]
161	–	–	107	41	84	–	–	41	44	[13]
–	–	–	–	–	–	25.3[b, a]	–	50[b]	–	[14]
–	–	–	–	–	–	–	–	–	–	[15]
2,452	3,188	561	494	–	99[b]	97[b]	94[b]	78–97[c, b]	–	[16]
44		39		0.58	94	–	66	–	81	[17]

[a] Calculated.
[b] Mass removal.
[c] Recirculation.

7.2 Treatment of Potato Washing Wastewater

Potatoes are used by many cultures across the planet and are processed to be eaten in many forms, such as potato chips, frozen French fries, dehydrated diced and mashed potatoes, potato flakes, starch and flour. Worldwide potato processing generates about 1.3×10^9 kg of waste in the United States alone [18]. Potato processing wastewater can cause pollution, and therefore needs to be treated before being released to the local waterways, in order to not disturb or degrade them.

The processing of potatoes involves several steps depending on the type of potato and the product desired. The first step in the process is to clean the potatoes after being harvested [19]. This is done by using a stream of water to remove the dirt and other organic particles attached to the potatoes. In addition, the removal of these particles reduces wear on machinery further down the production line. This step will produce around 1–3 kg per 1,000 kg potatoes [18].

After the cleaning, the potatoes are either peeled by abrasion, steaming or a lye treatment, depending on the final product [18]. Abrasion is the simplest form of the three treatments, using the least amount of water per pound of potatoes processed, and results in the lowest BOD in the effluent. However, this process results in the highest total suspended solids (TSS) levels of the three peeling processes. Steaming results in a slightly higher water usage and BOD levels than abrasion, but the exiting TSS is the lowest of the three. Lye treatment uses the most amount of water and creates the most BOD of the three, but TSS effluent levels are between the other two. Another consideration is that using lye, which is an alkaline compound, will have an effect on the effluent of the processing station that will need to be addressed in the wastewater treatment process [18].

Therefore, the necessity of potato processing (including cleaning) water treatment is due to its high level of pollutants in the water, mainly BOD, TSS and chemical oxygen demand (COD). In addition, the effluent from a potato processing plant will vary depending on the

Table 7.4 General characteristics of potato processing wastewater (adapted from Hung et al. [18]).

Parameter	Low levels	High levels
Water flow (m^3/ton of potatoes)	115	1,700
COD (mg/L)	1,100	12,582
BOD (mg/L)	2,307	7,420
TSS (mg/L)	280	62,444
pH	4.2	9
TP (mg/L)	2	60

potato used and the product being made. The levels of different measured components also vary. Table 7.4 shows common ranges for the most important parameters of potato processing wastewater.

The treatment of potato processing wastewater is done in four general steps. The exact treatment of each step can vary depending on the treatment needs of the water as well as the location of the plant. The first step is screening to remove the larger particles, followed by primary treatment to remove the remaining solids in the water. If the potato processing includes a pH lowering or raising like lye treatment, a neutralizing step is needed prior to the secondary, or biological treatment, which needs a pH level between 6.5 and 8.5 [18]. Otherwise, the pH will not be conducive to microbiological growth. The secondary treatment intents to the degradation of the soluble organic compounds through biological activity.

At the start of potato processing water treatment, from approximately 1950 to 1980, only a primary treatment was conducted [18]. Since CWs have the capacity to remove high levels of BOD, they are capable of treating potato-processing wastewater after primary treatment.

Treating potato wash water in the northern climates has become a new challenge for potato farmers. For example, some Canadian potato farmers must perform the potato washing themselves, before shipping the potatoes for processing. In addition, in the northern Canadian climate, the temperature can drop below $-20°C$ or $-30°C$ during the cold season. This might present a problem, since treatment may be slowed down, or even stopped [19].

In a case study by Bosak et al. [19], potato wash water with a BOD of 1,113 mg/L and TSS of 4,338 mg/L was treated by a system including a CW system. The treatment chain started with a sedimentation tank, followed by two aeration tanks, after which, the water was applied to a Horizontal Subsurface Flow (HSSF) CW.

The experiment lasted for two years. Results of the first year showed that, amongst other factors, unexpected high influent volumes led to lower results than expected. Therefore, it was decided to increase the size of the first three open-air ponds. In addition, the volumes of all three tanks were increased, which almost doubled their total volume. Hence, the HRT was quadrupled as well. Additionally, another aerator was added in the third tank. These modifications led to increased treatment efficiencies in the second year (Table 7.5) [19].

The increased dissolved oxygen availability, due to the added aerator, helped increase BOD$_5$ removal. Movement of the water from the sedimentation tank, to each of the subsequent aeration tanks, resulted in a general decreasing trend of concentrations for TSS, BOD$_5$ and NH$_3$-N. This was

Table 7.5 Treatment efficiencies (combined winter and spring/summer seasons) for each year of investigation [19].

Parameter	Year 1		Year 2	
	Pretreatment (%)	CW (%)	Pretreatment (%)	CW[a] (%)
BOD_5	84	29	43	65
TSS	98	21	98	36
TN	57	4	55	13
TP	26	13	83	5

[a] Winter months do not have wetland activity.

also the case for TN, but it must be noted that during the winter months, the ratio of NH_3-N/TN increased in the sedimentation tank. It was suggested that the organic nitrogen was settling out and mineralized to NH_3-N, resulting in the increased amount of NH_3-N [19].

In another study, Kato et al. [20] investigated the improvement of treatment of high concentration potato starch processing wastewater in the cold climate of northern Japan during two periods; winter and early/late fall. During the winter, two separate CWs were used to test their effectiveness in treating different potato processing wastewater:

- Wastewater stored over winter when it was too cold for the CW.
- Wastewater produced freshly from the potato processing factory.

The CW that was operational in the period from May to August treated the stored wastewater. The CW operating from September to November treated the wastewater coming directly from the potato processing factory. Both CWs treatment chains were the exactly the same:

- Pretreatment.
- Vertical Flow treatment bed (VF), with recirculation.
- Vertical Flow treatment bed (VF), with recirculation.
- Vertical Flow treatment bed (VF), with recirculation.
- Horizontal Sub-surface Flow (HSSF) bed.
- Vertical Flow treatment bed (VF).

In order to prevent clogging of the systems, a system of bypass pipes was built at the entrance of each of the recirculating VF beds. If the treatment bed is clogged, this setup allowed the influent to pass to the other, allowing the clogged bed to degrade the accumulated matter causing the clogging and allowing the wastewater to flow again in the treatment bed [20].

In addition, a section of the HSSF treatment bed incorporated a medium containing a phosphorus adsorption material, which can be reused as agricultural fertilizer [20].

COD removal efficiency was high with 92% in preserved treated wastewater, and 94% removal in freshly treated wastewater. Ammonia removal amounted to 74% in preserved wastewater and 72% in fresh wastewater. TN removal amounted to 82% in preserved wastewater and 90% in freshly treated wastewater [20].

Using four different VF and a HSSF (in the middle) in series, led to promising results, in spite of a harsh cold climate. In general, more pollutants were removed during the treatment of fresh wastewater.

Generally, the system located in the seasonally freezing climate of Canada showed good treatment performance. The changes after the first year led to slight increase in the removal of TSS, whereas the removal of BOD_5, TN and TP showed a large increase of removal. Looking at the removal efficiency of the treatment wetland, the results from the first and second year differed greatly; removal efficiencies increased with the exception of NH_3-N, which was almost the same 16% vs. 17%, and TP. The main difference was found in BOD_5 reduction, increasing from 29 to 65% removal, whereas a more modest increase was observed for TSS (21 to 36%) and TN (4 to 13%).

In the study based in Japan's seasonally freezing climate, potato starch processing wastewater was treated in a series of VF CWs with a HSSF CW. The study shows that in general, the treatment with the stored wastewater had a higher percentage removal of the studied parameters, with the exception of TN. That being said, the treatment of the unstored wastewater had an influent with a higher concentration of all parameters analyzed, and the actual mass removal of each parameter was higher, with the exception of TP, which had the same mass removal rate.

The two studies described above demonstrate the ability of CWs to remove a high level of pollutants from potato processing wastewater. Both studies were implemented in full-scale systems, and already showed promising results. It must be noted that these were operated in colder climates, which are more challenging for biological treatment processes. Hence, in warmer climates there could be a certain possibility of increased performance. Therefore, with the proper pretreatment, CWs have proven to be able to perform organic pollutant removal from potato processing wastewater, even in cold northern climates.

7.3 Treatment of Molasses Wastewater

Molasses is both a by-product of the sugar industry and also a substrate for the ethanol industry. Both industries generate wastewater that demands treatment. The global demand for ethanol is expected to rise above 120 billion liters by the year 2020 [21], for example in 2016 in Thailand, 1.3 billion liters were forecasted to be used [22]. The main sources for sugar production are sugar cane and beets. The effluents from these industries are both very high in COD, and BOD, as well as nitrogen, phosphorus, potassium and melanoidin [21]. In addition, the dark color of melanoidin can also inhibit photosynthesis [23]. These effluent characteristics can cause a strain on the ecosystems due to the degradation of water quality.

The general process of sugar production can be summarized as follows [24]:

- Sugarcane is crushed – releasing the sugar cane juices.
- Clarification by addition of substrate to induce coagulation of soluble and insoluble unwanted elements.
- Filtration to remove solids.
- Evaporation of excess water in remaining sugarcane juice.
- Crystallization of remaining liquid by addition of seed crystals.
- Centrifugation of crystals to remove the molasses wastewater.
- Recrystallization of sucrose to table sugar.

According to results by Sohsalam and Sirianuntapiboon [23], molasses wastewater contains high COD and BOD concentrations (1,600–21,000 mg BOD_5/L up to a distance of 8 km of distillery in the receiving water); however, there are also high concentrations of other nutrients present [23]:

- Nitrogen; 1,660–4,200 mg/L.
- Phosphorus; 225–3,038 mg/L.
- Potassium; 9,600–17,475 mg/L.

The processing of the molasses takes place in four main steps. First, the preparation of the ingredients, including yeast and molasses. This is followed by fermentation – here the molasses are metabolized by the yeast and carbon dioxide is released. This subsequent distillation process allows for the refinement of the alcohol to higher concentrations – here the molasses distillery wastewater is removed. The final product is then packaged [25].

The resulting effluent from the molasses processing plant is very high in pollutants. These values will vary depending on the type of molasses being processed. Molasses wastewater is a difficult wastewater to treat due to its high BOD, COD and melanoidin content. There has been a variety of treatments applied. One possibility is land and deep well disposal, which can only be used in certain areas due to the possible groundwater contamination and onsite geological limitations. More advanced types of treatments include anaerobic, aerobic and physico-chemical treatments [23].

Aerobic systems that have been used include single phase and biphase anaerobic systems, anaerobic lagoons and conventional anaerobic systems. High rate anaerobic systems used for treatment include anaerobic fixed film reactors, upflow anaerobic sludge blanket reactors and anaerobic fluidized bed reactors. Systems based on aerobic methods include bacterial systems and phytoremediation techniques, such as constructed wetlands and fungal systems. The last type of treatment falls under physico-chemical methods, including adsorption, coagulation and flocculation, as well as oxidation processes. Constructed wetlands have been used as a tertiary treatment step [23].

In a research conducted in Thailand, the characteristics of the used molasses wastewater were as shown in Table 7.6 [23].

In this study, Sohsalam and Sirianuntapiboon [23] used Surface Flow Constructed Wetlands in small-scale microcosms. Three different plants and five different organic loading rates (OLR) were compared. The systems were built in a way that the first part of the microcosm was used as a settling

Table 7.6 Characteristics of the used molasses wastewater in Thailand [23].

Parameter	Average value – measured in mg/L, except color density
SS	3,233
COD	57,935
BOD_5	614
TN	1,360
NH_4^+-N	639
NO_3^-	14.7
TP	77.2
Color density[a]	78.9

[a] Color density measured was measured by absorbency of wavelength of 475 nm.

tank, and the second half as the treatment wetland. The loading rates compared were 61.2, 69.6, 80.6, 92.9 and 121.3 g $BOD_5/m^2/d$ [23].

Plant biomass production for all three plant species decreased as the OLR was increased. In addition, in the case of the OLR of 121.3 g $BOD_5/m^2/d$, the plants died within 6 weeks. Nitrogen and phosphate accumulation within the plant material was similar in all plants, regardless of the level of OLR. There was also a reduction of biofilm observed in the wetland as the OLR increased [23]. These effects seemed to be linked to the decreasing Oxidation Reduction Potential (ORP) as the OLR increased, leading to more anaerobic conditions, reducing the overall aerobic bacteria numbers.

In contrast to the plant species, the ORP decrease had also an influence on the individual nitrogen species, resulting in a NH_4^+-N percentage removal drop from an average of 79% to an average of 12%, whereas the NO_3^--N removal rose from, an average of 95% to an average of 99% [23]. In general, the efficiency of the treatment by the constructed wetland decreased with the increase of OLR. TP showed the lowest variation, since it has less of a microbiologically driven removal process, and is more dependent on adsorption. Suspended solids removal changed only little in relation to the change between aerobic and anaerobic conditions, since this is physical removal process.

Table 7.7 presents the parameter and the average reduction in percentage of removal with an increased OLR.

This experiment was able to demonstrate the upper limits of the treatment of high concentration molasses wastewater. The CW system's performance was good at 61.2 g $BOD_5/m^2/d$, but slowly decreased as the OLR increased in the case of most parameters but more drastically in the case of COD and color intensity.

Another field-scale research project by Billore et al. [26] investigated the enhanced oxygen availability by incorporating a tidal flow system and a replaceable strip of broken bricks into a constructed wetland. The project was carried out because the current wastewater treatment system did not meet local minimum effluent standards. The CW received the effluent from the current wastewater treatment system, which was treating molasses based distillery effluent from ethanol production [26]. The system configuration was as follows [26]:

- Pretreatment.
- Unplanted anaerobic bed using gravel.
- Unplanted bed using inverted porous clay bowls – filled and emptied on a regular basis, creating a tidal affect.

Table 7.7 Average removal of pollutants with increased OLR [23].

Parameter	Pollutant reduction in percent with increased OLR
Color intensity	54
COD	52
BOD5	15
SS	3
TN	27
TP	13

- Treatment wetland using *Typha latifolia* – with added replaceable strip of broken brick pieces for added TP removal.
- Treatment wetland using *Typha latifolia*.

The Tidal Flow System implemented a low cost passive manner to introduce oxygen into the treatment bed, avoiding the extra need of electricity and pumps for artificial oxygen supply. Since the brick material was kept separate and was not mixed with the normal constructed wetland bed media, it can be removed and replaced when needed. This allows the system to improve TP removal rates and potentially extend the overall system's lifespan.

Common values found in the effluent of an ethanol production's distillery are as follows [27]:

- BOD; 50,000 – 60,000 mg/L.
- COD; 110,000 – 190,000 mg/L.
- Sulphates; 7,500 – 9,000 mg/L.
- Phosphates; 2,500 – 2,700 mg/L.
- Phenols; 8,000 – 10,000 mg/L.
- Total nitrogen; 5,000 – 7,000 mg/L.

Percentage of overall reduction for the studied parameters is shown in Table 7.8.

Ammonia showed an overall removal of 50%, whereby Billore et al. [26] suggested that this was partly because of the volatilization effect, due to the high outdoor temperatures of 56–62°C. It could also be seen that the TSS removal rate was double in planted in comparison to unplanted gravel wetlands, which indicates that a planted treatment bed is better at removing TSS. It is also noted that the first unplanted/unaerated treatment bed had a very low BOD removal rate (about 2%), whereas the other unplanted and the two planted treatment beds varied between 35–55% removal rate [26]. This difference can be attributed to the oxygen introduced through aeration.

Phosphorus showed overall removal rates of 80% in the whole treatment train. The wetland with the strip of broken bricks showed a 35% removal rate of phosphorus all by itself. The second planted treatment bed had a lower TP removal rate with about 28% [26].

In summary, molasses production is a food transformation industry that generates large volumes of wastewater and plays an important role in countries like India and Thailand. However, more improved solutions for wastewater treatment have to be found. Increased research in this area could help this industry reduce its impact on the environment, while also allowing companies to reduce costs, since traditional treatment systems tend to be relatively expensive. It can be seen that each

Table 7.8 Removal reduction of pollutants [26].

Parameter	Overall removal (%)
BOD	85
COD	65
NH_4-N	50
NO_3-N	60
TP	80

of the processes, producing molasses or converting it into ethanol, have high organic and nutrient loadings which could have a serious impact to our environment and health.

In both instances, the use of planted CWs improved the treatment efficiency by removing an average of 30% TP treating ethanol distillery effluent. Constructed wetlands also contributed to high percentage removal rates when treating molasses wastewater effluent from sugar refineries. However, the research also showed the limits of current CWs in relation to increased OLRs, which reduced removal rates from 75% to only 15%. These results allow for better design considerations when applying CWs for high load wastewaters. Possible future research could explore ways to treat high OLR by changing other parameters, such as bed depth and HRTs. In addition, it could be seen that the use of plants played a role in treatment efficiency. In this respect, further plant species could be investigated in terms of their abilities to increase treatment efficiencies and withstand high OLR.

The treatment of ethanol distillery effluent used two novel approaches in the field, including passive aeration, which had the desired effect of increasing the BOD removal. This was done without electricity need, a very useful solution for many parts of the world where power supply around the clock is not certain. The second novel approach was the addition of broken bricks to enhance TP reduction, resulting in a higher TP removal rate compared to the constructed wetland without broken bricks.

Both projects presented the adaptability of constructed wetlands for treatment of high strength wastewater. Considering that the research is relatively new and only few studies have been conducted for this type of wastewater, further studies are necessary in order to allow for further treatment improvement of these types of wastewaters by CWs. There might be more suitable plants, or maybe further improved passive aeration systems that will allow for an improved treatment.

In general, constructed wetlands are suitable to treat wastewater from sugar and ethanol production industry. However, like with most CW applications, a proper pretreatment is need. Systems were tested and applied successfully in the lab and field scale level, showing that treatment of high strength wastewater from the discussed industries is feasible.

7.4 Treatment of Effluents from Coffee Processing

Production and trade of coffee has a great global socioeconomic impact, while it constitutes a most beloved beverage consumed worldwide. The coffee plant is cultivated in tropic climates and has two main varieties, the Arabica (*Coffea arabica*) and the Robusta (*Coffea canephora*) between which the world production is about 75–80% and 20% respectively [28].

Coffee production, however, includes some processes relevant to the quality and taste of the final product, which produce effluents that cause severe environmental hazard when disposed into the water courses. The type of effluent produced by coffee industries is wastewater high in concentration of organic matter and requires treatment before its disposal to waterways. It has been reported that people residing close to coffee industrial plants and using stream water deriving from the coffee possessing wastewater (CPW) effluent for domestic reasons face serious health issues, including dizziness, nausea, skin and eye irritation, and respiratory problems [28].

Coffee processing can be achieved through the dry or the wet (washed) method. Each processing technique entails a different pollution extent and different quality of coffee. The dry coffee processing technique involves drying of coffee cherries directly once harvested, while the removal of the pericarp of the coffee seed follows. This technique is characterized by simplicity and by the least environmental burden [29]. The washed processing technique involves more steps, i.e. pre-sorting period, pulping of exocarp and mesocarp, fermentation period, and washing off. This method is applied to approximately 50% of the global coffee crop owing to the finer quality of coffee product [28].

Nevertheless, the downsides of the washed technique are the by-products of coffee pulp and pulping water discharged in water courses, while 40% of the mass of the fresh coffee cherry accounts for the pulp residuals [28]. The pulping water effluent consists of elevated organic matter, mainly of nitrogen, fermenting sugars, large protein supplies, and pectins [29]. Consequently, the amount of coffee wastewater discharge varies based on the processing method and the wastewater disposed into the water recipients has to comply with the purification standards. It has been reported that the amount of BOD required to subdivide organic matter from CPW reaches 20,000 mg/l for pulping wastewater discharges and 8,000 mg/l for fermentation outflows, while the BOD value ought to be below 200 mg/l prior to disposal into water courses [29]. Other pollutants that the CPW contains are toxic compounds such as tannins, caffeine and phenols [28, 29]. Furthermore, the resistant organic matter found in CPW – and particularly from coffee pulp – accounts for the 80% of pollution load, attaining higher than 50,000 mg/l COD [29]. It is therefore inferred that CPW effluents need to undergo proper treatment to achieve and comply with the health and safety water policy regulations.

The approaches applied to reduce organic material from CPW and to remove toxic chemicals incorporate chemical (i.e., activated carbon, upflow anaerobic sludge blanket (UASB), upflow anaerobic hybrid reactor (UAHR)), and/or biological/natural processes and systems, as there is a high interest and investigation in constructed wetlands (CW) performance on the removal of CPW pollutants because of the lower costs involved. Activated carbon, for instance, albeit efficient, incorporates additional expenditure for activated carbon production. CWs might have the potential to treat crude CPW and/or to post-treat effluent deriving from anaerobic systems. Anaerobic lagoons, for example, have been largely utilized in India for primary effluent treatment because under appropriate anaerobic conditions removal of organic and suspended sediment loads is high (70–80%) [28]; however, post-effluent treatment is still required to conform to the local authority environmental standards. To confront this post-treatment challenge and unite the removal efficiency with affordable costs, there is research conducted on combined systems, case studies of which are reported as follows.

Enden and Calvert [29] mention a pilot scale CPW treatment scheme implemented in Vietnam, consisting of an acidification pond, a neutralization tank, an UASB biogas digester, and a CW planted with common reeds (*Phragmites australis*) for biological treatment, while tertiary treatment was attained in a hyacinth pond. The primary treatment stage includes acidification and neutralization. Because the raw coffee pulp wastewater acidity is high, it stays in the water for 6 hours forming ultimately a floating crust, which is easily removed. Thereafter, the effluent undergoes neutralization, where the pH becomes 6.0 through the addition of natural limestone ($CaCO_3$). The effluent passes into the UASB biogas reactor being subjected to anaerobic digestion. This process achieved 70–90% abatement of BOD within a short period of 4 to 6 hours Hydraulic Residence Time (HRT). The following biological treatment is done through a subsurface flow CW, where beyond the dissolved oxygen (DO) supplied naturally via the root system, additional oxygen was provided to accelerate the aerobic breakdown. Ultimately, water was polished through a hyacinth water pond, which is known to remove microorganisms and heavy metals, and to attenuate organic loads.

Selvamurugan et al. [30] developed a combined treatment arrangement for wet CPW in India incorporating a UAHR for biomethanation with aeration, and a pilot scale vegetated subsurface flow CW for post-treatment. Two plant species were assessed; *Typha latifolia* and *Colacasia*. The reactor functioned for various HRTs, and the CW under different aerated conditions. The optimal efficiency was achieved at 18 h. For the post-treatment stage, oxygen is indispensable to further abate the organic loads. It was demonstrated that continuous aeration yielded greatest removal of BOD and COD compared to intermittent aeration. Furthermore, the longer aeration displayed higher TSS removal. Amongst the two assessed plant species, *Typha latifolia* presented higher reduction in BOD, COD and TSS than *Colacasia*. The overall integrated system performance evidenced acidity decrease (pH reached 7.5 from 4.1), and BOD, COD and TSS reduction of 98, 97, and 90% respectively. Importantly, the final effluent was within the conformity limits.

Fia et al. [31] used different organic loadings to assess the efficiency of anaerobic filters accompanied by pilot scale horizontal subsurface flow CWs, planted half-way with *Alternanthera phyloxeroides* and half-way with *Typha*. Results demonstrated the benefits of vegetation on the removal of nitrogen compared to the non-vegetated CW. Correction of the pH (close to 7.0) was realized using hydrated lime. The study used three different phases, in each of which different organic load was applied. Results reinforced the notion that increase in organic load does not promote nutrient and phenols abatement. The combination F1+CW1 (where F stands for filter), which received the lowest organic loading, exhibited the maximum efficiency, presenting nitrogen and phosphorus removal greater than 50% on average.

Research conducted by Fia et al. [32] on the performance of two winter plant species on the post-treatment of CPW, indicated that the plant species play a major role in the overall removal efficacy. The study took place in six pilot horizontal subsurface flow (HSSF) CWs, half of which were planted with ryegrass (*Lolium multiflorum*) and half with black oat (*Avena strigosa*). The assessment included the crude protein and the dry matter of the aforementioned forages. Results showed that black oat displayed descending productivity with applied organic load, in contrast with ryegrass whose productivity enhanced with increasing applied load. Ryegrass, in particular, managed to collect 3 to 25 times more crude protein than black oat, and overall, displayed better adaptability, suitability and removal efficiency.

As alluded to earlier, aeration is fundamental for the biological treatment in CWs, as they promote the nitrification process, and hence inadequate DO inhibits this activity. For this reason, Rossmann et al. [33] performed experiments to evaluate the contribution of artificial aeration and of plant removal efficiency of nutrients and phenols in CPW effluent. The aeration scheme included a submerged Sarlobetter S520 pump. The CWs were pilot scale, HSSF, vegetated with ryegrass, and followed the subsequent combinations: (i) aiCWc, aerated influent and planted; (ii) aiCW*, aerated influent and unplanted; (iii) CWc, non-aerated influent and planted; (iv) CW*, non-aerated influent and unplanted. The HRT was 12 days. The results consolidated the beneficial role of vegetation, indicating higher removal efficiencies in nitrogen (N), phosphorus (P), potassium (K) and phenols, compared to non-vegetated cases. In the case of artificial aeration in the absence of plants, it is acknowledged that the role of plants is superior and irreplaceable for the functions they serve and for the overall removal efficacy. However, the optimum removal was achieved with the combination of vegetation and artificial aeration, presenting the maximum removal rates of N, P, K and phenolic compounds. Therefore the use both of vegetation and artificial oxygen supply in CWs is encouraged, as this combination leads to optimum removal efficiency of nutrients and phenols despite the ancillary cost of artificial aeration. The authors further recommend the implementation of cascade aeration downstream of wastewater treatment plants, both as a cost-effective way of additional aeration of the final effluent, but also as a potential substitution of the artificial aeration in the CW unit.

Following the above-mentioned CWs study, Rossmann et al. [34] continued their tests and presented results evaluating the effects of vegetation presence and artificial aeration on organic load removal. It was proved that the CWs were very effective in removing BOD and COD, on average by

86% and 90% respectively. Interestingly, it was observed that artificial aeration did not account for the removal of organic loads in the CWs. TSS levels were reduced significantly, but this decrease was attributed possibly to lime insertion for pH correction.

Since the coffee pulp effluent is highly acidic and organic and the disposal volumes are significantly large, it is proposed by Padmapriya et al. [28] to be utilized as an organic fertilizer. This process involves three to twelve months piling of the pulp by-product, which is then converted into a rich, black humus compost. Alternatively, it is suggested that the coffee pulp is blended with cattle manure and is piled for a short time [28]. Either way, it offers a cost-effective and organic mode of fertilizing the lands.

This review described the problems of coffee processing wastewater and explored the potential of to-date subsurface flow constructed wetland studies either to treat fully or post-treat this highly organic effluent. Although to-date research is applied on pilot scale subsurface flow CW units, results are interesting and promising. Full-scale studies are required to prove or refute this potential. The literature recommended particular plant species that seem to be most appropriate for this particular effluent, such as ryegrass (*Lolium multiflorium*), and practices that should be followed or avoided; for example, aeration is fundamental for organic matter mitigation, and artificial aeration or cascade aeration is strongly recommended for organic loading removal, while vegetation presence is crucial because it provides a healthy root zone and space for microbial consortia activity.

References

1 WWAP (United Nations World Water Assessment Programme). The United Nations World Water Development Report 2015: Water for a Sustainable World. Paris: UNESCO; 2015.

2 Tilman D, Clark M. Global diets link environmental sustainability and human health. Nature. 2014; 515(7528): 518–522.

3 Bustillo-Lecompte CF, Mehrvar M. Slaughterhouse wastewater characteristics, treatment, and management in the meat processing industry: A review on trends and advances. J Environ Management. 2015; 161: 287–302.

4 Tritt WP, Schuchardt F. Materials flow and possibilities of treating liquid. Bioresour. 1992; 41(3): 235–245.

5 Kadlec R, Wallace S. Treatment Wetlands, 2nd edn. Boca Raton: CRC Press; 2009.

6 Van Oostrom AJ, Cooper RN. Meat processing effluent treatment in surface flow and gravel bed constructed wastewater wetlands. In: Cooper PF, Findlater BC, eds. Constructed Wetlands in Water Pollution Control. Oxford: Pergamon Press; 1990, pp. 321–332.

7 Finlayson CM, Chick AJ. Testing the potential of aquatic plants to treat abattoir wastewater. Water Res. 1983; 17(4): 415–422.

8 Van Oostrom AJ, Russel JM. Denitrification in constructed. Water Sci Technol. 1994; 29(4): 7–14.

9 Van Oostrom AJ. Nitrogen removal in constructed wetlands. Water Sci Technol. 1995; 32(3): 137–148.

10 Rivera F, Warren A, Curds CR, Robles E, Gutierrez A, Gallegos E, et al. The application of the Root Zone Method for the treatment and reuse of high-strength abattoir waste in Mexico. Water Sci Technol. 1997; 35(5): 271–278.

11 Lavigne RL, Jankiewicz J. Artificial wetland treatment technology and its use in the Amazon River forests of Ecuador. In Proceedings of the 7th International Conference on Wetland Systems for Water Pollution Control; 2000; Gainesville: University of Florida, pp. 813–820.

12 Gutierrez-Srabia A, Fernandez-Villagomez G, Martinez-Pereda P, Rinderknecht-Seijas N, Poggi-Varaldo HM. Slaughterhouse wastewater treatment in a full-scale system with constructed wetlands. Water Environ Res. 2004; 76(4): 334–343.

13 Gasiunas V, Strusevicius Z, Struseviciéne MS. Pollutant removal by horizontal subsurface flow constructed wetlands in Lithuania. J Environ Sci Health. 2005; 40A: 467–1478.

14 Tanner CC, Kadlec RH, Gibbs MM, Sukias JP, Nguyen ML. Nitrogen processing gradients in subsurface-flow treatment. Ecol Eng. 2002; 18: 499–520.

15 Kadlec RH, Tanner CC, Hally VM, Gibbs MM. Nitrogen spiraling in subsurface-flow treatment wetland. Ecol Eng. 2005; 25: 365–381.

16 Soroko M. Treatment of Wastewater from Small Slaughterhouse in Hybrid Constructed Wetlands system. In: Toczyłowska I, Guzowska G, eds. Proceedings of the Workshop Wastewater Treatment in Wetlands. Theoretical and Practical Aspects; 2007; Gdansk: Gdańsk University of Technology Printing Office, pp. 171–176.

17 Carreau R, Van Acker S, Van der Zaag AC, Madani A, Drizo A, Jamieson R, et al. Evaluation of a surface flow constructed wetland treating abattoir wastewater. Appl Eng Agric. 2012; 28(5): 757–766.

18 Hung YT, Lo HH, Awad A, Salman H. Potato wastewater treatment. In Wang LK, Hung YT, Lo HH, Yapijakis C, eds. Waste Treatment in the Food Processing Industry: CRC Press; 2005; pp. 193–254.

19 Bosak V, Vander Zaag A, Crolla A, Kinsley C, Gordon R. Performance of a Constructed Wetland and pretreatment system receiving potato farm washwater. Water. 2016; 8(5):183–196.

20 Kato K, Inoue T, Ietsugu H, Koba T, Sasaki H, Miyaki N, et al. Design and performance of hybrid reed bed systems for treating high content wastewater in the cold climate. In: 12th International Conference on Wetland Systems for Water Pollution Control; 2010; Venice, pp. 511–517.

21 Arimi MM, Zhang Y, Götz G, Geißen S. Treatment of melanoidin wastewater by anaerobic digestion and coagulation. Environ Technol. 2015; 36(19): 2410–2418.

22 USDA. Thailand Biofuels Annual 2016. Global Agricultural Information Network (GAIN) report. United States Department of Agriculture (USDA), Foreign Agricultural Service; 2016. Report No.: TH6075.

23 Sohsalam P, Sirianuntapiboon S. Feasibility of using constructed wetland treatment for molasses wastewater treatment. Bioresourc Technol. 2008: 5610–5616.

24 Taylor A. From raw sugar to raw materials. Chemical Innovation. 2000; 30: 45–48.

25 Satyawali Y, Balakrishnan M. Wastewater treatment in molasses-based alcohol distilleries for COD and color removal: A review. J Environ Manage. 2008; 86:481–497.

26 Billore SK, Singh N, Ram HK, Sharma JK, Singh VP, Nelson RM, et al. Treatment of a molasses based distillery effluent in a constructed wetland in central India. Wat Sci Technol. 2001: 441–448.

27 Mohana S, Acharya BK, Madamwar D. Distillery spent wash: Treatment technologies and potential applications. J Hazard Mater. 2009; (163):12–25.

28 Padmapriya R, Tharian JA, Thirunalasundari T. Coffee waste management – An overview. Int J Current Sci. 2013: 83–91.

29 Von Enden JC, Calvert KC, Sanh K, Hoa H, Tri Q. Review of coffee waste water characteristics and approaches to treatment. Project. German Technical Cooperation Agency (GTZ), Improvement of Coffee Quality and Sustainability of Coffee Production in Vietnam; 2002.

30 Selvamurugan M, Doraisamy P, Maheswari M. An integrated treatment system for coffee processing wastewater using anaerobic and aerobic process. Ecol Eng. 2010: 1686–1690.

31 Fia R, Matos AT, Fia FR. Biological systems combined for the treatment of coffee processing wastewater: II-Removal of nutrients and phenolic compounds. Acta Scientiarum. Technology. 2013: 451–456.

32 Fia R, Matos AT, Fia FR, Matos MP, Lambert TF, Nascimento FS. Performance of forage crops in wetlands used in the treatment of wastewater of coffee processing. Revista Brasileira de Engenharia Agrícola e Ambiental. 2010: 842–847.

33 Rossmann M, Matos AT, Abreau EC, Silva FF, Borges AC. Performance of constructed wetlands in the treatment of aerated coffee processing wastewater: Removal of nutrients and phenolic compounds. Ecol Eng. 2012: 264–269.

34 Rossmann M, Matos AT, Abreau EC, Silva FF, Borges AC. Effect of influent aeration on removal of organic matter from coffee processing wastewater in constructed wetlands. J Environ Manage. 2013: 912–919.

35 Preechajarn S, Prasertsri P. Biofuels Annual. Bangkok: USDA Foreign Agricultural Service; 2016. Report No.: TH6075.

Part III

Agro-Industrial Wastewater

8

Olive Mill Wastewater Treatment in Constructed Wetlands

F. Masi[1], A. Rizzo[1], R. Bresciani[1], Dimitrios V. Vayenas[2,8], C.S. Akratos[6], A.G. Tekerlekopoulou[3] and Alexandros I. Stefanakis[4,5,7]

[1]*Iridra Srl, Florence, Italy*
[2]*Department of Chemical Engineering, University of Patras, Patras, Greece*
[3]*Department of Environmental and Natural Resources Management, University of Patras, Agrinio, Greece*
[4]*Department of Engineering, German University of Technology in Oman, Athaibah, Oman*
[5]*Bauer Resources GmbH, BAUER-Strasse 1, Schrobenhausen, Germany*
[6]*Department of Civil Engineering, Democritus University of Thrace, Xanthi, Greece*
[7]*Bauer Nimr LLC, Muscat, Oman*
[8]*Institute of Chemical Engineering and High Temperature Chemical Processes (FORTH/ICE-HT), Platani, Patras, Greece*

8.1 Introduction

The olive oil industry is principally widespread in the Mediterranean region, playing an important role for the economies of countries such as Spain, Italy, Greece, Turkey, Syria, and Tunisia. Moreover, other countries are becoming emerging olive oil producers, including Argentina, Australia, and South Africa. According to the International Olive Council, world olive oil production was on average of 2,839,000 tonnes per year from 2009 to 2015, with 69.9% produced in Europe.

The sustainable management of the huge amount of olive mill wastewater (OMW) generated from the olive oil industry (more than 107 mm^3 per year [1, 2]) is still an open issue. Different technological approaches have been tested with success, such as incineration, aerobic and anaerobic biological treatment, nanofiltration and reverse osmosis, electro-oxidation, and electrocoagulation. However, these approaches were often too expensive and with issues such as the management of sludge and/or by-products [3], especially considering the small size of most of the mills [4]. Very often the preferred OMW management option is the direct discharge into the soil, considering the local availability of sufficient land to irrigate and the high costs of fuel for the transportation and dispersion that are increasing proportionally with the distance of the fields from the mill. Although constructed wetlands (CWs) could represent a low cost and easy management solution, the really high organic and suspended solid loads driven by OMW limited their application, due to the high risk of clogging and risk of poor performance. Indeed, prevalently pilot scale CWs, often in combination with other technologies, are available in the literature to date [3–9], with one recent full-scale system attempt [10].

This chapter aims to provide useful information to create a break-through for the use of CWs for OMW treatment. To this aim, the production and the characterization of the OMW are discussed, with particular attention to the most relevant issues to be managed during the design of CW treatment systems for this particular wastewater. In the following paragraphs, the long Greek

Constructed Wetlands for Industrial Wastewater Treatment, First Edition. Edited by Alexandros I. Stefanakis.
© 2018 John Wiley & Sons Ltd. Published 2018 by John Wiley & Sons Ltd.

Table 8.1 Olive mill wastewater characterization from the literature (adapted from Vymazal [3]).

		Min	Max	Common treatment targets
COD	mg/L	37,000	318,000	120–160
BOD$_5$	mg/L	10,000	150,000	20–40
TSS	mg/L	6,000	83,700	35–80
TP	mg/L	410	840	<2

experience on the use of CWs for OMW treatment is reviewed, as well as a successful Italian case study, an innovative pilot system that includes CWs. Finally, advantages, disadvantages, and lessons learnt from the described case studies are discussed.

8.2 Wastewater Production and Characterization

Olive mill wastewater (OMW) can be generated either in the two-phase olive oil extraction processes along with olive pomace, or in three-phase olive oil production as oil alone [11]. On average, 100 kg of olives produce 18 kg of virgin and extra virgin oil, 50 kg of pomace and 60 kg of wastewater, including vegetation water, contained inside the drupe, and water used to clean olives and machinery [4]. The OWM quality characterization encountered in the literature is summarized in Table 8.1, which clearly highlights that the main chemical parameters are orders of magnitude higher compared to common treatment targets requested for discharge in water bodies or on soil.

Indeed, OMWs are characterized by very high organic and suspended solid loads, typical acidic conditions (range of pH in between 4 to 5), high content of phosphorous and nitrogen, as well as high concentrations of phenols, oils, grease and fatty acids [11]. Moreover, OMWs are usually characterized by a quite low biodegradability, mainly due to several inhibitors such as polyphenols, lipids and organic acids [12]. An additional complexity to manage OMW is driven by the seasonality of oil production; indeed, the production phase is usually limited to only 2–3 months.

From the OMW characterization, it's clear that the treatment of OMW with CWs is quite a difficult task. Indeed, assuming the typical organic and suspended solid loads of domestic wastewater (130 g_{COD}/PE/d, 80 g_{TSS}/PE/d, and 200 l/PE/d; [13]) and the range of OMW quality reported in Table 8.1, oil industry produces a wastewater with an organic load 60–489 and a suspended solid load 15–200 higher than typical domestic wastewater. For these reasons, a relevant role for applying CWs for OMW treatment is driven by primary treatments, with different tested solutions reported in the literature [4, 7, 9, 14]. The direct application of raw OMW is rare, and should require dilution [7].

8.3 Applications and Configurations

In this chapter the wide Greek experience on the use of CWs for OWM treatment is reviewed. Moreover, an innovative Italian approach based on the coupling of CWs with an evaporator-condenser is also reported, as an example of potential benefits driven by the use of a technological solution with CWs for OWM management. The graphical schemes of all the systems presented in this section are reported in Figure 8.1.

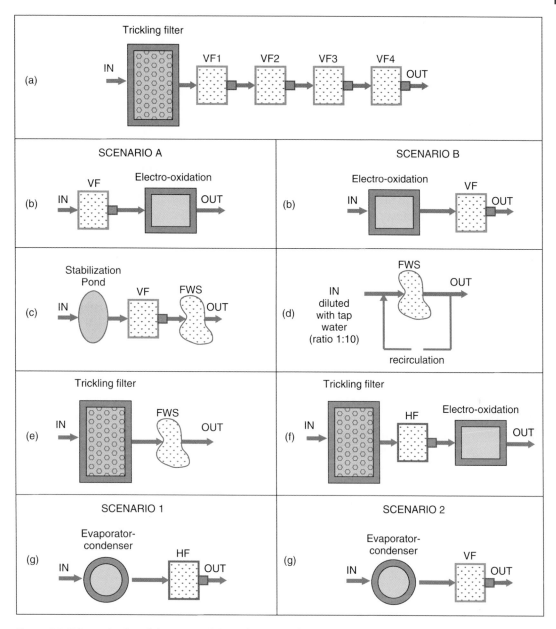

Figure 8.1 Schematization of the reviewed CWs schemes on the treatment of OMW: (a) Herouvim et al. [7]; (b) Grafias et al. [8]; (c) Gikas et al. [9]; (d) Kapellakis et al. [5]; (e) Michailides et al. [10]; (f) Tatoulis et al. [6]; (g) Masi et al. [4].

8.3.1 The Greek Experiences

Greece has the 3rd place in olive oil production worldwide, as it produces approximately 350,000–400,000 ton/year of olive oil [15]. In contrast to Spain and Italy, olive oil in Greece is produced mainly in three-phase (3-P) olive mills, which are scattered in rural areas. The estimated total number of olives mills in Greece is around 2,300 [16]. These olive mills are a serious environmental threat, as they produce large quantities of OMW, which is extremely hazardous when discharged untreated to surface water bodies. On the other hand, olive mills in Greece are usually small family businesses and they cannot afford investing in OMW treatment systems, which possess high capital and operational costs. In addition, 45,000 tons of black and 20,000 tons of green table olives are produced annually in approximately 75 factories in Greece [17] along with significant quantities of table olive processing wastewater (TOWW), which also pose a serious environmental threat due to the seasonal production and high organic and phenolic loads [18].

Due to their overall simplicity in construction and operation as well as their lower capital and operational costs, CWs are an excellent solution for either OMW or TOWW treatment. In fact, several attempts at OMW [5, 7–10] and TOWW treatment [6] have been reported in Greece (Table 8.2), which mainly concern pilot-scale experimental applications.

8.3.1.1 Free Water Surface CWs

Although FWS CWs appear to either have lower pollutant removal rates or require higher areas compared to other CWs types (Table 8.2), it seems to be the only CW type used for OMW in Greece in full or industrial scale applications. The first attempt at using FWS CWs for OMW treatment was by Kapellakis et al. [5], where OMW was diluted with fresh water at a ratio of 1/10 and then introduced in two series of FWS CWs (one operated under continuous flow and the other with 50% recirculation) with a hydraulic load of 3.15 m^3/d. Although the applied pollutant loads were well above the standard pollutants loads applied in CWs, pollutant removal rates were extremely high and reached almost 80% (Table 8.2). These rates were further improved when effluent recirculation was applied [5]. Phenols and COD seem to antagonize, as oxygen availability is limited and both are removed mainly through aerobic oxidation [5]. The relatively low phenol removal rates could be increased if the HRT, dilution ratio or recirculation ratio was further increased. Thus, it is clear that CWs could serve as a treatment stage for OMW, following an initial treatment stage, in which a significant portion of organic and phenol load should be removed. Furthermore, the use of only FWS CW limits nitrification, as anoxic conditions prevailed [5], which also enhanced the argument for the necessity of a pretreatment stage. Nevertheless, the ability of FWS CWs treating high strength wastewater as OMW, was proven.

Michailides et al. [10] presented the first full-scale FWS CW treating pretreated OMW. The FWS was located in the area of an olive mill near Amfilochia Aetoloakarnia, Greece and was the retrofitting of the olive mill's initial evaporation pond. The full-scale FWS CW was divided into five cells by soil barriers. To avoid OMW leakage, the base of the wetland was lined with clay. The wetland had a slope of 2% and the five cells covered a surface area of 400, 350, 700, 350 and 250 m^2, respectively (total area of 2,050 m^2). Cells 1 and 2, closest to the influent entry point, were kept unplanted because the high influent pollutant concentrations are toxic for plant growth. The wetland was operated under batch conditions for 60 days (HRT of 60 days).

CW was operated for only 60 days, as until end of May the majority of OMW stored inside the CW was lost due to evapotranspiration [10]. Parallel to the CW operation, photo-oxidation experiments were conducted in the laboratory, in order to fully assess pollutant photo-oxidation [10]. After

Table 8.2 CWs applications for OMW and TOWW in Greece.

Reference	CW type	CW area (m²)	Plant species	Pretreatment	Surface loads (g/m²/day)					Removal rates (%)				
					C	TKN	TP	TSS	d-phenols	C	TKN	TP	TSS	d-phenols
[7]	VF	0.1256	*Typha latifolia* *C. alternatifolius*	Biological filter	6589	175	20	–	997	73	75	88	–	75
[8]	VF		*Phragmites australis*	Electro-oxidation	15	–	–	–	–	86	–	–	–	–
[9]	VF- HSF	1.0–2.9	*Phragmites australis* *Typha latifolia*	Stabilization pond	2371	8	–	53	37.4	54	44	–	52	60
[5]	FWS	45.5	*Phragmites australis*	Dilution with tap water (ratio 1:10)	92.5	1.9	0.6	32.7	14.7	80	78	80	83	74
[10]	FWS	2050	*Phragmites australis*	Biological filter	285	8	2	–	50	94	98	95	–	95
[6]	HSF	0.24	*Phragmites australis*	Biological filter	77	–	–	–	1.7	70	–	–	–	78

HSF = horizontal subsurface load; VF = vertical flow; FWS = free water surface; C = organic matter; TKN = total Kjeldahl nitrogen; TP = total phosphorus; OP = orthophosphate; TSS= total suspended solids; d-phenols = diluted phenols.

the treatment through a biological trickling filter, OMW was introduced into the FWS CW, while its pollutant concentrations were 27,400 mg/L, 4,800 mg/L, 191 mg/L and 770 mg/L for COD, phenols, orthophosphate (OP) and total Kjeldahl nitrogen (TKN), respectively. After 60 days, pollutant concentrations were decreased to 1,584 mg/L, 248 mg/L, 5.2 mg/L and 18 mg/L for COD, phenols, OP and TKN, respectively (corrected values for days 60–10). Although pollutant removal rates were extremely high (above 95%), effluent concentrations remained above legislation limits for surface disposal [10]. A post-treatment stage is not imperative, since high evapotranspiration rates in FWS can retain the total OMW produced volume. These extremely high pollutant removal rates are mainly attributed to biological oxidation, as photo-oxidation was proved to remove only 18% of COD and 11% of phenols [10].

8.3.1.2 Horizontal Subsurface Flow CWs

Although horizontal subsurface flow (HF) CWs are considered to be more efficient than FWS, only one application of HF for TOWW is recorded [6]. This study included the operation of two identical pilot-scale HF CWs (72 cm long, 33 cm wide, 35 cm deep, surface area = 0.24 m^2) filled with fine gravel (D$_{50}$ = 6mm), while one unit was planted (OG-P) with common reeds (*Phragmites australis*) and the other was kept unplanted (OG-U) as the control. The units were operated for one year (from July 2014 to April 2015), while three HRTs (2, 4 and 8 days) were applied. HF CWs received pretreated TOWW in a biological trickling filter [6]. CWs effluent was post-treated by electro-oxidation. Batch electro-oxidation experiments were performed in a double-walled, cylindrical glass vessel with a liquid capacity of 0.2 L. A rectangular electrode (16 cm^2) made of boron-doped diamond was used as the anode and the cathode was made of stainless steel [6].

HF CWs managed to further reduce COD concentration from 2,750 $\pm$ 900 mg/L to 900 $\pm$ 800 mg/L and phenol concentrations from 200 mg/L to 50 mg/L [6]. Organic matter removal was not significantly affected by vegetation (ANOVA p value = 0.678 > 0.05), while HRT alterations significantly affected COD removal, as 2 days HRT leads to significantly lower ($p < 0.05$ for OG-P and OG-U) COD removal efficiencies (48% for OG-P and 40% OG-U). However, an HRT of 4 days is considered to be sufficient for COD removal (95% for OG-P and OG-U) [6]. On the contrary, the HRT did not significantly affect (OG-P, $p > 0.05$; OG-U, $p > 0.05$) the phenol removal [6]. Although HF CWs achieved high pollutant removal rates, pollutant effluent concentrations remained above legislation limits for discharge, thus, a post-treatment by electro-oxidation is imperative. In order to minimize electro-oxidation operational cost, CW's effluents (i.e., when operated with an HRT of 8 days) with relatively low pollutant concentrations (i.e., 450 mg/L for COD and 85 mg/L for phenols) were post-treated by electro-oxidation [6].

Electro-oxidation achieved complete removal of the residual pollutant loads after 90 min at 62.5–125 mA/cm^2. In order to just reach legislation limits for final discharge, only 45 min of electro-oxidation at 62.5 mA/cm^2 are necessary. Complete decolorization was also achieved [6]. The proposed hybrid treatment system (i.e., trickling filter, CW and electro-oxidation) can achieve full treatment of TOWW, while minimizing operational cost, as the operational cost of the biological filter per unit volume has been estimated at 0.09 Euro/m^3 [15] and operational cost for electro-oxidation is two orders of magnitude higher (i.e., 9.5 Euro/m^3) [6].

8.3.1.3 Vertical Flow CWs

Vertical subsurface flow (VF) CWs are able to remove higher organic loads compared to the other CWs types, therefore they are an excellent choice for OMW treatment. Nevertheless, until now only pilot-scale applications have been recorded in Greece. Herouvim et al. [7] have used twelve pilot-scale VF CWs, which consisted of three paralleled series (four VF CWs units per series), for the treatment of pretreated OMW. OMW pretreatment was implemented in a biological trickling filter. Each pilot-scale VF unit was 96 cm in length, 38.5 cm in width and 31 cm in depth, while two series (A and B) were planted with common reed (*Phragmites australis*) and the third (series C) was kept unplanted. Each series consisted of four VF stages; the first stage (units A1, B1 and C1) had a 7 cm deep drainage layer of cobbles (D_{50} = 60 mm). This layer was covered with 7 cm of medium gravel (D_{50} = 24.4 mm) and 17 cm of fine gravel (D_{50} = 6 mm). The second stage (units A2, B2 and C2) had a 9 cm deep drainage layer of cobbles (D_{50} = 60 mm), covered with a 6 cm deep layer of medium gravel (D_{50} = 24.4 mm), 9 cm of fine gravel (D_{50} = 6 mm), and then 6 cm of sand (D_{50} = 0.5 mm). Finally, the third (units A3, B3 and C3) and fourth stages (units A4, B4 and C4) had drainage layers of cobbles 5.5 cm deep (D_{50} = 60 mm) and on top of it layers of medium gravel (9.5 cm) (D_{50} = 24.4 mm), fine gravel (3.5 cm) (D_{50} = 6 mm), and sand (11.5 cm) (D_{50} = 0.5 mm) were placed.

All units were loaded with 820 L of OMW on a weekly basis, while loading period duration was 2 days and resting period duration was 5 days. VF CWs in this study managed to receive extremely high pollutant loads (i.e., 88–6589 g COD/m^2/d; 17–997 g d-phenols/m^2/d, 3.0–175 g TKN/d m^2; 3.0–20.0 g OP/m^2/d; [7]). Pollutant removal efficiencies (series A and B) were also high; 74% on average for COD, 75% for TKN and 87% for OP. This setup proved the overall ability of passive VFCWs to receive and remove high pollutant loads in multistage systems, as they were able to remove 4900 g COD/m^2/d, 130 g TKN/m^2/d, 15 g OP/m^2/d and 750 g Phenols/m^2/d.

Grafias et al. [8] also tested pilot-scale VF CWs for OMW treatment, which were circular tanks with a surface area of 0.24 m^2 and 0.65 m deep, planted with *Phragmites australis*. Pilot-scale units received OMW with organic load of 15 g COD/m^2/d, which was either raw OMW and after the CWs was post-treated by electro-oxidation (Scenario A), or was pretreated with electro-oxidation (Scenario B). Electro-oxidation experiments were conducted in a DiaCell (type 100) single compartment electrolytic flow-cell manufactured by Adamant Technologies (Switzerland) at 20 A, without the addition of supporting electrolytes and at ambient temperature [8]. VF CWs were again found to be rather effective in OMW treatment as they removed 86% of COD and 77% of color [8], while electro-oxidation as the post-treatment stage managed to further increase pollutant removal to 95% for COD and 94% for color. On the other hand, when OMW was treated first by electro-oxidation and afterwards by VFCWs, pollutant removal efficiencies were reduced to 81% for COD and 58 % for color [8].

8.3.1.4 Hybrid Wetland Systems

The only attempt at using hybrid CWs systems for OMW treatment recorded in Greece was presented by Gikas et al. [9], who used a cylindrical shape VFCW (0.82 m in diameter and 1.5 m high), followed by a rectangular FWS CW (3.4 m long and 0.85 m wide), which operated under an HRT of 13 days. Before OMW introduction to the hybrid CWs system, it remained for about 8 months in two open tanks, aiming at organic and suspended solid load reduction. Pollutant loads for both CWs were on average: 2,371 g COD/m^2/d, 53 g TSS/m^2/d, 8 g TKN/m^2/d and 233 g d-phenols/m^2/d for the VF unit and 382 g COD/m^2/d, 7.6 g TSS/m^2/d, 1.2 g TKN/m^2/d and 37.4 g d-phenols/m^2/d for the FWS unit, while overall pollutant removal efficiencies were 54% for COD, 44% for TKN, 54% for TSS and 60% for d-phenols.

8.4 Evaporation Plus Constructed Wetlands: An Italian Innovative Approach

The Italian case study describes a pilot study in which an evaporator-condenser (EC) has been coupled with CW to treat OMW. The system treats the OWM produced by the Santa Tea mill, located in Reggello municipality (Tuscany region), and is described in detail by Masi et al. [4]. The EC (model Alfa Flash, produced by Alfalaval) was designed to treat up to 48 m^3/d of OMW, approximately all the wastewater produced by the Santa Tea mill. The EC produced about 80% of condensate and 20% of concentrate. The concentrate was mixed with olive mill pomace and delivered to centralized treatment plants. The condensate was treated with two parallel CW beds to compare two different solutions: (i) HF bed, filled with gravel; (ii) vertical subsurface flow (VF), vertically filled with layers of different porous media (gravel and sand). Both HF and VF beds were designed to treat up to 500 L/d of condensate, and with a surface area of 10 m^2. The monitoring period lasted from October 2012 until January 2013.

The raw OMW was characterized by high concentrations of COD (138,000 ± 7,400 mg/L) and P (310 ± 22 mg/L), low pH (4.5 ± 0.3), and low biodegradability. The EC stage promoted a very effective reduction of OMW organic load (95–98.7% of COD removal), which is blocked in the concentrate state. As a consequence, the condensate proved to be more suitably treated by CWs, with the following characteristics: COD 2,423 ± 168 md/L; ammonium 0.17 ± 0.04 mg_{N-NH4}/L; phosphate 0.17 ± 0.04 mg_{P-PO4}/L; pH 3.8 ± 0.14 mg/L; VFA 274 ± 32 mg/L. Due to the very poor nutrient content of condensate, micronutrients were dosed within the CW beds. The performance of the CW beds in the treatment of OWM condensate was satisfactorily high, with slightly better COD removal efficiencies for the HF bed compared with the VF bed (92% and 88%, respectively). Both CW beds did not show evidence of clogging during the monitored period. The HF bed achieved a highly stable removal from the beginning of the operation, fitting in with the variations of the olive mill working periods.

8.5 Discussion and Conclusions

Some interesting key points can be learnt from the experiences so far of CW application for OMW treatment in Greece and Italy:

- Passive subsurface CWs (VF and HF) are more efficient than FWS CWs, as they have achieved high pollutant removal rates under the Mediterranean climatic conditions, while they received higher pollutant loads (Table 8.2).
- OMW pretreatment is necessary in order to reduce pollutant concentrations to non-toxic levels for wetland vegetation.
- OMW post-treatment is also necessary, as CWs cannot deliver effluent concentration within the permissible limits, while color reduction in all biological treatment systems is rather limited. Therefore, a physico-chemical treatment stage (e.g., electro-oxidation) could further reduce organic matter, phenols and color.

The Greek experience highlighted that CWs cannot treat OWM as a unique treatment stage, needing instead to be coupled with other techniques for meeting a very low standard for discharge. Usually, the additional solutions lead to very high costs, such as the use of electro-oxidation as a post-treatment [6]. Therefore novel solutions need to be studied to reduce the OMW treatment costs. One example is reported in this chapter, i.e., the novel approach tested in Italy, in which the costs of the OMW treatment are minimized as much as possible. In order to be applicable on a real scale, the efficiency of the proposed solutions need to fit with national regulations. As an example, the suitability of the ED+CW scheme is evaluated here considering the Italian law. In Italy, the current law regulates the discharge of wastewater produced by industrial activities either in sewers, water bodies or soil, which can be done after the release of a waiver by the public authorities. For the OMW case, only the possibility of discharge in sewer and in the water body are considered, due to stringent limits for discharge on soil. On this basis, three possibilities are available according to the ED+CW solution:

- discharge of the condensate to the public sewer (when present) after the granting of permission and the payment of a fee proportional to the load;
- discharge of the condensate after a treatment aimed at achieving the specific Italian limit for discharging into public sewers (COD 500 mg/L);
- treatment of the condensate to achieve the limit for discharge into surface water bodies (COD 160 mg/L).

During the monitoring phase, the minimum COD effluent concentrations from the ED+CW system were 217 mg/L and 285 mg/L for HF and VF beds, respectively. Therefore, a current upscaling of the ED+CW treatment plant seems to be feasible to respect the Italian limit for discharge in sewers. If a sewer is not available, the surface of the CW bed needs to be increased to further reduce the effluent COD concentration and respect the limit for discharge in the water body, increasing the hydraulic retention time to promote the degradation of the more recalcitrant residual organic matter.

References

1 Yalcuk A, Pakdil NB, Turan SY. Performance evaluation on the treatment of olive mill waste water in vertical subsurface flow constructed wetlands. Desalination. 2010; 262(1):209–214.

2 Benitez J, Beltran-Heredia J, Torregrosa J, Acero JL, Cercas V. Aerobic degradation of olive mill wastewaters. Appl Microbiol Biotechnol. 1997; 47(2):185–188.

3 Vymazal J. Constructed wetlands for treatment of industrial wastewaters: a review. Ecol Eng. 2014; 73:724–751.

4 Masi F, Bresciani R, Munz G, Lubello C. Evaporation–condensation of olive mill wastewater: Evaluation of condensate treatability through SBR and constructed Wetlands. Ecol Eng. 2015; 80:156–161.

5 Kapellakis IE, Paranychianakis NV, Tsagarakis KP, Angelakis AN. Treatment of olive mill wastewater with constructed wetlands. Water 2012; 4:260–271.

6 Tatoulis T, Stefanakis A, Frontistis Z, et al. Treatment of table olive washing waters using trickling filters, constructed wetlands and electrooxidation. Environ Sci Pollut Res. 2017; 24(2):1085–1092.

7 Herouvim E, Akratos CS, Tekerlekopoulou A, Vayenas DV. Treatment of olive mill wastewater in pilot-scale vertical flow constructed wetlands. Ecol Eng. 2011; 37(6):931–939.

8 Grafias P, Xekoukoulotakis NP, Mantzavinos D, Diamadopoulos E. Pilot treatment of olive pomace leachate by vertical-flow constructed wetland and electrochemical oxidation: an efficient hybrid process. Water Res. 2010; 44(9):2773–2780.

9 Gikas GD, Tsakmakis ID, Tsihrintzis VA. Treatment of olive mill wastewater in pilot-scale natural systems. 8th Int. Con. of EWRA "Water Resources Management in an Interdisciplinary and Changing Context", Porto, Portugal, 26–29 June 2013, paper #232, pp. 1207–1216.

10 Michailides M, Tatoulis T, Sultana M-Y, et al. Start-up of a free water surface constructed wetland for treating olive mill wastewater. Hemijskaindustrija 2015; 69.

11 Coskun T, Debik E, Demir NM. Treatment of olive mill wastewaters by nanofiltration and reverse osmosis membranes. Desalination. 2010; 259(1):65–70.

12 Filidei S, Masciandaro G, Ceccanti B. Anaerobic digestion of olive oil mill effluents: evaluation of wastewater organic load and phytotoxicity reduction. Water Air Soil Pollut. 2003; 145(1–4):79–94.

13 Masotti L. Depurazione delle acque. Tecniche ed impianti per il trattamento delle acque di rifiuto (Wastewater treatment. Technologies and plants for wastewater treatment). Calderini Editore – Il. Sole. 2011;24.

14 del Bubba M, Checchini L, Pifferi C, Zanieri L, Lepri L. Olive mill wastewater treatment by a pilot-scale subsurface horizontal flow (SSF-h) Constructed Wetland. Annali di Chimica. 2004; 94(12):875–887.

15 EC (European Commission). Prospects for the olive oil sector in Spain, Italy and Greece, 2012–2020, Brief N°2, July 2012.

16 Michailides M, Panagopoulos P, Akratos CS, Tekerlekopoulou AG, Vayenas DV. A full-scale system for aerobic biological treatment of olive mill wastewater. J Chem Technol Biotechnol. 2011; 86:888–892.

17 Kyriacou A, Lasaridi KE, Kotsou M, Balisa C, Pilidis G. Combined bioremediation and advanced oxidation of green table olive processing wastewater. Proc Bioch 2005; 40:1401–1408.

18 Kotsou M, Kyriakou A, Lasaridi K, Pilidis G. Integrated aerobic biological treatment and chemical oxidation with Fenton's reagent for the processing of green table olive wastewater. Proc Bioch. 2004; 39:1653–1660.

9

Dairy Wastewater Treatment with Constructed Wetlands: Experiences from Belgium, the Netherlands and Greece

C.S. Akratos[7], D. Van Oirschot[2], A.G. Tekerlekopoulou[1], Dimitrios V. Vayenas[3,4] and Alexandros I. Stefanakis[5,6,8]

[1] *Department of Environmental and Natural Resources Management, University of Patras, Agrinio, Greece*
[2] *RietLand bvba, Van Aertselaerstraat 70, Minderhout, Belgium*
[3] *Department of Chemical Engineering, University of Patras, Patras, Greece*
[4] *Institute of Chemical Engineering and High Temperature Chemical Processes (FORTH/ICE-HT), Platani, Patras, Greece*
[5] *Bauer Resources GmbH, BAUER-Strasse 1, Schrobenhausen, Germany*
[6] *Department of Engineering, German University of Technology in Oman, Athaibah, Sultanate of Oman*
[7] *Department of Civil Engineering, Democritus University of Thrace, Xanthi, Greece*
[8] *Bauer Nimr LLC, Muscat, Oman*

9.1   Introduction

Modern dairy farming involves automated milking, in so-called milking parlors. The cows are milked by a milking machine which extracts milk with a vacuum system. The milk is transported through a system of pipes called the milk line towards the milk cooling tank. The milk line is cleaned twice a day (after milking) and the milk cooling tank after each time the milk is collected by the dairy company.

The most widely applied cleaning method for milking installations uses three consecutive rinsing cycles. The first rinse uses clean water, then water with added detergent is used for the second rinse and the final rinse uses clean water again. The amount of water used for each rinsing cycle varies from 50–100 L, depending on the type of installation. The wastewater from the first rinsing cycle usually contains milk and can, thus, be highly loaded with organic matter. Therefore, in some situations it can be advisable to dispose of the wastewater from the first cycle in some other way (e.g., mixing it in the manure cellar). If the milk content of the first cycle is not too high, it can be treated with the rest of the wastewater.

Dairy wastewaters are generated by the milk processing units where the pasteurization and homogenization of fluid milk are processed. Large amounts of wastewaters are also generated from the production of dairy products such as butter, cheese, milk powder etc. Large volumes of water are used in cleaning the processing units, resulting in wastewater containing detergent, sanitizers, base, salts and organic matter, depending upon the sources [1], since the dairy industry generates large quantities of wastewater; 0.2–10 L of wastewater are produced per liter of processed milk [2]. Dairy wastewaters contain high concentrations of organic matter (e.g., fat, milk, protein, lactose, lactic acid), minerals and detergents. Typical dairy wastewater characteristics include 4,000–59,000 mg/L COD, 70–800 mg/L TSS, 100–1,400 mg/L TN and 25–450 mg/L TP [2].

Constructed Wetlands for Industrial Wastewater Treatment, First Edition. Edited by Alexandros I. Stefanakis.
© 2018 John Wiley & Sons Ltd. Published 2018 by John Wiley & Sons Ltd.

After the production of cheese or the removal of fat and casein (80% of the proteins) from milk, the remaining green-yellowish liquid is known as whey [3]. It is estimated that the worldwide production of whey is over 100 billion (100 thousand million) kg per year [3, 4]. Approximately half of this total whey production is produced in the European Union (EU) [5]. During the production of casein or fresh-cultured cheese, only 8% of the total produced whey is produced directly as a by-product from skimmed milk. Whey represents 85–95% of the milk volume and contains about 55% of the milk nutrients comprising milk sugar (lactose), serum proteins (whey proteins), minerals, a small amount of fat, and most of the water soluble minor nutrients from milk such as vitamins [6, 7]. About 85–95% of the milk volume is cheese whey and about 55% of the milk nutrients remain in this potential wastewater. It is a nutritious liquid, containing proteins, lactose, vitamins and minerals, but also enzymes, hormones and growth factors [7]. Around 39,000–60,000 mg/L of lactose is contained in cheese whey and this forms the main fraction of organic loads [8]. Aside from lactose, it also contains fats (0.99–10.58 kg/m^3, 990–$10,580$ mg/L), soluble protein (1.4–8 kg/m^3, $1,400$–$8,000$ mg/L), lipids (4–5 kg/m^3, $4,000$–$5,000$ mg/L) and mineral salts (8–10% of the dry extract) [9, 10]. The main components of the whey protein are true proteins, peptides and non-protein (NPN) components. Whey proteins are determined as the components which are soluble at pH 4.6 in their native form [11]. Characteristics of cheese whey also depend on the quality and type of the milk (cow, goat, sheep and buffalo) used for the cheese making [7], on the evaluated milk portion and other parameters such as mechanisms of cheese making, the acid used for coagulation, time and temperature of coagulation [9].

The liquid effluents of dairy industries and cheese dairies are major contributors to the worldwide industrial pollution issues [12]. Their direct disposal can affect the physical and chemical structure of soils, for which crop yields decrease. In addition, when cheese whey is discharged into water bodies, it affects the aquatic life by causing eutrophication of the receiving waters [13, 14]. The biological treatment of cheese whey depends on its organic load. If the ratio of BOD and COD in whey is higher than 0.5, then it is suitable for treatment by biological processes.

Dairy wastewaters, including cheese whey, are generally treated with various physico-chemical methods such as coagulation/flocculation by various inorganic and organic natural coagulants, and membrane processes such as nanofiltration (NF) and/or reverse osmosis (RO) [15]. Various biological methods are also used to treat cheese whey including activated sludge processes, aerated lagoons, trickling filters, sequencing batch reactors (SBR), anaerobic sludge blanket reactors (UASB), anaerobic filters, etc. [10, 16–18]. However, most of these technologies are connected with increased investment and operational costs, while they have higher maintenance demands. Thus, the sustainable technology of Constructed Wetlands increasingly appears as an attractive alternative. Therefore, the goal of this chapter is to combine and present the experiences from three different countries (Belgium, the Netherlands and Greece) on Constructed Wetlands systems applied for dairy wastewater treatment.

9.2 Brief Literature Review on Wetland Systems for Dairy Wastewater Treatment

Although CWs are mainly used for municipal wastewater treatment, they have been proved to be rather successful in dairy wastewater treatment as well (Table 9.1). Dairy wastewaters are characterized by their high organic load, therefore in most cases of CWs applications dairy wastewater is pretreated. The pretreatment stages aim mainly at removing suspended solids. Therefore, in the

Table 9.1 Applications of different Constructed Wetland systems for dairy wastewater treatment and related information (FWS: free water surface, HSF: horizontal subsurface flow).

Reference	CW Surface Area (m²)	Plant species	Pretreatment	Applied surface load (g/m²/d)				Removed surface load (g/m²/d)			
				C	TKN	TP	TSS	C	TKN	TP	TSS
FWS systems											
[24]	630	Typha domingensis, Scirpus validus, Phragmites australis	Solid separators, anaerobic lagoons, aerobic ponds	20.4	23.2	–	67.4	0.61	6.0	–	24.2
[26]	4265	Carex riparia, Typha latifolia, Phragmites australis, Sparganium erectum, Glyceria fluitans, Iris pseudacorus, Phalaris arundinaceae, Alisma plantago-aquatica	Three-chambered tank	6.8	–	0.05	2.55	6.22	–	0.04	2.52
[27]	100	Typha latifolia, Lemna spp.	Storage tank	6.5	0.8	0.12	4.8	6.37	0.64	0.11	4.61
[28]	500	Typha latifolia, Scirpus tabernaemontani, Litaneutria minor	Settling basin	–	–	–	–	–	–	–	–
Hybrid systems											
[19]	1990	Phragmites australis, Senecio sylvaticus, Urtica dioica, Typha latifolia, Sparganium erectum, Butomus umbellatus, Glyceria maxima	Settling tanks	1.28	0.74	0.10	1.96	1.2	0.68	0.09	1.92
[20]	50–1900	Phragmites australis	Septic tank, settling ponds	5.7–8.3	0.56–1.97	0.17–1.71	–	4.5–6.6	0.2–0.8	0.12–1.2	–

(Continued)

Table 9.1 (Continued)

Reference	CW Surface Area (m²)	Plant species	Pretreatment	Applied surface load (g/m²/d)				Removed surface load (g/m²/d)			
				C	TKN	TP	TSS	C	TKN	TP	TSS
[21]	892	Sotalia fluviatilis	Settling tank	68.5	–	0.6	16	–	–	–	–
[23]	398	Lemna sp., Pontederia cordata	Anaerobic/Facul-tative lagoon, Aerobic lagoon	29.4	3.4	2	8.4	11.2	1.45	0.5	4.9
[29]	1.87	Sotalia fluviatilis	–	173.5	–	2	22.3	156	–	1.6	20.1
[30]	80	Phragmites australis	–	50–1,500	50–1,500	1.5–40	20–400	45–1,350	33–975	0.8–21	19–376
[32]	1.63	Typha angustata	Secondary treatment in storage tank	8.6–34.5	1.3–4.9	OP: 0.3–1.1	0.2–0.7	5.2–20.7	0.78–2.94	OP: 0.09–0.33	0.16–0.56
[33]	600	Phragmites communis, Scirpus validus	Storage lagoons	–	–	–	–	–	–	–	–
[34]	Two beds (15 m² and 10 m²)	Phragmites australis, Stuckenia pectinata, Carex acutiformis, Glyceria maxima, Typha latifolia	Settling tank	27–45	–	–	–	24–41	–	–	–
[35]	72	Phragmites australis	Imhoff tank	110.1	5.84	1.15	62.3	101.3	2.83	0.70	56.6
[36]	20	–	SBR	–	–	–	–	–	–	–	–
HSF systems											
[20]	50–1900	Phragmites australis	Septic tank, settling in ponds	5.7–8.3	0.56–1.97	0.17–1.71	–	4.5–6.6	0.2–0.8	0.12–1.2	–
[21]	892	Sotalia fluviatilis	Settling tank	68.5	–	0.6	16	–	–	–	–
[22]	138.6 (each cell)	Typha angustifolia, Phragmites australis, Suillus pungens	Settling basin	51.3	1.96	0.5	24.6	38.9	0.55	0.23	22.1

Ref	Area	Plant species	System								
[23]	398	Lemna sp., Pontederia cordata	Anaerobic/Facul-tative lagoon, Aerobic lagoon	29.4	3.4	2	8.4	11.2	1.45	0.5	4.9
[25]	19	Scirpus validus	Oxidation pond	0.9–4.1	0.6–2.7	0.2–0.8	1.9–8.5	0.76–3.5	0.4–1.8	0.12–0.48	1.5–6.5
[31]	160	Phragmites australis	–	–	–	–	–	–	–	–	–
[32]	1.63	Typha angustata	Secondary treatment in storage tank	8.6–34.5	1.3–4.9	OP: 0.3–1.1	0.2–0.7	5.2–20.7	0.78–2.94	OP: 0.09–0.33	0.16–0.56
[33]	600	Phragmites communis, Scirpus validus	Storage lagoons	–	–	–	–	–	–	–	–
[34]	Two beds (15 m^2 and 10 m^2)	Phragmites australis, Stuckenia pectinata, Carex acutiformis, Glyceria maxima, Typha latifolia	Settling tank	27–45	–	–	–	24–41	–	–	–
[35]	72	Phragmites australis	Imhoff tank	110.1	5.84	1.15	62.3	101.3	2.83	0.70	56.6
[36]	20	–	SBR	–	–	–	–	–	–	–	–
[37]	7600	Typha latifolia, Carex riparia, Glyceria maxima, Carex riparia, Glyceria maxima, Phalaris arundinacea, Carex riparia	–	1.2	–	–	0.21	1.14	–	–	0.2
[38]	138.6	Typha angustifolia, Phragmites australis, Suillus pungens, Typha latifolia, Lythrum salicaria, Eleocharis obtusa, Litaneutria minor	Bulk tank	52.6	1.9	0.514	14.8	14.7	0.5	0.14	6.7
[39]	100	Typha latifolia	Heated storage tank	17	–	–	–	16.7	–	–	–

– Data not sufficient to calculate pollutant loads.

majority of experiments/applications (Table 9.1), the pretreatment stages include either simple settling basins [19–22] or a settling stage combined with biological treatment, such as aerobic and anaerobic lagoons [23, 24] or oxidation ponds [25].

One of the main issues in CW treatment systems is to identify the role of the plants in pollutant removal and to define their toxicity boundaries. In dairy wastewater treatment, where organic loads are very high, a variety of different plant species have been used (Table 9.1). As reported by Browne and Jenssen [19], *Phragmites australis, Scirpus sylvaticus* and *Urtica dioica* were able to grow in CWs treating dairy wastewater and did not exhibit any toxicity effects. The exact contribution of plants in nutrient removal is a controversial issue as almost all related studies give different removal efficiencies. Gottschall et al. [40] report that nutrient removal due to plant uptake was significantly lower in their study compared to previous studies [41, 42], which reported that plant uptake is responsible for 27–66% of nitrogen (N) removal and 47–65% of phosphorus (P) removal. In addition, Newman et al. [22] report that only 3% of N removal could be attributed to plant uptake. Mantovi et al. [35] also attribute most nutrient removal to biofilm biochemical oxidation and plant uptake. Tanner et al. [25] reported that planted wetlands showed greater removal efficiencies of N and P from dairy farm wastewaters than unplanted wetlands.

The unplanted wetland proved to be less efficient in removing both N and P with higher loading rates. Removal of NH_4-N increased with retention time in the planted horizontal subsurface flow (HSF) CW, whereas the unplanted wetland showed lower performance. Plant rhizosphere aeration may stimulate aerobic decomposition processes by increasing nitrification and subsequent gaseous losses of N through denitrification [43, 44] and by decreasing the relative levels of dissimilatory nitrate reduction to ammonium [45].

Regarding the effect of different plant species in the CW removal efficiency, Dipu et al. [46] examined various plant species i.e., *Typha* sp., *Eichhornia* sp., *Salvinia* sp., *Pistia* sp., *Azolla* sp. and *Lemna* sp. Generally, plants appeared to neutralize pH from the initial alkaline values recorded in dairy wastewaters. They also found that *Azolla* sp. and *Eichhornia* sp.-based CWs (83.2%) were more efficient in pollutant removal, followed by *Typha* sp.-based CWs (80%), however these differences were insignificant [46]. Ghosh and Gopal [32] examined the plant tolerance to dairy wastewater and found that young *Typha* plants yellowed when wastewater with high EC values was applied. They also mentioned that plant density and height were maximum near the CW's inlet and attributed this to the higher nutrient concentrations, which could promote plant growth [32].

Also, at 4 days HRT, plant growth was higher and pollutant removal efficiencies were also high [32]. According to Ibekwe et al. [33], treated dairy wastewater can be recycled and used for irrigation. The authors observed the existence of a diverse bacterial community (especially *Nitrosospira* sp.) that oxidizes ammonia, which could also be involved in the nitrification process in the wetlands and influence the final quality of effluent water. Munoz et al. [21] suggested that artificial aeration is needed to enhance the removal efficiency and the pore volumes. Schaafsma et al. [28] proposed that increasing plant density and wastewater recirculation will promote denitrification.

The type of wetland seems to be crucial for dairy wastewater treatment. As shown in Table 9.1, only four research groups have focused on dairy wastewater treatment using free water surface (FWS) CWs [24, 26–28]. As mentioned previously, dairy wastewater is characterized by high concentrations of organic matter, thus, a FWS CW cannot achieve efficient pollutant removal, since the high organic matter concentration creates anoxic or anaerobic conditions in the water column and reduces the amount of oxygen available for microbial organic matter oxidization. As shown in Table 9.1, when pollutant surface load is low in FWS CWs [24, 38], removal efficiencies are high for organic matter

(91–98%), N (80%), P (89–92%) and TSS (96–99%). According to Schaafsma et al. [28], organic matter and nutrient removal efficiencies are also significant, but in some cases, nitrate and nitrite concentrations increase in the effluent. In most experiments and/or applications using vertical flow (VF) CWs, poor phosphorus removal was observed. Dunne et al. [26], reported that the VFCW system did not show any significant reduction in pollutant concentration between the wetland's inlet and outlet, although BOD reduction was significant.

HSF CW systems, however, appear to be more efficient than the other two types for dairy wastewater treatment, and 17 experiments and/or applications employing HSF CWs have been found. HSF CWs are far more efficient bio-reactors than FWS CWs, as the removal efficiencies of the former are in the range of 28–99% for COD, 21–99% for N, 2–98% for P, and 45–95% for TSS. It should be mentioned, that these removal efficiencies were achieved with pollutant surface loads 10 times higher than those applied in FWS CWs. The most efficient CW system for dairy wastewater treatment appears to be a hybrid system of both VF and HSF stages. Hybrid systems have been tested by three research groups [19, 29, 30], and each demonstrated high removal efficiencies for all pollutants (83–96% for COD, 65–92% for N, 52–99% for P and 83–99% for TSS), even with higher pollutant surface loads than those applied to HSF systems.

Ghosh and Gopal [32] found that at 4 days hydraulic residence time (HRT), plant growth was higher and pollutant removal efficiencies were also high. According to Ibekwe et al. [33], treated dairy wastewater can be reused for irrigation. The authors observed the existence of a diverse bacterial community (especially *Nitrosospira* sp.) which can oxidize ammonia and which could also be involved in the nitrification process.

9.3 Experiences from the Netherlands and Belgium

In Dutch and Belgian rural areas where no sewerage system is available, untreated wastewater discharges in surface waters primarily originate from domestic dwellings. The second most important source is widespread dairy farms. The Belgian firm Rietland has been involved in 40 projects concerning wastewater from dairy farms in the Netherlands and Belgium, a number of which have been monitored for the first years after commissioning. Some systems in Belgium have been, however, monitored for a longer period of time. Thus monitoring results are available for CWs as much as 17 years old. The CW type used in most cases was a single stage, sand-filled VF type, but in one occasion, an intensified wetland in the form of a Forced Bed Aeration (FBA®) system was used. In some projects, additions to the substrate were tested for their phosphorus binding capacity. Also in some of the wetlands, the effect of effluent recirculation was tried out.

All wetlands treat a mix of the domestic wastewater of the family home, together with rinsing water from the milking machine and the cooling tank and often the cleaning water from the milking parlor. The domestic wastewater is pretreated in a septic tank. The wastewater from the milking parlor in a high volume degreaser (retention time > 5 d). Concrete degreasers lined with HDPE or with an epoxy coating on the inside were used, to prevent the concrete being dissolved by lactic acid that is produced as a result of the biodegradation of milk.

On the average, in a dairy farm in the Netherlands and Belgium about 50 cows are milked each day. Depending on the size of the family and the type and capacity of the milking machine, the wetlands were designed for an organic load varying between 10–40 PE. The most common size built was 60 m^2 (20 PE). Because dairy wastewater usually has higher BOD/COD concentrations compared to

Table 9.2 Some sample analyses of pure milk and first cycle wastewater at three dairy farms in Belgium (data courtesy of VLM/Lisec).

	Pure milk	Sample 1	Sample 2	Sample 3
BOD (mg/L)	117,000	1,990	3,190	317
COD (mg/L)	170,000	3,500	5,900	540
N-Kj (mg/L)	4,750	90.5	128	16.8
Tot-P (mg/L)	675	13.2	15.2	2.8
Liters/day		160	100	100
Milk content (L) (based on COD)	–	3.29	3.47	0.32

domestic wastewater, the hydraulic load of the wetlands at dairy farms is typically dimensioned at 25 $L/m^2/d$. This means that at dairy farms the hydraulic load is only 50% of what is normally used for treating domestic wastewater.

In the years 1994–2000, when the first CW systems for treating dairy wastewater were built in the Netherlands and Belgium, this was a rather new application. There was very little data available about the organic load of wastewater from the milking parlor. The data available also showed quite some variation in BOD and COD loads. The variation seemed to be caused by the method of cleaning and how much milk was left in the milking installation and cooling tank before cleaning commenced. In general, wastewater from the milking parlor shows a relative high BOD/COD content compared to domestic wastewater and relative low N-values due to the presence of milk (Table 9.2). However, there is significant variation in the values from day to day, as well as a lot of variation between individual dairy farms.

The samples shown in Table 9.2 lead to the conclusion that the daily milk discharge at those farms amounts 0.32–3.47 L. If we define a PE as 54 g BOD, 1 L of milk can be compared to 2.2 PE. The organic load from the first rinsing cycle alone on those farms ranges between 0.6–5.9 PE.

The detergents used for daily cleaning, contain sodium hydroxide and sodium hypochlorite (chlorine), to comply with food hygiene regulations. In addition to this, regular cleaning takes place with acid detergents, often containing phosphoric acid, which leads to occasional high levels of phosphorus in the wastewater to be treated. In addition to wastewater from cleaning the milking machine, other sources of wastewater may be present. This may include the cleaning water from the milking parlor itself, which may be contaminated with manure. The organic load from this source is very unpredictable. If the wetland is to be used to treat this wastewater as well, the wetland must be designed with a high safety margin.

9.3.1 Wetland System Description

The type of wetland used most often by RietLand has been the VFCW type with intermittent flow. This system exhibits a high purification efficiency and low maintenance costs. This type of wetland is also less prone to soil clogging than the HSF type. The basic design is a variation upon an original design by Klaus Bahlo from Germany [47]. The systems are typically single beds, laid out with a surface area of 3 m^2/PE. They consist of a main substrate layer of uniform, fine grained sand (0/1 mm) of 1 m thickness, in which additions of limestone ($CaCO_3$; lower layer) and sometimes iron (Fe; upper

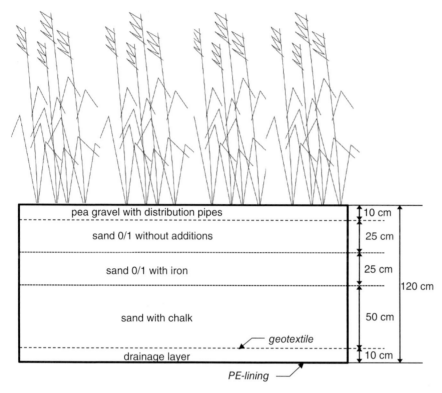

Figure 9.1 Cross-section with the various substrate layers in a VFCW used for dairy wastewater treatment by Rietland.

layer) are mixed (Figure 9.1). The purpose of the limestone addition is to stabilize the pH value of the wastewater and to remove some phosphorus. The purpose of the iron is to bind phosphorus.

A drainage layer is present at the bottom of the system, consisting of a layer of pea gravel 8/16 mm, of 10 cm thick in which drainage pipes (PE) of 80 mm are embedded. The drainage layer is covered with a geotextile to prevent sand from the substrate entering the layer of pea gravel. On top of the substrate layer, a second layer of pea gravel of the same grain size is laid out in which a system of distribution pipes is embedded. Typically those pipes are laid out in at distances of 1 m. The pipes are drilled with 6 mm holes at each m, facing downward. Thus a matrix of holes at every m² is formed. The diameter of the distribution pipes is calculated in such a way that the head loss along the length of the pipes is negligible in order to obtain an even distribution of wastewater across the top surface area of the wetland. Finally, the CW is planted with reeds (*Phragmites australis*) at a density of 6 plants/m².

9.3.2 Operation

Pre-settled wastewater (from a septic tank or degreaser) is collected in a concrete container with a capacity of the total expected wastewater production of one day. From this container, the wastewater is pumped to the CW twice a day. A submersible pump is used, with a capacity that guarantees quick flooding of the surface area to form a layer of several cm of wastewater in the top layer of gravel.

Typically this means the pump must have a capacity of about 5 L/min/m^2 of the wetland, at a total head of 4 m H$_2$O.

The pump is operated by means of a timer that is set to start the pump at times that allow roughly about the same amount of wastewater to be pumped to the reed bed twice a day. This typically means it is set to pump a couple of hours after the "morning peak" of wastewater discharge and again 12 h later, after the "evening peak". The timer switches the pump on for 15 min each time. If the container is emptied before that time (which is usually the case after about 5–10 min), the floating switch of the pump will automatically switch it off. This setup was chosen because most level switches proved unreliable in the long run and the timer/floating switch solution is very cost-efficient. In larger systems where multiple pumps are used, a small PLC is used to operate the pumps.

9.3.3 Results from the Netherlands

A summary of the treatment results of 13 CWs in the Netherlands is given in Table 9.3. These wetlands were constructed in the years 1994–1996. The sampling campaigns in general span a period of one year after commissioning, where monthly samples would be taken. The sampling and analysis was performed by the Dutch Water-boards who were involved in the projects.

These results show a very high removal of organic matter and organic nitrogen. There is also a significant removal of total phosphorus. Denitrification occurs for a little less than 50%, which is quite good for a one-stage VFCW and significantly better than results obtained with the same type of wetland which only treats domestic wastewater. This is probably due to the relatively high C/N ratio of the dairy wastewater. The limiting factor for denitrification in most VFCW seems to be a lack of carbon-sources in the lower, anaerobic zone.

9.3.3.1 Experimental Projects

Eight of the thirteen wetlands listed above were part of an experimental project, conducted in 1997 at dairy farms in Groningen. The experimental systems were built with slight differences in composition of the substrate layers. In two of the systems effluent recirculation was tried. The variations in the composition of the substrate layers and their results were as follows:

Table 9.3 Treatment results from 13 VFCWs treating wastewater from dairy farms in the Netherlands.

mg/L	n = 128 samples		
	Influent	Effluent	Removal (%)
BOD	592	24.3	97
COD	1125	84.6	93
TKN	62.6	10.4	84
NH$_4$-N	34.3	7.4	79
NO$_2$-N		0.3	
NO$_3$-N		20.3	
Tot-N		31.1	48
Tot-P	18.1	4.6	72

- In two wetlands, a layer of peat (20 cm) was added in the middle of the substrate layer. The purpose of this was to add a carbon-source for denitrification. After a short while this layer of peat turned out to compact too much and clogged, leading to ponding of the wetland. The layers of peat were removed. Addition of peat is therefore not further discussed in this paper.
- In six wetlands, a 30 cm layer consisting of a mixture of steel slag and sand was located at 20 to 50 cm below the substrate surface. This exact position was chosen because both deeper and shallower positions cause problems. In the top 20 cm of the substrate, the main oxidation of organic matter takes place, which would cause a bacterial film to form around the iron particles which in turn would prevent efficient binding of phosphorus. Adding the steel slag in deeper layers might lead to dissolution of iron under anaerobic conditions and subsequent re-release of bound phosphorus, which is then washed out of the wetland.
- In all thirteen filters in Table 9.3, some form of calcium carbonate was added.

9.3.3.2 Stimulation of Denitrification through Recirculation of Effluent

Theoretically, recirculation of nitrate-rich effluent from VFCWs can lead to denitrification processes under anaerobic circumstances. To this end, 50% of the effluent of two of the above wetlands was pumped back to the influent pump container. Regrettably samples of the influent were taken from this same influent pump container and thus no data is available of the raw influent. As such, the purification efficiency of the wetlands is difficult to compare to systems without recirculation. Because in most cases the influent would be diluted with an unknown amount of effluent, only the effluent concentration of Tot-N can be evaluated here and compared with wetlands in the same project without recirculation. The available data shows that one of the two filters has a significant lower Tot-N concentration in the effluent than the other wetlands. The Tot-N concentration of the second wetland with recirculation showed little difference to the other systems so the results here were inconclusive. Further investigation of the value of recirculation needs to be conducted. Another interesting possibility would be to recirculate effluent towards the pre-settling tank instead of the pump container, to obtain a longer residence time.

9.3.3.3 Phosphorus Removal

Phosphorus can be removed in wetlands by adding certain additions to the substrate that chemically bind phosphorus and thus retain it from the wastewater. Common additions include calcium and iron in various forms. Calcium-carbonate can also serve to stabilize the pH value of the effluent. Because the nitrification process lowers the pH, over the long-term low pH conditions could hamper the optimal efficiency of the wetland. In addition to calcium-carbonate, several types of iron-additions have been used, with varying levels of success (discussed below). Table 9.4 shows the results for CWs to which only calcium carbonate has been added in various forms.

In the system at Van Oirschot (no relation to the co-author) broken seashells were used. In the wetlands at Adema, Veltman and Cluster, a form of fine white limestone gravel (0.1–1.5 mm) called "Jura marble", supplied by Carmeuse, was used. This product is normally used in the drinking water purification process to stabilize the pH-value. Finally, at the wetland Wolters, another type of grey limestone gravel (0.1–1.5 mm) was used. Chemically the three products are virtually identical, consisting of mainly $CaCO_3$; however the phosphorus binding capacity varies a lot between the three products. Sea shells and grey limestone seem to lose their phosphorus binding capacity quite rapidly and the removal efficiency for phosphorus drops to 25–50% within the year and then stabilizes.

Table 9.4 Phosphorus concentrations in the effluent of 5 CWs at dairy farms, only with $CaCO_3$ addition.

Wetland facility	Mean Tot-P effluent per month (mg/L)				
	Van Oirschot	Adema	Veltman	Cluster	Wolters
Mean influent	14.6	6.9	13.3	21.4	7.3
1	0.4	1.1	0.9	1.3	2.4
2	1.0	1.4	1.2	3.6	–
3	0.3	1.7	1.4	5.2	–
4	0.7	0.8	2.3	5.1	2.8
5	2.2	2.0	2.7	8.1	–
6	6.8	1.6	2.2	6.5	–
7	7.7	1.7	2.2	7.7	–
8	8.4	3.0	3.4	4.5	3.0
9	10.0	1.4	2.2	11.0	–
10	9.7	0.4	2.2	7.6	–
11	11.0	–	–	5.1	5.9
12	10.0	0.7	1.6	–	–
13	9.5	1.2	–	1.5	–
14	8.4	1.2	2.8	4.0	5.5
15	9.9	–	–	5.2	–
16	10.0	–	–	–	–

At the systems where Jura marble is used, effluent phosphorus levels remain low throughout the sampling period and overall removal efficiency for phosphorus is as high as 75–80%. Comparing the three products, the main difference seems to be that Jura marble is softer and thus possibly less crystalline than the other two products. This could explain the difference in phosphorus binding capacity.

In six further CWs (Table 9.5), in addition to limestone, steel slag was used. Steel slag is a waste product from steel manufacturing and contains about 20% iron and 75% CaOH. The high CaOH content initially led to pH values as high as 11 in the effluent but this was only a temporary effect during the first few months. After that the pH value stabilized at around 7.5. A more problematic effect of the steel slag layer was that during the wastewater purification process the CaOH reacted with the organic compound of the wastewater to $CaCO_3$ and this led to binding with the sand to form a 20 cm thick hard crust in the substrate. This compromised the hydraulic conductivity of the substrate to the point that the whole system became clogged. In two of the systems (Onnes, ter Veer), the hard crust was broken using a small excavator and remixed in the top layer in November 1997. In the other four wetlands, the steel slag layer was removed entirely and replaced with sand. In those systems where the steel slag layer was only broken and not removed especially, very high phosphorus removal rates of up to 95% were reached, which proves the utility of using iron as a substrate addition to remove phosphorus. However, the steel slag form of iron seems not very suitable in practice.

Table 9.5 Phosphorus concentrations in the effluent of six CWs at dairy farms, with $CaCO_3$ and iron addition (steel slag).

Wetland facility	Mean influent	Mean Tot-P effluent (mg/L)				
		June '96	Aug '96	Dec '96	Mar '97	Jun '97
Bos	29.1	0.14	2.2	1.0	–	3.3
Hamming	26.5	0.27	0.49	2.5	6.1	7.1
Veening	8.8	0.27	0.84	6.5	4.3	8.8
Vrieling	13.4	0.33	0.16	1.9	1.0	2.0
Onnes	11.7	0.18	0.01	2.0	3.2	4.3
Ter Veer	23.3	0.39	0.9	–	–	2.0

9.3.4 Results from Belgium

In the period from 1996 to 2004, several CWs of the same type as above were built at Belgian dairy farms. These systems are of particular interest as they were monitored for a longer period than the systems in the Netherlands. So the results will be discussed, farm per farm in more detail.

9.3.4.1 System at Poppe, Eeklo

The system at the farm of Geert Poppe, was constructed in August 1999 and is still functioning today. The vertical flow wetland is laid out with 75 m^2 and treats the domestic wastewater from the house of the farmer (four persons) and the wastewater from the milking parlor. Additionally ice-cream specialties are produced at the farm that leads to a discharge of extra waste water from the manufacturing. The milk-lines are flushed twice per day with approximately 100 L of water. However the first flushing water (30 L) is discharged towards the manure cellar so is not treated in the wetland. The cleaning of the milk cooling tank is done three times a week with 240 L of water. For the ice-cream production, about 1,000 L are used, three times per week. Including the domestic wastewater, the total waste water production varies between 1,000 and 2,000 L/d. The domestic wastewater and the dairy wastewater are pretreated in separate septic tanks. The overflows of the septic tanks lead to a pump well from which the pretreated waste water is distributed over the surface of the wetland. In the wetland substrate, 0.4 ton of pure iron was mixed for binding phosphorus and 1.1 ton of limestone grit for pH stabilization.

The CW was monitored by the province of Oost-Vlaanderen (Eastern Flanders). Two samples of both influent and effluent were taken in 2008 when the system was 9 years old. In 2010, six effluent samples were taken and an additional sample of the effluent was taken in 2016, so almost 17 years after the commissioning of the system. The results are shown in Table 9.6.

Although the influent was only sampled twice, the BOD and COD values are in general much higher than the ones from the Dutch dairy farms above. This could be related to the ice-cream production. Based on these influent concentrations, the daily organic load towards the CW, was about 1.3 kg BOD and 3.4 kg COD. The specific surface load on the wetland was 17.6 g BOD/m^2/d and 45.6 g COD/m^2/d, which is a normal load for a VFCW of this type for BOD, but the COD surface load is high. On the other hand, the TN load towards the wetland is low compared to systems treating domestic waste water and the TP concentration in the waste water is relatively high.

Table 9.6 Monitoring data (influent, effluent) from the Geert Poppe wetland facility in Belgium.

Date	BOD In	BOD Eff	COD In	COD Eff	SS In	SS Eff	TKN In	TKN Eff	Tot-N In	Tot-N Eff	Tot-P In	Tot-P Eff
08/08/08	1,389	3	3,170	73	2,840	3.4	21.3	2.5	21.3	3.5	17.6	8.9
30/10/08	368	3	1,389	49	267	3.2	32.9	2.6	32.9	14.3	11.8	8.4
17/05/10		3		39		2.0		2.0		12.6		2.0
28/06/10		3		61		16.3		2.4		6.4		11.4
05/08/10		3		87		11.0		3.2		6.1		8.8
14/09/10		3		47		2.0		2.4		9.9		10.3
12/10/10		3		49		2.0		3.2		16.1		9.6
26/10/10		3		61		2.7		2.4		8.6		9.1
14/04/16		3		45		2.0		1.6		18.7		18.2
Aver	879	3	2,280	57	1,554	5.0	27.1	2.5		10.7	14.7	9.5
SD	755	0	1,259	15	1,819	5.1	8.2	0.5		5.1	4.1	4.2
%	99.7		97.5		99.7		90.9		60.6		35.2	

The effluent values show a consistent excellent quality with BOD-values below 3 mg/L and an average COD concentration of 57 mg O_2/L, even after having been in use for 17 years. Removal of BOD, COD, and SS are all high in the 90% range and TKN removal is very good with 90.9% removal. The Tot-N removal of 60.6% is good for a single stage system without recirculation. The Tot-P removal is poor and seems to be declining over the years, probably due to saturation of the added iron.

9.3.4.2 System at De Paep, Sint-Gillis Waas in Belgium

At the dairy farm of Jan de Paep, a VF sand-filled CW was constructed in June 2004. It has a total surface area of 60 m². To enhance denitrification, a 60 m² pond, planted with reeds was added as a polishing stage. The system treats the domestic waste water of the family (five persons) plus the rinsing water from the milklines (total 216 L/d), milk cooling tank (150 L, once every 3 days) and cleaning water of the milk parlor (approximately 100 L/day, possibly containing manure). The total daily waste water production varies between 1,250 and 1,400 L. The domestic and dairy wastewater are pretreated in septic tanks. The wetland substrate contains 1.3 tons of jura marble ($CaCO_3$) in the lower sand layers. No iron was added here.

The treatment wetland was monitored by the province of Oost-Vlaanderen. A first sampling took place in 2007 and in the next years, most years additional samples were taken until 2016. So the total sampling period spans 9 years, where the last sample was taken when the system was 12 years old. The results are shown in Table 9.7.

Compared to the average of the Dutch systems, the BOD and COD values are similar, but the TN and TP concentrations are higher. Possibly more manure enters the system with the cleaning water of the milking parlor. Also, in Belgium, as an acid cleaning agent, only phosphoric acid is legally permitted, which leads to relatively high P-loads in the waste water of Belgian dairy farms. The daily BOD and COD loads are 0.76 and 1.6 kg respectively which leads to a specific surface load of 12.6 g BOD/m²/d and 26 g COD/m²/d. The N-load is 126 g N/d and 2 g N/m²/d to the filter surface.

Table 9.7 Monitoring data (influent, effluent) from the Jan De Paep wetland facility in Belgium (n = 22, period 2007–2016).

	BOD			COD			SS		
	IN	Eff1	Eff2	IN	Eff1	Eff2	IN	Eff1	Eff2
Average	570	4	5	1,198	35	52	279	11.2	10.6
Sd	175	2	4	328	15	24	121	11.4	9.6
Removal (%)		99.4	99.1		97.1	95.7		96.0	96.2

	TKN			Tot-N			Tot-P		
	IN	Eff1	Eff2	IN	Eff1	Eff2	IN	Eff1	Eff2
Average	93.7	4.3	4.1	95.1	50.0	29.7	31.3	22.2	15.6
Sd	19.2	8.8	4.1	18.4	19.2	15.0	23.9	12.3	8.2
Removal (%)		95.4	95.6		47.5	68.8		29.1	50.4

Effluent samples were taken directly after the VFCW (Eff1) and after the polishing pond (Eff2). The removal rates of this system are similar to the previous one. Also here we see very consistent results and after 12 years, the BOD, COD, TKN and SS removal rates still reach excellent values of > 95% directly behind the VF wetland. The TN removal rate is at 47.5% a bit poorer than the system at Geert Poppe, but the influent TN concentration is higher. The total-P concentrations in the effluent vary quite a bit but on average nearly 30% is removed without any P-binding additions in the substrate.

The polishing pond has a clear effect on the overall removal rates. Especially for nutrients, the N and P concentrations are lowered substantially, probably due to plant uptake. A slight increase in the BOD and COD concentrations between inlet and outlet of the pond can however be observed.

9.3.4.3 System at PDLT, Geel in Belgium

In 1996 a first CW was constructed at PDLT (Provincial Agricultural Agency for the province of Antwerp) as a demonstration project for dairy farms in Flanders. The original constructed wetland was of the VF sand-filled type, like the ones described in the above, with a surface area of 75 m². In 2015, this system was replaced by an aerated wetland, as a consequence of a change in the milking parlor, where the traditional milking system was replaced by two milking robots. The new system needed to have a considerably larger capacity.

The system built in 1996 treated the wastewater from the flushing of the milk lines (200 L/d), the milk cooling tank (200 L/d), the cleaning water of the milking parlor (approx. 100 L/d and on average about 450 L/d of domestic wastewater from the farmers and visitor groups (mainly toilet use). In total about 1,500 L/d of wastewater was treated in the system. Pretreatment occurred in separate septic tanks for the dairy wastewater and the domestic wastewater.

The mixed waste water was pumped towards the treatment wetland. As a phosphorus binding agent, 5.4 tons of steel slag were added to the sand, which led to similar problems of formation of a hard crust in the substrate and consequently a lower hydraulic conductivity of the filter and ponding of waste water on the filter surface. The hard crust of the steel slag and sand mixture was broken here with a small excavator and put back into the filter substrate. After that the hydraulic conductivity was

Table 9.8 Monitoring data (influent, effluent) from the PDLT wetland facility in Belgium (n = 21, period 1999–2000).

	BOD		COD		TKN		NH$_4$-N		TN	
	IN	Eff	IN	Eff	IN	Eff	IN	Eff	IN	Eff
Average	499	5	830	38	38	3.0	20	0.3	38.2	21.8
Sd	298	6	516	29	12	1.5	12	0.5	12.0	15.0
Removal (%)		99.1		95.4		92.1		98.7		42.9

	NO$_3$-N		NO$_2$-N		TP		PO$_3$-P		SS	
	IN	Eff	IN	Eff	IN	Eff	IN	Eff	IN	Eff
Average	0.6	18.8	0.1	0.1	27.2	5.8	22.1	5.2	230	6
Sd	0.3	15.7	0.0	0.0	11.2	2.8	10.6	2.6	214	8
Removal (%)		–		4.5		78.8		76.5		97.4

restored sufficiently to avoid ponding. The effluent of the wetland was led to a splitting chamber where 50% of the effluent was recirculated to the influent pump well in order to enhance denitrification.

The monitoring of the system was done by the Provincial Institute for Hygiene (PIH) for the VMM (Flemish Environmental Agency). For the period, between April 1999 and April 2000, two-weekly samples were taken and analyzed, both of influent and effluent of the system. The influent concentrations however are diluted by the recirculated effluent. The results are shown in Table 9.8.

Compared to the systems described before, the influent concentrations are lower because of the dilution with effluent. In order to be able to calculate the organic loads to the wetland, it is necessary to multiply the concentrations with twice the influent flow as this is the real daily flow entering the wetland, including the recirculated water. This gives a daily BOD-load of 1.5 kg BOD/d, a COD-load of 2.5 kg COD/d and N-load of 115 g/day. The specific surface loads are then 20 g BOD/m^2/d, 33 g COD/m^2/d and 1.5 g N/m^2/d. So, the loads are in between the previous two systems.

The effluent samples after 3 years of operation, show excellent results again for BOD, COD and SS that can be compared to the previous two systems. The TN removal is good at 42.9%, but the recirculation did not have a significant positive effect, if compared to the system at De Paep that operates without recirculation and shows an even better TN-removal rate than this system. TP removal is however much better than in the other systems described, so apparently after 3 years the steel slag still shows a substantial P-binding capacity with a TP removal rate of almost 79%. This is however lower than the removal rates shown for the Dutch systems in their first year. The Dutch systems however also show some saturation effect, so on the long term, iron seems not to be an ideal P-binding material in CWs.

9.3.4.4 Aerated Wetland (FBA) at PDLT, Geel in Belgium

In 2015, the VFCW was replaced by a new wetland, as a consequence of expansion of the PDLT milking installation. A second stable and two milking robots were to be installed, which increased the organic load towards the wetland towards approximately 3–4 times the current load. In order to limit the space requirements and costs of a new wetland, a Forced Bed Aeration (FBA®) aerated wetland of 120 m^2 was installed, with a design BOD load of 4.8 kg BOD/d. The aeration concept makes it thus possible to reduce the required surface area with a factor 2–3 compared to standard VCWs.

FBA is a concept that was developed by Scott Wallace (Naturally Wallace Consulting) in 1997 and has been implemented in a variety of mostly large-scale wetland projects in the UK and in North America. FBA systems rely on the very uniform injection of small quantities of air at low pressure, in saturated SSF CWs, to increase oxygen transfer. This precise control over air flow rates is necessary since aeration-induced hydrodynamic mixing in gravel bed reactors (such as subsurface flow wetlands) is much less than in open-tank systems such as activated sludge.

Rietland installed a specific variant of this concept, based on a two-stage concept, where the first stage has 2/3 of the total surface area and is of the saturated VFCW type and the second stage at 1/3 of the total surface area is of the HSF CW type (Figure 9.2). Both beds have a fixed water table with a height of 1 m inside the substrate layer of 1.1 m. The substrate consists of LECA (brand name Argex) with a grain size of 8–16 mm. Both cells are aerated by means of a grid of perforated pipes at the bottom of the wetland cells. The air is forced into the aeration grid by means of two regenerative blowers. The blowers run 50% of the time and are switched by means of a timer, to warrant both aerobic and anaerobic circumstances in the cells. As a post-treatment stage for phosphorus removal, three basins filled with apatite were installed.

At the time of writing, the stables at PDLT have not been finished completely yet, so only one of the two milking robots is in use. Next to the dairy waste water from this milking robot and the cleaning water of the milking parlor, the sanitary wastewater is discharged to the system as before. Once the second milking robot will be installed, a sampling campaign will be started. One preliminary sample has already been taken after the system was in operation for three months. The results are shown in Table 9.9.

The aerated wetland shows excellent results so far with removal rates for all values above 93%. Especially the TN removal rate is noteworthy and probably a consequence of aerobic and anaerobic

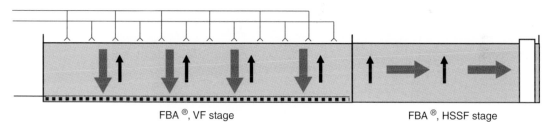

FBA ®, VF stage   FBA ®, HSSF stage

Figure 9.2 Two-stage FBA® system. Blue arrows indicate the water flow and the red arrows the air flow.

Table 9.9 First results of the aerated wetland at PDLT on 15/12/2015.

	BOD			COD			TKN			TN	
IN	Eff	%	IN	Eff	%	IN	Eff	%	IN	Eff	%
1,370	3	99.8	2,190	10	99.8	68	2	97.1	68.6	4.2	93.9
	NO$_3$-N			NO$_2$-N			SS			TP	
IN	Eff	%	IN	Eff	%	IN	Eff	%	IN	Eff	%
0.5	2.1	–	0.1	0.1	–	230	2	99.1	34.2	0.15	99.6

operation in combination with a relatively high C/N ratio that is common for dairy waste water. As the wetland is not loaded yet to the full capacity, the results in the above are however to be treated with caution.

9.4 Experiences from Greece

9.4.1 First Experimental Project

The first experimental investigation in Greece relates to four pilot-scale HSF CWs (Figure 9.3), which were constructed using plastic trapezoidal tanks (units) with dimensions 1.26 m length, 0.68 m width (upper base), 0.73 m depth, and a total volume of 0.62 m^3 [48–50]. The units were filled with fine

Figure 9.3 Pilot-scale HSF CWs for the treatment of Cr(VI) and SCW (upper two photos); perforated inflow pipes (diffuser) (bottom left photo); outlet tube of the CWs (bottom right photo).

gravel (D_{50} = 6mm). The four pilot-scale CWs units operated for two years (1st operational period) in order to treat either SCW or tap water spiked with Cr(VI). Two of the pilot-scale HSF CWs were used during 1st operational period to treat SCW. One was planted (SCW-P) with common reeds (*Phragmites australis*) and the other was kept unplanted (SCW-U). A series of HRTs (i.e., 8, 4, 2 and 1 day) were studied to examine the effect of HRT COD removal. Influent COD concentrations ranged from 1,200 to 7,200 mg/L in order to also examine the effect of pollutant load. The other two pilot-scale HSF CWs were used during the 1st operational period to treat tap water spiked Cr(VI) and, again, one was planted (Cr-P) with common reeds (*Phragmites australis*) and the other was kept unplanted (Cr-U).

During the 2nd operational period, tap water enriched with Cr(VI) and SCW was introduced into the four CW units. HRTs and influent COD and Cr(VI) concentrations were similar to those used in 1st operational period [48, 50]. Two HRTs were applied (i.e., 4 and 8 days). Influent COD concentrations ranged between 1,300–4,100 mg/L in the SCHW-P and SCHW-U units, and 1,800–4,000 mg/L in the Cr-P and Cr-U units. In addition, Cr(VI) influent concentrations ranged between 0.4–5 mg/L for the SCHW-P and SCHW-U units and 2.2–5.5 mg/L for the Cr-P and Cr-U units. The four pilot-scale units operated co-treating SCHW and Cr(VI) for seven months. Evapotranspiration (ET) was assessed on a daily basis. Precipitation levels and influent and effluent volumes were measured daily. During days with high solar radiation and temperatures, ET exceeded the wastewater influent volumes, leading to a reduction in the units' water level, as ET values reached 15.5 L/d. Table 9.10 presents the experimental results for the two operational periods.

Pilot-scale CWs during the 1st operational period removed substantial organic matter load, as COD removal efficiencies were on average 76.4% and 71% for the SCW-P and SCW-U units, respectively, although organic load was high (19.4–770 g/m²/d for the SCW-U and from 11.4–620 g/m²/d for the SCW-P) [50]. The pilot-scale CWs had a stable performance during the whole year and only during HRT alterations or vegetation harvesting the COD removal was decreased [50].

One-way ANOVA showed no significant effect on COD removal for HRTs between 2 to 8 days ($p > 0.05$ for both SCW-P and SWH-U units). When the HRT of 1 day was applied, COD removal efficiency was significantly reduced ($p < 0.05$ for both SCW-P and SWH-U units) [50]. Nevertheless, the pilot-scale CWs showed extremely high COD removal efficiencies (up to 100%), while receiving one of the highest COD influent concentrations (1,200–7,200 mg/L) and operating under the lowest HRT (i.e., 2 days) ever reported in the literature [50].

The plant effect on CWs performance is a controversial issue, as many researchers argue that vegetation is not significant. Small differences occurred between the SCW-P and SCW-U units, as mean COD removal efficiencies were 83% for the SCW-P and 70% for the SCW-U [50]. A paired *t*-test, performed to determine statistical significance at the 95% confidence level, showed that vegetation significantly affected COD removal [50].

During the 2nd experimental period, all four pilot-scale units co-treated SCW and Cr(VI). Concerning Cr(VI), all pilot-scale CW units achieved constantly complete removal (100% removal rate). On the contrary Cr-U unit during the 1st experimental period achieved extremely low Cr(VI) removal rates (approximately 21%) [49]. This indicates that Cr(VI) removal was achieved not only through absorption by plant biomass but also by microbial activity, since the microorganisms developed adequately due to the presence of nutrients in the SCW. Moreover, the acidic nature of the SCW [16] promoted the growth of filamentous biomass [8]. Most microorganisms grow best at neutral pH (7.0) [51]. Bacterial metabolic activity increases when pH range is between 3.5 and 5.5 [52], but it is reported [53] that bacterial growth generally occurs in neutral to slightly acidic conditions. In

Table 9.10 Physico-chemical characteristics of co-treated wastewaters (mixed) cheese whey and Cr(VI) solution [48–50].

		Cr-P	Cr-U	SCW-P	SCW-U
1st operational period					
pH	Influent	7.49 ± 0.19	7.46 ± 0.20	5.29 ± 1.1	5.41 ± 1.1
	Effluent	6.76 ± 0.23	7.46 ± 0.23	6.26 ± 0.2	6.35 ± 0.3
EC (μS/cm)	Influent	356 ± 42	360 ± 49	894 ± 315	901 ± 321
	Effluent	446 ± 118	357 ± 53	$1{,}907 \pm 383$	$1{,}885 \pm 620$
COD (mg/L)	Influent	–	–	$3{,}273 \pm 958$	$3{,}300 \pm 1{,}060$
	Effluent	–	–	516 ± 668	$1{,}132 \pm 1{,}334$
Cr (VI) (mg/L)	Influent	2.9 ± 2.2	2.9 ± 2.1	–	–
	Effluent	0.3 ± 0.6	2.4 ± 1.8	–	–
2nd operational period					
pH	Influent	7.04 ± 0.44	7.04 ± 0.45	6.9 ± 0.53	6.95 ± 0.52
	Effluent	6.66 ± 0.47	6.98 ± 0.42	6.93 ± 0.49	6.5 ± 0.45
EC (μS/cm)	Influent	612 ± 426	618 ± 410	651 ± 80	660 ± 69
	Effluent	564 ± 452	470 ± 351	724 ± 456	$1{,}054 \pm 422$
COD (mg/L)	Influent	$2{,}910 \pm 630$	$2{,}950 \pm 660$	$3{,}240 \pm 550$	$3{,}250 \pm 540$
	Effluent	990 ± 570	$1{,}240 \pm 790$	955 ± 620	$1{,}660 \pm 890$
Cr (VI) (mg/L)	Influent	2.5 ± 5.5	2.5 ± 5.5	0.5 ± 5	0.5 ± 5
	Effluent	0 ± 0	0 ± 0	0 ± 0	0 ± 0

this study, the mean pH value was found to be near neutral, therefore the pH promoted microbial growth and participated in Cr(VI) removal mechanisms. The gravel bed system also assisted Cr(VI) removal by various mechanisms including co-precipitation, sorption and ion exchange [54]. Additionally, it is reported that gravel media dominates Cr removal through sorption to iron oxides and co-precipitation of iron phases with available Cr [55].

On the other hand, COD removal rates (Cr-P: 66%; Cr-U: 58%; SCW-P: 71%; SCW-U: 48%) were lower than those of Cr(VI) in all units. To compare COD removal rates between the two operational periods, specific time periods were selected when the average temperature, the HRT and the inlet COD concentration were almost identical. One-way ANOVA was performed for the SCW-P and SCW-U units, which showed that COD removals between the two periods were not statistically significant ($p > 0.05$) for both pilot units. This result suggests that the combined wastewater of cheese whey and Cr(VI) can be treated in CWs and that organic matter removal rates are substantial, even in the presence of Cr(VI). Although there were no significant variations between the two experimental periods, the COD removal rate was reduced in both units. This reduction might be due to the addition of Cr(VI) to the SCW solution. As a result, the effluent COD values were relatively high, but the removal rates remained satisfactory in the systems fed with the combined wastewater.

The toxicity of Cr(VI) in the dairy wastewater could affect the COD removal efficiency. An increased concentration of Cr(VI) $\geq$ 5 mg/L decreases the COD removal [56]. It is reported that 1–3 mg/L of Cr(VI) did not have any significant impact on COD removal, but reductions in removal were observed when Cr(VI) concentrations were greater than 5 mg/L [56]. In this case, the reduction in COD removal was less than 10% at concentrations of 5 mg Cr(VI)/L. COD removal did not change under shock loading conditions with a concentration of 5 mg/L of Cr(VI) [56]. Moreover, little to no impact on microorganisms was observed during the first day of shock loading with 5 mg/L and 10 mg/L Cr(VI) [56]. Shock loading of 5 mg/L and 10 mg/L Cr(VI) also decreased COD removal efficiency in the second day, and the CWs required 3 and 5 days, respectively, to recover to a steady state condition [57]. As the concentration of Cr(VI) increased, a more pronounced effect of Cr(VI) on microorganisms was observed in this study.

Heavy metals like Cr(VI) are highly toxic substances and have toxic effects on microorganisms, because they tend to adhere to microorganisms and produce complexes with bacteria [58]. Metal ion complexion with heavy metals affects the bacterial growth and can lead to their death. Cr and Ni were found to inhibit microbial activity and negatively affect COD removal efficiency at concentrations above 10 mg/L [59, 60]. Thus, results showed that wastewater containing trace levels of heavy metal ions within the effluent discharge limits may affect COD removal efficiency at short HRTs. Moreover, at trace amounts, heavy metal ions act as a stronger competitor and reduce COD adsorption on the bioflocs [58].

The average COD removal efficiency was 71% for the SCW-P and 48% for the SCW-U (Table 9.2). SCW-P showed over 50% removal rate of COD, when the input COD concentration ranged between 2,000–5,000 mg/L for both HRTs (8 and 4 days). Both SCW-P and SCW-U units presented slightly higher COD removal rates (74.8% for the SCW-P and 51.5% for the SCW-U) during the first operational period [50]. These differences in COD removal could be attributed to the possible toxic effects of Cr(VI) to the microorganisms that removed the organic matter. Cr-P and Cr-U presented lower COD removal rates compared to SCW-P and SCW-U (66% for the Cr-P and 58% for the Cr-U). One-way ANOVA showed that the COD removal rates for the SCW-P and SCW-U units were not significantly different ($p > 0.05$), which was also the case for the Cr-P and Cr-U units ($p > 0.05$).

Furthermore, no significant variation in COD removal between the planted and unplanted units even in the presence of Cr(VI) was found, as COD removal in all units was around 65%. However, in the first experimental period [48–50], the SCW-P and SCW-U units showed significantly higher COD removal rates (almost 100%) when treated only with SCW wastewater. In the second experimental period, Cr(VI) was added to the SCW wastewater and the SCW-P and SCW-U units showed lower COD removal rates, even with moderate Cr(VI) loadings. In the Cr-P and Cr-U units, a reasonable COD removal rate was observed when SCW was first introduced into these two units.

To assess the effect of HRT on COD removal, two different HRTs were applied in the pilot units, i.e., 8 and 4 days. In addition, two different COD concentrations were used as influent (2,000 to 3,000 mg/L) in order to examine the removal of COD. The different HRTs and influent COD concentrations affected the COD surface loading rate (SLR), which ranged from 34.2–128.3 g/m^2/d for the SCW-P, 38.5–120.2 g/m^2/d for the SCW-U, 31.6–122.8 g/m^2/d for the SCW-P and 24.2–87.8 g/m^2/d for the SCW-U units. Significant difference in COD removal was recorded in all units with moderate surface COD loading at both HRTs (i.e., 8 and 4 days). Thus, it can be concluded that HSF CWs can treat the combined wastewater by removing organic matter at low residence times, i.e. up to 4 days, even in the presence of heavy metals such as Cr(VI).

9.4.2 Second Experimental Project

The second experiment involved four pilot-scale HSF CWs, constructed and placed outdoors in Western Greece [61]. Each unit was a HDPE tank (72 cm length, 33 cm width, 35 cm depth and surface area of 0.24 m^2). Two-thirds of the length of two units (i.e., 48 cm) were filled with igneous fine gravel obtained from a local river bed (D$_{50}$ = 6mm, specific surface area = 3,075 m^2/m^3, porosity = 35%) and the remaining one third of their length (i.e., 24 cm) was filled with natural zeolite (D$_{50}$ = 4 mm, specific surface area = 4,370 m^2/m^3, porosity = 25%). One of these units (GZ-P) was planted with common reeds (*Phragmites australis*) obtained from local streams and the other was kept unplanted (GZ-U). Two thirds of the length (i.e., 48 cm) of the other two pilot-scale units were filled with random type high density polyethylene (HDPE), with a specific area of 188 m^2/m^3, porosity of 95%, specific weight of 47 kg/m^3 and piece dimensions 5.7 × 5.5 mm (Figure 9.4). The remaining one third (i.e., 24 cm) of the length was also filled with natural zeolite (D$_{50}$ = 6mm). One unit was kept unplanted (PZ-U) and the other was planted (PZ-P) with common reeds (*Phragmites australis*). To achieve rapid vegetation growth, three reed stems were initially planted (i.e., 12 reeds/m^2).

Figure 9.4 Photo of the HDPE substrate material in the pilot wetland units.

SCW was used as the influent source, obtained from the effluent of a full-scale biological filter, operated in a nearby dairy factory. The raw SCW's COD concentration ranged between 40,000–50,000 g/L, while COD concentration in the effluent of the biological filter (i.e., influent to the CWs) varied between 1,000–5,000 mg/L [18, 62]. The four CW units operated one year with a HRT of 4 days and for another year with a HRT of 2 days. Organic surface load rates (SLR) were on average 66.6, 70.1, 231 and 229 $g/m^2/d$ for the GZ-P, GZ-U, PZ-P and PZ-U units, respectively [61].

Influent COD concentrations to the CWs varied between 120–6,135 mg/L, since the quality of the pretreated SCW was not stable. Mean COD removal efficiencies for the two-year operation period were 83%, 79%, 79% and 72% for the GZ-P, GZ-U, PZ-P and PZ-U units, respectively [61]. The majority of organic matter (i.e., 60–70%) was removed within the first 2/3 of the length of all pilot CWs. This is in agreement with results from other studies [63, 65, 66]. Organic matter removal further continued in the last part of the units containing zeolite, indicating the ability of zeolite to remove organics [67, 68]. The units with gravel and the plastic material showed similar COD removal rates, although the ones with the plastic material received much higher organic loads.

The first section of the CW units containing either gravel or plastic material removed only a small portion of NH_4+-N, with respective removal efficiencies of 39%, 48%, 28% and 29% for the GZ-P, GZ-U, PZ-P and PZ-U units. Ammonia nitrogen was further removed in the last section of the unit containing zeolite (up to 75%) [61]. Phosphorus was also gradually removed along all pilot-scale units (approximately 65% in all units) [61].

ANOVA showed that HRTs of 4 and 2 days did not significantly affect COD removal ($p > 0.05$ for all units) [61]. This result confirms previous findings [49]. It should be highlighted that the CWs units in this experiment, although receiving one of the highest surface organic loads reported in the literature for HSF CWs [64], managed to achieve very high COD removal efficiencies (up to 100%). Ammonium and phosphorus removal were also not affected by HRT ($p > 0.05$).

The main novelty of the experimental setup is the CW design and the use of the plastic material as substrate. This specific material was chosen to overcome the limitations of common CW substrate, such as low porosity and clogging. The main goal is to increase the porosity (95% for the plastic media, while only 35% for the gravel) [61]. Results showed that, although the CW units containing the plastic material received three times higher HLR, they showed similar or even lower effluent COD concentrations than the units containing gravel. ANOVA confirmed this finding, since removal rates were not significantly affected ($p > 0.05$) by the substrate material.

In summary, the units with the plastic substrate material treated three times higher hydraulic load and four times higher organic load, compared to the units with the gravel media. This means that four times smaller area is required for the CWs containing the plastic medium without change in their overall performance. Thus, the use of plastic material as substrate reduces the required footprint of the wetland by 75% and limits the clogging potential, providing an additional advantage to the use of wetland technology for wastewater treatment [61].

9.5 Conclusions

The treatment of dairy wastewater represents itself a challenge due to the contained high pollutants loads. Experiences from both experimental and full-scale applications from three different countries (Belgium, the Netherlands and Greece) show that the green technology of Constructed Wetlands can be effectively applied for the treatment of these agro-industrial wastewaters.

In Belgium and the Netherlands, sand-filled VFCWs have proven to be highly efficient wastewater purification systems, with very little maintenance demands and relatively little surface area required. A design loading rate of 20 g BOD/m^2/d is recommended. At this loading rate, the systems show excellent removal rates for BOD/COD and organic N with percentages well above 90%, that stay consistently high over a period of at least up to 17 years as demonstrated. When used for dairy wastewater, the VFCWs also show reasonable TN removal of 40–60%. Adding a pond or FWS system can enhance the TN removal. The TP removal, especially when adding iron to the substrate, is initially high, but decreased over time due to saturation of the iron. Aerated wetlands seem to be a promising, economical alternative, but the available data on treatment of dairy waste water with aerated wetlands is limited so far.

Research activities from Greece show the good overall efficiency of HSF CW systems for the treatment of cheese whey, when used in combination with upstream biological filters. The use of a recycled plastic material for the first time in a CW system seems a novel design modification that can reduce the required footprint by up to 75%, without altering the performance of the system. In this way, a three times higher hydraulic load and a four times higher organic load can be applied to the HSF CW, compared to a traditional CW system containing gravel media.

References

1 Belyea RL, Williams JE, Gieseka L, Clevenger TE, Brown JR. Evaluation of dairy waste water solids as a feed Ingredient. J Dairy Sci. 1990; 73:1864–1871.
2 Vourch M, Balannec B, Chaufer B, Dorange G. Treatment of dairy industry wastewater by reverse osmosis for water reuse. Desalination. 2008; 219:190–202.
3 Smithers GW. Whey and whey proteins – From "gutter-to-gold". Int Dairy J. 2008; 18:695–704.
4 OECD–FAO. Agricultural Outlook 2008–2017. Highlights. Paris. Accessed on 5 September 2012. Available at www.agrioutlook.org/dataoecd/ 54/15/40715381.pdf.
5 EC. Communication from the European Commission to the Council 2009/385/EC of 22 July 2009 on Dairy Market Situation, 2009.
6 Farizoglu B, Keskinler B, Yildiz E, Nuhoglu A. Cheese whey treatment performance of an aerobic jet loop membrane bioreactor. Process Biochem. 2004; 39:2283–2291.
7 Wit JN de. Lecturer's Handbook on whey and whey products. European Whey Products Association (EWPA) 14, Rue Montoyer 1000 Brussels, Belgium 2001.
8 Ghaly AE, Kamal MA. Submerged yeast fermentation of acid cheese whey for protein production and pollution potential reduction. Water Res. 2004; 38:631–644.
9 Chatzipaschali AA, Stamatis AG. Biotechnological utilization with a focus on anaerobic treatment of cheese whey: current status and prospects. Energies. 2012; 5:3492–3525.
10 Prazeres AR, Carvalho F, Rivas J. Cheese whey management: A review. J Environ Manage. 2012; 110:48–68.
11 Fox PF. Milk proteins: general and historic aspects. Fox PF, Mc Sweeney PLH. Advanced Dairy Chemistry – 1, Proteins, 3rd edn. New York: Kluwer Academic/Plenum Publishers; 2003, pp. 5–55.
12 Papachristou E, Lafazanis CT. Application of membrane technology in the pretreatment of cheese dairies wastes and co-treatment in a municipal conventional biological unit. Water Sci Technol. 1997; 32:361–367.

13 Panesar PS, Kennedy JF, Gandhi DN, Bunko K. Bioutilisation of whey for lactic acid production. Food Chemistry. 2007; 105:1–14.

14 Siso MIG. The biotechnological utilization of cheese-whey: A review. Bioresour Technol. 1996; 57:1–11.

15 Kushwaha JP, Srivastava VC, Mall ID. An overview of various technologies for the treatment of dairy wastewaters. Crit Rev Food Sci. 2011; 51:442–452.

16 Carvalho F, Prazeres AR, Rivas J. Review, Cheese whey wastewater: Characterization and treatment. Sci Total Environ. 2013; 445–446:385–396.

17 Demirel B, Yenigun O, Onay TT. Anaerobic treatment of dairy wastewaters: a review. Process Biochem. 2005; 40:2583–2595.

18 Tatoulis TI, Tekerlekopoulou AG, Akratos CS, Pavlou S, Vayenas DV. Aerobic biological treatment of second cheese whey in suspended and attached growth reactors. J Chem Technol Biot. 2015; 90:2040–2049.

19 Browne W, Jenssen PD. Exceeding tertiary standards with a pond/reed bed system in Norway. Water Sci Technol. 2005; 51:299–306.

20 Gasiunas V, Strusevicius Z, Struseviciene MS. Pollutant removal by horizontal subsurface flow constructed wetlands in Lithuania. J Environ Sci Health. 2005; 40:1467–1478.

21 Munoz P, Drizo A, Hession WC. Flow patterns of dairy wastewater constructed wetlands in a cold climate. Water Res. 2006; 40:3209–3218.

22 Newman J, Clausen J, Neafsey J. Seasonal performance of a wetland constructed to process diary milk house wastewater in Connecticut. Ecol Eng. 2000; 14:181–198.

23 Moreira VR, LeBlanc BD, Achberger EC, Frederick DG, Leonardi C, Design and evaluation of a sequential biological treatment system for dairy parlor wastewater in southeastern Louisiana. App Eng Agric. 2010; 26:125–136.

24 Shamir E, Thompson TL, Karpiscak MM, Freitas RJ, Zauderer J. Nitrogen accumulation in a constructed wetland for dairy wastewater treatment. J Am Water Resour Ass. 2001; 37:317–325.

25 Tanner CC, Clayton JS, Upsdell MP. Effect of loading rate and planting on treatment of dairy farm wastewaters in constructed wetlands – I. removal of oxygen demand, suspended solids and faecal coliforms. Water Res. 1995; 29:17–26.

26 Dunne EJ, Culleton N, Donovan GO, Harrington R, Olsen AE. An integrated constructed wetland to treat contaminants and nutrients from dairy farmyard dirty water. Ecol Eng. 2005; 24:221–234.

27 Jamieson R, Gordon R, Wheeler N, Smith E, Stratton G, Madani A. Determination of first order rate constants for wetland treating livestock wastewater in cold climates. J Env Eng Sci. 2007; 6:65–72.

28 Schaafsma JA, Baldwin AH, Streb CA. An evaluation of a constructed wetland to treat wastewater from a dairy farm in Maryland, USA. Ecol Eng. 2000; 14:199–206.

29 Lee MS, Drizo A, Rizzo DM, Druschel G, Hayden N, Twohig E. Evaluating the efficiency and temporal variation of pilot-scale constructed wetlands and steel slag phosphorus removing filters for treating dairy wastewater. Water Res. 2010; 44:4077–4086.

30 Rousseau D, Vanrolleghem PL, Pauw ND. Constructed wetlands in Flanders: a performance analysis. Ecol Eng. 2004; 23:151–163.

31 Farnet AM, Prudent P, Ziarelli F, Domeizel M, Gros R. Solid-state 13C NMR to assess organic matter transformation in a subsurface wetland under cheese–dairy farm effluents. Bioresourc Technol. 2009; 100:4899–4902.

32 Ghosh D, Gopal B. Effect of hydraulic retention time on the treatment of secondary effluent in a subsurface flow constructed wetland. Ecol Eng. 2010; 36:1044–1051.

33 Ibekwe AM, Grieve CM, Lyon SR. Characterization of microbial communities and composition in constructed dairy wetland wastewater effluent. Appl Environ Microbiol. 2003; 69:5060–5069.

34 Kern J, Idler C, Carlow G. Removal of fecal coliforms and organic matter from dairy farm wastewater in a constructed wetland under changing climate conditions. J Environ Sci Health. 2000; 35:1445–1461.

35 Mantovi P, Marmiroli M, Maestri E, Tagliavini S, Piccinini S. Application of horizontal sub-surface flow constructed wetland on treatment of dairy parlor wastewater. Bioresourc Technol. 2003; 88:85–94.

36 Moir SE, Svoboda I, Sym G, Clark J, McGechan MB, Castle K. An experimental plant for testing methods of treating dilute farm effluents and dirty water. Biosyst Eng. 2005; 90:349–355.

37 Mustafa A, Scholz M, Harrington R, Carroll P. Long-term performance of a representative integrated constructed wetland treating farmyard runoff. Ecol Eng. 2009; 35:779–790.

38 Newman JM, Clausen JC. Seasonal effectiveness of a constructed wetland for processing milk-house wastewater. Wetlands. 1997; 17:375–382.

39 Smith E, Gordon R, Madani A, Stratton G. Cold climate hydrological flow characteristics of constructed wetlands. Canadian Biosyst Eng. 2005; 47:1–7.

40 Gottschall R, Boutin, C, Crolla A, Kinsley C, Champagne P. The role of plants in the removal of nutrients at a constructed wetland treating agricultural (dairy) wastewater Ontario, Canada. Ecol Eng. 2007; 29:154–163.

41 Comin FA, Romero JA, Astorga V, Garcia G. Nitrogen removal and cycling in restored wetlands used as filters of nutrients for agricultural runoff. Water Sci Technol. 1997; 35:255–261.

42 Greenway M, Woolley A. Changes in plant biomass and nutrient removal over 3 years in a constructed free water surface flow wetland in Cairns Australia. 7th International Conference on Wetland Systems for Water Pollution Control. Lake Buena Vista, Florida, 2000, pp. 707–718.

43 Hansen JI, Anderson FO. Effects of *Phragmites australis* roots and rhizomes on redox potentials, nitrification and bacterial numbers in the sediment. 9th Nordic Symposium, Malma, Sweden, 1981, pp. 72–88.

44 Reddy KR, Patrick WH, Lindau CW. Nitrification–denitrification at the plant root–sediment interface in wetlands. Limnol Oceanogr. 1989; 34;1004–1013.

45 Tiedje JM. Ecology of denitrification and dissimilatory nitrate reduction to ammonium In: Zehnder AJB, ed. Biology of Anaerobic Microorganisms. New York: John Wiley & Sons; 1988, pp. 179–244.

46 Dipu S, Kumar AA, Thanga VSG. Phytoremediation of dairy effluent by constructed wetland technology. Environmentalist. 2011; 31:263–278.

47 Bahlo K, Wach G. Naturnahe Abwasserreinigung, Planung und Bau von Pflanzenkläranlagen, Ökobuch, Verlag, 1992, ISBN 3-922964-52-4.

48 Sultana M-Y, Chowdhury AKMMB, Michailides MK, Akratos CS, Tekerlekopoulou AG, Vayenas DV. Integrated Cr(VI) removal using constructed wetlands and composting. J Haz Mat. 2015; 281:95–105.

49 Sultana M–Y, Tatoulis T, Akratos CS, Tekerlekopoulou AG, Vayenas DV. Effect of operational parameters on the performance of a horizontal subsurface flow constructed wetland treating secondary cheese whey and Cr(VI) wastewater. Int J Civ Str Eng. 2015; 2:286–289.

50 Sultana M-Y, Mourti C, Tatoulis T, Akratos CS, Tekerlekopoulou AG, Vayenas DV. Effect of hydraulic retention time, temperature, and organic load on a horizontal subsurface flow constructed wetland treating cheese whey wastewater, J. Chem. Tech. Biot. 2016; 91:726–732.

51 Hamad SH. Factors Affecting the Growth of Microorganisms. In: Bhat R, Karim A, Paliyath G, eds. Food. In Progress in Food Preservation. Oxford: Wiley-Blackwell; 2012.

52 Cimino G, Caristi C. Acute toxicity of heavy metals to aerobic digestion of waste cheese whey. Biol Wastes. 1990; 33:201–210.

53 Beaubien A, Jolicoeur C. The toxicity of various heavy metal salts, alcohols and surfactants to microorganisms in a biodegradation process: a flow microcalorimetry investigation. In: Liu D, Dutka BJ, eds. Toxicity Screening Procedures Using Bacterial Systems. New York: Marcel Dekker; 1984, pp. 261–281.

54 Marchand L, Mench M, Jacob DL, Otte ML. Metal and metalloid removal in constructed wetlands, with emphasis on the importance of plants and standardized measurements: A review. Environ Pollut. 2010; 158:3447–3461.

55 Dotro G, Palazolo P, Larsen D. Chromium fate in constructed wetlands treating tannery wastewaters. Water Environ Res. 2009; 81:617–625.

56 Stasinakis AS, Thomaidis NS, Mamais D, Papanikolaou EC, Tsakon A, Lekkas TD. Effects of chromium (VI) addition on the activated sludge process. Water Res. 2003; 37:2140–2148.

57 Ertugrul T, Berktay A, Nas B. Influence of salt and Cr(VI) Shock loadings on oxygen utilization and COD removal in SBR. Env Eng Sci. 2006; 23:1055–1064.

58 Sin SN, Chua H, Lo W, Yu PHF. Effects of trace levels of copper, chromium, and zinc ions on the performance of activated sludge. Appl Biochem Biotechnol. 2000; 84–86:487–500.

59 Suthirak S, Sherrard JH. Activated sludge Nickel toxicity studies. J Water Poll Control Federation. 1981; 53:1314–1332.

60 Dilek FB, Gokcay CF, Yetis U. Combined effects of Ni(II) and Cr(VI) on activated sludge. Water Res. 1998; 32:303–312.

61 Tatoulis T, Akratos CS, Tekerlekopoulou AG, Vayenas DV, Stefanakis AI. A novel horizontal subsurface flow Constructed Wetland: reducing area requirements and clogging risk. Chemosphere. 2017; 186:257–268, doi: 10.1016/j.chemosphere.2017.07.151

62 Tatoulis TI, Tekerlekopoulou AG, Akratos CS, Vayenas DV. Genesis and diaspora of the dairy process: Aerobic biological treatment of its wastewaters. In: Proceedings, IWA Regional Symposium on Water, Wastewater and Environment: Traditions and Culture, 21–24 March, Patras, 2014.

63 Akratos CS, Tsihrintzis VA. Effect of temperature, HRT, vegetation and substrate material on removal efficiency of pilot-scale horizontal subsurface flow constructed wetlands. Ecol Eng. 2007; 29:173–191.

64 Sultana MY, Akratos CS, Pavlou S, Vayenas DV. Constructed wetlands in the treatment of agro-industrial wastewater: A review. Hemijska Industrija. 2015; 69:127–142.

65 Herouvim E, Akratos CS, Tekerlekopoulou AG, Vayenas DV. Treatment of olive mill wastewater in pilot-scale vertical flow constructed wetlands. Ecol Eng. 2011; 37:931–939.

66 Stefanakis AI, Seeger E, Dorer C, Sinke A, Thullner M. Performance of pilot-scale horizontal subsurface flow constructed wetlands treating groundwater contaminated with phenols and petroleum derivatives. Ecol Eng. 2016; 95:514–526.

67 Stefanakis AI, Akratos CS, Gikas GD, Tsihrintzis VA. Effluent quality improvement of two pilot-scale, horizontal subsurface flow constructed wetlands using natural zeolite (clinoptilolite). Microp Mesop Mat. 2009; 12:131–143.

68 Stefanakis AI, Akratos CS, Tsihrintzis VA. Vertical Flow Constructed Wetlands: Eco-engineering systems for wastewater and sludge treatment, 1st edn. Oxford: Elsevier; 2014.

10

The Performance of Constructed Wetlands for Treating Swine Wastewater under Different Operating Conditions

Gladys Vidal, Catalina Plaza de Los Reyes and Oliver Sáez

Engineering and Biotechnology Environmental Group (GIBA-UDEC), Environmental Science Faculty and Center EULA–Chile, University of Concepción, Concepción-Chile

10.1 Introduction

10.1.1 The Swine Sector and the Generation of Slurries

Pork production is led globally by China, the European Union and the United States, with Chile accounting for less than 1% of global production. The high demand for pork in Asian countries (South Korea, Japan and China) continues to be most important focus for exports, including from Chile. Because of this, the pork sector in Chile is becoming more important. Chile now ranks sixth among producing countries globally and second among countries in Latin America after Brazil, according to the US Department of Agriculture (USDA). This has resulted in increased quantity of swine slurries, which is the main residue of the pork industry. Slurry management varies from the productive process to final disposition through a system of appropriate treatment.

10.1.2 Characterization of Slurries

Swine slurries are characterized by a mixture of feces and urine. The physiochemical composition of slurries is heterogeneous owing to the high degree of variation in diet, stabling systems, wash water management, accumulation time and type of storage [1]. Table 10.1 shows the generation and characterization of pig slurries considering 40 parameters, 125 citations, and close to 7,000 to 14,000 data entries, taking into consideration principally data for swine at the fattening stage [2].

The characterization highlights the generation of urine (43.0 ± 15.0 kg/day/1000 kg live weight) and feces (89 ± 30 kg/day/1000 kg live weight). As well, slurries have high content of solids (e.g., total: 12.6–42.7 g/L; volatiles: 7.8–23.9 g/L), organic matter (e.g., chemical oxygen demand or COD: 16.1–56.2 g/L; biochemical oxygen demand or BOD_5: 3.1–26.3 g/L), electrical conductivity (8.4–18.7 mS/cm) and nutrients (e.g., total nitrogen (TN): 1.5–5.2 g TN/L; ammonium (NH_4^+-N): 0.9–4.3 g NH_4^+-N/L) and total phosphorous (TP): 0.5–1.3 g TP/L), among others [3–5].

Other specific compounds, such as heavy metals (e.g., mean Cu^{+2} values in swine slurries are between 30 and 40 mg Cu^{+2}/L, and Zn^{+2} with 60 mg Zn^{+2}/L), hormones, antibiotics and others, which can have different environmental effects depending on the management and final disposal of these residues [3, 5].

Table 10.1 Physico-chemical characterization of pig slurry (n: number of studies evaluated by parameter) (modified from [2]).

	Data reporting mechanism							
	Generation rate (g/day/1000 kg live weight)				ppm wet basis (mg/kg wet total manure)			
Analyzed parameters	Median	Mean	SD	n	Median	Average	SD	n
Urine	45,000	43,000	15,000	10	–	–	–	–
Total waste	85,000	89,000	30,000	74	–	–	–	–
TS	8,900	9,600	4,600	50	100,000	120,000	55,000	71
VS	5,600	6,000	2,400	28	87,000	100,000	40,000	48
COD	7,500	7,700	2,900	38	89,000	99,000	34,000	46
TOC	2,200	2,100	160.0	4	25,000	39,000	35,000	7
BOD_5	2,200	2,400	1,200	24	32,000	34,000	9,600	31
Volatile acids	360	360	–	1	4,200	3,600	1,000	4
Alkalinity	ND	ND	ND	–	250	250	–	1
TSS	4,800	5,000	960	5	70,000	100,000	52,000	7
VSS	4,300	4,300	300	2	52,000	52,000	1,300	2
TKN	450	450	140	44	5,700	5,700	2,000	55
Ammonia-N	220	230	62	10	3,300	3,600	1,100	14
TP	140	150	56	38	1,700	1,900	740	48
Potassium	300	290	120	37	3,700	3,600	1,600	48
Calcium	360	380	230	10	2,600	3,400	1,400	21
Magnesium	64	64	25	11	800	800	170	22
Sodium	38	57	34	3	530	800	630	11
Sulfur	120	120	52	7	800	870	480	13
Chloride	ND	ND	ND	–	3,100	3,100	620	2
pH	–	–	–	–	7.3[a]	7.3[a]	0.8	2
Iron	25	22	9.5	8	180	200	78	14
Manganese	1.8	1.6	0.6	5	19	21	9.1	12
Boron	3.9	4	0.5	4	42	41	5.6	6
Molybdenum	0.05	0.05	0.04	2	0.2	0.33	0.4	5
Aluminum	4.6	4.6	–	1	49	45	5.4	3
Zinc	5.1	4.9	1.6	11	56	54	7.8	20
Copper	1.7	2.3	1.6	11	11.0	15	11	20
Cadmium	0.008	0.008	–	1	0.09	0.33	0.3	3
Lead	0.1	0.1	–	1	1.1	1	0.1	3
Cobalt	0.05	0.05	–	1	0.53	0.45	0.1	4
Arsenic	0.8	0.8	–	1	8.4	8.4	–	1
Total coliforms[b]	6.4E + 05	7.4E + 05	6.0E + 05	4.0E + 00	2.5E + 03	4.2E + 03	4.1E+03	4
Fecal coliforms[b]	4.3E + 05	4.3E + 05	2.3E + 05	2.0E + 00	2.8E + 03	9.6E + 03	1.4E+04	4
Fecal Streptococci[b]	1.8E + 06	1.8E + 06	1.5E + 06	2.0E + 00	8.4E + 04	5.1E + 04	4.1E+04	5
Total Streptococci[b]	9.0E + 09	9.0E + 09	–	1.0E + 00	2.3E + 08	2.3E + 08	–	1
Total Enterococcus[b]	4.5E + 08	4.5E + 08	–	1.0E + 00	5.5E + 03	4.7E + 03	3.4E + 03	3
Escherichia coli[b]	ND	ND	ND	–	1.0E + 02	1.0E + 02	–	1

[a] Standard units.
[b] ND = not determined.

10.1.3   Environmental Effects of the Application of Slurry in Soils

One of the most common ways of disposing of crude slurry is direct application to the soil. This generates a series of negative impacts on water, air and soil quality owing to concentrations of organic matter, nutrients, mineral salts, heavy metals and antibiotics, among others [6]. Compounds generated during animal production can enter the aquatic environment by lixiviation from slurry accumulation systems with low levels of impermeability, from overflows from during heavy rains, atmospheric deposition and from runoff from irrigated fields where slurry has been applied [7].

Slurries are rich in nitrogen and can generate problems of toxicity because of nitrates (NO_3^--N) in ground waters used for human consumption. One consequence is the blue baby syndrome or methemoglobinemia, which affects infants (under six months of age) that have consumed water with nitrate concentrations in excess of 10 mg/L [8]. It is estimated that 7% of wells for drinking water in the USA have been shut down because of nitrate contamination owing to agricultural activities, with an estimated 44,000 children being at risk [9]. There have been no cases of this disease reported in Chile to date. Nevertheless, nitrate concentrations in the range of 0.02–10.67 mg NO_3^--N/L have been found in wells in the Bio Bio Region, a highly agricultural zone.

There are also risks of contamination involving the transmission of diseases to human beings by parasites, like that produced by larvae of *Eristalis tenax*, which is classified as an agent of accidental myiasis [10]. Intensive swine production and the disposal of slurries in soil affect air quality and the atmosphere. Bad odors are the product of the diffusion of gases like NH_3, CO_2, H_2S, CH_4, N_2O, CO and COVs (amines, amides, carbonyls) and by bacterial action on different components of animal waste and the uncontrolled fermentation of residues [11].

10.1.4   Integrated Management for Treating Swine Slurry

The impact of slurries on the environment constitutes one of the main challenges for agriculture worldwide. Once dominated by small and medium scale operations as part of traditional agriculture, pig production is becoming highly concentrated. Animal production is generally separated from crop production, because of which the quantity of slurry produced often exceeds the local demand for its use as a fertilizer for crop production [12].

Proper slurry management seeks to use it as a nutritional source and as an amendment for crop soils. Treatment can be improved with biological, chemical and physical methodologies, above all in combination, as part of holistic systems that (1) are integrated to meet the needs of other agricultural activities; and (2) maximize the value of slurries through the production of energy and other beneficial sub-products, the concentration of nutrients, recycling, and the reduction of greenhouse gases. The challenge for many countries is the form of applying of these technologies in an economically feasible manner and at a wider scale [12].

Figure 10.1 shows developing technologies applied for the treatment of swine slurries. The figure shows the three main alternative foci for managing swine slurry. The main focus is the development of dry systems, such as warm beds in which fresh manure is mixed with filler or the use of inclined belts under the floor grating to separate solid waste from urine so that all the manure coming out of the system is in a solid state.

The second focus consists of improving or adapting existing liquid treatment systems so that volatile solids and organic nutrients can be separated from fresh manure in order to transport and/or treat manure with a variety of technologies to obtain products with added value. Solid-liquid separation

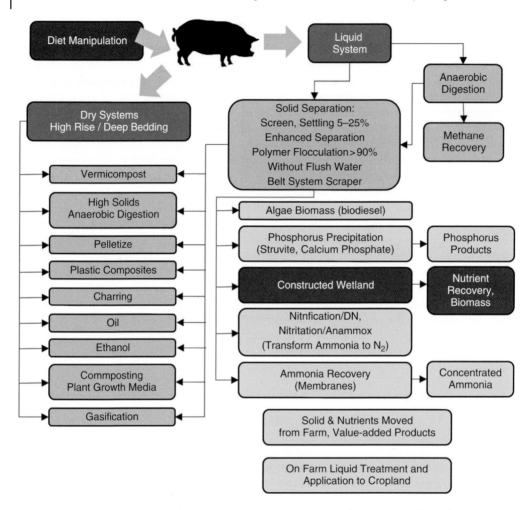

Figure 10.1 Treatment systems applied in the integrated management of pig slurry (modified from [12] and [13]).

of crude slurry increases the possible applications and widens the scope for decisions about its use. The initial separation allows for the recovery of organic compounds that can be used in producing compost, energy and other added-value products. These products include stabilized substitutes for peat, humus, biomass, organic manure, soil amendments and energy. The remaining liquid must be treated at the farm. A variety of biological, physical and chemical processes can be used to manage specific nutrient content and comply with environmental standards.

A third focus is the use of anaerobic digesters to recover carbon-based methane and energy in swine slurry [13]. Biogas recover systems obtain methane from slurry for subsequent burning to generate electricity or heat. Biogas production from manure using digesters is projected to have major importance globally. However, complementary treatment systems need to be developed in areas with intensive swine production to deal with excess nitrogen and to recover phosphorous from the effluents of anaerobic digesters in a form that can be removed from the environment.

10.1.5 Primary Treatment (Solids Removal)

Primary treatment of swine slurry refers to a physical process for separating solids from liquid manure to generate two distinct fractions, one solid and the other liquid. The solid fraction has a much higher concentration of solids (937.8 g TS/kg) than the original manure (62.2 g TS/kg), given that 93% of the composition is water, resulting in a liquid fraction with a solid concentration of 20.4 g TS/L [14].

Traditional systems for separating solids in suspension in slurries using sieves are not very efficient (0.0–50%). The best results have been obtained with filtering materials like sand or with centrifuge decanters (20–60%) [14]. However, the most efficient approaches to date involve the use of organic flocculants (polyacrylamide polymer) prior to mechanical separation. Successful systems have been developed with this technology to obtain solids with more than 50% dry matter and liquid with over 97% reduction in suspended solids and 84% reduction in BOD_5 [15, 16].

10.1.6 Secondary Treatment (Organic Matter Removal)

While the technologies applied during primary treatment can be highly efficient in removing suspended solids (40–60%), organic matter loads can remain high (3,000 mg COD/L) [17]. Organic matter can be removed by aerobic and anaerobic systems, but anaerobic systems are preferred in the pork industry because aerobic systems involve high energy costs for large-scale aeration required in removing high levels of organic matter and nutrients.

10.1.6.1 Anaerobic Treatment Systems

Several configurations have been developed for anaerobic digestion of swine slurry. Over several decades anaerobic digestion applied to swine slurry has proven to be technically viable and versatile in adapting to different working conditions: in large-scale operations with centralized management; plants in individual farms; simple gas recovery operations in covered pools [18]; treatment of the liquid fraction or of the solid fraction [19, 20]. The generic advantages of the process are well-known and the lines of research and development are directed at having a better understanding of the process at the microbiological level to increase the velocity of the process and biogas production and to improve control over the process, the energy balance and the economic balance and to integrate the process in a thorough treatment [21].

Anaerobic digestion is highly applicable to swine slurries owing to the high organic load (0.5 and 15.0 kg $COD/m^3/d$) [3, 22]. With anaerobic digestion it is possible to obtain energy, methane, stabilized residues, liquid fertilizers and soil conditioners. However, as Bonmati and Flotats [23] have noted, the economic viability of this technology is reduced by the high concentrations of ammonia nitrogen in swine slurries, which reduce hydrolysis and biogas production.

Table 10.2 shows the different anaerobic configurations used for this type of residue and their operational parameters. The yield of converted organic matter can be 77 to 86% of volatile solids, with methane production of 66.7 to 77%, depending on the configuration used.

10.2 Removal of Nutrients by Constructed Wetlands

Anaerobic processes have an efficiency of close to 80% in removing organic matter, depending on the type of system used [25]. However, the resulting effluent still has high nutrient concentration, in

Table 10.2 Anaerobic settings applied in the treatment of pig manure.

System (volume)	Influent	Temperature (T°)	OLR (kg VS/m³/d)	HRT (d)	VS or COD removal (%)	Production CH₄ (m³/kg VS added)	CH₄ (%)	Ref.
AL (7,200 m^3)	Slurry	13.5	0.0125	343	86.0% (VS)	0.26	66.7	[18]
UASB (2.6 L)	Manure from the pig	35	12.39	0.9–3.6	75.0% (COD)	–	77.0	[20]
CSRT (4.5 L)	Manure from the pig	37, 45, 55, 60	–	15	–	0.19, 0.14, 0.07, 0.02	–	[19]
UASB (0.2 L)	Slurry	37	–	10	82.0% (COD)	–	–	[24]

AL: Anaerobic Lagoon; UASB: Upflow Anaerobic Sludge Bed; CSRT: Continuous Stirred-Tank Reactor; OLR: Organic Loading Rate.

particular nitrogen. Consequently, a follow-up system is necessary to reduce nitrogen and recalcitrant high-molecular-weight organic matter still present.

Campos et al. [26, 27] noted that the COD/N ratio in the effluent determines which would be the most appropriate process to reduce nitrogen content. For a COD/N ratio > 20 the best method for nitrogen reduction is assimilation by heterotrophic bacteria. For a 20 > COD/N ratio of > 5 indicates that the mean route of nitrogen reduction is bacterial assimilation through conventional nitrification/denitrification. Finally, a COD/N ratio < 5 indicates that the nitrogen reduction is by means of nitrification/partial denitrification or nitrification/partial anammox.

Technically complex systems to reduce nitrogen are not viable at the small and medium scale owing to installation and maintenance. Consequently, unconventional alternative systems have been proposed for treating swine slurries, including constructed wetlands as an efficient and viable alternative at the small and medium scale [28].

10.2.1 Constructed Wetland (CW)

This type of system can be defined as an area that is permanently inundated by shallow or deep water with dense vegetation adapted to this technology [29, 30]. CWs can be considered complex bioreactors because of the varied physical, chemical and biological processes that occur among the microbial communities, aquatic macrophytes, soils and sediments [31, 32].

Constructed wetlands can have free-water surface (FWS), horizontal subsurface flow (HSSF) or vertical subsurface flow (VSSF) [33] as is showed in Table 10.3. The FWS system has been used for tertiary slurry treatment and the subject of several studies [34–38]. Elimination efficiencies are often 35–51% for TSS loads of 17–116 kg TSS/ha/d, 30–50% for organic matter loads (expressed as COD) of 34–291 kg COD/ha/d, 37–51% for TN loads of 2–51 kg TN/ha/d, and 13–26% for TP loads of 3–22 kg TP/ha/d [39].

Horizontal subsurface flow (HSSF) is the main mechanism for removing nitrogen in constructed wetlands via nitrification/denitrification. However, studies have shown that oxygen in the rhizome section prevent complete nitrification. Volatilization, adsorption and incorporation by plants play minor roles in eliminating nitrogen [40]. Vymazal [41] analyzed the applicability of this technology for different types of liquid residues and determined the mean percentages of TN eliminated from

Table 10.3 Operational characteristics for different systems of artificial wetlands.

Type	Pretreatment	Dimensions	Support medium	Macrophyte	NLR (kgTN/ha/d)	Efficiency TN (%)	Ref.
HS	AL	4 cells 3.6 × 33.5 × 1.0 m	Clay (0.3 m) Loamy sand (0.25 m)	*Sc, Ju, Ty*	3–40	>50%	[28]
HSS	AD	6 cells 4.0 × 1.0 × 1.0 m	Gravel (0.5 m)	*Ph, Ty, Pi, Ei*	22.4	74–78	[46]
HSS	ST	1.2 × 1.2 × 1.2 m	Gravel (0.1 m)	*Cy*	90	97–98	[47]
VS			Loamy sand (0.7 m)				

HS: Horizontal Surface Flow; HSS: Horizontal Subsurface Flow; VS: Vertical Subsurface Flow; AL: Anaerobic Lagoon; AD: Anaerobic Digester; ST: Sedimentation Tank; NLR: Nitrogen Loading Rate; TN: Total Nitrogen; *Sc: Scirpus* spp.; *Ju: Juncus* spp.; *Ty: Typha* spp.; *Ph: Phragmites* spp.; *Pi: Pistia* spp.; *Ei:Eichhornia* spp.; *Cy: Cyperus* spp.

domestic wastewater (39.4%), industrial wastewater (18%), agricultural runoff (51.3%), and lixiviated landfills (33%). Nitrogen is eliminated in the VSSF system through adsorption/harvests, nitrification/denitrification, volatilization and ionic exchange, the latter two processes making the least contribution for this type of wetland [42]. The main advantages of this system are the prevention of disease vectors, bad odors and the risk of public contact with partially treated water. This type of system is still in the development stage for treating swine slurries [43–46], with elimination efficiencies of 28–38% for organic matter loads of 904 to 3,900 kg COD/ha/d, 4–15% for ammonium loads of 37–114 kg NH_4^+-N/ha/d, and 37–49% for TSS loads of 170–667 kg TSS/ha/d [45].

10.2.1.1 Macrophyte Species Used in Constructed Wetlands

Several studies have indicated the importance of macrophyte species in constructed wetland systems for nitrogen treatment [49, 50]. The rhizome and roots provide surfaces and oxygen for the growth of microorganisms that can carry out nitrification [51]. The rhizome area is also a source of carbon based on root exudates, optimizing denitrification and the elimination of organic substances in the system [29, 52, 53].

The most commonly used macrophyte species are the genera *Schoenoplectus* and *Typha*, usually in combination in surface flow wetlands fed with a nitrogen loading rate (NLR) of 3–40 kg TN/ha/d. Studies from the United States should be highlighted as they represent close to 90% of referenced works. The majority of the studies have been based on surface flow systems, highlighting the studies by Lee et al. [43] and Tapia et al. [54], who have operated this type of system with high organic loading rate (OLR) levels using *Eichhornia crassipes* as a monoculture or a combination of emerging species (*Typha latifolia; Fimbristylis spadicea; Eleocharis interstincta; Arundinella berteroniana; Cladium jamaicensis*) respectively.

10.2.1.2 Nitrogen Elimination Mechanisms in Constructed Wetlands

CWs have proven to be a profitable and efficient alternative for treating different types of effluents [34, 39, 55, 56]. The use of these systems has been increasing in recent decades owing to their efficiency in eliminating excessive nutrient loads [57]. They can eliminate 70–95% of TN for a NLR of 3–36 kg TN/ha/d [58, 59].

Nitrogen elimination mechanisms include denitrification, ammonium volatilization, incorporation to plant tissue, ammonium adsorption, anommox processes and organic nitrogen mineralization [32]. Other processes like ammonification and nitrification intervene in converting nitrogen to more simple compounds [41, 58, 61, 62]. Nitrification is the limiting microbial mechanism in eliminating nitrogen given the elimination of larger quantities of TN is associated with denitrification [32]. Figure 10.2 shows the processes related to nitrogen elimination in constructed wetlands, where denitrification (anoxic process) is limited to oxic processes of nitrate formation.

Denitrification is considered the main way to eliminate nitrogen in constructed wetlands. Denitrification is defined as the process by which NO_3^- is converted into N_2 via an intermediary nitrite (NO_2^-), nitric oxide (NO) and nitrous oxide (N_2O) in the absence of oxygen [33, 64–66]. Denitrification is done through heterotrophic bacteria that can use molecular oxygen of nitrites or nitrates as final acceptors of alternative electrons during cellular respiration [67].

The environmental factors that influence denitrification rates include the absence of O_2, redox potential, soil moisture, temperature, pH level, denitrifying bacteria, soil type organic matter, nitrate concentration and the water level [60]. Paul and Clark [68] indicated that the optimal pH range is between 6 and 8. At a pH <5 denitrification slows down but can still be significant, with denitrification by organotrophs being negligible or nonexistent at a pH <4. Denitrification is also highly dependent on temperature. Maximum denitrification rates occur in a range of 60 to 75°C and decrease significantly at higher temperatures [68].

Denitrification requires 2.3 g of organic matter (BOD_5) per gram of NO_3^--N as a carbon energy source. In the absence of this or another equivalent source of carbon denitrification is inhibited [33]. Denitrification also increases alkalinity at an approximate ratio of 3 g $CaCO_3$ for every g NO_3^--N reduced. Increased alkalinity translates into higher pH in the wetland surface. Estimations in the literature of denitrification rates are highly variable and in the range of 0.03–10.2 kg TN/ha/d, while TN elimination by denitrification is between 60 and 90% [32]. Table 10.4 shows the percentages of TN elimination via denitrification in surface flow constructed wetlands for treating swine slurries and associated physiochemical parameters.

The loads applied are in the range of 3–40 kg TN/ha/d. The most often used macrophyte species are *Typha* spp. and *Schoenoplectus* spp. Hunt et al. [64] assessed the effect of macrophyte species in denitrification and found that system inoculated with *Juncus* spp. *and Schoenoplectus* spp. has elimination rates of 44% and provide a better environment for this process than *Typha* spp. *and Sparganium* spp., the latter with an elimination rate of only 18.1%. Subsequently, Hunt et al. [35] analyzed the efficiency of the system with the combination of *Typha* spp. and *Schoenoplectus* spp. and found that under this combination denitrification can reach 70%, because of which they concluded that the macrophyte species that most favors denitrification is *Schoenoplectus* spp.

The influent feed in all the studies had a concentration of DO <1.7 mg O_2/L, with a redox potential with anoxic characteristics, which favors denitrification, mainly during summer when temperatures range between 21.7 and 28.7°C. Hunt et al. [35] found that elimination via denitrification is in the range of 18.1–91%.

However, studies by Plaza de los Reyes et al. [69] indicate that elimination by denitrification only reaches 0.3–5.6% because of the lack of bioavailable organic matter for denitrification given that the influent comes from an anaerobic bioreactor that can eliminate close to 80% of COD, leaving an influent with an elevated concentration of recalcitrant organic matter of high molecular weight

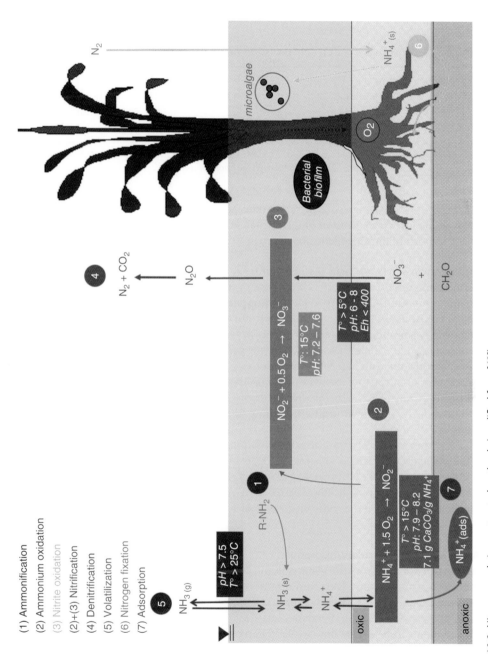

(1) Ammonification
(2) Ammonium oxidation
(3) Nitrite oxidation
(2)+(3) Nitrification
(4) Denitrification
(5) Volatilization
(6) Nitrogen fixation
(7) Adsorption

Figure 10.2 Nitrogen cycle in constructed wetlands (modified from [63]).

Table 10.4 Elimination of TN systems via denitrification in surface flow constructed wetlands for the treatment of pig manure.

Pretreatment	NLR (kg TN/ ha/d)	Plant Species	T (°C)	pH	DO (mg O_2/L)	Eh (mV)	Elimination via denitrification (%)	Ref
AL	3–40	*Ju, Sc*	4.2–28.7	7–7.5	0.3–1.5	39–147	44.4	[64]
		Ty, Sp	4.2–28.7	6.9–7.2	0.3–1.7	−2–175	18.1	
AL	4–35	*Ty, Sc*	–	6.8–7.5	–	36–61	71–91	[35]
UASB	2–30.2	*Ty*	11–21.7	7.4–8.3	0.3–0.7	−139.3–54.2	0.3–5.6	[69]

Anaerobic Lagoon; UASB: Upflow Anaerobic Sludge Bed; NLR: Nitrogen Loading Rate; DO: Dissolved Oxygen; Eh: Redox Potential; *Ju: Juncus* spp.; *Sc: Schoenoplectus* spp.; *Ty: Typha* spp.; *Sp: Sparganium* spp.

(1,000–10,000 Da), as well as a high concentration of N-NH_4^+ (810–1,700 mg NH_4^+-N/L) that cannot be nitrified given the low concentration of OD, which limits the formation of nitrates in the system. Nitrate is more important limiting factor for denitrification than organic carbon [35].

10.2.1.3 Incorporation into Plant Tissue (Assimilation)

Nitrogen assimilation refers to a variety of biological processes that convert inorganic forms of nitrogen into organic compounds that serve as basic components for plant cells and tissue. The two forms of nitrogen that can be assimilated via plants are nitrate and ammonium [32, 56]. However, the main source of nitrogen is ammonia nitrogen (NH_4^+-N), given it has lower energy state [30]. In addition to direct incorporation for cellular growth, NH4+-N can easily be transformed into amino acids by a wide range of autotrophic and heterotrophic microorganisms [41].

The incorporation of nitrogen depends mainly on the seasonal cycles of plant species like *Typha* spp. and *Phragmites australis*, with maximum incorporation in spring and summer (maximum plant growth) and decreasing in autumn until falling to zero incorporation in winter owing to plant senescence [30].

Plaza de los Reyes et al. [70, 71] and Szögi et al. [72] indicate that the incorporation of nitrogen by plants is in the range of 0.7 to 2.6 kg TN/ha/d. Szögi et al. [72] noted that more than 80% of nitrogen is stored in leaves, with a range of 0.8 to 0.9 kg TN/ha/d. TN accumulation in the roots is on the order of 0.1 to 0.2 kg TN/ha/d. Conversely, Plaza de los Reyes et al. [69, 71] indicated that over 60% of TN incorporation in *Typha* sp. is in the roots, in a range of 1.1–2.2 kg TN/ha/d and 0.4–0.5 kg TN/ha/d, respectively.

10.2.1.4 Ammonium Sedimentation/Adsorption

Ammonium can be adsorbed from the water column through a cationic exchange reaction with detritus, inorganic sediments or soils. The ammonium ion (NH_4^+) is generally absorbed by clays as an exchangeable ion, chemically adsorbed by humic substances or fixed within the clay pores, with these reactions possibly occurring simultaneously [32]. The quantity of adsorbed ammonium present in detritus and sediments in a surface flow wetland can exceed 20 g TN/m^2 [33]. These processes are influenced by diverse factors like the nature and quantity of clays, periods of flooding and drought, the nature and quantity of organic matter in the soil, the saturation period, the presence of vegetation and the age of the wetland.

Ammonium volatilization. The loss of ammonium (NH_4^+) to the atmosphere by volatilization is a complex process mediated by a combination of physical, chemical and biological factors. Ammonia exchange (NH_3) between the water column, soil and the atmosphere plays an important role in the nitrogen cycle in wetlands [33]. The conversion between NH_3 and ammonium ions is highly dependent on factors like ammonium concentration, pH and temperature. Conversion decreases significantly at low pH levels and low temperature [39, 73]. Reddy et al. [74] indicated that NH_3 loss by volatilization in inundated soils and sediments are not significant at pH levels below 7.5. Under normal temperature and pH conditions (25°C; pH of 7.0), un-ionized ammonium represents only 0.6% of the total ammonium present [39]. At a pH of 9.5 and a temperature of 30°C, the percentage of un-ionized ammonium increases to 75% of total ammonium [39, 73]. Among the biological factors that affect ammonium volatilization in wetlands are total microbial respiration and microalgae photosynthesis, the latter increases the pH level during the day [41].

10.2.1.5 Anammox (or Anaerobic Ammonia Oxidation)

There are concrete tests for the elimination of nitrites using ammonium, known as anaerobic ammonium oxidation, or anammox [33]. Anammox is an autotrophic process (because of which it does not require carbon) in which bacteria convert nitrite and ammonium into nitrogenous gas (N_2). It is strictly anaerobic and carried out by bacteria of the order *Planctomycetes* [61]. The process requires 1.94 g O_2/g NH_4^+ [33, 41]. The presence of anammox bacteria can be expected in artificial wetlands maintained under optimal conditions for their growth (e.g. pH: 7.5; T: 30°C) [75, 76].

10.3 Removal of Nutrients by Constructed Wetlands using Biological Pretreatments

The nitrogen elimination routes in artificial wetland systems are directly related to environmental factors like pH, temperature, and dissolved oxygen, and to the operational strategies applied, which in turn are related to the presence of organic carbon, the hydraulic load, the feeding mode, retention time, the contaminant load, recirculation, and the harvesting of macrophyte species.

The pH level plays a preponderant role in nitrification because alkalinity is reduced [30]. This can cause substantial falls in the pH level, which in turn can hinder and denitrification. However, Vymazal [41] indicated that denitrification can occur at pH levels lower than 5. Brix et al. [77] found that the incorporation of nitrogenized compounds (NH_4^+-N; NO_3^--N) is completely inhibited at a pH of 3.5 in *Typha latifolia*.

The optimal temperature range for nitrification in artificial wetlands is 16.5–32°C, with low levels of nitrification at temperature below 5–6°C and over 40°C. Similarly, it has been found that denitrification occurs more slowly at low temperatures (~5.0°C) and in the absence of inhibiting factors increases exponentially with increases in temperature up to 20–25°C. Studies of wetland efficiency under different climatic conditions have shown the effect of temperature [39]. The effect of temperature on TN treatment has also been reported by Reddy et al. [78], who found TN elimination rates between 70–77%, with a NLR of 16 kg TN/ha/d during the summer (19–21.9°C), while in winter (4–6.4°C) the rate did not exceed 41%.

Oxygen availability is a limiting environmental parameter in constructed wetlands for treating swine slurries. Ro et al. [79] found that oxygen flow was 38.9 kg O_2/ha/d (equivalent to 3.9 g O_2/m^2/d) in a surface flow wetland inoculated with *Schonoplectus americanus* and *Typha latifolia* to treat swine slurry. However, Wu et al. [80] indicated that this value only represents 0.02 to 0.04 g $O_2/m^2/d$. Chen et al. [46] identified variations in the redox potential (Eh) between 0.1 and 0.4 m below the surface of 18–39 mV and –42–63 mV, respectively, which indicates simultaneous nitrification and denitrification in the system.

The effect of organic matter in eliminating nitrogen in artificial wetlands has been a matter of discussion. Ding et al. [81] and Wu et al. [82] noted that NO_2^--N and NO_3^--N accumulate under a C/N ratio of 0, with nitrification under aerobic conditions predominating, given that the major part of BOD_5 is consumed by heterotrophic organisms. Consequently, the necessary concentration for denitrification is not obtained (0.93–1.07g BOD_5/ NO_3^--N reduced) [83]. With a C/N ratio of 2/5 nitrification and denitrification occur simultaneously, resulting in a decrease in COD and NH_4^+-N and NO_3^--N content, while with a C/N ratio of 5–9, NO_2^--N and NO_3^--N contents decrease significantly, with the highest levels of efficiency of the system at this C/N ratio. Finally, a C/N ratio between entre 10 and 20 under anoxic conditions result in an N_2O production rate of 5,590.6 µg $N_2O/m^2/h$ [82], considering that N_2O is more harmful than CO_2.

Xu et al. [84] indicated that COD concentrations of 600–800 mg/L inhibit photosynthesis and consequently nutrient incorporation in *Typha angustifolia*. Reddy and DeLaune [67] stated that anaerobic soils cause stress symptoms in macrophytes similar to those of hydric stress, among them stomatic closure and reduced photosynthesis. Moreover, low Eh values inhibit radicular growth and development (elongation), with complete inhibition of root elongation of *Spartina patens* in a range of redox potential of -50 to 70 mV.

NT elimination efficiencies of over 70.0% have been reported for constructed wetlands used to treat swine slurries, with a NLR ranging between 2–50 kg TN/ha/d [39, 58]. However, given that swine slurries have high concentrations of NH_4^+-N, phytotoxic effects on macrophyte species have been reported [71, 85]. Clarke and Baldwin [86] indicated that concentrations of NH_4^+-N/L of between 200–400 mg have phytotoxic effects on *Typha* sp., inhibiting growth by 75.0% and biomass production by and 80.0% (from 9.1–1.5 tons in dry weight/ha/yr) [41, 67]. Growth inhibition translates into variations in the incorporation of nutrients of 0.3–4.2 kg TN/ha/d [67, 87, 88]. Hunt et al. [58] assessed nitrogen assimilation by plants in surface flow wetlands to treat swine slurry and found that in a system with a VCN below 9 kg TN/ha/d, assimilation by plants represents 30% of eliminated TN, while this falls to less than 3% when NLR exceeds 10 kg TN/ha/d.

Harvesting the macrophyte species can also improve the efficiency of TN elimination in wetlands [89]. Studies by Hunt et al. [58] and Szögi et al. [72] indicate that TN via plants presents a range between 11.4 and 59.5 g/m^2, with a NLR between 3–40 kg TN/ha/d for a surface flow wetland to treat swine slurry. In contrast, removing the plants can have a negative impact on the microbial population that lives on the plant stalk, thus decreasing the efficiency of the system.

Table 10.5 summarizes the purification efficiencies of different configurations of shallow constructed wetlands as tertiary treatment systems [69, 90, 91]. There are secondary treatments in all the cases shown in Table 10.5 that involve eliminating organic matter by aerobic or anaerobic biological systems. Efficiency in eliminating nutrients ranges broadly between 22–96% for N and 13–86% for P. N elimination is favored by prior nitrification and/or aerobic treatment of the wetland [59, 92]. From the table it is also evident that the mode of operating the system strongly affects elimination efficiency. It can be observed that N (28–47%) and P (24-39%) elimination decreases at lower levels of hydraulic retention time (HRT) (2–4 d) [46]. The nutrient load velocity also affects the efficiency of nutrient elimination, such that nitrogen and phosphorous load velocities of 2–36 kgN/ha/d and 1-2 kgP/ha/d can result in efficiency levels of 78–89% and 56–66%, respectively [93].

Table 10.5 Operational characteristics of constructed wetlands used in the pig sector.

Species	Type	Pretreatment	HRT (d)	Loading rate		Efficiency (%)		References
				kg N/ha/d	kg P/ha/d	N	P	
Tl, Sa	6 (HS-Lagoon-HS)	Anaerobic lagoon	18	7–40	3–22	37–51	13–31	[78, 94]
Sa, Sc, Sv, Je, Spa, Ta, Tl	4HS	Anaerobic lagoon Nitrification unit	11–13	4.8–27.2 44–51	0.9–6 7–9	50–84 78–88	25–38 <10	[58, 59, 95]
Ec	1HS	Activated sludge	28	69–262	15–47	10–24	47–59	[43]
Pc, Pa, Tl	Hybrid (2VS + 1HSS)	Anaerobic lagoon + Sand filter	–	214	30	50	42	[96]
		Pig slurry treated recirculated (25 a 100%)	–	–	–	54–67	47–49	
Pc, To, Ps, Ec	1HS	Aerobic unit Anaerobic unit	2–4 5–7	–	–	28–47 74–78	24–39 57–63	[46]
Gm, Ga, Gs, Mm, Poa	16HS	Anaerobic digester	–	11–36	1.1–1.8	78–90	56–66	[93]
Cdp, Ap, Tl	2HSS	Filtration tank	5	93	22	–	–	[97]
Ci, So, Pa	Hybrid (3 VS + 1 HSS)	Aerated lagoon	4–5	76	2	64	61	[92]
Ta	HS	Anaerobic digester Storage lagoon	20	2–30	–	37–72 22–51	–	[69, 91]
Me, Ap, Ec	HS	–	30	–	–	47–96	–	[98]
Pa	VS	Raw swine wastewater diluted with tap water	1	176	24	44–61	85–86	[99]

Acorus calamis: Ac; *Alisma plantago-aquatica*: Apa; *Alternanthera philoxeroides*: Ap; *Canna indica*: Ci; *Carex pseudocyperus*: Cp; *Carex acutiformis*: Ca; *Cynodon dactylon* Pers: Cdp; *Elodea cyemsa*: Eca; *Eichhomia crassipes*: Ec; *Filipendula ulmaria*: Fu; *Glyceria aquatic*: Ga; *Glyceria máxima*: Gm; *Iris pseudacorus*: Ip; *Juncus effusus*: Je; *Lemna minor*: Lm; *Lythrum salicaria*: Ls; *Mentha aquatica*: Ma; *Myriophyllum elatinoides*: Me; *Myriophyllum spicatum*: Msp; *Molinia máxima*: Mm; *Phragmites australis*: Pa; *Phragmites communis*: Pc; *Pistia stratiotes*: Ps; *Poa populus* spp.: P; *Ranunculus lingua*: Rl; *Salix* spp.: S; *Scirpus lacustris*: Sl; *Scirpus maritimus*: Sm; *Scirpus validus*: Sv; *Scirpus cyperinus*: Sc; *Schenoplectus amercanus*: Sc; *Sparganium erectum*: Se; *Sparganium americanum*: Spa; *Stratiotes aloides*: Sta; *Symphytum officinale*: So; *Typha angustifolia*: Ta; *Typha latifolia*: Tl; *Typha orintalis*: To; *Horizontal Surface Flow*: HS; *Horizontal Subsurface Flow*: HSS; *Vertical Subsurface Flow*: VS.

Acknowledgements

This work was partially supported by the INNNOVA BIO BIO Project N° 13.3327-IN.IIP and FONDAP/CONICYT/15130015.

References

1 Hjorth M, Christensen KV, Christensen ML, Sommer SG. 2010. Solid–liquid separation of animal slurry in theory and practice. A review. Agron Sustainable Dev. 2010; 30(1):153–180.

2 Barker JC, Overcash MR. Technical Note: Swine Waste Characterization: A Review. T ASABE. 2007; 50(2):651–657.

3 Choi E. Piggery Waste Management: Towards a Sustainable Future. London: IWA Publishing; 2007.

4 Vanotti MB, Szögi AA, Hunt PG, Millner PD, Humenik FJ. Development of environmentally superior treatment system to replace anaerobic swine lagoons in the USA. Bioresource Technol. 2007; 98(17):3184–3194.

5 Moral R, Perez-Murcia MD, Perez-Espinosa A, Moreno-Caselles J, Paredes C, Rufete B. Salinity, organic content, micronutrients and heavy metals in pig slurries from South eastern Spain. Waste Manage. 2008; 28(2):367–371.

6 Burkholder J, Libra B, Weyer P, Heathcote S, Kolpin D, Thome PS, Wichman M. Impacts of waste from concentrated animal feeding operations on water quality. Environ Health Persp. 2007; 115(2):308–312.

7 Aneja VP, Nelson DR, Roelle PA, Walker JT, Battye W. Agricultural ammonia emissions and ammonium concentrations associated with aerosols and precipitation in the southeast United States. J Geophys Res-Atmos. 2003; 108: 1–11.

8 Majumdar D. The blue baby syndrome. Resonance. 2003; 8(10):20–30.

9 Gebremariam SY, Beutel MW. Nitrate removal and DO levels in batch wetland mesocosms: Cattail (*Typha* spp.) versus bulrush (*Scirpus* spp.). Ecol Eng. 2008; 34(1):1–6.

10 González M, Comte G, Monárdez J, Díaz de Valdés M, Matamala I. Miasis genital accidental por *Eristalis tenax*. Rev Chil Infectol. 2009; 26(3):270–272.

11 Babot D, de la Peña L, Chávez E. Techniques of Environmental Management in Swine Production. Lleida: Fundation Catalana of Cooperation; 2004.

12 Vanotti M. Development of clean technologies for management of wastes from pig production and their environmental benefits. In: Proceedings of the VI Congress of Pig Production of MERCOSUR (CPPM)/XI National Congress of Pig Production of Argentina (CNPP). Salta City; 2012.

13 U.S. EPA. Anaerobic Digestion of Manure. [Internet]. U.S. Environmental Protection Agency: AG-Star Program; 2009. [cited 2013 May] Available from: www.epa.gov/agstar/anaerobic/ad101/index.html.

14 Campos E, Illá J, Magrí A, Palatsi C, Solé F, Flotats X. Guide treatment of cattle manure [Internet]. Lleida; 2004 [cited 2013 May] Available from: www.arc cat.net/altres/purins/guia/pdf/guia_dejeccions.pdf.

15 Vanotti MB, Rashash DMC, Hunt PG. Solid-liquid separation of flushed swine manure with PAM: effect of wastewater strength. T ASAE. 2002; 45(6):1959–1969.

16 García MC, Szogi AA, Vanotti MB, Chastain JP. 2007. Solid-liquid separation of dairy manure with PAM and chitosan polymers. In: International Symposium on Air Quality and Waste Management for Agriculture, 16–19 September 2007, Broomfield, Colorado, 2007.

17 Tousignant E, Fankhauser O, Hurd S. Guidance manual for the design, construction and operations of constructed wetlands for rural applications in Ontario [Internet]. Ontario; 1999 [cited 2013 May] Available from: http://agrienvarchive.ca/bioenergy/download/wetlands_manual.pdf.

18 Heubeck S, Craggs RJ. Biogas recovery from a temperate climate covered anaerobic pond. Water Sci Technol. 2010; 61(4):1019–1026.

19 Hansen KH, Angelidaki I, Ahring BK. Anaerobic digestion of swine manure: inhibition by ammonia. Water Res. 1998; 32(1):5–12.

20 Kalyuzhnyi S, Sklyar V, Fedorovich V, Kovalev A, Nozhevnikova A, Klapwijk A. The development of biological methods for utilisation and treatment of diluted manure streams. Water Sci Technol. 1999; 40(1):223–229.

21 Flotats X, Campos E, Palatsi J, Bonmatí X. Anaerobic digestion of pig manure and co-digestion residues from the food industry. Porci. 2001; 65:51–65.

22 Nishio N, Nakashimada Y. Recent development of anaerobic digestion processes for energy recovery from wastes. J Biosci Bioeng. 2007; 103(2):105–112.

23 Bonmati A, Flotats X. Air stripping of ammonia from pig slurry: characterisation and feasibility as a pre- or post-treatment to mesophilic anaerobic digestion. Waste Manage. 2003; 23(3):261–272.

24 Rodríguez DC, Belmonte M, Penuela G, Campos JL, Vidal G. Behaviour of molecular weight distribution for the liquid fraction of pig slurry treated by anaerobic digestion. Environ Technol. 2011; 32(4):419–425.

25 Chamy R, Carrera J, Jeison D, Ruíz G. Advances in environmental biotechnology: treatment of liquid and solid waste, 2nd edn. Valparaiso: University Editions of Valparaiso; 2003.

26 Campos J, Figueroa M, Fernández I, Mosquera-Corral A, Méndez R. Anammox: The future process for removing ammonia from effluent liquid manure digesters. In: Bonmati A, Palatsi J, Prenafeta F, Fernández B, Flotats X, eds. II Spanish Congress on Management of Livestock Manure. Barcelona; 2010a, pp. 165–178.

27 Campos J, Vázquez-Padín J, Figueroa M, Fajardo C, Mosquera-Corral A, Méndez R. Novel biological nitrogen-removal processes: applications and perspectives. In: Fluid Waste Disposal. Canton KW; 2010b, pp. 155–180.

28 Hunt PG, Szögi, AA, Humenik FJ, Rice JM, Matheny TA, Stone KC. Constructed wetlands for treatment of swine wastewater from an anaerobic lagoon. T ASAE. 2002; 45(3):639–647.

29 Brix H. Do macrophytes play a role in constructed treatment wetlands?. Water Sci Technol. 1997; 35(5):11–17.

30 Kadlec RH, Knight RL. Treatment Wetlands. CRC Press, Boca Raton, FL, 1996, 893 pp.

31 Vymazal J, Brix H, Cooper P, Haberl R, Perfler R, Laber J. Removal mechanisms and types of constructed wetlands. In: Vymazal, J, Brix, H, Cooper P, Green M, Haberl R, eds. Constructed wetlands for wastewater treatment. Leiden: Backhuys Publisher; 1998, pp. 17–66.

32 Lee CG, Fletcher TD, Sun G. Nitrogen removal in constructed wetland systems. Eng Life Sci. 2009; 9(1):11–22.

33 Kadlec RH, Wallace S. Treatment Wetlands, 2nd edn. Boca Raton: Taylor & Francis Group; 2008.

34 Hunt P, Poach ME. State of the art for animal wastewater treatment in constructed wetlands. Water Sci Technol. 2001; 44(11–12):19–25.

35 Hunt PG, Poach ME, Matheny TA, Reddy GB, Stone KC. Denitrification in marsh-pond-marsh constructed wetlands treating swine wastewater at different loading rates. Soil Sci Soc Am J. 2006; 70(2):487–493.

36 Poach ME, Hunt PG, Reddy GB, Stone KC, Johnson MH, Grubbs A. Effect of intermittent drainage on swine wastewater treatment by marsh-pond-marsh constructed wetlands. Ecol Eng. 2007; 30(1):43–50.

37 Stone KC, Poach ME, Hunt PG, Reddy GB. Marsh-pond-marsh constructed wetland design analysis for swine lagoon wastewater treatment. Ecol Eng. 2004; 23(2):127–133.

38 Szögi AA, Hunt PG, Sadler EJ, Evans DE. Characterization of oxidation reduction processes in constructed wetlands for swine wastewater treatment. Appl Eng Agric. 2004; 20(2):189–200.

39 Poach ME, Hunt PG, Reddy GB, Stone KC, Johnson MH, Grubbs A. Swine wastewater treatment by marsh-pond-marsh constructed wetlands under varying nitrogen loads. Ecol Eng. 2004; 23(3):65–175.

40 Vymazal J, Kröpfelová L. Wastewater treatment in constructed wetlands with horizontal sub-surface flow. Dordrecht: Springer Science & Business Media; 2008.

41 Vymazal J. Removal of nutrients in various types of constructed wetlands. Sci Total Environ. 2007; 380(1):48–65.

42 U.S. EPA. Wastewater Technology Fact Sheet: Free Water Surface Wetlands. U.S. Environmental Protection Agency: Office of Water, Washington, D.C., EPA 832-F-00-024, 2000.

43 Lee CY, Lee CC, Lee FY, Tseng SK, Liao CJ. Performance of subsurface flow constructed wetland taking pretreated swine effluent under heavy loads. Bioresource Technol. 2004; 92(2):173–179.

44 Zhao YQ, Sun G, Allen SJ. Anti-sized reed bed system for animal wastewater treatment: a comparative study. Water Res. 2004; 38(12):2907–2917.

45 Sun G, Zhao YQ, Allen SJ. An alternative arrangement of gravel media in tidal flow reed beds treating pig farm wastewater. Water Air Soil Poll. 2007; 182(1–4):13–19.

46 Chen SW, Kao CM, Jou CR, Fu YT, Chang YI. Use of a constructed wetland for post-treatment of swine wastewater. Environ Eng Sci. 2008; 25(3):407–418.

47 Kantawanichkul S, Neamkam P, Shutes RBE. Nitrogen removal in a combined system: vertical vegetated bed over horizontal flow sand bed. Water Sci Technol. 2001; 44(11–12):137–142.

48 Sun ZG, Liu JS. Nitrogen cycling of atmosphere-plant-soil system in the typical *Calamagrostis angustifolia* wetland in the Sanjiang Plain, Northeast China. J Environ Sci. 2007; 19(8):986–995.

49 Leverenz HL, Haunschild K, Hopes G, Tchobanoglous G, Darby JL. Anoxic treatment wetlands for denitrification. Ecol Eng. 2010; 36(11):1544–1551.

50 Herouvim E, Akratos CS, Tekerlekopoulou A, Vayenas DV. Treatment of olive mill wastewater in pilot-scale vertical flow constructed wetlands. Ecol Eng. 2011; 37(6): 931–939.

51 Cui L, Ouyang Y, Lou Q, Yang F, Chen Y, Zhu W, Luo S. Removal of nutrients from wastewater with *Canna indica* L. under different vertical-flow constructed wetland conditions. Ecol Eng. 2010; 36(8):1083–1088.

52 Bialowiec A, Janczukowicz W, Randerson PF. Nitrogen removal from wastewater in vertical flow constructed wetlands containing LWA/gravel layers and reed vegetation. Ecol Eng. 2011; 37(6):897–902.

53 Wang R, Baldy V, Périssol C, Korboulewsky N. Influence of plants on microbial activity in a vertical-downflow wetland system treating waste activated sludge with high organic matter concentrations. J Environ Manage. 2012; 95:S158–S164.

54 Tapia González F, Gíacoman Vallejos G, Herrera Silveira J, Quintal Franco C, García J, Puigagut J. Treatment of swine wastewater with subsurface-flow constructed wetlands in Yucatán, Mexico: Influence of plant species and contact time. Water SA. 2009; 35(3):335–342.

55 Poach ME, Hunt PG, Vanotti MB, Stone KC, Matheny TA, Johnson MH, Sadler EJ. Improved nitrogen treatment by constructed wetlands receiving partially nitrified liquid swine manure. Ecol Eng. 2003; 20(2):183–197.

56 Neubauer ME, Plaza de los Reyes C, Pozo G, Villamar CA, Vidal G. Growth and nutrient uptake by *Schoenoplectus californicus* (C.A. Méyer) Sójak in a constructed wetland fed with swine slurry. J Soil Sci Plant Nutr. 2012; 12(3):421–430.

57 Vymazal J. 2005 Natural and constructed wetlands: Nutrients, metals and management. Leiden Netherlands: Backhuys Publishers; 2005.

58 Hunt PG, Matheny TA, Szögi AA. Denitrification in constructed wetlands used for treatment of swine wastewater. J Environ Qual. 2003; 32(2):727–735.

59 Hunt PG, Stone KC, Matheny TA, Poach ME, Vanotti MB, Ducey TF. Denitrification of nitrified and non-nitrified swine lagoon wastewater in the suspended sludge layer of treatment wetlands. Ecol Eng. 2009; 35(10):1514–1522.

60 Hernández ME, Mitsch WJ. Denitrification in created riverine wetlands: Influence of hydrology and season. Ecol Eng. 2007; 30(1):78–88.

61 Erler DV, Eyre BD, Davison L. The contribution of anammox and denitrification to sediment N_2 production in a surface flow constructed wetland. Environ Sci Technol. 2008; 42(24):9144–9150.

62 Sasikala S, Tanaka N, Wah HW, Jinadasa KBSN. Effects of water level fluctuation on radial oxygen loss, root porosity, and nitrogen removal in subsurface vertical flow wetland mesocosms. Ecol Eng. 2009; 35(3):410–417.

63 Plaza de los Reyes C, Vera L, Salvato M, Borin M, Vidal G. Considerations for nitrogen removal in constructed wetlands. J Water Technol (In Spanish). 2011; 330: 40–49.

64 Hunt PG, Szögi AA, Humenik FJ, Rice JM, Matheny TA, Stone KC. Constructed wetlands for treatment of swine wastewater from an anaerobic lagoon. Transactions ASAE. 2002; 45(3):639–647.

65 Li YX, Wei LI, Juan WU, Xu LC, Su QH, Xiong XI. Contribution of additives Cu to its accumulation in pig feces: study in Beijing and Fuxin of China. J Environ Sci. 2007; 19(5):610–615.

66 Zaman M, Nguyen ML, Gold AJ, Groffman PM, Kellogg DQ, Wilcock RJ. Nitrous oxide generation, denitrification, and nitrate removal in a seepage wetland intercepting surface and subsurface flows from a grazed dairy catchment. Aust J Soil Res. 2008; 46(7):565–577.

67 Reddy KR, DeLaune RD. Biogeochemistry of wetlands: science and applications. Boca Raton: CRC Press; 2008.

68 Paul E, Clark F. Soil microbiology and biochemistry, 2nd edn. San Diego: Academic Press; 1996.

69 Plaza de los Reyes C, Pozo G, Vidal G. Nitrogen behavior in a free water surface constructed wetland used as post-treatment for anaerobically treated swine wastewater effluent. J Environ Sci Health A Tox Hazard Subst Environ Eng. 2014; 49(2):218–227.

70 Plaza de los Reyes C, Villamar CA, Neubauer ME, Pozo G, Vidal G. Behavior of *Typha angustifolia* L. in a free water surface constructed wetlands for the treatment of swine wastewater. J Environ Sci Health A Tox Hazard Subst Environ Eng. 2013; 48(10):1216–1224.

71 Plaza de los Reyes C, Villamar C, Neubauer M, Pozo G, Vidal G. Nitrogen behavior in a constructed wetland for the treatment of swine wastewater. J Environ Sci Health A Tox Hazard Subst Environ Eng. 2013; 48:1–9.

72 Szögi AA, Hunt PG, Humenik FJ. Nitrogen distribution in soils of constructed wetlands treating lagoon wastewater. Soil Sci Soc Am J. 2003; 67(6):1943–1951.

73 Bustamante MA, Paredes C, Marhuenda-Egea FC, Pérez-Espinosa A, Bernal MP, Moral R. Co-composting of distillery wastes with animal manures: Carbon and nitrogen transformations in the evaluation of compost stability. Chemosphere 2008; 72(4):551–557.

74 Reddy KR, Patrick WH, Broadbent FE. Nitrogen transformations and loss in flooded soils and sediments. Crit Rev Env Sci Tec. 1984; 13(4):273–309.

75 Jianlong W, Ning Y. Partial nitrification under limited dissolved oxygen conditions. Process Biochem. 2004; 39(10):1223–1229.

76 Shipin OV, Lee SH, Chiemchaisri C, Wiwattanakom W, Ghosh GC, Anceno AJ, Stevens WF. Piggery wastewater treatment in a tropical climate: biological and chemical treatment options. Environ Technol. 2007; 28(3):329–337.

77 Brix H, Dyhr-Jensen K, Lorenzen B. Root-zone acidity and nitrogen source affects *Typha latifolia* L. growth and uptake kinetics of ammonium and nitrate. J Exp Bot. 2002; 53(379):2441–2450.

78 Reddy GB, Hunt PG, Stone K, Grubbs A. Treatment of swine wastewater in marsh-pond-marsh constructed wetlands. Water Sci Technol. 2001; 44(11–12):545–550.

79 Ro KS, Hunt PG, Johnson MH, Matheny TA, Forbes D, Reddy GB. Oxygen transfer in marsh-pond-marsh constructed wetlands treating swine wastewater. J Environ Sci Health A Tox Hazard Subst Environ Eng. 2010; 45(3):377–382.

80 Wu MY, Franz EH, Chen S. Oxygen fluxes and ammonia removal efficiencies in constructed treatment wetlands. Water Environ Res. 2001; 73(6):661–666.

81 Ding Y, Song X, Wang Y, Yan D. Effects of dissolved oxygen and influent COD/N ratios on nitrogen removal in horizontal subsurface flow constructed wetland. Ecol Eng. 2012; 46:107–111.

82 Wu J, Zhang J, Jia W, Xie H, Gu RR, Li C, Gao B. Impact of COD/N ratio on nitrous oxide emission from microcosm wetlands and their performance in removing nitrogen from wastewater. Bioresource Technol. 2009; 100(12):2910–2917.

83 Sun G, Austin D. Completely autotrophic nitrogen-removal over nitrite in lab-scale constructed wetlands: Evidence from a mass balance study. Chemosphere. 2007; 68(6):1120–1128.

84 Xu J, Li C, Yang F, Dong Z, Zhang J, Zhao Y, Qi P, Hu Z. *Typha angustifolia* stress tolerance to wastewater with different levels of chemical oxygen demand. Desalination. 2011; 280(1):58–62.

85 Li C, Zhang B, Zhang J, Wu H, Xie H, Xu J, Qi P. Physiological responses of three plant species exposed to excess ammonia in constructed wetland. Desalin Water Treat. 2011; 32(1–3):271–276.

86 Clarke E, Baldwin AH. Responses of wetland plants to ammonia and water level. Ecol Eng. 2002; 18(3):257–264.

87 Shamir E, Thompson T, Karpiscak M, Freitas R, Zauderer J. Nitrogen accumulation in a constructed wetland for dairy wastewater treatment. J Am Water Resour As. 2001; 37(2):315–325.

88 Maddison M, Mauring T, Remm K, Lesta M, Mander Ü. Dynamics of *Typha latifolia* L. populations in treatment wetlands in Estonia. Ecol Eng. 2009; 35(2):258–264.

89 Sawaittayothin V, Polprasert C. Nitrogen mass balance and microbial analysis of constructed wetlands treating municipal landfill leachate. Bioresource Technol. 2007; 98(3):565–570.

90 Villamar A. Incidence of the anaerobic digestion on the nutrient and metals treatment by constructed wetlands [Ph.D. Thesis]. University of Concepcion; 2014.

91 Plaza de los Reyes C, Vidal G. Effect of variations in the nitrogen loading rate and seasonality on the operation of a free water surface constructed wetland for treatment of swine wastewater. J Environ Sci Health A Tox Hazard Subst Environ Eng. 2015; 50(13):1324–1332.

92 Borin M, Politeo M, De Stefani G. Performance of a hybrid constructed wetland treating piggery wastewater. Ecol Eng. 2013; 51:229–236.

93 Harrington C, Scholz M. Assessment of pre-digested piggery wastewater treatment operations with surface flow integrates constructed wetland systems. Bioresource Technol. 2010; 101(20):7713–7723.

94 Poach ME, Hunt PG, Sadler EJ, Matheny TA, Johnson MH, Stone KC, Humenik FJ, Rice JM. Ammonia volatilization from constructed wetlands that treat swine wastewater. T ASAE. 2002; 45(3):619–627.

95 Stone KC, Hunt PG, Szögi AA, Humenik FJ, Rice JM. Constructed wetlands desing and performance for swine lagoon wastewater treatment. T ASAE. 2002; 45(3):723–730.

96 Lian-sheng H, Hong-liang L, Bei-dou X, Ying-bo Z. Enhancing treatment efficiency of swine wastewater by effluent recirculation in vertical-flow constructed wetland. J Environ Sci. 2006; 18(2):221–226.

97 Matos AT, Freitas S, Martinez MA, Tótola MR, Azevedo AA. Tifton grass yield on constructed wetland used for swine wastewater treatment. Rev Bras Eng Agríc Ambient. 2010; 14(5):510–516.

98 Zhang S, Liu F, Xiao R, Li Y, Zhou J, Wu J. Emissions of NO and N_2O in wetland microcosms for swine wastewater treatment. Environ Sci Pollut R. 2015; 22(24):19933–19939.

99 Doherty L, Zhao Y, Zhao X, Wang W. Nutrient and organics removal from swine slurry with simultaneous electricity generation in an alum sludge-based constructed wetland incorporating microbial fuel cell technology. Chem Eng J. 2015; 266:74–81.

Part IV

Mine Drainage and Leachate Treatment

11

Constructed Wetlands for Metals: Removal Mechanism and Analytical Challenges

Adam Sochacki[1,2,3], Asheesh K. Yadav[4], Pratiksha Srivastava[5], Naresh Kumar[6], Mark Wellington Fitch[7] and Ashirbad Mohanty[8]

[1]*Environmental Biotechnology Department, Faculty of Power and Environmental Engineering, Silesian University of Technology, Gliwice, Poland*
[2]*Centre for Biotechnology, Silesian University of Technology, Gliwice, Poland*
[3]*Department of Applied Ecology, Faculty of Environmental Sciences, Czech University of Life Sciences Prague, Kamýcká, Czech Republic*
[4]*Department of Environment and Sustainability, CSIR – Institute for Minerals and Materials Technology, Bhubaneswar, India*
[5]*Australian Maritime College (AMC), University of Tasmania, Launceston, Australia*
[6]*Department of Geological and Environmental Sciences, Stanford University, Stanford, USA*
[7]*Missouri University of Science and Technology, Civil Engineering, Rolla, USA*
[8]*Environmental Biotechnology, Division Helmholtz-Zentrum für Umweltforschung - UFZ, Leipzig, Germany*

11.1 Sources of Metal Pollution and Rationale for Using Constructed Wetlands to Treat Metal-Laden Wastewater

Metal contamination of soils, sediments and waters occurs from diverse geogenic and anthropogenic sources. The anthropogenic sources of metal-impacted waters include various industries that discharge significant amounts of metals and metalloids in their waste streams, particularly electroplating, tanning and mining. For metal-impacted waters, due to limited financial budgets, simple methods for wastewater treatment like settling ponds or constructed wetlands (CWs) are often preferred.

CWs are artificial wetlands, designed specifically for anthropogenic discharge and use natural processes to improve water quality. Wetlands have much larger footprints than chemical treatment processes, but have low operating costs as they do not require any chemical addition, mechanical machinery or skilled manpower to operate, nor in most cases, any powered equipment. However, they are not always an alternative to conventional physical–chemical methods of industrial wastewater treatment. For example, high concentrations of toxic pollutants in an industrial effluent will preclude direct application of biological methods for treatment. CWs may, though, serve as a polishing step after the conventional treatment, and may also act as a buffer volume in case of failure of the upstream treatment facilities. In the long-term perspective, CWs can remove trace concentrations of metals that are challenging to remove by classical physical–chemical methods [1].

The concept of using CWs for cost- and energy-efficient treatment of industrial wastewaters in the world has been demonstrated with a high degree of success (see other chapters). CWs have a high potential to remove metals and other pollutants, but the metals in the wetlands may result in a highly contaminated substrate which would be regarded as a hazardous material [2–4]. The major advantages of using CWs for the treatment of industrial wastewater as compared to active chemical treatment are: (i) that they are more cost-effective, particularly in the long-term operation; (ii) that they employ natural biological processes; and (iii) that they do not produce bulky oxidized sludges that require further management (11.5).

11.2 Removal Mechanisms

Designers of CWs for metal removal often focus on sulfide generation and assume removal is due to the formation of metal sulfides. However, the removal of metals in CWs is generally dominated by mechanisms like adsorption, filtration and sedimentation, association with metal oxides and hydroxides, precipitation as sulfides, and plant uptake. The other removal mechanisms that might also play a role in CWs are chelation and complexation with organic matter, precipitation as carbonates and phosphates, and microbial transformations. Another important geochemical process in CWs is the redox transitions found in near-surface environment. Change in redox potential along with the presence of organic matter create biogeochemical conditions suitable for biotic/abiotic speciation changes, particularly in metalloids like arsenic (As) and selenium (Se). Redox speciation is critical in metal mobility. Reducing conditions also enhance reductive dissolution of Fe-oxides, which are important scavengers for metals, and may cause metal release.

11.2.1 Adsorption

In the wetland substrates, metals can be adsorbed by cation exchange (physical adsorption) or chemisorption. Chemisorption is a stronger and more permanent form of binding than cation exchange. The metal removal mechanism is dependent on the adsorption process especially in the initial period of a CW's operational lifetime before sorption sites are not saturated [6]. The adsorption of metals onto organic matter is mediated by the carboxyl and phenolic hydroxyl residues of molecules (e.g., humic acids) produced or exposed during the decomposition of plants. Further degradation of this detrital matter may lead to more stable chelate complexes. Copper (Cu) and nickel (Ni) have been shown to have particular affinity to organic matter and to be retained by adsorption in wetlands [7]. It was observed that metals may bind to functional groups of organic matter adsorbed on Fe-hydr(oxides), in particular in acidic conditions [8]. Adsorption depends on pH as functional groups responsible for metal sorption are deprotonated and become available for binding dissolved metals [9]. The adsorption process can be enhanced using specific natural or synthetic sorbents.

11.2.2 Filtration and Sedimentation

Filtration and sedimentation are particularly effective processes when metals are in particulate or colloidal form [6]. The means of enhancing these processes is the use of fine filtering medium or to use systems with free water surface with low flow velocities and preferably dense vegetation to trap metallic particulates. Filtration and sedimentation are responsible for the removal of precipitates of external origin and those that are formed in CWs. For example, residual metal hydroxides present in the pretreated industrial wastewater may be entrapped in the bed media of CWs. Also the efficiency of metal removal as sulfides or (oxy)hydroxides depends to some extent on the filtration and sedimentation processes [9].

11.2.3 Association with Metal Oxides and Hydroxides

Aluminium (Al), iron (Fe) and manganese (Mn) may form insoluble compounds in the CWs by oxidation (abiotic or bacterially-mediated) or hydrolysis according to the simplified reactions [7]:

$$Al^{3+} + 3H_2O \rightarrow Al(OH)_3 + 3H^+ \tag{11.1}$$

$$2Fe^{2+} + 0.5O_2 + 2H^+ \rightarrow 2Fe^{3+} + H_2O \tag{11.2}$$

$$Fe^{3+} + 2H_2O \rightarrow FeOOH + 3H^+ \tag{11.3}$$

$$Mn^{2+} + 0.25O_2 + 1.5H_2O \rightarrow MnOOH + 2H^+ \tag{11.4}$$

The most common reaction for Al removal is precipitation as hydroxide [10]. If dissolved sulfate is present Al may also precipitate as hydroxysulfate [4]. Solubility of Al is dependent only on the pH of water and is unaffected by the oxidation and reduction processes that govern the mobility of other metals [7].

In oxygenated environment ferric iron (Fe^{III}) is usually present as insoluble (oxy)hydroxides, denoted as FeOOH [4]. In a reduced environment, the presence of Fe^{II} causes reduction of oxidized manganese precipitates (e.g., MnO_2 and MnOOH) [7]. Oxidation of Mn is much slower than for Fe and can be markedly accelerated by bacteria [6]. Iron and Mn oxides and (oxy)hydroxides are considered excellent scavengers for various metals such as Cu, Ni, Pb and Zn [4, 11]. Other metals may become associated with these Fe or Mn oxyhydroxides due to adsorption or co-precipitation [7]. For example, 97% of Zn was found to be bound to hydroxide phase in the top layer of wetland sediments [4, 12].

The major removal mechanisms of Mn are precipitation and co-precipitation as or with oxides and (oxy)hydroxides. It is noteworthy that manganese sulfide is stable only under very basic pH conditions.

The fate of both Fe and Mn in CWs depends on redox potential. These metals precipitate as oxides, (oxy)hydroxides, or hydroxides in the oxidizing (micro)environment of the wetland system. When wetland substrate becomes reducing, the metals bound to Fe and Mn oxides and (oxy)hydroxides can be remobilized, resulting in massive release. Because it can act as an electron acceptor, presence of oxidized Fe can inhibit bacterial sulfate reduction, with bacteria capable of reducing ferric iron outcompeting sulfate-reducing bacteria for electron donors. The inhibition of sulfate reduction stems also from the fact that the redox conditions in the substrate are buffered above the range required for this process [13]. Wetland plants introduce oxygen into the subsurface, so reduced sediment may be oxidizing at the root; rhizoconcretions on plant roots, commonly known as iron plaque, can be highly efficient sinks of trace metals. These structures may contain 5–10-fold more trace metals then the adjacent substrate.

11.2.4 Precipitation as Sulfides

Much of the research on metal removal in CWs has focused on biological reduction of sulfate resulting in metal sulfides. Most metal sulfides are insoluble and remain stable in reduced conditions, resulting in low effluent metal concentrations.

11.2.4.1 Mechanism of the Process

Sulfate-reducing bacteria (SRB) mediate dissolved sulfate reduction to hydrogen sulfide and metals are immobilized by the resulting sulfide according to the simplified reactions:

$$SO_4^{2-} + 2CH_2O(org) \Rightarrow H_2S + 2HCO_3^- \tag{11.5}$$

$$H_2S + Me^{2+} = MeS(s) + 2H^+ \tag{11.6}$$

where Me is a divalent metal cation. The bicarbonate produced in reaction (11.5) balances the acidity produced by reaction [6, 14]:

$$SO_4^{2-} + Me^{2+} + 2CH_2O(org) = MeS(s) + 2H_2CO_3(aq) \tag{11.7}$$

The thermodynamic equilibria of metal sulfide precipitation are as follows [15]:

$$H_2S \Longleftrightarrow HS^- + H^+ K_{p1} = \frac{[HS^-][H^+]}{[H_2S]}, pK_1 = 6.99 \tag{11.8}$$

$$HS^- \Longleftrightarrow S^{2-} + H^+ K_{p2} = \frac{[S^{2-}][H^+]}{[HS^-]}, pK_2 = 17.44 \tag{11.9}$$

$$Me^{2+} + S^{2-} \Longleftrightarrow MS(s) \tag{11.10}$$

$$Me^{2+} + HS^- \Longleftrightarrow MeS(s) + H^+ \tag{11.11}$$

$$2Me^+ + HS^- \Longleftrightarrow Me_2S(s) + H^+ \tag{11.12}$$

These equations are simplified and may not be true for non-stoichiometric forms and diagenetic MeS$_2$ forms, such as the Fe sulfides greigite and pyrite. It is noteworthy that sulfide precipitation is a multistage process, with initially non-stoichiometric amorphous phases undergoing diagenetic reorganization to crystalline forms. Sulfides are (meta)stable under anaerobic conditions but undergo oxidation at the oxic–anoxic interface, which leads to the release of metals [16]. In general, metal sulfides are less soluble than their carbonate or hydroxide counterparts, achieving more complete precipitation and stability over a broader pH range (Table 11.1) [17].

Sulfate is often present in industrial wastewater at much greater concentration than metals, and the production of excessive H$_2$S (or HS$^-$) may not only adversely affect the environment due to toxicity and odor, but also may increase metal mobility. The latter stems from the fact that HS$^-$ may form stable aqueous complexes with metals. For example, if HS$^-$ concentration is above 0.1 mol/L the dominant copper species is CuS(HS)$_3^{3-}$, and the lower solubility CuS(HS)$_2^{2-}$ does not become significant until the HS$^-$ is below 0.2 mol/L [15]. In general, the re-dissolution of precipitated metal sulfides by excessive sulfide is described as [15]:

$$MeS(s) + HS^-(aq) \Longleftrightarrow MeS(HS)^-(aq) \tag{11.13}$$

Removal of metals in sulfate-reducing wetlands was reported to follow two mechanisms: the adsorption of metal onto organic matter and the formation of metal-sulfide precipitates. Although equilibrium favors sulfide formation, adsorption onto organic matter appears to occur much more rapidly than sulfide precipitation. Metals adsorbed to organic matter apparently convert slowly to sulfide precipitates [18, 19]. Removal not due to *via* other than sulfides may also occur because the SRB biofilm is able to bind metals by several mechanisms as mentioned in Section 11.2.5. The key requirements for sulfate-reducing systems are: anaerobic conditions (oxidation-reduction potential below −100 mV), electron donors (simple organic compounds), microbial groups capable of utilizing inorganic sulfur compounds as electron acceptors (dissolved sulfate) [10, 13, 20].

Table 11.1 Solubility products of selected metal sulfides, hydroxides and carbonates (adapted from [17]).

Element	Sulfide		Hydroxide		Carbonate	
	Form	log K_s[a]	Form	log K_s	Form	log K_s
Al			$Al(OH)_3$ alpha	−33.5		
Cu	Cu_2S	−47.6	$Cu(OH)$	−14.7	$CuCO_3$	−11.5
	CuS	−35.2	$Cu(OH)_2$	−18.6		
Fe	FeS	−17.2	$Fe(OH)_2$	−15.1	$FeCO_3$	−10.2
	Fe_2S_3	−85.0	$Fe(OH)_3$	−37.4		
Mn	MnS pink	−9.6	$Mn(OH)_2$	−12.7	$MnCO_3$	−10.6
	MnS green	12.6				
Ni	NiS alpha	−18.5	$Ni(OH)_2$	−14.7	$NiCO_3$	−6.9
	NiS gamma	−25.7				
Pb	PbS	−27.0	$Pb(OH)_2$	−16.1	$PbCO_3$	−13.1
Zn	ZnS sphalerite	−23.0	$Zn(OH)_2$	−15.7	$ZnCO_3$	−10.0
	ZnS wurtzite	−24.3			$ZnCO_3 \cdot H_2O$	−10.3

[a] K_s solubility product constant.

In freshwater wetlands, sulfate reduction rates are generally limited by the amount of sulfates. In CWs, however, sulfate concentration is usually high and sulfate reduction rates are mostly dependent on the supply of electron donor. Therefore, a major factor limiting the application of microbial sulfate reduction in wetland systems is the availability of carbon and energy sources to drive the process. For sulfate reduction to be effective for treating such wastewater, a key requirement is the availability of a carbon source as an electron donor [11]. This aspect is discussed in Section 11.2.4.2. Some of the metals (or metalloids) which may react with hydrogen sulfide to form highly insoluble metal sulfides are: As, Cu, Fe, Ni, Pb, and Zn [18]. Precipitation of metals with sulfides has several advantages when compared with precipitation by hydroxides: (i) the residual metal concentrations in the effluent are lower; (ii) the interference of the chelation agents in wastewater is less problematic; (iii) precipitation offers increased selectivity; (iv) and higher reaction rates result from lower hydraulic retention time [17].

There are several processes that may increase removal of metals related to bacterial sulfate reduction but not directly due to sulfide formation. Some SRB have broad-specificity enzymes able to change the oxidation state of metals or metalloids, thereby directly reducing solubility. Cells and associated extracellular polymers are capable of biosorption and bio-concentration of soluble ions, nucleation of precipitate at biopolymer surfaces, and electrostatic entrapment of particulates. These processes thus may contribute directly to metal removal but may also enhance the efficiency of sulfide precipitation [16]. The chemical reduction of metals may occur abiotically in sufficiently reducing conditions, but also can be attributed to enzymatic action. For example, *Desulfovibrio* are able to reduce Fe^{III} to Fe^{II}, and Cu^{II} is reduced to Cu^I in sulfidic solutions [15, 16]. Interestingly, Cu may also be reduced to its metallic form under slightly acidic and reduced conditions when the amount of sulfide is low [7].

Not all metals noticeably precipitate as sulfides, particularly Al and Mn. It is suspected that Al may precipitate in reduced conditions (as alunite) according to the reaction:

$$3Al^{3+} + K^+ + 6H_2O + 2SO_4^{2-} \rightarrow KAl_3(OH)_6(SO_4)_2 + 6H^+ \tag{11.14}$$

Not only are some metals not precipitated as sulfides, metalloids like As, can be mobilized by microbial sulfate reduction. Dissolved sulfide produced by SRB can reduce As^V to the more mobile As^{III}, causing release into the aqueous environment [21]. If sulfide is available in excess, As can also form *thio*-arsenic species, which has a low sorption affinity towards Fe-S and Fe-oxides and consequently, is more mobile [22, 23].

11.2.4.2 Bacterial Sulfate Reduction in Constructed Wetlands

Sulfate reduction may occur through either assimilatory or dissimilatory pathways. In the assimilatory pathway, oxidized forms of sulfur are reduced to generate sulfide, which is eventually transferred to amino acids as sulfhydryl groups (R-SH). Most of the reduced sulfur is fixed within the cell and only a small portion is released. This process occurs in both aerobic and anaerobic conditions. Upon death and decay, organisms release the sulfur in the reduced state, which can result in the accumulation of high levels of hydrogen sulfide in wetland sediments or which can be reoxidized to elemental sulfur (S^0) and sulfate in anaerobic and aerobic environments, respectively [4, 13, 24].

Dissimilatory sulfate reduction is mediated by SRB; although referred to as sulfate reducing bacteria, both bacteria and archaea can use sulfate as an electron acceptor [25]. The majority of the SRB fall into one of the four phylogenetic lineages: (i) the mesophilic δ-proteobacteria *Desulfovibrio, Desulfobacterium, Desulfobacter,* and *Desulfobulbus*; (ii) the thermophilic Gram-negative bacteria *Thermodesulfovibrio*; (iii) the Gram-positive bacteria *Desulfotomaculum*; and (iv) the archaeal *Euryarchaeota* with the genus *Archaeoglobus* [26]. Existence of a fifth, novel lineage of SRB, the *Thermodesulfobiaceae*, was reported by Mori et al. [27]. Many SRB are versatile, using different electron acceptors than sulfate, including elemental sulfur, fumarate, nitrate, dimethysulfoxide, Fe^{III}, and/or Mn^{IV}. Sulfate reducers include heterotrophic and autotrophic bacteria. The latter use H_2 as an electron donor [25]. The composition of the SRB community may have an effect on the efficiency of metal sulfide precipitation. Thus the choice of inoculum may be a key design aspect but has not been studied in detail [28].

SRB occurring in CWs may interact with other groups of bacteria. For example, methanogens and SRB share many ecological and physiological similarities and they often coexist in the reduced environment. When the source of electron donor is not limited, both methanogens and SRB can thrive [29]. Under substrate-limiting conditions, both groups of bacteria compete for acetate and H_2. There are several factors that govern this competition, e.g. pH, COD/SO_4^{2-} ratio, substrate type, and temperature. Methanogens and SRB may also be synergistically related, with mutual reliance for metabolic activities. Methanogens may utilize acetate or H_2 produced by SRB, and SRB may oxidize methane produced by methanogens [29].

In addition to competition or interaction for electron donors, SRB in CWs are often dependent on others to provide electron donor when only complex organic matter is available (e.g. wood, peat, or compost). The volatile fatty acids utilized by SRB are the products of fermentation [30]. Fermentative bacteria utilize simple organic compounds obtained by enzymatic hydrolysis of detrital matter composed of cellulose, hemicellulose, proteins, lipids, waxes, and/or lignin [13]. Thus the fermenters degrade complex organics and generate fatty acids, the organic matter used by SRBs.

11.2.4.3 Carbon Source for Sulfate-Reducing Bacteria

The source of organic matter is a key requirement for the treatment of sulfate-containing wastewater by biological sulfate reduction. Carbon-deficiency is typical for sulfate-rich wastewaters such as mine drainage or electroplating wastewater. Electron donors should be added to such wastewaters to ensure reduction of sulfate and, indirectly, precipitation of metal sulfides [31]. That electron donor can be a solid material mixed with the filtering medium or an aqueous solution added to the wastewater (liquid carbon source). Cost effective, widely available and highly effective organic substrates are desirable in both cases.

11.2.4.3.1 Solid Carbon Sources

Natural organic substrates tested to date include a wide range of roughage and organic wastes including cow manure, rice stalks, straw, molasses, ryegrass, oak chips, spent oak from shiitake farms, spent mushroom compost, sewage sludge, organic-rich soil, activated sludge, compost of various origin, leaf mulch, mixtures of poultry manure, wood chips, and peat [32]. Also more defined compounds such as polymers from lactic acid have been investigated [16, 33]. It is believed that high sulfate-reduction rates can be achieved when a mixture of at least two materials is used. This is especially the case when relatively readily biodegradable material (e.g., manure) is mixed with more recalcitrant (e.g., saw dust) [33]. Some of these substrates may induce clogging, thus reducing treatment performance and leading to system failure. Using a substrate with good hydraulic properties, such as gravel, and amending the wastewater with a soluble carbon source has therefore been suggested [34, 35].

11.2.4.3.2 Liquid Carbon Sources

A drawback of many of the solid substrates noted above is a low rate of transformation into electron donor for SRBs. It is possible to use synthetic carbon sources in the liquid state such as acetate, butyrate, lactate, propionate and pyruvate to supply that electron donor in a more controlled manner. Ethanol and other alcohols can also be used to this end. Molasses was also reported to be used as an external carbon source due to the high content of sucrose [31]. These soluble sources can be pure solutions or waste products containing these substances, e.g., dairy wastewater.

SRB compete with methane producing bacteria for electron donors [36]. The advantage in this competition depends, amongst other factors, on the amount of carbon available and the ratio between electron donors and acceptors, expressed for example as COD/SO_4^{2-}. This ratio is adjusted according to the goals of the process: stimulation of methane production or production of sulfides. The stoichiometry of the sulfate reduction half-reaction is:

$$8e + 8H^+ + SO_4^{2-} \rightarrow S^{2-} + 4H_2O \qquad (11.15)$$

The theoretical minimum COD/SO_4^{2-} ratio required for achieving sulfate reduction is therefore 0.67 [36].

The disadvantage of using organic substrates such as lactate rather than acetate is that they may not be fully oxidized by SRB if in excess, thus causing elevated COD levels in the effluent [31]. Lactate is the most energetically advantageous substrate for SRB in terms of biomass produced and thus may be a good liquid substrate choice. The main drawback of using lactate is that only certain species of SRB can directly oxidize it to CO_2 (e.g., *Desulfotomaculum*), whereas other species can only partly oxidize it to acetate (e.g., *Desulfovibrio*) and thus acetate might build up and COD in the effluent could be high [33].

11.2.5 Microbial Removal Processes

Some microbial processes have direct or indirect impact on metal removal in CWs. Kosolapov et al. [11] categorized these processes into (i) biosorption; (ii) microbial oxidation-reduction of metals; (iii) methylation; and (iv) biogeochemical cycle-assisted metal removal. Biosorption is the passive uptake of metal, which is common on dead or living microbial biomass [37]. Sorption sites become saturated, so this is not a long-term approach if the metal-loaded biomass is not harvested [11]. Biosorption involves several functional groups including carboxyl, sulfonate, phosphate, hydroxyl, and amino moieties. The non-specific term biosorption is used because it can involve mechanisms and interactions like ion exchange, chelation, adsorption and entrapment, many of which are poorly understood [38, 39]. As noted in Sections 11.2.3 and 11.2.4, microorganisms which govern biogeochemical cycles can immobilize metals due to sulfate reduction or Fe- or Mn-oxidation, leading respectively to the (co-) precipitation of metals with sulfides and Fe- or Mn-(oxy-) hydroxides [35]. Anaerobic metal-reducing bacteria use metals as terminal electron acceptors in respiration, and the reduced metals may be less or more soluble, depending on the specific elemental chemistry. Microbial reduction of Cr (VI) to Cr (III) is an example of this process [11]. There are some microbial oxidation reactions in non-reduced zones of CWs, like the oxidation of dissolved iron (II) and manganese (II), which precipitate the metal as metal (oxy)hydroxide.

Some metals, including Hg, As and Se, can be biomethylated by aerobic and anaerobic bacteria of CWs, resulting in the production of volatile derivatives such as dimethylmercury, dimethylselenide or trimethylarsine [11, 40]. Although the metalloid is removed from water, these volatile derivatives are generally more toxic than the inorganic forms, because of their lipophilic character [35].

11.2.6 Plant Uptake of Metals in Constructed Wetlands

Plant uptake may play a major role in the removal of metals in CWs. Some plants act as accumulators, but most plants exclude metals [41]. Accumulator plants generally transform metals into a non-reactive form which is stored in their tissues with minimum harm. Metals may be micronutrients, an efficient mechanism works in plants to take up micronutrients from the environment. Plant roots, aided by plant-produced chelating agents and plant-driven pH changes and redox reactions, are able to solubilize and take up micronutrients not only from very low concentration levels but from almost insoluble precipitates present in the soil [42].

11.2.6.1 Metal Uptake by Aquatic Macrophytes

Aquatic macrophytes usually depend on the micronutrients, which may include Fe, Mn, Zn, Cu, Mo and Ni, available in the water or substrate [43]. There are other metals like Cd, Cr, Pb, Co, Ag, Se and Hg which are not known for specific biological function in the plant, but can also accumulate in some plants [40]. Depending on the nature of the plants, uptake routes of metals are different; emergent and rooted submerged plants uptake through roots, whereas floating plants take up metals through leaves or shoots [44–46]. In general, the metal uptake rate per unit area of the wetland is often much higher for herbaceous plants, or macrophytes such as cattails. Depending on plant growth rate and background concentration of the metals in plant tissue, the uptake rate of metal varies largely. In the case of foliar absorption of metals, the metal in aqueous phase moves through cracks in the cuticle or through the stomata to the cell wall and then to plasmalemma in passive mode [47, 48]. Grill et al. [49] recognized these sites in the cell wall and termed them as phytochelatins. Phytochelatins are metal-complexing peptides which are involved in detoxification and in homeostatic balance. Metals also may be bound to the cell wall by a process called metathiolate formation through mercaptide complexes.

11.2.6.2 Metal Uptake by the Roots

Plant roots can mobilize metals by various mechanisms including (i) secretion of metal-solubilizing and metal-chelating molecules (phytosiderophores); (ii) reduction of metal ions by metal-reductases; and (iii) the secretion of protons [40, 50]. Once metal is solubilized, it may enter the roots by extracellular (apoplastic) or intracellular (symplastic) pathways. The apoplast provides extracellular space into which water molecules and dissolved low molecular mass substances diffuse. The symplastic compartment consists of a variety of cells connected via plasmodesmata. The apoplast plays an important role in the binding, transport, and distribution of ions and in response to environmental stress. Uptake occurs either by passive diffusion through the cell membrane or more commonly by active, carrier-mediated transport. These carriers can be complexing agents, such as organic acids or proteins that bind to the metal species [51]. After entering the roots, metals are either stored or transported to the shoots. Transport of metals to the shoots probably occurs in the xylem and metals can be re-distributed in the shoots via the phloem. Metals in the xylem and phloem may be transported in a complexed/chelated form with organic acids, phytochelatins or metallothioneins [40, 52].

Metal present in soil or water is often not readily available for uptake by plants; some metals are more mobile and available for uptake than others [53]. The mobility of metal ions mainly depends on the pH, redox, and presence of chelating agents. Many other factors such as root size, external metal concentrations, temperature, metal interaction, addition of nutrients, and salinity also play a minor role in determining mobility [54]. Thus, such parameters of the CW substrates have impact on metal uptake by aquatic macrophytes [55, 56].

11.2.6.3 Metal Uptake by the Shoots

When shoots are in water, they can also become the entry point for metals. A few studies on submerged and floating aquatic plants establish the relative importance of shoot/root uptake of metals, as a good correlation exists between metal concentrations in leaves and the surrounding water [57, 58]. However, the direct uptake of metal by shoots as compared to roots is still an issue of debate [57, 59].

11.2.6.4 Indirect Assistance in Metal Removal by Plants

There are some indirect roles of aquatic macrophytes in metal removal in CWs due to interaction with the substrate and/or micro-organisms [60, 61]. The notable indirect roles of plant in metal removal are (i) the secretion of exudates; (ii) release of oxygen in the root zone; and (iii) decomposing plant material as a carbon source.

11.2.6.4.1 Excretion of Exudates

Some studies indicate plant roots can excrete protons and exudates (organic matter) in the rhizosphere which can acidify and mobilize metals [40, 62]. Plant-derived exudates provide sites for metal sorption, as well as carbon sources for bacterial metabolism, thus promoting long-term functioning [63–66]. Root exudates can also enhance metal removal in CWs by affecting the redox potential, such as demonstrated by Buddhawong et al. [67]. However, the quantity of exudates is very low and may only affect redox potential in carbon-limited wastewater like mining and industrial effluents [68].

Roots also may exude organic acids such as citrate, oxalate, malate, malonate, fumarate, and acetate [69]. These anions can chelate metallic ions to varying degrees, thus decreasing phytotoxicity.

Tricarboxylate anions (e.g., citrate) chelate metallic cations more strongly than dicarboxylate anions (e.g., oxalate), which in turn chelate more strongly than monocarboxylate anions (e.g., acetate) [70].

11.2.6.4.2 Release of Oxygen in Root Zone

Wetland plants provide transport oxygen to the roots, generally through aerenchyma [71, 72]. Some part of this oxygen leaks from the roots to the rhizosphere, termed radial oxygen loss. Such oxygen transport in the presence of dissolved Fe causes an oxidative protective film, also called "plaque", to form on the root surface and mainly composed of Fe oxyhydroxides. This type of environment is particularly significant in horizontal-flow CWs as it generates an aerobic micro-environment near the root zone [68]. There are conflicting reports as to whether the presence of plaques reduces or increases the uptake of metals by the plants [62, 73, 74].

11.2.6.4.3 Decomposing Plant Material as Carbon Source

When there is no harvesting of plant biomass in CWs, large amounts of organic matter are ultimately recycled due to decomposition [75]. Such decomposition of biomass may release or be a sink for metals. A few studies in natural wetland systems demonstrated that metal concentrations in plant litter generally increase with time, although different mechanisms for this enrichment have been reported, such as adsorption of metals from the sediment, accumulation of fine particulates in the litter, and active uptake by the microbial community [62]. Metals are also removed by small invertebrates and vertebrates present in CWs and grazing on plant matter, however, the degree to which metals contained within decomposing macrophytes enter invertebrates and vertebrates is still not well known. Jackson [56] suggested that macrophytes act as carriers for metals between sediments and food webs.

11.2.6.5 Role of Plants in Removing Metals from Industrial Wastewater

There is general understanding in the scientific literature that substrate acts as the primary sink for metals [62]. However, the role of plant in removal of metals is debatable. Some authors conclude that metal uptake by vegetation is significant [76–81], while some consider them of minor importance [82–86].

Calheiros et al. [87] have explored the application of different plant species in CWs receiving tannery wastewater. It was found that *T. latifolia* and *P. australis* were the species better adapted to tannery wastewater in terms of survival and propagation where inflow wastewater was having the Cr concentration of 0.01 mg/L. Hadad et al. [82] studied the possibility of treating industrial wastewater within CW in Argentina and found average metal removal efficiencies of 83, 82, 69, and 55% for Fe, Cr, Ni, and Zn, respectively. Besides, metal concentration in macrophyte tissues increased significantly where metal concentration in the roots was 2–3 times higher than in leaves. However, only a small amount of metal accumulated (7, 2, and 4% of Cr, Ni, and Zn, respectively) in the wetland was stored in the macrophyte tissue.

Aslam et al. [88] investigated the compost-based and the gravel-based vertical flow constructed wetland to treat wastewater of an oil refinery in Pakistan. The results indicated that the treatment efficiency was low at the beginning but it improved gradually with the growth of plants and biofilm. A significant removal of metals including Fe^{2+}, Cu^{2+}, and Zn^{2+} was observed. Total amounts in whole *Phragmites* plants (roots, rhizomes, and culms), were 35.5 and 27.6 mg/m^2 for copper, 82.4 and 58.9 mg/m^2 for Fe, and 18.5 and 13.6 mg/m^2 for Zn in the compost-based and gravel-based CWs, respectively. Plant tissues took up significant amounts of metals (35–56%) in this study.

Mantovi et al. [74] studied two horizontal subsurface flow reed beds treating dairy parlor effluent and domestic sewage in an Italian rural settlement. Copper and Zn content were evaluated in

P. australis tissues and showed that plant tissues took up only 1–2% of metals entering the treatment systems. Lesage et al. [89, 90] studied the submerged plant *Myriophyllum spicatum* for the treatment of metal-contaminated industrial wastewater. The study focused on the sorption/desorption characteristics of the surface of *M. spicatum* for Co, Cu, Ni, and Zn. The experimental results showed that the metal removal process by plants involves a combination of rapid sorption on the surface and slow accumulation and translocation in the biomass. The Langmuir model nicely described the sorption of Co, Ni, and Zn, whereas sorption of Cu was better described by the Freundlich model. In the study it was found that the biomass had the highest affinity for Cu and Zn. The Langmuir sorption maximum of Co, Ni, and Zn were found to be 2.3, 3.0, and 6.8 mg/g, respectively. At the highest initial concentration of 100 mg/L, a maximum of 29 mg/g of Cu was sorbed onto the surface of the biomass.

Sochacki et al. [91] focusing on the polishing of carbon-deficient electroplating wastewater observed that the removal of Cu, Ni, Zn in lab-scale saturated vertical-flow CWs was only slightly (<10%) affected by vegetation (*Phragmites australis* (Cav.) Trin. ex Steud and *Phalaris arundinacea* L.). Interestingly, the effect of vegetation was negative in the CWs filled with gravel, but positive in the CWs with peat.

11.2.7 Other Processes

The other processes that may play a role in the removal of metals can be: formation of carbonates, metals hydrolysis (catalysed by bacteria under acidic conditions), reduction to non-mobile forms (biologically or chemically), biological methylation (and subsequent volatilization; especially for mercury) [7, 92]. Metals could be complexed by organic ligands from plants (mentioned in 11.2.6.4), from organic in wastewater, or microbial sources. This phenomenon, however, cannot occur under typical conditions prevailing in wetlands [14]. Aluminium and some other metals may precipitate with phosphate [4].

One example of the very complex metal chemistry in natural systems is observed for Pb, which apart from sulfides, carbonates and hydroxides may form oxide (PbO), sulfate ($PbSO_4$), charged hydroxide ($PbOH^+$), and biplumbite ion ($HPbO_2^-$), and if phosphate is present, pyromorphite ($Pb_5(PO_4)_3Cl$) [13, 15]. Metals such as Pb in wetlands complex, but numerous mechanisms, both biological and abiotic, may play a role in the removal of metals. Although studied to some extent as single processes, the interplay between various biotic and abiotic processes remains a topic for further investigation [14]. Based on the solubility product constant (K_s) of different metal species (Table 11.1), it is assumed that the removal of metals as sulfides is dominant, as in general sulfides have much lower solubility than hydroxides and carbonates. However, the mobility of metals in sulfidic systems is often underestimated as the solubility of freshly precipitated amorphous forms of metal sulfides can be several times higher than for the crystalline form (which are shown in Table 11.1) [14]. Thus, straightforward application of solubility data cannot explain trace metals levels, nor speciation, in natural systems [13].

11.3 Analytical Challenges

11.3.1 Background and Overview of Methods

There are at least two definitions of chemical speciation. It is either "the active process of identification and quantification of the different species forms or phases in which an element occurs in a material" or "the description of the amounts and kinds of species, forms or phases present in the material", as proposed by Ure [93]. The same author suggested that speciation be divided into three classes:

- classical, which refers to specific chemical compounds or oxidation states;
- functional, which refers to the specific behavior of the element in a particular environment or in specific conditions, and is characterized as 'plant available' or 'mobile' species; and
- operational, which refers to the situation where the reagent used to extract the sample defines the species, e.g. "acetic acid soluble" species.

Direct determination of the chemical form of metal in solid-phase environmental samples can be achieved by means of various instrumental techniques [94, 95]. The scanning electron microscopy (SEM) approach coupled with X-ray microanalysis proved successful for identifying sulfides in reactive mixtures from CWs [8, 34]. Neculita et al. [96] and Gammons and Frandsen [14] used scanning electron microscopy – energy-dispersive X-ray spectroscopy (SEM-EDS) and XRD techniques to evaluate mineralogy of substrate from anaerobic microcosm and full-scale constructed wetland, respectively. Raman spectroscopy and infrared spectroscopy were used to study the residues after a sequential extraction procedure [97], for identification of both organic biomolecules and inorganic mineral material in the same samples [98], to search for small quantities of pyrite and amorphous iron sulfides or metal sulfides in anoxic river sediments [99], to confirm the presence of vivianite ($Fe^{II}_2(PO_4)\cdot 2H_2O$) in the sediments [100], together with Fourier transform infrared spectroscopy (FTIR) to analyze ochreous sediments of CWs [101]. The other method, which proved to be useful in the analysis of sediments was X-ray photoelectron spectroscopy (XPS) because it gives quantitative information on all the elements on the surface (except hydrogen and helium) and, in addition, gives information on the chemical bonding [102]. The application of XPS allows analysis of unaltered environmental samples when fast-frozen samples are analyzed and can be used to study colloidal and amorphous forms of sulfides [103].

The distribution, mobility and bioavailability of metals in the environment depend on their concentration and also (to even larger extent) on the form in which they are bound to solid phase. The short-term or long-term changes of the chemical–physical conditions in the environment may increase the mobility of metals. The bioavailability of metals is associated with many variables such as characteristics of the particle surface, on the type of binding, and properties of the solution in contact with the solid phase [104]. Another crucial factor deciding whether an element will be incorporated into an organism is its physiology. As stated by Bacon and Davidson [95], functional speciation depends strongly on the context and aim of the experiment so that the same metal pool could be considered as "bioavailable", when plant uptake is of interest but "mobile" or "labile" in leaching studies.

The complexity of possible reactions and often unknown reaction kinetics in CWs often restrict studies of metal species distribution in solid phases to operationally defined analytical procedures such as sequential extraction procedures [105]. Sequential extraction only divides the content of a sample into portions soluble in particular reagents under particular conditions. These reagents are often recommended in a given protocol to target well-defined mineral phases, but their specificity cannot be assured. Thus, the results of only sequential extraction are insufficient to elucidate binding characteristics of trace metals to specific minerals. Reliable interpretation can be made only if additional analytical techniques are employed (e.g., XRD, XPS, SEM-EDS and Raman spectroscopy) to analyze the samples and the residues after each stage of the extraction [95, 106]. Nonetheless, the results of a sequential extraction procedure may furnish useful information on the mobility and fractionation of metals in the substrate especially when combined with the determination of acid volatile sulfides and simultaneously extracted metals (AVS-SEM). Metals are believed to be stable or potentially mobile if the SEM/AVS molar ratio is less than one or greater than one, respectively [96].

11.3.2 Sequential Extraction Procedures and their Applicability to Wetland Substrates

Sequential extraction procedures (SEPs) use a series of reagents added to the same sample to sub-divide the total metal content. Each step of the procedure tends to be more vigorous then the previous one, starting with mild conditions (e.g., shaking with water, a salt solution or dilute acetic acid) to end with highly reactive reagents (e.g., hot mineral acid). Generally, the elements extracted early during the procedure are those most weakly bound to the solid phase. They have thus greater potential mobility and environmental impact than those extracted later [95]. A large number of different protocols has been reported, however, the Tessier [107] and BCR [108, 109] schemes are among the most commonly applied [95]. In an IUPAC report Hlavay et al. [110] stated: "Despite some drawbacks, the sequential extraction method can provide a valuable tool to distinguishing among trace element fractions of different solubility related to mineralogical phases. The understanding of the speciation of trace elements in solid samples is still rather unsatisfactory because the appropriate techniques are operationally defined". The results of only sequential extraction are insufficient to elucidate binding characteristics of metals to specific minerals. Reliable interpretation can be made only if additional analytical techniques are employed (e.g., X-ray diffraction or SEM-EDS) to analyze the residues after each stage of the extraction. A SEP is not advisable for absolute studies in which distribution of metals between specific soil phases is to be identified [95].

The considerable advantage of SEPs is ease of use, and that they can be adjusted to a wide variety of environmental samples. However, SEPs have significant limitations that may compromise their validity and must be borne in mind. The most important of these limitations is the potential introduction of artifacts that obscure the true metal speciation [111]. These artifacts may be introduced by two routes: (i) redistribution of leached metals into other sediment phases; or (ii) partial or complete dissolution of sediment phases prior to their targeted extraction step. In the first case, extracted metal precipitates and will be quantified in a later fraction, whereas in the second case later fractions (e.g., reducible) will be underestimated [106].

The latter problem becomes particularly important when anoxic sediments such as most CWs contain are studied. As discussed above, metal-sulfide precipitates may be a major sink for metals, but these sulfide phases are amenable to premature dissolution during SEP. Several studies demonstrated that hydroxylammonium hydrochloride may leach sulfide-bound metals so they can be construed as being bound to Fe-Mn (oxy)hydroxides [106, 112, 113]. This reagent is used in the most commonly applied procedures, mainly those based on the Tessier protocol [107] and the original and modified BCR protocols [108, 109]. Peltier et al. [106] examined the speciation of Zn and Pb in anoxic wetland sediments using SEP based on a modified Tessier protocol and simultaneously performing X-ray absorption spectroscopy analysis to check the accuracy of the extraction results and found that the sequential extraction method resulted in misidentification of approximately 50% of Zn bound to reducible Fe-Mn oxides, rather than sulfides. This early solubilization of metals from sulfide phases was observed mostly during the step when hydroxylammonium hydrochloride is used. As mentioned above, this reagent is commonly used in many sequential extraction methods and routinely used to examine metal speciation in sediments, thus its low selectivity may reduce applicability of these methods for anoxic sediments. It was suggested that the content of metals bound with sulfides would be underestimated particularly for samples with freshly formed metal sulfides (amorphous), such as those in wetland sediments. In samples where metal sulfides occur as primary minerals, such as sphalerite (ZnS) and galena (PbS), the results of sequential extraction seem to correspond well to metal speciation determined with other methods [106].

11.3.3 State-of-the-Art Instrumental Methods

As metals cannot be degraded as such, and can only be transformed to different mineral phases, the post-entrapment fate of metals is of particular interest in environmental remediation. It has now been shown that precipitated metals can also re-mobilize if environmental conditions such as pH or redox change over time [114, 115]. Thus, it is imperative to understand the form in which metals are precipitated in the solid phase to potentially design better remediation wetlands. Though analytical capabilities are already well developed for metal quantification and speciation in aqueous medium, speciation in solid phase (precipitates, sediments) is rather difficult. Though extraction methods are extensively used, as described in the previous section the results are not only operational but suffer that at least some reduced phases are transformed during SEP.

Metal analysis in wetland sediments thus often fails to give a full picture of metal behavior in order to predict the mobility/stability of metals in the sediment. Traditional non-destructive bench top solid phase analysis techniques like X-ray diffraction (XRD) and X-ray Fluorescence (XRF) are very effective and now easily available. Bench top XRF measurements are very efficient in getting direct total metal concentration without using the destructive methods like sequential extraction, maintaining sample integrity. XRF methods have many advantages over conventional methods beyond the non-destructive nature of the assay, especially (11.1) that advanced XRF instruments, make possible the elemental mapping with high resolution (in microns) and from such results to establish the possible interaction between different elements, providing indirect information on mineral phases; and also (11.2) with advances in analytical designs, XRF units are now available in hand-held size and can be used to determine elemental composition of samples directly in the field. The main limitation of XRF methods is that they typically lack the sensitivity of inductively coupled plasma (ICP), so the detection limits are higher and thick samples can have X-ray self-absorption making quantification more difficult.

The XRD on the other hand, is a very effective technique for mineral identification, but for environmental samples like wetland sediments, the detection limits are relatively higher and the presence of silica and quartz makes it more difficult to identify the relevant phases due to many diffraction peaks generated by quartz. XRD is useful only for crystalline minerals, consequently amorphous and less crystalline minerals remain undetected. Recently, environmental scientists have started to use more sophisticated analytical techniques, e.g. high-energy X-ray synchrotron analysis like X-ray Absorbance Spectroscopy (XAS), which is very effective for both crystalline and amorphous phases even at low concentration with little or no interferences as the energy range is element specific.

With advanced facilities, it is possible now to determine elemental speciation of metals at concentrations as low as 20 ppm. This becomes even more critical while dealing with redox sensitive metals and metalloids like arsenic (As), where mobility is directly linked to the elemental speciation. It is by now well established that reduced species of As (As^{3+}) are more mobile than the oxidized counterparts [115]. So, to assess the risk, it is critical to determine different reduced species in the solid phase as well. The understanding of solid phase speciation of contaminants in wetland sediments is also essential in terms of environmental management. Many researchers interested in wetland biogeochemistry now use a combinatory approach for techniques such as XRF, XRD and XAS. In addition, in wetland systems, accumulation of micronutrients and contaminants in plants, e.g. speciation and localization of metals and metalloids in hyper and non-hyper accumulator plants, studies of the rhizosphere are also of great interest and with improved bench top XRF and synchrotron based XRF imaging gives access to this information by directly mapping elemental speciation in plants parts.

Overall, the mobility and successful remediation of most metals and particularly metalloids (As, Se etc.) in wetland systems, highly depends on their elemental speciation. Similarly, the elemental speciation of metals will also control the process of surface adsorption or structural incorporation in Fe/Fe-Mn oxides. The complexity and inter-dependence of redox reactions in wetland sediments often restrict understanding of metal behavior and distribution, if only traditional methods like sequential extraction are used.

11.3.4 Advanced Analytical Techniques

Recent advances in technology and access to the synchrotron platforms for environmental scientists have expanded the scope of understanding metal and metalloid behavior in sediments. Synchrotron radiation facilities have already became an important and widely used tool for probing the structure, composition, and bonding of metals in all forms, including those of interest in the environmental sciences. It is now possible to quantify and determine elemental speciation of metals and metalloids in sediment and aquifer samples without any pre-treatment. This is particularly interesting in reducing environments such as in wetlands. High energy X-rays with a wide range of energy (1–70,000 eV), up to a billion times as bright as those produced by conventional X-ray tubes used in XRD are now available. This remarkable increase in X-ray brightness (number of photons per unit solid angle) has immensely extended the range of structural and chemical investigations possible, allowing studies at low concentrations, extremely small sample size, and sample which are redox sensitive or phenomenon of very short duration. Experiments that were considered impractical a few years ago (e.g., structural compositions of molecular complexes on metal surfaces) are now very common.

Synchrotron radiation is a specific electromagnetic radiation generated in storage rings from high energetic electrons. Electromagnetic radiation emitted when electrons moving at velocities close to the speed of light, are forced to change direction under the action of a magnetic field. Today more than 50 synchrotron facilities are in operation and available to scientific community worldwide [116]. The large penetration depth of high energetic X-rays in matter, allows one to investigate the interior of an object without destructive sample preparation. Chemical speciation analysis through energy scanning around the K or L edge at high-energy resolution is now possible at the spatial resolution level in the nanometer range. The fine structure of X-ray absorption edges in the X-ray absorption spectroscopy (XAS) depends on the local chemical environment and state of the excited atom. Measurement in the transmission or the fluorescence modes provides this chemical information with two complementary methods, X-ray absorption near edge structure (XANES) and extended X-ray absorption fine structure analysis (EXAFS).

Sample collection and preservation methods are now well developed and analysis at N_2 or He temperatures gives noise free signals for XANES and EXAFS analysis for elemental concentrations as low as 20 ppm. It is also possible to establish the first and second neighbors of the element of interest to determine the phase and structure using EXAFS spectra interpretation with freely available computer programs. For more information and general details on synchrotron radiations and analysis for environmental scientist, readers will benefit from general reviews like of Lombi and Susini [117].

Figure 11.1 shows XANES spectra of different arsenic species in solid phase analyzed at synchrotron beamline at As K-edge. It is evident that different species can be identified easily in the sample with different peak positions. The higher energy part of a XAS spectrum above the

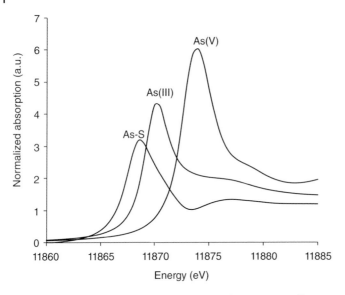

Figure 11.1 As K-edge XANES spectra recorded for Arsenite (As^{III}), Arsenate (As^V) and Arsenic-sulfide species.

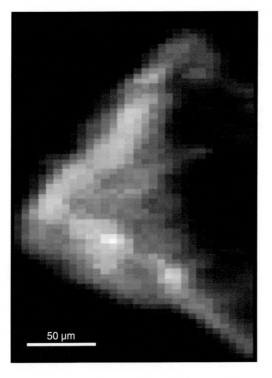

Figure 11.2 An elemental map of a sediment grain showing co-existence of Arsenic (light gray) and Manganese (dark gray) obtained using μ-XRF imaging at a synchrotron beamline.

edge is the EXAFS region, which provides oscillation generated by the constructive and destructive interferences between the outgoing and backscattered photoelectron wave and provides information on characteristics of the neighboring atoms and their co-ordination number.

Another approach to determine elemental speciation and quantification in sediment samples is using synchrotron based μ-XRF imaging. In this method, an area of interest is scanned at high resolution to get elemental mapping of that area. It is also possible now with advance synchrotrons, to take XAS measurements along with the μ-XRF mapping at selected points with spatial resolution of few hundred nanometers to few microns. Figure 11.2 shows an μ-XRF image from selected sediment grains showing co-precipitation of As and Mn in the sample. This technique is very often used to determine metal speciation and accumulation in plant parts (root, stem and leaves) [118].

References

1 Diels L, Spaans PH, Van Roy S, Hooyberghs L, Ryngaert A, Wouters H, Walter E, Winters J, Macaskie L, Finlay J, Pernfuss B, Woebking H, Puempel T, Tsezos M. Heavy metals removal by sand filters inoculated with metal sorbing and precipitating bacteria. Hydrometallurgy. 2003; 71:235–241.

2 Eger P, Lapakko K. Use of wetlands to remove nickel and copper from mine drainage. In: Constructed Wetlands for Wastewater Treatment; Municipal, Industrial, and Agricultural, Hammer D.A. (ed.) Lewis Publishers: Boca Raton, Florida; 1989, pp. 780–787.

3 Gessner TP, Kadlec RH, Reaves RP. Wetland remediation of cyanide and hydrocarbons. Ecol Eng. 2005; 25:457–469.

4 Kadlec RH, Wallace S. Treatment wetlands, 2nd edn. Boca Raton, FL, United States: CRC Press; 2009.

5 Johnson DB, Hallberg KB. Pitfalls of passive mine water treatment. Rev Environ Sci Biotechnol. 2002; 1:335–343.

6 Sheoran AS, Sheoran V. Heavy metal removal mechanism of acid mine drainage in wetlands: A critical review. Miner Eng. 2006; 19:105–116.

7 Sobolewski A. A review of processes responsible for metal removal in wetlands treating contaminated mine drainage. Int J Phytoremediation. 1999; 1(1):19–51.

8 Song Y. Mechanisms of lead and zinc removal from lead mine drainage in constructed wetland. PhD Dissertation. Civil Engineering Department, Faculty of Graduate School, Univ. Missouri-Rolla, Rolla, MO, USA, 2003.

9 Sheoran AS, Sheoran V, Choudhary RP. Bioremediation of acid-rock drainage by sulphate-reducing prokaryotes: A review. Miner Eng. 2010; 23:1073–1100.

10 PIRAMID Consortium. Engineering guidelines for the passive remediation of acidic and/or metalliferrous mine drainage and similar wastewaters, European Commission 5th Framework RTD Project no. EVK-CT-1999-000021 "Passive In-situ Remediation of Acidic Mine/Industrial Drainage" (PIRAMID), University of Newcastle Upon Tyne: Newcastle Upon Tyne, United Kingdom, 2003.

11 Kosolapov DB, Kuschk P, Vainshtein MB, Vatsourina AV, Wiessner A, Kaestner M, Mueller, R.A. Microbial processes of heavy metal removal from carbon-deficient effluents in constructed wetlands. Eng Life Sci. 2004; 4:403–411.

12 Bostick BC, Hansel CM, Fendorf S. Seasonal Fluctuations in Zinc Speciation within a Contaminated Wetland. Environ Sci Technol. 2001; 35: 3823–3829.

13 Reddy KR, DeLaune R. Biogeochemistry of Wetlands: Science and Applications. CRC Press: Boca Raton, Florida, 2008.

14 Gammons C, Frandsen A. Fate and transport of metals in H_2S-rich waters at a treatment wetland. Geochem Trans. 2001; 2,1.

15 Lewis AE. Review of metal sulphide precipitation. Hydrometallurgy. 2010; 104:222–234.

16 Lens PNL, Vallero M, Esposito G. Bioprocess engineering of sulphate reduction for environmental technology. In: Barton LL, Hamilton WA, eds. Sulphate-Reducing Bacteria: Environmental and Engineered Systems. Cambridge University Press; 2007, pp. 285–295.

17 Blais JF, Djedidi Z, Cheikh RB, Tyagi RD, Mercier G. Metals Precipitation from Effluents: Review. Pract Period Hazard Toxic Radioact Waste Manag. 2008; 12:135–149.

18 Wallace SD, Knight RL. Small-Scale Constructed Wetland Treatment Systems: Feasibility, Design Criteria, and O&M Requirements. Water Environment Research Foundation, 2006.

19 Fitch M, Burken J. Constructed Wetlands for Metals Removal: Design for Neutral Waters and AMD Remediation. Proposal to the Midwest Hazardous Substance Research Center, 2003.

20 Younger PL, Banwart SA, Hedin R. Mine Water: Hydrology, Pollution, Remediation. Kluwer Academic Publishers: London, United Kingdom, 2002.

21 Kumar N, Couture R-M, Millot R, Battaglia-Brunet F, Rose J. Microbial Sulfate Reduction Enhances Arsenic Mobility Downstream of Zerovalent-Iron-Based Permeable Reactive Barrier. Environ. Sci. Technol. 2016; 50(14): 7610–7617.

22 Couture RM, Rose J, Kumar N, Mitchell K, Wallschlager D, Van Cappellen P. Sorption of arsenite, arsenate, and thioarsenates to iron oxides and iron sulphides: A kinetic and spectroscopic investigation. Environ Sci Technol. 2013; 47: 5652–5659.

23 Stucker VK, Silverman DR, Williams KH, Sharp JO, Ranville JF. Thioarsenic species associated with increased arsenic release during biostimulated subsurface sulfate reduction. Environ Sci Technol. 2014; 48:13367–13375.

24 Tang K, Baskaran V, Nemati M. Bacteria of the sulphur cycle: an overview of microbiology, biokinetics and their role in petroleum and mining industries. Biochem Eng J. 2009; 44:73–94.

25 Thauer RK, Stackebrandt E, Allan Hamilton AW. Energy metabolism and phylogenetic diversity of sulphate-reducing bacteria. In: Barton LL, Hamilton WA, eds. Sulphate-Reducing Bacteria: Environmental and Engineered Systems. Cambridge University Press. 2007, pp. 1–37.

26 Castro HF, Williams NH, Ogram A. Phylogeny of sulphate-reducing bacteria. FEMS Microbiol Ecol. 2000; 31:1–9.

27 Mori K, Kim H, Kakegawa T, Hanada S. A novel lineage of sulphate-reducing microorganisms: *Thermodesulfobiaceae* fam. nov., *Thermodesulfobium narugense*, gen. nov., sp nov., a new thermophilic isolate from a hot spring. Extremophiles. 2003; 7:283–290.

28 Webb JS, McGinness S, Lappin-Scott HM. Metal removal by sulphate-reducing bacteria from natural and constructed wetlands. J Appl Microbiol. 1998; 84:240–248.

29 Khanal S. Anaerobic Biotechnology for Bioenergy Production: Principles and Applications. John Wiley & Sons; 2008.

30 Sturman PJ, Stein OR, Vymazal J, Kröpfelová L. Sulfur Cycling in Constructed Wetlands. In: Vymazal J, editor. Wastewater Treatment, Plant Dynamics and Management in Constructed and Natural Wetlands. Springer Netherlands, Dordrecht. 2008, pp. 329–344.

31 Liamleam W, Annachhatre AP. Electron donors for biological sulfate reduction. Biotechnol Adv. 2007; 25:452–463.

32 Gibert O, De Pablo J, Cortina JL, Ayora C. Municipal compost-based mixture for acid mine drainage bioremediation: Metal retention mechanisms. Appl Geochem. 2005; 20:1648–1657.

33 Neculita CM, Zagury GJ, Bussière B. Passive treatment of acid mine drainage in bioreactors using sulfate-reducing bacteria – critical review and research needs. J Environ Qual. 2007; 36:1–16.

34 Machemer SD, Wildeman TR. Adsorption compared with sulfide precipitation as metal removal processes from acid mine drainage in a constructed wetland. J Contam Hydrol. 1992; 9:115–131.

35 Lesage E. Behaviour of heavy metals in constructed treatment wetlands. PhD thesis. Faculty of Bioscience Engineering. Ghent: Belgium Ghent University. 2006.

36 Song YC, Piak BC, Shin HS, La SJ. Influence of electron donor and toxic materials on the activity of sulfate reducing bacteria for the treatment of electroplating wastewater. Water Sci Technol. 1998; 38:187–194.

37 Unz RF, Shuttleworth KL. Microbial mobilization and immobilization of heavy metals. Curr Opin Biotechnol. 1996; 7(3):307–310.

38 Schiewer S, Volesky B. Biosorption process for heavy metal removal. In: DR Lovley, ed. Environmental Microbe–Metal Interactions. Washington: ASM Press. 2000, pp. 329–362.

39 Gadd GM. Bioremedial potential of microbial mechanisms of metal mobilization and immobilization. Curr Opin Biotechnol. 2000; 11:271–279.

40 Salt DE, Blaylock M, Kumar NPBA, Dushenkov V, Ensley BD, Chet I. Raskin I. Phytoremediation: A novel strategy for the removal of toxic metals from the environment using plants. Biotechnology (NY). 1995; 13:468–475.

41 Sinha RK, Herat S, Tandon PK. Phytoremediation: Role of plants in contaminated site management. In: Environmental Bioremediation Technologies. Springer Berlin Heidelberg; 2007, pp. 315–330.

42 US Department of Energy. Plume Focus Area, December. Mechanisms of plant uptake, translocation, and storage of toxic elements. Summary Report of a workshop on phytoremediation research needs. 1994. Available from: www.osti.gov/scitech/servlets/purl/10109412 [Accessed: 30 November 2017].

43 Welch RM. Micronutrient nutrition of plants. CRC Crit Rev Plant Sci. 1995; 14(1):49–82.

44 Cowgill VM. The hydrogeochemical of Linsley Pond, North Branford. Part 2. The chemical composition of the aquatic macrophytes. Arch Hydrobiol. 1974; 45(1):1–119.

45 Matagi SV, Swai D, Mugabe R. A review of heavy metal removal mechanisms in wetlands. Afr J Trop Hydrobiol Fish. 1998; 8:23–35.

46 Sriyaraj K, Shutes RBE. An assessment of the impact of motorway runoff on a pond, wetland and stream. Environ Int. 2001; 26(5–6): 433–439.

47 Price C.E. Penetration and transactions of herbicides and fungicide in plants. In: McFarlane NR, ed. Herbicides and Fungicides. The Chemical Society Special Publication, Burlington House London WIV OBN, 29. 1977, pp. 42–66.

48 Everard M, Denny P. Flux of lead in submerged plants and its relevance to a fresh water system. Aquat Bot. 1995; 21:181–193.

49 Grill E, Winnacker EL, Zenk MH. Phytochelatins: The principal heavy-metal complexing peptides of higher plants. Science. 1985; 230:674–676.

50 Meers E, Ruttens A, Hopgood M, Lesage E, Tack FMG. Potential of *Brassica rapa, Cannabis sativa, Helianthus annuus* and *Zea mays* for phytoextraction of heavy metals from calcareous dredged sediment derived soils. Chemosphere. 2005; 61(4):561–572.

51 Fergusson JE. The heavy metals: chemistry, environmental impact and health effects. Oxford: Pergamon Press; 1990, pp. 382–388.

52 Raven PH, Evert RF, Eichhorn SE. Biology of Plants. New York: Worth Publishers, 1992.

53 Lasat MM. Phytoextraction of metals from contaminated soil: a review of plant/soil/metal interaction and assessment of pertinent agronomic issues. J of Hazard Subst Res. 2000; 2(5):1–25.

54 Rieuwerts JS, Thornton I, Farago ME, Ashmore MR. Factors influencing metal bioavailability in soils: preliminary investigations for the development of a critical loads approach for metals. Chem Spec Bioavailab. 1998; 10(2): 61–75.

55 Gambrell RP. Trace and toxic metals in wetlands – A review. J Environ Qual. 1994; 23:883–891.

56 Jackson LJ. Paradigms of metal accumulation in rooted aquatic vascular plants. Sci Total Environ. 1998; 219:223–231.

57 Guilizzoni P. The role of heavy metals and toxic materials in the physiological ecology of submersed macrophytes. Aquat Bot. 1991; 41(1–3):87–109.

58 Keskinkan O. Investigation of heavy metal removal by a submerged aquatic plant (*Myriophyllum spicatum*) in a batch system. Asian J Chem. 2005; 17(3):1507–1517.

59 Maine MA, Suñé NL, Lagger SC. Chromium bioaccumulation: comparison of the capacity of two floating aquatic macrophytes. Water Res. 2004; 38:1494–1501.

60 Dunbabin JS, Bowmer KH. Potential use of constructed wetlands for treatment of industrial wastewaters containing metals. Science Total Environ. 1992; 111:151–168.

61 Williams JB. Phytoremediation in wetland ecosystems: progress, problems, and potential. CRC Crit Rev Plant Sci. 2002; 21(6):607–635.

62 Weis JS, Weis P. Metal uptake, transport and release by wetland plants: implications for phytoremediation and restoration. Environ Int. 2004; 30:685–700.

63 Beining BA, Otte ML. Retention of metals and longevity of a wetland receiving mine leachate. In: Proceedings of 1997 National Meeting of the American Society for Surface Mining and Reclamation, Austin, Texas, May 10–16; 1997, pp. 43–46.

64 Beining BA, Otte ML. Retention of metals originating from an abandoned lead-zinc mine by a wetland at Glendalough, Co. Wicklow. Biol Environ. 1996; 96B:117–126.

65 Jacob DL, Otte ML. Conflicting processes in the wetland plant rhizosphere: metal retention or mobilization? Water Air Soil Pollut Focus. 2003; 3:91–104.

66 Jacob DL, Otte ML. Long-term effects of submergence and wetland vegetation on metals in a 90-year old abandoned Pb-Zn mine tailings pond. Environ Pollut. 2004; 130:337–345.

67 Buddhawong S, Kuschk P, Mattusch J, Wiessner A, Stottmeister U. Removal of arsenic and zinc using different laboratory model wetland systems. Eng Life Sci. 2005; 5(3): 247–252.

68 Stottmeister U, Wießner A, Kuschk P, Kappelmeyer U, Kästner M, Bederski O, Müller RA, Moormann H. Effect of plants and microorganisms in constructed wetlands for wastewater treatment. Biotechnol Adv. 2003; 22:93–117.

69 Ryan PR, Delhaize E, Jones DL. Function of mechanism of organic acid exudation from plant roots. Annu Rev Plant Physiol Plant Mol Biol. 2001; 52:527–560.

70 Chaudhry Q, Blom-Zandstra M, Gupta S, Joner EJ. Utilising the synergy between plants and rhizosphere microorganisms to enhance breakdown of organic pollutants in the environment. Environ Sci Pollut Res. 2005; 12:34–48.

71 Armstrong J, Armstrong W, Beckett PM. *Phragmites australis*: venture- and humidity-induced pressure flows enhance rhizome aeration and rhizosphere oxidation. New Phytol. 1992; 120:197–207.

72 Armstrong J, Armstrong W, Beckett PM, Halder JE, Lythe S, Holt R, Sinclair A. Pathways of aeration and the mechanisms and beneficial effect of humidity- and venture-induced convections in Phragmites australis (Cav.) Trin Ex Steud Aquat Bot. 1996; 54:177–197.

73 Peverly JH, Surface JM, Wang T. Growth and trace metal absorption by *Phragmites australis* in wetlands constructed for landfill leachate treatment. Ecol Eng. 1995; 5:21–35.

74 Mantovi P, Marmiroli M, Maestri E, Tagliavini S, Piccinini S, Marmiroli N. Application of a horizontal subsurface flow constructed wetlands on treatment of dairy parlor wastewater. Bioresour Technol. 2003; 88:85–94.

75 Leuridan, I. Complexation of copper by decomposition residuals of *Phragmites australis* in surface flow constructed wetlands. MSc thesis. Faculty of Agricultural and Applied Biological Sciences. Ghent: Belgium Ghent University, 2004.

76 Bragato C, Brix H, Malagoli M. Accumulation of nutrients and heavy metals in *Phragmites australis* (Cav.) Trin. ex Steudel and *Bolboschoenus maritimus* (L.) Palla in a constructed wetland of the Venice lagoon watershed. Environ Pollut. 2006; 144:967–975.

77 Cheng S, Grosse W, Karrenbrock F, Thoennessen M. Efficiency of constructed wetlands in decontamination of water polluted by heavy metals. Ecol Eng. 2002; 18:317–325.

78 Collins BS, Sharitz RR, Coughlin DP. Elemental composition of native wetland plants in constructed mesocosm treatment wetlands. Bioresour Technol. 2005; 96:937–48.

79 Grisey E, Laffray X, Contoz O, Cavalli E, Mudry J, Aleya L. The bioaccumulation performance of reeds and cattails in a constructed treatment wetland for removal of heavy metals in landfill leachate treatment (Etueffont, France). Water Air Soil Pollut. 2012; 223:1723–1741.

80 Khan S, Ahmad I, Shah MT, Rehman S, Khaliq A. Use of constructed wetland for the removal of heavy metals from industrial wastewater. J Environ Manage. 2009; 90:3451–3457.

81 Maine MA, Suñe N, Hadad H, Sánchez G, Bonetto C. Influence of vegetation on the removal of heavy metals and nutrients in a constructed wetland. J Environ Manage. 2009; 90:355–363.

82 Hadad HR, Maine MA, Bonetto CA. Macrophyte growth in a pilot-scale constructed wetland for industrial wastewater treatment. Chemosphere. 2006; 63:1744–1753.

83 Manios T, Stentiford EI, Millner P. Removal of heavy metals from a metaliferous water solution by *Typha latifolia* plants and sewage sludge compost. Chemosphere. 2003; 53:487–494.

84 Nyquist J, Greger M. A field study of constructed wetlands for preventing and treating acid mine drainage. Ecol Eng. 2009; 35:630–642.

85 Yang B, Lan CY, Yang CS, Liao WB, Chang H, Shu WS. Long-term efficiency and stability of wetlands for treating wastewater of a lead/zinc mine and the concurrent ecosystem development. Environ Pollut. 2006; 143:499–512.

86 Marchand L, Mench M, Jacob DL, Otte ML. Metal and metalloid removal in constructed wetlands, with emphasis on the importance of plants and standardized measurements: A review. Environ Pollut. 2010; 158:3477–3461.

87 Calheiros CSC, Rangel AOSS, Castro PML. Constructed wetland systems vegetated with different plants applied to the treatment of tannery wastewater. Water Res. 2007; 41:1790–1798.

88 Aslam MM, Malik M, Baig MA, Qazi IA, Iqbal J. Treatment performances of compost-based and gravel-based vertical flow wetlands operated identically for refinery wastewater treatment in Pakistan. Ecol Eng. 2007; 30:34–42.

89 Lesage E, Mundia C, Rousseau DPL, Van de Moortel AMK, Du Laing G, Meers E, Tack FMG, De Pauw N, Verloo MG. Sorption of Co, Cu, Ni and Zn from industrial effluents by the submerged aquatic macrophyte *Myriophyllum spicatum* L. Ecol Eng. 2007a; 30:320–325.

90 Lesage E, Rousseau DPL, Meers E, Tack FMG, Pauw ND. Accumulation of metals in a horizontal subsurface flow constructed wetland treating domestic wastewater in Flanders, Belgium. Sci Total Environ. 2007b; 380: 102–115.

91 Sochacki A, Surmacz-Górska J, Faure O, Guy B. Polishing of synthetic electroplating wastewater in microcosm upflow constructed wetlands: Effect of operating conditions. Chem Eng J. 2014; 237:250–258.

92 Obarska-Pempkowiak H, Gajewska M, Wojciechowska E. Hydrofitowe oczyszczanie wód i ścieków. [Treatment of water and wastewater by hydrophytes]. Warszawa, Poland: Wydawnictwo Naukowe PWN. 2010 (in Polish).

93 Ure AM. Trace element speciation in soils, soil extracts and solutions, Microchim Acta. 1991; 104:49–57.

94 D'Amore JJ, Al-Abed SR, Scheckel KG, Ryan JA. Methods for speciation of metals in soils: a review. J Environ Qual. 2005; 34(5):1707–1745.

95 Bacon JR, Davidson CM. Is there a future for sequential chemical extraction? Analyst. 2008; 133:25–46.

96 Neculita CM, Zagury GJ, Bussière B. Effectiveness of sulfate-reducing passive bioreactors for treating highly contaminated acid mine drainage: II. Metal removal mechanisms and potential mobility. Appl Geochem. 2008; 23:3545–3560.

97 Kierczak J, Neel C, Aleksander-Kwaterczak U, Helios-Rybicka E, Bril H, Puziewicz J. Solid speciation and mobility of potentially toxic elements from natural and contaminated soils: a combined approach. Chemosphere. 2008: 73(5):776–784.

98 Edwards HG, Jorge Villar SE, Bishop JL, Bloomfield M. Raman spectroscopy of sediments from the Antarctic Dry Valleys; an analogue study for exploration of potential paleolakes on Mars. J Raman Spectrosc. 2004; 35(6):458–462.

99 Villanueva U, Raposo JC, Castro K, de Diego A, Arana G, Madariaga JM. Raman spectroscopy speciation of natural and anthropogenic solid phases in river and estuarine sediments with appreciable amount of clay and organic matter. J Raman Spectrosc. 2008; 39(9):1195–1203.

100 Taylor KG, Hudson-Edwards KA, Bennett AJ, Vishnyakov V. Early diagenetic vivianite $[Fe_3(PO_4)_2 \cdot 8H_2O]$ in a contaminated freshwater sediment and insights into zinc uptake: A μ-EXAFS, μ-XANES and Raman study. Appl Geochem. 2008; 23(6):1623–1633.

101 Gagliano WB, Brill MR, Bigham JM, Jones FS, Traina SJ. Chemistry and mineralogy of ochreous sediments in a constructed mine drainage wetland. Geochim Cosmochim Acta. 2004; 68(9):2119–2128.

102 Arnarson TS, Keil RG. Organic–mineral interactions in marine sediments studied using density fractionation and X-ray photoelectron spectroscopy. Org Geochem. 2001; 32(12):1401–1415.

103 Shchukarev A, Gälman V, Rydberg J, Sjöberg S, Renberg I. Speciation of iron and sulphur in seasonal layers of varved lake sediment: An XPS study. Surf Interface Anal. 2008; 40(3–4):354–357.

104 Filgueiras AV, Lavilla I., Bendicho C. Chemical sequential extraction for metal partitioning in environmental solid samples. J Environ Monit. 2002; 4:823–857.

105 Tack FMG, Verloo MG. Chemical speciation and fractionation in soil and sediment heavy metal analysis: A review. Int J Environ Anal Chem. 1995; 59:225–238.

106 Peltier E, Dahl AL, Gaillard JF. Metal speciation in anoxic sediments: When sulfides can be construed as oxides. Environ Sci Technol. 2005. 39:311–316.

107 Tessier A, Campbell PGC, Bisson M. Sequential extraction procedure for the speciation of particulate trace metals. Anal Chem. 1979; 51:844–851.

108 Quevauviller P, Rauret G, Muntau H, Ure AM, Rubio R, López-Sánchez JF, Fiedler HD, Griepink B. Evaluation of a sequential extraction procedure for the determination of extractable trace metal contents in sediments. Fresenius Jof Anal Chem. 1994; 349:808–814.

109 Rauret G, López-Sánchez JF, Sahuquillo A, Rubio R, Davidson C, Ure A, Quevauviller P. Improvement of the BCR three step sequential extraction procedure prior to the certification of new sediment and soil reference materials. J Environ Monit. 1999; 1:57–61.

110 Hlavay J, Prohaska T, Weisz M, Wenzel WW, Stingeder GJ. Determination of trace elements bound to soil and sediment fractions (IUPAC Technical Report). Pure Appl Chem. 2004; 76:415–442.

111 Tipping E, Hetherington NB, Hilton J, Thompson DW, Howles E, Hamilton-Taylor J. Artifacts in the use of selective chemical-extraction to determine distributions of metals between oxides of manganese and iron. Anal Chem. 1985; 57:1944–1946.

112 Lacal J, Da Silva MP, García R, Sevilla MT, Procopio JR, Hernández L. Study of fractionation and potential mobility of metal in sludge from pyrite mining and affected river sediments: changes in mobility over time and use of artificial ageing as a tool in environmental impact assessment. Environ Pollut. 2003; 124:291–305.

113 Burton ED, Phillips IR, Hawker DW. Factors controlling the geochemical partitioning of trace metals in estuarine sediments. Soil Sediment Contam. 2006; 15: 253–276.

114 Kumar N, Chaurand P, Rose J, Diels L, Bastiaens L. Synergistic effects of sulfate reducing bacteria and zero valent iron on zinc removal and stability in aquifer sediment. Chem Engin J. 2015; 260:83–89.

115 Kumar N, Couture RM, Millot R, Battaglia-Brunet F, Rose J. Microbial sulfate reduction enhances arsenic mobility downstream of zero valent based permeable reactive barrier. Environ Sci Technol. 2016; 50(14):7610–7617.

116 Lightsources. Available from: www.lightsources.org.

117 Lombi E, Susini J. Synchrotron-based techniques for plant and soil science: opportunities, challenges and future perspective. Plant Soil. 2009; 320:1–35.

118 Vijayan P, Willick IR, Lahlali R, Karunakaran C, Tanino KK. Synchrotron radiation shed fresh light on plant research: The use of powerful techniques to probe structure and composition of plants. Plant Cell Physiol. 2015; 56(7):1252–1263.

12

A Review on the Use of Constructed Wetlands for the Treatment of Acid Mine Drainage

C. Sheridan[1], A. Akcil[2], U. Kappelmeyer[3] and I. Moodley[1]

[1] *Industrial Mine Water Research Unit (IMWaRU), Centre in Water Research Development (CIWaRD), School of Chemical and Metallurgical Engineering (CHMT), University of the Witwatersrand, Johannesburg, South Africa*
[2] *Mineral-Metal Recovery and Recycling (MMR&R) Research Group, Mineral Processing Division, Department of Mining Engineering, Suleyman Demiral University, Isparta, Turkey*
[3] *Helmholtz Centre for Environmental Research-UFZ, Department of Environmental Biotechnology, Leipzig, Germany*

12.1   What is Acid Mine Drainage?

Acid mine drainage (AMD) or acid rock drainage (ARD) is an environmental pollutant that is associated with mining activities. The production of AMD is actually a natural occurrence but mining activities have, however, accelerated this process, resulting in negative impacts to the environment [1–3].

In a natural environment small amounts of acid are produced when sulfide ore is exposed to an oxidizing environment and this occurs through the natural weathering of rocks that enclose sulfide. Subsequently, naturally occurring processes neutralize the acid, with surrounding alkaline rocks providing the most alkalinity [1]. These neutralization processes (shown below) include dissolution of carbonate substrates like calcite and dolomite [equations (12.1), (12.2)]; reduction of iron hydroxides within sediment [equation (12.3)]; and dissimilatory sulfate reduction [equation (12.4)]. All of these processes collectively consume protons and produce alkalinity, thereby reducing AMD impact. Mining of sulfide ore has fractured large amounts of these sulfide-containing rocks and this has led to the exposure of roots-rock surfaces. This consequently leads to the production of large quantities of AMD. In addition to this, these naturally occurring processes have been become unbalanced and as such, natural buffering sources become less and less available [2].

$$CaCO_3 + 2\,H^+ \rightarrow Ca^{2+} + H_2O + CO_2 \tag{12.1}$$

$$CaCO_3 + H_2SO_4 \rightarrow Ca^{2+} + HCO_3^- \tag{12.2}$$

$$3\,Fe(OH)_3 + H^+ \rightarrow Fe_3(OH)_8 + H_2O \tag{12.3}$$

$$SO_4^{2-} + 2\,CH_2O \rightarrow H_2S + 2\,HCO_3^- \tag{12.4}$$

In a mining environment, AMD generation begins with the mining of sulfide ore and continues to generate, even after the mine is closed or abandoned [4, 5]. Initially AMD is formed in the groundwater of an active mine; however the rate of AMD generation is slow as the water level is kept to a minimum for dewatering, through the use of pumps. When the mine is shut down however, pumping operations cease, leading to groundwater levels rising and eventual flooding of the mine [6]. The rate of AMD generation thereby rapidly increases and the groundwater eventually leaches into surface waters and waterways [7].

Chemically, AMD is formed from the exposure of sulfide ore minerals to water and oxygen. Oxidizing bacteria also play a role in this generation and typically act as catalysts. Once the ore is exposed, sulfate and heavy metals such iron, copper, lead, nickel, manganese, cadmium, aluminium and zinc are leached and solubilize into the water [4, 6, 8]. In Johannesburg, the ore body is uraniferous, and this has resulted in the mobilization of uranium into the aquatic environment. Tutu et al. [9] found uranium in open, accessible surface waters in excess of 70 mg/L in certain instances. In general, the concentration of sulfate and other contaminants in the water is highly variable as a consequence of ore body heterogeneity. Thus, depending on the characteristics of the mine, the drainage generally has a low pH, high salinity and high concentrations of heavy metals. It should be noted that mine drainage can also be neutral or alkaline depending on the pH and that these waters may also have significant environmental impacts. They are not considered further here.

There are four well-known equations that describe the formation of AMD and they are expressed in terms of pyrite, as it is the most common sulfide mineral mined. For other sulfide-containing minerals the stoichiometry and reaction rates differ slightly. The generation of AMD begins through the oxidation of pyrite to ferrous iron and sulfate [equation (12.5)]. Through the action of iron-oxidizing bacteria, ferrous iron is oxidized to ferric iron [equation (12.6)]. This oxidation can also occur when Fe^{2+} travels to surface waters such as rivers and dams, where the pH is relatively higher i.e. pH > 5 [1]. Ferric iron then spontaneously reacts with water to form ferric hydroxide [equation (12.7)] and this is the orange-red precipitate seen in AMD, also known as yellow boy. Excess Fe^{3+} acts as an additional and secondary reducing agent for pyrite [equation (12.8)]. Equations (12.5), (12.7) and (12.8) release a substantial amount of acidity into the water, which characterizes AMD as acidic. The overall process results in sulfuric acid and ferric hydroxide being produced [equation (12.9)].

$$FeS_2 + \tfrac{7}{2} O_2 + H_2O \rightarrow Fe^{2+} + 2\,SO_4^{2-} + 2\,H^+ \tag{12.5}$$

$$Fe^{2+} \tfrac{1}{4} O_2 + H^+ \rightarrow Fe^{3+} + \tfrac{1}{2}\,H_2O \tag{12.6}$$

$$Fe^{3+} + 3\,H_2O \rightarrow Fe(OH)_3 + 3\,H^+ \tag{12.7}$$

$$FeS_2 + 14\,Fe^{3+} + 8\,H_2O \rightarrow 15\,Fe^{2+} + 2\,SO_4^{2-} + 16\,H^+ \tag{12.8}$$

$$FeS_2 + 3.75\,O_2 + 3.5\,H_2O \leftrightarrow Fe(OH)_3 + 2\,SO_4^{2-} + 4\,H^+ \tag{12.9}$$

(Iron sulfide + Oxygen + Water ↔ Ferric hydroxide + aqueous sulfuric acid)

The initial oxidation step of other common sulfide minerals such as arsenopyrite [equation (12.10)], chalcopyrite [equation (12.11)] and sphalerite [equation (12.12)] are shown below:

$$4FeAsS + 13O_2 + 6H_2O \rightarrow 4Fe^{2+} + 4SO_4^{2-} + 4H_2AsO_4^- + 4H^+ \tag{12.10}$$

$$2CuFeS_2 + 4O_2 \rightarrow 2Cu^{2+} + Fe^{2+} + SO_4^{2-} \tag{12.11}$$

$$ZnS + 2O_2 \rightarrow Zn^{2+} + SO_4^{2-} \tag{12.12}$$

12.2 Sources of AMD

AMD is produced within and around mining areas. These typically include abandoned and active mines, both of which can be open-pit or underground [5, 10]. Secondary sources of AMD include

Table 12.1 Mineral sources of sulfide which can contribute to acidic mine water

Mineral	Chemical formula
Arsenopyrite	$FeS_2.FeAs$
Bornite	$CuFeS_4$
Chalcocite	Cu_2S
Chalcopyrite	$CuFeS$
Covellite	CuS
Galena	PbS
Gold sulfide	Au_2S
Millerite	NiS
Mobybdenite	MoS_2
Pyrite	FeS_2
Pyrrhhdite	$Fe_{11}S_{12}$
Sphalerite	ZnS

mine waste dumps, ore stockpiles, tailings dumps, tailings dams, haul roads, quarries and sludge ponds [6, 10, 11]. The nature of AMD from these secondary sources may be more aggressive than the drainage water due to factors such as particle size, surface area, amorphosity, homogeneity and disaggregation [12].

Any sulfide-containing mineral is classified as a potential source of AMD [4, 11]. Generally speaking, the most common metal sulfide associated with AMD is pyrite. Metal concentrations, however vary according to the characteristics of the mine and thus other metals may be more significant [1]. Particularly for South Africa where the coal and gold mining industries are extensive, pyrite especially contributes massively to the formation of AMD. A list of other sulfide based minerals, all of which could potentially contribute to AMD is presented in Table 12.1.

12.3 Environmental and Social Impacts of AMD

12.3.1 Environmental Impacts

There has been increased public awareness on the environmental hazards associated with AMD. AMD is globally considered to be one of the most hazardous forms of water pollution and has significant environmental impact [4, 11]. The Environmental Protection Agency (EPA) considers it to be "second only to global warming and ozone depletion" in terms of ecological risk.

The impacts of AMD on the environment and ecosystems have been widely reported [13–16]. It is a multifactor pollutant that affects the environment chemically, biologically, ecologically and physically (Figure 12.1). The toxicity of metals, high sulfate concentrations (and consequent high salinity), total dissolved solids, low pH and high acidity of AMD all play a role in negatively impacting the environment as well as life. Owing to this multifarious nature of AMD, food chains and ecosystems collapse at much higher rates than if it were a single pollutant. The ultimate result is the irreversible destruction of habitats and the death of organisms in both terrestrial and aquatic habitats [11].

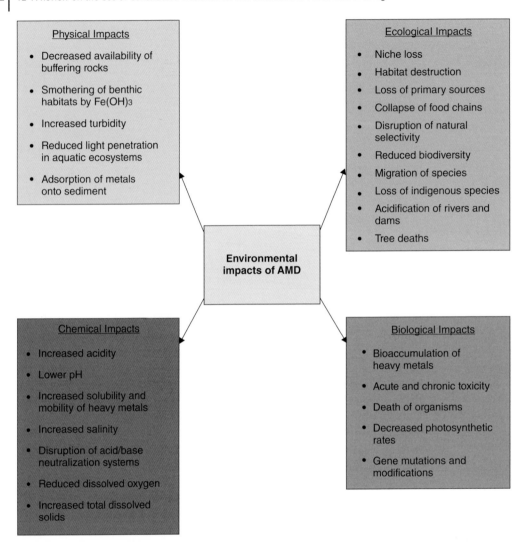

Figure 12.1 The physical, chemical, biological and ecological impacts of AMD.

Among the negative environmental impacts presented in Figure 12.1, the most significant and visible impact of AMD is the smothering of metal precipitates on the water surface. This poses significant impacts to water habitats in that it reduces the penetration of sunlight, creates turbidity and lessens the diffusion of oxygen. This makes it difficult for photosynthetic organisms to thrive and can lead to the loss of autotrophic sources, thereby disrupting food chains. In benthic habitats soil is smothered with precipitates and heavy metals adsorb onto the sediment. All of these limits vegetation speciation and reduces biodiversity.

Acid/base neutralizing processes also become disturbed as buffering sources such as carbonate rocks become less available owing to their excessive neutralization of AMD. In particular bicarbonate and carbonate sources get broken down to carbon dioxide and water as a result of the low pH. This also affects photosynthesis as many plants use bicarbonate as an inorganic carbon source [17]. The heavy metals in AMD can also have disastrous effects on organisms. Depending on the concentration and toxicity of the metal, organisms may bioaccumulate the metal [18], which can cause acute and chronic toxicity or at the worst, sudden death can occur.

12.3.2 Social Impacts of AMD

AMD is discharged or leached into dams and rivers, which can further contaminate public water systems. Communities residing in the vicinity of sulfidic ore mines are most affected and threatened by AMD. AMD contaminated water cannot be utilized for potable or domestic purposes as it does not meet the required water quality standards [7]. AMD thus poses a significant threat to humans, not only in terms of toxicity but also in terms of water supply. In developing countries, it is estimated that billions of people will face water scarcity by the year 2050 [19] and if AMD is not controlled or treated, this risk will be exacerbated.

Economically, AMD also has significant impacts. Owing to the acidic nature of AMD, it is highly corrosive and can corrode most iron-based structures [20]. In functional mines AMD-contaminated water cannot be re-used as it increases the level of toxicity in a mine, which can lead to disastrous effects. As such the operating costs of pumping and using potable water are high. AMD also limits the use of downstream waterways, affecting industries such as fishing, agriculture, irrigation and recreation [7]. In addition, Johannesburg, the economic hub of Africa is the site of an estimated AMD discharge which will exceed 250 mega liters per day, which has to be treated. This costs the South African fiscus in excess of US50 million per year, just for pH neutralization using lime. It is estimated that this cost will increase to treat the AMD to legislative limits [21]. Thus, effective, cheap remedial techniques are required to prevent the economic benefits from mining being eroded through protracted clean-up.

12.4 Remediation of AMD

There are various physical, chemical and biological mechanisms by which AMD can be treated. These are summarized in Table 12.2. They are categorized as physical, chemical or biological. All three of these mechanisms can be employed individually or in combination in both active or passive systems.

Table 12.2 Physical, chemical and biological mechanisms of AMD treatment (modified from [20])

Physical	Chemical	Biological
Adsorption	pH control	Dissimilatory sulfate reduction
Absorption	Complexation	Action of oxidizing bacteria
Sedimentation	Oxidation/reduction	
Flocculation	Chelation	
Crystallization	Electrochemical	
Filtration Nanofiltration	Ion exchange Precipitation	

AMD remediation is generally classified as active or passive. Active systems are highly effective and can treat the AMD to almost any desired purity. Active processes include using alkali precipitation, reverse osmosis, ion exchange and active bioremediation (typically within reactors). Most active processes have a fairly small surface footprint, but are instead complex, costly, require continuous chemical feed and rely on well trained technical personnel to operate and maintain. In addition, as the purity of the treated AMD is increased, the level of technology required increases exponentially.

Whilst highly technical solutions might well be applicable in many locations, especially in the developed world or at active mine sites, following closure of a mine site the feasibility of an active processes diminishes rapidly as skilled personnel move elsewhere. In the developing world, in particular, there may be no skilled technical expertise anywhere following closure of the mine, and thus if ongoing treatment is required, it is generally preferable to utilize passive treatment techniques. Constructed wetlands (CWs) are obvious passive remediation technologies.

12.4.1 Constructed Wetlands

CWs were first described and documented in the early 1960s where the quality of wastewater was improved through the use of natural sphagnum moss wetlands [22, 23] and this eventually stimulated the idea of wetlands being used to neutralize AMD [24]. Considerable research expanded on this idea and after five decades, CWs have now been globally implemented at many mine sites for the treatment of AMD [25].

12.4.1.1 Constructed Wetland Configuration Types

CWs are classified in terms of hydrology, although some sources do classify them according to macrophyte type [26]. In terms of hydrology there are two types of CWs:

1) Surface flow wetlands
2) Sub-surface flow wetlands which include:
   - Horizontal sub-surface flow
   - Vertical sub-surface flow.

Since AMD characteristics and CW designs and configurations are highly variable, the use of CWs for treating AMD is presented chronologically from the 1990s to the present. The use of CWs for treating AMD prior to 1990 is excluded from this analysis as it can be found elsewhere in literature. Also, the work described here is not necessarily exhaustive, and there may be documented use of CWs for treating AMD which are not discussed here. In general, however, the mechanisms described will be similar.

12.4.1.2 Mechanism by which CWs Remediate Most AMD/ARD

There are three main components to AMD: acidity (and pH), sulfate, and (heavy) metals. Sulfate is primarily removed through the process of dissimilatory sulfate reduction, which occurs in the anaerobic regions of the CW. In the anaerobic zone, a hydrocarbon, together with the sulfate are reacted with a resultant sulfide and carbonate [27–29]. The reaction is shown as Equation (12.13):

$$2CH_2O + SO_4^{2-} \rightarrow H_2S + 2HCO_3^- \tag{12.13}$$

In this reaction CH_2O represents organic matter. Thus, the pH starts to increase as alkalinity is generated. The CW may also contain a source of alkalinity, such as being constructed from dolomite, which may also increase the pH. The rise in pH is due to the consumption of protons within the system. This can be due to a reaction that consumes protons (such as oxidation). Indeed, if the initial pH is especially low, fermentations that produce weak acids are also able to increase the pH. As the pH increases, many of the metals in solution precipitate, provided the pH does not increase too far (in which case some may solubilize again).

The solubility of most metals and the chemistry of their precipitates at different pH values are shown in their unique Pourbaix diagrams, which are freely available in literature. For example, iron has a very well described Pourbaix diagram, which clearly shows how by pH 7 that most forms of iron are insoluble. Thus, heavy and other metals tend to precipitate. In the presence of H_2S and free alkalinity, the soluble metals are also able to precipitate as metal sulfides, which are insoluble. In this instance, the cycle will be complete, and the cause of the AMD, which is typically a sulfide mineral, will be present again. Provided it remains within an anaerobic environment, it will remain in this stable, benign form. The formula for this reaction is given by equation (12.14) [27].

$$M^{2+} + H_2S + 2CH_2O \rightarrow MS + 2H_2O + 2CO_2 \tag{12.14}$$

Whilst this description is accurate on a macro-scale, there is still very little known about what happens at the micro-scale level, and more specifically about the interaction between the roots, the microbes within the roots and the CW matrix, and the AMD. Notwithstanding this, CWs have been successfully used for treating AMD for more than 30 years. In the next section, the performance of these wetlands is discussed. The section will be presented chronologically, since the lessons learned at some sites were incorporated in later systems.

12.4.1.3 Constructed Wetlands for Treating AMD Prior to 2000

Prior to 2000, a number of CWs were constructed for the treatment of AMD, and those described here were constructed primarily in North America and the UK. In the UK, a pilot-scale CW was designed and constructed for the Wheal Jane Mine in Cornwall [27]. Wheal Jane was formally abandoned in 1991, and in 1992, an adit plug failed, causing a large-scale release of AMD. Passive systems were considered appropriate and the final system included an aerobic reed bed, primarily designed for iron and arsenic removal. Hamilton et al. [27] describe their results on a pilot-scale system, with flow rates of between 0.1 and 0.6 L/s. These aerobic reed beds were successful in removing iron; however during ferric iron hydrolysis, acidity was generated, and this caused a decrease in the pH, despite the purpose of wetlands being to neutralize pH and remove iron.

Machemer et al. [29] described the design and optimization of the wetland constructed at the Big Five Tunnel, in Idaho Springs, Colorado. The CW consisted of five cells, some of which were vegetated, some not. Significant quantities of organic matter such as mushroom compost and animal manure were placed on the cells. The system overall was able to raise the pH from 3 to 6, and effectively removed iron, nickel and nitrate. It was, however, not successful at removing much more than 15% of the sulfate.

Mitsch and Wise [30] describe the use of a 0.39 ha CW for the treatment of AMD in a stream in South Eastern Ohio, USA. The CW was constructed as a 9-cell system, with plants grown in organic matrices such as mushroom compost and manure. There was a combination of surface and subsurface flow systems described. This system had little effect on acidity or pH, although 80% of the iron was removed, and 33% of the aluminium was removed. They also found very little sulfate reduction within the CW. An important finding was that plants have little to do with the retention of metals, despite the fact that they have elevated levels of metals within their tissues. Groudeva and

Groudev [31] conducted research on remediating AMD from a uranium mine using a passive system which included an aerobic wetland. Their AMD feed had a pH of 3.2, sulfate approaching 1.9 g/L, 700 g/L of iron, 44 g/L of copper, uranium and radium with a combined β-radioactivity of 3.85B q/L. The aerobic CW portion of this passive system removed 28% of the sulfate, 40% of the iron, 20% of the copper, 60% of the zinc and almost 40% of the radioactivity.

The predominant trends in this period before 2000, were thus towards mimicking natural wetland systems which were known to remediate AMD somewhat and to apply "ecological engineering approaches to decommissioning" [32]. The fundamental mechanisms of remediation were known, and if sulfate reduction was required a carbon source was added to the CW.

12.4.1.4 Constructed Wetlands for Treating AMD Between 2001 and 2010

The literature gathered for this period can be broadly divided into two categories:

1) Papers describing the performance of constructed wetlands and exploring the mechanism of remediation in greater detail; and
2) Reflexive papers considering the limitations of existing CWs treating AMD.

A selection of these findings is discussed here.

12.4.1.4.1 Descriptive Papers

There was considerable effort towards understanding the microbial processes occurring within CWs during this period. Kosolapov et al. [33] describe the contribution of microbial processes to removing metals in carbon-deficient effluents being treated by CWs. They discuss the role of biosorption, specifically that it is a mechanism by which metals are able to be removed; however that unless the biomass is removed, the effect is short-term due to the short life-cycle of microorganisms. They describe the main process being through precipitation as a metal sulfide and further discuss the role of reduction or oxidation of different metals into insoluble states.

Reduction is an important mechanism for some priority metals such as U(VI), Cr(VI) and others, which can be reduced to a valencies of IV or III. This mechanism is in some instances very important, since trivalent chromium is much less toxic than hexavalent chromium, and U(IV) is much less mobile. The removal of Fe(II) in contrast was described as being removed through abiotic processes through the oxidation of Fe(II) to Fe(III) which reacts with water to form a hydroxide which is insoluble. In this work, they also discuss that CWs cause biological methylation of certain metals such as mercury. This is an anaerobic process and occurs naturally within CWs. In certain locations, such as Johannesburg, there is considerable mercury contamination of the environment due to informal gold mining [34] and this must be considered as part of the design of any system.

Nelson et al. [35] discussed the removal of metals by CW systems treating metal rich water in North Carolina, and specifically in this case they had elevated levels of mercury, copper, zinc and lead. Whilst at this site, the water being treated was not AMD, and was therefore not acidic; it was able to remove more than 80% of copper and mercury, and they noted that the systems were shallow and primarily aerobic, which explains the high removal of mercury (rather than methylation). Manganese, as described by Hallberg and Johnson [7], is also difficult to remove within a wetland system. It is difficult to oxidize Mn(II) to Mn(IV) which is the insoluble form. This typically requires pH > 8 and does not typically form an insoluble sulfide. Thus, CWs may be less useful as a treatment technology if Mn removal is a priority.

Nyquist and Gregor [36] discuss a CW used to treat Kristineberg Mine tailings at very high latitude. In their study system, AMD entered into the CW with a pH of 2.65 and the CW was found to have almost no impact on the AMD, except for approximately 30% removal of the copper. The pH was unchanged as were most other metals including iron. Unfortunately, there is no satisfactory reason in this paper as to why this wetland failed; however the residence time according to the authors would range from 60 to 120 hours and it is may be that this is insufficient for treating AMD, especially within a cold climate environment. Notwithstanding these limitations, CWs are effective sinks for many metal species.

During this period, considerable attention was paid to the microbiology of sulfate reduction within CWs. Hallberg and Johnson [37] continued research on the Wheal Jane CW with a microbiological study on the CW ecosystem. They found acidophilic iron oxidizing bacteria, and over the course of their study, heterotrophic acidophiles emerged as the dominant group of bacteria, sometimes accounting for as much as 25% of the total enumerable microbial population. Riefler et al. [38] reported that in the CW that they were studying, in Ohio, USA, they found that the sludge within their system contained more than 6×10^6 sulfate-reducing bacteria cells (SRBs) per gram of sludge. The system was also effective at removing iron and aluminium.

Kuschk et al. [39] conducted field research to assess the effectiveness of CWs treating the acidity within AMD. They compared subsurface flow, surface flow and hydroponic systems with unplanted systems. Their findings showed that the planted surface flow appeared most effective, removing 80% of acidity, and increasing the pH from 3.3 to greater than 4.5. Using the same experimental apparatus, Wiessner et al. [40] assessed the effectiveness of the different CWs to remove iron and zinc. They found that a maximum of 97% of the iron and 77% of the zinc was removed, but that rainfall events remobilized some of the metals accumulated within the systems.

Koschorreck [41] prepared a mini-review of microbial sulfate reduction at low pH, typically the conditions to be found within CWs treating AMD. It was established that DSR can occur at pH < 5 and that the inhibitory effects on this process included H_2S and certain organic acids. Metal sulfide precipitation was also noted to have an inhibitory effect, especially if in competition with iron-reducing bacteria. He concludes that there is not always a reasonable explanation for why sulfate reduction does not occur in low pH environments, especially one like that described by Nyquist and Gregor [36].

12.4.1.4.2 Reflexive Papers

Following the emergence of CWs as a treatment technology thought to be suitable for post-mining AMD treatment (as a result of their low maintenance requirements) in the 1980s and 1990s, by the next decade there were a number of researchers who reflected on the efficacy of CWs as a remediation technology. Hallberg and Johnson [42], following much research on the Wheal Jane wetland, noted the following limitations with the use of CWs for treating AMD.

- Passive AMD remediation techniques allow for very little system control and are subject to seasonal and other variabilities (such as rainfall).
- There were insufficient data for accurate assessment of sizing and performance of CWs, and this is a problem which is still relevant today [43].
- CWs are sometimes ineffective when used in isolation.
- CWs sometimes fail catastrophically.
- The role of the macrophytes (at that time) was not fully resolved.
- Manganese removal was not always effective.

Further, they importantly conclude that the biogeochemistry and microbiology of the systems is highly complex and that there is limited understanding of these processes.

Wiseman and Edwards [44] considered the use of CWs for treating coalmine AMD in Wales in the 1990s. The systems they monitored were effectively able to remove in excess of 80% of the iron in the feed; however, the authors are clear to note that the system required maintenance, which must be costed for in the design phase. Whitehead and Neal [45] briefly discuss the long-term conclusions arising from the Wheal Jane wetland. They point out that the sludges which accumulate within the CW are toxic and hazardous; thus care needs to be taken during maintenance or refurbishment of any CW treating AMD for disposal of wastes. In environments where AMD is enriched in uranium, this problem is further amplified as a host of international regulations considering the handling and transport of radioactive substances would need to be considered.

12.4.1.5 Constructed Wetlands for Treating AMD from 2010 to the Present

In this period, the reporting of CWs treating AMD became global with the technologies having spread to Asia, Australia, Africa and the rest of Europe. De Matos and Zhang [46] report on the use of CWs for treating AMD to allow final discharge to natural ecosystems. The study was located in Spain and described two systems – a horizontal flow compost wetland followed by an anoxic limestone drain and a second system comprising a layered vertical flow system (with free water, organic material and limestone) followed by a series of aerobic wetlands. These systems treated different AMD streams, the first being designed for a more acidic AMD. In the first system, the pH was increased from 4 to 6.8; 90% of iron was removed but only 30% on manganese was removed. The second system removed 71% of iron, 80% of the aluminium present and 60% of the manganese.

Gikas et al. [47] considered the use of CWs for removing trace quantities of heavy metals. Using horizontal sub-surface flow CWs, with an artificially prepared feed of 2 mg/L of Cr, Pb and Fe, they obtained removals of greater than 87% for all three metals of concern. Guittonny-Philippe et al. [48] considered the use of CWs in removing pollution from industrial catchments. Specifically, the authors discuss the role of plants within CWs reporting a number of papers with conflicting results in this regard.

Sheridan et al. [21] reported on the use of a CW utilizing a charcoal matrix and one using a basic oxygen furnace slag matrix for treating AMD. In their study, they used two artificial AMD solutions both containing 6000 ppm of sulfate, 2000 ppm of iron and one with a pH of 4 and the with a pH of 4 and another with a pH of 1.35. Both the charcoal and the BOF CWs removed more than 75% of the sulfate and almost all iron. The pH was increased significantly in all cases. The authors point out that the mechanism of remediation, especially in the charcoal system was unknown.

Ji et al. [49] describe the use of a passive mine water treatment system which included a CW for treating AMD from the Ho-Nam Coal mine in Korea. Their findings show that the system was effective overall, but did not perform as designed or expected. Iron was removed prior to entering the CW. Other metals that were removed include aluminium, manganese, calcium and magnesium. Specifically, approximately 60% of the Mn was removed. This result is likely due to this CW being aerobic.

Panda et al. [50] prepared a mini-review on the bioremediation of acidic mine effluents with a specific focus on the role of the sulfidogenic microorganisms. They report that over 30 different bacteria (or strains thereof) and more than 25 archaea have been identified within AMD ecosystems, including CWs. The importance of understanding the microbiology and microbial ecosystems is explicitly described since the microbes (bacteria and archaea) are responsible for a large percentage of the removal of sulfate.

12.5 Summary

It is tempting for consultants, consulting engineers, legislators and mines to use CWs as remediation technologies for AMD treatment or mine closure remediation. This is primarily due to the CWs looking pretty and natural (which is vastly different to most active mine sites); a lack of long-term costs (critically important given the longevity of most AMD sources) and the lack of continued ability to treat the AMD (often the mine has closed and only the legacy remains). The past 30 years of research quite clearly shows that it is still very difficult to design a CW to effect a specified remediation. There are rough rules of thumb which can be applied for design purposes. However the use of these rules of thumb in no way guarantees success. Even worse, the construction of a pilot-scale CW also does not guarantee success.

The heterogeneous nature of AMD also makes it very difficult to understand how a CW would react given a specific feed. It is much simpler to design CWs for urban/domestic effluent since the chemical constituents leaving any given wastewater treatment facility do not vary significantly. AMD, on the other hand, is as unique as the geochemical fingerprint associated with the ore body. Whilst most AMD contains iron and sulfate, the pH, the concentrations, the ratio of the concentrations and the presence of any other constituents will be vastly different. The synergistic and antagonistic effects of this chemical melange are not possible to predict, and thus research at each given site is required. Furthermore, to design a CW for treating AMD requires that the designer has an in-depth understanding of inorganic, redox and aquatic chemistry. As shown, some metals require oxidation to be immobilized, others require reduction, and these redox reactions may require very different pH values to occur. Thus, for adequate design, a CW may need a low pH zone, a high pH zone, an aerobic zone and an oxidative zone. Failure to understand the intersection of chemistry, redox and pH could take a dangerous AMD and cause extreme environmental toxicity and consequent damage (as in the case of the transformation of mercury to methylated mercury).

However, if careful analysis of the AMD is conducted, if the consulting engineers are able to communicate effectively in collaborative teams with geologists, microbiologists and chemists to understand the interplay between the system, the biogeochemistry and the microbiology, it is possible to design CWs for treating almost any AMD effectively to a specified limit.

References

1 Skousen, R.A., Geidel, G., Foreman, J., Evans, R., Hellier, W., 1998. A handbook of technologies for avoidance and remediation of acid mine drainage. The National Mine Land Reclamation Centre, West Virginia, USA.
2 Mayes, W.M., Batty, L.C., Younger, P.L., Jarvis, A.P., Kõiv, M., Vohla, C., Mander, U., 2009. Wetland treatment at extremes of pH: A review. Sci. Total Environ. 407:3944–3957.
3 McCarthy, T.S., 2011. The impact of acid mine drainage in South Africa. South Afr. J. Sci. 107. doi:10.4102/sajs.v107i5/6.712.
4 Akcil, A., Koldas, S., 2006. Acid Mine Drainage (AMD): causes, treatment and case studies. J. Clean. Prod. 14:1139–1145.
5 Skousen, J., Rose, A., Geidel, G., Foreman, J., Evans, R., Hellier, W. Handbook of technologies for avoidance and remediation of acid mine drainage. West Virginia University, Morgantown, West Virginia; The National Mine Land Development Center, 1998.

6 Johnson, D.B., Hallberg, K.B., 2005. Acid mine drainage remediation options: a review. Sci. Total Environ. 338:3–14.

7 Hallberg, K.B., Johnson, D.B., 2005. Microbiology of a wetland ecosystem constructed to remediate mine drainage from a heavy metal mine. Sci. Total Environ. 338:53–66.

8 Taylor, J., Pape, S., Murphy, N., 2005. A summary of passive and active treatment technologies for acid and metalliferous drainage (AMD). In: 5th Australian Workshop on Acid Mine Drainage, Fremantle, Australia.

9 Tutu, H., McCarthy, T.S. and Cukrowska, E., 2008. The chemical characteristics of acid mine drainage with particular reference to sources, distribution and remediation: The Witwatersrand Basin, South Africa as a case study. Applied Geochem. 23(12):3666–3684.

10 Udayabhanu, S., Prasad, B., 2010. Studies on environmental impact of acid mine drainage generation and its treatment: an appraisal. IJEP 30:953–967.

11 Gray, N.F., 1997. Environmental impact and remediation of acid mine drainage: a management problem. Environ. Geol. 30:62–71.

12 Kuyucak, N., 1999. Acid Mine Drainage Prevention and Control options. IMWA Proceedings.

13 Baird, C., Cann, M., 2005. Environmental chemistry. Basingstoke: Macmillan.

14 Bell, F.G., Hälbich, T.F.J., Bullock, S.E.T., 2001. The effects of acid mine drainage from an old mine in the Witbank Coalfield, South Africa. Q. J. Eng. Geol. Hydrogeol. 35:265–278. doi:10.1144/1470-9236/00121.

15 Evangelou, V., Zhang, Y.L., 2009. A review: Pyrite oxidation mechanisms and acid mine drainage prevention. In: Critical Reviews in Environmental Science and Technology.

16 Simate, G.S., Ndlovu, S., 2014. Acid mine drainage: Challenges and opportunities. J. Environ. Chem. Eng. 2:1785–1803. doi:10.1016/j.jece.2014.07.021.

17 Lambers, H., Chapin III,, F.S. and Pons, T.L., 2008. Plant Physiological Ecology. Photosynthesis Respiration and Long Distance Transport. Springer Science and Business Media, 2nd edn, pp. 10–12.

18 UXL Encyclopedia of Water Science, 2005. Bioaccumulation of heavy metals. The Gale Group, Inc. Environmental Protection Agency, Vienna, Austria, 101 pp.

19 Mthembu, M.S., Swalaha, F.M., Bux, F., Odinga, C.A., 2013. Constructed wetlands: A future alternative wastewater treatment technology.

20 Taylor, J., Page, S., Murphy, N., 2005. A summary of passive and active treatment technologies for acid and metalliferous drainage (AMD). Australian Centre for Minerals Extension and Research (ACMER), 2005.

21 Sheridan, C., 2013. Paying the price. TCE The Chemical Engineer (867):30–32.

22 Seidel, K., 1966. Reinigung von Gewassern durch hohere Pflanzen. Deutsche Naturwissenschaft.

23 Weider, R.K., Lang, G.E., Whitehouse, A.E., 1985. Metal removal in a Sphagnum-dominated wetlands. In: Brooks, R.P., et al. (eds). Wetlands and water management on mined lands. Process of a Conference, October 1985. Pennsylvania State University.

24 Gazea, B., Adam, K., Kontopoulos, A., 1996. A review of passive systems for the treatment of acid mine drainage. Miner. Eng. 9:23–42. doi:10.1016/0892-6875(95)00129-8.

25 Zhi, W., Ji, G., 2012. Constructed wetlands, 1991–2011: A review of research development, current trends, and future directions. Sci. Total Environ. 441:19–27. doi:10.1016/j.scitotenv.2012.09.064.

26 Vymazal, J., 2009. The use constructed wetlands with horizontal sub-surface flow for various types of wastewater. Ecol. Eng. 35:1–17. doi:10.1016/j.ecoleng.2008.08.016.

27 Hamilton, Q.U.I., Lamb, H.M., Hallett, C., Proctor, J.A., 1999. Passive treatment systems for the remediation of acid mine drainage at Wheal Jane, Cornwall. J. CIWEM. 13, 93–103.

28 Benner, S.G., Blowes, D.W., Ptacek, C.J., 1997. A full-scale porous reactive wall for prevention of acid mine drainage. Ground Water Monit. Remediat. 17:99–107. doi:10.1111/j.1745-6592.1997.tb01269.x.

29 Machemer, S.D., Reynolds, J.S., Laudon, L.S., Wildeman, T.R., 1993. Balance of S in a constructed wetland built to treat acid mine drainage, Idaho Springs, Colorado, U.S.A. Appl. Geochem. 8(6):587–603.

30 Mitsch, W.I., Wise, K.M., 1998. Water quality, fate of metals, and predictive model validation of a constructed wetland treating acid mine drainage. Water Res. 32(6):1888–1900.

31 Groudeva, V.I., Groudev, S.N., 1998. Cleaning of acid mine drainage from a uranium mine by means of a passive treatment System. Min. Pro. Ext. Met. Rev. 19:89–95.

32 Kalin, M., 2001. Biogeochemical and ecological considerations in designing wetland treatment systems in post-mining landscapes. Waste Manage. 21:191–196.

33 Kosolapov, D.B., Kuschk, P., Vainshtein, M.B., et al., 2004. Microbial processes of heavy metal removal from carbon-deficient effluents in constructed wetlands. Eng. Life. Sci. 4(5):403–411.

34 Lusilao-Makiese, J.G., Cukrowska, E.M., Tessier, E., Amouroux, D., Weiersbye, I., 2013. The impact of post gold mining on mercury pollution in the West Rand region, Gauteng, South Africa. J. Geochem. Exploration. 134:111–119.

35 Nelson, E., Specht, W., Knox, S., 2006. Metal removal from water discharges by a constructed treatment wetland. Eng. Life Sci. 6:26–30.

36 Nyquist, J., Gregor, M., 2008. A field study of constructed wetlands for preventing and treating acid mine drainage. Ecol. Eng. 35:630–642.

37 Hallberg, K.B., Johnson, D.B., 2005. Biological manganese removal from acid mine drainage in constructed wetlands and prototype bioreactors. Sci. Total Environ. 338:113–124.

38 Riefler, R.G., Krohn, J., Stuart, B., Socotch, C., 2008. Role of sulfur-reducing bacteria in a wetland system treating acid mine drainage. Sci. Total Environment. 394:222–229.

39 Kuschk, P., Wiessner, A., Buddhawong, S., Stottmeister, U., Kästner, M., 2006. Effectiveness of differently designed small-scale constructed wetlands to decrease the acidity of acid mine drainage under field conditions. Eng. Life. Sci. 6(4):394–398.

40 Wiessner, A., Kuschk, P., Buddhawong, S., Stottmeister, U., Mattusch, J., Kästner, M., 2006. Effectiveness of various small-scale constructed wetland designs for the removal of iron and zinc from acid mine drainage under field conditions. Eng. Life. Sci. 6(6):584–592.

41 Koschorreck M., 2008. Microbial sulphate reduction at a low pH. FEMS Microbiol Ecol. 64(3):329–342.

42 Hallberg, K.B., Johnson, D.B., 2003. Passive mine water treatment at the former Wheal Jane Tin Mine, Cornwall: important biogeochemical and microbiological lessons. Land Contamin. Reclam. 11(2):213–220.

43 Sheridan, C., Hildebrand, D., Glasser, D., 2014. Turning wine (waste) into water: toward technological advances in the use of constructed wetlands for winery effluent treatment. AIChE J. 60 (2):420–431.

44 Wiseman, I.M., Edwards, P.J., 2004. Constructed wetlands for minewater treatment: performance and sustainability. Water Environ. J. 18(3):127–131.

45 Whitehead, P.G., Neal, C., 2005. The Wheal Jane wetland remediation system study: some general conclusions. Sci. Total Environ. 338:155–157.

46 de Matos, M.C.C.F., Zhang, Z., 2011. Mining acid rock passive treatment for closure of a lignite mine in Spain: Achieving necessary water quality for discharge into lakes and drinking water reservoirs. Lakes and Reservoirs: Res. Manage. 16:195–204.

47 Gikas, P., Ranieri, E., Tchobanoglous, G., 2013. Removal of iron, chromium and lead from wastewater by horizontal subsurface flow constructed wetlands. J. Chem. Technol. Biotechnol. 88:1906–1912

48 Guittony-Philippe, A., Masotti, V., Höhener, P., Boudenne, J., Viglione, J.; Laffont-Schwob, I., 2013. Constructed wetlands to reduce metal pollution from industrial catchments in aquatic Mediterranean ecosystems: A review to overcome obstacles and suggest potential solutions. Environ. Internat. 64:1–16.

49 Ji, S.-W., Lim, G.-J., Cheong, Y.-W., Yoo, K., 2012. Geosystem Engineering. 15(1), 27–32.

50 Panda, S., Mishra, S., Akcil, A., 2016. Bioremediation of acidic mine effluents and the role of sulfidogenic biosystems: a mini-review. Euro-Mediterranean J. Environ. Integ. 1(1):1–8.

13

Solid Waste (SW) Leachate Treatment using Constructed Wetland Systems

K.B.S.N. Jinadasa[1], T.A.O.K. Meetiyagoda[2] and Wun Jern Ng[3]

[1] *Department of Civil Engineering, University of Peradeniya, Sri Lanka*
[2] *R&D Unit, CETEC Pvt Ltd, Kandy, Sri Lanka*
[3] *Nanyang Environment and Water Research Institute, and School of Civil and Environmental Engineering, Nanyang Technological University, Singapore*

13.1 The Nature of Solid Waste (SW) and SW Leachate

Solid wastes are any discarded or abandoned materials and can be solid, semi-solid, or containerized gaseous and liquid materials. In the urban context, Municipal Solid Waste (MSW), commonly referred to as trash or garbage, comprises everyday items used and then discarded. Such items include product packaging, grass clippings, furniture, clothing, bottles, food scraps, newspapers, appliances, paint, and batteries. These can originate from homes, schools, hospitals, and commercial premises. A typical MSW gross composition is given in Table 13.1, where the largest component is organic materials. MSW composition varies even at a particular site; Tables 13.2 and 13.3 show such a variation at a site in Punjab, India. In Table 13.2, the largest component is the compostable materials and water, while nitrogen, phosphorus and potassium percentages are relatively low (Table 13.3)[1, 2].

Solid wastes have become a sanitation threat in much of the developing world. The most common method of solid waste disposal is landfilling. Proper sanitary landfill sites have liners and covers, and operations such as compaction, leachate management, and gas collection to protect human health and the environment. Unfortunately, not all landfills have been appropriately constructed and operated, and in South and Southeast Asia more than 90% of all landfills are non-engineered disposal facilities [3].

Leachate is water, which has passed through the filled material and leached (i.e., extracted) material such as dissolved and suspended organic and inorganic matter from the landfill. The water is often the result of precipitation infiltrating the landfills. In addition, the water originally present in the wastes can be released and so contributes to leachate generation [1].

The quantity of leachate production and leachate composition are affected by the climatic factors at the landfill site. Tropical countries have year round temperatures of above 18°C. There is also often high precipitation and hence humidity. Therefore, during rainy seasons (especially in monsoon seasons), leachate production can be expected to increase. While during the hot and dry season,

Constructed Wetlands for Industrial Wastewater Treatment, First Edition. Edited by Alexandros I. Stefanakis.
© 2018 John Wiley & Sons Ltd. Published 2018 by John Wiley & Sons Ltd.

Table 13.1 Typical MSW gross composition [1].

Component	Percentage (wt. %)
Organic materials	40
Unrecyclable plastics	10
Unrecyclable materials	30
Agriculture waste	20

Table 13.2 Composition of MSW at Punjab site, India [2].

Category	Item	%
Recyclable material	Paper, plastic, rags	3–5
	Leather, rubber, synthetic	1–3
	Glass, ceramics	0.5–1
	Metals	0.2–2
Compostable material	Food articles, fodder, dung, Leaves, Organic material	40–60
Inert material	Ash, dust, sand, building material	20–50
Moisture		40–80
Density		250–500 kg/m^3

Table 13.3 Chemical Composition of MSW at a Punjab site, India [2].

Item	%
Nitrogen	0.56–0.71
Phosphorus	0.52–0.82
Potassium	0.52–0.83
C/N	21–30

accelerated degradation of wastes can occur although leachate volume is reduced. For example, leachate is formed mostly in the period from November to April in Poland, with the maximum leachate volume occurring in December. During dry periods, leachate would not be noted from May to October [4]. In the tropical environment where rain can occur year round, such distinct periods of leachate production may not occur.

The volume of leachate requiring management would very much depend on the waste's initial moisture content and composition, biochemical and physical transformations, and inflow of water from outside the landfill [5, 6].

13.2 Characteristics of SW Leachate in Tropical Developing Countries

Tropical regions, in the lowlands, have mean annual temperatures exceeding 20°C and these can even exceed 25°C. There are tropical regions which are affected by distinctly seasonal rainfalls, the monsoons. Therefore, leachate production can have distinct variations in the year.

Many landfills in developing countries are in effect open dumpsites where the solid wastes are dumped in an uncontrolled manner. Composition of landfill leachates is influenced by internal and external factors. External factors include solid waste composition, operational mode of landfill, climate, and hydrological conditions. The internal factors are biochemical activities, moisture, temperature, pH and the age of the landfill [5]. MSW leachates are concentrated and complex effluents, which contain inorganic and organic compounds including possibly toxic and hazardous chemicals [7–9]. The organics can include aromatic compounds, halogenated compounds, phenols, pesticides and nitrogenous compounds [10, 11].

Properties of a leachate are often categorized as physical, chemical and biological. Physical properties include solids content, turbidity, temperature, conductivity, color, and density. Chemical properties can be further divided into two categories: inorganic and organic constituents. Chloride, pH, alkalinity, gases, metals, nitrate, phosphate, ammonium are among the inorganic constituents. Proteins, carbohydrates, and oil and grease are among the organic chemical constituents. Pathogens and other microorganisms are among the biological constituents.

The amount of organic pollutants in landfill leachate can be measured in terms of biochemical oxygen demand (BOD_5), chemical oxygen demand (COD), total organic carbon (TOC), inorganic carbon (IC) and total carbon (TC). Leachate quality is often also defined with parameters such as dissolved oxygen (DO), pH, oxidation reduction potential (ORP), electrical conductivity (EC), suspended solids, total nitrogen (TN) and total phosphorous (TP).

The composition of landfill leachate at different landfills in Sri Lanka is given in Table 13.4. The highest BOD_5 and COD values were recorded at the Kolonnawa site.

Table 13.4 Leachate composition at Sri Lankan solid waste dumpsites [12].

Sample	BOD_5 (mg/L)	COD (mg/L)	BOD_5/COD	TN (mg/L)	TP (mg/L)
Matale	71	4236	0.02	1549	43
Hambantota	49	2475	0.02	224	12
Kataragama	38	1047	0.04	180	20
Bandargama	938	9279	0.1	305	24
Kolonnawa	49600	82577	0.6	664	11
Gampola	41	1249	0.03	1212	6
Gohagoda	19	714	0.03	684	8
Wennappuwa	86	2405	0.04	361	18
Rathnapura	3757	13951	0.27	558	35
Negombo	294	18328	0.02	703	86

Leachate characteristics at three landfilling sites in Ludhiana City, Punjab (India) are shown in Table 13.5. The variability needs to be noted and this is also related to age of the site. The concentrations of leachate contaminants at Jamalpur and Noorpur belt are higher than that at Jainpur, which is the oldest one.

The age of a landfill significantly affects the quantity and quality of the leachate produced. The water retaining capacity is affected due to mineralization of organic substances [13]. The relationship between leachate properties and landfill age is given in Table 13.6 [11, 14, 15] and [16].

Table 13.5 Leachate characteristics at sites in Ludhiana City, Punjab, India [2].

Parameters	Jainpur	Jamalpur	Noorpur Belt
Appearance	Brownish	Brownish	Brownish
Odor	Sewage smell	Sewage smell	Sewage smell
pH (–)	9.3	9.8	9.5
TS (mg/L)	5,963	7,695	6,579
SS (mg/L)	615	1,132	886
TDS (mg/L)	5,348	6,563	5,693
Turbidity (NTU)	43	79	68
Hardness (mg/L)	585	638	621
BOD_5 (mg/L)	329	495	406
COD (mg/L)	1,335	2,535	2,018
BOD_5/COD	0.24	0.19	0.20
Chloride (mg/L)	1,448	1,836	1,653
Nitrate (mg/L)	12.5	18.6	15.9
Total phosphorus (mg/L)	52.8	83.5	64.3
Sulphate (mg/L)	48.7	65.1	53.8

Table 13.6 Leachate characteristics at various landfill ages [13].

Age (years)	BOD (mg/L)	COD (mg/L)	BOD: COD	NH_3-N (mg/L)	SS (mg/L)	pH	Reference
Young (<5)	>2000	>10000	>0.3	–	–	6.5	[11, 14]
Intermediate (5–10)	150–2000	4000–10000	0.1–0.3	–	–	6.5–7.5	
Old (>10)	<150	<4000	<0.1	–	–	>7.5	
Young (5)	2031.62	3641.2	0.31	288.6	–	6.52	[15]
Old (15)	196.83	875.44	0.20	260.03	–	5.72	
Young (<5)	32790	41507	0.79	1896	1873	6.6	[16]
Intermediate (5–10)	2684	5348	0.50	1826	143	7.9	
Old (>10)	145	1367	0.11	892	17.2	8.2	

Landfilled organic materials undergo anaerobic degradation and this goes through two phases of biological transformations – the acidogenic phase and methanogenic phase. Old landfills would have largely passed the acidogenic phase and so release methanogenic leachate, while producing methane and carbon dioxide [17]. In the acidogenic phase, acidogens convert soluble organic material mainly to acetate, propionate, butyrate, hydrogen, and carbon dioxide.

Less than 5 year-old landfills are largely acidogenic and so can have leachate pH of 3.7–6.5, reflecting the presence of carboxylic acids and bicarbonate ions. As the landfills age, leachate becomes neutral or weakly alkaline (pH of 7.0–7.6) and eventually alkaline (pH 8.0–8.5) [18].

13.3   Treatment Methods for SW Leachate

Commonly, leachate is characterized by high values of COD, pH, ammonia nitrogen and heavy metals, as well as strong color and odor. However, these values are time dependent in relation to a landfill's age [19, 20]. It is not only the quantities that change with age, but also how a particular leachate component is formed can also change with age. Ammonia nitrogen in the leachate from young landfills results from the deamination of amino acids during destruction of organic compounds [21], but high concentration of ammonia nitrogen can also be found in leachate of older landfills due to hydrolysis and fermentation of the nitrogenous fractions of biodegradable substrates [22]. Such behavior makes leachate treatment potentially difficult.

Inanc et al. [23] identified methods to treat landfill leachate including:

- Aerobic treatment such as aerated lagoons, and activated sludge.
- Anaerobic treatment such as anaerobic lagoons.
- Physicochemical treatment such as air stripping, pH adjustment, chemical precipitation, oxidation, and reduction.
- Coagulation using lime, alum, and ferric chloride.
- Advanced techniques such as carbon adsorption, and ion exchange.

Biological processes are often used to treat landfill leachate because of their overall reasonable costs, and especially their operational costs (compared to chemical methods). Biological methods can serve as "pretreatment" to reduce high concentrations of BOD, COD and ammonium before polishing with physico-chemical methods. Biological treatments are, therefore, suitable for application on young leachates (<5 years), which would have higher organic concentrations and BOD/COD ratio. However, mature landfill leachates not only have lower organic concentrations, but the biodegradable organic fraction declines. Therefore biological treatment is potentially less effective for old landfill leachate [11].

Where the activated sludge method is used, leachate is aerated in an open tank with diffusers or mechanical aerators. This is a suspended growth process [24, 25]. Aerobic treatment in its most basic engineered configuration is the oxidation lagoon. This is a naturally aerated pond. However, to achieve satisfactory treatment results, oxidation lagoons require a large amount of land area to ensure sufficient oxygen transfer and, hence, avoid septic conditions. Unlike the activated sludge process, which is dependent on flocculated biomass, the active biomass in the oxidation lagoon largely accumulates as a biofilm on the lagoon bottom and sides.

Landfill leachate has been effectively treated with the rotating biological contactor (RBC). RBC consists of large disks supporting biofilms and with radial and concentric passages slowly rotating in a trough. While rotating, 40% of the support media surface area is in the leachate. The continuous rotation and alternating exposure to air and leachate allows microorganisms to metabolize substrates aerobically [24, 25].

The Sequencing Batch Reactor (SBR) is a complete-mix batch activated sludge system without a secondary clarifier. The SBR differs from the conventional activated sludge plant in its temporal-frame process arrangement in contrast with the spatial-frame arrangement in the latter [26]. The SBR operating cycle typically comprises fill, react, settle, draw and idle.

It should be noted that aerobic processes, as those described above, are unlikely to be able to degrade all the types of organic compounds found in leachate, as these may be persistent in the face of aerobic metabolic activity. Consequently, the treated leachate can still have a significant residual organic content. Often the aerobic process is preceded by an anaerobic process, which then serves as a pretreatment stage to reduce organic strength. Such an approach is often necessary to reduce both the capital and operating costs of an aerobic process only system. A high degree of waste stabilization is possible with low production of excess biological sludge. The anaerobic process also has lower nutrient requirements (making nutrients supplementation unnecessary), no oxygen requirements, and the process by-product of methane is a useful energy source.

There are many types of anaerobic systems incorporating suspended or attached growth, fixed and moving film systems, and combinations of these. In the absence of oxygen (free and combined), organic matter is converted to carbon dioxide and methane gas (Figure 13.1). Since such metabolism yields relatively little energy to the microorganisms, their growth rates and cell yields are lower than those in aerobic processes.

Volatile fatty acid intermediates are primarily acetic and propionic acid. During acidogenesis, pH can be reduced due to fatty acid production if there is inadequate buffering capacity. Low pH has the potential to inhibit further acidogenesis and would inhibit methanogenesis. Typically pH values are maintained between 6.4 and 7.2, to allow both acidogenesis and methanogenesis to proceed satisfactorily.

In engineering terms, the simplest of the anaerobic systems to construct would be anaerobic ponds. These are designed deep (deeper than aerobic ponds) and operated with high organic loadings, and such high loadings are used to maintain anaerobic conditions in the pond beneath the top liquid layer. Anaerobic ponds are typically operated with long hydraulic detention times and would have no induced mixing. Consequently the design and placement of the inlet and outlet works must be such that short-circuiting is avoided.

The upflow anaerobic sludge blanket (UASB) process is a sludge blanket system. UASB technology has been applied in tropical countries where the relatively high ambient temperatures make it easy to operate a mesophilic process. In the UASB, the leachate enters the reactor from the bottom and flows upward through an anaerobic sludge blanket. The three key engineered elements in the reactor

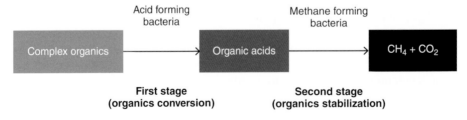

Figure 13.1 Organic matter conversion in the absence of oxygen [1].

are the influent distribution system, the three phase (gas, liquid, and solids) separator, and effluent withdrawal system.

Ion exchange resins, synthetic and natural, have had wide application in water and wastewater treatment [27] and can also be expected to potentially applicable in leachate treatment.

Ammonia dissolved in the leachate can be removed as a gas, using air stripping. Air stripping can be carried out in tanks or packed towers, or in counter-current, multi-stage reactors. The pH value is important and when it is lower than 7, ammonia tends to remain in the solution. However, when pH is raised to 11, ammonia tends to come out of the solution and enters the gas phase. Therefore, in a gas stripper, pH is adjusted to 11–12 by adding NaOH.

Coagulation can be used to remove suspended and colloidal matter, and some dissolved matter as well. Its efficiency can, however, be influenced by the appropriateness of the coagulant in relation to the target pollutants, dosage of coagulant, pH, mixing speed and duration, temperature and dissolved solids content [28]. Inorganic coagulants such as aluminum sulfate (Alum) and ferric chloride ($FeCl_3$) are most commonly used [29]. Poly aluminum chloride (PAC) has also been found an effective coagulant and can be used in place of alum [30]. Typically PAC dosages are lower than alum and this would result in lower quantities of aluminum sludge produced, which would require subsequent disposal.

Residual organics remaining after biological and chemical treatments can be removed with activated carbon adsorption. The disadvantage of this method is its associated operating costs, since the activated carbon should be replaced (or regenerated) at regular intervals [25].

Given the complex nature of leachate, a single unit treatment process is unlikely to be adequate. Leachate treatment systems would typically begin with an anaerobic process, followed by aerobic, chemical, and final physical processes (e.g., sorption).

Constructed wetlands (CWs) do inherently include many of the processes described above. CW systems have aerobic and anaerobic processes. These are mediated by macrophytes and microphytes. Additionally, the soil on which the macrophytes grow, can also mediate physical and chemical processes.

13.3.1 Advantages of Constructed Wetlands for Leachate Treatment Under Tropical Climate

Constructed wetlands (CWs) are engineered systems that are used as a secondary or tertiary treatment process for wastewater treatment. Compared to other treatment systems, CWs are low cost, easily operated and maintained, and have a strong potential for application in developing countries [31].

There are two general types of wetlands: surface flow systems and subsurface flow systems [31]. Subsurface flow CWs are further divided into two types according to flow direction: vertical subsurface flow (VSSF CWs) and horizontal subsurface flow (HSSF CWs) [31]. Surface flow CWs are further divided into three types based on the plants used (Figure 13.2). Hybrid constructed wetlands combine the different types of CWs to enhance treatment efficiency. Hybrid CWs tend to have higher pollutant removal efficiency than the single type CW systems [31]. Hybrid CWs can be seen as analogs of the anaerobic–aerobic treatment systems discussed earlier. For example, the VSSF CW would act as an aerobic system, while the HSSF CW system would act as an anaerobic system. Constructed wetlands have ecological conditions similar to natural wetlands [32].

Microorganisms in the wetlands system are in suspension, adhering to the plants, and in the soil. These then use the pollutants in the leachate as substrates. Aerobic heterotrophic bacteria degrade BOD_5 and COD, while nitrogen compounds are converted by autotrophic bacteria.

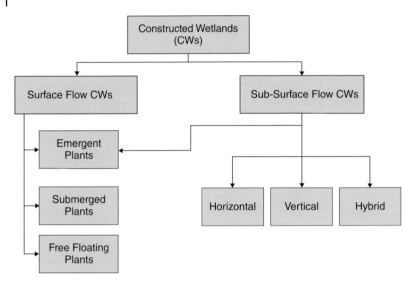

Figure 13.2 Constructed Wetlands classification [31].

The processes that affect nitrogen conversion include ammonia volatilization, nitrification, denitrification, nitrogen fixation, plant and microbial uptake, mineralization (ammonification), anaerobic ammonia oxidation, and, sorption [31, 33]. CWs have ability to remove heavy metals in the leachate by binding to soil or sediments due to positive charge of the heavy metal cations, precipitation as insoluble salts (CO_3^{2-}, HCO_3^-, and OH^-), and uptake by bacteria and plants. Various processes are also involved in the removal of pathogens in CWs and these include adhering to particulates and sedimentation, filtration, UV radiation, unfavorable water chemistry, temperature, predation, and antibiotics that are produced by the plants [31].

CWs are more sustainable in terms of energy requirements compared to mechanical systems such the activated sludge, since the primary energy requirement for macrophyte maintenance would be fulfilled by solar energy [31]. CWs can also be aesthetically landscaped and have often served as wildlife sanctuaries. Moreover, treated effluent from CWs can be used as irrigation water for agriculture.

13.4 Experimental Methodology for Plant Species and CW Performance Evaluation

Evaluation of CW performance with different plant species can be performed at three levels. Analysis can be initiated with batch tests to assess leachate treatment efficiency. For this, laboratory-scale sub-surface flow CW units have been used with three wetland plant species: narrow leaf cattail (*Typha angustifolia*), green bulrush (*Scirpus atrovirens*) and umbrella palm (*Cyperus alternifolius*). Replicates of each plant species are fed with various dilutions leachate, e.g., 25%, 50%, 75% and 100% leachate and with the challenges lasting 7 days. This allows for quick screening and determination of suitable plant species to use.

Figure 13.3 Plate 1: Pilot scale hybrid constructed wetland units planted with *Typha angustifolia*.

The second level of tests could be a pilot scale experiment in the field. For example, subsurface flow hybrid wetlands with 7 days HRT have been used with each CW test unit comprising a VSSF followed by a HSSF sub-unit and planted with *Typha angustifolia* (Plate 1; Figure 13.3). The length, width, and height of the VSSF sub-unit are 1 m, 0.7 m and 0.6 m, respectively. The effective height was 0.5 m. The configuration of the HSSF sub-unit is 1 m × 0.7 m × 0.45 m.

In the VSSF sub-unit, the bottom layer (5 cm) comprised 32 mm coarse gravel to disperse the flow and to allow for drainage of the treated leachate. The next 20 cm of the bed is filled with 19–20 mm medium gravel, and on top of that is 20 cm of 6–8 mm fine gravel. The top layer is a 5 cm soil layer (Figure 13.4).

A 45 cm high layer of 32 mm coarse gravel is used to distribute the influent flow into the HSSF sub-unit and to collect the treated effluent at the other end of the system. The middle layer is filled with 6–8 mm fine gravel up to 40 cm in height and the remaining 5 cm is filled with soil (Figure 13.5). If the HSSF sub-units are not receiving effluent from the VSSF sub-units, then the leachate may be applied at the following dilutions: 10%, 25%, and 50% of leachate.

While second level tests may be conducted with synthetic leachate so that consistent test conditions may be obtained, third level tests should be performed with the real leachate. The example shown in Plate 1 (Figure 13.3) had performance analysis based on real leachate collected from the "Gohagoda dumpsite" in Kandy, Sri Lanka (Plate 2; Figure 13.6) and with three different wetland plant species (Plates 3–5; Figure 13.7): narrow leaf cattail (*Typha angustifolia*) (Plate 3; Figure 13.7a), green bulrush (*Scirpus atrovirens*) (Plate 4; Figure 13.7b) and umbrella palm (*Cyperus alternifolius*) (Plate 5; Figure 13.7c). In such tests, the plants were challenged with 5% and 25% dilutions of the real

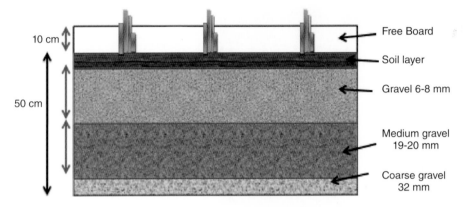

Figure 13.4 Layer arrangement of the pilot-scale VSSF CWs.

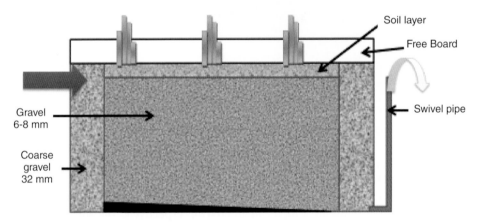

Figure 13.5 Layer arrangement of the pilot-scale HSSF CWs.

Figure 13.6 Plate 2: Gohagoda dumpsite, Kandy, Sri Lanka.

(a)   (b)   (c)

Figure 13.7 (a) Plate 3: *Typha angustifolia*; (b) Plate 4: *Scirpus atrovirens* and (c) Plate 5: *Cyperus alternifolius*.

leachate. Narrow leaf cattail (*Typha angustifolia*) (Plate 3; Figure 13.7c) belongs to the *Typhaceae* family. This is a commonly used plant for wastewater treatment. This plant is a helophyte and can survive in marshy conditions. There have been reports in the literature of this plant having been used for wastewater and leachate treatment. Green bulrush (*Scirpus atrovirens*) (Plate 4; Figure 13.7b) belongs to the *Cyperaceae* family. This plant has also been used in CWs for wastewater treatment. This perennial plant is about 2½–4' tall, un-branched and more or less erect. The culm is green, glabrous, and terete. It can survive flooding and drought conditions. Umbrella palm (*Cyperus alternifolius*) (Plate 5; Figure 13.7c) belongs to the *Cyperaceae* family and is commonly grown as a pond plant especially in shallow water. A favorable condition for growth is when plant is submerged up to 10 cm and so this plant has a higher tolerance to flooded conditions.

The Gohagoda dumpsite (Plate 2; Figure 13.6) is otherwise known as the Thekkawatta Landfill and is located in Kandy district at 7°18'47.33"N, 80°37'17.19"E Sri Lanka. This area receives annual rainfall of 2,500 mm. Gohagoda dumpsite has been in operation since the 1970s. The Mahaweli River flows 200 m below the dumpsite and there is a threat of leachate contamination.

13.5 Effect of Plant Species on Leachate Components

13.5.1 Effect on Organic Compounds

Organic compounds in landfill leachate include fatty acids, humic acids and fulvic-like substances [34]. The molecular weights of these three groups are low, high and intermediate respectively. The biodegradable fraction of the organics is measured by the BOD, while the non-biodegradable fraction is measured in terms of the difference between the BOD and COD values [35]. Aerobic and anaerobic degradation of soluble organic substances are responsible for removal of BOD. Hydrolysis and catabolic activities of autotrophic and heterotrophic bacteria contribute to the biodegradation of these compounds. The following reaction shows the biodegradation of organic compounds involving aerobic heterotrophic organisms:

$$(CH_2O) + O_2 \rightarrow CO_2 + H_2O$$

Anaerobic degradation of organic compounds can also occur in CWs, going deeper into the soil supporting the macrophytes, and in thick biofilms occurring elsewhere on the macrophytes and suspended particles. This process has different stages of reactions involving facultative and obligate anaerobic heterotrophic microorganisms. The first step is fermentation performed by facultative microorganisms.

$$C_6H_{12}O_6 \rightarrow 3\ CH_3COOH$$
$$C_6H_{12}O_6 \rightarrow 2\ CH_2CHOHCOOH + H_2$$
$$C_6H_{12}O_6 \rightarrow 2\ CO_2 + 2\ C_2H_5OH$$

Then, anaerobic microbes use the above end-products and convert these into methane and carbon dioxide. Where there are oxidized species present such as nitrates and sulphates, then there can also be end-products such as nitrogen and hydrogen sulphide.

$$CH_3COOH + H_2SO_4 \rightarrow 2\ CO_2 + 2H_2O + H_2S$$
$$CH_3COOH + 4\ H_2 \rightarrow 2\ CH_4 + 2\ H_2O$$

The BOD_5/COD ratio indicates the biodegradability of a leachate. At young landfills (landfilled wastes not older than 3–5 years), the BOD_5/COD ratio is high (0.7), indicating high biodegradability of organics in the leachate. In mature landfills (5–0 years) the BOD_5/COD ratio decreases to 0.5–0.3 and the proportion of easily biodegradable organic matters decreases proportionately, while the non-biodegradable matter would have remained intact. Landfills over 10 years are considered as old landfills and these typically have low BOD_5/COD ratios of possibly less than 0.1. As the BOD_5/COD ratios decline, biological treatment processes become less effective [36]. Portions of the Gohagoda dumpsite may be considered an intermediate stage landfill site (given the continuous use) and its BOD_5/COD is about 0.3. This would suggest dilution at 25% leachate would be appropriate for the tests. Examples of test results at various organic (in terms of BOD and COD) loading rates on CWs are shown in Table 13.7.

The BOD, COD, and TC removal efficiencies by the three plant species tested are given in the Table 13.8. All three plant species performed well for BOD, COD, and TC removal with 25% leachate

Table 13.7 Organic loading rates of synthetic leachate applied to CWs.

Dilution at X% Synthetic leachate	Organic loading rate (g/m³/d)	
	BOD	COD
100%	63.24	371.2
75%	41.08	280.5
50%	23.34	190.8
25%	11.69	90.7

Table 13.8 BOD$_5$ Removal efficiency (%).

Plant type	100%			75%			50%			25%		
	BOD$_5$	COD	TC	BOD$_5$	COD	TC	BOD$_5$	COD	TC	BOD$_5$	COD	TC
Narrow leaf cattail	86.4	83.1	77.3	86.3	88.7	84.7	89.0	96.2	91.2	89.5	99.1	90.3
Green bulrush	85.3	79.7	68.0	79.8	85.6	72.0	89.9	87.8	82.8	92.9	98.2	77.7
Umbrella palm	87.9	89.6	82.7	93.4	96.5	88.7	97.7	98.7	94.0	99.3	99.6	97.3

Table 13.9 Organic loading rates (OLR) and hydraulic loading rates (HLR) of different leachate-fed CW systems.

Dilution	Type of CW	OLR(COD g/m^3/d)	HLR(m/day)
10% Leachate feeding system	VSSF CW	17.6	0.071
	HSSF CW	0.9	0.064
25% Leachate feeding system	VSSF CW	35.4	0.071
	HSSF CW	9.6	0.064
50% Leachate feeding system	VSSF CW	99.1	0.071
	HSSF CW	43.1	0.064

Table 13.10 COD concentration profile in the leachate-fed CW systems.

Stage	10% system	25% system	50% system
Inlet (mg/L)	246.4	496	1386.3
Effluent of VSSF CWs (mg/L)	13.7	142	635.7
Effluent of HSSF CWs (mg/L)	7.4	86	160.1

and Umbrella Palm was the most effective. The pilot-scale tests results (using synthetic leachate) are given in Table 13.9. The good performances indicated by the laboratory tests were reproduced in the pilot tests. In addition, Table 13.10 shows the average COD concentration profiles at various stages in the CWs.

COD removal efficiencies of the hybrid systems with the 10%, 25% and 50% synthetic leachate feeding rates were 97.0, 82.6 and 88.5%, respectively, while the values for the VSSF alone CWs were 94.4, 71.3 and 54.1%, respectively.

HLRs for the VSSF CWs and HSSF CWs were 0.071 and 0.064 m/d, respectively (Table 13.9). When the hydraulic loading rate was < 0.01 m/d, organic matter removal was reduced, although removal of other pollutants had increased with the increase in HRT. Increased loading rates reduced the pollutant removal capacity of the CWs. For Hybrid CWs (VSSF-HSSF), an HLR < 0.045 m/d and OLR 100 g/m^3/d were found effective for leachate treatment. Pilot-scale experiments using hybrid CWs and real leachate also showed high BOD, COD and TC removals. Mean influent BOD$_5$ was mg/L (for 5% leachate) and 381 mg/L (for 25% leachate). Standard deviation was ±36.4 and ±162.8 for 5% and 25% leachate, respectively. Mean influent COD was 564 mg/L and 1,135 mg/L for the 5%

Table 13.11 Organic loading rates (OLR) with 5% and 25% real leachate.

Leachate concentration	OLR(BOD g/m³/d)	OLR(COD g/m³/d)
5% Leachate	7.3	16.4
25% Leachate	24.2	48.7

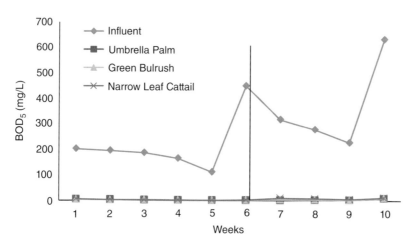

Figure 13.8 BOD$_5$ profiles with respect to time for hybrid CWs planted with four different plant species.

and 25% leachate, respectively. Total carbon (TC) was 184 mg/L and 368 mg/L for the 5% and 25% leachate, respectively, with respective standard deviations ±31.8 and ±70.6. Table 13.11 shows the organic loading rates based on 5% and 25% leachate.

COD and BOD$_5$ removal efficiency of the VSSF CWs was above 94% for all plant species. Supradata [37] reported oxygen in the media could be supplied by the plant roots. The oxygen is a by-product of photosynthesis. Aerobic conditions around the plants' root system could therefore be maintained and hence contributing to the high organic removals.

Figures 13.8 and 13.9 show the residual BOD and COD profiles with respect to time in the hybrid CWs planted with the various macrophytes.

13.5.2 Effect on Removal and Transformation of Nitrogen Compounds

Common nitrogen forms in landfill leachate are ammonia, organic nitrogen and possibly some nitrite (NO_2^-) and nitrate (NO_3^-). Cyclic processes that occur in "conceptual compartments" such as the water column, sediments, plant roots, biofilms, and stem and leaves can participate in the removal of the nitrogenous compounds from the leachate.

Nitrification (13.1) and denitrification (13.3) are processes that are important in nitrogen removal [31, 38].

$$NH_4^+ + O_2 \rightarrow NO_2^- + 2H^+ + H_2 (Nitrosomonas) \tag{13.1}$$
$$NO_2^- + O_2 \rightarrow 2\,NO_3^- (Nitrobacter) \tag{13.2}$$

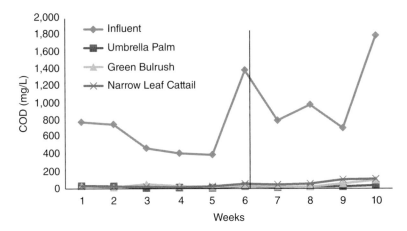

Figure 13.9 COD profiles with respect to time for hybrid CWs planted with with four different plant species.

$$NO_3^- + \text{Organic C} \rightarrow N_2(NO \text{ and } N_2O) + CO_2 + H_2O \ (\textit{Pseudomonas}) \qquad (13.3)$$

The different biochemical transformations of nitrogen in wetlands are given in Table 13.12.

The Hybrid Constructed Wetlands can have more stable removal rate of nitrogen in comparison to that of one-stage systems. The average removal rate was 7.8 kg N_{tot}/ha/d [39]. One stage system would not be able to effectively achieve both nitrification and denitrification processes in the same compartment. Pilot-scale experiments for leachate treatment using hybrid CWs with natural leachate have clearly revealed the transformation and removal of nitrogen compounds within wetland beds with the help of plant species. Ammonia nitrogen is one of the major components in the leachate. Average influent concentrations are 73.24 mg/L and 131.5 mg/L for 5% and 25% diluted real leachate.

VSSF CWs are dominated by aerobic conditions that enhance the aerobic microbial activities. Aerobic bacteria convert NH_4^+-N into NO_3^--N by nitrification process. Therefore considerable reduction of NH_4^+-N can be observed in VSSF CWs. The best average NH_4^+-N removal efficiency of 88.7% was

Table 13.12 Biogeochemical transformation of nitrogen in wetlands [31, 33].

Process	Transformation
Volatilization	$NH_3(aq) \rightarrow NH_3 (g)$
Ammonification	$\text{Organic-N (aq)} \rightarrow NH_3 (aq)$
Nitritation	$2NH_4^+(aq) + 3O_2 \rightarrow 2NO_2^-(aq) +2H_2O (aq) + 4H^+(aq)$
Nitrification	$2NO_2^-(aq) + O_2 \rightarrow 2NO_3^-(aq)$
Denitrification	$2NO_3^- (aq) \rightarrow 2NO_2^-(aq) \rightarrow 2NO (g) \rightarrow N_2O (g) \rightarrow N_2 (g)$
Dissimilatory nitrate reduction	$2NO_3^- (aq) \rightarrow NH_3 (aq)$
N_2 fixation	$N_2 (g) \rightarrow \text{Organic-N (aq)}, NH_3 (aq)$
Biological assimilation	$NH_3 (aq), NO_2^-(aq), NO_3^- (aq) \rightarrow \text{Organic-N (aq)}$
Ammonia adsorption	$NH_3 (aq) \rightarrow NH_3 (s)$
ANAMMOX	$NH_3 (aq) + NO_2^-(aq) \rightarrow N_2 (g)$

found for the green bulrush plant and 87.3%, 79.9% for umbrella palm and narrow leaf cattail respectively. Vymazal [33] has also mentioned that NH_4^+-N removal in VSSF CWs was 84.2% and 48.3% in HSSF CWs.

Intermittent feeding can be helpful in nitrogen removal as substrate pores can then be periodically re-supplied with oxygen (in the entrained air) and this could supplement the oxygen supplied by the roots in the plant rhizosphere, hence stimulating nitrification [31]. Thus, purposefully pulse loaded VSSF wetlands have been noted to promote NH_4^+-N oxidation [31, 40, 41].

In HSSF CWs normally saturated with water, anaerobic conditions occur. This therefore would restrict the nitrification process and enhance the denitrification process. Low average NH_4^+-N removal efficiencies can be observed in HSSF CWs [31]. HSSF CWs can remove NO_3^--N by the denitrification process. The anoxic conditions in the saturated HSSF CWs enhance the anaerobic microbial activities. The Umbrella palm in the HSSF CW model has shown the best average removal efficiency of 61%.

Influent NO_3^--N concentrations were very low in the leachate at 0.64 mg/L and 1.18 mg/L for 5% leachate and 25% leachate, respectively. But NO_3^--N concentrations in VSSF CWs had increased due to the nitrification process. Umbrella palm VSSF CWs model showed the best average NO_3^--N production of 18.3 mg/L and 34 mg/L for 5% and 25% leachate, respectively.

Average influent TN concentrations are 93.9 mg/L and 171.3 mg/L for 5% and 25% diluted leachate, respectively. Average total efficiency was also highest in Umbrella palm (74.6%). According to Vymazal [33], TN removal efficiencies were 44.6% and 42.3% for VSSF CWs and HSSF CWs, respectively. Laboratory-scale synthetic leachate treatment models have revealed high removal efficiency for TN (Table 13.13). The Umbrella palm was the most effective.

Influent concentrations of NH_4^+-N in the hybrid CWs ranged from 22 mg/L to 102 mg/L when using synthetic leachate. The average removal efficiency of NH_4^+-N in VSSF beds was better (up to 56% removal) than in HF beds (at 29% removal). In a hybrid system, the combined removal was 69% in the VSSF-HSSF CWs.

NO_3^--N concentrations in the outlet of the hybrid systems (15.4, 18.0, and 47.2 mg/L, respectively) are generally higher than that in the inlet (1.6, 2.9, and 10.0 mg/L, respectively) under different feeding loading rates (Table 13.14). This would suggest inadequate denitrification.

Negative rates of nitrate removal can occur, implying high nitrification potential in the VSSF systems and inadequate denitrification in the HSSF systems. A higher TN removal efficiency can be observed in hybrid systems, compared to individual VSSF or HSSF CW. The TN removal efficiency at 10% synthetic leachate concentration is 70.1%, while the values are 30.6 and 58.4% in VF CW and HF CW respectively.

Table 13.13 TN removal efficiency for laboratory-scale CW batch experiment.

Plant type	100%	75%	50%	25%
Narrow leaf cattail	73.5	83.3	90.6	94.9
Green bulrush	69.3	71.3	89.6	84.8
Umbrella palm	76.4	87.4	93.5	98.8

Table 13.14 Nitrate profile at various synthetic leachate dilutions applied to the CW.

	Leachate concentration		
	10%	25%	50%
Inlet	1.62 ($\pm$0.4) mg/L	2.9 ($\pm$1.5) mg/L	10 ($\pm$5.9) mg/L
VSSF effluent	28.2 ($\pm$20.2) mg/L	43.5 ($\pm$32.7) mg/L	89.3 ($\pm$46.5) mg/L
Outlet	15.4 ($\pm$9.9) mg/L	18.0 ($\pm$17.8) mg/L	47.2 ($\pm$24.0) mg/L

13.6 Summary

When considering the organic compounds removal (i.e., BOD, COD, and TC) at different influent concentrations, the batch experiment with all three plant species showed higher performance with 25% leachate. In addition, among the three selected plants, Umbrella palm had the highest efficiency. Pilot-scale experiments of the hybrid system cited better performance compared to the values of VSSF CWs. Furthermore, the experiments with real leachate showed similarly good performance with the hybrid system being better than the VSSF CW and HSSF CW applied individually.

When considering the removal and transformation of nitrogen compounds with different concentrations, removal efficiency for TN was recorded highest in all concentration groups. Similar to organic removal, Umbrella palm showed the highest average removal efficiency among the three selected plant species. Apart from that, the average removal efficiency of NH_4^+-N and NO_3^--N in VSSF-HSSF CWs systems showed an improvement in the efficiency figures compared to VSSF and HSSF CWs. Field scale experiments for leachate treatment also revealed that nitrogen removal and transformations within the hybrid CWs are comparably higher compared to other treatment systems.

The three plant species selected are tolerant to the concentrations of COD, BOD and NH_4^+-N present in the diluted leachate. Among these, the green bulrush plant shows the best removal efficiencies for many of the parameters in one-stage systems. However, the highest overall performance in hybrid systems is achieved with the Umbrella palm.

Hybrid constructed wetlands are capable of more stable removal of organics and nitrogen in comparison to the one-stage systems, when treating landfill leachate. The average removal rates of organics and nitrogen compounds are higher than those in the conventional systems. This would be due to the sequence of aerobic and anaerobic conditions in the hybrid CW system.

References

1 Raghab SM, El Meguid AMA, Hegazi HA. Treatment of leachate from municipal solid waste landfill. J HBRC 2013; 9(2):187–192.
2 Bhalla B, Saini MS, Jha MK. Characterization of leachate from Municipal Solid Waste (MSW) Landfilling Sites of Ludhiana, India: A Comparative Study. Int J Eng Res Applic. 2012; 2(6):732–745.

3 Visvanathan C. Solid Waste Management in Asian Perspectives. Environmental Management Tools 2006, Bangkok, pp. 1–7.

4 Jędrczak A. Quantity and chemical composition of landfill leachates (In Polish). Mat. III Konferencji Szkoleniowej. Budowa bezpiecznych składowisk. Fundacja PUK, Wisła, 1993.

5 Johansen OJ, Carlson DA. Characterization of sanitary landfill leachates. Water Res. 1976; 10(12):1129–1134.

6 Jędrczak A, Haziak K. Quantity, chemical composition, and treatment of landfill leachates. In: Mat. Konferencyjne IV Konferencji Szkoleniowej nt. Budowa bezpiecznych składowisk odpadów. Fundacja PUK, Poznań, 1994.

7 Plotkin S, Ram NM. Multiple bioassays to assess the toxicity of a sanitary landfill leachate. Arch Environ Cont Toxicol. 1984; 13(2):197–206.

8 Rojíčková-Padrtová R, Maršálek B, Holoubek I. Evaluation of alternative and standard toxicity assays for screening of environmental samples: selection of an optimal test battery. Chemosphere 1998; 37(3):495–507.

9 Ward ML, Bitton G, Townsend T, Booth M. Determining toxicity of leachates from Florida municipal solid waste landfills using a battery-of-tests approach. Environ Toxicol. 2002; 17(3):258–266.

10 Oman CB, Junestedt C. Chemical characterization of landfill leachates – 400 parameters and compounds. Waste Manage. 2008; 28:1876–1891.

11 Renou S, Givaudan JG, Poulain S, Dirassouyan F, Moulin P. Landfill leachate treatment: Review and opportunity. J Hazard Mater. 2008; 150(3):468–493.

12 Sewwandi B, Takahiro K, Kawamoto K, Hamamoto S, Asamoto S, Sato H. Characterization of landfill leachate from municipal solid wastes landfills in Sri Lanka. In: ICSBE. [online], 2012. Available at: www.civil.mrt.ac.lk/conference/ICSBE2012/SBE-12-236.pdf [Accessed 1 December 2017].

13 Szpadt R. Characteristics and treatment methods of municipal landfill leachates. Municipal Rev. 2006; 60–66.

14 Bhalla B, Saini MS, Jha MK. Effect of age and seasonal variations on leachate characteristics of municipal solid waste landfill. Int J Res Eng Technol. 2013; 2:223–232.

15 Lee AH, Nikraz H, Hung YT. Influence of waste age on landfill leachate quality. Internat J Environ Sci Develop. 2010; 1(4):347.

16 Kang KH, Shin HS, Park H. Characterization of humic substances present in landfill leachates with different landfill ages and its implications. Water Res. 2002; 36(16):4023–4032.

17 McCarty PL. Anaerobic waste treatment fundamentals. Public works. 1964; 95(9):107–112.

18 Tałałaj I. Quality of underground waters in the vicinity of municipal landfills. In Jakość wód gruntowych wokół wysypisk odpadów komunalnych. W: II Forum Inżynierii Ekologicznej "Monitoring środowiska"red. Wiatr I, Marczak H. Nałęczów (p. 640), 1998.

19 Malina J. Design of anaerobic processes for treatment of industrial and municipal waste (Vol. 7). CRC Press, 1992.

20 Im J-H, Woo H-J, Choi M-W, Han K-B, Kim C-W. Simultaneous organic and nitrogen removal from municipal landfill leachate using an anaerobic-aerobic system. Water Res. 2001; 35(10):2403–2410.

21 Klimiuk E, Kulikowska D, Koc-Jurczyk J. Biological removal of organics and nitrogen from landfill leachates – A review. Management of pollutant emission from landfills and sludge. Taylor & Francis Group, London, 2007, pp. 187–204.

22 Tatsi AA, Zouboulis AI. A field investigation of the quantity and quality of leachate from a municipal solid waste landfill in a Mediterranean climate (Thessaloniki, Greece). Adv Environ Res. 2002; 6(3):207–219.

23 Inanc B, Calli B, Saatci A. Characterization and anaerobic treatment of the sanitary landfill leachate in Istanbul. Water Sci Technol. 2000; 41(3):223–230.

24 Goorany O, Oztürk I. Soluble microbial product formation during biological treatment of fermentation industry effluent. Water Sci Technol. 2000; 42(1-2):111–116.

25 Aquino SF, Stuckey DC. Soluble microbial products formation in anaerobic chemostats in the presence of toxic compounds. Water Res. 2004; 38(2):255–266.

26 Poltak RF. Sequencing Batch Reactor Design and Operational Considerations. New England Interstate Water Pollution Control Commission, 2005.

27 Wang S, Peng Y. Natural zeolites as effective adsorbents in water and wastewater treatment. Chem Eng J. 2010; 156(1):11–24.

28 Wang JP, Chen YZ, Ge XW, Yu HQ. Optimization of coagulation–flocculation process for a paper-recycling wastewater treatment using response surface methodology. Colloids and Surfaces A: Physicochem Eng Aspects. 2007; 302(1):204–210.

29 Amokrane A, Comel C, Veron J. Landfill leachates pretreatment by coagulation-flocculation. Water Res. 1997; 31(11):2775–2782.

30 Mamlook R, Badran O, Abu-Khader MM, Holdo A, Dales J. Fuzzy sets analysis for ballast water treatment systems: best available control technology. Clean Technol Environ Policy. 2008; 10(4):397–407.

31 Stefanakis AI, Akratos CS, Tsihrintzis VA. Vertical Flow Constructed Wetlands: Eco-engineering Systems for Wastewater and Sludge Treatment. Elsevier Science: Oxford, 2014.

32 Sayadi MH, Kargar R, Doosti MR, Salehi H. Hybrid constructed wetlands for wastewater treatment: A worldwide review. Internat Acad Ecol Environ Sci. 2012; 2(4):204.

33 Vymazal J. Removal of nutrients in various types of constructed wetlands. Sci Total Environ. 2007; 380(1):48–65.

34 Bricken EC. Constructed Wetlands as an Appropriate Treatment of Landfill Leachate. Doctoral dissertation, University of Natal, 2003.

35 Tchobanoglous G, Burton FL, Stensel HD. Wastewater Engineering Treatment and Reuse, 4th edn. McGraw-Hill: New York, 2003.

36 Wojciechowska E, Gajewska M, Obarska-Pempkowiak H. Treatment of landfill leachate by constructed wetlands: three case studies. Polish J Environ Stud. 2010; 19(3):643–650.

37 Supradata. Domestic Waste Water Treatment by using *Cyperus alternifolius*, L. with Subsurface Flow Wetland System (SSF-Wetlands). Master Program of Environmental Science, Diponegoro University, Semarang, 2005.

38 Redmond ED, Just CL, Parkin GF. Nitrogen removal from wastewater by an aerated subsurface-flow constructed wetland in cold climates. Water Environment Research. 2014; 86(4):305–313.

39 Pempkowiak HO, Gajewska M. The removal of nitrogen compounds in Constructed Wetlands in Poland. Polish J Environ Studies. 2003; 12(6):739–746.

40 Cooper P. The performance of vertical flow constructed wetland systems with special reference to the significance of oxygen transfer and hydraulic loading rates. Water Science and Technology. 2005; 51(9):81–90.

41 Yalcuk A, Ugurlu A. Comparison of horizontal and vertical constructed wetland systems for landfill leachate treatment. Bioresource Technol. 2009; 100(9):2521–2526.

Part V

Wood and Leather Processing Industry

14

Cork Boiling Wastewater Treatment in Pilot Constructed Wetlands

Arlindo C. Gomes[1], Alexandros I. Stefanakis[2,3,4], António Albuquerque[1] and Rogério Simões[1]

[1] *Universidade da Beira Interior, Materiais Fibrosos e Tecnologias Ambientais (FibEnTech-UBI), Rua Marquês d'Ávila e Bolama, Covilhã, Portugal*
[2] *Bauer Resources GmbH, BAUER-Strasse 1, Schrobenhausen, Germany*
[3] *Department of Engineering, German University of Technology in Oman, Athaibah, Oman*
[4] *Bauer Nimr LLC, Muscat, Oman*

14.1 Introduction

14.1.1 Cork Production and Manufacture

Cork is the natural and renewable material extracted from the out bark of the producer tree (*Quercus suber* L.) by traditional procedures after growth cycles of nine years, from trees older than 20 years. This activity is strictly regulated and intends to ensure a lifespan production over 200 years and ultimately contribute to the preservation of these natural forest ecosystems, typical of the Mediterranean region. Apart from cork production, these forests are also characterized by their high level of biodiversity, their contribution to hydrology regulation, prevention of desertification and carbon dioxide fixing (18 kg per kg of cork extracted) [1]. Thus, cork production is a model of sustainability between human activity and natural resources preservation [2–4]. Moreover, since 1993 the ecological importance of cork oak forests was also taken into account by the European Union and they were classified as protected habitats under the framework of the Natura 2000 Network [2].

Besides the environmental importance for the Iberian countries, where 51% of this natural ecosystem on earth is located (i.e., about 1,289,000 ha), the economic and social aspects have also been taken into consideration, given that Portugal and Spain produce more than 80% of the world cork annual production, estimated to surpass 200 thousand tons per year. Portugal alone has a market share of 70% for cork product exports, mostly wine stoppers and agglomerates for thermal and acoustic insulation [1]. The economic and social outlook of production and manufacture activities are critically dependent on wine consumers' preference for one-piece stoppers made with cork in compared with those produced with synthetic materials, because this application uses only 15% of the cork produced, but accounts for approximately 75% of the industry revenues [5]. Thus, if the market share of cork stoppers declines, the profit from cork extraction activity is also reduced and the maintenance of these ecosystems of high ecological value would probably decline.

The industrial processing of the cork slabs after extraction from trees starts with maturing and drying stages followed by the boiling process, a low technology operation intended to improve the cork texture and properties, making this material more homogeneous, flat, elastic and mostly free

Constructed Wetlands for Industrial Wastewater Treatment, First Edition. Edited by Alexandros I. Stefanakis.
© 2018 John Wiley & Sons Ltd. Published 2018 by John Wiley & Sons Ltd.

of biological and chemical contamination. The traditional procedure continues to be used in most industries and includes the immersion of the corkwood in boiling water for 1.0–1.5 h. Due to the critical importance of organic contamination removal, especially if the raw material is intent to produce stoppers, the reuse of hot water has to be limited and ranges from 6 up to 30 loads of raw cork according to factory procedures and quality standards [6–8]. The high-quality requirements for the water used in the boiling stage contribute to increment the fresh water consumption and consequently the volume of effluent produced, which ranges from 140 to 1,200 L/ton of corkwood processed [7, 9, 10]. Thus, this stage of the industrial cork processing raises some environmental concerns due to the high specific water consumption and the organic load of bio-recalcitrant nature of the effluent.

Hitherto, most studies published on the topic of cork boiling wastewater (CBW) treatment or valorization used physico-chemical treatment options rather than biological processes. However, the related costs are high and above those of the biological treatment alternative for wastewaters with similar organic loads but from different sources [11, 12]. Thus, constructed wetland systems (CWs) can be an alternative to conventional biological treatment systems, namely to activated sludge systems, with the prominent virtues of low construction and operation costs [13–15]. CWs are engineered to take advantage of several mechanisms to remove pollutants, namely physical processes (precipitation, filtration, sedimentation and volatilization) and biochemical processes induced by wetland plants and microorganisms [13, 16, 17]. Additionally, the landscape value of wetlands is perfectly compatible with cork oak forests.

Until today, there is no study or research published for the treatment of CBW using CW systems. Thus, this investigation is a novelty and included the monitoring over a period of four years of the operation of a horizontal subsurface flow constructed wetland (HSF CW) microcosm-scale system planted with common reeds (*Phragmites australis*) and filled with light expanded clay aggregates (LECA), as support media for the plants and for biofilm development. The contribution of plants to the treatment was assessed by the comparison with an identical wetland unit without plantation (control bed). After this extended period of operation intended to maximize biomass development and acclimatization, which included stepwise increase of the organic load rate (OLR) up to 8.9 g COD/m^2/d, the assessment of the treatment capacity of the system was done by doubling the OLR to 16.4 g COD/m^2/d during 200 days.

14.1.2 Cork Boiling Wastewater Characteristics

CBW has an intense dark color, high concentration of organic pollutants, namely polysaccharides, phenols, polyphenolic compounds and cork extracts including polymerized tannins. Some of these compounds have high molecular weight (MW), as it is the case of tannins with 500 to 3,000 Da and when polymerized can reach up to 40 kDa [18]. The combination of these features results in poor bioavailability and biodegradability, increases the toxicity and ultimately restrains the feasibility of conventional biological processes for wastewater treatment discharge or reuse [10, 18, 19].

The most complete and recent published CBW characterization from different studies is presented in Table 14.1. These results show a great diversity of values for all parameters analyzed due to variations in specific water consumption, differences in the contamination levels and/or composition of the raw corkwood processed [1, 3–5]. For instance, the chemical oxygen demand (COD) and biological oxygen demand for 5-days incubation (BOD$_5$) concentrations range from 1,240–11,500 mg/L and 320–3,500 mg/L, respectively.

Table 14.1 Main characteristics of cork boiling wastewater from recent published literature.

Parameter[a] (units)	Vilara et al. [23]	Teixeira et al. [22]	Bernardo et al. [10]	Pintor et al. [29]	Santos et al. [21]	Gomes et al. [28]	De Torres-Socías et al. [28]	Marques et al. [11]	Fernandes et al. [30]
pH	7.50	4.6–6.2	4.70	5.0–6.5	5.42	5.81	7.2	5.8 ± 0.0	6.5 ± 0.5
Conductivity (mS/cm)	2.90		0.935	1.2			1.1	1.5 ± 0.1	0.70 ± 0.07
Turbidity (NTU)	30.3			58–84			163		
Abs 254 nm				0.97–1.17	0.395[b]	0.562[b]			0.43 ± 0.09[b]
Color	11,300[c]	Visible[d]	7,100[e]		0.118[f]	0.221[f]			0.08 ± 0.01[f]
SS (mg/L)		65–900		126	970		290		
COD (mg/L)	4,692	2,260–11,500	2,604	1,786–2,403	1,536	1,878	1,240	6,500 ± 100	2,041 ± 114
DOC (mg/L)	1,448			763–892	595	498	586		800 ± 71
BOD_5 (mg/L)	750	500–3,500	900	320–456	407	684			260 ± 52
BOD_{20} (mg/L)			1,225		554				474 ± 62
BOD_5/COD			0.35	0.18–0.19	0.26	0.27			0.13 ± 0.02
BOD_{20}/COD			0.47		0.36	0.36			0.23 ± 0.03
TN (mg/L)	308.8	60–200		21–58	15.33	17.02	27.8	40	20 ± 3
TP (mg/L)	10.3	20–60		17.7–19.9	6.46	5.85			
TPh (mg/L)	740[g]	1,000–3,500[h]	410[i]		110.3[i]	523[i]		1,200[g]	140 ± 20[i]
Tannins (mg/L)		850–1,700[i]	270[i]			399[i]			

[a] Abs: Absorbance, SS: suspended solids, COD: chemical oxygen demand, DOC: dissolved organic carbon, BOD_5: biological oxygen demand after 5 days incubation, BOD_{20}: biological oxygen demand after 20 days incubation, TN: total nitrogen, TP: total phosphorus, TPh: total phenols;
[b] for dilutions (1:50);
[c] Pt-Co units (at 400 nm);
[d] dilution (1:20);
[e] Hazen Units;
[f] Absorbance at 580 nm for dilutions (1:5);
[g] expressed as caffeic acid equivalents;
[h] expressed as gallic acid equivalents;
[i] expressed as tannic acid equivalents.

Beside the differences in the methodologies used to quantify color (Table 14.1), in all cases CBW samples had dark brownish color due to high concentration of corkwood extracts with high molecular size, namely of tannins. The quantification of these pollutants was made through the total phenols (TPh) analysis and less frequently includes tannin determination by the gravimetric method described by Makkar et al. (1993) [20]. The methodologies reported for TPh quantification involve several variations on the original procedure published by Folin and Ciocalteu (1927), namely employ the Folin–Ciocalteu reagent solution, absorbance measurements at 765 nm, use of different compounds for calibration and then to express the overall concentration, namely gallic, caffeic and tannic acids [11, 21–25]. The major drawbacks of these methods are the low selectivity of the reagent solution, which reacts with any reducing substances in addition to phenols [25].

The majority of the research results published includes bioassays to assess biodegradability and, less frequently, the acute toxicity before and after the treatment to evaluate the environmental impact of the effluent discharge or, in the case of pretreatment stages, to evaluate the feasibility of the subsequent biological treatment. The most common biodegradability indices used are based on the ratio between the BOD and COD; the rapidly bioavailable fraction of the organics in the wastewater corresponds to the BOD_5 using non-acclimatized aerobic biomass or extended incubation up to 20-days (BOD_{20}) to account for the overall biodegradable fraction [10, 21]. According to some authors, values of BOD_5/COD ratios higher than 0.40–0.50 are required for effective biological treatment [26, 27]. Less frequently, the Zahn–Wellens test protocol (with 28-days incubation) was used to assess the ultimate biodegradability and the determination of the oxygen uptake rate (OUR) for the short-term biodegradability assessment [23, 28, 29].

To date, only one publication reported the number of total heterotrophic bacteria in CBW using serially diluted, spread on plate count agar with 48 hours incubation at 25°C and 50°C. These temperatures were selected taking into consideration the cooling of CBW from 100°C to ambient temperature. The bacterial enumeration at 25°C ranged from 3.8×10^4–1.8×10^7 colony-forming units (CFU) per mL and at 50°C were $<1 \times 10^1$–2.1×10^4 CFU/mL [7]. In any case, CBW has also low bacterial biodiversity, which is further reduced in the case of the isolation of cultures grown using phenolic compounds as selective carbon sources [7, 30].

Hitherto, the quantification of the organic pollutants in CBW includes more than 50 compounds with concentration ranging from 0.22–2.00 mg/L for the syringic acid up to 238.7 mg/L for the ellagic acid. However, research results published for CBW treatment (in mg/L) were limited to phenolic acids, namely gallic (2.46–103.70), protocatechuic (0.38–71.10), caffeic (2.14), vanillic (2.00–8.00), *p*-cumaric (4.10), ferulic (6.00–7.70) and ellagic (2.92–238.70) [11, 21, 31, 32]. Some of these compounds are also present in other agro-industrial wastewaters, namely from olive mills (OMW), wine-distillery, wood debarking and coffee processing [11, 33]. The potential for recovery and valorization of these by-products as promising sources of renewable chemicals is high due to their bioactive properties and extended range of potential uses, namely as antioxidants, anti-inflammatories, anti-carcinogenics and inhibitors of enzymes with activity related to several diseases, but this has not yet been explored [25, 34].

The anaerobic digestion of CBW raises the opportunity to recover energy and resources through biogas production and concentration increase of compounds from the benzoic acid family (i.e., gallic, protocatechuic, vanillic and syringic acids), highlighting the potential for added value to CBWs [11]. The use of ionizing radiation (gamma radiation) also increased the concentration of phenolic compounds and antioxidant capacity [34]. Another opportunity for CBW valorization is the replacement of the vegetable extracts used in the tanning industry by nanofiltration (NF) concentrates [19, 22]. Overall, all of these approaches have the potential to enhance the environmental and economic sustainability of the cork manufacture industry but until today they have not yet been implemented on a real scale.

14.2 Cork Boiling Wastewater Treatment

Despite the significant investment in innovation to create new cork products and to improve the quality of those already sold, especially to ensure close to complete elimination of organic contaminants, namely of the chlorophenols (such as 2,4,6-trichloroanisole) to prevent damage of the organoleptic quality of the wine during the storage period, in most factories CBW treatment still limited. In fact, the CBW disposal has only progressed from direct discharge into public water-courses without any treatment, or after retention basins to allow for equalization, homogenization and partial evaporation, to the application of coagulation–flocculation as a pretreatment stage, followed by discharge into municipal wastewater treatment plants [7, 28, 35–37]. However, the acceptance of cork products and the profitability of production and transformation activities can be increased if they are viewed as products with an environmental and sustainable character [1]. Therefore, technologies allowing for the reduction of water consumption and the mitigation of pollution discharged to water bodies are necessary and can contribute to this goal.

The literature review showed a vast array of possibilities for CBW treatment, including physico-chemical processes, membrane separation technologies, biological treatments and treatment sequences. In most cases, several pretreatments options are applied to reduce the suspended solids (SS) concentration (between 23–59%), in addition to COD removal (between 43–70%) [19, 31, 32, 38–40]. Besides the removal of gross solids, which is critical for membrane technology performance, oxidation is often used to increase the biodegradability and reduce the toxicity, so that subsequent biological treatment can be successfully applied and, in one case, to enhance the performance of ultrafiltration (UF) membranes in a wide range of molecular weight cut-offs (MWCOs), from 4 to 98 kDa, through fouling reduction [6, 21, 24].

However, it should be taken into account that sequential treatments, through combination of physico-chemical processes or of chemical and biological oxidation, present an additional difficulty for an efficient operation of a large-scale plant. Moreover, it is necessary to ensure that the chemical oxidant and the biological culture do not come in undue contact with each other [12, 28].

Next, the results published for CBW treatments are revised and categorized according to the methodology used, i.e., physico-chemical processes, membrane separation, biological treatment and sequential treatments (Tables 14.2–14.5).

14.2.1 Physico-Chemical Treatment

The physico-chemical processes applied to CBW can be divided into those requiring the consumption of reactants, namely coagulation–flocculation, chemical oxidation and anodic oxidation, and those using membrane technologies (MTs). However, the large variations of CBW characteristics (Table 14.1) make difficult the direct comparisons.

As summarized in Table 14.2, coagulation–flocculation processes consume reactants to adjust the pH and to promote the formation of flocs, allowing the transformation of non-settleable solutes into

Table 14.2 Literature review results for treatment of CBW with physico-chemical processes.

Treatment method	pH IN	pH OUT	Color IN	Color Rem (%)	COD IN (mg/L)	COD Rem (%)	TPh IN (mg/L)	TPh Rem (%)	Reference
Coagulation-Flocculation									
(200 mg Fe^{+3}/L)	5.6	7.2			2,280	53	380[d]	89	[41]
(166 mg Al^{+3}/L)	5.0[a]				3,047	54	381[d]	82	[8]
(100 mg Chitosan/L)	2.9[a]				2,469[b]	17	958[e]	26	[32]
(200 mg Chitosan/L)	3.0[a]					25		45	
(Fe^{3+} added up to 20 mg/L)	7.2	3.1			1,170[c]	43			[42]
(Fe^{+3} added up to 20 mg/L)	6.6	2.6			1,780[c]	70			
(Fe^{+3} added up to 20 mg/L)	2.8[a]				1,240	61			[9]
(Fe^{+2} added to 20 mg/L)						45			
Fenton-Oxidation									
(10.6 g H_2O_2/L; H_2O_2:Fe^{2+} weight ratio of 1:5)	3.2[a]				5,000[c]	87			[29]
Photo-Fenton									
(13.64 g H_2O_2/L; H_2O_2:Fe^b = 1:8.3 weight ratio; Irradiation time = 10 min)	3.2[a]				2,100[b]	66			[43]
Solar Photo-Fenton									
(20-80 mg Fe^{3+}/L; 4.36 g H_2O_2/L)[g]	2.6–2.8[a]	2.8			1,786–2,404	65	352[d]	95	[29]
		2.9				91		81	
([Fe^{3+}] set to 20 mg/L; 750 mg H_2O_2/L; irradiation time 377 min)	2.6–2.8[a]				1,240	59			[28]
([Fe^{3+}] set to 20 mg/L; 780 mg H_2O_2/L; irradiation time 435 min)					480[c]	52			

Ozonation

$(O_{3,app}/COD_i = 0.32\text{–}1.64)$	~5.3	2.7–3.6			~1,900	19–48	~290[d]	65–80	[42]
$(O_{3,app}/COD_i = 2.90\text{–}3.60)$	4.8				~1,600	42–69	~305[d]	80–94	[35]
$O_{3,app}/COD_i = 0.27\text{–}2.63$	3.3[a]	2.2–2.7	0.554[h]	70–87	1,878	15–53	523[d]	66–83	[24]
	5.8	3.3–5.4		57–91		16–59		60–82	
	10.0[a]	4.0–6.3		59–92		25–62		38–75	

AOP

50 mg O_3/L + 0.34–3.40 g H_2O_2/L	4.8				~1,600	76–80	305[d]	95–98	[35]
(0.82 g O_3/L + 0.6 g H_2O_2/L; Oxidation Time = 11 h)	7.0[a]	5.1			480[c]	14			[28]
	10.0[a]	6.4				33			

Anodic Oxidation

BDD electrodes Na_2SO_4 = 0–1,5 g/L; current density of 30 mA/cm[b]; Oxidation time up to 8 h	6.5		0.080[h]	>50	~2,000	>74	140[f]	>80	[44]

[a] pH after adjustment;
[b] values for total organic carbon (TOC) concentration;
[c] after CBW pre-treatment;
[d] expressed as caffeic acid equivalents;
[e] expressed as gallic acid equivalents;
[f] expressed as tannic acid equivalents;
[g] using cork bleaching WW as source for H_2O_2 (at 7.7 g H_2O_2/L) results for end of solar-photo-treatment;
[h] Abs at 580 nm for samples dilutions of 1:10;

Table 14.3 Literature review results for treatment of CBW with membrane separation technology.

Membrane(pore size or MWCO)	Operational conditions[1]	Color		COD		TPh		Reference
		IN	Rem (%)	IN (mg/L)	Rem (%)	IN (mg/L)	Rem (%)	
MF								
DUR-0.65 (0.65 μm)	ΔP = 0.45–2.3 bar; Lp = 860 L/m^b/h/bar;	0.114^b	47–83	4,290	17–39	761^c	21–48	[39]
DUR-0.10 (0.10 μm)	ΔP = 0.45–2.3 bar; Lp = 248 L/m^b/h/bar;		60–85		19–41		26–48	
UF								
Bio-300K (300 kDa)	ΔP = 0.35–1.8 bar; Lp = 769 L/m^b/h/bar;	0.114^b	66–83	4,290	24–39	761^c	26–49	[39]
Bio-300K (300 kDa)	ΔP = 0.75 bar; Lp = 769 L/m^b/h/bar;	0.020^b	83	2,670	7	197^c	16	[45]
Bio-10K (100 kDa)	ΔP = 0.75–1.80 bar; $J_{v(initial)}$ = 86–147 L/m^b/h;		95		29		41	
GR40PP (100 kDa)	ΔP = 1–3 bar; VRF = 4; Lp = 60 L/m^b/h/bar	0.059^b	64	1,536	45	110.3^d	37	[21]
– (91 kDa)	ΔP = 3 bar; Lp = 106 L/m^b/h/bar	5,700^e	66	2,285	45	360^d	46	[10]
GR51PP (50 kDa)[6]	ΔP = 1–3 bar; VRF=4; Lp = 52.5 L/m^b/h/bar	0.022^b	24	851	45	69.4^d	48	[21]
– (45 kDa)	ΔP = 3 bar; Lp = 56 L/m^b/h/bar	5,700^e	91	2,285	52	360^d	65	[10]
– (25 kDa)	ΔP = 3 bar; Lp = 37.8 L/m^b/h/bar	7,100^e	93	2,604	62	410^d	68	
GR61PP (20 kDa)^g	ΔP = 1–3 bar; VRF=4; Lp = 26.6 L/m^b/h/bar	0.016^b	50	686	24	31.6^d	20	[21]
– (13.6 kDa)	ΔP = 3 bar; Lp = 34.8 L/m^b/h/bar	5,700^e	95	2,285	68	360^d	75	[10]

Membrane	Conditions							Ref.
GR81PP (10 kDa)[h]	ΔP = 1–3 bar; VRF=4; Lp = 13.5 L/m[b]/h/bar	13	0.008[b]	520	26	28.8[d]	42	[18]
– (3.8 kDa)	ΔP = 3 bar; Lp = 2.5 L/m[b]/h/bar	97	5,700[e]	2,285	74	360[d]	82	[10]
– (1.2 kDa)	ΔP = 3 bar; Lp = 1.4 L/m[b]/h/bar	99			90		90	[10]
NF								
NF270 (400 Da)	ΔP = 3 bar; VRF>7; Lp = 11 L/m[b]/h/bar; $J_{v(initial)}$ = 3.6 L/m[b]/h	>99		2,260–11,500	>96	1,000–3,000[i]	>96	[22]
CK (150–300 Da)	ΔP = 10–30 bar; VRF = 2; Lp = 6.5 L/m[b]/h/bar; $J_{v(initial)}$ = 21–61 L/m[b]/h	99	0.114[b]	4,290	96	760[(3)]	98	[45]
DK (150–300 Da)	ΔP = 30 bar; VRF = 2; Lp = 4.6 L/m[b]/h/bar; $J_{v(initial)}$ = 73.1 L/m[b]/h	98			93		96	
NF90 (~150Da)	ΔP = 15 bar; VRF>7; Lp = 7 L/m[b]/h/bar; $J_{v(initial)}$ = 2.3 L/m[b]/h	>99		2,260–11,500	>99	1,000–3,000[i]	>99	[22]
– (125 kDa)	ΔP = 3 bar; Lp = 5.2 L/m[b]/h/bar	~100	5700[e]	2,285	95	360[d]	92	[10]

[a] ΔP = transmembrane pressure, Lp = permeability to pure water and $J_{v(initial)}$ = initial permeate flux with CBW;
[b] Abs at 580 nm for samples dilutions of 1:10;
[c] tannic content;
[d] expressed as tannic acid equivalent;
[e] Hazen Units;
[f] results for operation with the permeate from GR40PP (MWCO of 100 kDa);
[g] operation with the permeate from GR51PP (MWCO of 50 kDa);
[h] operation with the permeate from GR61PP (MWCO of 20 kDa);
[i] expressed as gallic acid equivalent.

Table 14.4 Literature review results for treatment of CBW with biological processes.

Treatment method (Main operation conditions)	pH		Color		COD (mg/L)		TPh (mg/L)		Reference
	IN	Final	IN	Rem (%)	IN	Rem (%)	IN	Rem (%)	
Activated-sludge (HRT=24-96 h; $T=20°C$)	5.40 ± 0.20	7.89–8.55			~1,810	13–37	290 ± 60[a]	20–32	[6]
Fungal strains[b] (48 h batch incubation; 2 g/L of dry weight fungal biomass; $T = 25°C$)				−43–47	7,500 ± 500	49.5–58.9			[39]
Enriched cultures[c] (aeration rate = 0.2 (V/V); 15 days incubation; $T = 25°C$)	4.7–5.7				1,670[d]	~25	730[e]	~18	[7]
Self-bioremediation (bacterial immobilization into residual cork particles)								60–80	[30]
Anaerobic digestion (Batch incubation during 15–44 d with inoculum from OMW treatment at of 5 g VSS/L; $T = 37°C$)	7.20	7.67			3,000	40	553[a]	−63	[11]
HSF CW[f] ($T = 18 ± 5$ °C; HRT = 6.1 ± 0.6 d; HLR = 6.2 ± 0.6 L/m[b]/day; OLR = 16.5 ± 2.3 g COD/m[b]/day)	7.04 ± 0.25[g]	8.02 ± 0.35	0.109 ± 0.090 [h]	−111 ± 33	2,675 ± 289	62 ± 10[i]	116.8 ± 13.0 [e]	58 ± 7[i]	Present study

[a] Expressed as caffeic acid equivalent;
[b] 4 strains isolated from outer bark of cork oak trees and 2 for their ability to degrade lignin and to grow on phenol;
[c] selected for their ability to use tannic acid as single carbon and energy sources;
[d] values for TOC concentration;
[e] expressed as tannic acid equivalent;
[f] HSF – horizontal subsurface flow, CW – constructed wetland, HRT – hydraulic retention time, HLR – hydraulic loading rate and OLR – organic loading rate and the values presented corresponds to average ± standard deviation for $n = 22$;
[g] after pH adjustment;
[h] Abs at 580 nm for samples dilutions of 1:10;
[i] percentage of removals calculated through mass balance.

Table 14.5 Literature review results for treatment of CBW with sequential processes.

Treatment sequence	CBW			Removal (%)			Reference
	Color	COD(mg/L)	TPh(mg/L)	Color	COD	TPh	
O_3 (0.91 g/g COD; HRT = 6 h) → **Biodegradation** (Activated sludge with acclimatized biomass; HRT = 48 h)		1,900	290[a]		65	94	[6]
Biodegradation (Activated sludge with acclimatized biomass; HRT = 48 h) → O_3 (0.85 g/g COD; HRT = 3 h)		1,800	250[a]		77	92	
Fenton → **coagulation-flocculation** (reactants for combined process Iron 0.06–11.2 g/L + H_2O_2 2–34 g/L)		3,500–3,900	480–580[a]		22–85	4–98	[9]
Fenton ($[Fe^{2+}]_{initial}$ = 1 g/l;$[H_2O_2]_{initial}$ = 5 g/L) → **Biodegradation** (Aerobes selected using tannic acid as sole carbon and energy source)		2,300 – 4,600[b]	660 – 780[c]		>90		[7]
Fenton ($[Fe^{2+}]_{initial}$ = 56–11,200 mg/L; $[H_2O_2]_{initial}$ = 2–34 g/L) → **Coagulation-flocculation** (NaOH added up to pH>10, 1 day sedimentation)		3,500 – 3,900	480 – 580[a]		22–85	4–98	[9]
Coagulation-Flocculation+Filtration (pH = 2.8; addition of Fe^{2+} or Fe^{3+} to have **dissolved concentration of 20 mg/L + 75 μm pore size**) → **Solar photo-Fenton** (1.9–2.4 g H_2O_2/L; irradiation time of 500–750 min)		1,170			45–90		[29]
		1,780					
Coagulation-Flocculation+Filtration (pH = 2.8; addition of Fe^{3+} to have **dissolved concentration of 20 mg/L + 75 μm pore size**) → **Solar photo-Fenton**(780 mg H_2O_2/L; irradiation time of 435 min)		1,240			52		[23]
Solar Photo-Fenton (pH = 2.6–2.8; 20–80 mg Fe^{2+}/L; H_2O_2 consumption 2.14–2.70 g/L provide from cork bleaching WW) → **Biodegradation** (Zahn-Wellens test with supply of nutrients, 28 days)		1,786 – 2,403	304 – 399[a]		65		[28]
O_3(0.05 g/g TOC) → **UF** (MWCO = 300 kDa; ΔP = 0.75 bar)	0.157[d]	4,400	897[e]	83	7	16	[39]
O_3(0.05 g/g TOC) → **UF** (MWCO = 10 kDa; ΔP = 1.25 bar)				95	29	41	
O_3(0.05 g/g TOC) → **UF** (MWCO = 5 kDa; ΔP = 0.75 bar)				91	24	31	

(Continued)

Table 14.5 (Continued)

Treatment sequence	CBW			Removal (%)			Reference
	Color	COD(mg/L)	TPh(mg/L)	Color	COD	TPh	
Sequences for the treatment of permeates							
MF(Pore size = 0.65 µm; ΔP = 0.75 bar) → $\mathbf{O_{3,appl}}$(0.05 g/g TOC)	0.157^d	4,400	897^e	98	82	94	[39]
MF(Pore size = 0.10 µm; ΔP = 0.75 bar) → $\mathbf{O_{3,appl}}$(0.05 g/g TOC)				96	83	97	
UF(MWCO = 300 kDa; ΔP = 0.75 bar) → $\mathbf{O_{3,appl}}$(0.05 g/g TOC)				96	87	98	
UF(MWCO = 300 kDa; ΔP = 0.75 bar) → **UV**				40	81	43	
UF(MWCO = 300 kDa; ΔP = 0.75 bar) → $\mathbf{O_{3,appl}}$(0.05 g/g TOC)+**UV**				98	97	100	
UF(MWCO = 300 kDa; ΔP = 0.75 bar) → **AOP** ($O_{3,appl}$ = 0.05 g/g TOC; $[H_2O_2]_{initia}l$ = 34 mg/L)				99	98	100	
Sequences for the treatment of concentrates							
UF (MWCO = 100 kDa; ΔP = 1–3 bar; VRF = 4) → $\mathbf{O_3}$ ($O_{3,appl}$ = 0.7–1.3 g/L)	0.171^d	3436	233^c	90–97	59–69	88–92	[21]

[a] Expressed as caffeic acid equivalent;
[b] values for TOC concentration;
[c] expressed as tannic acid equivalent;
[d] Abs at 580 nm for samples dilutions of 1:10;
[e] tannic content.

coagulated settleable particles next removed as sludge. COD and TPh removals up to 70% and 89% were reported, but these results are strongly influenced by the type and dose of coagulant, coagulation mixing time and stirring rate, pH, ionic strength, nature and concentration of the organic compounds present in the solution [42]. Nevertheless, these processes are not an effective way for pollutant elimination, but they promote the concentration of the organics into chemical sludge, thus, further treatment is required before the final disposal (which is usually not considered).

Variations of Fenton reaction, ozonation and advanced oxidation processes (AOP) are the most studied physico-chemical techniques for CBW treatment (Table 14.2). In this case, a mixture of oxidants, O_3 and H_2O_2, are applied alone or in combination with UV radiation to increase the amount of the highly reactive hydroxyl radicals ($OH^\bullet$). However, due to the lack of selectivity of oxidants, which is minimal for $OH^\bullet$, it is difficult to avoid their consumption in reaction with biodegradable organics or radical scavengers; the biocompatibility of the produced compounds is also difficult to estimate. Thus, most frequently, the oxidation extent is only the minimal to ensure the viability of the subsequent biological treatment. Moreover, the oxidant efficiency decreases with the dose; thus, it is necessary to carefully set the optimum oxidant dosage and to optimize the operational conditions (reactor configuration, increase of the mass transfer, temperature, reaction time, etc.) [12].

Fenton oxidation also takes advantage of the production of $OH^\bullet$ with high oxidant potential (2.08 V) resulted from the reaction of H_2O_2 with the catalyst (Fe^{2+}), which can be increased by exposure to UV radiation [12, 29]. The results reported were dependent on organic load (COD of 480–5,000 mg/L), oxidant doses (0.75–13.6 g H_2O_2/L), Fe^{2+} concentration, source of radiation and irradiation time. Despite the fact that most of the published results ignore the characteristics and amounts of sludge produced in Fenton oxidation, it is important to notice their magnitude, which in one case accounts for 176 mL/L of CBW with a water content of 96% [43, 45]. These processes allow an increase of the biodegradability from 0.18–0.27 up to 0.28–0.70, using the BOD_5/COD ratio for evaluation, and from 0.13 to 0.70 with the Zahn–Wellens test [28, 29, 43, 45]. Taking into account the process economics, the most interesting results were reported with 8.5–12.5 hours of solar irradiation and use of cork bleaching wastewater as source of H_2O_2 (at 7.7 g/L), which allows for COD and TPh removal of 65–91% and 81–91%, respectively. However, the strong color of the samples impaired any additional increases through irradiation and raised reaction times [29].

In ozonation trials, amounts of O_3 equivalent to $O_{3(appl)}$/COD_0 ratios of 0.27–3.60 were used to achieve color, COD and TPh removal of 59–92%, 15–69% and 38–94%, respectively [24, 35, 38]. Despite the pH of the sample and O_3 dose being critical parameters requiring optimization, the results reported show that ozonation is the best option for decolorization and that sample pH influences the extent of oxidation by molecular O_3 and by $OH^\bullet$, as for values below 5.5 the $OH^\bullet$ formed by ozone self-decomposition is suppressed [12]. The effect of pH (set at 3 and 10) and of O_3 dose ($O_{3(appl)}$/COD_0 ratios of 0.27–2.68) affected the BOD_5/COD ratio increase from 0.27 up to 0.44 and the toxicity reduction from 3.08 to 1.28 toxicity units (TU) (for Microtox) [24]. Other authors also reported significant increases of biodegradability from 0.13–0.60 up to 0.59–0.93 after ozonation [6, 35]. These results showed also that mineralization extension is not correlated with biocompatibility augmentation, because for close oxidant doses the COD removals were maximum at alkaline pH, but it was the large removal of TPh at acid pH that led to the highest increase of BOD_5/COD ratios and toxicity reduction [24].

The use of boron-doped diamond (BDD) anodes allowed the reduction of COD, dissolved organic carbon (DOC), TPh and color greater than 90%, and increase of the BOD_5/COD and BOD_{20}/COD ratios from 0.13 and 0.23 up to 0.59 and 0.72, respectively, after 8 h oxidation at current density of

30 mA/cm^2. However, to achieve conductivities that enabled anodic oxidation at the highest current intensities applied and to minimize the specific energy consumptions, addition of supporting electrolyte was required at 0.75 g/L of Na_2SO_4 [44].

Although MTs have a wide range of selectivity, from microfiltration (MF) having pore sizes ranging from 0.1 to 2 μm and operate at pressures below 5 bar to reverse osmosis (RO) requiring pressures in the range from 50 to 100 bar (seawater desalination) or of 15 to 50 bar (other applications), the presence of vegetal extracts covering a wide range of MW, as is the case of polyphenolic compounds having colloidal behavior, promotes severe membrane fouling, leading to a drastic permeate flux decline with operation time [10, 22, 31, 32, 42]. The other major drawback of MTs is the produced concentrates containing the pollutants rejected, requiring further treatment before final discharge; until now, the potential opportunities for valorization were limited to the reuse of NF retentates by the leather industry [19, 22]. Thus, for the remaining studies the alternative solutions for the concentrates were not presented.

Besides the differences in CBW characteristics and in the pretreatments used to remove gross SS, the comparison between the results presented in Table 14.3 is also difficult due to differences in experimental set-ups, membrane materials, module configuration, hydrodynamic and operational conditions. However, it is possible to assert that permeate flux and quality are correlated with membrane selectivity, usually reported in terms of MWCO. In any case, the results show that permeates obtained with lower MWCO membranes are less colored, have lower COD and TPh concentrations and are more readily biodegradable, which confirms that large MW compounds cause the strong color of CBW and are less available to undergo biodegradation [10, 21, 22].

All published results for operations with UF membranes having MWCOs of 4 up to 300 kDa, transmembrane pressures from 1 to 3 bar were performed after CBW pretreatment, which was not enough to prevent hydrophobic compounds, namely ellagic acid, to develop an adsorbed layer that increases membrane fouling and consequently leads to severe drops in permeate flux [10, 31, 32, 42]. The increase of selectivity with NF membranes raised the rejection of organic compounds, TPh and color above 92%, indicating that NF membranes have a very good ability to concentrate the pollutants and produce a permeate stream with characteristics close to those required for discharge or water reuse [10, 19, 22, 27]. In one case, it was suggested that about 86% of the CBW volume can be recovered for reuse through NF membrane (MWCO of 125 Da). However, this wastewater was collected after the patented Symbios cork boiling process instead of the traditional procedure, which reduces the TCA formation [22].

14.2.2 Biological Treatment

Biological treatment options either in aerobic, anaerobic or sequential conditions can be successfully applied to the treatment of agro-industrial effluents with high concentration of phenolic compounds, thus with characteristics similar to CBW [12]. Beside the aforementioned CBW features restraining biological treatment, the unbalanced composition (C:N:P ratio) should be noticed and the fact that 56% of the organic load comes from compounds with MW above 100 kDa [11, 21]. Thus, the results published for biological treatment (Table 14.4) are limited to conventional activated sludge inoculated with acclimatized biomass [6], use of enriched cultures of aerobic bacteria isolated from CBW samples with tannic acid as single carbon and energy source [7], fungal strains isolated from the outer bark of cork oak trees [46] and, more recently, anaerobic digestion [11]. In this last case, the authors claimed an effective contribution to the economic viability of the treatment process through reduction of the amount of sludge produced and recovery of the energy potential through methane production.

The limitations of biological treatment were well illustrated by the results reported for an activated sludge system operated at hydraulic retention times (HRT) of 24–96 h, which allowed only limited removals of 13–37% and 20–32% for COD and TPh, respectively [6]. Even the use of enriched microbial cultures taken from retention basins used to collect CBW, did not increase the bioavailability of the organic compounds, namely of the polyphenols. Therefore, the authors concluded that biodegradation needs to be preceded by Fenton oxidation to increase the total organic carbon (TOC) removal up to 90% [7]. Four fungal strains isolated from cork bark and another two selected due to their ability to degrade a large variety of persistent environmental pollutants, were incubated with CBW and after 5 days incubation the reductions of COD and color were of 48–62% and up to 47%, respectively. In any case, the values of EC_{50}-5 min for the Microtox bioassay showed a decline of toxicity ranging from a ten-fold decay up to complete loss [46]. Aerobic bacteria collected from CBW storage pounds and selected for their tolerance against phenol and chlorophenols were immobilized onto residual cork particles. The results reported for an overall of 16 strains isolated were limited to the removal of TPh (60–80%) and very dependent on experimental conditions (pH, temperature, nutrient addition, aeration, etc.) [30].

Regarding the anaerobic treatment of CBW, the absence of lag phase with biomass adapted to OMW, led to the conclusion that organic compound bioavailability limitations prevailed over the microbial consortium inhibition or toxicity, because colored compounds in CBW are not biodegradable but their presence does not affect the removal of the colorless fraction [11]. After 11–15 days' incubation period, the COD removals were 36–40% for CBW with 3 and 6 g COD/L. Besides the limited COD reduction, the methane yield was of 0.126–0.142 L/g COD added or 0.315 to 0.394 L/g COD removed. In addition, the authors suggested that the phenolic fraction, the dark colour and the remaining COD can only be removed through electrochemical processes, which will reduce the energy outcome of the combined treatment [11]. According to previous research, several genera of aerobic and anaerobic microorganisms can degrade phenolic compounds but chlorophenols are hardly degraded due to the high stability of the carbon-halogen bond [31].

In summary, in most cases, one single treatment stage is not enough to ensure the fulfillment of the environmental requirements for wastewater discharge or reuse and sequential treatments are required.

14.2.3 Sequential Treatment

The remediation of industrial wastewaters with a wide variety of pollutants and of concentrations, which is the case with CBW, is a complex problem. Thus, when the possibilities and capabilities of the available conventional treatments are not adequate to achieve the desired reclamation, sequential approaches are required (Table 14.5). The most common objective of the first stage of treatment based on chemical oxidation is the increase of the bioavailability and biodegradability of the pollutants, rather than mineralization [6, 7, 24].

The sequences of chemical oxidation followed by biodegradation are interesting because they limit the amount of reactants below those used for single stage treatment, allowing for the reduction of operational costs for COD removal of 65–90% [12]. In these cases, the percentage of mineralization should be minimal, to avoid unnecessary expenditure of chemicals and energy, but excellent process optimization is required. Most AOPs lower the pH of the wastewater due to the generation of organic acids, or the best results are achieved at pH values out of the range necessary for biodegradation (i.e., pH of 6.5–7.5), as it is the case of pH around 3 for Fenton or 9 for ozone, being necessary neutralization before biological treatment [12].

The results reported in Table 14.5 for treatment sequences including MTs can be divided in those using oxidation to improve the performance of MF (0.1 and 0.65 µm) and UF (MWCO of 5 to 300 kDa) membranes, through complete removal of the ellagic acid, previously associated with permeate flow decay with time of operation, or to improve permeate quality [32, 39, 40]. In the case of a 10 kDa membrane, the initial permeate flux increased 25% with pre-ozonation, which also yielded removal of color and tannins in the range of 80–90% and of COD around 40%. However, permeates post-ozonation procedures were more efficient in terms of the overall removal of COD, color and tannins, which were above 82% with both MF membranes [40]. Only for one study combining a 100 kDa membrane and ozonation, the characteristics and volumes of permeate and concentrates were provided. Taking into consideration the volumetric reduction factor (VRF = 4), 25% of the CBW volume process corresponds to a concentrate stream with a COD of 3,436 mg/L, TPh of 233 mg/L and Abs at 580 nm of 0.171 (dilution 1:10). The remaining volume was a permeate stream with COD of 851 mg/L, TPh of 69 mg/L and an absorbance at 580 nm of 0.042 (color); which corresponded to reductions of 45%, 37% and 64%, respectively [21].

The results for sequences of physico-chemical treatments still depend on the amount of reactants and energy consumed, which can be reduced using the solar photo-Fenton process already successfully implemented with other wastewaters, namely from wineries, pesticide and dyeing industries. In this context, the most interesting results were reported for the integration of solar photo-Fenton with biodegradation using cork bleaching wastewater as source for H_2O_2, with COD removal up to 65% [29].

14.3 Constructed Wetland Technology

CWs is an established technological solution for domestic and municipal wastewater treatment [13, 14]. They are considered internationally as an attractive, green alternative to conventional wastewater treatment methods for small and medium size communities [15]. The success of these eco-tech systems in providing high efficiency rates for the treatment of these wastewaters, enabled the investigation of their use for various industrial effluents. Today, a wide range of advances in wetland technology relates to industrial applications, such as effluents from the textile industry, food processing industry, wineries, oil and gas industry, etc. Especially wastewater generated from small-scale industrial facilities in rural areas represent a technical and economic challenge. Usually these contain high concentrations of organics, nutrients and heavy metals. Effective treatment with conventional methods, which include combinations of chemical, physical and biological processes is not always economically feasible. Moreover, the operation of such conventional treatment facilities can also be challenging and costly. From this point of view, CWs can be not only an effective but also a cost-effective treatment solution [14].

Wetland technology has already been introduced in Portugal. Various design configurations have been applied in the country, but the most widely used system in Portugal is the Horizontal Subsurface Flow (HSF) CW, as in most European and Mediterranean countries, for applications of domestic wastewater treatment. The substrate medium is essential for plant establishment, growth and development of the biofilm. Problems with substrate clogging are common, and its causes are still under investigation, while the search for new, appropriate materials is still ongoing. It is assumed that clogging is related to the characteristics of the substrate medium, the excessive growth of biomass, the retention and accumulation of organic solids, the precipitate formation and the development of rhizomes and roots [13, 47]. Light-expanded clay aggregates (LECA) have been presented as alternative substrate media to reduce the clogging problem and increase the treatment capacity, since they present both higher porosity and specific surface area, which allow for a good biofilm adhesion and a high hydraulic conductivity [48].

CWs have not been tested yet for CBW treatment, which means that – to the best of the authors' knowledge – the project presented in this chapter is the first attempt to investigate the feasibility and the potential use of wetland systems for this agro-industrial wastewater source. However, CWs have been tested in effluents with a similar composition to that of CBW, mainly in wastewater containing a high organic load and phenol concentration. Such wastewater sources are olive mills [49–52], pulp and paper industry [53, 54], wineries [55, 56], coffee processing [57] and contaminated groundwater with refinery effluents and petrochemical industry effluent [58]. As reported in a review paper by Stefanakis and Thullner [59], various designs and wetland types have been tested with wastewater containing phenolic compounds. Although it has been proved that CWs have the ability to remove phenols from water, even at high concentrations, it is not yet clear whether discharge standards can be reached. However, the feasibility of CW systems in phenol removal seems to be a proven case, despite the need for further studies to evaluate the removal capacity. Hence, the present project aims at investigating the efficiency of microcosm-scale wetland units for the treatment of CWB, which also contains high levels of phenolic compounds.

14.3.1 Experimental Setup of Microcosm-Scale Constructed Wetlands

A long-term investigation was conducted to assess the performance of a microcosm-scale CW for the treatment of CBW. Two HSF CWs were used in the laboratory. Each wetland unit was a rectangular PVC container (15.0 cm wide, 34.8 cm long and 14.3 cm deep), filled with LECA (Figure 14.1). The porosity of LECA was initially measured and found to be 38.3% One unit was planted with common reeds (*Phragmites australis*) (CWP) and the other was left unplanted (CWC) and used as control unit. The planting took place in May 2012, followed by a commissioning period, during which synthetic wastewater with 300 mg COD/L was introduced to the two beds before the start of loadings with CBW. The maximum water level was at 9.8 cm, corresponding to an initial void volume of 1.961 L.

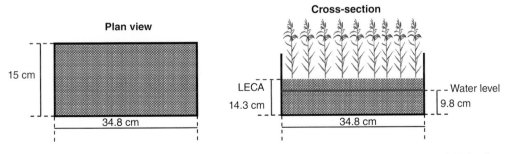

Figure 14.1 Plan view and cross-section of the lab-scale Horizontal Subsurface Flow Constructed Wetlands.

For more than four years the adaptation and acclimatization of the plants and biofilm development were carried out through six gradual increases of the OLRs from 2.55 to 8.87 g COD/m^2/d. Each phase ran until steady results for COD and TPh mass removal rates (MRR) were achieved; thus, the length of each phase varied according to the OLR increase, season and CBW characteristics. This long and carefully conducted acclimatization process was intended to overcome the difficulties resulting from the CBW complex composition, with low bacterial enumeration and metabolic biodiversity [7, 30].

The operating conditions of both bioreactors were set taking into consideration the COD of the CBW collected in an industrial plant located in the Portalegre district (Portugal) at the outlet of an equalization and homogenization tank, and by adjusting the feed flow rate supplied by two peristaltic pumps working in intermittent mode (15 min each operation hour). The characteristics of the CBW samples were in the range of those previously presented in Table 14.1, namely 2,028–3,237 mg COD/L and 89.5–135.1 mg TPh/L. The pH of the raw wastewater was set to 6.5–7.5 (with a mean value of 7.04 ± 0.25). The unbalanced C:N:P ratio of 100:1:0.4 was not adjusted to the reference value of 100:5:1 for aerobic biodegradation, i.e., the lack of nitrogen and phosphorus was not corrected. The feed flow rate of both pumps was measured weekly at least twice and adjustments were made when necessary to ensure steady values for the OLR. Both bioreactors were placed indoor in a laboratory of the Chemical Department of the University of Beira Interior (Covilhã-Portugal), in order to avoid excessive temperature variations, but ensuring at the same time plenty of natural light. The inside temperatures were maintained near 20°C through the regulation of a thermostatic device for a room temperature of 18 ± 5°C.

14.3.2 Experimental Results

The results reported in Table 14.4 (last row) were obtained with a mean OLR of 16.5 g COD/m^2/d during an operation period of 200 days (from 25 January to 11 August). According to CBW COD concentration, an HRT close to 6 days was ensure through adjustment of the inflow rates; thus resulting in the variation of the hydraulic loading rate (HLR) between 5.8–7.8 L/m^2/d. This high HRT was required because of the positive correlation between HRT and treatment efficiency and operation stability, highly recommended for wastewaters of bio-recalcitrant nature as it is the case of CBW (Table 14.1).

Due to the small size of the CWs used, the evaporation (EV) and evapotranspiration (ET) were important in all seasons. In the CWP, the ET values of 2.22 ± 0.84 L/m^2/d were above the values obtained with the CWC (1.17 ± 0.32 L/m^2/d) and these differences could be correlated with plant activity. Despite the contribution of plants to water losses, thus increasing the HRT, the significant development of plant roots biomass during successive growth cycles over four years resulted in the rise of the bed height from the initial value of 14.3 cm up to 20 cm; which corresponds to reduction of the void volume and offset the ET influence on operational conditions. Since the CWs were placed indoors, plants were not harvested and they maintained green leaves even during cold seasons, but after spring and until autumn the plants' height and leaves presented the highest development. It is worth mentioning that after several growth cycles with CBW, the plants never showed any toxicity signs and the expansion of roots also confirms their excellent physiological condition.

The removal efficiencies of COD and TPh were calculated through mass balance and expressed in percentages. This option was considered best suited to evaluate performance removals than the calculations based on concentration, because of the significant values of the EV and ET. However, the values of water loss calculated through mass balance and presented here for the planted unit of 35.7 ±

13.6% are commonly reported for operations with small-scale CWs under warm climatic conditions, being a consequence of the so-called "oasis" effect which restrains the outflow volumes [48, 60]. The color removals reported in Table 14.4 are for the difference between the inflow and outflow values. In this case the correction of the water loss allows decolorization improvement from $-111 \pm 33\%$ up to $-28 \pm 32\%$, which means that the color is higher in the outflow than in the inflow. This drawback of the biological treatments with CBW was also previously reported, namely for anaerobic and aerobic conditions, and it was correlated with the high MW of the pollutants contributing to color, which have also low bioavailability [6, 10, 11, 21]. On the other hand, a very high variability of this parameter was observed; this can be due to the physical process contribution to decolorization, namely sedimentation, precipitation and adsorption, which are reversible processes sensitive to the variation of operational conditions. Similar variability was observed for the CWC, with $-25 \pm 21\%$.

Overall, if the water loss of 14.4 ± 7.5 is reported by Albuquerque et al. (2009) [60] for a real-scale HSF CW operation (surface area of 773 m^2) with similar characteristics in a very close location and temperature ($18.4 \pm 9.2°C$), it is possible to anticipate a great improvement of the outflow color, COD and TPh concentration. For this case scenario, the average characteristics of the outflow can be estimated to be: a pH close to neutrality, color of 0.196 for the Abs at 580 nm (dilution 1:10), COD of 1,357 mg/L and TPh of 68.4 mg/L; which clearly exceeds the legal requirements for discharge and indicates the need for post-treatment or the reduction of the OLRs.

The occurrence of environmental conditions appropriate to anaerobic or aerobic metabolism can be assessed through measurements of redox potential (Eh) in the middle of the beds; oxidative conditions require values bigger then +100 mV, reductive conditions for values less than −100 mV and the intermediary values can be classified as anoxic conditions [61, 62]. According to the wetland design, the oxygen required for aerobic degradation can be provided by diffusion, convection and oxygen emission from the macrophyte roots into the rhizosphere. However, it was this last source of oxygen that made the difference between CWC and CWP for the availability of oxygen, considering that for *P. australis* plants the reported value of oxygen input rate found in the literature is 0.02–12 g/m^2/d [61]. Thus, the increase of COD and TPh removals in the CWP system corresponds to additional available oxygen and to high metabolic diversity.

The measurements (in the middle of the beds) of Eh in the CWP were always above those of the CWC, with measurements between −61 to +56 and −117 to −75 mV, respectively. Thus, these conditions allowed for enhanced aerobic biodegradation, which increased close to the exit of the CWP where the Eh measurements were closer to oxidative conditions with −35 to 154 mV. However, the oxygen availability was still limited compared to the organic load in both units. In any case, the oxygen availability at the planted unit was always more than those of the control unit. It is also important to mention that root growth followed plant development, and also had a positive contribution to the performance by providing additional surface area for biofilm development with different characteristics than the LECA material and with variable oxygen availability.

Previous results reported for CWs systems suggest that a multitude of bacterial metabolisms can take place at the same time in different locations of the wetlands according to oxygen concentration and availability, namely aerobic respiration, denitrification, sulfate reduction and methanogenesis, etc. [49, 63]. The amount of oxygen released by plants into their immediate root environment is still a controversial issue, and subject of debate because multiple factors are involved in its calculation, namely the HRT. The assessments of oxygen transfer rate are lower than that reported in literature for estimations based on mass balances and theoretical stoichiometric calculations (from 5.4 to 22 g O_2/m^2/d) [17, 61].

As expected from past experience, the treatment performance of the CWP was always higher than the unplanted unit [15, 17, 58]. The COD mean reduction was of 62% (values in the range from 44 up to 79%), which corresponded to an MRR of 10.2 ± 2.6 g COD/m^2/d. The average values for the CWC were of $48 \pm 9\%$ and 6.9 ± 2.1g COD/m^2/d, respectively. The removal efficiencies for the TPh were of $41 \pm 6\%$ and $58 \pm 11\%$ for MRRs of 0.25 ± 0.05 and 0.41 ± 0.10 g TPh/m^2/d for the unplanted and planted units, respectively. The results obtained by the CWP system are compared with other biological treatment alternatives in Table 14.4. With the exception of decolorization, the removals of COD and TPh in the CWP system are clearly higher than those reported for other methods. Moreover, the HRT of 6 days set for the CWP operation is in between the values of the activated sludge system and those of the anaerobic digestion of 1–4 and 15 days, respectively [6, 11].

The assessment and comparison of the phenolic compound removals reported here for the CWP unit is relevant but problematic to achieve, because it was difficult to find wastewaters with characteristics close to CBW and CW operations with values in the range of the OLR experienced by us. Stefanakis et al. [58] studied the application of HSF CW to the treatment of groundwater contaminated with phenol and m-cresol, corresponding to OLRs of 0.62 and 0.08 g/m^2/d respectively, and reported removals higher than 84%. For CWs with the same configuration but a more close context, with aerated and non-aerated coffee processing wastewater, for OLR of 0.52 and 0.66 g TPh/m^2/d and HRT of 12 days, removals were reported of 72 and 66%, respectively; which correspond to MRRs in the range of 0.37–0.44 g TPh/m^2/d; thus in agreement with our results [57].

The extended range of microbial metabolisms after a long period of acclimatization to wastewater characteristics had a positive contribution to TPh removal. However, even taking into consideration the water loss through ET, the COD mean concentration was only reduced from 2,675 to 1,009 mg/L (ranging from 670–1,545 mg/L) and the TPh concentration from 116.8 to 47.9 (ranging from 29.6–74.5 mg/L), which is still above the legal requirements for water discharge or reuse. Thus, further downstream treatment is necessary to reduce the remaining organic load, which has low bioavailability and biodegradability.

The most frequently mentioned constraints of HSF CWs operations, namely the significant land requirements, the lack of performance under cold climatic conditions and the limited removals of N and P [14, 61, 63], are anticipated to be overcome because most of the cork processing units are located in the Iberian Peninsula, thus under the warm climate typical of the Mediterranean region and large land areas are available. Finally, the less efficient removal of N and P by comparison with COD and BOD are not relevant issues in this effluent because of their low initial content and plant uptake.

14.4 Conclusions

The preservation of the cork oak natural ecosystem forest is closely dependent on the valorization of cork products, namely the stoppers used in wine bottles, which has a significant contribution for the economic viability of the cork supply chain. However, to ensure the consumer's preference for cork, it is necessary to enhance the sustainability and eco-efficiency of the manufacture stages, especially by reducing the amount of water and pollution generated during the transformation process. The application of Constructed Wetland systems in the treatment of CBW was tested for the first time, using microcosm-scale beds. After a long-term experimental run under varying pollutant loads, it was found that CWs can contribute to the reduction of the impacts resulting from post-harvest stages as an environmentally friendly technology, which can be easily integrated into the urban or agricultural landscape, in addition to low construction and operation costs. The CW systems with the horizontal subsurface flow configuration showed good treatment performance, namely for COD and TPh removal, which were higher compared to respective results of other biological treatment methods. This project is the first positive outcome regarding the use of wetland technology for CBW treatment, which enables the scaling up for the implementation of a full-scale wetland facility.

Acknowledgements

The authors are grateful to FCT, COMPETE and FEDER for financing the PTDC/AGR-AAM/102042/2008 research project.

References

1 APCOR. APCOR's Cork Yearbook 2015. Available at: /www.apcor.pt/wp-content/uploads/2015/12/APCOR-Boletim-Estatistico.pdf.

2 Council Directive 92/43 EEC of 21 May 1992 on the Conservation of Natural Habitats and Wild Fauna and Flora. O.J. Eur. Comm. No. L206/7.

3 Leite C, Pereira H. Cork-Containing Barks – A Review. Front Mater. 2017; 3(63. doi:10.3389/fmats.2016.00063.

4 Rives J, Fernandez-Rodriguez I, Rieradevall J, Gabarrell X. Environmental analysis of raw cork extraction in cork oak forests in southern Europe (Catalonia e Spain). J Environ Manage. 2012; 110:236–245.

5 Mazzoleni SG, di Pasquale G, Mulligan M, di Martino P, Rego FC, Recent dynamics of the Mediterranean vegetation and landscape. Chichester, UK: John Wiley & Sons; 2004.

6 Benitez FJ, Acero JL, Garcia J, Leal AI. Purification of cork processing wastewaters by ozone, by activated sludge, and by their two sequential applications. Water Res. 2003; 37(17):4081–4090.

7 Dias-Machado M, Madeira LM, Nogales B, Nunes OC, Manaia CM. Treatment of cork boiling wastewater using chemical oxidation and biodegradation. Chemosphere. 2006; 64(3): 455–461.

8 Domínguez JR, González T, García HM, Sánchez-Lavado F, de Heredia JB. Aluminium sulfate as coagulant for highly polluted cork processing wastewaters: Removal of organic matter. J Hazard Mater. 2007; 148(1–2):15–21.

9 De Heredia JB, Domínguez JR, López R. Treatment of Cork Process Wastewater by a Successive Chemical–Physical Method. J Agric Food Chem. 2004; 52(14):4501–4507.

10 Bernardo M, Santos A, Cantinho P, Minhalma M. Cork industry wastewater partition by ultra/nanofiltration: A biodegradation and valorisation study. Water Res. 2011; 45(2):904–912.

11 Marques IP, Gil L, La Cara F. Energetic and biochemical valorization of cork boiling wastewater by anaerobic digestion. Biotechnol Biofuels. 2014; 7: Available at: www.biotechnologyforbiofuels.com/content/7/1/67.

12 Oller I, Malato S, Sánchez-Pérez JA. Combination of Advanced Oxidation Processes and biological treatments for wastewater decontamination – A review. Sci Total Environ. 2011; 409(20):4141–4166.

13 Stefanakis AI, Akratos CS, Tsihrintzis VA. Vertical Flow Constructed Wetlands: Eco-engineering Systems for Wastewater and Sludge Treatment. New York: Elsevier Science; 2014.

14 Stefanakis AI. Constructed Wetlands: description and benefits of an eco-tech water treatment system. In: E. McKeown and G. Bugyi (eds) Impact of Water Pollution on Human Health and Environmental Sustainability. Hershey – Pennsylvania: Information Science Reference (IGI Global); 2015, pp. 281–303.

15 Wu H, Zhang J, Ngo HH, Guo W, Hu Z, Liang S, Fan J, Liu H. A review on the sustainability of constructed wetlands for wastewater treatment: Design and operation. Bioresour Technol. 2015; 175:594–601.

16 Tyroller L, Rousseau DPL, Santa S, García J. Application of the gas tracer method for measuring oxygen transfer rates in subsurface flow constructed wetlands. Water Res. 2010; 44(14):4217–4225.

17 García J, Rousseau DPL, Morato J, Lesage E, Matamoros V, Bayona JM. Contaminant removal processes in subsurface-flow constructed wetlands: a review. Crit Rev Environ Sci Technol. 2010; 40:561–661.

18 Aguilera-Carbo A, Augur C, Prado-Barragan LA, Favela-Torres E, Aguilar CN. Microbial production of ellagic acid and biodegradation of ellagitannins. Appl Microbiol Biotechnol. 2008;78: Available from: http://link.springer.com/article/10.1007/s00253-007-1276-2/fulltext.html

19 Geraldes V, Minhalma M, de Pinho MN, Anil A, Ozgunay H, Bitlisli BO, Sari O. Nanofiltration of cork wastewaters and their possible use in leather industry as tanning agents. Polish J Environ Stud. 2009; 18(3):353–357.

20 Makkar HPS, Blummel M, Borowy M, Becker NK. Gravimetric determination of tannins and their correlations with chemical and protein precipitation methods. J Sci Food Agric. 1993; 61(2):161–165.

21 Santos DC, Silva L, Albuquerque A, Simões R, Gomes AC. Biodegradability enhancement and detoxification of cork processing wastewater molecular size fractions by ozone. Bioresour Technol. 2013; 147:143–151.

22 Teixeira ARS, Santos JLC, Crespo JG. Sustainable membrane-based process for valorisation of cork boiling wastewaters. Sep Purif Technol. 2009; 66(1):35–44.

23 Vilara VJP, Maldonado MI, Oller I, Malato S, Boaventura RAR. Solar treatment of cork boiling and bleaching wastewaters in a pilot plant. Water Res. 2009; 43(16):4050–4062.

24 Gomes AC, Silva L, Simões R, Canto N, Albuquerque A. Toxicity reduction and biodegradability enhancement of cork processing wastewaters by ozonation. Water Sci Technol. 2013; 68(10):2214–2219.

25 Guiberteau-Cabanillas A, Godoy-Cancho B, Bernalte E,Tena-Villares M, Guiberteau Cabanillas CG, Martínez-Canas MA. Electroanalytical behavior of gallic and ellagic acid using graphene modified screen-printed electrodes. method for the determination of total low oxidation potential phenolic compounds content in cork boiling waters. Electroanalysis. 2015; 27(1):177–184.

26 Tchobanoglous G, Burton FL, Stensel HD. Wastewater engineering treatment and reuse. Metcalf and Eddy, McGraw-Hill Science; 2003.

27 Oliveira J, Nunes M, Santos P, Cantinho P, Minhalma M. Cork processing wastewater treatment/valorization by nanofiltration. Desalination Water Treat. 2009; 11:224–228.

28 De Torres-Socías E, Fernández-Calderero I, Oller I, Trinidad-Lozano MJ, Yuste FJ, Malato S. Cork boiling wastewater treatment at pilot plant scale: Comparison of solar photo-Fenton and ozone $(O_3, O_3/H_2O_2)$. Toxicity and biodegradability assessment. Chem Eng J. 2013; 234:232–239.

29 Pintor AMA, Vilar VJP, Boaventura RAR. Decontamination of cork wastewaters by solar-photo-Fenton process using cork bleaching wastewater as H_2O_2 source. Sol Energy. 2011; 85(3):579–587.

30 Fernandes A, Santos D, Pacheco MJ, Ciríaco L, Simões R, Gomes AC, Lopes A. Electrochemical treatment of cork boiling wastewater with a boron-doped diamond anode. Environ Technol. 2015; 36(1):26–35.

31 Del Castillo I, Hernández P, Lafuente A, Rodríguez-Llorente ID, Caviedes MA, Pajuelo E. Self-bioremediation of cork-processing wastewaters by (chloro)phenol-degrading bacteria immobilized onto residual cork particles. Water Res. 2012; 46(6):1723–1734.

32 Minhalma M, de Pinho MN. Tannic-membrane interactions on ultrafiltration of cork processing wastewaters. Sep Purif Technol. 2001; 22–23:479–488.

33 García-Ballesteros S, Mora M, Vicente R, Sabater C, Castillo MA, Arques A, Amat AM. Gaining further insight into photo-Fenton treatment of phenolic compounds commonly found in food processing industry. Chem Eng J. 2016; 288(15):126–136.

34 Custódio L, Patarra J, Alberício F, Neng NR, Nogueira JMF, Romano A. Phenolic composition, antioxidant potential and in vitro inhibitory activity of leaves and acorns of *Quercus suber* on key enzymes relevant for hyperglycemia and Alzheimer's disease. Ind Crops Prod. 2015; 64:45–51.

35 Acero JL, Benitez FJ, Beltran de Heredia J, Leal AI. Chemical treatment of cork-processing wastewaters for potential reuse. J Chem Technol Biotechnol. 2004; 79:1065–1072.

36 Madureira J, Melo R, Botelho ML, Leal JP, Fonseca IM. Effect of ionizing radiation on antioxidant compounds present in cork wastewater. Water Sci Technol. 2013; 67(2):374–379.

37 Bianchi F, Careri M, Mangia A, Musci M. Optimization of headspace sampling using solid-phase microextraction for chloroanisoles in cork stoppers and gas chromatography–ion-trap tandem mass spectrometric analysis. J Sep Sci. 2003; 26(5):369–375.

38 De Torres-Socías E, Cabrera-Reina A, Trinidad MJ, Yuste FJ, Oller I, Malato S. Dynamic modelling for cork boiling wastewater treatment at pilot plant scale. Environ Sci Pollut Res. 2014; 21:12182–12189.

39 Benítez FJ, Acero JL, Leal AI. Application of microfiltration and ultrafiltration processes to cork processing wastewaters and assessment of the membrane fouling. Sep Purif Technol. 2006; 50(3):354–364.

40 Benítez FJ, Acero JL, Leal AI, González M. The use of ultrafiltration and nanofiltration membranes for the purification of cork processing wastewater. J Hazard Mater. 2009; 162(2–3):1438–1445.

41 Domínguez JR, de Heredia JB, González T, Sanchez-Lavado F. Evaluation of ferric chloride as a coagulant for cork processing wastewaters. Influence of the operating conditions on the removal of organic matter and settleability parameters. Ind Eng Chem Res. 2005, 44(17):6539–6548.

42 Minhalma M, Domínguez, JR, de Pinho MN. Cork processing wastewaters treatment by an ozonization/ultrafiltration integrated process. Desalination. 2006; 191:148–152.

43 Silva CA, Madeira LM, Boaventura RA, Costa CA. Photo-oxidation of cork manufacturing wastewater. Chemosphere. 2004; 55(1):19–26.

44 Guedes AM, Madeira LP, Boaventura RR, Costa AV. Fenton oxidation of cork cooking wastewater – overall kinetic analysis. Water Res. 2003; 37(13):3061–3069.

45 Benítez FJ, Acero JL, Leal AI, Real FJ. Ozone and membrane filtration based strategies for the treatment of cork processing wastewaters. J Hazard Mater. 2008; 152(1):373–380.

46 Mendonça E, Pereira P, Martins A, Anselmo AM. Fungal Biodegradation and Detoxification of Cork Boiling Wastewaters. Eng Life Sci. 2004; 4(2):144–149.

47 Nivala J, Knowles P, Dotro G, García J, Wallace S. Clogging in subsurface-flow treatment wetlands: Measurement, modeling and management. Water Res. 2012; 46:1625–1640.

48 Albuquerque A, Oliveira J, Semitela S, Amaral L. Evaluation of the effectiveness of horizontal subsurface flow constructed wetlands for different media. J Env Sci. 2012; 22:820.

49 Del Bubba M, Checchini L, Pifferi C, Zanieri L, Lepri L. Olive mill wastewater treatment by a pilot-scale horizontal flow (SSF-h) constructed wetland. Ann Chim. 2004; 94:875–887.

50 Kappelakis IE, Paranychianakis NV, Tsagarakis KP, Angelakis AN. Treatment of olive mill wastewater with constructed wetlands. Water 2012; 4:260–271.

51 Sultana MY, Akratos CS, Vayenas DV, Pavlou S. Constructed wetlands in the treatment of agro-industrial wastewater: A review. Hem Ind. 2015; 69(2):127–142.

52 Tatoulis T, Stefanakis AI, Frontistis Z, Akratos CS, Tekerlekopoulou AG, Mantzavinos D, Vayenas DV. Treatment of table olive washing water using trickling filters, constructed wetlands and electrooxidation. Environ Sci Poll Res. 2016;1-8: DOI 10.1007/s11356-016-7058-62016.

53 Abira MA, Van Bruggen JJA, Denny P. Potential of a tropical subsurface constructed wetland to remove phenol from pre-treated pulp and papermill wastewater. Water Sci Techol. 2005;51(9):173–176.

54 Choudhary AK, Kumar S, Sharma C. Removal of chlorophenolics from pulp and paper mill wastewater through Constructed Wetland. Water Environ Res. 2013;85(1):54–62.

55 Shepherd HL, Grismer ME, Tchobanoglous G. Treatment of high-strength winery wastewater using a subsurface flow constructed wetland. Water Environ Res. 2001;73(4):394–403.

56 Arienzo M, Christen EW, Quayle WC. Phytotoxicity testing of winery wastewater for constructed wetland treatment. J Hazard Mater. 2009; 169:94–99.

57 Rossmann M, de Matos AT, Abreu EC, de Silva FF, Borges AC. Performance of constructed wetlands in the treatment of aerated coffee processing wastewater: removal of nutrients and phenolic compounds. Ecol Eng. 2012; 49:264–69.

58 Stefanakis AI, Seeger E, Dorer C, Sinke A, Thullner M. Performance of pilot-scale horizontal subsurface flow constructed wetlands treating groundwater contaminated with phenols and petroleum derivatives. Ecol Eng. 2016; 95:514–526.

59 Stefanakis AI, Thullner M. Fate of Phenolic Compounds in Constructed Wetlands Treating Contaminated Water. In: AA Ansari, SS Gill, R Gill, G Lanza, L Newman (eds) Phytoremediation – Management of Environmental Contaminants Volume 4. Switzerland: Springer International Publishing; 2016, pp. 311–325.

60 Albuquerque A, Arendacz M, Obarska-Pempkowiak H, Borges M, Correia M. Simultaneous removal of organic and solid matter and nitrogen in SSHF constructed wetlands in temperate Mediterranean climate. KKU Res J. 2009; 13(10):1117–1127.

61 Brix H. Do macrophytes play a role in constructed treatment wetlands? Water Sci. Technol. 1997; 35(5):11–17.

62 Garcia J, Capel V, Castro A, Ruíz R, Soto M. Anaerobic biodegradation tests and gas emissions from subsurface flow constructed wetlands. Bioresour Technol. 2007; 98(16):3044–3052.

63 Truu M, Juhanson J, Truu J. Microbial biomass, activity and community composition in constructed wetlands. Sci Total Environ. 2009; 407(13):3958–3971.

15

Constructed Wetland Technology for Pulp and Paper Mill Wastewater Treatment

Satish Kumar[1] and Ashutosh Kumar Choudhary[2]

[1] *Department of Paper Technology, Indian Institute of Technology Roorkee, Saharanpur Campus, Saharanpur, India*
[2] *Department of Applied Sciences and Humanities, HSET, Swami Rama Himalayan University, Dehradun, India*

15.1 Introduction

The pulp and paper industry is a growing sector, which mainly depends on natural resources, i.e., water, wood, agro-residues and fossil fuels for the production of paper. The worldwide paper and paperboard production (about 403 million tons in 2013) is dominated by North America, Europe, and Asia. The global demand is projected to grow by about 3% yearly, reaching an expected 490 million tons by 2020 [1].

Pulp and paper mills generate large quantities of wastewater with high organic load and color that may cause serious environmental impacts to the receiving water bodies. Wood preparation, pulping, bleaching, washing, and coating operations are the main sources of pollution [2]. A number of chemical pulping processes are used to convert ligno-cellulosic materials to obtain fibers for papermaking. The pulping process results in lignin network degradation and removal of its soluble fractions from the plant tissue producing unbleached pulp (cellulose 80–90%, hemicelluloses 10–15%, and residual lignin 2.5–4%). The residual lignin is responsible for the unwanted dark color and photo yellowing of the pulp [3]. The chemical pulping processes leave behind small amounts of modified lignin, which is removed subsequently by various bleaching processes with chlorine, chlorine based chemicals or chlorine free bleaching chemicals. Normally, bleaching is done to brighten the pulp to obtain final target brightness.

Traditionally hard woods and softwoods have been used as raw materials for paper making. Reduction of forest resources has encouraged industries to use agricultural residues for production of paper. In developing countries, medium sized paper mills are using agro-residues as raw materials and commonly practice soda pulping and conventional bleaching processes for the production of bleached grade pulps and are discharging very high pollution loads into the environment. While by switching over to new fiber lines (involving modern technologies in pulping, bleaching, pulp washing etc.), the mills in developed countries have reduced the pollution level below the prescribed limits by switching over to oxygen delignification and chlorine dioxide bleaching; a number of mills are also adopting chlorine free bleaching. The mills in developing countries continue using elemental chlorine for bleaching for techno-economic reasons. The bleaching process is essentially chlorination (C) and/or chlorine dioxide (D) followed by alkali extraction (E) and calcium hypochloride (H). Bleaching sequences like CEH, CE$_p$H, D/CEH, CEHH, or D/CE$_p$HH are common in developing countries.

Constructed Wetlands for Industrial Wastewater Treatment, First Edition. Edited by Alexandros I. Stefanakis.
© 2018 John Wiley & Sons Ltd. Published 2018 by John Wiley & Sons Ltd.

As a consequence, the bleach plants have become a major source of environmental pollution contributing to high chloro-organic release with high biochemical oxygen demand (BOD), chemical oxygen demand (COD), adsorbable organic halides (AOX) and color loads.

The presence of toxic chemicals has been detected in paper mill effluent samples, sludge, sediments and pulp and paper samples. The levels are found to be lower for mechanical pulping processes than chemical pulping processes. Internationally, it has been recognized that chlorinated compounds are generated during the chlorination step of pulp bleaching. Thus, in order to reduce the generation of chloro-organics, the mills have switched over to elemental chlorine free (ECF) bleaching using chlorine dioxide (D stage), which reduces the chloro-organic generation up to 20%. Total chlorine free (TCF) bleaching practically eliminates the formation of these compounds.

15.2   Pulp and Paper Mill Wastewater Characteristics

The characteristics of the wastewater generated by the pulp and paper mill depends upon the pulping and bleaching process operations, type of the wood material, i.e., soft wood, hard wood or agro-residue and practices adopted for chemical recovery. The pulp and paper mill generates wastewater (per ton of OD pulp) with high BOD (10–40 kg), COD (20–200 kg), suspended solids (10–50 kg), AOX (0–4 kg) loads, color, toxicity, and high concentration of nutrients (phosphorus and nitrogen), which cause eutrophication in receiving water bodies [4]. Bleach plant discharges account for 60–70% of BOD and 80–90% of color load of the entire mill having chemical recovery [5]. The typical characteristics of pulp and paper mill wastewater after primary treatment are shown in Table 15.1.

Pulp and paper mills are discharging a large number of organic compounds formed during the various operations of paper manufacture. Among these operations, pulping and pulp washing contributes

Table 15.1 Characteristics of pulp and paper mill wastewater after primary treatment [6–11].

Parameter	Value
pH	4.7–10.0
Suspended solids (mg/L)	620–1,120
COD (mg/L)	399–2,035
BOD (mg/L)	110–582
Color (Pt-Co unit)	959–5,830
AOX (mg/L)	12–37.5
Phenol (mg/L)	0.43–1.7
Chlorophenolics (µg/L)	12–239
Chloro-resin and fatty acids (µg/L)	37–157
2,3,7,8-TCDD (ng/L)	0–4.2
Kjeldahl nitrogen (mg/L)	3–13
P total (mg/L)	0.5–1.8
Zn (mg/L)	1.3
Pb (mg/L)	1.4

maximum concentrations of BOD, COD, color, and toxicity. The principal compounds are degraded lignin fragments, carbohydrates and extractives [3, 12, 13]. With the installation of chemical recovery, the contribution from pulping section is largely reduced.

The next largest contributing section is pulp bleaching. The bleaching is carried out by chlorine or chlorine based compounds. More than 500 different chlorinated organic compounds have been identified including chlorate, chloroform, chlorophenolics (chlorophenols, chlorocatechols, chloroguaiacols, chlorosyringols, chlorosyringaldehydes and chlorovanillins), chlorinated resin and fatty acids (RFA), chlorinated hydrocarbons, dioxins and furans, etc. [14, 15]. In wastewater, these compounds are estimated collectively as AOX. Many of these organo-chlorine compounds are toxic, mutagenic and are slowly degraded in the environment [3, 16, 17]. In addition to these chemicals, dyes and biocides are also present.

Chlorophenolics are primarily formed as a result of chlorination of the lignin remaining in the pulp after pulping [17]. These compounds bio-accumulate in aquatic organisms due to their hydrophobic nature. Resin and fatty acids (RFAs) are compounds that are present in kraft pulping liquor. The amount of these acids in natural and chlorinated form that ends up in the bleach plant effluent depends on the wood species and on the degree of washing of unbleached pulp. Among RFAs, chlorinated RFAs are the major contributors to toxicity of pulp and paper mill wastewater to aquatic organisms. Chlorinated dioxins and furans have also been detected in bleaching effluent and paper mill sludge [18, 19]. There are many possible chlorinated isomers of dioxin and furan but the two isomers (2,3,7,8-tetrachlorodibenzo-p-dioxin and 2,3,7,8-tetrachlorodibenzo-p-furan) are very toxic.

One of the key effluent quality parameter for monitoring the level of organo-chlorine compounds is AOX. Lower value of AOX means lower concentration of organo-chlorine compounds and consequently lower wastewater toxicity. Toxicity is the property of an individual compound. Some compounds are highly toxic, i.e., show toxicity at trace levels, whereas others show toxicity at appreciable concentration. An accident in the operation of the bleach plant can cause the concentration of toxic substances in the diluted wastewater to become too high. Chloro-organics are toxins, bio-recalcitrant, and tend to persist in the environment, and hence are known as persistent organic pollutants (POPs) [20]. Discharge of these POPs to the environment may cause the ecological imbalance.

Many authors have reported the presence of toxic pollutants in fish and their toxic effect on fish such as mutagenicity, respiratory stress, liver damage, genotoxicity, and lethal effects when exposed to pulp and paper mill wastewater [21]. 2,4-dichlorophenol, 2,4,5-trichlorophenol, pentachlorophenol, chlorinated dioxin, and chloroform are carcinogenic, whereas chlorocatechols are strongly mutagenic [22]. Resin acids have been reported to cause teratogenicity in fish larvae [23] and genotoxicity to mussels [24]. The nature and concentration of compounds formed/present in the wastewater depends upon the raw material used and the paper-making process variables. The removal efficiency of these compounds is also dependent upon the biochemical treatment process used. Thus, the nature and concentration of various toxic compounds varies from mill to mill.

15.3 Remediation of Pulp and Paper Mill Wastewater Pollution

The remediation of paper mills wastewater can be done either by in-plant modification or by wastewater treatment. In-plant mainly includes extended delignification, oxygen delignification, enzyme pretreatment of pulp, chlorine stage modification, ECF bleaching, H_2O_2 reinforced alkaline extraction and TCF bleaching. The general treatment processes adopted by the pulp and paper mills include two stages of treatment. Primary treatment reduces the suspended solids (SS), whereas secondary treatment reduces the BOD and is essentially a biological treatment (activated sludge process (ASP), aerated lagoons trickling filter etc.) involving the use of mixed bacterial population. The reduction in COD is modest and color reduction is very marginal, while reduction of toxic compounds is also small. Only lower chlorine substituted compounds are reduced, whereas higher chlorine substituted compounds largely responsible for wastewater toxicity are not eliminated. Thus, treated paper mill wastewater still contains sufficient quantities of toxic chlorinated compounds.

The different treatment systems developed for reducing the wastewater color and toxicity are termed as tertiary processes. As color and toxicity mainly come from first alkaline extraction stage (E_1), chlorination stage (C) and/or chlorine dioxide stage (D_1), these methods have been selectively tested for treating D/C and E_1 stage effluents, considering the high treatment cost of these methods. The various tertiary treatment processes are based on ion-exchange/adsorption, flocculants, membrane and advanced oxidation processes (AOPs). Different flocculants such as ferric chloride, magnesium chloride, and poly-aluminum chloride, have been tested on paper mill wastewater. All these methods give good color reduction (90–95%), modest COD reduction and good removal of chlorinated organics. These processes require the treatment of excess sludge obtained. Membrane methods such as ultrafiltration are basically a concentration step. The organic compounds are concentrated into a small volume from initial large volume. This process gives frequent clogging problems and requires wastewater absolutely free from suspended solids. Different AOPs such as the Fenton/photo-Fenton process, wet air oxidation, electro-chemical oxidation, photo-catalysis, ozonation etc., have been used to remove both low and high concentrations of organic compounds from pulp and paper mill wastewater. These processes, although they often have high capital and operating costs, are the only viable treatment methods for wastewaters containing refractory, toxic, and non-biodegradable materials.

15.4 Constructed Wetlands

Constructed wetlands (CWs) are engineered natural systems and have a great potential for the bio-remediation of industrial wastewaters. In the last few decades, these systems have been constructed to treat wastewaters originating from different sources. These systems have more aesthetic appearance with lower maintenance and running costs than traditional wastewater treatment systems [25]. CWs efficiently reduce/remove the contaminants including color, inorganic matter, organic matter, toxic compounds, metals and pathogens from wastewaters. Reduction or removal of contaminants is accomplished by combination of physical, chemical and biological treatment mechanisms. The treatment efficiency of these systems depends mainly on the CW design, hydraulic loading rate (HLR), type of contaminant, root zone interactions and the climatic factors. There are so many plant species that could be used for wastewater treatment in CWs, but till now few plant species have been tested for pulp and paper mill wastewater treatment as shown in Table 15.2.

15.4.1 Performance of CWs for Pulp and Paper Mill Wastewater Treatment

CWs have been successfully utilized for the treatment of pulp and paper mills wastewater with promising results. Abira et al. [6] observed the effectiveness of a pilot CW with ten cells to remove the contaminants from pre-treated pulp and paper mill wastewater under varying hydraulic retention

Table 15.2 Plant species used in CWs for pulp and paper mill wastewater treatment.

Botanical name	Common name
Canna indica	Canna lily
Colocasia esculenta	Taro, Elephant ear
Cyperus immensus	Papyrus sedge
Cyperus pangorei	Galingale, Korai grass
Cyperus papyrus	Papyrus sedge
Phragmites australis	Common reed
Phragmites mauritianus	Reed grass
Typha angustifolia	Bulrush, Narrow leaf cattail
Typha domingensis	Southern cattail
Typha latifolia	Bulrush, broadleaf cattail

time (HRT) of 3 to 6 days. The study was carried out at the premises of the PANPAPER Mills, in Webuye, Kenya. This mill produced large amounts of wastewater (32,000–37,000 m^3/d) and discharged into the Nzoia River after primary clarification and secondary treatment by aerated lagoons. Four pairs of subsurface flow rectangular-shaped cells were constructed with dimensions of 1.2 m (width) × 3.2 m (length) × 0.8 m (depth) covering a total surface area of 30.7 m^2. These cells were operated in parallel and planted with *Typha domingensis, Phragmites mauritianus, Cyperus immensus,* and *Cyperus papyrus.* The focus of the project was on the investigation of the removal of nutrients (nitrogen and phosphorus), organic matter (BOD, COD), phenols and total suspended solids (TSS) from the wastewater. The average inlet concentrations of BOD, COD, TSS, nitrogen, phosphorous and phenol were 45, 394, 52, 2.56, 0.76, and 0.64 mg/L, respectively. Mean removal efficiency for nitrogen ranged from 49–75% for the planted cells and 42–49% for the unplanted ones. During batch operation mode, nitrogen removal efficiency ranged from 32–68% for the planted cells and 29–35% for the unplanted ones. In the planted cells phosphorus removal was in the range of 30–60% and 4–38% in the unplanted cells under continuous flow, compared to 10–58% in the planted and 9–12% in the unplanted cells under batch operation.

In CWs, nitrogen removal takes place by different processes that include adsorption, ammonia volatilization, nitrification and denitrification, nitrogen fixation by plant uptake and microbes. But in most of the wetland treatment systems the process of nitrification–denitrification plays an important role compared to other processes [25]. Phosphorus transformation in CWs includes sedimentation, adsorption, precipitation, dissolution, plant and microbial uptake [25]. The removal of BOD and TSS was up to 90% and 81%, respectively, from the wastewater to concentrations below the national guidelines in Kenya. However, COD removal was low (up to 52%) and the guideline value of 100 mg/L was not achieved in this study. All wetland cells showed good removal efficiency for phenol. However, the *Phragmites*-planted wetland cells showed higher removal efficiency (95.7 ± 0.7%) than the *Papyrus* cells (92.8 ± 1.4%). Phenol removal efficiency was higher when the macrophytes were in active growth phase compared to the steady growth phase for all species. Phenol removal takes place by a combination of processes involving adsorption, biodegradation, plant uptake and volatilization. According to Eckenfelder [26], biodegradation is the major process.

In surface flow CWs, removal by volatilization can be up to 30% [27]. In this study, phenol removal could have been achieved by both biodegradation and adsorption onto the sediment particles.

Prabu and Udayasoorian [28] reported the treatment of pulp and paper mill wastewater by three microcosm-scale CWs planted with *Cyperus pangorei, Phragmitis australis,* and *Typha latifolia,* individually. Bench-scale wetlands were constructed in plastic tubs measuring 48 cm in diameter and 50 cm in height. Hydrosoil height (combination of pebbles, gravels, coarse sand, fine sand, and soil) in wetland containers was 35 cm. The entire study was carried out at an HRT of 48 h. After the treatment, the BOD of the pulp and paper mill wastewater decreased by 77%, 74% and 64%, respectively, in *Phragmitis australis, Typha latifolia* and *Cyperus pangorei* planted wetlands and COD decreased by 62%, 55% and 49%, respectively. The overall chlorinated phenol degradation of *Phragmitis australis, Cyperus pangorei* and *Typha latifolia* planted tubs was 87%, 81% and 72%, respectively.

Rani et al. (29) evaluated the effectiveness of horizontal subsurface flow constructed wetland (HSSF-CW) for pulp and paper mill wastewater treatment with plant species *Canna indica* and *Typha angustifolia* at different HRTs of 1.5, 3.5 and 6.5 days. Results showed that the CW system could effectively reduce the total solids ($87.6 \pm 1.1\%$), COD ($86.6 \pm 2.0\%$), BOD_5 ($80.01 \pm 0.1\%$), and color ($89.4 \pm 0.6\%$) during summer at a HRT of 3.5 days. The study showed that plants species *Typha angustifolia* and *Canna indica* could survive and propagate well in pulp and paper mill wastewater. Removal efficiency for all pollutants also improved at higher temperatures. Therefore, the results obtained showed the higher removal efficiencies during the summer as compared to winter.

In India, recently a detailed laboratory study on the application of CW for treatment of pulp and paper industry wastewater has been reported by Choudhary et al. [30–34]. The study was carried out for two years, i.e., 2010 and 2011. The main focus of this study was to detect and remove the different chloro-organic compounds (i.e., AOX, chlorophenolics, and cRFAs) from wastewater. This study was carried out at different HRTs of 4.0, 5.9 and 8.6 days. Additional objectives of the study were to analyze the soil and plant material for chloro-organic accumulation and to study the mass balance of chlorophenolics in the treatment system. The study was performed in two phases. Years 2010 and 2011 were considered as Phase I and Phase II, respectively. In Phase I and Phase II plant species, i.e., *Canna indica* and *Colocasia esculenta* were selected, respectively (Figure 15.1). Two HSSF-CW cells, named as Cell A and Cell B, were constructed with surface area 5.25 m^2 each. Cell A was used for control experiments, i.e., only fresh water was used as feed. Cell B was used for treatment studies. The dimensions of each cell were 3.5 m in length, 1.5 m in width, and effective depth of 0.28 m. The wastewater samples (after primary treatment) were collected from a pulp and paper mill in India.

During Phase I, the first three trials (i.e., HRTs of 4.0, 5.9, and 8.6 days) were conducted to find out the optimum HRT for the treatment of wastewater. The treatment efficiency of the HSSF-CW was examined by monitoring wastewater quality parameters (COD and color) in the inlet and outlet at different hydraulic conditions. The average decrease in pH was 0.1 at a HRT of 4 days, 0.3 at a HRT of 5.9 days and 0.4 at a HRT of 8.6 days. The decrease in pH increased with increase in HRT value. The decrease in pH value from inlet to outlet may be due to the formation of some acidic components by the microbial degradation of lignin and its derivatives present in the wastewater.

The average COD removal efficiencies of the HSSF-CW planted with *Canna indica* were 78%, 88% and 91% at HRTs of 4, 5.9 and 8.6 days, respectively. Results show that as the HRT was increased (i.e., hydraulic load decreases), COD removal efficiency also increased. It was also observed that as the HRT increased from 5.9 days to 8.6 days, there was only marginal increase in COD removal efficiency (3%). It was concluded that the HRT of 5.9 days was adequate for the removal of organic load from the pulp and paper mill wastewater and taken as optimum value for further treatment of wastewater.

(a)   (b)

Figure 15.1 View of the experimental HSSF-CW. (a) CW planted with *Canna indica*, (b) CW planted with *Colocasia esculenta*.

Table 15.3 Characteristics of wastewater at the inlet of the HSSF-CW (for a HRT of 5.9 d).

| | Plant species | | | |
| | *Canna indica* | | *Colocasia esculenta* | |
Parameter	At inlet[a]	% Removal	At inlet[a]	% Removal
pH	7.7 ± 0.17	–	7.9 ± 0.31	–
COD (mg/L)	1,011 ± 82	88	1,084 ± 191	84
Color (Pt-Co mg/L)	2,553 ± 238	96	2,691 ± 364	94
BOD$_5$ (mg/L)	248 ± 15	93	294 ± 53	90
AOX (mg/L)	16.5 ± 3.50	89	17.6 ± 1.29	87
Chlorophenolics (µg/L)	40.0 ± 5.98	90	76.5 ± 9.91	87
cRFAs (µg/L)	83.8 ± 13.42	94	112.6 ± 17.21	93

[a] Average of four observations.

The characteristics of the wastewater collected from the inflow of HSSF-CW during the Phase-I and Phase-II at the optimum HRT of 5.9 days are shown in Table 15.3 and concentrations of individual chlorophenolics and cRFA in Tables 15.4 and 15.5, respectively.

Results showed that the removal efficiency for BOD (90–93%) was comparatively higher than COD (84–88%). The removal was due to the microbial interactions that take place in the root zone of the CW. Microorganisms easily decompose the biodegradable matter than recalcitrant compounds that are the part of COD. The results with both plant species showed good removal efficiency for AOX (87–89%), chlorophenolics (87–90%), and cRFAs (93–94%).

The untreated wastewater characteristics showed the presence of 26 chlorophenolic compounds belonging to six categories (chlorophenols (CP), chlorocatechols (CC), chloroguaiacols (CG),

Table 15.4 Removal of chlorophenolics by HSSF-CW (for a HRT of 5.9 d).

| | Plant species | | | |
| | Canna indica | | Colocasia esculenta | |
Name of compound	Inlet concentration (µg/L)	Removal (%)	Inlet concentration (µg/L)	Removal (%)
3-CP	3.90 ± 1.87	92	4.90 ± 2.07	91
4-CP	1.60 ± 0.77	93	2.90 ± 1.34	90
2,6-DCP	2.79 ± 1.10	87	5.68 ± 3.51	87
2,5-DCP	0.75 ± 0.36	88	8.88 ± 3.02	87
2,4-DCP	3.17 ± 0.90	94	4.63 ± 0.32	89
2,3-DCP	0.02 ± 0.01	ND	ND	ND
3,4-DCP	0.15 ± 0.04	ND	0.21 ± 0.07	ND
4-CG	0.79 ± 0.24	89	10.25 ± 7.01	89
2,4,5-TCP	11.28 ± 4.29	84	20.39 ± 11.57	81
2,3,6-TCP	0.06 ± 0.03	67	0.02 ± 0.01	ND
2,3,5-TCP	0.43 ± 0.19	ND	0.05 ± 0.03	ND
2,4,6-TCP	5.01 ± 1.23	96	0.01 ± 0.01	ND
4,5-DCG	2.58 ± 0.86	96	9.39 ± 3.58	86
2,3,4-TCP	0.26 ± 0.17	77	0.94 ± 0.27	91
4,6-DCG	0.21 ± 0.13	91	0.35 ± 0.23	ND
3,6-DCC	0.33 ± 0.17	ND	2.96 ± 1.85	92
3,5-DCC	0.08 ± 0.02	ND	2.36 ± 2.08	93
3,4,6-TCG	0.27 ± 0.07	85	0.17 ± 0.02	ND
3,4,5-TCG	0.04 ± 0.01	ND	0.12 ± 0.07	88
4,5,6-TCG	0.01 ± 0.006	ND	0.13 ± 0.04	ND
5,6-DCV	1.61 ± 0.66	94	ND	ND
PCP	1.10 ± 0.44	89	ND	ND
TeCG	0.22 ± 0.04	77	0.55 ± 0.23	88
TCS	0.27 ± 0.05	93	0.42 ± 0.12	92
TeCC	0.93 ± 0.17	84	ND	ND
2,6-DCSA	2.14 ± 0.85	94	1.18 ± 0.48	87
Total	**40.0 ± 5.98**	**90**	**76.5 ± 9.91**	**87**

(ND= not detected; D-di; T-tri; Te-tetra; P-penta; CP-chlorophenol; CG-chloroguaiacol; CC-chlorocatechol; CV-chlorovanillin; CSA-chlorosyringaldehyde).

chlorosyringol (CS), chlorosyringaldehyde (CSA), and chlorovanillin (CV)) (Table 15.4). Among the individual chlorophenolic compounds 2,4,5-TCP contributed the highest concentration (11.28–20.39 µg/L) followed by 4-CG, 4,5-DCG, 2,4,6-TCP, 3-CP, 2,4-DCP, 2,5-DCP, 2,6-DCP, 4,5-DCG, 2,6-DCSA, 3-CP, and 4-CP. Other chlorophenolics were present in minor quantities. Among different categories of chlorophenolics compounds, CP contributed the highest share

Table 15.5 Removal of cRFAs by HSSF-CW (for a HRT of 5.9 d).

Name of compound	Plant species			
	Canna indica		*Colocasia esculenta*	
	Inlet concentration (µg/L)	Removal (%)	Inlet concentration (µg/L)	Removal (%)
DCSA	39.56 ± 7.43	96	40.35 ± 5.42	96
CDAA	19.62 ± 3.21	93	31.32 ± 8.10	91
DCDAA	6.75 ± 2.26	95	21.46 ± 4.31	92
TCSA	17.90 ± 4.81	92	19.50 ± 9.74	92
Total	**83.8 ± 13.42**	**94**	**112.6 ± 17.21**	**93**

(30.52–48.59 µg/L) followed by CG (4.12–20.96 µg/L), CC (1.34–5.32 µg/L), CSA (1.18–2.14 µg/L), CV (1.61 µg/L), and CS (0.27–0.42 µg/L). On the basis of chlorine atom substitution, dichlorophenolics (DCP) contributed highest share (23.83–35.64 µg/L) followed by trichlorophenolics (TCP) (17.63–22.23 µg/L), monochlorophenolics (MCP) (6.29–18.05 µg/L), tetrachlorophenolics (TeCP) (0.55–1.15 µg/L), and pentachlorophenolics (PCP) (1.10 µg/L). The results further indicated that 94–99% of the identified compounds in the untreated wastewater are MCP, DCP and TCP compounds. The quantity of various chlorophenolics present in the wastewater mainly depends on the bleaching process variables (bleach chemical dose, bleaching sequence, and temperature for bleaching), wood species and type of pulp used in bleaching by the paper mill.

After treatment with *Canna indica*, 67–100% removal of chlorophenolics was achieved (Table 15.4). The chlorophenolics, i.e., 2,3-DCP, 3,4-DCP, 2,3,5-TCP, 2,4,6-TCP, 3,5-DCC, 3,6-DCC, and 4,5,6-TCG were not detected after treatment by *Canna indica*. After treatment with *Colocasia esculenta*, 81–100% removal of chlorophenolics was achieved (Table 15.4). Figure 15.2 shows the removal of different categories of chlorophenolics by HSSF-CW planted with Canna *indica* and *Colocasia esculenta*. The results showed very high removal of chlorinated phenolics. Removal values were between 86–94%. The treatment by *Colocasia esculenta* resulted in very high removal of chlorinated phenolics but was slightly lower than *Canna indica* (86–92%). Figure 15.3 shows the

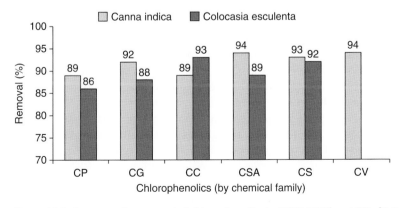

Figure 15.2 Category wise removal of chlorophenolics by HSSF-CW (for a HRT of 5.9 days).

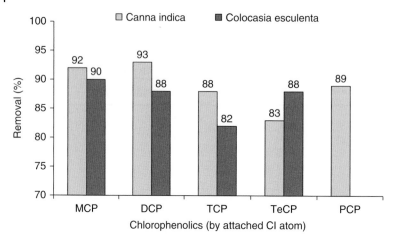

Figure 15.3 Removal chlorophenolics by HSSF-CW (for a HRT of 5.9 days).

removal of chlorophenolics (according to Cl atom attached) by the CW treatment. Similar findings were observed on the removal of chlorophenolics by the two species *Canna indica* and *Colocasia esculenta*, in terms of attached chlorine atom.

The removal of CP was low as compared to other categories of chlorophenolics when treated with *Canna indica*. The reason for low CP removal was the comparatively higher concentration of CP in the untreated wastewater. DCP were removed to the higher extent followed by MCP, PCP, TCP, and TeCP. The higher removal of PCP as compared to TCP may be due to very low initial share of PCP (1.10 µg/L) in the untreated wastewater as compared to TCP (17.63 µg/L). The share of TeCP (1.15 µg/L) and PCP (1.10 µg/L) in the untreated wastewater was nearly the same, but the removal of PCP was slightly higher than the TeCP. This may be due to the conversion of PCP into TeCP and other lower CP. In the case of *Colocasia esculenta*, the quantity of CP and CG was comparatively higher in the untreated wastewater, which gave comparatively lower removal of CP and CG. Treatment by *Colocasia esculenta* gave the highest removal of MCP followed by DCP, TeCP and TCP. The higher removal of TeCP as compared to TCP may be due to very low initial share of TeCP (0.55 µg/L) in the untreated wastewater as compared TCP (22.23 µg/L). The removal of lower chlorinated chlorophenolics was higher as compared to highly chlorinated chlorophenolics, as highly halogenated organic compounds are extremely resistant to decomposition, due to their lower solubility in water and the lack of a structural site for enzyme attachment for biotransformation or degradation.

The quantities of various cRFAs detected in the CW inflow wastewater during Phase-I and Phase-II are shown in Table 15.5. Four cRFAs, i.e. chlorodehydroabietic acid (CDAA), 12,14-dichlorodehydroabietic acid (DCDAA), 12,13-dichlorostearic acid (DCSA), and 9,10,12,13-tetrachlorostearic acid (TCSA) were detected in the untreated wastewater of the HSSF-CW. The average inflow concentration of DCSA was highest (39.56 µg/L) followed by CDAA (19.62 µg/L), TCSA (17.90 µg/L), and DCDAA (6.75 µg/L) during the Phase-I of the treatment.

During the second phase of CW operation, the average inflow concentration of all cRFA was comparatively higher than the Phase-I. The removal efficiency of different cRFA was high between 91–96% for both *Canna indica* and *Colocasia esculenta* (Table 15.5). Thus, the CW can almost completely remove the chlorinated resin and fatty acids. The molecular size of cRFAs is large as compared to

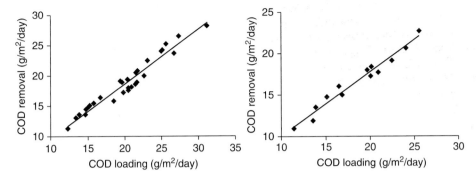

Figure 15.4 Plot of mass removal rate and mass loadings for (a) *Canna indica* and (b) *Colocasia esculenta*.

chlorophenolics, and due to this it takes more time to pass from inlet to outlet, i.e., more retention time is needed for the treatment. Hence, the removal efficiency was higher for cRFAs in comparison to chlorophenolics.

During the study, the mass removal of COD was proportional to the inflow load. A linear correlation between COD mass loading and mass removal was reported in the mass loading range studied, as shown in Figure 15.4. For *Canna indica* and *Colocasia esculenta*, the COD mass loading varied from 12.3–31.2 g/m^2/d and 11.4–25.5 g/m^2/d, respectively. The COD mass loading removal reported during the treatment process was up to 28.3 and 22.8 g/m^2/d for *Canna indica* and *Colocasia esculenta*, respectively.

The removal rate was directly proportional to the inflow mass quantities and, thus, followed first-order kinetics. The first-order area based removal rate constants (k) for COD removal by HSSF-CW were estimated by assuming exponential removal to non-zero background concentrations. Fitted values of k were derived from the equation (15.1) [25]:

$$ln\frac{C_o - C^*}{C_i - C^*} = -k/q \tag{15.1}$$

where, k is the area based first-order removal rate constant (m/d), q is the hydraulic loading rate (m/d), C_i is the inlet concentration (mg/L), C_o is the outlet concentration (mg/L), and C^* is the irreducible background wetland concentration (mg/L). The inlet and outlet concentrations of COD were calculated for Phase-I and Phase-II at different hydraulic loadings and the data was fitted to k-C* model by using first-order kinetics equation (15.1). The reported first order rate constant (k) values are 0.06 and 0.052 m/d for *Canna indica* and *Colocasia esculenta*, respectively. The background concentrations for COD in the k-C* model were 75 and 90 mg/L for *Canna indica* and *Colocasia esculenta*, respectively, which confirmed good removal of COD in the treatment system, since the typical background concentration of COD was in the range 30–100 mg/L [25].

Choudhary et al. [32–34] also reported that plant uptake play a direct role in the removal of chloro-organic compounds (chlorophenolics) from pulp and paper mill wastewater. Plants indirectly provide aerobic conditions to the microorganisms in the root zone for the degradation of chlorophenolics. After completion of the treatment studies (Phase-I), biomass of plant species was harvested from the CW and analyzed for the chlorophenolics. Results showed that 14 chlorophenolic compounds of four different categories (CP, CG, CV, and CSA) were detected in the plant biomass, in which CV contributed the highest share of 53.4% followed by CSA (19.0%), CG (14.3%) and

CP (13.4%). In the biomass of *Colocasia esculenta*, six chlorophenolic compounds of three different categories (CP, CG, and CSA) were detected. Among the identified chlorophenolics, 2,6-DCSA contributed the highest accumulation in the plant biomass (115.1 µg/kg) followed by 4-CG (61.1 µg/kg) and 4,5-DCG (36.7 µg/kg). Other chlorophenolics were found to accumulate in minor quantities. On the basis of chlorine atom substitution, DCP contributed the highest share of 77% and 63% followed by MCP (18% and 34%) and TCP (5% and 3%) for *Canna indica* and *Colocasia esculenta*, respectively. In both the plant species, DCP contributed the highest share and TCP contributed the lowest. MCP and DCP contributed share was 95% in *Canna indica* and 97% in *Colocasia esculenta*.

The uptake of any organic compound into plant tissue is predominantly affected by the lipophilic nature, which can be characterized by the octanol-water partition coefficient (Kow) [35]. Generally, organic compounds with a log K_{OW} value between 0.5 and 3 are taken up the best by plants [36]. Organic compounds with a log Kow > 4 are believed not to be significantly taken up by the plant unit membrane system because of retention within the root epidermis [37]. The results further showed that only mono, di, and tri-chlorophenolics were detected in the plant biomass. It may be due to the reason that for mono, di, and tri-chlorophenolics the log K_{OW} values were < 4 and for tetra and penta-chlorophenolics the log K_{OW} values were > 4 [38, 39]. Plant biomass was also analyzed for cRFAs content but no cRFAs were detected in the plant biomass of *Canna indica* and *Colocasia esculenta*. The log K_{OW} values for cRFAs were considerably high (>7); that's why these compounds were not taken up by the plant species.

In CWs, one can easily estimate the degradation of pollutants through mass balance studies. Choudhary et al. [32–34] reported that large fraction of the total chloro-organics loaded to the HSSF-CW became degraded at an HRT of 5.9 days. The degradation of chlorophenolics and cRFAs were 68–71% and 78–79%, respectively, in the CW cell. The accumulation of chlorophenolics was about ten times more in plant species *Canna indica* (0.2%) as compared to *Colocasia esculenta* (0.02%). Thus *Canna indica* has more accumulative potential for chlorophenolics. The kinetics of different physical, chemical and biological processes defines the actual degradation or accumulation of organic compounds in the CW treatment system. Plant species *Canna indica* and *Colocasia esculenta* survived and grew well under the specific experimental conditions. Besides this, *Canna indica* produced a vegetation cover with flowers that gave the wastewater treatment system a nice appearance. The ornamental plants can also be used to increase the aesthetic appearance of CWs.

The AOX, chlorophenolics, and cRFA were removed by the number of physio-chemical and biological processes when these compounds pass through the substrate of CWs. In order to check effectiveness of soil adsorption/absorption and microbial degradation for the removal of chlorophenolics, studies with some model chlorophenolics were also carried out in the laboratory by Choudhary et al. [33]. These studies indicated that soil adsorption/absorption and microbial degradation effectively removed the chlorophenolics from the solution. The removal varied from 36.5 to 77.9% by soil adsorption/absorption process. This process proved to be more effective for highly chlorinated chlorophenolics as compared to lower ones. The microbial degradation studies showed 19.9 to 28.8% and 30.6 to 43.3% removal of chlorophenolics after the time interval of 48 and 96 h, respectively [33].

The degree of sorption depends on the characteristics of the organic compounds (such as organic carbon partition coefficient (K_{OC})) and the solid surface (plants, substrate and litter) [40, 41]. The microorganisms attached to the root zone of the plants play a very important role in the degradation of organic matter, whereas oxygen is supplied by the plant roots [42]. The extent of microbial degradation of organic compounds within a CW is also expected to strongly depend on the physico-chemical properties (such as concentration, water solubility) of the contaminant [40].

A variety of microorganisms (e.g., *Pseudomonas* sp., *Sphingomonas* sp., *Flavobacterium* sp., *Novosphingobium lentum, Rhodococcus chlorophenolicum, Sphingomonas chlorophenolica, Desulfitobacterium hafniense,* and *Mycobacterium* sp.) have the ability to interact with the substances leading to structural changes or complete degradation of the chlorophenolic compounds [43, 44]. These compounds become soluble after some initial reductive dechlorination steps, and thus can be easily degraded by microbial interactions (aerobic/anaerobic). Both aerobic and anaerobic microbial degradation processes simultaneously takes place in the CWs. The rate of aerobic degradation is relatively slower for highly chlorinated compounds in comparison to low chlorinated compounds [45]. The biodegradation of pentachlorophenol is faster under anaerobic than under aerobic conditions. Under anaerobic conditions, dechlorination of pentachlorophenol leads to the formation of a mixture of di, tri, and tetra-chlorophenols [46]. Under aerobic conditions, highly chlorinated phenolics are resistant to biodegradation because chlorine atoms interfere with the action of oxygenase enzymes, which generally initiate the degradation of aromatic rings [47]. The cRFAs are almost insoluble in water and may also be removed by a number of separation processes including adsorption/absorption, when they pass through the substrate of CW. The degree of sorption and its rate are dependent on the characteristics of both the organic matter and the solid surface [48]. Being organic in nature, cRFAs could be used as food for microorganisms, e.g., *Pseudomonas* sp., *Sphingomonas* sp. [49, 50]. These compounds become soluble after some initial reductive dechlorination steps, and thus can be easily degraded by microbial interactions.

The most common biological treatments used in the pulp and paper mill include: activated sludge process (ASP), aerated stabilization basin (ASB), oxidation lagoon, anaerobic processes and biological processes such as trickling filter, rotatory disc etc. These secondary treatment methods are not capable of reducing toxicity and color of the pulp and paper mill wastewater. So, the wastewater discharged is colored and expected to contain large amount of pollutants contributing to AOX. More than 80% of biological wastewater treatment systems are based on the principle of activated sludge process, in which microorganisms oxidize the organic compounds [53]. On the other hand, the use of CWs for the treatment of pulp and paper mill wastewater is limited, generally due to the lack of experience and awareness in developing countries. The performances achieved by activated sludge plants and CWs used for the treatment of wastewaters are shown in Table 15.6. These data show that

Table 15.6 Efficiencies of constructed wetland and activated sludge processes [7, 51, 52].

Parameters	Removal efficiency (%)		
	Activated sludge process (ASP)	Surface flow CW	Sub-surface flow CW
BOD	85–93	85–95	90–98
COD	50–85	60–80	75–94
Color	50–72	75–85	90–97
AOX	30–42	–	87–89
Chlorophenolics	28–59	–	87–90
cRFAs	30–37	–	93–94
Nitrogen (N)	30–40	20–80	30–65
Phosphorus (P)	30–45	30–50	40–87

ASP is not efficient for the removal of nutrients (N and P), AOX and chloro-organic compounds from pulp and paper mills. On the other hand, CWs proved to be an efficient treatment facility.

15.5 Conclusions

Pulp and paper mill wastewater contains a large number of chloro-organic compounds. Among chlorophenolics, MCP, DCP, and TCP are the major compounds present in wastewater. CWs have a great potential to treat contaminated wastewater generated by pulp and paper mills. An HRT of about 6 days is sufficient for the removal of the majority of pollutants and color from paper mill wastewater. Through CWs, wastewater can be effectively treated and reach values below the toxic range with very little remaining color. Plants and microorganisms are the active agents in the treatment process. Soil adsorption/absorption and biosorption/microbial decomposition plays an important role in the removal of chloro-organic compounds from the wastewater in CWs. HSSF-CW has higher removal efficiency for lower chlorinated phenolics compared to highly chlorinated phenolics. Major fractions of the toxic chloro-organic compounds can be degraded in the CW. The removal of BOD by CWs from pulp and paper mill wastewater is comparable to the existing conventional biological treatment systems (e.g., activated sludge process), but CWs treatment system is a good option for the removal of COD, high load of color, and toxic chloro-organic compounds from pulp and paper mill wastewater. Additionally, these systems give a pleasant appearance and aesthetic value to the local environment.

References

1 Paperage [Internet]. Cohasset, USA: Paperage Magazine; 2006 December. Available from: www.paperage.com/issues/nov_dec2006/11_2006ofinterest.html

2 Catalkaya EC, Kargi F. Color, TOC and AOX removals from pulp mill effluent by advanced oxidation processes: A comparative study. J Hazard Mat B. 2007 Jan; 139(2): 244–253.

3 Dence CW, Reeve DW. Pulp Bleaching – Principles and Practice. Atlanta, Georgia: TAPPI Press; 1996.

4 Pokhrel D, Viraraghavan T. Treatment of pulp and paper mill wastewater a review. Sci Tot Env. 2004 Oct; 333(1–3):37–58.

5 Rao NJ. Approaches to cleaner production in pulp and paper industry. IPPTA J Con Iss. 1997; 135–155.

6 Abira MA, Van Bruggen JJA, Denny P. Potential of a tropical subsurface constructed wetland to remove phenol from pre-treated pulp and paper mill wastewater. Wat Sci Tech. 2005 May; 51(9):173–176.

7 Choudhary AK, Kumar S, Sharma C. Monitoring of chloro-organic compounds from Indian paper mills wastewater. IPPTA J. 2014 Apr; 26(2):104–115.

8 Latorre A, Rigol A, Lacorte S, Barceló D. Organic Compounds in Paper Mill Wastewaters: In: Hutzinger O, ed. The Handbook of Environmental Chemistry. Vol. 5 Part O. Berlin, Germany: Springer Verlag; 2005, pp. 25–51.

9 Schnell A, Steel P, Melcer H, Hodson PV, Carey JH. Enhanced biological treatment of bleached kraft mill effluents: I. Removal of chlorinated organic compounds and toxicity. Wat Res. 2000 Feb; 34(2):493–500.

10 Subrahmanyam PVR. Waste management in pulp and paper industry. J Ind Assoc Env Manag. 1990; 17:79–94.

11 Verma VK, Gupta RK, Rai JPN. Biosorption of Pb and Zn from pulp and paper industry effluent by water hyacinth (*Eichhornia crassipes*). J Sci Ind Res. 2005 Oct; 64(10):778–781.

12 Singh RP, ed. The bleaching of pulp. USA: Tappi Press; 1979.

13 Catalkaya EC, Kargi F. Advanced oxidation treatment of pulp mill effluent for TOC and toxicity removals. J Env Manag. 2008 May; 87(3):396–404.

14 Kringstad KP, Lindstrom K. Spent Liquors from pulp bleaching. Env Sci Tech. 1984 Aug; 18(8):236–247.

15 Freire CSR, Silvestre AJD, Neto CP. Carbohydrate derived chlorinated compounds in ECF bleaching of hardwood pulps: Formation, degradation and contribution to AOX in a bleached kraft pulp mill. Env Sci Tech. 2003 Jan; 37(4):811–814.

16 Smook GA. Handbook for Pulp and Paper Technologists. Atlanta: Tappi Press; 1989.

17 Savant DV, Rahman AR, Ranade DR. Anaerobic degradation of adsorbable organic halides (AOX) from pulp and paper industry wastewater. Bio Tech. 2006 Jun; 97(9):1092–1104.

18 *Cincinnati*, Ohio. United States Environmental Protection Agency (USEPA). The National Dioxin Study. EPA 420/4-87-003 office of water regulation & standards; 1987.

19 Kumar S, Gupta A, Singh M. Minimising the formation of 2378 TCDD and 2378 TCDF in bleaching Effluents – A Laboratory Study. IPPTA Con Iss. 1999; 49–59.

20 United States Environmental Protection Agency (USEPA) and U.S. Department of Agriculture. Cincinnati, Ohio. Clean Water Action Plan: Restoring and Protecting America's Waters. EPA-840-R-98-001; 1998.

21 Erisction G, Larsson A. DNA-Adducts in perch (*Perca fluviatillis*) in coastal water pollution with bleachening pulp mill effluents. Ecotox Env Saf. 2000 June; 46(2):167–173.

22 Savant DV, Ranade DR. Biodegradation of AOX from paper and pulp industry wastewater. Interaction Meet on Environmental Impact of Toxic Substances Released in Pulp and Paper Industry. New Delhi. 2002 December 10; 41–43. Available from: www.dcpulppaper.org/gifs/report47.pdf

23 Brinkworth LC, Hodson PV, Tabash S, Lee P. CYP1A induction and blue sac disease in early developmental stages of rainbow trout (*Oncorhynchus mykiss*) exposed to retene. J Tox Env Hea: A. 2003; 66(7):627–646.

24 Gravato C, Oliveira M, Santos MA. Genotoxic effects and oxidative stress responses induced by retene in marine mussels (*Mytilus galiprovincialis*). Fre Env Bul. 2004; 13:795–800.

25 Kadlec RH, Knight RL. Treatment Wetlands. New York: Lewis Publishers; 1996.

26 Eckenfelder WW Jr., Industrial Water Pollution Control, 3rd edn. New York USA: McGraw-Hill; 2000.

27 Polprasert C, Dan NP, Thayalakumaran N. Application of constructed wetlands to treat some toxic wastewaters under tropical conditions. Wat Sci Tech. 1996 Dec; 34(11):165–171.

28 Prabhu PC, Udayasoorian C. Treatment of pulp and paper mill effluent using constructed wetland, EJEAFChe. 2007 Jan; 6(1):1689–1701.

29 Rani N, Maheshwari RC, Kumar V, Vijay VK. Purification of pulp and paper mill effluent through *Typha* and *Canna* using constructed wetlands technology. J Wat Reu Des. 2011 Dec; 1(4): 237–242.

30 Choudhary AK, Kumar S, Sharma C, Kumar P. Performance of constructed wetland for the treatment of pulp and paper mill wastewater. Proceedings of World Environmental and Water

Resources Congress 2011: Bearing Knowledge for Sustainability. California, USA. 2011 May 22–26; pp. 4856–4865.

31 Choudhary AK, Kumar S, Sharma C. Organic load removal from paper mill wastewater using green technology. Proceedings of Vth World Aqua Congress, 2011. New Delhi, India. 2011 November 16–18; pp. 103–109.

32 Choudhary AK, Kumar S, Sharma C. Removal of chlorinated resin and fatty acids from paper mill wastewater through constructed wetland. Wor Acad Sci Eng Tech. 2011 Aug; 5(8):61–65.

33 Choudhary AK, Kumar S, Sharma C. Removal of chlorophenolics from pulp and paper mill wastewater through constructed wetland. Wat Env Res. 2013 Jan; 85(1):54–62. Available from: 10.2175/106143012X13415215907419

34 Choudhary AK, Kumar S, Sharma S, Kumar V. Green technology for removal of chloro-organics from pulp and paper mill wastewater. Wat Env Res. 2015 Jul; 87(7):660–669. Available from: 10.2175/106143014X14182397986688

35 Ryan JA, Bell RM, O'Connor GA. Plant uptake of non-ionic organic chemicals from soils. Chemo. 1988; 17(12):2299–2323.

36 Trapp S, Karlson U. Aspects of phytoremediation of organic pollutants. J Soils Sed. 2001 Mar; 1(1):37–43.

37 Trapp S. Model for uptake of xenobiotics into plants. In Trapp S, McFarlane JC, editors. Plant Contamination: Modeling and Simulation of Organic Chemical Processes. Boca Raton USA: Lewis Publishers; 1995, pp. 107–152.

38 Shiu WY, Ma KC, Varhaníčková D, Mackay D. Chlorophenols and Alkylphenols: A review and correlation of environmentally relevant properties and fate in an evaluative environment. Chemo. 1995 Sept; 29(6):1155–1224.

39 Xie TM, Hulthe B, Folestad S. determination of partition coefficients of chlorinated phenols, guaiacols and catechols by shake-flask GC and HPLC. Chemo. 1984; 13(3):445–459.

40 Imfeld G, Braeckevelt M, Kuschk P, Richnow HH. Monitoring and assessing processes of organic chemicals removal in constructed wetlands. Chemo. 2009 Jan; 74(3):349–362.

41 Karickhoff SW. Semi-empirical estimation of sorption of hydrophobic pollutants on natural sediments and soils. Chemo. 1981; 10(8):833–846.

42 Vymazal J, Brix H, Cooper PF, Green MB, Haberl R. Constructed wetlands for wastewater treatment in Europe. Netherlands: Backhuys; 1998.

43 Czaplicka M. Sources and transformations of chlorophenols in the natural environment. Sci Tot Env. 2004 Apr; 322(1–3):21–39.

44 Field JA, Alvarez RS. Microbial degradation of chlorinated phenols. Rev Env Sci Biotech. 2008 Sept; 7(3):211–241.

45 Amon JP, Agrawal A, Shelley ML, Opperman BC, Enright MP, Clemmer ND. Development of a wetland constructed for the treatment of groundwater contaminated by chlorinated ethenes. Eco Eng. 2007 May; 30(1):51–66.

46 D'Angelo EMD, Reddy KR. Aerobic and anaerobic transformations of pentachlorophenol in wetland soils. Soil Sci Soc Am J. 2000 May; 64(3):933–943.

47 Copley SD. Diverse mechanistic approaches to difficult chemical transformations: Microbial dehalogenation of chlorinated aromatic compounds. Chem Bio. 1997 Mar; 4(3):169–174.

48 United States Environmental Protection Agency (USEPA). Cincinnati, Ohio. Design Manual: Constructed Wetlands Treatment of Municipal Wastewaters, EPA/625/R-99/010; 2000.

49 Liu HW, Lo SN, Lavallée HC. Mechanisms of removing resin and fatty acids in CTMP effluent during aerobic biological treatment. Tappi. 1996; 79(5):145–154.

50 William WM, Gordon RS. Bacterial metabolism of chlorinated dehydroabietic acids occurring in pulp and paper mill effluents. App Env Micro. 1997 Aug; 63(8): 3014–3020.

51 Sirianuntapiboon S, Jitvimolnimit S. Effect of plantation pattern on the efficiency of subsurface flow constructed wetland (SFCW) for sewage treatment. Afr J Agri Res. 2007 Sept; 2(9):447–454.

52 VonSperling M. Comparison among the most frequently used systems for wastewater treatment in developing countries. Wat Sci Tech. 1996 Feb; 33(3):59–72.

53 Tizghadam M, Dagot C, Baudu C. Wastewater treatment in a hybrid activated sludge baffled reactor. J Hazar Mat. 2008 Jun; 154(1–3):550–557.

16

Treatment of Wastewater from Tanneries and the Textile Industry using Constructed Wetland Systems

Christos S. Akratos[4], A.G. Tekerlekopoulou[1] and Dimitrios V. Vayenas[2,3]

[1] *Department of Environmental and Natural Resources Management, University of Patras, Agrinio, Greece*
[2] *Department of Chemical Engineering, University of Patras, Patras, Greece*
[3] *Institute of Chemical Engineering and High Temperature Chemical Processes (FORTH/ ICE-HT), Platani, Patras, Greece*
[4] *Department of Civil Engineering, Democritus University of Thrace, Xanthi, Greece*

16.1   Introduction

The production and use of chemical compounds has increased tremendously worldwide and many of these compounds are biologically non-degradable. Effluents discharged from industrial and agricultural activities are high in volume and contain high concentrations of heavy metals and organic compounds. However, wastewaters are not treated properly and this negatively affects water and soil quality [1, 2]. Therefore, the major concern is to treat these wastewaters before they are discharged into the environment.

Wastewater is categorized and defined according to its sources of origin. Wastewaters discharged principally from residential sources and generated by such activities as food preparation, laundry, cleaning and personal hygiene, are termed as "domestic wastewater". Industrial/commercial wastewaters originate and are released from manufacturing processing industries and commercial activities such as printing, textile, steel, food and beverage processing [3].

Various industries are major sources of pollution. Based on the type of industry, various levels of pollutants can be discharged into the environment directly or indirectly through public sewer lines. Industrial wastewater can also include sanitary wastes, manufacturing wastes and relatively uncontaminated water from heating and cooling operations [4]. One of the major problems of industrial wastewaters is their high content of heavy metals, solvents, dyes, pesticides, etc. [5], which can have toxic effects [6, 7].

16.1.1   Tannery Wastewaters

Tanning is the chemical process through which animal hides and skin are converted into leather and related products. The conversion of hides into leather is usually achieved by various tanning agents and the process generates a highly turbid, colored and foul-smelling wastewater. During the tanning process, about 300 kg of chemicals are added per ton of hides [15] and, after processing, about 30–35 L of processed wastewaters are generated per kg of skin with variable pH values and high pollutant concentrations [16]. Untreated tannery wastewaters contain high COD (4,000 mg/L), trivalent chromium (Cr(III)) (150 mg/L), sulfides (160 mg/L), sodium chloride (NaCl) (5,000 mg/L), TKN (160 mg/L), sulfate (1,400 mg/L), calcium (Ca), magnesium (Mg), organics, toxic ingredients, as well as large quantities of solid waste and suspended solids such as animal hair [8]. Therefore, when these untreated wastewaters are released directly into natural water bodies, they effect ecosystem flora and fauna and increase the health risk to human beings [9–14]. The chemicals used in the tanning process can also cause diseases in tannery workers such as ophthalmological conditions, skin irritations, kidney failure and gastrointestinal problems [15]. Finally, certain tannery wastewater streams contain high salinity, which can pollute the environment and jeopardize the biological treatment of this wastewater [16].

Chromium salts are used during the tanning process and these generate two forms of chromium: hexavalent chromium (Cr(VI)) and trivalent chromium (Cr(III)). Of these two forms, Cr(VI) is highly toxic to living organisms even at low concentrations causing carcinogenic effects. It also creates toxicity in anaerobic digestion by accumulating metal in the intracellular fraction of biomass. Environmental protection regulations have imposed strict limits for Cr emission. Due to the high toxicity of Cr(VI), the US Environmental Protection Agency (EPA) and the European Union (EU) regulated total Cr, both Cr(III) and Cr(VI), concentrations in surface waters to below 0.05 mg/L. Thus, removal of chromium from tannery wastewater is essential.

Although several physico-chemical and biological methods have been tested to treat tannery wastewaters, their high pollutant concentrations lead to either insufficient treatment or high operational cost [8]. Therefore hybrid treatment systems containing a constructed wetland (CW) polishing stage could provide viable and cost-effective solutions. Pretreatment methods used for tannery wastewaters before they enter CWs (Table 16.1), include the use of primary settlement tanks [17] or primary settlement after treatment with alkali [18]. Both these pretreatment technologies aim to reduce suspended solid concentrations in the wastewater before it enters the CWs, thus avoiding clogging phenomena. The above pretreatment technologies also reduce Cr concentrations. The use of settlement alone to reduce Cr concentrations is common, as Cr can form coagulants with other pollutants.

Various studies on treatment efficiency with or without vegetation in CWs have concluded that pollutant removal rates are higher when vegetation is present [26]. The ability of CWs to uptake pollutants depends on the type of the pollutant. From the results of all the currently available literature presented in Table 16.1, no significant variations are observed in the treatment efficiency of CWs, depending on the plant species used. CWs using different plant species demonstrate comparable removal efficiencies for Cr. All the species tested appear to be rather tolerant to Cr concentrations, as they can survive and grow even with Cr concentrations of 30 mg/L [18, 25].

According to Table 16.1, the main type of CW tested for tannery wastewater treatment has been the horizontal subsurface flow (HSF). Additionally, HSF CWs have also been mainly used for Cr removal from wastewater and have demonstrated very high removal efficiencies. Specifically, Aguilar et al. [18] report removal efficiencies for COD, TKN and Cr (96–94%, 92–94% and 99%, respectively) using high surface loads for COD, TKN and Cr (955.5 g/m^2/d), 12.2 g/m^2/d and 1.7 g/m^2/d, respectively; Table 16.1). They also report that sulphate reduction bacteria (SRB) and sulphide oxidizing bacteria (SOB) played a significant role in the cycling of carbon, nitrogen and sulphates. High removal efficiencies were also reported by Calheiros et al. [17, 19, 21] in a series of experiments, in which different plant species, substrates and hydraulic loadings were tested. The treatment efficiency achieved

Table 16.1 Experiments and applications of Constructed Wetlands treating tannery wastewaters.

Reference	CW surface area (m²)	Plant species	Pretreatment	Surface load (g/m²/d)					Removal (%)				
				C	TN	TP	TSS	Cr	C	TN	TP	TSS	Cr
HSF													
[18]	405	Schoenoplectus americanus, Typha sp.	Treatment with alkali and sedimentation	878–1,246	8–19	–	–	1.5–2.2	96–98	92–94	–	–	99
[17]	1.2	Phragmites australis, Typha latifolia	Primary treatment	47.9–146.6	3.7–12.3	0.009–0.028	2.4–15.1	–	62–84	53–60	–	88–90	30–100
[19]	36	Arundo donax, Sarcocornia sp.	–	24.2–60.5	1.3–3.2	0.088–0.22	11–27	–	60	60	63	67	–
[20]	1.2	Typha latifolia, Phragmites australis	–	95.9–293.2	7.32–14	0.025–0.6	4.8–30.2	0.022–0.66	31–66	29–39	16–33	69–85	55–97
[21]	1.2	Phragmites australis, Typha latifolia, Stenotaphrum secundatum, Iris pseudacorus	–	59–125.6	0.43–7.6	0.009–0.015	2.25–4.7	Total Cr: 0.0005–0.015 Cr(VI): 0.00003–0.0003	55–65	23–27	–	70–75	Total Cr: 53–95 Cr(VI): 80
[22]	72	Arundo donax, Sarcocornia sp.	Biological treatment	3.8–23.6	0.5–0.7	0.02–0.09	–	–	51–80	41–90	40–93	–	–
[23]	378	Phragmites australis	Control bed	9.53	–	OP: 0.16	–	0.007	12–23	–	OP: 20	–	43–55
[24]	0.31	Typha spp.	n.a.	18.9	–	–	–	0.18	93	–	–	–	98
Hybrid													
[25]	–	Pennisetum purpureum, Brachiaria decumbens, Phragmites australis	–	–	–	–	–	–	–	–	–	–	97–99.6

n.a.: data not available.

for COD, TKN, TSS, total Cr and Cr(VI) was 55–65%, 23–27%, 70–75%, 53–95% and 80%, respectively. The surface loads of the above contaminants were 59–125.6 $g/m^2/d$ for COD, 0.43–7.6 $g/m^2/d$ for TKN, 0.009–0.015 $g/m^2/d$ for TP, 2.25–4.74 $g/m^2/d$ for TSS, 0.0005-0.015 $g/m^2/d$ for Cr, and 0.00003–0.0003 $g/m^2/d$ for Cr(VI). Similar results were also reported by Kucuk et al. [23]. From the experimental results it appears that COD removal efficiency was not affected by the different plant species used (*Schoenoplectus americanus, Typha* sp., *Phragmites australis, Arundo donax, Sarcocornia* sp., *Stenotaphrum secundatum, Iris pseudacorus, Pennisetum purpureum, Brachiaria decumbens*), therefore the main removal mechanism of organic matter in HSF CWs is microbial oxidation, as also reported by other researchers [26–28]. Calheiros et al. [21] stated that the nutrient removal efficiencies observed were lower than expected for a HSF CW, and this was attributed to the relatively low hydraulic residence time (HRTs) (3.4 to 6.8 days) applied to these CWs, while it is elsewhere reported that for sufficient removal of nitrogen and phosphorus, HRTs of more than 14 days are necessary [26]. Mant et al. [25] used a hybrid CW system that consisted of HSF and vertical flow (VF) CWs to treat tannery wastewaters. Their system resulted in high Cr removal efficiency ranging from 97 to 99.6%.

All the above-mentioned research showed significant Cr removal when using CWs to treat either tannery wastewater or artificial wastewater with Cr. In general, the main mechanism of Cr removal in CWs was Cr accumulation by the plants, specifically by their root systems [29]. Removal efficiencies of Cr reached almost 100%, thus establishing CWs as a very attractive treatment method, as they are a low-cost technology and produce only small amounts of by-products. Basically, the only by-product produced during treatment in CWs is plant biomass, which should be removed once a year by harvesting. Harvested plants derived from CWs treating wastewater with Cr should be handled as hazardous waste. Although this by-product is a hazardous material, it can be handled easily as the quantities produced are very low compared to other physico-chemical or biological methods that produce large quantities of hazardous sludge.

16.1.2 Azo Dye and Textile Industries

Among the waste producing industries, textile dyeing processes are the most environmentally unfriendly, because they produce colored wastewaters that are heavily polluted with dyes, textile auxiliaries and chemicals [30]. In the textile industry, the presence of color in discharges is one of the major problems. The characteristics of textile wastewater depend on the production, technology and chemicals used [31]. Usually, textile industries consume huge quantities of water and generate large volumes of wastewater through various steps in the dyeing and finishing processes and the released wastewater is a complex mixture of different polluting substances such as inorganic, organic, elemental and polymeric products [32]. Dye wastes are the most predominant substances in textile wastewater and these substances are not only toxic to the biological world, but also their dark color blocks sunlight and creates acute problems in ecosystems [33].

A wide range of chemicals and dyestuffs are required for dyeing and finishing and these are generally organic compounds of complex structure. Not all of these compounds are contained in the final product, but they are all found in waste and cause disposal problems [34]. Major pollutants in textile wastewaters include high suspended solids (SS) (26,200 mg/L), chemical oxygen demand (COD) (12,000–50,000 mg/L), chromium (Cr) (50 mg/L), heat, color (4,750 mg/L), acidity, and other soluble substances [35–37]. In addition, only 47–87% of dyestuffs are biodegradable [38]. More than 100,000 dyes are available today (of which azo dyes represent about 70% on weight basis), and over one million tons dyes are produced per year, of which 50% are textile dyes [39].

Textile wastewater treatment processes include both biological and physico-chemical methods. Biological treatment usually includes a combination of anaerobic and aerobic processes that are carried out successively to effectively degrade azo-dye effluents. Anaerobic microorganisms participate in textile wastewater treatment by breaking the azo bonds and decolorizing aromatic amines. As a next step, aerobic processes mineralize the aromatic amine residues and effectively purify the textile wastewater of organic matter [40].

Textile industry wastewaters are usually treated by biological (aerobically and anaerobically) and physico-chemical methods; however, biological treatment cannot successfully remove color, and physico-chemical methods require high capital and operational costs [8]. Thus, hybrid treatment systems including a CW stage are attractive in terms of removal efficiency and overall treatment cost. Available studies using CWs for textile wastewater treatment are presented in Table 16.2. None of these studies included pretreatment of the wastewater before it entered the CW stage. This lack of pretreatment is explained by the use of artificial wastewater in each study, therefore eliminating the need to remove suspended solids or lower the pollutant load before introducing the wastewater into the CW systems.

Only three attempts to treat textile wastewater with CWs have been published (Table 16.2). In two experiments, VF CWs were used explicitly, and in the third a hybrid system comprising a VF and two HSF CWs was used. Once again, VF and HSF systems are chosen when wastewater has high pollutant loads or contains rather toxic compounds, as HSF and VF CWs are more effective bioreactors than FWS CWs and can degrade even the most toxic pollutants. The most interesting result arising from these three studies is not only the rather high removal rates of organic matter and nitrogen observed, but rather the high removal rates of color, which range from 60 to 97% (Table 16.2). Color removal from textile wastewater is noteworthy because it is the only pollutant that other biological methods cannot remove efficiently. In these experiments it was also observed that pH values decreased and sulfates were removed. Although the removal of color is rather promising for the future application of CWs in textile wastewater treatment, only a few experiments have been performed to date, and the exact mechanism of color removal by CWs has not yet been fully investigated. Therefore, further

Table 16.2 Experiments and applications of Constructed Wetlands treating textile wastewaters.

Reference	CW type	CW surface area (m^2)	Plant species	Surface load (g/m^2/d)					Removal (%)				
				C	TN	TP	TSS	Color	C	TN	TP	TSS	Color
[41]	VF,HSF	80	*Phragmites australis*	58-140	5-7.4	_	129 mg/L	_	70-84	28-52	_	80-93	90
[40]	VF	0.0254	*Phragmites australis*	22.7	3.3	0.40	_	_	86	67	26	_	97
[42]	VF	0.48	*Phragmites australis, Typha latifolia*	46.8	_	_	_	_	59.6	_	_	_	60

research should be conducted using control experiments to identify the role of biodegradation, plant accumulation and porous media absorbance of azo-dyes.

16.2 Discussion

Studies of tannery wastewater treatment using CWs showed that microorganisms were mainly responsible for removing organic matter, nitrogen and sulfur, while the wetland vegetation was the main contributor to Cr removal. Even though the use of different plant species did not show any significant differences in Cr removal, the choice of plant species used should be made after examining different parameters such as root depth, plant growth/biomass production, local climate, and tolerance to high pollutant loads [43]. Species used in CW systems for tannery wastewater treatment showed high tolerance to Cr toxicity. Another parameter that enhances the removal efficiencies of CWs treating tannery wastewaters is the use of multiple CW stages in series, or hybrid CW systems mainly comprising both HSF and VF configurations.

The key factor in treating textile wastewater is color removal. CWs show high color removal abilities due mainly to azo-dye biotransformation. Although most plant species tested were influenced by the toxic effect of the azo-dyes, *Phragmites australis* was the most tolerant species and contributes greatly to azo-dye removal. As color removal is the key factor in textile wastewater treatment, CWs could be used as the main component in a hybrid treatment system.

Published experiments show that CWs achieve high pollutant removal efficiencies when treating pretreated tannery and textile wastewaters. These high removal efficiencies in combination with the extremely low construction and operation cost, establish CWs as a very attractive treatment method for owners of small industrial units. However, effluent pollutant concentrations still remain above EU recommended limits and prohibit its direct disposal or use for irrigation. This insufficient treatment could be easily overcome by using hybrid CW systems.

Chromium is present in various forms in tannery wastewaters and other industrial effluents. Therefore chromium removal is an important issue for industrial wastewater treatment. The following case study presents the ability of CWs to safely remove Cr(VI).

16.3 Constructed Wetlands for Cr(VI) Removal: A Case Study

Two pilot-scale HSF CWs (Figure 16.1) were constructed using plastic trapezoidal tanks (units) with dimensions 1.26 m long, 0.68 m wide (upper base), 0.73 m deep, and a total volume of 0.62 m^3 [44]. All units were filled with fine gravel ($D_{50} = 6$ mm) and placed in an open-air facility. One pilot-scale unit was planted (Cr-P) with six stems (7 reeds/m^2) of common reeds (*Phragmites australis*), and the other was kept unplanted (Cr-U). Wastewater flow took place through a perforated plastic pipe (diffuser) placed across the width of the tank at the upstream side. The outlet structure of the units was an orifice (1/4 inch diameter) at the base of the downstream end of each unit which was connected to a U pipe. Overflow wastewater was collected in a 35 L plastic tank for proper disposal.

Mean values of the physico-chemical characteristics are presented in Table 16.3. Figure 16.2 presents a times series charts for Cr(VI) removal rate in the Cr-P and Cr-U pilot-scale units. During the first year of operation, with a HRT of 8 days, Cr(VI) removal efficiency did not show great variations between the two units [45]. Moreover, with the exception of the initial commissioning phase,

Figure 16.1 Pictures of the pilot-scale constructed wetland units.

Table 16.3 Physico-chemical characteristics of the Cr(VI) solution [44].

		Cr-P	Cr-U
pH	Influent	7.49 ± 0.19	7.46 ± 0.20
	Effluent	6.76 ± 0.23	7.46 ± 0.23
EC (μS/cm)	Influent	356 ± 42	360 ± 49
	Effluent	446 ± 118	357 ± 53
Cr(VI) (mg/L)	Influent	2.9 ± 2.2	2.9 ± 2.1
	Effluent	0.3 ± 0.6	2.4 ± 1.8

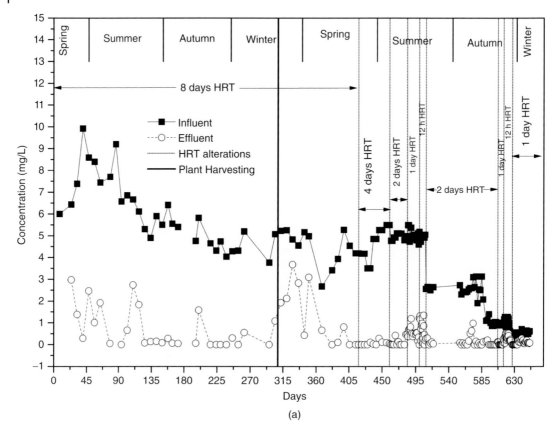

Figure 16.2 Time series charts for Cr(VI) removal rate in (a) Cr-P and (b) Cr-U pilot-scale units. [44].

Cr-U showed high concentrations of Cr(VI) in the effluent (mean value 3.84 mg/L), which were not affected by temperature fluctuations. To assess temperature effect on Cr(VI) removal, Sultana et al. [44] performed one-way ANOVA in two different experimental sets (one for temperatures above and one for temperatures below 15°C). Although both units seemed to be affected by temperature, the Cr-P unit was most probably affected by the reeds' annual growth cycle, as in low temperatures common reeds limit their growth and usually decay.

pH is a key factor that influences metal chemistry and mobility and has a significant impact on the uptake of heavy metals [46]. The solubility, mobility and bioavailability of metals increase when the pH decreases [47]. Hydroxide minerals of metals are less soluble under pH conditions of natural water. As hydroxide ion activity is directly related to pH, the solubility of metal hydroxide minerals increases with decreasing pH, and they become potentially available for biological processes as pH decreases [48]. During the operational period, pH values were monitored at the inlet and outlet points of both the Cr-P and Cr-U units. The mean influent and effluent pH values in the Cr-U unit were identical (pH 7.46), while maximum and minimum influent values of pH in Cr-U were 7.96 and 7.04, respectively, and 7.91 and 5.77, respectively, at the Cr-U outlet.

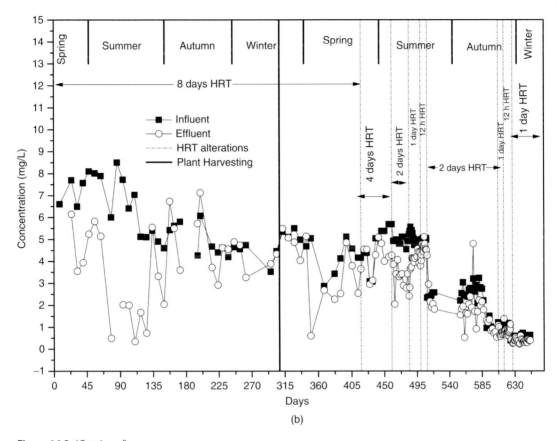

Figure 16.2 (*Continued*)

As stated above, metal mobility is influenced by pH value. Heavy metals are co-precipitated in ferrous-iron oxides and the oxiferric hydroxide surfaces are positively charged in acidic pH conditions and negatively charged in alkaline pH conditions [49]. Thus, to enhance the adsorption and removal of metal oxyanions, iron co-precipitation should occur under acidic pH conditions [50]. Furthermore, as pH increases competition occurs between OH^- and chromate ions (CrO_4^{2-}) and these become the dominant species at higher pH values (<6), although adsorption ceases at pH >9 [51]. Kongroy et al. [52] stated that Cr accumulation in sediment could be due to precipitation under high pH conditions (pH = 8–9). Lesage et al. [53] reported that too low and too high pH values are unfavorable for metal sorption. Mant et al. [25] came to the same conclusion, although they did not define a specific pH value. In the above case study, the mean pH value of the Cr-U unit was found to be 7.46, therefore it could be concluded that this pH value may enhance the precipitation of Cr(VI) in the unplanted unit.

During the first year of operation, the two CW units showed significant differences in Cr(VI) removal, with the exception of the first one hundred days. During these first one hundred days, both units showed a similar fluctuation in removal efficiency that was attributed to the CW commissioning phase [45]. In the first year of treatment, the wetlands were treated with mean Cr(VI) influent concentrations of 5.61 and 5.47 mg/L for CW-P and CW-U, respectively and an HRT of 8 days was used. The mean Cr(VI) effluent concentrations were recorded as 0.86 and 3.84 mg/L for CW-P and CW-U, respectively, and the mean removal efficiency of Cr was recorded as 85% and 26% for CW-P and CW-U, respectively. From the results of the first year, it is clear that Cr(VI) removal in CW-P was higher than in CW-U with an HRT of 8 days. Therefore, the difference in removal efficiencies between the two units could only be attributed to the presence of the reed vegetation.

To ensure the most effective pollutant removal within the wetlands, the plant vegetation should be harvested in winter when the reed's life cycle is complete. Following plant harvesting, the effluent concentrations of the CW-P increased, enhancing this way the assumption that the main Cr(VI) removal mechanism is plant biomass accumulation. With the exception of the initial 100 days, the Cr-U showed high Cr(VI) effluent concentrations most of the year. These high effluent concentrations were not affected by temperature fluctuations, thus implying that microbial accumulation of Cr(VI) is minimal and that during the initial 50 days of operation, Cr(VI) was mainly removed via porous media absorption. The ability of reeds to accumulate Cr(VI) in their roots and transfer it to stems and leaves has been proved in previous experiments by Aguilar et al. [18].

In the second year of the experiment, following reed plant regeneration, both the wetlands were treated with different concentrations of Cr(VI) with different HRTs. Throughout the operational period, the two units showed significant differences in Cr(VI) removal, with the exception of the first 40 days as the plants regrew. During these 40 days, both units showed a similar fluctuation in removal efficiency. These differences in Cr(VI) removal efficiencies were not due to temperature fluctuations, as the environmental temperature rose in the same period. However, both Cr-P and Cr-U presented clear differences in Cr(VI) removal, as the mean removal rate in Cr-P was 87.47%, but only 21% in Cr-U.

Cr(VI) concentrations decreased slightly in the Cr-U unit. In this unit, the main mechanisms of Cr(VI) removal were probably accumulation by microorganisms and absorption by the porous media. Although the removal rate was low, it may also be explained as some chromium hydroxide precipitation is dependent on pH. When pH decreases, precipitation of chromium hydroxide increases within a certain range of pH (5.5–8) [54, 55]. This limited Cr(VI) removal in the Cr-U unit can also be attributed to the fact that microorganisms received water contaminated with Cr(VI) and not wastewater. The contaminated water has lower concentrations of organic carbon and nutrients and, therefore, limits the growth of microorganisms. Furthermore, the porous media used in these pilot-scale units was river gravel which does not have high absorption ability.

On the contrary, the Cr-P unit removed almost all the Cr(VI), as its removal efficiency was close to 100% for most of the operating period. In this unit, Cr(VI) removal efficiency showed variations in specific time periods. In the winter period when the plants were regenerating, Cr(VI) removal efficiency in the Cr-P showed some fluctuation but still performed better than Cr-U. This reinforces the argument that the main removal mechanism of Cr(VI) is accumulation in reed plants throughout the year. The other occasion where Cr-P showed lower Cr(VI) removal efficiency, was when the HRTs were altered. This is because the Cr-P unit required an adjustment period when higher Cr(VI) loads were introduced. Otherwise, a high removal efficiency of Cr(VI) was observed throughout the experimental period in the Cr-P unit, which suggests that the Cr(VI) removal was performed by the plants. In addition to this, hydroponic gravel bed systems contain an extensive variety of micro-environments near the plant root surfaces and within the biofilms that form on root surfaces and gravel [56]. Nevertheless, if no biofilm develops on the plant roots, roots still serve as ground for metal binding, due to the presence of organic ligands such as carboxyl, hydroxyl, and phenolic functional groups [25].

Although the Cr-P was subjected to several variations of HRT and Cr(VI) feed concentration, the plants did not show any signs of injury and continued to grow normally.

To examine the effect of HRT on Cr(VI) removal efficiency, Sultana et al. [44] used different HRTs (i.e., 8, 4, 2, 1, 0.5 days), while mean influent surface loading rate (SLR) ranged from 0.03 to 0.55 $g/m^2/d$ for Cr-U and from 0.03 to 0.52 $g/m^2/d$ for Cr-P. One-way ANOVA statistical analysis showed that [44]:

- 0.5 days HRT was insufficient for Cr(VI) removal, as both units showed great variations probably due to the diurnal cycles of the plant vegetation.
- Under steady state conditions, the Cr-P unit managed to completely remove Cr(VI) (100% removal rate), thus proving that CWs can operate under extremely low HRTs (i.e., 1 day) and still achieve complete Cr(VI) removal.
- On the other hand, the Cr-U unit was also not affected by HRT alterations, although Cr(VI) removal rates were extremely low throughout the entire operational period, thus proving the beneficial role of vegetation in CWs.

The by-product of CWs is plant biomass and when the treated wastewater contains high concentrations of heavy metals, plant biomass is also considered as hazardous waste. Therefore, a significant issue when using CWs for removing heavy metals is plant biomass management. Sultana et al. [44] proposed to co-compost plant biomass with other agricultural wastes (i.e., olive leaves and olive pomace). The final product of this experiment was a soil amendment of excellent quality [44]:

- Final pH values of the compost were approximately 8, while electrical conductivity values were around 700 µS/cm, which is below the Greek standard upper limits (4 S/cm) [57].
- Final Total Kjeldahl Nitrogen (TKN) was 3.5%, which qualifies the present compost to be used as nitrogenous fertilizer [58].
- The final C/N ratio was around 14, which is below the value of 20 that is the limit used to characterize mature compost.
- The concentration of Cr within the compost (i.e., 10 mg/kg dry mass) did not vary during composting, as expected, and is below the EU limit (70 mg/kg dry mass) for organic farming applications [59].
- Germination index values of the experimental compost reached 157%, thus indicating that the compost favors plant growth.

16.4 Conclusions

Although CWs have been used mainly to treat municipal wastewater, a significant number of experiments and applications have proved their ability to also treat wastewaters originating from tanning and dyeing processes. Based on the main findings from all currently available literature sources, it can be concluded that:

- CWs demonstrate significantly high pollutant removal efficiencies when applied to treat tannery and textile wastewaters.

- Despite the fact that all experimental CWs received very high pollutant loads, none showed signs of phytotoxicity. Additionally, CW vegetation does not seem to be affected by the toxic substances contained in wastewaters.
- CWs can provide suitable environments, in which heavy metals can be biodegraded.
- In most cases, a pretreatment stage is necessary to remove SS and to lower pollutant loads.
- Although the specific plant species used had no significant impact on pollutant removal, the vegetation does increase the removal of all pollutants. Plant vegetation is the main removal mechanism for heavy metals.
- Although CWs show significantly high removal efficiencies, effluent pollutant concentrations remain above legal limits. Therefore, a combination of CW systems with other conventional treatment systems is imperative to further lower pollutant levels.

References

1 UNESCAP. Integrating environmental consideration into economic policy making: institutional issues. 2000 (ST/ESCAP/1990).
2 UNESCAP. Integrating environmental consideration into economic policy making process: background readings, vol 1. Intuitional arrangement and mechanism at national level, 1999 (ST/ESCAP/1944).
3 MedCities and ISR EWC. Guidelines for municipal solid waste management in the Mediterranean region, 2003. Barcelona: MedCities.
4 Emongor V, Nkegbe E, Kealotswe B, Koorapetse I, Sankwasa S, Keikanetswe S. Pollution Indicators in Gaborone Industrial Effluent. J Appl Sci. 2005; 5:147–150.
5 Oller I, Malato S, Sánchez-Pérez JA, Combination of Advanced Oxidation Processes and biological treatments for wastewater decontamination – a review. Sci Total Environ. 2011; 409:4141–4166.
6 Nies DH. Microbial heavy-metal resistance. Appl Microbiol Biotechnol. 1999; 51:730–750.
7 Duncan JR, Stoll A, Wilhelmi B, Zhao M, Van Hille R. The use of algal and yeast biomass to accumulate toxic and valuable heavy metals from wastewater. Final report to the water research commission: 2003, Rhodes University.
8 Stefanakis AI, Akratos CS, Tsihrintzis VA. Vertical flow constructed wetlands: Eco-engineering systems for wastewater and sludge treatment, first edn. 2014, Elsevier, Burlington, USA.
9 Kolomaznik K, Adamek M, Andel I, Uhlirova M. Leather waste – potential threat to human health, and a new technology of its treatment. J Hazard Mater. 2008; 160:514–520.
10 Chattopadhyay B, Gupta R, Chatterjee A, Mukhopadhyay SK. Characterization and eco-toxicity of tannery wastes envisaging environmental impact assessment. J Am Leather Chem As. 1999; 94:337–346.
11 Reemtsma T, Jekel M. Dissolved organics in tannery wastewater and their alteration by a combined anaerobic and aerobic treatment. Water Res. 1997; 31:1035–1046.
12 Reemtsma T, Zywicki B, Stueber M, Kloepfer A, Jekel M. Removal of sulfur-organic polar micropollutants in a membrane bioreactor treating industrial wastewater. Environ Sci Technol. 2002; 36:1102–1106.
13 Verheijen LAHM, Werseema D, Hwshoffpol LW, Dewit J. Livestock and the environment: finding a balance management of waste from animal product processing. International Agriculture Centre, Wageningen, 1996, The Netherlands.

14 Nandy T, Kaul SN, Shastry S, Manivel W, Deshpande CV. Wastewater management in cluster of tanneries in Tamilnandu through implementation of common treatment plants. J Sci Ind Res. 1999; 58:475–516.

15 Midha V, Dey A. Biological treatment of tannery wastewater for sulfide removal. International. J Chem Sci. 2008; 6:472–486.

16 Lefebvre O, Moletta R. Treatment of organic pollution in industrial saline wastewater: A literature review. Water Res. 2006; 40:3671–3682.

17 Calheiros CSC, Rangel AOSS, Castro PML. Treatment of industrial wastewater with two-stage constructed wetlands planted with *Typha latifolia* and *Phragmites australis*. Bioresour Technol. 2009; 100:3205–3213.

18 Aguilar JRP, Cabriales JJP, Vega MM. Identification and characterization of sulfur-oxidizing bacteria in an artificial wetland that treats wastewater from a tannery. Int J Phytorem. 2008; 10:359–370.

19 Calheiros CSC, Teixeira A, Pires C, Franco AR, Duque AF, Crispim LFC, Moura SC, Castro PML. Bacterial community dynamics in horizontal flow constructed wetlands with different plants for high salinity industrial wastewater polishing. Water Res. 2010; 44:5032–5038.

20 Calheiros CSC, Rangel AOSS, Castro PML. Evaluation of different substrates to support the growth of *Typha latifolia* in constructed wetlands treating tannery wastewater over long-term operation. Bioresour Technol. 2008; 99:6866–6877.

21 Calheiros CSC, Rangel AOSS, Castro PML. Constructed wetland systems vegetated with different plants applied to the treatment of tannery wastewater. Water Res. 2007; 41:1790–1798.

22 Calheiros CSC, Quitério PVB, Silva G, Crispim LFC, Brix H, Moura SC, Castro PML. Use of constructed wetland systems with *Arundo* and *Sarcocornia* for polishing high salinity tannery wastewater. J Environ Manage. 2012; 95:66–71.

23 Kucuk OS, Sengul F, Kapdan IK. Removal of ammonia from tannery effluents from a red bed constructed wetland. Water Sci Technol. 2003; 48:179-186.

24 Dotro G, Larsen D, Palazolo P. Treatment of chromium-bearing wastewaters with constructed wetlands. Water Environ J. 2011; 25:241–249.

25 Mant C, Costa S, Williams J, Tambourgi E. Phytoremediation of chromium by model constructed wetland. Bioresour Technol. 2006; 97:1767–1772.

26 Akratos CS, Tsihrintzis VA. Effect of temperature, HRT, vegetation and porous media on removal efficiency of pilot-scale horizontal subsurface flow constructed wetlands. Ecol Eng. 2007; 29:173–191.

27 Greenway M, Woolley A. Changes in plant biomass and nutrient removal over 3 years in a constructed free water surface flow wetland in Cairns Australia. 7th International Conference on Wetland Systems for Water Pollution Control. 2000. Lake Buena Vista, Florida, pp. 707–718.

28 Vymazal J. The use of sub-surface constructed wetlands for wastewater treatment in the Czech Republic: 10 years experience. Ecol Eng. 2002; 18:633–646.

29 Sultana M-Y, Akratos CS, Pavlou S, Vayenas DV. Chromium removal in constructed wetlands: A review. Int Biodeter Biodegr. 2014; 96:181–190.

30 Roussy J, Chastellan P, Van Vooren M, Guibal E. Treatment of ink-containing wastewater by coagulation/flocculation using biopolymers. Water SA. 2005; 31:369–376.

31 Awomeso JA, Taiwo AM, Gbadebo AM, Adenowo JA. Studies on the pollution of water body by textile industry effluents in Lagos, Nigeria. J Appl Sci Environ Sanitation. 2010; 5:353–359.

32 Brown D, Laboureur P. The aerobic biodegradability of primary aromatic amines. Chemosphere. 1983; 12: 405–414.

33 Choi JW, Song HK, Lee W, Koo KK, Han C, Na BK. Reduction of COD and colour of acid and reactive dyestuff wastewater using ozone. Korean J Chem Eng. 2004; 21:398–403.

34 Savin II, Butnaru R. Wastewater characteristics in textile finishing mills. Environ Eng Manag J. 2008; 7:859–864.

35 Bisschops I, Spanjers H. Literature review on textile wastewater characterization. Environ Technol. 2003; 24:1399–1411.

36 Venceslau MC, Tom S, Simon JJ. Characterization of textile wastewaters a review. Environ Technol. 1994; 15:917–929.

37 World Bank. Environmental, Health, and Safety guidelines for textile manufacturing, International Finance Corporation, 2007. World Bank Group, On line at: www.ifc.org/ifcext/sustainability.nsf/ Attachments.

38 Pagga U, Brown D. The degradation of dyestuffs: Part II. Behaviour of dyestuffs in aerobic biodegradation tests. Chemosphere. 1986; 15:479–491.

39 Boyter HA. Environmental legislations USA. In: Christie RM, ed. Environmental aspects of textile dyeing. Cambridge, England: Woodhead, 2007.

40 Ong SA, Uchiyama K, Inadama D, Yamagiwa K. Simultaneous removal of color, organic compounds and nutrients in azo dye-containing wastewater using up-flow constructed wetland. J Hazard Mater. 2009; 165:696–703.

41 Bulc TG, Ojstrsek A. The use of constructed wetland for dye-rich textile wastewater treatment. J Hazard Mater. 2008; 155:76–82.

42 Nilratnisakorn S. Thiravetyan P, Nakbanpote W. A constructed wetland model for synthetic reactive dye wastewater treatment by narrow-leaved cattails (*Typha angustifolia* Linn.). Water Sci Technol. 2009; 60:1565–1574.

43 Brix H. Functions of macrophytes in constructed wetlands. Water Sci Technol. 1994; 29:71–78.

44 Sultana M-Y, Chowdhury AKMMB, Michailides, MK, Akratos CS, Tekerlekopoulou AG, Vayenas DV. Integrated Cr(VI) removal using constructed wetlands and composting. J Hazard Mater. 2015; 281:106–113.

45 Michailides MK, Sultana M-Y, Tekerlekopoulou AG, Akratos CS, Vayenas DV. Biological Cr(VI) removal using bio-filters and constructed wetlands, Water Sci Technol. 2013; 68:2228–2233.

46 Zeng F, Ali S, Zhang H, Ouyang Y, Qiu B, Wu F, Zhang G. The influence of pH and organic matter content in paddy soil on heavy metal availability and their uptake by rice plants. Environ Pollut. 2011; 159:84–91.

47 Gambrell RP. Trace and toxic metals in wetlands – a review. J Environ Qual. 1994; 23:883–891.

48 Salomons W. Environmental impact of metals derived from mining activities: processes, predictions, prevention. J Geochem Explor. 1995; 52:5–23.

49 Sheoran AS, Sheoran V. Heavy metal removal mechanism of acid mine drainage in wetlands: a critical review. Miner Eng. 2006; 19:105–116.

50 Brix H. Wastewater treatment in Constructed Wetlands: system design, removal processes, and treatment performance. Constructed Wetlands for Water Quality Improvement. G.A. Moshiri. Florida, America, CRC Press, 1993, pp. 9–22.

51 Baral SS, Dasa SN, Rath P. Hexavalent chromium removal from aqueous solution by adsorption on treated sawdust. Biochem Eng J. 2006; 31:216–222.

52 Kongroy P, Tantemsapya N, Lin YF, Jing SR, Wirojanagud W. Spatial distribution of metals in the Constructed Wetlands. International Journal of Phytoremediation. 2012; 14: 128–141; doi: 10.1080/15226514.2011.573825.

53 Lesage E, Mundia C, Rousseau DPL, Van de Moortel AMK, du Laing G, Meers E, Tack FMG, De Pauw N, Verloo MG. Sorption of Co, Cu, Ni and Zn from industrial effluents by the submerged aquatic macrophyte *Myriophyllum spicatum* L. Ecol Eng. 2007; 30:320–325.

54 Karale RS, Wadkar DV, Nangare PB. Removal and recovery of hexavalent chromium from industrial waste Water by precipitation with due consideration to cost optimization. J Environ Res Dev. 2007; 2:209–216.

55 Alves MM, González Beça CG, Carvalho RG, de Castanheira JM, Pereira MCS, Vasconcelos LAT. Chromium removal in tannery wastewaters "polishing" by *Pinus sylvestris* bark. Water Res. 1993; 27:1333–1338.

56 Williams JB. Microbial factors affecting the design and operation of a gravel bed hydroponic sewage treatment system. Ph.D. thesis. Department of Civil Engineering, 1993. University of Portsmouth.

57 Lasaridi K, Protopapa I, Kotsou M, Pilidis G, Manios T, Kyriacou A. Quality assessment of composts in the Greek market: the need for standards and quality assurance. J Environ Manage. 2006; 80: 58–65.

58 Legislativo D. Riordino e revisione della disciplina in materia di fertilizzanti, a norma dell'articolo, In: Legge, D. (ed.), 2010. 88. Gazzetta Ufficiale.

59 EU. Heavy metals and organic compounds from wastes used as organic fertilizers. 2004. Ref. Nr.: TEND/AML/2001/07/20.

Part VI

Pharmaceuticals and Cosmetics Industry

17

Removal Processes of Pharmaceuticals in Constructed Wetlands

A. Dordio[1,2] and A.J.P. Carvalho[1,3]

[1] *Chemistry Department, School of Sciences and Technology, University of Évora, Évora, Portugal*
[2] *MARE – Marine and Environmental Research Centre, University of Évora, Évora, Portugal*
[3] *CQE – Évora Chemistry Centre, University of Évora, Évora, Portugal*

17.1   Introduction

Pharmaceuticals are substances that are developed and used to perform a specific biochemical function in the diagnosis, prophylaxis, or therapy of a disease, disorder or abnormal physical state, or its symptoms in human beings or animals. Increasing amounts of pharmacologically active substances are consumed yearly in human and veterinary medicine and, as a side-effect of its widespread use, they are frequently present in wastewaters.

There are many aspects in which pharmaceuticals differentiate from conventional industrial chemical pollutants [1–4]:

- Pharmaceuticals usually are chemically complex molecules that vary widely in molecular weight (ranging typically from 200 to 1000 Dalton), structure, functionality, and shape; due to the variety of functions that pharmaceuticals must perform, this class of compounds spans very different physicochemical properties and biological functionalities.
- In general, pharmaceuticals are polar molecules with several ionizable groups, and the degree of ionization and its properties depend on the pH of the medium, with many molecules exhibiting an amphoteric behavior; most notably they can be characterized by at least some moderate solubility in water, as they commonly must take effect in aqueous media, but some of them present some lipophilicity as well.
- Some pharmaceuticals have a year-long persistence in the environment (e.g., erythromycin, cyclophosphamide, naproxen, and sulfamethoxazole) or even longer (e.g., clofibric acid, carbamazepine) and become biologically active through accumulation.
- After their administration to humans or animals, pharmaceutical molecules are absorbed, transported, distributed by the organism, and therein subjected to metabolic reactions that may modify their chemical structure; frequently, these transformations yield metabolites with a reduced biodegradability.

In summary, pharmaceuticals are usually resistant to degradation, highly persistent in aqueous medium, and potentially able to produce adverse effects in aquatic organisms, as well as having negative impacts to human health.

Over the last few years, the detection of pharmaceutical residues in environmental samples (mostly collected from aquatic environment sources) has become a hot topic. Pharmaceuticals have been one of the most important emerging classes of pollutants to be detected worldwide, albeit usually at very low concentrations (in the μg/L to ng/L levels), in raw and treated wastewater, biosolids and sediments, receiving water bodies, groundwaters and even drinking water [3, 5–26]. Despite the only recent public awareness, pharmaceuticals, their metabolites and transformation products most certainly have been entering the environment for many decades already. However, the actual extent of this problem has only been exposed by more recent in-depth environmental monitoring studies using the latest advances in chemical analysis methodologies and instrumentation (in particular chromatographic separation techniques coupled to highly sensitive detection methods such as mass spectrometry, which allowed to significantly lower the detection and quantification limits for analyses of organic compounds in complex environmental matrices).

In numerous monitoring studies that have been carried out during the last decades, the most frequently detected pharmaceutical residues belong to the pharmaco-therapeutic classes of lipid regulating drugs, analgesics and anti-inflammatory drugs, antibiotics, hormones, antidiabetics, neuroactive drugs and beta-blockers [3, 5–27]. It is important to note that the classification of pharmaceuticals by their active substances, within groups of pharmaceuticals whose defining criteria may even be non-chemical (e.g., therapeutic), does not imply that families of pharmaceuticals follow a certain common chemical behavior. Even small changes in chemical structure may have significant effects on solubility, polarity and other properties, which in turn will create some difficulty to predict environmental distribution in air, sediments and soils, water and biota [4].

While the presence of organic micropollutants in the aquatic environment has for long been associated with numerous harmful effects, including short-term and long-term toxicity and endocrine disrupting effects, there is less data available concerning specific effects of pharmaceuticals, which has been produced only recently [5, 9, 11, 23, 26, 28–38]. Therefore, the risks for the environment posed by pharmaceutical pollutants cannot be fully assessed currently, although, considering the nature and functions of this type of chemicals, potential for similarly hazardous effects as other organic micropollutants may be expected. Even though the trace concentration levels, at which most of these pollutants are detected in the environment, may be seen as a mitigation factor in respect to its hazardous potential, a concomitant issue is raised by the huge variety of substances that increasingly are being identified in environmental monitoring studies. In fact, given that pharmaceuticals share, by design, a common purpose of performing a biochemical effect, there is a significant potential for cumulative or synergistic effects between several different drugs (especially if they were designed for the same or similar targets), which is still largely unpredictable and undocumented. Therefore, an environmental risk must be considered even for concentrations lower than usual in comparison with other types of pollutants.

In this chapter, the application of Constructed Wetlands to deal with the environmental contamination caused by pharmaceutical residues is addressed. In the following section, a presentation is made of the several sources of pharmaceutical pollutants, the various pathways of their environmental fate and potential effects to living organisms (including humans). Excretion of metabolites or non-metabolized parent drugs, or improper disposal of unused or expired products are the main sources of introduction of pharmaceuticals in the sewage system. Many of these pollutants do not receive efficient treatment in wastewater treatment plants (WWTPs), because conventional WWTPs are designed to reduction of bulk pollutant loads and are generally inefficient in removing micropollutants such as pharmaceuticals.

The rising concern with emergent pollutants has been highlighted several times by specialists [9, 39–43]. Awareness of this problem resulted over the last decade in a move by environmental agencies worldwide towards electing some of these new substances as priority pollutants and requiring new environmental risk assessments to be carried out as part of the process of approving new substances for public use [41, 42, 44, 45]. In turn, this situation led to a progressive reform of WWTPs, with adaptations of existing conventional plants and consideration of these new requirements in the design of new plants. Consequently, there has been a growing need for an optimization of conventional wastewater treatment processes (e.g., increasing hydraulic and solid retention times) or the development and implementation of alternative/complementary wastewater treatment processes that may achieve higher efficiencies at removing pharmaceuticals from wastewater, at reasonable cost of operation and maintenance.

Several advanced technologies (e.g., advanced oxidation processes or membrane processes) [14, 46–53] have been evaluated but not always the cost-effectiveness of these solutions at a large scale is sufficiently attractive, with the high efficiency of some technologies being sometimes offset by the high costs of its implementation and/or operation [5, 12, 14, 15, 54, 55].

Phytotechnologies have gained a good reputation as an interesting low-cost wastewater treatment technology for non-conventional pollutants such as heavy metals and organic xenobiotics. In particular, constructed wetlands systems (CWS) are increasingly being used to provide a form of secondary or tertiary treatment for wastewaters [56–65]. In the removal of pharmaceuticals from wastewater, the use of CWS have already shown in studies both at pilot and at full scale some interestingly high efficiencies for some compounds, including some that are notoriously recalcitrant to conventional treatment methods [60, 66, 66–84].

In section 17.3, the topic of the specific use of CWS for the removal of pharmaceuticals in wastewater treatment is addressed. The section begins with a description of the state of the art in this subject, presenting a sample of the variety of studies conducted to evaluate the efficiencies of these systems to treat this type of pollutants. Then, a discussion is presented of the most relevant processes that are involved in the removal of pharmaceuticals in CWS and the role played by each CWS component in the efficiency of the system.

It is not easy to accomplish a detailed and comprehensive description of such processes and for long these wastewater treatment systems have been operated as "black boxes". In the past, much of the CWS design was done with little knowledge of (or consideration for) the role played by CWS components and how their effects could be enhanced and optimized. However, over the years the interest is growing over mechanistic studies of CWS processes and attempts to optimize CWS design (layout of the systems, selection of CWS components, i.e. plant species, support matrix materials and microorganisms strains, etc.) and operation (type and rate of flow, hydraulic retention time) [61, 65, 68, 75, 82, 85–94].

This chapter presents a current overview of the use of CWS for treating wastewaters contaminated with pharmaceuticals, showcasing some successful applications reported in the literature and higlighting the potential of this phytotechnology, and pointing out aspects that have until now received less attention and present opportunities for further study and improvement of this type of systems.

17.2 Pharmaceutical Compounds in the Environment: Sources, Fate and Environmental Effects

The detection of pharmaceutical residues in the environment has become a relevant subject lately, causing a growing concern in the scientific community and the general public. The research on this theme has established, so far, a large and sometimes unexpected variety of routes through which pharmaceuticals cross (and are distributed to) various environmental compartments. Figure 17.1 shows a representation of possible pathways for pharmaceuticals (and metabolites, and other transformation/degradation products) in the environment.

Wastewater point sources are widely regarded as the primary route of entry for pharmaceuticals (and metabolites or transformation products) in the environment [3, 5–7, 9, 10, 12, 14, 97]. They are introduced in the sewage systems as they are excreted by the organism, either in non-metabolized form (sometimes amounting to the majority of the initial ingested substance) or in the form of their

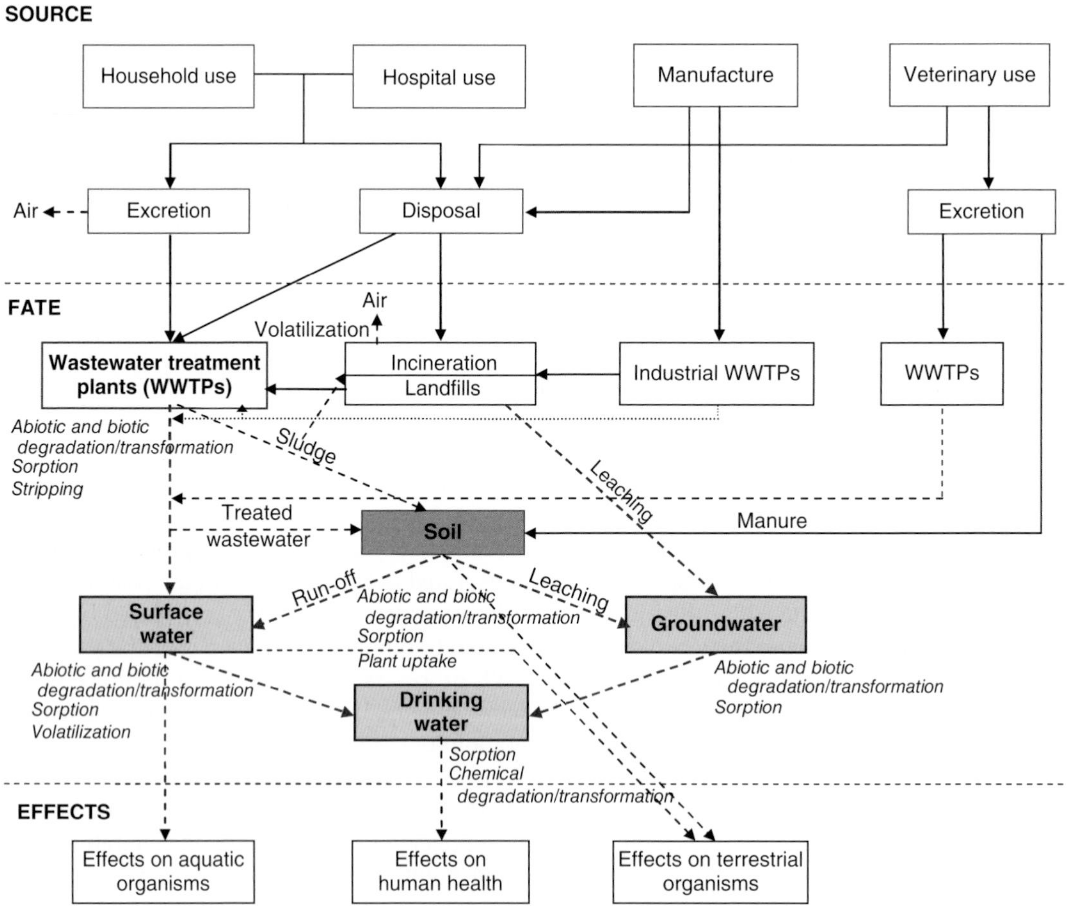

Figure 17.1 Sources, pathways and impacts of pharmaceuticals in the environment (sources: [8, 29, 95, 96]).

various metabolites. Frequently there is also an additional source of expired medications that are improperly disposed of directly in the sewage systems. Meanwhile, the fate of these substances in WWTPs in many cases corresponds to only some minor removal or transformation. In general, conventional WWTPs are designed to treat bulk pollutants and the wastewater treatment processes used in such plants are mostly inefficient to remove organic micropollutants such as pharmaceuticals [7, 10, 14, 15, 43]. Therefore, these contaminants are usually still present in effluents from WWTPs, which represent the most important source point for aquatic exposure to pharmaceuticals [6, 7, 10, 14, 15, 32, 43].

Other important sources of aquatic contamination with pharmaceuticals originate from wastewaters of hospitals [8, 46, 98–100], the pharmaceutical industry, livestock and aquacultures [11, 100–103], as well as landfill leachates [8, 104–107].

On the other hand, the contamination of soils (especially agricultural ones) with pharmaceuticals, originating both from human consumption and from veterinary use, are mainly caused by irrigation with reclaimed water and from the use of sewage sludge (biosolids) and manure as fertilizer or compost [13, 18–21]. The presence of pharmaceuticals in the soil may then leads to the contamination of surface water by run-off or of groundwater by leaching [8, 11, 16, 17, 97, 108, 109]. Furthermore, as some of these substances have the potential to be taken up by plants being grown on contaminated soil, there is a risk that crops can also become contaminated, thus constituting a public health issue [110–117]. In general, there is a broader risk that such vegetation, taking up and accumulating pharmaceuticals, will take part of the diet of herbivores and, subsequently, be passed along the food chain.

Pharmaceuticals in the environment are potentially subjected to the same type of transport, transfer and transformation/degradation processes as other organic pollutants, namely involving sorption, hydrolysis, biological transformation/degradation, redox reactions, photodegradation, volatilization and precipitation/dissolution (Figure 17.1) [2–4, 6, 8, 16, 24, 29, 118–120].

These processes occur continuously in the environment and influence the presence and mobility of pharmaceuticals in it. Their behavior, under any of these pathways ultimately leading to partitioning, degradation or transformation, may contribute to reduce their concentrations in the environment (or their availability, through their stabilization in inert forms) or even to eliminating them entirely and, thereby, to lower their potential to harm human health and aquatic life. However, some pharmaceutical metabolites and transformation products from these processes can be more persistent and/or more toxic than the parent compound [6, 14, 15, 32, 34, 121, 122].

The pharmaceuticals with highest potential to reach steady-state levels in the environment and to be detected in surface and ground waters and in drinking water supplies are those with high consumptions that also combine the properties of significant water solubility and high resistance to degradation both by biotic and abiotic processes [2, 3, 123].

However, major differences between pharmaceuticals and other "traditional" organic pollutants (such as, for example, pesticides, PCBs, PAHs or explosives) is that pharmaceutical substances, in general, are designed into a sufficiently water soluble form because usually they are supposed to function in that medium. In addition, before they reach the environment they have already passed through the digestive tracts of humans or animals and then, in most cases, also through some wastewater treatment processes. Two consequences of this pre-exposure to a special biotic environment and to biochemical metabolism are that: (i) many pharmaceuticals will enter the aquatic environment in a modified form that is more stable in regard to biotic transformation or degradation; and (ii) those pharmaceuticals still remaining unaltered at the end of this path are probably highly resistant to biotic transformation or degradation. These realizations allow certain inferences to be made regarding the importance of abiotic processes for the fate of pharmaceutical compounds in the aquatic environment [119]. Given the water solubility of many pharmaceuticals, the abiotic processes that are most likely to transform these pollutants and to more permanently remove them from the aquatic environment include hydrolysis and photodegradation [2, 3, 29, 118, 124]. However, considering the previous exposure of pharmaceuticals to the digestive tract and their relatively long-residence time in aqueous environments within the WWTPs, hydrolysis reactions likely play a less important role in the aquatic fate of many pharmaceuticals after they reach the environment [119]. Conversely, direct photodegradation by sunlight may be an important elimination process for those pharmaceuticals that have significant absorbances in the 290–800 nm region [124–127].

In any case, both pharmaceutical properties and environmental conditions influence the extent to which the various abiotic and biotic processes may affect the short-term behavior and long-term fate of a pharmaceutical in the environment. Some of their most important physical and chemical properties that may affect their environmental fate are molecular structure, polarity, ionization constant, water solubility, octanol-water partition coefficient, sorption distribution coefficient and the compound's half-life. In addition, environmental conditions with most relevance with regard to these pollutants' environmental fate include temperature, sunlight, pH, content of organic matter in soils and sediments and redox conditions.

Under ordinary conditions the physicochemical properties of many pharmaceuticals makes them in many cases refractory to degradation and transformation, according to evidence accumulated over the latest years [2–5, 95, 97]. Consequently, they do indeed have the potential to reach and persist in the environment. However, relatively little is known about the impending human or ecological hazards that can arise from the cumulative exposure to the extensively varied blend of pharmaceuticals and metabolites present in the different environmental compartments (notwithstanding the low concentrations at which they are observed to occur). Pharmaceuticals are designed to target specific biochemical pathways and, as a side-effect, when introduced in the environment, they may intervene in analogous pathways in other organisms having identical or similar target organs, tissues, cells or biomolecules. Even in organisms that lack or have modified receptors for a particular pharmaceutical molecule, the drug may still cause a disruptive effect through an alternative mode of action. It is important to realize that, for many pharmaceuticals, their specific modes of action are not well described and often not only one but many different modes of actions occur. Therefore, the ecotoxicity of most pharmaceuticals is difficult to assess or predict [5]. In addition, metabolites and transformation products of pharmaceuticals can also be of concern, because many of them have a toxicity which in many cases may be similar to or even higher than that of the parent compounds [5, 7, 14, 34, 47, 128].

The current literature about ecotoxicological effects of human pharmaceuticals deals mainly with the acute toxicity resulting from short-term exposure, as evaluated through standardized tests focused on aquatic organisms. Acute toxicity values are determined in the mg/L dose range for most of the pharmaceuticals detected in the environment [95], but concentration levels reported for surface water are at least three orders of magnitude below [3, 5, 97]. On the other hand, it is more relevant but also much more difficult to assess whether long-term chronic effects have any environmental significance as these toxicity data is generally lacking [5]. Nonetheless, some primary effects can be identified which derive from the presence of pharmaceuticals and related substances in the environment including cumulative impacts, endocrine disruption and genotoxic effects [5, 9, 11, 23, 26, 28–36, 38, 117]. In particular, the extensive use of antibiotics and consequent continual exposure of bacteria to antibiotic residues in the aquatic environment has contributed to the development of antibiotic resistance genes and bacteria, reducing the therapeutic potential against human and animal pathogens [12, 31, 35, 129–132].

In addition to toxicity, the aspect of persistence is of particular importance when considering the effects of pharmaceuticals in the environment. Some pharmaceuticals are so-called "pseudo persistent pollutants", which means that, even if their properties do not favor their bioaccumulation, they may still be persistent due to their continuous introduction in significant amounts in the environment. While not consisting of persistence in terms of a long half-life, this still results in a long-term exposure for the aquatic ecosystem.

Overall, although the ecotoxicity of pharmaceuticals is not clearly characterized in most cases, the possibility of negative impacts is noteworthy and a number of researchers are trying to quantify the risk posed by various pharmaceuticals [1, 34, 38, 40, 42, 43, 121, 133–136, 136–138].

The risks presented by the disseminated presence of pharmaceuticals in the environment point out to the urgent need for finding ways of retaining and removing these pollutants at point sources such as WWTPs, before they reach the receiving water bodies which are the main environmental receptacles for these pollutants. In the attempt to enhance the efficiency of removal of these pollutants by the technologies currently in use, optimizations of the conventional wastewater treatment processes and of WWTP operating parameters have been explored, for example, by increasing hydraulic and solid retention times [10, 14, 139–143]. In recent years, a variety of advanced treatment technologies for removal of pharmaceutical compounds from water have been evaluated. Among them activated carbon adsorption, membrane treatment using both biological (membrane bioreactors) and non-biological processes (reversed osmosis, ultrafiltration, nanofiltration), and advanced oxidation processes have been the most frequently studied at lab experiments and at pilot scale for removal of emerging pollutants at trace concentrations [5, 7, 10, 12, 14, 15, 48–52, 54, 144–154].

However, despite the high removal efficiencies featured in many cases, sometimes even reducing pharmaceutical concentrations to levels below detection limits [5, 48, 50, 149, 150, 153, 155–158], the high cost of implementation and/or operation is a major drawback of most advanced treatment technologies currently available. In most contexts, these processes are considered as non cost-effective to be used on a large scale in wastewater treatment [4, 5, 12, 14, 54, 55, 148, 158]. Moreover, when complete decomposition (mineralization) is not achieved, some of these treatments may also originate transformation products, which in some cases may be more persistent or toxic than the parent compounds [6, 7, 15, 29, 48, 128, 159–161]. Consequently, the development of alternative or complementary wastewater treatment processes that are effective at removing pharmaceuticals within reasonable costs of operation/maintenance remains a priority.

As a more economical alternative, the use of aquatic plant-based technologies has been increasingly explored in the latest years and successfully used for removal of many organic xenobiotics from contaminated soils, water and wastewater [162–169]. In fact, the implementation of phytotechnologies such as constructed wetland systems (CWS) is becoming an increasingly popular option lately for removing several organic pollutants of various classes [56–65], including pharmaceuticals [60, 66, 66–68, 68–84, 170–173]. These systems are gradually becoming a frequent option for secondary wastewater treatment systems or as treatment units for polishing secondary effluent from WWTPs. In addition to low cost, simple operation and maintenance (there by not requiring highly skilled labour), and environmental friendliness are some of their most attractive characteristics.

17.3 Pharmaceuticals Removal in Constructed Wetlands

For a long time, natural wetlands have been credited with the capacity to depurate the water that inundated such areas. Constructed wetlands systems (CWS) are engineered systems that take advantage of many of the same processes that occur in natural wetlands, but do so within a more controlled environment [57, 174, 175]. CWS can be optimized to attain greater efficiencies by enhancing the concerted action of all the components (support matrix, vegetation and microbial population) through a variety of interdependent chemical, physical and biological processes (sorption, microbial degradation, plant uptake, etc.).

In the past, CWS have been used mainly as an alternative to the conventional wastewater treatment systems or as a complementary system for domestic WWTPs of small communities and, thus, have been mostly applied for the removal of bulk wastewater pollutants such as suspended solids, organic matter, excess of nutrients and pathogens. The use of these systems has subsequently evolved for also treating a broader variety of wastewater types and many examples can now be found of applications in the treatment of municipal, industrial or agricultural wastewaters as well as urban runoffs [57, 58, 62, 63, 85, 174–178]. As a consequence of this enlargement of the CWS' domain of applications, a wider range of pollutant types to be removed as well as a focus of the treatments into more specific targets such as pharmaceuticals, has been increasingly under research in lab and pilot scale and implemented in full scale [59, 60, 66, 66–68, 68–84, 170–173, 179–183]. In most current cases, the targets of treatment in CWS have been especially those pollutants, which are more recalcitrant to conventional wastewater treatment processes. In many of such studies, CWS have been proving to be efficient and cost-effective solutions for the removal of some pharmaceuticals [59, 59, 65, 68, 75, 82, 183].

17.3.1 Removal Efficiency of Pharmaceuticals in CWS

Over the latest years, there has been a growing interest in the use of CWS for removal of pharmaceuticals from wastewater. In fact, a wide variety of pharmaceuticals from different therapeutic classes (with varied chemical structures and physico-chemical properties) have been studied in regard to their removal in CWS (including the use of CWS as a secondary wastewater treatment system or as a polishing stage). Studies have been performed on different types of CWS (Surface Flow Constructed Wetlands (SF), Horizontal Subsurface Flow Constructed Wetlands (HSSF), Vertical Subsurface Flow Constructed Wetlands (VSSF) and hybrid constructed wetlands (hybrid CWs) at microcosm scale, mesocosm (or pilot) scale and in full scale. Various types of plants and support matrices were also investigated as well different operating modes. So far, many of the studies conducted on this topic have shown a noteworthy potential of these systems to remove a wide variety of pharmaceutical compounds (Table 17.1).

A sample of recent research on pharmaceuticals removal in CWS is presented in Table 17.1. From these examples, it can be seen that the therapeutic classes with most relevant coverage in CWS studies are the analgesics/anti-inflammatory drugs, antibiotics, beta-blockers, diuretics, lipid regulators, psychiatric drugs and stimulants/psychoactive drugs. Within these families of pharmaceuticals, the substances that have been most commonly evaluated were diclofenac, ibuprofen, ketoprofen, naproxen, salicylic acid, sulfamethoxazole, atenolol, clofibric acid, carbamazepine and caffeine.

In the setup of the assessed treatment systems, the use of two plants stands out: *Phragmites* spp. (by far the most commonly used plant species in these CWS studies) and *Typha* spp.; additionally a

Table 17.1 Pharmaceuticals removal in different types of CWS.

Organic compound	Physico-chemical properties[a] S_W (25°C) (mg/L)	$\log K_{ow}$ (25°C)	pK_a	Type of CWS	Type of substrate/plant	% Removed	Removal processes suggested by authors	Reference
Analgesic/anti-inflammatory								
Acetaminophen	14,000	0.46	9.38	Lagoons + SF (*full*)	*Hydrocotle* spp. + *Phragmites australis*	100	n.d.	[184]
				HSSF (*pilot*)	Soil: stones:gravel/*Phragmites australis* or *Typha latifolia*	47–99	Biodegradation	[179]
				HSSF (*pilot*)	Gravel/*Phragmites australis*	95	Biodegradation	[185]
Diclofenac	2.4	4.51	4.15	HSSF (*pilot*)	Gravel/*Phragmites australis*	15	Sorption	[84]
				VSSF (*pilot*)	Gravel/*Phragmites australis*	73	Sorption	[186]
				SF (*full*)	*Typha* spp. + *Phragmites australis*	73 (winter) 96 (summer)	Sorption	[173]
				HSSF (*full*)	*Phragmites australis*	21 ± 18	Sorption	[187]
				HSSF (*pilot*)	Gravel/*Phragmites australis*	99	Biodegradation	[188]
				HSSF (*mesocosms*)	Gravel/*Phragmites australis*	n.r.-40	Biotic processes (plants and microorganisms)	[66]
				SF (*mesocosms*)	*Phragmites australis* or *Typha* sp.	n.r.-55	Biotic processes (plants and microorganisms)	
				HSSF (*mesocosms*)	Gravel/*Typha angustifolia*	55	Sorption	[170]
				SF (*microcosms*)	*Salvinia molesta, Lemna minor, Ceratophyllum demersum, Elodea canadensis*	99	Photodegradation	[189]
				HSSF (*mesocosms*)	Gravel/*Phragmites australis* or *Typha angustifolia*	n.r.-75	Photodegradation	[180]
				SF (*mesocosms*)	*Phragmites australis* or *Typha angustifolia*	n.r.-90	Photodegradation	[180]
				HSSF (*mesocosms*)	Gravel/*Typha angustifolia*	55	n.d.	[190]

(Continued)

Table 17.1 (Continued)

Organic compound	Physico-chemical properties[a]			Type of CWS	Type of substrate/plant	% Removed	Removal processes suggested by authors	Reference
	S_W (25°C) (mg/L)	log K_{ow} (25°C)	pK_a					
				HSSF (mesocosms)	Gravel/Typha angustifolia	44	Biotic processes (plants and microorganisms)	[191]
				HSSF (mesocosms)	Gravel/Phragmites australis	32–70	Biodegradation	[185]
				VSSF (full)	Sand/Phragmites australis and Lemna	35 ± 11, 20 ± 19 (winter)	Biodegradation	[182]
				SF (full)	Iris pseudacorus, Scirpus sp. and Carex sp. and Lemna	71 ± 5 (summer)	Photodegradation, biodegradation	[192]
				Down flow CW (microcosms)	Sand: gravel: soil/Phalaris arundinacea L. var. picta L.	51 ± 6	Biodegradation	[181]
				SF (mesocosms)	Cyperus alternifolius	69.3	Plant uptake and other biotic and abiotic processes	
Ibuprofen	21	3.97	4.9	HSSF (pilot)	Gravel/Phragmites australis	48 (deep) 81 (shallow)	Biodegradation, sorption	[193]
				HSSF (pilot)	Gravel/Phragmites australis	71	Biodegradation, sorption	[84]
				VSSF (pilot)	Gravel/Phragmites australis	99	Biodegradation, sorption	[186]
				Lagoons + SF (full)	Hydrocotle spp. + Phragmites australis	>99	n. d.	[184]
				SF (full)	Typha spp. + Phragmites australis	95 (Winter) 96 (Summer)	Biodegradation, sorption	[173]
				HSSF (microcosms)	Gravel/Phragmites australis	52	Biodegradation, sorption	[194]
				HSSF (full)	Phragmites australis	65 ± 23	Biodegradation, sorption	[187]
				VSSF (full)	Phragmites australis	89 ± 8	Biodegradation, sorption	

				System	Substrate/Plant	Removal (%)	Removal process	References
				HSSF (mesocosms)	Gravel/Phragmites australis	98–99	Biodegradation	[188]
				VSSF (microcosms)	Expanded clay/Typha spp.	82 (winter) 96 (summer)	Biodegradation, plant uptake, sorption	[70]
				HSSF (mesocosms)	Gravel/Phragmites australis or Typha angustifolia	38–75	Biotic processes (plants and microorganisms)	[66]
				SF (mesocosms)	Phragmites australis or Typha sp.	54–98	Biotic processes (plants and microorganisms)	[170]
				HSSF (mesocosms)	Gravel/Typha angustifolia	78	Biotic processes (plants and microorganisms)	[180]
				HSSF (mesocosms)	Gravel/Phragmites australis or Typha angustifolia	n.r.–95	Biodegradation	
				SF (mesocosms)	Phragmites australis or Typha angustifolia	n.r.–98	Biodegradation	
				HSSF (mesocosms)	Gravel/Phragmites australis	52–85	Biodegradation	[185]
Ketoprofen	51	3.12	4.45	HSSF (pilot)	Gravel/Phragmites australis	38	Sorption	[84]
				SF (full)	Typha spp. + Phragmites australis	97 (winter) 99 (summer)	Photodegradation	[173]
				HSSF (full)	Phragmites australis	90	Sorption	[187]
				VSSF (full)	Phragmites australis	n.r.	Sorption	
				HSSF (mesocosms)	Gravel/Phragmites australis or Typha angustifolia	22–55	Biotic processes (plants and microorganisms)	[66]
				SF (mesocosms)	Phragmites australis or Typha sp.	36–50	Biotic processes (plants and microorganisms)	[66]
				HSSF (mesocosms)	Gravel/Phragmites australis or Typha angustifolia	n.r.–83	Photodegradation	[180]
				SF (mesocosms)	Phragmites australis or Typha angustifolia	n.r.–95	Photodegradation	
				HSSF (mesocosms)	Gravel/Typha angustifolia	95	Biodegradation, photodegradation	[190]
				HSSF (mesocosms)	Gravel/Typha angustifolia	96	Biotic processes (plants, microorganisms)	[191]

(Continued)

Table 17.1 (Continued)

Organic compound	Physico-chemical properties[a]			Type of CWS	Type of substrate/plant	% Removed	Removal processes suggested by authors	Reference
	S_W (25°C) (mg/L)	log K_{ow} (25°C)	pK_a					
Naproxen	15.9	3.18	4.15	HSSF (pilot)	Gravel/Phragmites australis	85	Biodegradation, sorption	[84]
				VSSF (pilot)	Gravel/Phragmites australis	89	Biodegradation, sorption	[186]
				Lagoons + SF (full)	Hydrocotle spp. + Phragmites australis	99	n.d.	[184]
				SF (full)	Typha spp. + Phragmites australis	52 (winter) 92 (summer)	Biodegradation, sorption	[173]
				HSSF (full)	Phragmites australis	45 ± 52	Biodegradation, sorption	[187]
				VSSF (full)	Phragmites australis	92 ± 2	Biodegradation, sorption	
				HSSF (pilot)	Phragmites australis	99	Biodegradation	[188]
				HSSF (mesocosms)	Gravel/Phragmites australis or Typha angustifolia	43–90	Biotic processes (plants and microorganisms)	[66]
				SF (mesocosms)	Phragmites australis or Typha sp.	30–70	Biotic processes (plants and microorganisms)	
				HSSF (mesocosms)	Gravel/Typha angustifolia	91	Biotic processes (plants and microorganisms)	[170]
				SF (microcosms)	Salvinia molesta, Lemna minor, Elodea canadensis, Ceratophyllum demersum	40–53	Photodegradation	[189]
				HSSF (mesocosms)	Gravel/Typha angustifolia	91	Biodegradation	[190]
				HSSF (mesocosms)	Gravel/Typha angustifolia	93	Biotic processes (plants, microorganisms)	[191]
				HSSF (mesocosms)	Gravel/Phragmites australis or Typha angustifolia	n.r.–82	Photodegradation	[180]
				SF (mesocosms)	Phragmites australis or Typha angustifolia	22–92	Photodegradation	

Salicylic acid	2,240	2.26	2.97	HSSF (pilot)	Gravel/Phragmites australis	96	Biodegradation, sorption	[84]
				VSSF (pilot)	Gravel/Phragmites australis	98	Biodegradation, sorption	[186]
				HSSF (full)	Phragmites australis	95 ± 4	Biodegradation, sorption	[187]
				VSSF (full)	Phragmites australis	87 ± 8	Biodegradation, sorption	
				HSSF (mesocosms)	Gravel/Phragmites australis or Typha angustifolia	62–90	Biotic processes (plants and microorganisms)	[66]
				SF (mesocosms)	Phragmites australis or Typha sp.	35–90	Biotic processes (plants and microorganisms)	
				HSSF (mesocosms)	Gravel/Phragmites australis or Typha angustifolia	60–99	Biotic processes (microorganisms and plants)	[180]
				SF (mesocosms)	Phragmites australis or Typha sp.	55–99	Biotic processes (plants and microorganisms)	
Antibiotics								
Amoxicillin	3,433	0.87	n.a.	HSSF (mesocosms)	Phragmites australis or Typha angustifolia	21–45	Biotic processes	[172]
Ciprofloxacin	30,000	0.28	6.09	VSSF (mesocosms)	Red soil : volcanic rock or zeolite: gravel/Hybrid pennisetum	82–85	Adsorption	[195]
Clarithromycin	0.342	3.16	8.99	HSSF (mesocosms)	Phragmites australis or Typha angustifolia	31–39	Photodegradation, sorption, biodegradation	[172]
				VSSF (full)	Sand/Phragmites australis and Lemna	17 ± 28 (summer) 89 ± 3 (winter)	Biodegradation	[182]
				SF (full)	Iris pseudacorus, Scirpus sp. and Carex sp. and Lemna	70 ± 10 (summer)	Photodegradation, biodegradation	
Doxycycline	n.a.	n.a.	n.a.	HSSF (mesocosms)	Gravel/Phragmites australis or Typha angustifolia	71–75	Adsorption/retention, biotic processes (plants and microorganisms)	[172]

(Continued)

Table 17.1 (Continued)

Organic compound	Physico-chemical properties[a] S_W (25°C) (mg/L)	$\log K_{ow}$ (25°C)	pK_a	Type of CWS	Type of substrate/plant	% Removed	Removal processes suggested by authors	Reference
Enrofloxacin	3,397	0.7	n.a.	VSSF (microcosms)	Lava rock: gravel/Phragmites australis	98	Adsorption, biodegradation	[79]
				VSSF (microcosms)	Lava rock : gravel/Phragmites australis	98	Biodegradation	[78]
Erythromycin	1,437	3.06	8.88	HSSF (mesocosms)	Gravel/Phragmites australis or Typha angustifolia	n.r.–64	Biotic processes	[172]
				VSSF (full)	Sand/Phragmites australis and Lemna	>92 (summer) 86 ± 3 (winter)	Biodegradation	[182]
				SF (full)	Iris pseudacorus, Scirpus sp. and Carex sp. and Lemna	80 ± 7	Photodegradation, biodegradation	
Ofloxacin	28,267	−0.39	n.a.	SSF (microcosms)	Ceramsite and gravel/Cyperus alternifolius	90.8 ± 1	Plant uptake, adsorption, biodegradation	[196]
Oxytetracycline	313	−0.9	3.27	VSSF (microcosms)	Gravel, Expanded clay/Phragmites australis	>97	Adsorption, biotic processes (plants and microorganisms)	[71]
				VSSF (mesocosms)	Red soil:volcanic rock or zeolite: gravel/Hybrid pennisetum	91–95	Adsorption	[195]
Sulfamethazine	1,500	0.19	7.65	VSSF (mesocosms)	Red soil:volcanic rock or zeolite: Gravel/Hybrid pennisetum	68–73	Adsorption	[195]
Sulfamethoxazole	610	0.89	n.a.	Lagoons + SF (full)	Hydrocotyle spp. + Phragmites australis	91	n.d.	[184]
				FWS (pilot)	Acorus + Typha spp.	30	n.d.	[197]
				HSSF (mesocosms)	Gravel/Phragmites australis or Typha angustifolia	80–87	Biodegradation	[172]
				VSSF (pilot)	Gravel/Thalia dealbata	24 ± 31 (summer) 34 ± 9 (winter)	Biodegradation	[77]

Compound				System	Substrate/Plant	Removal (%)	Mechanism	Reference
				VSSF (pilot)	Zeolite/Arundo donax var.	15 ± 18 (winter)	Biodegradation	[182]
				Up-ward VSSF (pilot)	Vesuvianite/Thalia dealbata	58± 18 (summer) 19± 46 (winter)	Biodegradation	
				VSSF (full)	Sand/Phragmites australis and Lemna	63 ± 8 (summer) − 74 ± 43 (winter)	Biodegradation	[192]
				SF (full)	Iris pseudacorus, Scirpus sp. and Carex sp. and Lemna	49 ± 11 (summer)	Photodegradation, biodegradation	
				Down flow CW (microcosms)	Sand: gravel: soil/Phalaris arundinacea L. var. picta L.	24 ± 15	Biodegradation	[196]
				SSF (microcosms)	Ceramsite and gravel/Cyperus alternifolius	77 ± 4	Plant uptake, adsorption, biodegradation	
Tetracycline	231.1	−1.3	3.3	VSSF (microcosms)	Lava rock : gravel/Phragmites australis	94	Adsorption, biodegradation	[79]
				VSSF (microcosms)	Lava rock : gravel/Phragmites australis	94	Biodegradation	[78]
Trimethoprim	400	0.91	7.12	HSSF (mesocosms)	Gravel/Phragmites australis or Typha angustifolia	92-99	Biodegradation	[172]
				SF(mesocosms)	Phragmites australis or Typha angustifolia	65-88	Biodegradation	
				VSSF (pilot)	Gravel/Thalia dealbata	31 ± 37 (summer)	Biodegradation	[77]
				VSSF (pilot)	Zeolite/Arundo donax var.	38 ± 10 (summer)	Biodegradation	
				Up-ward VSSF (pilot)	Vesuvianite/Thalia dealbata	87 ± 3 (summer) 78 ± 9 (winter)	Biodegradation	

(Continued)

Table 17.1 (Continued)

Organic compound	Physico-chemical properties[a]			Type of CWS	Type of substrate/plant	% Removed	Removal processes suggested by authors	Reference
	S_W (25°C) (mg/L)	log K_{ow} (25°C)	pK_a					
				VSSF (full)	Sand/Phragmites australis and Lemna	>95 (summer) 89 ± 3 (winter)	Biodegradation	[182]
				SF (full)	Iris pseudacorus, Scirpus sp. and Carex sp. and Lemna	>95 (summer)	Photodegradation, biodegradation	[182]
Beta-blockers								
Atenolol	13,300	0.16	9.6	Lagoons + SF (full)	Hydrocottle spp. + Phragmites australis	>99	n.d.	[184]
				SF (pilot)	Acorus + Typha spp.	97	n.d	[197]
				VSSF (microcosms)	Expanded clay/Typha spp. or Phragmites australis	>92	Sorption, plant uptake	[171]
				VSSF (full)	Sand/Phragmites australis and Lemna	>99 (summer) 88 ± 5 (winter)	Biodegradation	[182]
				SF (full)	Iris pseudacorus, Scirpus sp. and Carex sp. and Lemna	96 ± 2 (summer)	Photodegradation, biodegradation	[184]
Metoprolol	16,900	1.88	9.6	Lagoons + SF (full)	Hydrocottle spp. + Phragmites australis	92	n.d.	[184]
				VSSF (full)	Sand/Phragmites australis and Lemna,	72 ± 5 (summer) 80 ± 8 (winter)	Biodegradation	[182]
				SF (full)	Iris pseudacorus, Scirpus sp. and Carex sp. and Lemna	75 ± 4 (summer)	Photodegradation, biodegradation	[184]
Sotalol	5,510	0.24	n.a.	Lagoons + SF (full)	Hydrocottle spp. + Phragmites australis	30	n.d.	[184]

						Biodegradation	
Diuretics								
Furosamide	73.1	2.03	3.8	HSSF (full)	Phragmites australis	71	Biodegradation	[187]
Lipid regulators								
Clofibric acid	583	2.57	3.18	HSSF (pilot)	Gravel/Phragmites australis	n.r.	n.d	[193]
				SF (full)	Typha spp. + Phragmites australis	32 (winter) 36 (summer)	n.d	[173]
				SF (microcosms)	Salvinia molesta, Lemna minor, Ceratophyllum demersum, Elodea canadensis	16–23	n.d.	[189]
				VSSF (microcosms)	Expanded clay/Typha spp	48 (winter) 75 (summer)	Sorption, plant uptake	[70]
				HSSF (mesocosms)	Gravel/Typha angustifolia	34	Sorption, plant uptake	[190]
				HSSF (mesocosms)	Gravel/Typha angustifolia	39	n.d.	[191]
Gemfibrozil	10.9	4.77	n.a.	Lagoons + SF (full)	Hydrocotle spp. + Phragmites australis	64	n.d.	[184]
Psychiatric drugs								
Carbamazepine	17.7	2.45	14	HSSF (pilot)	Gravel/Phragmites australis	26 (deep) 16 (shallow)	Sorption	[193]
				HSSF (pilot)	Gravel/Phragmites australis	16	Sorption	[84]
				VSSF (pilot)	Gravel/Phragmites australis	26	Sorption	[186]
				HSSF (microcosms)	Gravel/Phragmites australis	5	Sorption	[194]
				SF (full)	Typha spp. + Phragmites australis	30–47	Sorption	[173]
				HSSF (full)	Phragmites australis	38	Sorption	[187]
				SF (pilot)	Acorus + Typha spp.	65	Plant uptake	[197]
				VSSF (microcosms)	Expanded clay/Typha spp.	88 (winter) 97 (summer)	Sorption, plant uptake	[70]
				HSSF (mesocosms)	Gravel/Phragmites australis or Typha angustifolia	22–50	Sorption	[66]
				SF (mesocosms)	Phragmites australis or Typha sp.	10–45	Sorption	[170]
				HSSF (mesocosms)	Gravel/Typha angustifolia	27	Sorption	[170]
				HSSF (mesocosms)	Gravel/Phragmites australis or Typha angustifolia	n.r.–70	Plant uptake	[180]

(Continued)

Table 17.1 (Continued)

Organic compound	Physico-chemical properties[a]			Type of CWS	Type of substrate/plant	% Removed	Removal processes suggested by authors	Reference
	S_W (25°C) (mg/L)	log K_{ow} (25°C)	pK_a					
				SF (mesocosms)	Phragmites australis/Typha angustifolia	n.r.–95	Plant uptake	
				HSSF (mesocosms)	Gravel/Typha angustifolia	27	Adsorption	[190]
				HSSF (mesocosms)	Gravel/Typha angustifolia	28	Biotic processes (plants and microorganisms)	[191]
				SSF (microcosms)	Ceramsite and gravel/Cyperus alternifolius	64 ± 2	Plant uptake, adsorption, biodegradation	[196]
Oxazepam	20	2.24	n.a.	VSSF (full)	Sand/Phragmites australis and Lemna	34 ± 11 24 ± 25	Biodegradation	[182]
				SF (full)	Iris pseudacorus, Scirpus sp. and Carex sp. and Lemna	37 ± 8	Photodegradation, biodegradation	
Stimulants/Psychoactive drug								
Caffeine	21,600	–0.07	10.4	HSSF (pilot)	Gravel/Phragmites australis	97	Biodegradation, sorption	[84]
				VSSF (pilot)	Gravel/Phragmites australis	99	Biodegradation, sorption	[186]
				Lagoons + SF (full)	Hydrocotyle spp. + Phragmites australis	>99	n.d.	[184]
				HSSF (full)	Phragmites australis	97	Biodegradation, sorption	[187]
				VSSF (full)	Phragmites australis	99	Biodegradation, sorption.	
				HSSF (mesocosms)	Gravel/Phragmites australis or Typha angustifolia	57–99	Biodegradation	[66]
				SF (mesocosms)	Phragmites australis or Typha sp.	20–99		

System type	Substrate / plant species	Removal (%)	Process	Ref.
SF (microcosms)	Salvinia molesta, Lemna minor, Ceratophyllum demersum, Elodea canadensis	81–99	Biodegradation, plant uptake	[189]
HSSF (mesocosms)	Gravel/Phragmites australis or Typha angustifolia	45–99	Biotic processes (microorganisms and plants)	[180]
SF (mesocosms)	Phragmites australis or Typha sp.	20–99	Biotic processes (plants and microorganisms)	[190]
HSSF (mesocosms)	Gravel/Typha angustifolia	93	Biodegradation	[191]
HSSF (mesocosms)	Gravel/Typha angustifolia	90	Biotic processes (plants and microorganisms)	
HSSF (mesocosms)	River gravel: fine volcanic gravel or coarse volcanic gravel/ Typha latifolia, Phragmites australis or Cyperus papyrus	89 ± 4	Biotic processes (microorganisms, plants), adsorption	[198]

a Source: PHYSPROP [199].

Notes: S_w =Water solubility; K_{ow} =Octanol-water partition coefficient; CWS: Constructed wetlands system; HSSF: Horizontal subsurface flow; VSSF: Vertical subsurface flow; SF: Surface flow; n.a.: not available; n.d.: not detailed; n.r.: not removed.

still relevant use of *Lemna* spp. can be noted and, then, only a minor use of a variety of other species is seen is these studies.

The most frequently studied types of CWS are of the sub-surface flow (SSF) kind, although the number of surface flow (SF) systems is also relevant. Within SSF systems, the horizontal type (HSSF) is significantly more common than the vertical one (VSSF). Most studies are conducted at a pilot scale/mesocosm level, but an also significant number of assessments have been performed on full-scale systems, many of which are systems in operation. Studies on lab-scale microcosm systems are relatively less frequent, probably because they are regarded as being too small to adequately represent and model the large complexity of a CWS.

With regard to the types of materials used for the support matrices of SSF systems, the vast majority of the studied SSF-CWS employed gravel as bed substrate. Soil, sand or rock (especially volcanic rock) are also sometimes used, but only in very few studies have other materials (such as zeolites or expanded clay) been tested.

Most studies attempt to provide an explanation for the behavior of the selected pharmaceuticals in the systems under study. Photodegradation, sorption/adsorption, biodegradation and plant uptake are the major processes which are advanced as being responsible for pharmaceuticals removal in CWS. Among these, biodegradation is the most common suggestion for the major pharmaceutical removal mechanism. Together with the second most frequently proposed process, sorption/adsorption, pharmaceuticals removal in CWS is generally regarded as being due to either one of the two processes (or to both, in concurrency). However, part of these proposals consists merely of hypotheses, as this topic is not always investigated thoroughly and not enough data is presented to support those claims.

In particular, biodegradation is sometimes advanced as an obvious "default" explanation for pharmaceuticals removal, seen as inherent to the functioning of biological systems and, therefore, whose effectiveness does not require experimental confirmation. On the other hand, uptake of pharmaceuticals by plants is suggested as a relevant removal process in a minority of cases, either presented just as a possibility among biotic processes or actually confirmed by the detection of the pharmaceuticals within plant tissues. Although still scarce, studies of plant uptake of pharmaceuticals (and, more broadly, of other organic micropollutants) is an evolution in the perceived role of plants in the removal of these pollutants, as plant uptake was traditionally seen as relevant only in the case of heavy metals and was neglected in the case of organic xenobiotic compounds. Studies to detect these molecules in plant tissues and characterize their metabolization by plants' enzymatic machinery or plants' endophytic bacteria, however, are gaining importance.

In regard to abiotic removal processes, photodegradation appears as the second most commonly reported mechanism, after sorption/adsorption phenomena.

The relative importance of a particular process can vary significantly, depending on the characteristics of the pharmaceutical substance, the CWS type (e.g., SF, HSSF or VSSF), operational parameters (e.g., retention time), the environmental conditions, the type of vegetation within the system, as well as the type of support matrix materials. In fact, in a complex treatment such as CWS, several removal processes of pharmaceuticals may occur simultaneously, cooperatively or concurrently. Ultimately, CWS optimization for the removal of specific targets such as pharmaceuticals requires a basic knowledge of the processes involved in the removal of the pollutants and the relationships between those processes and the characteristics of the CWS components.

Considering the significantly more frequent use of the SSF type of CWS, as Table 17.1 indicates, and the fact that SSF contain the same components as a SF-CWS and additionally the support matrix component that a SF-CWS does not include, in the following section, the components and processes occurring at a SSF-CWS will be described, which is a more comprehensive description that is still relevant for the case of a SF-CWS.

17.3.2 Main Removal Processes for Pharmaceuticals in SSF-CWS

The removal of pharmaceutical pollutants in CWS is attained through essentially the same variety of processes that are effective for other kinds of organic pollutants. Organic xenobiotics removal by SSF-CWS involves several inter-dependent processes which, to some extent, may have an enhanced overall effect in pollutant removal as their concomitant action may lead to cooperative or synergistic effects.

Under one of the possible classification criteria, pollutant removal processes may be distinguished as biotic (carried out by living organisms such as plants and microorganisms) or abiotic (physical or chemical). Many of these processes are fundamentally the same occurring in natural wetlands and are also identical to those responsible for the environmental fate of xenobiotics that contaminate the water bodies [200]. The way in which CWS differ from a natural system such as a natural wetland is that the effects of some of these processes may be enhanced (at least, for a specific set of contaminants) through an optimization of the most influential conditions within the controlled environment of a CWS.

The primary abiotic and biotic processes that can participate in removing organic xenobiotics such as pharmaceuticals from contaminated water in a SSF-CWS are described in Table 17.2.

17.3.2.1 Abiotic Processes

The abiotic removal of contaminants in SSF-CWS involves a range of physical and chemical processes. However, sorption is usually the most relevant of the abiotic processes occurring in SSF-CWS, at the surface of plants roots and of the solid media or at biofilms that developed at those surfaces [57, 59, 61, 65, 68, 85, 175, 200, 201]. Sorption results in a short-term retention of the contaminants, but in the long term the sorbed compounds should be slowly desorbed and released for a continuous feed of pollutants to the other much slower processes such as plant uptake or biodegradation. If on the contrary, contaminants are irreversibly sorbed, sorbents are expected to eventually become saturated and sorption should progressively be less effective over time. Other effects such as the development of biofilms on the sorbent media additionally contribute to the alteration of sorption efficiency with time. Overall, sorption processes cannot be seen as a single solution for the removal of pollutants, but in cooperation with other processes in a SSF-CWS the former may certainly contribute to the enhancement of the efficiency of the latter.

The type of materials composing the support matrix and their chemical characteristics will obviously have a strong influence over the occurrence of sorption phenomena [65, 75, 200, 202], but the extent of the retention by these processes is also a function of several wastewater characteristics (e.g., dissolved organic matter content, pH and electrolyte composition), as well as of the physical and chemical properties of the pollutants themselves (e.g., pK_a, solubility, log K_{ow}, etc.). In addition, a uniform contact of the wastewater with the SSF-CWS media is crucial for the efficiency of the system, which depends on a good hydraulic conductivity of the support matrix that avoids the occurrence of overland flows and preferential channeling [57, 175].

Other common abiotic processes such as hydrolysis, photodegradation, redox reactions and volatilization (Table 17.2) may also contribute, in varied extents, to the removal of some particular

Table 17.2 Abiotic and biotic processes involved in organic xenobiotics removal in SSF-CWS [162, 200, 201].

Processes	Description
Abiotic	
Sorption	The physical-chemical processes occurring at the surface of plants roots and of the supporting solid matrix which result in a short-term retention or long-term immobilization of the xenobiotics.
	The term refers broadly and without clear distinction to outer surface effects (adsorption) as well as diffusion into the inner surface of porous materials and incorporation in the sorbent (absorption).
Hydrolysis	The chemical breakdown of organics by the action of water, a process which usually is pH-dependent.
Photodegradation	Decomposition of the pollutant molecules by the action of sunlight.
Redox reactions	Modification, which sometimes may be quite substantial, of the pollutant due to the action of oxidizing or reducing agents. Redox reactions are also frequently brought about by biotic agents (e.g. bacteria), or enzymatically catalyzed.
Precipitation	For those compounds which can exist in several forms having distinct water solubilities, this corresponds to the conversion into the most insoluble forms. This removal process actually is just a transference between environmental compartments (i.e. liquid to solid).
Filtration	Removal of particulate matter and suspended solids. It may lead to the removal of some dissolved organic pollutants as they are adsorbed to the filtered particles.
Volatilization	Release of some organic xenobiotics (or smaller molecules resulting from their prior decomposition through photodegradation, etc.) as vapors, which occurs when these compounds have significantly high vapor pressures.
Biotic	
Aerobic/anaerobic biodegradation	Decomposition of organic substances by the metabolic processes of microorganisms, which usually play a significant role in the removal of organic compounds in CWS.
Rhizodegradation	Enhancement of biodegradation of some organic xenobiotics through the stimulation of microorganisms provided by substances (such as sugars, organic acids, enzymes, etc.) released in roots exudates.
Phytodegradation/ Phytotransformation	Direct uptake of organic compounds by plants and ensuing decomposition inside both root and shoot tissues. In addition to plant metabolism, endophytic microorganisms may be involved in the degradation of some compounds. Degradation products may be further subjected to the sequestration/lignification or undergo complete decomposition (mineralization). This process is usually much faster than biodegradation or rhizodegradation.
Phytovolatilization	Uptake and transpiration of volatile organics through the aerial plant parts.
Phytostabilization	Set of processes contributing to reduce the mobility of the organic pollutants and to attain their long-term immobilization in the rhizosphere.

classes of compounds, depending on their specific properties [57, 59, 61, 68, 85]. However, these are not, in general, major removal processes for most organic compounds such as pharmaceuticals, because either the conditions in SSF-CWS or the properties of the compounds themselves are not suitable. For example, in SSF-CWS the water level is typically below the surface of the solid matrix and, furthermore, most of its surface in under a moderate shade produced by the plants leaves. Therefore, in these conditions, exposure of the pollutants to sunlight is very limited and photodegradation does not occur in appreciable extent. The process of volatilization is also of modest importance for substances with low volatility, as is the case of most pharmaceuticals. Where SSF-CWS are used as a tertiary treatment stage after conventional secondary treatment in a WWTP, organic compounds do not suffer, in most cases, appreciable hydrolysis either, as they have already been subjected to such type of processes during secondary treatment stage. Therefore, the organic xenobiotics still present in the influents to the SSF-CWS stage are those that have resisted exposure to such hydrolysis processes or they are the transformation products of those substances that did not resist it.

17.3.2.2 Biotic Processes

SSF-CWS are biological systems and, as such, biological processes play in them a major role in the removal of pollutants. The two biotic components responsible in SSF-CWS for the biological contribution to the removal of organic pollutants are the wetland vegetation and the microbial populations. Biotic removal of organic pollutants by SSF-CWS can take place through plant uptake and ensuing metabolization, sequestration/lignification or mineralization/volatilization (phytodegradation and phytovolatilization); through the action of plant exudates or enzymes, either directly by catalyzing the pollutants' chemical degradation or indirectly by stimulating the action of rhizosphere microorganisms (rhizodegradation); or by biodegradation through the normal activity of microorganisms [61, 68, 162, 163, 165, 203–205]. In this complex phytotechnology, biotic processes can all occur simultaneously and contribute in varying degrees to the overall effect of the biotic components on the removal of pollutants. Their efficiency depends on the compounds' properties and their interaction with the plant system [162, 165, 204, 205].

Direct uptake of pollutants by plants is more widely recognized as a prominent removal process in the case of inorganic substances (heavy metals). However, in the case of various organic pollutants, recent studies have shown that plant uptake may also play a significant role among biotic processes [61, 65, 68, 75, 110, 112, 162, 164, 206, 206–209]. Uptake and translocation of organic contaminants by the whole plant system are dependent on the contaminant concentration, the physical and chemical properties, and the plant type.

Many pharmaceuticals are synthetic substances that are xenobiotic to the plants. As a consequence, there are generally no transporters for such molecules in plant cell membranes and, therefore, their movement into (uptake) and within (translocation) plant tissues is a passive process, i.e., it is driven by simple diffusion. Because this movement has a physical rather than a biological nature, it is fairly predictable across plant species [162] and it is related to the flow rate of the evapotranspiration stream.

In order to be taken up by a plant, an organic compound must have favorable chemical characteristics (e.g., suitable solubilities, polarity and partitioning coefficients) to be mobile across cell membranes [210]. It is generally considered that contaminants with log K_{ow} (the logarithm of the octanol-water partition coefficient, frequently used as a measure of a compound's hydrophobicity) between around 0.5 and 3.5, i.e., with a moderate hydrophobicity, are able to move inside the plant. Conversely, compounds with log K_{ow} out of this range are either too hydrophilic or too hydrophobic to be able to easily move in and out between aqueous and lipidic media, therefore being restricted to one type of environment with a reduced mobility [162, 164, 210]. As an example, compounds such as carbamazepine (log K_{ow} = 2.45), are more likely to be taken up by plants [207], as they have a moderate hydrophobicity that allows them to move across the roots' cell walls, but are also sufficiently water soluble to travel through cell fluids.

Some studies, however, indicate that the log K_{ow} value of an organic pollutant may not be the sole factor determining its tendency to be taken up and some compounds have been shown to be able to penetrate plant membranes despite a low log K_{ow} [211]. The capacity of a compound to be removed from water by a given plant, may also depend on other factors such as initial pollutant concentration, the anatomy and the root system of the plant [212]. In addition, very hydrophobic chemicals (log K_{ow} > 3.5) are also candidates for phytostabilization and/or rhizosphere bioremediation by virtue of their long residence times in the root zone [162, 164, 213].

Non-charged compounds (e.g., caffeine) are more easily taken up by aquatic plants, whereas some negatively charged compounds such as diclofenac and naproxen are not. In the latter case, the effect of pK_a and pH of ionizable compounds seems to be more relevant than lipophilicity [214], as electrostatic repulsions between the negative charges of the compounds and of the plants' biomembranes are an obstacle to the mobility of the compounds across the membranes (189, 215).

Inside the plants, organic xenobiotics such as pharmaceuticals are subjected to a set of metabolic transformations aiming their detoxification to plants. These processes should ultimately (and hopefully) lead to their transformation into less toxic compounds, either by a more or less extensive decomposition into smaller (possibly volatile) molecules or by conjugation/stabilization and subsequent sequestration in plant cell vacuoles or lignification and incorporation into cell walls. The metabolic processes operating in the phytotransformation of organic xenobiotics in plants will be briefly presented in section 17.3.3.1, but are also presented in detail in some reviews on phytoremediation [162, 164].

Notwithstanding the significant role that is currently recognized to be played by pollutants uptake by plants in the overall removal of organic pollutants by SSF-CWS, the action of microorganisms is still generally regarded as the major route for the elimination of organic xenobiotics in CWS [57, 75, 85, 86, 143]. Even so, in microbial processes plants still hold a relevant contribution as the efficiency of biodegradation may be further enhanced by the synergistic interactions between plants (in particular, the exudates release by their roots) and the microorganisms [162, 163, 165, 216].

Microorganisms may attain a measurable amount of organic pollutants uptake and storage, but it is the machinery of their metabolic processes that can play the most significant role in the decomposition of organic compounds, transforming the most complex molecules into simpler ones [200]. Biodegradation generally provides an important biological mechanism for removal of a wide variety of organic compounds. However, in the case of pharmaceuticals (a loose chemical family of compounds comprising widely diverse chemical structures and properties) the efficiency and rate of degradation of this class of pollutants by microorganisms is highly variable, ranging from extremely efficient as in the case of ibuprofen to nearly negligible as in the case of carbamazepine.

Microbial transformations can involve more than one type of mechanism and under different conditions several degradation products can be derived from the same initial compound depending on the environmental conditions. Additionally, transformations can be mediated by one organism or through combined effects of several organisms [86, 200].

One of the major distinctions is whether biodegradation occurs under aerobic or anaerobic conditions. Depending on the oxygen input by macrophytes in the rhizosphere and the availability of other electron acceptors, pharmaceuticals can be metabolized in various ways. For example, biodegradation of most organic xenobiotics is enhanced under aerobic conditions, and highly reduced conditions are favorable for the degradation of some chlorinated organics [200]. The efficiency and rate of organic xenobiotic degradation and the extent of microbial growth during degradation is also heavily influenced by the chemical structure of the xenobiotic [217]. Structurally simple pharmaceuticals with high water solubility and low adsorptivity are usually more similar to the naturally occurring substances which are usually used as energy sources by the microorganisms and are easily degraded. In contrast, pharmaceuticals with chemical structures very different from the naturally occurring compounds are often degraded slowly since microorganisms do not possess suitable degrading genes. In these cases degradation by non-specific enzymes may still occur but at a slower rate by non-specific reactions, which do not support microbial growth (co-metabolism) [218].

Xenobiotic degradation by the microorganisms is also strongly influenced by the support medium where the microbial populations grow. Temperature, pH, oxygen levels, presence of toxic substances and nutrients availability within the CWS are expected to play a very important role in the removal efficiency [86, 87, 200, 219, 220].

While the location of microorganisms' action is frequently attributed exclusively to the region of the rhizosphere, the role of endophytic microorganisms inside plants to biodegrade the pollutants that plants have taken up may also be relevant. This is another contribution of biodegradation that may be concurrent, complementary or even synergistic to plant's metabolic processes, but have until now been scarcely studied [221–223].

In summary, subjected to the effects of the processes enumerated above, pharmaceuticals may be removed from water through storage in the wetland solid matrix and in the biomass (plants and microorganisms), or through losses to the atmosphere (Figure 17.2).

The degree to which any process will contribute to the overall removal of pharmaceuticals from a wastewater in a CWS will depend on many of the same factors that affect processes responsible for the fate of pharmaceuticals in the environment. Such factors are related with the physical and chemical properties of the pharmaceuticals (molecular structure, polarity, ionization constant,

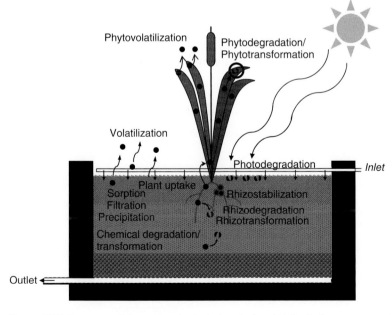

Figure 17.2 Summary of the major physical, chemical and biological processes controlling pollutant removal in a SSF-CWS (adapted from Dordio *et al.* [224]).

water solubility, hydrophobicity, sorption affinity, volatility, persistence) and with the environmental conditions (temperature, rainfall, sunlight). A few additional factors are particular to SSF-CWS, namely the wastewater characteristics (pH, organic matter content, redox conditions), the SSF-CWS design (type of solid matrix and biotia, layout, type of flow, rate of flow) and operation parameters (hydraulic retention time, dissolved oxygen levels, etc.).

A basic knowledge of how these pharmaceutical removal processes inter-operate in wetlands is extremely helpful for assessing the potential applications, benefits and limitations of SSF-CWS. Furthermore, the design of the system, comprising the selection of its components (plant species, materials for support matrix, and possibly inoculation of selected microorganisms strains), is better guided by a good understanding of the several roles played by each SSF-CWS component and also of their inter-dependences.

17.3.3   The Role of SSF-CWS Components in Pharmaceuticals Removal

A SSF-CWS is basically composed of three main components: the support matrix (i.e., the substrate in the planted bed, which supports the plants roots), the vegetation and the microbial populations. It is the concerted action of these inter-dependent components (through the variety of chemical, physical and biological processes that have been enumerated and described before) that is responsible for the depuration of wastewaters in these wastewater treatment systems (Figure 17.3).

A detailed characterization of the roles played by each SSF-CWS component in the overall pharmaceuticals removal efficiency, related with the properties of each substance involved, is necessary to guide the selection of each component for an optimal design of the system.

17.3.3.1   The Role of Biotic Components (Plants and Microorganisms) in Pharmaceuticals Removal
Plants play several important roles in a SSF-CWS with both direct and indirect influence in the system's efficiency and, therefore, a careful selection of the more suitable species is an important design decision. Major role of plants in CWS include the filtration of large debris; provision of surface area for microorganisms development and release of exudates by roots (normally including

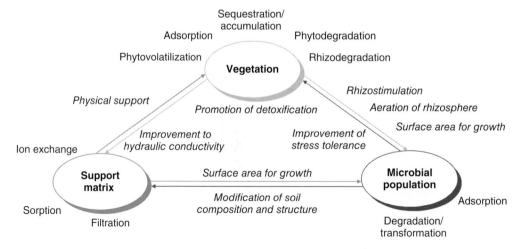

Figure 17.3 Overview of the pharmaceutical removal processes occurring within the domain of each SSF-CWS component and of the interactions and inter-dependences established between the components.

organic acids, sugars, amino acids, vitamins and enzymes [163, 166]) that stimulate their growth; enhancement of the hydraulic conductivity of the support matrix (roots and rhizomes growth help to prevent clogging in the support matrix); transport and release of oxygen through the roots (which increases aerobic degradation and nitrification); and contribution to the immobilization or elimination of pollutants (by adsorption, phytostabilization, and/or uptake of pollutants followed by their phytodegradation) [57, 58, 85, 88, 136, 204].

One of the primary criteria in the selection of plant species is their tolerance to the pollutants' toxicity and the extent of their role in lowering the pollutants concentrations in wastewater. Additionally, other desirable characteristics of plants are [57, 85, 174, 225]: ecological adaptability (the chosen plants should pose no risk to the natural ecosystem surroundings); tolerance to local conditions in terms of climate, pests and diseases; tolerance to the hypertrophic water-logged conditions of wetlands; ready propagation, rapid establishment and growth, good root system development and perennial duration rather than annual. Furthermore, other aspects such as economic, aesthetic, recreational, etc. may also have to be considered in the selection of the plant species [85, 225].

In addition to the selection of species, other aspects related to the CWS design, such as plant density and layout of the specimens (e.g., the way specimens of different species may be intermixed when planted in the beds), may be important and probably should be carefully planned as their influence may range from subtle differences in the system behavior to more pronounced impacts in the global efficiency [57, 90]. In particular, the cycles of vegetative activity of some species in addition to variations of climate conditions may lead to significant seasonal changes in the system efficiency, which in some cases may be mitigated by using polycultures of vegetation [57].

The plants growing in natural wetlands (often called wetland plants or macrophytes), are also typically the plant species used in CWS as these are well adapted to the water saturated conditions found in these systems [57, 88]. Aquatic macrophytes seem to be especially resistant to the toxicity of a great variety of organic pollutants at concentrations normally encountered in typical wastewater compositions. In addition, numerous studies have shown the capability of many macrophyte species to reduce the aqueous concentrations of various organic xenobiotics (e.g., explosives, petroleum hydrocarbons, and pesticides) and, more recently, of various pharmaceuticals too (see Table 17.3 further ahead).

Several different species of macrophytes can be used in CWS depending on the type of CWS design (SF or SSF) and its mode of operation (continuous or batch), loading rate, wastewater characteristics and environmental conditions. However, despite the fact that the dominant species of macrophyte varies locally, the most commonly used in SSF-CWS in temperate zones are the common reeds (*Phragmites australis*) and the cattails (*Typha* spp.) (Table 17.3). In Europe, *Phragmites australis* are the preferred plants for these systems. *Phragmites* has several advantages since it is a fast growing hardy plant and is not a food source for animals or birds [174]. However, in some parts of the U.S. the use of *Phragmites* is not permitted because it is an aggressive plant and there are concerns that it might infest natural wetlands. In these cases cattails or bulrush (*Scirpus* spp.) may be used [213].

The role of plants in the direct degradation and uptake of organic pollutants, in particular xenobiotics of anthropogenic origin, is not as important as the one provided by the action of microorganisms, but in many cases it may still represent a very significant contribution especially because some organic xenobiotics, including many pharmaceuticals, are resistant to biodegradation. In fact, an increasingly important role is being attributed to plants in removing those poorly or non-biodegradable organic xenobiotics through their capacity to sorb them in their roots and/or uptake them and sequester/transform them in their tissues [162–164, 226, 227].

Table 17.3 Removal of selected pharmaceuticals by plants in hydroponic conditions (adapted from Carvalho et al., 2014 [89]).

Plant	Pharmaceutical concentration	% Removal (exposure time)	Plant uptake	Reference
Armoratia rusticana	Diclofenac (31.8, 63.7 mg/L)	60–70 (8 d)	n.d.	[238]
	Ibuprofen (20.6, 41.3 mg/L)	40–94 (8 d)		
	Acetaminophen (15.1, 30.2 mg/L)	100 (8 d)		
Azzola filicuolides (aquatic plant)	Sulfadimethoxine (50–450 mg/L)	56.3–88.5 (35 d)	In the order of $\mu g/g$ - mg/g	[239]
Cabomba caroliniana	Codeine (24 ng/L, wastewater)	25 (4 d)	n.d.	[240]
	Methamphetamine (30 ng/L, wastewater)	70.7 (4 d)		
	Benzoylecgonine (18 ng/L, wastewater)	27.8 (4 d)		
	Tramadol (710 ng/L, wastewater)	48 (4 d)		
	Venlafaxine (259 ng/L, wastewater)	40.9 (4 d)		
	Oxazepam (83 ng/L, wastewater)	8.4 (4 d)		
	Citalopram (84 ng/L, wastewater)	64.3 (4 d)		
	Buprenorphine (11 ng/L, wastewater)	10 (4 d)		
	Methadone (18 ng/L, wastewater)	70.6 (4 d)		
Ceratophyllum demersum (aquatic plant)	Diclofenac (10 $\mu g/L$)	99 ± 1 (38 d)	Mainly photodegradation	[189]
	Naproxen (10 $\mu g/L$)	53 ± 5 (38 d)		
	Ibuprofen (10 $\mu g/L$)	52 ± 2 (38 d)	n.d.	
	Caffeine (10 $\mu g/L$)	81 ± 2 (38 d)		
	Clofibric acid (10 $\mu g/L$)	17 ± 3 (38 d)		
Chrysopogon zizanioides (vetiver grass)	Tetracycline (5–15 mg/L)	100 (40 d)	In the order of $\mu g/g$	[241]
Eichhornia crassipe	Carbamazepine (0.8 mg/L)	36.2 ± 7.2 (12 d)	n.d.	[242]
	Ibuprofen (0.8 mg/L)	93.7 ± 4.8 (12 d)		
	Sulfadiazine (0.8 mg/L)	46.6 ± 3.5 (12 d)		
	Sulfamethoxazole (0.8 mg/L)	42 ± 7.2 (12 d)		
	Sulfamethazone (0.8 mg/L)	42.5 ± 6.1 (12 d)		

(Continued)

Table 17.3 (Continued)

Plant	Pharmaceutical concentration	% Removal (exposure time)	Plant uptake	Reference
Elodea canadensis (aquatic plant)	Diclofenac (10 µg/L)	99 ± 1 (38 d)	Mainly photodegradation	[189]
	Naproxen (10 µg/L)	53 ± 3 (38 d)		
	Ibuprofen (10 µg/L)	77 ± 2 (38 d)	n.d.	
	Caffeine (10 µg/L)	94 ± 2 (38 d)		
	Clofibric acid (10 µg/L)	23 ± 2 (38 d)		
Hordeum vulgare L. (barley)	Sulfadimethoxine (12 mg/L)	75 (2 d)	In the order of µg/g (accumulated in roots)	[243]
	Sulfamethazine (11 mg/L)	40 (2 d)		
	Diclofenac (31.8, 63.7 mg/L)	25–43 (8 d)	n.d.	[238]
	Ibuprofen (20.6, 41.3 mg/L)	17–58 (8 d)		
	Acetaminophen (15.1, 30.2 mg/L)	42–83 (8 d)		
Iris pseudacorus L.	Ibuprofen (10mg/L)	~100 (24 d)	Taken up by roots and translocated to the aerial tissues	[244]
	Ihexol (10 mg/L)	13 (24 d)		
	Codeine (24 ng/L, wastewater)	>92.9 (4 d)	n.d.	[240]
	Methamphetamine (30 ng/L, wastewater)	85.3 (4 d)		
	Benzoylecgonine (18 ng/L, wastewater)	81.7 (4 d)		
	Tramadol (710 ng/L, wastewater)	70 (4 d)		
	Venlafaxine (259 ng/L, wastewater)	69.5 (4 d)		
	Oxazepam (83 ng/L, wastewater)	57.8 (4 d)		
	Citalopram (84 ng/L, wastewater)	93.5 (4 d)		
	Buprenorphine (11 ng/L, wastewater)	>70 (4 d)		
	Methadone (18 ng/L, wastewater)	>87.2 (4 d)		
Juncus acutus L.	Ciprofloxacin (0.1, 10 mg/L)	92–95 (28 d)	n.d.	[245]
	Sulfamethoxazole (50 mg/L)	61 (28 d)		
Juncus effusus L.	Ibuprofen (10 mg/L)	~100 (24 d)	Taken up by the roots and translocated to the aerial tissues	[244]
	Iohexol (10 mg/L)	31 (24 d)		
Landoltia punctata and *Lemna minor* (duckweeds)	Ibuprofen (2 mg/L)	47.5 ± 3.9 (9 d)	n.d.	[246]
	Fluoxetine (3 mg/L)	100 ± 3.9 (4 d)		
	Clofibric acid (2 mg/L)	0 (9 d)		

(Continued)

Table 17.3 (Continued)

Plant	Pharmaceutical concentration	% Removal (exposure time)	Plant uptake	Reference
Lemna gibba (aquatic plant)	Acetaminophen (1–1000 µg/L)	66.12 ± 1.4–92.2 ± 1.9 (10 d)	n.d.	[247]
	Diclofenac (1–1000 µg/L)	47.50 ± 2.0–83.51 ± 3.0 (10 d)		
	Progesterone (1–1000 µg/L)	66.50 ± 1.7–95.25 ± 3 (10 d)		
Lemna minor (aquatic plant)	Flumequine (0.050–1 mg/L)	96 (35 d) (16–26 % were attributed to the plant)	0.72–13.93 µg/g	[248]
	Diclofenac (10 µg/L)	99 ± 1 (38 d)	Mainly photodegradation	[189]
	Naproxen (10 µg/L)	40 ± 5 (38 d)		
	Ibuprofen (10 µg/L)	44 ± 3 (38 d)	n.d.	
	Caffeine (10 µg/L)	99 ± 1 (38 d)		
	Clofibric acid (10 µg/L)	16 ± 4 (38 d)		
Limnophila sessiliflora	Codeine (24 ng/L, wastewater)	33.3 (4 d)	n.d.	[240]
	Methamphetamine (30 ng/L, wastewater)	76 (4 d)		
	Benzoylecgonine (18 ng/L, wastewater)	33.3 (4 d)		
	Tramadol (710 ng/L, wastewater)	58.7 (4 d)		
	Venlafaxine (259 ng/L, wastewater)	61.8 (4 d)		
	Oxazepam (83 ng/L, wastewater)	9.6 (4 d)		
	Citalopram (84 ng/L, wastewater)	75 (4 d)		
	Buprenorphine (11 ng/L, wastewater)	12.7 (4 d)		
	Methadone (18 ng/L, wastewater)	63.3 (4 d)		
Linum usitatissimum	Diclofenac (63.6 mg/L)	100 (8 d)	n.d	[238]
	Ibuprofen (41.3 mg/L)	100 (8 d)		
	Acetaminophen (30.2 mg/L)	50 (8 d)		
Lupinus luteolus (pale yellow lupine)	Diclofenac (31.8, 63.7 mg/L)	3–7 (8 d)		
	Ibuprofen (20.6, 41.3 mg/L)	4–11 (8 d)		
	Acetaminophen (15.1, 30.2 mg/L)	100 (8 d)		

(Continued)

Table 17.3 (Continued)

Plant	Pharmaceutical concentration	% Removal (exposure time)	Plant uptake	Reference
Lythrum salicaria L (aquatic weed)	Flumequine (100 mg/L)	n.d. (10–30 d)	In the order of μg/g	[249]
	Flumequine (0.05–5 mg/L)	n.d. (35 d)		
Myriophyllum aquaticum (parrot feather)	Tetracycine (5.0 mg/L)	70 (15 d)	n.d.	[250]
	Oxytetracycline (5.0 mg/L)	>90 (15 d)		
Phragmites australis	Diclofenac (31.8, 63.7 mg/L)	18–33 (8 d)	n.d.	[238]
	Ibuprofen (20.6, 41.3 mg/L)	49–60 (8 d)		
	Acetaminophen (15.1, 30.2 mg/L)	0–16 (8 d)		
	Ciprofloxacin (0.1–1000 μg/L)	n.d (62d)	In order of ng/g$_{DW}$	[251]
	Oxytetracycline (0.1–1000 μg/L)			
	Sulfamethazine (0.1–1000 μg/L)			
	Metformin (64.6,129.2, 258.3 mg/L)	n.d.	Taken up by the roots and potential for subsequent translocation	[252]
	Albendazole (2.65 mg/L)	n.d.	Taken up (in order of pmol/g – nmol/g) and biotransformation	[253]
	Flubendazole (3.13 mg/L)			
	Ibuprofen (10 mg/L)	~100 (24 d)	Taken up by the roots and translocated to the aerial tissues	[244]
	Iohexol (10 mg/L)	80 (24 d)		
	Carbamazepine (5 mg/L)	35 (1 d)	n.d.	[254]
		66 (4 d)		
		90 (9 d)		
	Enrofloxacin (10, 100 μg/L)	67–91 (7 d)	n.d.	[255]
	Tetracycline (10, 100 μg/L)	60–98 (7 d)		
	Enrofloxacin (100 μg/L, wastewater)	94 (7 d)		
	Tetracycline (100 μg/L, wastewater)	75 (7 d)		
Pistia stratiotes (water lettuce)	Tetracycine (5.0 mg/L)	>90 (6 d)	n.d.	[250]
	Oxytetracycline (5.0 mg/L)	>90 (4 d)		

(Continued)

Table 17.3 (Continued)

Plant	Pharmaceutical concentration	% Removal (exposure time)	Plant uptake	Reference
Pistia stratiotes	Carbamazepine (0.8 mg/L)	34.3 ± 5.9 (12 d)	n.d.	[242]
	Ibuprofen (0.8 mg/L)	99.8 ± 0.6 (12 d)		
	Sulfadiazine (0.8 mg/L)	77.3 ± 8.1 (12 d)		
	Sulfamethoxazole (0.8 mg/L)	65.7 ± 5.3 (12 d)		
	Sulfamethazone (0.8 mg/L)	88.7 ± 5.4 (12 d)		
Populus nigra L. (callus cultures)	Ibuprofen (0.03, 3, 30 mg/L)	100 (21 d)	n.d.	[256]
Salvinia molesta (aquatic plant)	Diclofenac (10 µg/L)	99 ± 1 (38 d)	Mainly photodegradation	[189]
	Naproxen (10 µg/L)	43 ± 2 (38 d)		
	Ibuprofen (10 µg/L)	48 ± 5 (38 d)	n.d.	
	Caffeine (10 µg/L)	99 ± 1 (38 d)		
	Clofibric acid (10 µg/L)	18 ± 4 (38 d)		
Salix alba L. (clone SS5)	Ibuprofen (3, 30 mg/L)	37.1–81.3 (14 d)	n.d.	[257]
Salix alba L. (clone SP3)	Ibuprofen (3, 30 mg/L)	29.5–77.1 (14 d)		
Salix fragilis (willow)	Sulfadiazine (0–620 mg/L)	n.d. (25 d)	Detected in roots but not in the epigeal part of plants	[258]
Scirpus validus	Diclofenac (0.5–2 mg/L)	84–87 (7 d) >98 (21 d)	In the order of $\mu g/g_{FW}$; root uptake:0.47–4.94%; shoot uptake:0.28–2.07%	[259]
	Carbamazepine (0.5 –2 mg/L)	74 (21 d)	In the order of $\mu g/g$; 22% Carb. <5% naproxen	[214]
	Naproxen (0.5–2 mg/L)	98 (21 d)		
	Clofibric acid 0.5 –2 mg/L	41 ± 3–56 ± 2 (7 d) 65 ± 3–78 ± 1(21 d)	Taken up and translocation In roots:5.4–26.8 $\mu g/g_{FW}$ (21d) In shoots:7.2–34.6 $\mu g/g_{FW}$ (21 d)	[260]
Typha spp.	Clofibric acid (20 µg/L)	50% (within 1–2 d) 80% (21 d)	n.d.	[261]
	Ibuprofen (20 µg/L)	60% (within 1 d) 99% (21 d)	n.d.	[262]
	Carbamazepine (0.5–2 mg/L)	28 ± 2–52 ±3 (7 d) 39 ± 3–66 ± 2 (14 d) 56 ± 2–82 ± 2 (21 d)	Taken up, detected in leaves	[207]

(Continued)

Table 17.3 (Continued)

Plant	Pharmaceutical concentration	% Removal (exposure time)	Plant uptake	Reference
	Metformin (6.5, 19.4, 32.3 mg/L)	74 ± 4.1–81.1 ± 3.3 (28 d)	Taken up in the order of $\mu mol/g_{FW}$), translocation to shoots was restricted	[263]
	Metformin (64.6,129.2, 258.3 mg/L)	n.d.	Taken up by the roots and has the potential for subsequent translocation	[252]
	Diclofenac (1 mg/L)	n.d. (7 d)	Taken up (in the order of $\mu g/g_{FW}$) and rapid metabolization	[264]
Zannichellia palustris (aquatic plant)	Sulfathiazole (200 $\mu g/L$, wastewater)	35 ± 12 (20 d)	n.d.	[265]
	Sulfamethazine (200 $\mu g/L$, wastewater)	8 ± 6 (20 d)		
	Sulfamethoxazole (200 $\mu g/L$, wastewater)	20 ± 2 (20 d)		
	Sulfapyridine (200 $\mu g/L$, wastewater)	45 ± 3 (20 d)		
	Tetracycline (200 $\mu g/L$, wastewater)	83 ± 5 (20 d)	Mainly photodegradation	
	Oxytetracycline (200 $\mu g/L$, wastewater)	91 ± 1 (20 d)		

d: days; n.d. not detailed; FW: fresh weight; DW: dry weight.

Plants thus play an important role in the biotic processes of pharmaceuticals removal in CWS, involving several of the processes described previously, of which many details still remain to be characterized thoroughly. Subjected to the direct or indirect action of plants, pharmaceuticals can be stabilized or degraded in the rhizosphere, adsorbed or accumulated in the roots and transported to the aerial parts, volatilized or degraded inside the plant tissues (Figure 17.4).

Many pharmaceuticals can be readily taken up by plants but, as consequence of being xenobiotic, there are no specific transporters for these compounds in the plant membranes. Therefore, they move into and within plant tissues via diffusion (or passive uptake) [162, 164, 227]. The tendency of an organic compound to move into plant roots from an external solution is expressed as the root concentration factor (*RCF*, corresponding to the ratio between the equilibrium concentration in roots and the equilibrium concentration in external solution). The flux is driven by the water potential gradient created throughout the plant during transpiration, which depends on the plants characteristics and the CWS environmental conditions. Translocation of the compounds is highly dependent on their concentrations and physicochemical properties, such as water solubility, log K_{ow} (optimal at ~ 0.5–3.5) and pK_a [162, 164, 166, 213, 226].

Uptake has primary control over translocation, metabolism and phytotoxic action of the pharmaceutical, because the total amount of the substance available for these processes is determined

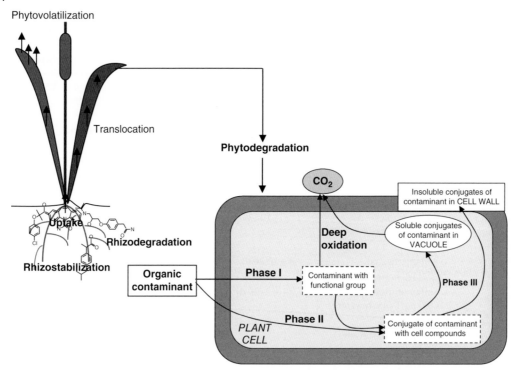

Figure 17.4 Major removal processes and transformation pathways of organic xenobiotics (including pharmaceuticals) in plants (adapted from Kvesitadze *et al.* [228] and Abhilash *et al.* [229]).

by the amount absorbed by the plant. Metabolism influences both pharmaceutical uptake and their phytotoxic action by either rendering the compound less or more active [164, 228].

Pharmaceuticals taken up by roots are translocated into different organs of the plants as a result of the physiological processes involved in the transport of nutrients. The main forces involved in this transport are related to the transpiration stream, i.e., transport of water and dissolved substances from roots to shoots, passing through vessels and tracheids located in the xylem [228, 230].

The importance of the transpiration stream for the uptake and translocation of organics by plants is expressed in the following equation [164, 231]:

$$U = (TSCF) \ (T) \ (C) \tag{17.1}$$

where U is the rate of organic compound assimilation (mg/d); T, the rate of plant transpiration, (L/d); C, the organic compound concentration in the water phase (mg/L); $TSCF$, the transpiration stream concentration factor, is a dimensionless ratio between the concentration of the organic compound in the liquid of the transpiration stream (xylem sap) and the bulk concentration in the root zone solution [164, 228, 230, 232].

Pharmaceuticals that penetrate the plant cells are exposed to plant's metabolic transformations that may lead to their partial or complete degradation, through which they may be transformed in less toxic compounds and bound in plant tissues [226, 228].

Metabolism of xenobiotic compounds in plant systems is generally considered to be a "detoxification" process that is similar to the metabolism of xenobiotic compounds in humans, hence the name "green liver" that is used to refer to these systems [233]. Once an organic xenobiotic is taken up and translocated, it undergoes one or several phases of metabolic transformation, as is illustrated by the diagram in Figure 17.4.

Three possible phases of metabolic transformation of organic compounds in higher plants can be identified [233]:

- *Phase I – Functionalization:* involves a conversion/activation (oxidation, reduction or hydrolysis) of lipophilic xenobiotic compounds [167, 234]; in this phase the molecules of the hydrophobic compound acquire a hydrophilic functional group (e.g., hydroxyl, amine, carboxyl, sulphydryl) through enzymatic transformations. The polarity and water solubility of the compound increase as a result of these processes, which also causes an increased affinity to enzymes catalyzing further transformation (conjugation or deep oxidation [226, 228]) by the addition or exposure of the appropriate functional groups. In the case of a low concentration, oxidative degradation of some xenobiotics to common metabolites of the cell and CO_2 may take place. Following this pathway, a plant cell not only detoxifies the compound but also assimilates the resulting carbon atoms for cell needs. In case of a high concentration, full detoxification is not achieved and the contaminant is exposed to conjugation [226].

  During this phase several different groups of enzymes are known to play an important role [163, 167, 235, 236]. In plants, oxidative metabolism of the xenobiotics is mediated mainly by cytochrome P450 monooxygenase, which is of crucial importance in the oxidative processes to bioactivate the xenobiotics into chemically reactive electrophilic compounds, which subsequently form conjugates during Phase II. Peroxidases are another important group of enzymes, which help in the conversion of some of the xenobiotics. Peroxygenases may also be involved in the oxidation of some compounds. Nitroreductase is needed for the degradation of nitroaromatics and laccase for breaking up aromatic ring structures.

  Phase I reactions are the first step needed to ultimately make a xenobiotic less toxic; those reactions modify the molecule to be ready for Phase II and Phase III reactions which further detoxify the chemical. However, if it already has a functional group suitable for Phase II metabolism, the compound can directly be used for Phase II without entering Phase I.

- *Phase II – Conjugation:* involves conjugation of xenobiotic metabolites of Phase I (or the xenobiotics themselves when they already contain appropriate functional groups) to endogenous molecules (proteins, peptides, amino acids, organic acids, mono-, oligo- and polysaccharides, pectins, lignin, etc.) [164, 167, 206, 226, 237]; as result of conjugation, compounds of higher molecular weight are formed with greatly reduced biological activity and usually reduced mobility. The end products of Phase II are usually less toxic than the original molecules or Phase I derivatives.

  Conjugation is catalyzed by transferases. Enzymes such as glutathione-S-transferases, glucosyl transferase and N-malonyl transferases are associated with Phase II [167].

  Conjugation of Phase I metabolites takes place in the cytosol, but it is harmful to accumulate these compounds in cytosol [167].

- *Phase III – Compartmentalization:* involves modified xenobiotics getting compartmentalized in vacuoles or getting bound to cell wall components such as lignin or hemicellulose [164, 167, 237].

In this phase (a potential final step in the non-oxidative utilization of xenobiotics), the conjugates are removed from vulnerable sites in cytosol and transported to sites where they may not interfere with cellular metabolism: soluble conjugates (with peptides, sugars, amino acids, etc.) are accumulated in vacuoles, whereas insoluble conjugates (coupled with pectin, lignin, xylan and other polysaccharide) are taken out of the cell and accumulated in plant cell walls [167, 235, 236].

Phase III reactions are unique to plants because they do not excrete xenobiotics as animals do. Plants therefore, need to somehow remove the xenobiotic within their own system. ATP driven vacuolar transporters are the main enzymes involved in this phase and further processing of conjugates may take place in the vacuolar matrix [167]. It is assumed that Phase III products are no longer toxic; however, this area of xenobiotic fate in plants is poorly understood, especially with reference to the identity of the sequestered products and any subsequent fate in herbivores who might consume those plants.

In order to illustrate the active role that may be played by plants in removing pharmaceuticals from water, a sample of recent studies is presented in Table 17.3 providing data obtained in hydroponic experiments using various types of plants exposed to the most common pharmaceuticals dissolved in their growth media (water, wastewater or nutrients solution).

The observed reductions in pharmaceuticals concentrations are reported, as well as indications wherever plant uptake of the pharmaceuticals has been confirmed in the studies. In addition, some studies are also carried out with the aim of assessing phytotoxicity effects on the tested plant species. Similar studies conducted within CWS are still scarce, using hydroponic conditions which have been almost exclusively used in this line of research until now. The rationale may be to keep the systems simpler, with minimal perturbations from the other components, when the goal is to address specifically plant–pharmaceutical interactions. Conversely, in CWS studies, the determination of the pharmaceutical fate in plants requires harvesting plant specimens and performing chemical analysis on their tissues, and such tasks have rarely been included in the experimental protocols of this kind of study.

The majority of the studies were performed on a single plant species and a single pharmaceutical compound. A much wider variety of plants species have been studied in these hydroponic studies than, later on, are typically employed in CWS. Table 17.3 also includes some studies on crop plant species, whose use in CWS is advised against because of the danger of being ingested by herbivores and potentially contaminating the food chain with pollutants taken up by those plants.

These studies clearly show that plants have in general a significant capacity to contribute to the removal of pharmaceuticals from contaminated waters. Although the contribution of plant uptake is not always verified, the studies where such data is available provide an indication that this process may be relevant in many cases.

Interesting results have also been found with regard to biochemical processes occurring with pharmaceuticals in plants, which indicate that plants possess mechanisms to cope with pharmaceutical compound toxicity and can either eliminate them or stabilize them and sequester them in their tissues. However, studies focusing on the metabolism of pharmaceutical substances in plants are still scarce until now [207, 208, 253, 263, 264, 266–269], in spite of the great interest that such data represents for phytoremediation applications such as CWS.

As relevant as the role of plants may be, a fundamental role in the removal of biodegradable pollutants is unquestionably played by the microorganisms in the rhizosphere, through their metabolic machinery [85].

Populations of microorganisms develop naturally in a CWS and are very diverse. A variety of algae, bacteria, fungi, protozoa, etc. can be found among the microorganisms that thrive in wetlands, an environment which provides adequate conditions for their growth and development. Additionally, the characteristics of these populations are also pronouncedly influenced by the type of vegetation and the support matrix materials [57, 85–87, 219]. This diversity potentiates the capacity of microbial biofilms to carry through an assortment of microbial activities which are essential in the performance of a CWS. In particular, a diverse mixture of both aerobic and anaerobic bacteria is useful in the degradation of organic matter, including organic xenobiotics such as pharmaceuticals [61, 182, 185, 188, 244, 270]. Therefore, there may be some interest in the characterization of microbial communities in CWS. The biodegradation of organic compounds is primarily attributed to heterotrophic bacteria and certain autotrophic bacteria, fungi including basidiomycetes and yeasts, and specific protozoa [57, 200]. Unfortunately, limited information is available to date concerning the characteristics of the microbial communities developing in CWS support media as well as on the influences due to the wetland vegetation [86, 87, 219, 271]. In addition, significantly missing from the literature is the influence of pharmaceuticals on microbial structure and the response of the microbial community to the presence of pharmaceuticals in CWS [78, 254, 272, 273].

Microbial populations may additionally be improved by inoculation of CWS with strands that are more suitable for removal of the target pollutants, namely strands that are selected (or genetically modified) to more efficiently biodegrade a given set of pharmaceuticals. However, not much work has been carried out to date on these possibilities and this is an optimization opportunity which is still largely unexplored.

Although the action of microorganisms is usually regarded to be undertaken almost exclusively within the rhizosphere, a potentially very significant role (that may have been long underestimated) may be played by microorganisms that are endophytic to plants, i.e., microorganisms that are present inside plants [86, 221–223, 254, 274]. In fact, at the moment it is not clearly understood what is the contribution of endophytic microorganisms to the degradation of organic xenobiotics inside plants (of course, much less of pharmaceuticals in particular) in relation to the plants' phytodegradation processes, as well as any cooperative or synergistic interplay between the two types of organism. This topic, too, is open for further research and ample debate.

One feature of the action of the biotic components (microorganisms and plants) is that their activities typically go through high and low cycles throughout the year and therefore they introduce undesirable seasonal variations in the CWS performance. To some extent, this may possibly be reduced by the action of some processes occurring at the abiotic component (support matrix).

17.3.3.2 The Role of the Support Matrix in Pharmaceuticals Removal

The primary role of the support matrix in a SSF-CWS is to provide a physical support for anchoring the plant root systems and a surface for microorganisms' attachment and development [65, 85, 85, 175]. In addition, plant roots and microorganisms can find in this component their supplies of water, air and nutrients. The support matrix also may provide a relevant service in the moderation of environmental conditions, such as temperature or pH, which may influence the biotic components' development. Furthermore, adequate characteristics of the support matrix such as its hydraulic conductivity are crucial for an even infiltration and flow of the wastewater through the SSF-CWS, avoiding the occurrence of overland flows and preferential channeling and, thus, assuring a good contact of the wastewater with the CWS media. It should be evident that the major importance of this component is for the proper function and efficiency of the SSF-CWS.

Concomitant to this supportive role, this component may play an active role in wastewater cleanup. It acts as a filter medium that can trap particulate pollutants and additionally may retain through sorption some types of dissolved pollutants. Some mineral materials also have ion exchange capacity, which may provide for some supplementary ability to retain polar or ionic pollutants.

The support matrix is a vital component in a SSF-CWS, which provides the link between all the components and the main treatment processes occurring within the system. The type of materials that compose the solid matrix will, in fact, have a strong influence over the occurrence of many of the removal processes and the proper functioning of the other components.

The support matrix may be composed of several different materials, each with its own particular characteristics such as particle size distribution, hydraulic conductivity, mineralogical composition, acid–base and surface charge properties, organic matter contents, sorptive properties, among others. These characteristics may influence the development of the biota which the substrate provides support for. In fact, it has been observed that some plant species and microorganism strains seem to prefer particular types of media, exhibiting an enhanced development in specific types of materials [86, 87, 219, 220]. Additionally, the chemical nature of the materials will determine the type and strength of the surface interactions between the support matrix and the pollutant molecules, and thereby regulates the extent to which the support matrix may contribute, through sorption processes, to the retention of the pollutants in the solid media. For different types of support matrices, the correspondingly different partitioning of the pollutant between the water and the solids compartments will influence the pollutant's behavior in the CWS and the processes to which it will be submitted.

The selection of the support matrix materials to be used in a SSF-CWS is an opportunity to optimize and significantly enhance the role of the support matrix in the removal of the pollutants, which may contribute to an overall improvement of the system's treatment efficiency. For example, sorption processes are usually seen as only providing a short-time retention of pollutants until sorption capacity of the support matrix becomes saturated. However, it has been observed [61, 70, 275] that these processes may be important to moderate the seasonal efficiency variations of a CWS that result from the seasonal variations in the activities of the biotic components. The support matrix may, thus, contribute to a more stable performance of the systems throughout the whole year.

Recently, the increasing use of CWS for more specific targets, such as pharmaceuticals, has spurred numerous attempts to optimize each component's contribution to the overall pollutants removal. Specifically in the selection of the support matrix components, a distinction between those materials that effectively interact with the target pollutants from those that are inert to them is a useful guidance to endow the support matrix with a more active role in the removal of the target pollutants.

A variety of materials have already been studied relative to their capacity to sorb pharmaceuticals, as is presented in Table 17.4. However, most experimental work carried out to date consists of batch sorption studies and in only a few cases have the studies been performed at CWS beds.

From the sample of published works presented in Table 17.4, it is clear that a large amount of research has been produced on the adsorption of pharmaceuticals over the last years. Different materials have different characteristics, and thus different pollutant retention properties. Activated carbon is one of the oldest and most popular choices for filter systems, but it has rarely been applied in SSF-CWS as it is generally considered to be an expensive material whose effectiveness, although high, does not justify its also rather high cost. Other options mainly consisted of natural materials, either in raw state or subjected to relatively inexpensive treatments (chemical modification or thermal and mechanical processing). One of the main criteria used in the selection a particular material is, in fact, its cost and wide availability. Materials tested range from a variety of mineral sources (especially clays) and, recently, the class of agricultural wastes and by-products has also been gaining

Table 17.4 Studies of pharmaceuticals sorption by several natural, modified and synthetic materials and agricultural wastes and by-products.

Material[a]	Pharmaceutical	Study type (assay type/mixture type/solvent)[b]	Removal efficiency/ Sorption capacity	Reference
Activated carbon				
"	Ibuprofen	BSS/SCS/PW	139.2–393.4 mg/g (298K)	[276]
"	Paracetamol, Acetylsalicylic acid	BSS/SCS/PW	236–354 mg/g	[277]
"	Clofibric acid	BSS/SCS/1:9 methanol: water	139–295 mg/g	[278]
"	Ibuprofen	BSS/SCS/PW	28.5 mg/g	[279]
"	Acetaminophen, Diclofenac, Sulfamethoxazole	BSS/SCS/PW	1.3–3.82 mmol/g	[280]
Clay-based materials				
Bentonite	Amoxicillin	BSS/SCS/PW	47 mg/g	[281]
Expanded clay	Clofibric acid	BSS/SCS/PW	0.0045–0.0123 mg/g	[282]
"	Clofibric acid, Carbamazepine, Ibuprofen	BSS/SCS/PW, BSS/MCS/PW, BSS/MCS/WW	58%–95% (SCS/PW), 51%–93% (MCS/WW) 0.023–0.033 mg/g (SCS/PW), 0.014–0.030 mg/g (MCS/PW), 0.011–0.027 mg/g (MCS/WW)	[283]
"	Atenolol	CWS/SCS/WW	82.0% (unplanted beds)	[171]
"	Ibuprofen, Carbamazepine, Clofibric acid	CWS/MCS/WW	43%–91% (unplanted beds, summer) 41%–87% (unplanted beds, winter)	[70]
"	Gemfibrozil, Mefenamic acid, Naproxen	BSS/SCS/PW	46.9%–98% 0.016–0.024 mg/g	[284]
Kaolinite (110 mM)	Oxytetracycline	BSS/MCS/PW	6.0 mg/g	[285]
Kaolinite	Ibuprofen	BSS/SCS/PW	3.1 mg/g	[279]
Montmorillonite	Chlortetracycline	BSS/MCS/PW	Poor fit of the Langmuir model	[285]
"	Oxytetracycline	BSS/MCS/PW	Poor fit of the Langmuir model	[285]
"	Tetracycline	BSS/MCS/PW	Poor fit of the Langmuir model	[285]
Montmorillonite (cation-saturated and with surfactants)	Carbamazepine	BSS/SCS/PW	$K_d = 0.048–0.073$ L/g (K^+ and Ca^{2+} saturated), $K_d = 0.705$ L/g (with surfactant)	[286]
Montmorillonite	Diphenhydramine	BSS/SCS/PW	0.88 mmol/g	[287]
"	Ibuprofen	BSS/SCS/PW	6.1 mg/g	[279]
Sepiolite	Clofibric acid	BSS/SCS/PW	36%–43%	[283]

(Continued)

Table 17.4 (Continued)

Material[a]	Pharmaceutical	Study type (assay type/mixture type/solvent)[b]	Removal efficiency/ Sorption capacity	Reference
Expanded vermiculite	Clofibric acid	BSS/SCS/PW	51%–60%	[283]
"	Gemfibrozil, Mefenamic acid, Naproxen	BSS/SCS/PW	26%–80% 0.066 mg/g (Naproxen)	[284]
Palygorskite	Tetracycline	BSS/SCS/PW	99 mg/g	[288]
Zeolites[a]				
Natural zeolite	Enrofloxacin	BSS/SCS/PW	19.3 mg/g (pH 7, 28°C)	[289]
Zeolite Y	Sulfonamide	BSS/MCS/PW	>90% (saturated solutions)	[290]
Natural (80% clinoptilolite) with surfactant	Diclofenac	BSS/SCS/PW	0.07–0.16 mmol/g	[291]
Y, mordenite, ZSM-5	Erythromycin, Carbamazepine, Levofloxacin	BSS/SCS/PW and BSS/MCS/WW (in Y)	42–100 mg/g (Y-PW), 26–32 mg/g (mor.-PW), 16–26 mg/g (ZSM-5-PW), 96–100 % (Y-WW)	[292]
Natural (clinoptilolite), Beta, ZSM-5	Nicotine	BSS/SCS/PW	1.0 mmol/g (Beta), 0.07 mmol/g (ZSM-5), 0.04 mmol/g (cli.)	[293]
Other siliceous materials				
Expanded perlite	Clofibric acid	BSS/SCS/PW	<5%	[282]
Agricultural wastes and by products				
Almond shell	Amoxicillin	BSS/SCS/PW	2.5 mg/g	[294]
Cork (granulated)	Ibuprofen, Carbamazepine Clofibric acid	BSS/SCS and MCS/PW and WW	0.06056–0.3668 mg/g (SCS, PW) 8–63% (MCS,WW)	[295]
Cork bark	Paracetamol	BSS/SCS/PW	0.99 mg/g	[296]

[a] Y, Beta and ZSM-5 are common abbreviations/designations for certain types of zeolites
[b] BSS – Batch sorption studies; CWS – constructed wetlands systems; SCS – single-component solutions; MCS – multi-component solutions; PW – pure water; WW – wastewater.

considerable popularity. Agro-wastes, due to their high carbon content, have also been increasingly used as low-cost precursors for the production of cheaper types of activated carbons.

A shortcoming of many of these studies is that most of them have been conducted in very ideal conditions. Typically, many studies have been conducted as batch assays under controlled laboratory conditions, like controlled temperature, focusing on single-component solutions prepared in pure water or, at best, using synthetically prepared wastewater.

The difference between batch feed and continuous flow hydraulics is an essential one as effects such as clogging of the substrate cannot be properly assessed from a batch assay, and the kinetics of the adsorption process may also substantially differ in the two conditions. Evaluation of multicomponent adsorption is also very important since wastewaters are typically very complex in their composition and, in such systems, the adsorbed amounts of a particular substance will unavoidably depend on the equilibrium between adsorption competition from all other substances. In addition, pollutant concentrations in these assays are frequently well above those typical of real environmental samples. Other far-from-real conditions include the stirring of the solutions (whereas in realistic systems flow is very slow) and the very low ratio of adsorbent mass-to-volume of solution in comparison to typical ratios of real systems.

In many studies a single substance is often used to represent or model the behavior of a whole family of chemically related substances (or a big part of it). While in some cases such a behavior may be observed and assumed, in general the validity of this assumption should be assessed as some traditional "family" denomination (e.g., pharmaceutical) refer in fact to substances that vary considerably in structure and properties. Even when the chemical structures of the compounds present some similarity, this does not correspond to similar behavior in a particular process such as adsorption.

These batch experiments are useful, nevertheless, to highlight the potential applicability and selectivity of the materials, determine their adsorptive capacities and characterize the mechanisms, but, once these are established, research work must be conducted at larger scale under more realistic conditions to verify their viability in fully operational CWS.

For filter systems, one of the important factors in assessing their viability is the aspect of the lifetime of the adsorbent and the option between disposal or regeneration of the exhausted material. In CWS, the frequent disposal and renewal of the support matrix is practically unfeasible, which means that either desorption can be easily induced or the support matrix will, after a period of the system's operation, saturate its adsorption capacity and lose this specific function in the CWS.

Sorption studies conducted so far have revealed various options for materials (of very diverse nature and characteristics) with very similar adsorption capacities for the main groups of organic pollutants. Within a set of equivalently performing sorbents, it should be recalled, however, that the selection of materials is also based on other criteria as well, as has already been enunciated, namely the hydraulic properties, adequacy for supporting the development of plants and microorganisms, ability for desorption/regeneration, cost of the materials, etc. Cost is actually an important parameter for comparison of adsorbent materials. Recently, numerous approaches have been studied for the development of cheaper and effective adsorbents obtained from industrial and agricultural wastes and by-products.

17.4 Final Remarks

Phytotechnologies such as CWS present several advantages, among which the most salient, relative to other technologies, are the relatively lower implementation and maintenance costs as well as lower requirements for skilled labor. Other less prominent features (but not without importance) are the aesthetically pleasant installations and the usually good public acceptance. Among the limitations of these systems are the generally larger area requirements for their implementation, the typically long periods required for achieving the treatment goals, and a performance that usually is not uniform throughout the year, having typically lower efficiencies during winter seasons. Furthermore, as the wastewater treatment in these systems is undertaken essentially by living components (plants and microorganisms), the response of the system to the types and amounts of pollutant loads and how efficiently they cope with it are more difficult to predict. Therefore, more extensive preparation and study of the behavior of CWS with a given type of wastewater is necessary in order to assess its potential usefulness for treating that sort of pollutants.

The topic of environmental contamination with pharmaceuticals is relatively recent and, therefore, as alternative wastewater treatment technologies are still being evaluated and developed, studies on the applications of CWS to specifically remove pharmaceuticals are not yet as abundant as for other types of pollutants, including other organic xenobiotics such as pesticides, for example. However, the data collected on the research conducted so far does look promising as these systems display in many cases sufficiently high efficiencies to be considered as interesting options.

The complexity of these systems, while responsible for the difficulty to attain a thorough and comprehensive understanding of all the mechanisms involved in the removal of pollutants underlying wastewater treatment, also presents further opportunities to tweak and optimize its operation. Therefore, it is worthwhile to invest in a better knowledge of many of the roles played by each CWS component and of the pollutant removal processes occurring within it. However, especially in regard to the CWS behavior towards pharmaceuticals, there are still several aspects of CWS processes and components that are still scarcely studied, and research on many of these matters may still be considered as currently pioneering. This subject, nonetheless, is currently an active field of research and the situation will evolve with further studies and the production of more scientific data.

Among the topics that may be more relevant for the future, these may be emphasized:

- Further studies need to be carried out on plant uptake of pharmaceuticals and their metabolization in plant tissues in order to understand better the role of plants and characterize the fate of pharmaceuticals inside plants. Among other aims, the study of the latter topic is important for assessing the risk posed by decaying plants and plant debris of CWS, and in carefully harvesting and adequately disposing of them, so as to avoid reintroduction of the pharmaceutical metabolites in the treated wastewater.
- Populations of microorganisms in CWS (including plants' endophytes) should be better characterized, as well as their processes affecting pharmaceuticals; the characteristics of these populations may eventually be modified and improved.
- Finally, evaluations of alternative materials to use as an SSF-CWS support matrix should be pursued with the intent of enhancing the temporary retention of pharmaceuticals at the solid media, while keeping them bioavailable for the slower biotic removal processes provided by plants and microorganisms. The role of this component may be useful for moderating environmental conditions and mitigating the lower activities of the biotic components during winter seasons. Research of low cost materials may also contribute to the economic appeal of SSF-CWS solutions as the support matrix may easily be the most expensive of the components in the construction phase.

References

1 Enick OV, Moore MM. Assessing the assessments: pharmaceuticals in the environment. Environ Impact Assess Rev. 2007; 27(8):707–729.
2 Petrovic M, Barceló D. Analysis. Fate and removal of pharmaceuticals in the water cycle. Amsterdam: Elsevier; 2007.

3 Aga DS. Fate of pharmaceuticals in the environment and in water treatment systems. Boca Raton: CRC Press; 2008.

4 Kümmerer K. The presence of pharmaceuticals in the environment due to human use – present knowledge and future challenges. J Environ Manage. 2009; 90(8):2354–2366.

5 Fent K, Weston AA, Caminada D. Ecotoxicology of human pharmaceuticals. Aquat Toxicol. 2006; 76(2):122–159.

6 Noguera-Oviedo K, Aga DS. Lessons Learned from more than two decades of research on emerging contaminants in the environment. J Hazard Mater. 2016; 316(5):242–251.

7 Evgenidou EN, Konstantinou IK, Lambropoulou DA. Occurrence and removal of transformation products of PPCPs and illicit drugs in wastewaters: A review. Sci Total Environ. 2015; 505:905–926.

8 Lapworth DJ, Baran N, Stuart ME, Ward RS. Emerging organic contaminants in groundwater: A review of sources, fate and occurrence. Environ Pollut. 2012; 163:287–303.

9 Bolong N, Ismail AF, Salim MR, Matsuura T. A review of the effects of emerging contaminants in wastewater and options for their removal. Desalination. 2009; 239(1–3):229–246.

10 Verlicchi P, Al Aukidy M, Zambello E. Occurrence of pharmaceutical compounds in urban wastewater: removal, mass load and environmental risk after a secondary treatment – a review. Sci Total Environ. 2012; 429:123–155.

11 Du LF, Liu WK. Occurrence, fate, and ecotoxicity of antibiotics in agro-ecosystems: a review. Agron Sustain Dev. 2012; 32(2):309–327.

12 Michael I, Rizzo L, McArdell CS, Manaia CM, Merlin C, Schwartz T, et al. Urban wastewater treatment plants as hotspots for the release of antibiotics in the environment: a review. Water Res. 2013; 47(3):957–995.

13 García-Santiago X, Franco-Uría A, Omil F, Lema JM. Risk assessment of persistent pharmaceuticals in biosolids: Dealing with uncertainty. J Hazard Mater. 2016; 302:72–81.

14 Luo Y, Guo W, Ngo HH, Nghiem LD, Hai FI, Zhang J, et al. A review on the occurrence of micropollutants in the aquatic environment and their fate and removal during wastewater treatment. Sci Total Environ. 2014; 473–474(0):619–641.

15 Fatta-Kassinos D, Meric S, Nikolaou A. Pharmaceutical residues in environmental waters and wastewater: current state of knowledge and future research. Anal Bioanal Chem. 2011; 399(1):251–275.

16 Li WC. Occurrence, sources, and fate of pharmaceuticals in aquatic environment and soil. Environ Pollut. 2014; 187:193–201.

17 Sui Q, Cao X, Lu S, Zhao W, Qiu Z, Yu G. Occurrence, sources and fate of pharmaceuticals and personal care products in the groundwater: A review. Emerg Contam. 2015; 1(1):14–24.

18 Topp E, Hendel JG, Lapen DR, Chapman R. Fate of the nonsteroidal anti-inflammatory drug naproxen in agricultural soil receiving liquid municipal biosolids. Environ Toxicol Chem. 2008; 27(10):2005–2010.

19 Gottschall N, Topp E, Edwards M, Payne M, Kleywegt S, Russell P, et al. Hormones, sterols, and fecal indicator bacteria in groundwater, soil, and subsurface drainage following a high single application of municipal biosolids to a field. Chemosphere. 2013; 91(3):275–286.

20 Verlicchi P, Zambello E. Pharmaceuticals and personal care products in untreated and treated sewage sludge: occurrence and environmental risk in the case of application on soil – a critical review. Sci Total Environ. 2015; 538:750–767.

21 Gottschall N, Topp E, Metcalfe C, Edwards M, Payne M, Kleywegt S, et al. Pharmaceutical and personal care products in groundwater, subsurface drainage, soil, and wheat grain, following a high single application of municipal biosolids to a field. Chemosphere. 2012; 87(2):194–203.

22 Gothwal R, Shashidhar T. Antibiotic pollution in the environment: a review. Clean-Soil Air Water. 2015; 43(4):479–489.

23 Calisto V, Esteves VI. Psychiatric pharmaceuticals in the environment. Chemosphere. 2009; 77(10):1257–1274.

24 Caliman FA, Gavrilescu M. Pharmaceuticals, personal care products and endocrine disrupting agents in the environment – a review. Clean-Soil Air Water. 2009; 37(4–5):277–303.

25 Rivera-Utrilla J, Sánchez-Polo M, Ferro-García MÁ, Prados-Joya G, Ocampo-Pérez R. Pharmaceuticals as emerging contaminants and their removal from water: a review. Chemosphere. 2013; 93(7):1268–1287.

26 Boxall ABA. The environmental side-effects of medication – How are human and veterinary medicines in soils and water bodies affecting human and environmental health? EMBO Rep. 2004; 5(12):1110–1116.

27 Miège C, Choubert JM, Ribeiro L, Eusèbe M, Coquery M. Fate of pharmaceuticals and personal care products in wastewater treatment plants – Conception of a database and first results. Environ Pollut. 2009; 157(5):1721–1726.

28 Bendz D, Paxéus NA, Ginn TR, Loge FJ. Occurrence and fate of pharmaceutically active compounds in the environment, a case study: Höje River in Sweden. J Hazard Mater. 2005; 122(3):195–204.

29 Farré Ml, Pérez S, Kantiani L, Barceló D. Fate and toxicity of emerging pollutants, their metabolites and transformation products in the aquatic environment. Trac-Trends Anal Chem. 2008; 27(11):991–1007.

30 Zounkova R, Odraska P, Dolezalova L, Hilscherova K, Marsalek B, Blaha L. Ecotoxicity and genotoxicity assessment of cytostatic pharmaceuticals. Environ Toxicol Chem. 2007; 26(10):2208–2214.

31 Kümmerer K. Antibiotics in the aquatic environment – a review – Part II. Chemosphere. 2009; 75(4):435–441.

32 Barra Caracciolo A, Topp E, Grenni P. Pharmaceuticals in the environment: Biodegradation and effects on natural microbial communities. A review. J Pharm Biomed Anal. 2015; 106:25–36.

33 Santos LHML, Araújo AN, Fachini A, Pena A, Delerue-Matos C, Montenegro MCBS. Ecotoxicological aspects related to the presence of pharmaceuticals in the aquatic environment. J Hazard Mater. 2010; 175(1–3):45–95.

34 Escher BI, Fenner K. Recent advances in Environmental Risk Assessment of transformation products. Environ Sci Technol. 2011; 45(9):3835–3847.

35 Kumar RR, Lee JT, Cho JY. Fate, occurrence, and toxicity of veterinary antibiotics in environment. J Korean Soc Appl Biol Chem. 2012; 55(6):701–709.

36 Arnold KE, Brown AR, Ankley GT, Sumpter JP. Medicating the environment: assessing risks of pharmaceuticals to wildlife and ecosystems. Philos Trans R Soc B – Biol Sci. 2014; 369(1656).

37 Pal A, Gin KY-H, Lin AY-C, Reinhard M. Impacts of emerging organic contaminants on freshwater resources: review of recent occurrences, sources, fate and effects. Sci Total Environ. 2010; 408(24):6062–6069.

38 Cooper ER, Siewicki TC, Phillips K. Preliminary risk assessment database and risk ranking of pharmaceuticals in the environment. Sci Total Environ. 2008; 398(1–3):26–33.

39 Robinson I, Junqua G, Van Coillie R, Thomas O. Trends in the detection of pharmaceutical products, and their impact and mitigation in water and wastewater in North America. Anal Bioanal Chem. 2007; 387(4):1143–1151.

40 Küster A, Adler N. Pharmaceuticals in the environment: scientific evidence of risks and its regulation. Philos Trans R Soc B – Biol Sci. 2014; 369(1656).

41 Carere M, Polesello S, Kase R, Gawlik BM. The Emerging Contaminants in the context of the EU Water Framework Directive. In: Petrovic M, Sabater S, Elosegi A, Barceló D, eds. Emerging Contaminants in River Ecosystems: Occurrence and Effects Under Multiple Stress Conditions. Cham: Springer International Publishing; 2016, pp. 197–215.

42 Ågerstrand M, Berg C, Björlenius B, Breitholtz M, Brunström B, Fick J, et al. Improving Environmental Risk Assessment of human pharmaceuticals. Environ Sci Technol. 2015; 49(9):5336–5345.

43 Gavrilescu M, Demnerová K, Aamand J, Agathos S, Fava F. Emerging pollutants in the environment: present and future challenges in biomonitoring, ecological risks and bioremediation. New Biotech. 2015; 32(1):147–156.

44 Kot-Wasik A, Debska J, Namiesnik J. Analytical techniques in studies of the environmental fate of pharmaceuticals and personal-care products. Trac-Trends Anal Chem. 2007; 26(6):557–568.

45 Lolic A, Paíga P, Santos LHML, Ramos S, Correia M, Delerue-Matos C. Assessment of non-steroidal anti-inflammatory and analgesic pharmaceuticals in seawaters of North of Portugal: Occurrence and environmental risk. Sci Total Environ. 2015; 508:240–250.

46 Verlicchi P, Al Aukidy M, Galletti A, Petrovic M, Barceló D. Hospital effluent: Investigation of the concentrations and distribution of pharmaceuticals and environmental risk assessment. Sci Total Environ. 2012; 430(0):109–118.

47 Fatta-Kassinos D, Vasquez MI, Kümmerer K. Transformation products of pharmaceuticals in surface waters and wastewater formed during photolysis and advanced oxidation processes – Degradation, elucidation of byproducts and assessment of their biological potency. Chemosphere. 2011; 85(5):693–709.

48 Feng L, van Hullebusch ED, Rodrigo MA, Esposito G, Oturan MA. Removal of residual anti-inflammatory and analgesic pharmaceuticals from aqueous systems by electrochemical advanced oxidation processes. A review. Chem Eng J. 2013; 228:944–964.

49 Cincinelli A, Martellini T, Coppini E, Fibbi D, Katsoyiannis A. Nanotechnologies for removal of pharmaceuticals and personal care products from water and wastewater. A review. J Nanosci Nanotechnol. 2015; 15(5):3333–3347.

50 Esplugas S, Bila DM, Krause LG, Dezotti M. Ozonation and advanced oxidation technologies to remove endocrine disrupting chemicals (EDCs) and pharmaceuticals and personal care products (PPCPs) in water effluents. J Hazard Mater. 2007; 149(3):631–642.

51 Taheran M, Brar SK, Verma M, Surampalli RY, Zhang TC, Valero JR. Membrane processes for removal of pharmaceutically active compounds (PhACs) from water and wastewaters. Sci Total Environ. 2016; 547:60–77.

52 Ganiyu SO, van Hullebusch ED, Cretin M, Esposito G, Oturan MA. Coupling of membrane filtration and advanced oxidation processes for removal of pharmaceutical residues: A critical review. Sep Purif Technol. 2015; 156(3):891–914.

53 Liu PX, Zhang HM, Feng YJ, Yang FL, Zhang JP. Removal of trace antibiotics from wastewater: A systematic study of nanofiltration combined with ozone-based advanced oxidation processes. Chem Eng J. 2014; 240:211–220.

54 Ganzenko O, Huguenot D, van Hullebusch ED, Esposito G, Oturan MA. Electrochemical advanced oxidation and biological processes for wastewater treatment: a review of the combined approaches. Environ Sci Pollut Res. 2014; 21(14):8493–8524.

55 Plumlee MH, Stanford BD, Debroux JF, Hopkins DC, Snyder SA. Costs of advanced treatment in water reclamation. Ozone – Sci Eng. 2014; 36(5):485–495.

56 Yu J, Chen WQ, Shui YH, Liu JQ, Ho WT, Zhang SY. Research on wastewater treatment through integrated constructed wetlands. Hangzhou: ICMT 2011 – International Conference on Multimedia Technology; 2011.

57 Kadlec RH, Wallace SD. Treatment wetlands, 2nd edn. Boca Raton: CRC Press; 2009.

58 Haberl R, Grego S, Langergraber G, Kadlec RH, Cicalini AR, Martins-Dias S, et al. Constructed wetlands for the treatment of organic pollutants. J Soils Sediments. 2003; 3(2):109–124.

59 Garcia-Rodriguez A, Matamoros V, Fontas C, Salvado V. The ability of biologically based wastewater treatment systems to remove emerging organic contaminants – a review. Environ Sci Pollut Res. 2014; 21(20):11708–11728.

60 García J, Rousseau D, Morató J, Lesage E, Matamoros V, Bayona J. Contaminant removal processes in subsurface-flow constructed wetlands: A review. Crit Rev Environ Sci Technol. 2010; 40(7):561–661.

61 Imfeld G, Braeckevelt M, Kuschk P, Richnow HH. Monitoring and assessing processes of organic chemicals removal in constructed wetlands. Chemosphere. 2009; 74(3):349–362.

62 Vymazal J, Kröpfelová L. Removal of organics in constructed wetlands with horizontal sub-surface flow: A review of the field experience. Sci Total Environ. 2009; 407(13):3911–3922.

63 Vymazal J, Brezinová T. The use of constructed wetlands for removal of pesticides from agricultural runoff and drainage: A review. Environ Int. 2015; 75:11–20.

64 Olette R, Couderchet M, Biagianti S, Eullaffroy P. Toxicity and removal of pesticides by selected aquatic plants. Chemosphere. 2008; 70(8):1414–1421.

65 Dordio AV, Carvalho AJP. Organic xenobiotics removal in constructed wetlands, with emphasis on the importance of the support matrix. J Hazard Mater. 2013; 252–253(0):272–292.

66 Hijosa-Valsero M, Matamoros V, Sidrach-Cardona R, Martín-Villacorta J, Bécares E, Bayona JM. Comprehensive assessment of the design configuration of constructed wetlands for the removal of pharmaceuticals and personal care products from urban wastewaters. Water Res. 2010; 44(12):3669–3678.

67 Zhang DQ, Gersberg RM, Hua T, Zhu JF, Nguyen AT, Law WK, et al. Effect of feeding strategies on pharmaceutical removal by Subsurface Flow Constructed Wetlands. J Environ Qual. 2012; 41(5):1674–1680.

68 Zhang DQ, Gersberg RM, Ng WJ, Tan SK. Removal of pharmaceuticals and personal care products in aquatic plant-based systems: A review. Environ Pollut. 2014; 184:620–639.

69 Avila C, Matamoros V, Reyes-Contreras C, Pina B, Casado M, Mita L, et al. Attenuation of emerging organic contaminants in a hybrid constructed wetland system under different hydraulic loading rates and their associated toxicological effects in wastewater. Sci Total Environ. 2014; 470:1272–1280.

70 Dordio A, Carvalho AJP, Teixeira DM, Dias CB, Pinto AP. Removal of pharmaceuticals in microcosm constructed wetlands using *Typha* spp. and LECA. Bioresour Technol. 2010; 101(3):886–892.

71 Dordio A, Carvalho AJP. Constructed wetlands with light expanded clay aggregates for agricultural wastewater treatment. Sci Total Environ. 2013; 463:454–461.

72 Hijosa-Valsero M, Reyes-Contreras C, Domínguez C, Bécares E, Bayona JM. Behaviour of pharmaceuticals and personal care products in constructed wetland compartments: Influent, effluent, pore water, substrate and plant roots. Chemosphere. 2016; 145:508–517.

73 Verlicchi P, Galletti A, Petrovic M, Barcelo D, Al Aukidy M, Zambello E. Removal of selected pharmaceuticals from domestic wastewater in an activated sludge system followed by a horizontal subsurface flow bed – Analysis of their respective contributions. Sci Total Environ. 2013; 454:411–425.

74 Zhang DQ, Ni WD, Gersberg RM, Ng WJ, Tan SK. Performance characterization of pharmaceutical removal by Horizontal Subsurface Flow Constructed Wetlands using multivariate analysis. Clean-Soil Air Water. 2015; 43(8):1181–1189.

75 Li YF, Zhu GB, Ng WJ, Tan SK. A review on removing pharmaceutical contaminants from wastewater by constructed wetlands: Design, performance and mechanism. Sci Total Environ. 2014; 468:908–932.

76 Berglund B, Khan GA, Weisner SEB, Ehde PM, Fick J, Lindgren PE. Efficient removal of antibiotics in surface-flow constructed wetlands, with no observed impact on antibiotic resistance genes. Sci Total Environ. 2014; 476:29–37.

77 Dan A, Yang Y, Dai YN, Chen CX, Wang SY, Tao R. Removal and factors influencing removal of sulfonamides and trimethoprim from domestic sewage in constructed wetlands. Bioresour Technol. 2013; 146:363–370.

78 Fernandes JP, Almeida CM, Pereira AC, Ribeiro IL, Reis I, Carvalho P, et al. Microbial community dynamics associated with veterinary antibiotics removal in constructed wetlands microcosms. Bioresour Technol. 2015; 182:26–33.

79 Carvalho PN, Araújo JL, Mucha AP, Basto MC, Almeida CM. Potential of constructed wetlands microcosms for the removal of veterinary pharmaceuticals from livestock wastewater. Bioresour Technol. 2013; 134:412–416.

80 Ávila C, Bayona JM, Martín I, Salas JJ, García J. Emerging organic contaminant removal in a full-scale hybrid constructed wetland system for wastewater treatment and reuse. Ecol Eng. 2015; 80:108–116.

81 Zhu SC, Chen H. The fate and risk of selected pharmaceutical and personal care products in wastewater treatment plants and a pilot-scale multistage constructed wetland system. Environ Sci Pollut Res. 2014; 21(2):1466–1479.

82 Verlicchi P, Zambello E. How efficient are constructed wetlands in removing pharmaceuticals from untreated and treated urban wastewaters? A review. Sci Total Environ. 2014; 470–471:1281–1306.

83 Hijosa-Valsero M, Matamoros V, Sidrach-Cardona R, Pedescoll A, Martin-Villacorta J, Garcia J, et al. Influence of design, physico-chemical and environmental parameters on pharmaceuticals and fragrances removal by constructed wetlands. Water Sci Technol. 2011; 63(11):2527–2534.

84 Matamoros V, Bayona JM. Elimination of pharmaceuticals and personal care products in subsurface flow constructed wetlands. Environ Sci Technol. 2006; 40(18):5811–5816.

85 Sundaravadivel M, Vigneswaran S. Constructed wetlands for wastewater treatment. Crit Rev Environ Sci Technol. 2001; 31(4):351–409.

86 Truu M, Juhanson J, Truu J. Microbial biomass, activity and community composition in constructed wetlands. Sci Total Environ. 2009; 407(13):3958–3971.

87 Calheiros CSC, Duque AF, Moura A, Henriques IS, Correia A, Rangel AOSS, et al. Substrate effect on bacterial communities from constructed wetlands planted with *Typha latifolia* treating industrial wastewater. Ecol Eng. 2009; 35(5):744–753.

88 Brix H. Do macrophytes play a role in constructed treatment wetlands? Water Sci Technol. 1997; 35(5):11–17.

89 Carvalho PN, Basto MC, Almeida CM, Brix H. A review of plant–pharmaceutical interactions: from uptake and effects in crop plants to phytoremediation in constructed wetlands. Environ Sci Pollut Res. 2014; 21(20):11729–11763.

90 Brisson J, Chazarenc F. Maximizing pollutant removal in constructed wetlands: Should we pay more attention to macrophyte species selection? Sci Total Environ. 2009; 407(13):3923–3930.

91 Albuquerque A, Oliveira J, Semitela S, Amaral L. Evaluation of the effectiveness of horizontal subsurface flow constructed wetlands for different media. J Environ Sci. 2010; 22(6):820–825.

92 Akratos CS, Tsihrintzis VA. Effect of temperature, HRT, vegetation and porous media on removal efficiency of pilot-scale horizontal subsurface flow constructed wetlands. Ecol Eng. 2007; 29(2):173–191.

93 Vymazal J. Constructed wetlands for treatment of industrial wastewaters: A review. Ecol Eng. 2014; 73:724–751.

94 Vymazal J. Plants used in constructed wetlands with horizontal subsurface flow: a review. Hydrobiologia. 2011; 674(1):133–156.

95 Halling–Sørensen B, Nors Nielsen S, Lanzky PF, Ingerslev F, Holten Lützhøft HC, Jørgensen SE. Occurrence, fate and effects of pharmaceutical substances in the environment – a review. Chemosphere. 1998; 36(2):357–393.

96 Heberer T. Occurrence, fate, and removal of pharmaceutical residues in the aquatic environment: a review of recent research data. Toxicol Lett. 2002; 131(1–2):5–17.

97 Nikolaou A, Meric S, Fatta D. Occurrence patterns of pharmaceuticals in water and wastewater environments. Anal Bioanal Chem. 2007; 387(4):1225–1234.

98 Mendoza A, Aceña J, Pérez S, López de Alda M, Barceló D, Gil A, et al. Pharmaceuticals and iodinated contrast media in a hospital wastewater: A case study to analyse their presence and characterise their environmental risk and hazard. Environ Res. 2015; 140:225–241.

99 Frédéric O, Yves P. Pharmaceuticals in hospital wastewater: Their ecotoxicity and contribution to the environmental hazard of the effluent. Chemosphere. 2014; 115:31–39.

100 Sim WJ, Lee JW, Lee ES, Shin SK, Hwang SR, Oh JE. Occurrence and distribution of pharmaceuticals in wastewater from households, livestock farms, hospitals and pharmaceutical manufactures. Chemosphere. 2011; 82(2):179–186.

101 Mo WY, Chen Z, Leung HM, Leung AOW. Application of veterinary antibiotics in China's aquaculture industry and their potential human health risks. Environ Sci Pollut Res. 2015;1–12.

102 Kim KR, Owens G, Kwon SI, So KH, Lee DB, Ok YS. Occurrence and environmental fate of veterinary antibiotics in the terrestrial environment. Water Air Soil Pollut. 2011; 214(1–4):163–174.

103 Rico A, Van den Brink PJ. Probabilistic risk assessment of veterinary medicines applied to four major aquaculture species produced in Asia. Sci Total Environ. 2014; 468–469:630–641.

104 Lu MC, Chen YY, Chiou MR, Chen MY, Fan HJ. Occurrence and treatment efficiency of pharmaceuticals in landfill leachates. Waste Manage. 2016; 55:257–264.

105 Eggen T, Moeder M, Arukwe A. Municipal landfill leachates: a significant source for new and emerging pollutants. Sci Total Environ. 2010; 408(21):5147–5157.

106 Ramakrishnan A, Blaney L, Kao J, Tyagi RD, Zhang TC, Surampalli RY. Emerging contaminants in landfill leachate and their sustainable management. Environ Earth Sci. 2015; 73(3):1357–1368.

107 Masoner JR, Kolpin DW, Furlong ET, Cozzarelli IM, Gray JL. Landfill leachate as a mirror of today's disposable society: pharmaceuticals and other contaminants of emerging concern in final leachate from landfills in the conterminous United States. Environ Toxicol Chem. 2016; 35(4):1–13.

108 Sabourin L, Beck A, Duenk PW, Kleywegt S, Lapen DR, Li H, et al. Runoff of pharmaceuticals and personal care products following application of dewatered municipal biosolids to an agricultural field. Sci Total Environ. 2009; 407(16):4596–4604.

109 Bottoni P, Caroli S, Caracciolo AB. Pharmaceuticals as priority water contaminants. Toxicol Environ Chem. 2010; 92(3):549–565.

110 Wu CX, Spongberg AL, Witter JD, Fang M, Czajkowski KP. Uptake of pharmaceutical and personal care products by soybean plants from soils applied with biosolids and irrigated with contaminated water. Environ Sci Technol. 2010; 44(16):6157–6161.

111 Goldstein M, Shenker M, Chefetz B. Insights into the uptake processes of wastewater-borne pharmaceuticals by vegetables. Environ Sci Technol. 2014; 48(10):5593–5600.

112 Wu X, Dodgen LK, Conkle JL, Gan J. Plant uptake of pharmaceutical and personal care products from recycled water and biosolids: a review. Sci Total Environ. 2015; 536:655–666.

113 Sabourin L, Duenk P, Bonte-Gelok S, Payne M, Lapen DR, Topp E. Uptake of pharmaceuticals, hormones and parabens into vegetables grown in soil fertilized with municipal biosolids. Sci Total Environ. 2012; 431:233–236.

114 Calderón-Preciado D, Matamoros V, Bayona JM. Occurrence and potential crop uptake of emerging contaminants and related compounds in an agricultural irrigation network. Sci Total Environ. 2011; 412–413:14–19.

115 Mohapatra DP, Cledon M, Brar SK, Surampalli RY. Application of wastewater and biosolids in soil: occurrence and fate of emerging contaminants. Water Air Soil Pollut. 2016; 227(3).

116 Prosser RS, Sibley PK. Human health risk assessment of pharmaceuticals and personal care products in plant tissue due to biosolids and manure amendments, and wastewater irrigation. Environ Int. 2015; 75:223–233.

117 Daughton CG, Ternes TA. Pharmaceuticals and personal care products in the environment: agents of subtle change? Environ Health Perspect. 1999; 107:907–938.

118 Kümmerer K. Pharmaceuticals in the environment: sources, fate, effects and risks. Berlin: Springer-Verlag; 2008.

119 Arnold WA, McNeill K. Transformation of pharmaceuticals in the environment: photolysis and other abiotic processes. In: Petrovic M, Barceló D, eds. Analysis, fate and removal of pharmaceuticals in the water cycle. Amsterdam: Elsevier; 2007, pp. 361–385.

120 Tijani JO, Fatoba OO, Petrik LF. A Review of pharmaceuticals and endocrine-disrupting compounds: sources, effects, removal, and detections. Water Air Soil Pollut. 2013; 224(11).

121 Celiz MD, Tso J, Aga DS. Pharmaceutical metabolites in the environment: analytical challenges and ecological risks. Environ Toxicol Chem. 2009; 28(12):2473–2484.

122 Bletsou AA, Jeon J, Hollender J, Archontaki E, Thomaidis NS. Targeted and non-targeted liquid chromatography-mass spectrometric workflows for identification of transformation products of emerging pollutants in the aquatic environment. Trac-Trends Anal Chem. 2015; 66:32–44.

123 Jjemba PK. Excretion and ecotoxicity of pharmaceutical and personal care products in the environment. Ecotox Environ Safe. 2006; 63(1):113–130.

124 Boreen AL, Arnold WA, McNeill K. Photodegradation of pharmaceuticals in the aquatic environment: A review. Aquat Sci. 2003; 65(4):320–341.

125 Challis JK, Hanson ML, Friesen KJ, Wong CS. A critical assessment of the photodegradation of pharmaceuticals in aquatic environments: defining our current understanding and identifying knowledge gaps. Environ Sci –Process Impacts. 2014; 16(4):672–696.

126 Andreozzi R, Raffaele M, Nicklas P. Pharmaceuticals in STP effluents and their solar photodegradation in aquatic environment. Chemosphere. 2003; 50(10):1319–1330.

127 Velagaleti R. Behavior of pharmaceutical drugs (human and animal health) in the environment. Drug Inf J. 1997; 31:715–722.

128 Hübner U, von Gunten U, Jekel M. Evaluation of the persistence of transformation products from ozonation of trace organic compounds – A critical review. Water Res. 2015; 68:150–170.

129 Amador PP, Fernandes RM, Prudencio MC, Barreto MP, Duarte IM. Antibiotic resistance in wastewater: Occurrence and fate of Enterobacteriaceae producers of Class A and Class C beta-lactamases. J Environ Sci Health Part A – Toxic/Hazard Subst Environ Eng. 2015; 50(1):26–39.

130 Borghi AA, Palma MA. Tetracycline: production, waste treatment and environmental impact assessment. Braz J Pharm Sci. 2014; 50(1):25–40.

131 Milic N, Milanovic M, Letic NG, Sekulic MT, Radonic J, Mihajlovic I, et al. Occurrence of antibiotics as emerging contaminant substances in aquatic environment. Int J Environ Health Res. 2013; 23(4):296–310.

132 Rizzo L, Manaia C, Merlin C, Schwartz T, Dagot C, Ploy MC, et al. Urban wastewater treatment plants as hotspots for antibiotic resistant bacteria and genes spread into the environment: A review. Sci Total Environ. 2013; 447:345–360.

133 Hernando MD, Mezcua M, Fernández-Alba AR, Barceló D. Environmental risk assessment of pharmaceutical residues in wastewater effluents, surface waters and sediments. Talanta. 2006; 69(2):334–342.

134 Emblidge JP, DeLorenzo ME. Preliminary risk assessment of the lipid-regulating pharmaceutical clofibric acid, for three estuarine species. Environ Res. 2006; 100(2):216–226.

135 Cunningham VL, Binks SP, Olson MJ. Human health risk assessment from the presence of human pharmaceuticals in the aquatic environment. Regul Toxicol Pharmacol. 2009; 53(1):39–45.

136 de Garcia SAO, Pinto GP, Garcia-Encina PA, Irusta-Mata R. Ecotoxicity and environmental risk assessment of pharmaceuticals and personal care products in aquatic environments and wastewater treatment plants. Ecotoxicology. 2014; 23(8):1517–1533.

137 Godoy AA, Kummrow F, Pamplin PA. Occurrence, ecotoxicological effects and risk assessment of antihypertensive pharmaceutical residues in the aquatic environment – A review. Chemosphere. 2015; 138:281–291.

138 Crane M, Watts C, Boucard T. Chronic aquatic environmental risks from exposure to human pharmaceuticals. Sci Total Environ. 2006; 367(1):23–41.

139 Kim S, Eichhorn P, Jensen JN, Weber AS, Aga DS. Removal of antibiotics in wastewater: Effect of hydraulic and solid retention times on the fate of tetracycline in the activated sludge process. Environ Sci Technol. 2005; 39(15):5816–5823.

140 Vieno N, Sillanpaa M. Fate of diclofenac in municipal wastewater treatment plant – A review. Environ Int. 2014; 69:28–39.

141 Clara M, Kreuzinger N, Strenn B, Gans O, Kroiss H. The solids retention time-suitable design parameter to evaluate the capacity of wastewater treatment plants to remove micropollutants. Water Res. 2005; 39(1):97–106.

142 Maurer M, Escher BI, Richle P, Schaffner C, Alder AC. Elimination of beta–blockers in sewage treatment plants. Water Res. 2007; 41(7):1614–1622.

143 Onesios K, Yu J, Bouwer E. Biodegradation and removal of pharmaceuticals and personal care products in treatment systems: a review. Biodegradation. 2009; 20(4):441–466.

144 Miralles-Cuevas S, Oller I, Agüera A, Llorca M, Sánchez Pérez JA, Malato S. Combination of nanofiltration and ozonation for the remediation of real municipal wastewater effluents: Acute and chronic toxicity assessment. J Hazard Mater. 2016; 323(A):442–451.

145 Li CC, Cabassud C, Guigui C. Evaluation of membrane bioreactor on removal of pharmaceutical micropollutants: a review. Desalin Water Treat. 2015; 55(4):845–858.

146 Wang J, Chu L. Irradiation treatment of pharmaceutical and personal care products (PPCPs) in water and wastewater: An overview. Radiat Phys Chem. 2016; 125:56–64.

147 Rahim Pouran S, Abdul Aziz AR, Wan Daud WMA. Review on the main advances in photo-Fenton oxidation system for recalcitrant wastewaters. J Ind Eng Chem. 2015; 21:53–69.

148 Tahar A, Choubert JM, Coquery M. Xenobiotics removal by adsorption in the context of tertiary treatment: a mini review. Environ Sci Pollut Res. 2013; 20(8):5085–5095.

149 Ek M, Baresel C, Magner J, Bergstrom R, Harding M. Activated carbon for the removal of pharmaceutical residues from treated wastewater. Water Sci Technol. 2014; 69(11):2372–2380.

150 Snyder SA, Adham S, Redding AM, Cannon FS, DeCarolis J, Oppenheimer J, et al. Role of membranes and activated carbon in the removal of endocrine disruptors and pharmaceuticals. Desalination. 2007; 202(1–3):156–181.

151 Cirja M, Ivashechkin P, Schäffer A, Corvini P. Factors affecting the removal of organic microp-ollutants from wastewater in conventional treatment plants (CTP) and membrane bioreactors (MBR). Rev Environ Sci Biotechnol. 2008; 7(1):61–78.

152 Dolar D, Gros M, Rodriguez-Mozaz S, Moreno J, Comas J, Rodriguez-Roda I, et al. Removal of emerging contaminants from municipal wastewater with an integrated membrane system, MBR–RO. J Hazard Mater. 2012; 239:64–69.

153 Rodriguez-Mozaz S, Ricart M, Kock-Schulmeyer M, Guasch H, Bonnineau C, Proia L, et al. Pharmaceuticals and pesticides in reclaimed water: Efficiency assessment of a microfiltration–reverse osmosis (MF–RO) pilot plant. J Hazard Mater. 2015; 282:165–173.

154 Secondes MFN, Naddeo V, Belgiorno V, Ballesteros F. Removal of emerging contaminants by simultaneous application of membrane ultrafiltration, activated carbon adsorption, and ultra-sound irradiation. J Hazard Mater. 2014; 264:342–349.

155 Rosal R, Rodriguez A, Perdigon-Melon JA, Petre A, Garcia-Calvo E, Gomez MJ, et al. Occurrence of emerging pollutants in urban wastewater and their removal through biological treatment followed by ozonation. Water Res. 2010; 44(2):578–588.

156 Rizzo L, Fiorentino A, Grassi M, Attanasio D, Guida M. Advanced treatment of urban wastewater by sand filtration and graphene adsorption for wastewater reuse: Effect on a mixture of pharmaceuticals and toxicity. J Environ Chem Eng. 2015; 3(1):122–128.

157 Yu JG, Zhao XH, Yang H, Chen XH, Yang Q, Yu LY, et al. Aqueous adsorption and removal of organic contaminants by carbon nanotubes. Sci Total Environ. 2014; 482–483:241–251.

158 Klavarioti M, Mantzavinos D, Kassinos D. Removal of residual pharmaceuticals from aqueous systems by advanced oxidation processes. Environ Int. 2009; 35(2):402–417.

159 Mathon B, Choubert JM, Miege C, Coquery M. A review of the photodegradability and transformation products of 13 pharmaceuticals and pesticides relevant to sewage polishing treatment. Sci Total Environ. 2016; 551–552:712–724.

160 Postigo C, Richardson SD. Transformation of pharmaceuticals during oxidation/disinfection processes in drinking water treatment. J Hazard Mater. 2014; 279:461–475.

161 Wang XH, Lin AY-C. Is the phototransformation of pharmaceuticals a natural purification process that decreases ecological and human health risks? Environ Pollut. 2014; 186:203–215.

162 Pilon-Smits E. Phytoremediation. Annu Rev Plant Biol. 2005; 56:15–39.

163 Macek T, Mackova M, Kas J. Exploitation of plants for the removal of organics in environmental remediation. Biotechnol Adv. 2000; 18(1):23–34.

164 Dietz AC, Schnoor JL. Advances in phytoremediation. Environ Health Perspect. 2001; 109:163–168.

165 Susarla S, Medina VF, McCutcheon SC. Phytoremediation: an ecological solution to organic chemical contamination. Ecol Eng. 2002; 18(5):647–658.

166 Alkorta I, Garbisu C. Phytoremediation of organic contaminants in soils. Bioresour Technol. 2001; 79(3):273–276.

167 Eapen S, Singh S, D'Souza SF. Advances in development of transgenic plants for remediation of xenobiotic pollutants. Biotechnol Adv. 2007; 25(5):442–451.

168 Gan S, Lau EV, Ng HK. Remediation of soils contaminated with polycyclic aromatic hydrocarbons (PAHs). J Hazard Mater. 2009; 172(2–3):532–549.

169 Schwitzguébel JP. Phytoremediation of soils contaminated by organic compounds: hype, hope and facts. J Soils Sediments. 2015;1–11.

170 Zhang DQ, Tan SK, Gersberg RM, Sadreddini S, Zhu JF, Nguyen AT. Removal of pharmaceutical compounds in tropical constructed wetlands. Ecol Eng. 2011; 37(3):460–464.

171 Dordio A, Pinto J, Pinto AP, da Costa CT, Carvalho A, Teixeira DM. Atenolol removal in microcosm constructed wetlands. Intern J Environ Anal Chem. 2009; 89(8–12):835–848.

172 Hijosa-Valsero M, Fink G, Schlusener MP, Sidrach-Cardona R, Martin-Villacorta J, Ternes T, et al. Removal of antibiotics from urban wastewater by constructed wetland optimization. Chemosphere. 2011; 83(5):713–719.

173 Matamoros V, Garcia J, Bayona JM. Organic micropollutant removal in a full-scale surface flow constructed wetland fed with secondary effluent. Water Res. 2008; 42(3):653–660.

174 Cooper PF, Job GD, Green MB, Shutes RBE. Reed beds and constructed wetlands for wastewater treatment. Medmenham: WRc Publications; 1996.

175 Vymazal J, Brix H, Cooper PF, Green MB, Haberl R. Constructed wetlands for wastewater treatment in Europe. Leiden: Backhuys Publishers; 1998.

176 Wu S, Wallace S, Brix H, Kuschk P, Kirui WK, Masi F, et al. Treatment of industrial effluents in constructed wetlands: challenges, operational strategies and overall performance. Environ Pollut. 2015; 201:107–120.

177 Masi F, Rochereau J, Troesch S, Ruiz I, Soto M. Wineries wastewater treatment by constructed wetlands: a review. Water Sci Technol. 2015; 71(8):1113–1127.

178 Calheiros CSC, Rangel AOSS, Castro PML. Constructed Wetlands for tannery wastewater treatment in Portugal: ten years of experience. Int J Phytoremediat. 2014; 16(9):859–870.

179 Ranieri E, Verlicchi P, Young TM. Paracetamol removal in subsurface flow constructed wetlands. J Hydrol. 2011; 404(3–4):130–135.

180 Reyes-Contreras C, Hijosa-Valsero M, Sidrach-Cardona R, Bayona JM, Bécares E. Temporal evolution in PPCP removal from urban wastewater by constructed wetlands of different configuration: A medium-term study. Chemosphere. 2012; 88(2):161–167.

181 Zhai J, Rahaman M, Ji J, Luo Z, Wang Q, Xiao H, et al. Plant uptake of diclofenac in a mesocosm-scale free water surface constructed wetland by *Cyperus alternlifolius*. Water Sci Technol. 2016; 73(12):3008–3016.

182 Rühmland S, Wick A, Ternes TA, Barjenbruch M. Fate of pharmaceuticals in a subsurface flow constructed wetland and two ponds. Ecol Eng. 2015; 80:125–139.

183 Wu H, Zhang J, Ngo HH, Guo W, Hu Z, Liang S, et al. A review on the sustainability of constructed wetlands for wastewater treatment: design and operation. Bioresour Technol. 2015; 175:594–601.

184 Conkle JL, White JR, Metcalfe CD. Reduction of pharmaceutically active compounds by a lagoon wetland wastewater treatment system in Southeast Louisiana. Chemosphere. 2008; 73(11):1741–1748.

185 Ávila C, Reyes C, Bayona JM, García J. Emerging organic contaminant removal depending on primary treatment and operational strategy in horizontal subsurface flow constructed wetlands: Influence of redox. Water Res. 2013; 47(1):315–325.

186 Matamoros V, Arias C, Brix H, Bayona JM. Removal of pharmaceuticals and personal care products (PPCPs) from urban wastewater in a pilot Vertical Flow Constructed Wetland and a Sand Filter. Environ Sci Technol. 2007; 41(23):8171–8177.

187 Matamoros V, Arias C, Brix H, Bayona JM. Preliminary screening of small-scale domestic wastewater treatment systems for removal of pharmaceutical and personal care products. Water Res. 2009; 43(1):55–62.

188 Ávila C, Pedescoll A, Matamoros V, Bayona JM, García J. Capacity of a horizontal subsurface flow constructed wetland system for the removal of emerging pollutants: An injection experiment. Chemosphere. 2010; 81(9):1137–1142.

189 Matamoros V, Nguyen LX, Arias CA, Salvadó V, Brix H. Evaluation of aquatic plants for removing polar microcontaminants: A microcosm experiment. Chemosphere. 2012; 88(10):1257–1264.

190 Zhang DQ, Gersberg RM, Zhu J, Hua T, Jinadasa KBSN, Tan SK. Batch versus continuous feeding strategies for pharmaceutical removal by subsurface flow constructed wetland. Environ Pollut. 2012; 167:124–131.

191 Zhang DQ, Gersberg RM, Hua T, Zhu J, Tuan NA, Tan SK. Pharmaceutical removal in tropical subsurface flow constructed wetlands at varying hydraulic loading rates. Chemosphere. 2012; 87(3):273–277.

192 Nowrotek M, Sochacki A, Felis E, Miksch K. Removal of diclofenac and sulfamethoxazole from synthetic municipal waste water in microcosm downflow constructed wetlands: Start-up results. Int J Phytoremediat. 2016; 18(2):157–163.

193 Matamoros V, Garcia J, Bayona JM. Behavior of selected pharmaceuticals in subsurface flow constructed wetlands: A pilot-scale study. Environ Sci Technol. 2005; 39(14):5449–5454.

194 Matamoros V, Caselles-Osorio A, García J, Bayona JM. Behaviour of pharmaceutical products and biodegradation intermediates in horizontal subsurface flow constructed wetland. A microcosm experiment. Sci Total Environ. 2008; 394(1):171–176.

195 Liu L, Liu C, Zheng J, Huang X, Wang Z, Liu Y, et al. Elimination of veterinary antibiotics and antibiotic resistance genes from swine wastewater in the vertical flow constructed wetlands. Chemosphere. 2013; 91(8):1088–1093.

196 Yan Q, Feng G, Gao X, Sun C, Guo Js, Zhu Z. Removal of pharmaceutically active compounds (PhACs) and toxicological response of *Cyperus alternifolius* exposed to PhACs in microcosm constructed wetlands. J Hazard Mater. 2016; 301:566–575.

197 Park N, Vanderford BJ, Snyder SA, Sarp S, Kim SD, Cho J. Effective controls of micropollutants included in wastewater effluent using constructed wetlands under anoxic condition. Ecol Eng. 2009; 35(3):418–423.

198 Herrera-Cárdenas J, Navarro AE, Torres E. Effects of porous media, macrophyte type and hydraulic retention time on the removal of organic load and micropollutants in constructed wetlands. J Environ Sci Health Part A – Toxic/Hazard Subst Environ Eng. 2016; 51(5):380–388.

199 SRC. SRC PhysProp Database [Internet]. 2016 [cited 2016 Jul 14]. Available from: www.srcinc .com/what-we-do/databaseforms.aspx?id=386

200 Reddy KR, DeLaune RD. Toxic organic compounds. Biogeochemistry of wetlands: Science and applications. Boca Raton: CRC Press; 2008, pp. 507–536.

201 Bulak P, Walkiewicz A, Brzezinska M. Plant growth regulators-assisted phytoextraction. Biol Plant. 2014; 58(1):1–8.

202 Muller K, Magesan GN, Bolan NS. A critical review of the influence of effluent irrigation on the fate of pesticides in soil. Agric Ecosyst Environ. 2007; 120(2–4):93–116.

203 Evans GM, Furlong JC. Environmental Biotechnology: Theory and Application. Chichester: Wiley; 2003.

204 Zhang BY, Zheng JS, Sharp RG. Phytoremediation in engineered wetlands: mechanisms and applications. Procedia Environ Sci. 2010; 2(0):1315–1325.

205 Suresh B, Ravishankar GA. Phytoremediation – a novel and promising approach for environmental clean-up. Crit Rev Biotechnol. 2004; 24(2–3):97–124.

206 Schroder P, Collins C. Conjugating enzymes involved in xenobiotic metabolism of organic xenobiotics in plants. Int J Phytoremediat. 2002; 4(4):247–265.

207 Dordio AV, Belo M, Teixeira DM, Carvalho AJP, Dias CMB, Picó Y, et al. Evaluation of carbamazepine uptake and metabolization by *Typha* spp., a plant with potential use in phytotreatment. Bioresour Technol. 2011; 102(17):7827–7834.

208 Huber C, Bartha B, Schröder P. Metabolism of diclofenac in plants – hydroxylation is followed by glucose conjugation. J Hazard Mater. 2012; 243:250–256.

209 Herklotz PA, Gurung P, Vanden Heuvel B, Kinney CA. Uptake of human pharmaceuticals by plants grown under hydroponic conditions. Chemosphere. 2010; 78(11):1416–1421.

210 Tsao DT. Overview of phytotechnologies. Berlin: Springer Berlin Heidelberg; 2003.

211 Renner R. The KOW Controversy. Environ Sci Technol. 2002; 36(21):410A–413A.

212 Chaudhry Q, Schröder P, Werck-Reichhart D, Grajek W, Marecik R. Prospects and limitations of phytoremediation for the removal of persistent pesticides in the environment. Environ Sci Pollut Res. 2002; 9(1):4–17.

213 USEPA. Wastewater technology fact sheet wetlands: Subsurface flow. EPA 832-F-00-023. Washington: Office of Wastewater Management; 2000.

214 Zhang DQ, Hua T, Gersberg RM, Zhu J, Ng WJ, Tan SK. Carbamazepine and naproxen: Fate in wetland mesocosms planted with *Scirpus validus*. Chemosphere. 2013; 91(1):14–21.

215 Trapp S. Bioaccumulation of polar and ionizable compounds in plants. Boston: Springer US; 2009.

216 Singer AC, Crowley DE, Thompson IP. Secondary plant metabolites in phytoremediation and biotransformation. Trends Biotechnol. 2003; 21(3):123–130.

217 Dua M, Singh A, Sethunathan N, Johri A. Biotechnology and bioremediation: successes and limitations. Appl Microbiol Biotechnol. 2002; 59(2):143–152.

218 Seffernick JL, Wackett LP. Rapid evolution of bacterial catabolic enzymes: a case study with atrazine chlorohydrolase. Biochemistry. 2001; 40(43):12747–12753.

219 Li M, Zhou Q, Tao M, Wang Y, Jiang L, Wu Z. Comparative study of microbial community structure in different filter media of constructed wetland. J Environ Sci. 2010; 22(1):127–133.

220 Salomo S, Münch C, Röske I. Evaluation of the metabolic diversity of microbial communities in four different filter layers of a constructed wetland with vertical flow by Biolog™ analysis. Water Res. 2009; 43(18):4569–4578.

221 Afzal M, Khan QM, Sessitsch A. Endophytic bacteria: prospects and applications for the phytoremediation of organic pollutants. Chemosphere. 2014; 117:232–242.

222 Ijaz A, Imran A, Anwar ul Haq M, Khan QM, Afzal M. Phytoremediation: recent advances in plant–endophytic synergistic interactions. Plant Soil. 2015;1–17.

223 Weyens N, van der Lelie D, Taghavi S, Vangronsveld J. Phytoremediation: plant – endophyte partnerships take the challenge. Curr Opin Biotechnol. 2009; 20(2):248–254.

224 Dordio A, Carvalho AJP, Pinto AP. Wetlands: Water "living filters"? In: Russo RE, ed. Wetlands: Ecology, Conservation and Restoration. Hauppauge: Nova Science Publishers; 2008, pp. 15–72.

225 Scholz M, Lee B. Constructed wetlands: a review. Int J Environ Stud. 2005; 62(4):421–447.

226 Korte F, Kvesitadze G, Ugrekhelidze D, Gordeziani M, Khatisashvili G, Buadze O, et al. Organic toxicants and plants. Ecotox Environ Safe. 2000; 47(1):1–26.

227 Collins C, Fryer M, Grosso A. Plant uptake of non-ionic organic chemicals. Environ Sci Technol. 2006; 40(1):45–52.

228 Kvesitadze G, Khatisashvili G, Sadunishvili T, Ramsden JJ. The fate of organic contaminants in the plant cell. In: Kvesitadze G, Khatisashvili G, Sadunishvili T, Ramsden JJ, eds. Biochemical Mechanisms of Detoxification in Higher Plants. Berlin: Springer; 2006, pp. 103–165.

229 Abhilash PC, Jamil S, Singh N. Transgenic plants for enhanced biodegradation and phytoremediation of organic xenobiotics. Biotechnol Adv. 2009; 27(4):474–488.

230 Dodgen LK, Ueda A, Wu X, Parker DR, Gan J. Effect of transpiration on plant accumulation and translocation of PPCP/EDCs. Environ Pollut. 2015; 198:144–153.

231 Briggs GG, Bromilow RH, Evans AA, Williams M. Relationships between lipophilicity and the distribution of non-ionized chemicals in barley shoots following uptake by the roots. Pestic Sci. 1983; 14(5):492–500.

232 Doucette WJ, Chard JK, Moore BJ, Staudt WJ, Headley JV. Uptake of sulfolane and diisopropanolamine (DIPA) by cattails (*Typha latifolia*). Microchem J. 2005; 81(1):41–49.

233 Sanderson H, Johnson DJ, Reitsma T, Brain RA, Wilson CJ, Solomon KR. Ranking and prioritization of environmental risks of pharmaceuticals in surface waters. Regul Toxicol Pharmacol. 2004; 39(2):158–183.

234 Komives T, Gullner G. Phase I xenobiotic metabolic systems in plants. Z Naturforsch (C). 2005; 60(3–4):179–185.

235 Sandermann H. Plant-metabolism of xenobiotics. Trends Biochem Sci. 1992; 17(2):82–84.

236 Sandermann H. Higher-plant metabolism of xenobiotics – the green liver concept. Pharmacogenetics. 1994; 4(5):225–241.

237 Coleman JOD, BlakeKalff MMA, Davies TGE. Detoxification of xenobiotics by plants: Chemical modification and vacuolar compartmentation. Trends Plant Sci. 1997; 2(4):144–151.

238 Kotyza J, Soudek P, Kafka Z, Vanek T. Phytoremediation of Pharmaceuticals – Preliminary Study. Int J Phytoremediat. 2010; 12(3):306–316.

239 Forni C, Cascone A, Fiori M, Migliore L. Sulphadimethoxine and *Azolla filiculoides* Lam.: a model for drug remediation. Water Res. 2002; 36(13):3398–3403.

240 Mackuľak T, Mosný M, Škubák J, Grabic R, Birošová L. Fate of psychoactive compounds in wastewater treatment plant and the possibility of their degradation using aquatic plants. Environ Toxicol Pharmacol. 2015; 39(2):969–973.

241 Datta R, Das P, Smith S, Punamiya P, Ramanathan DM, Reddy R, et al. Phytoremediation potential of vetiver grass [*Chrysopogon zizanioides* (L.)] For tetracycline. Int J Phytoremediat. 2013; 15(4):343–351.

242 Lin YL, Li BK. Removal of pharmaceuticals and personal care products by *Eichhornia crassipe* and *Pistia stratiotes.* J Taiwan Inst Chem Eng. 2016; 58:318–323.

243 Ferro S, Trentin AR, Caffieri S, Ghisi R. Antibacterial sulfonamides: accumulation and effects in barley plants. Fresenius Environ Bull. 2010; 19(9B):2094–2099.

244 Zhang Y, Lv T, Carvalho PN, Arias CA, Chen Z, Brix H. Removal of the pharmaceuticals ibuprofen and iohexol by four wetland plant species in hydroponic culture: plant uptake and microbial degradation. Environ Sci Pollut Res. 2016; 23(3):2890–2898.

245 Christofilopoulos S, Syranidou E, Gkavrou G, Manousaki E, Kalogerakis N. The role of halophyte *Juncus acutus* L. in the remediation of mixed contamination in a hydroponic greenhouse experiment. J Chem Technol Biotechnol. 2016; 91(6):1665–1674.

246 Reinhold D, Vishwanathan S, Park JJ, Oh D, Michael Saunders F. Assessment of plant-driven removal of emerging organic pollutants by duckweed. Chemosphere. 2010; 80(7):687–692.

247 Allam A, Tawfik A, Negm A, Yoshimura C, Fleifle A. Treatment of drainage water containing pharmaceuticals using duckweed (*Lemna gibba*). Energy Procedia. 2015; 74:973–980.

248 Cascone A, Forni C, Migliore L. Flumequine uptake and the aquatic duckweed, *Lemna minor* L. Water Air Soil Pollut. 2004; 156(1):241–249.

249 Migliore L, Cozzolino S, Fiori M. Phytotoxicity to and uptake of flumequine used in intensive aquaculture on the aquatic weed, *Lythrum salicaria* L. Chemosphere. 2000; 40(7):741–750.

250 Gujarathi NP, Haney BJ, Linden JC. Phytoremediation potential of *Myriophyllum aquaticum* and *Pistia stratiotes* to modify antibiotic growth promoters, tetracycline, and oxytetracycline, in aqueous wastewater systems. Int J Phytoremediat. 2005; 7(2):99–112.

251 Liu L, Liu Yh, Liu Cx, Wang Z, Dong J, Zhu Gf, et al. Potential effect and accumulation of veterinary antibiotics in *Phragmites australis* under hydroponic conditions. Ecol Eng. 2013; 53:138–143.

252 Cui H, Hense BA, Müller J, Schröder P. Short term uptake and transport process for metformin in roots of *Phragmites australis* and *Typha latifolia*. Chemosphere. 2015; 134:307–312.

253 Podlipná R, Skálová L, Seidlová H, Szotáková B, Kubíček V, Stuchláková L, et al. Biotransformation of benzimidazole anthelmintics in reed (*Phragmites australis*) as a potential tool for their detoxification in environment. Bioresour Technol. 2013; 144:216–224.

254 Sauvetre A, Schrüder P. Uptake of Carbamazepine by rhizomes and endophytic bacteria of *Phragmites australis*. Front Plant Sci. 2015; 6.

255 Carvalho PN, Basto MC, Almeida CM. Potential of *Phragmites australis* for the removal of veterinary pharmaceuticals from aquatic media. Bioresour Technol. 2012; 116:497–501.

256 Iori V, Pietrini F, Zacchini M. Assessment of ibuprofen tolerance and removal capability in *Populus nigra* L. by in vitro culture. J Hazard Mater. 2012; 229–230:217–223.

257 Iori V, Zacchini M, Pietrini F. Growth, physiological response and phytoremoval capability of two willow clones exposed to ibuprofen under hydroponic culture. J Hazard Mater. 2013; 262:796–804.

258 Michelini L, Reichel R, Werner W, Ghisi R, Thiele-Bruhn S. Sulfadiazine uptake and effects on *Salix fragilis* L. and *Zea mays* L. plants. Water Air Soil Pollut. 2012; 223(8):5243–5257.

259 Zhang DQ, Hua T, Gersberg RM, Zhu J, Ng WJ, Tan SK. Fate of diclofenac in wetland mesocosms planted with *Scirpus validus*. Ecol Eng. 2012; 49:59–64.

260 Zhang DQ, Gersberg RM, Hua T, Zhu J, Ng WJ, Tan SK. Assessment of plant-driven uptake and translocation of clofibric acid by *Scirpus validus*. Environ Sci Pollut Res. 2013; 20(7):4612–4620.

261 Dordio AV, Duarte C, Barreiros M, Carvalho AJP, Pinto AP, da Costa CT. Toxicity and removal efficiency of pharmaceutical metabolite clofibric acid by *Typha* spp. – Potential use for phytoremediation? Bioresour Technol. 2009; 100(3):1156–1161.

262 Dordio A, Ferro R, Teixeira D, Carvalho AJP, Pinto AP, Dias CMB. Study on the use of *Typha* spp. for the phytotreatment of water contaminated with ibuprofen. Intern J Environ Anal Chem. 2011; 91(7–8):654–667.

263 Cui H, Schrüder P. Uptake, translocation and possible biodegradation of the antidiabetic agent metformin by hydroponically grown *Typha latifolia*. J Hazard Mater. 2016; 308:355–361.

264 Bartha B, Huber C, Schröder P. Uptake and metabolism of diclofenac in *Typha latifolia* – How plants cope with human pharmaceutical pollution. Plant Sci. 2014; 227:12–20.

265 Garcia-Rodríguez A, Matamoros V, Fontàs C, Salvadò V. The influence of light exposure, water quality and vegetation on the removal of sulfonamides and tetracyclines: A laboratory-scale study. Chemosphere. 2013; 90(8):2297–2302.

266 Bartha B, Huber C, Harpaintner R, Schröder P. Effects of acetaminophen in *Brassica juncea* L. Czern.: investigation of uptake, translocation, detoxification, and the induced defense pathways. Environ Sci Pollut Res. 2010; 17(9):1553–1562.

267 Pietrini F, Di Baccio D, Acena J, Pérez S, Barceló D, Zacchini M. Ibuprofen exposure in *Lemna gibba* L.: Evaluation of growth and phytotoxic indicators, detection of ibuprofen and identification of its metabolites in plant and in the medium. J Hazard Mater. 2015; 300:189–193.

268 Wu X, Fu Q, Gan J. Metabolism of pharmaceutical and personal care products by carrot cell cultures. Environ Pollut. 2016; 211:141–147.

269 Huber C, Preis M, Harvey PJ, Grosse S, Letzel T, Schröder P. Emerging pollutants and plants – Metabolic activation of diclofenac by peroxidases. Chemosphere. 2016; 146:435–441.

270 Conkle JL, Gan J, Anderson MA. Degradation and sorption of commonly detected PPCPs in wetland sediments under aerobic and anaerobic conditions. J Soils Sediments. 2012; 12(7):1164–1173.

271 Stottmeister U, Wiessner A, Kuschk P, Kappelmeyer U, Kastner M, Bederski O, et al. Effects of plants and microorganisms in constructed wetlands for wastewater treatment. Biotechnol Adv. 2003; 22(1–2):93–117.

272 Zhang D, Luo J, Lee ZMP, Maspolim Y, Gersberg RM, Liu Y, et al. Characterization of bacterial communities in wetland mesocosms receiving pharmaceutical-enriched wastewater. Ecol Eng. 2016; 90:215–224.

273 Li Y, Wu B, Zhu G, Liu Y, Ng WJ, Appan A, et al. High-throughput pyrosequencing analysis of bacteria relevant to cometabolic and metabolic degradation of ibuprofen in horizontal subsurface flow constructed wetlands. Sci Total Environ. 2016; 562:604–613.

274 Syranidou E, Christofilopoulos S, Gkavrou G, Thijs S, Weyens N, Vangronsveld J, et al. Exploitation of endophytic bacteria to enhance the phytoremediation potential of the wetland helophyte *Juncus acutus*. Front Microbiol. 2016; 7.

275 Wen Y, Xu C, Liu G, Chen Y, Zhou Q. Enhanced nitrogen removal reliability and efficiency in integrated constructed wetland microcosms using zeolite. Front Environ Sci Eng China. 2012; 6(1):140–147.

276 Mestre AS, Pires J, Nogueira JMF, Carvalho AP. Activated carbons for the adsorption of ibuprofen. Carbon. 2007; 45(10):1979–1988.

277 Beninati S, Semeraro D, Mastragostino M. Adsorption of paracetamol and acetylsalicylic acid onto commercial activated carbons. Adsorpt Sci Technol. 2008; 26(9):721–734.

278 Mestre AS, Pinto ML, Pires J, Nogueira JMF, Carvalho AP. Effect of solution pH on the removal of clofibric acid by cork-based activated carbons. Carbon. 2010; 48(4):972–980.

279 Behera SK, Oh SY, Park HS. Sorptive removal of ibuprofen from water using selected soil minerals and activated carbon. Int J Environ Sci Technol. 2012; 9(1):85–94.

280 Chang EE, Wan JC, Kim H, Liang CH, Dai YD, Chiang PC. Adsorption of selected pharmaceutical compounds onto activated carbon in dilute aqueous solutions exemplified by acetaminophen, diclofenac, and sulfamethoxazole. Sci World J. 2015; 2015:11.

281 Putra EK, Pranowo R, Sunarso J, Indraswati N, Ismadji S. Performance of activated carbon and bentonite for adsorption of amoxicillin from wastewater: Mechanisms, isotherms and kinetics. Water Res. 2009; 43(9):2419–2430.

282 Dordio AV, Teimao J, Ramalho I, Carvalho AJP, Candeias AJE. Selection of a support matrix for the removal of some phenoxyacetic compounds in constructed wetlands systems. Sci Total Environ. 2007; 380(1–3):237–246.

283 Dordio AV, Candeias AJE, Pinto AP, da Costa CT, Carvalho AJP. Preliminary media screening for application in the removal of clofibric acid, carbamazepine and ibuprofen by SSF-constructed wetlands. Ecol Eng. 2009; 35(2):290–302.

284 Dordio AV, Miranda S, Prates Ramalho JP, Carvalho AJP. Mechanisms of removal of three widespread pharmaceuticals by two clay materials. J Hazard Mater. 2016; 323(A):575–583.

285 Figueroa RA, Leonard A, Mackay AA. Modeling tetracycline antibiotic sorption to clays. Environ Sci Technol. 2004; 38(2):476–483.

286 Zhang WH, Ding YJ, Boyd SA, Teppen BJ, Li H. Sorption and desorption of carbamazepine from water by smectite clays. Chemosphere. 2010; 81(7):954–960.

287 Li ZH, Chang PH, Jiang WT, Jean JS, Hong HL, Liao LB. Removal of diphenhydramine from water by swelling clay minerals. J Colloid Interface Sci. 2011; 360(1):227–232.

288 Chang PH, Li Z, Yu TL, Munkhbayer S, Kuo TH, Hung YC, et al. Sorptive removal of tetracycline from water by palygorskite. J Hazard Mater. 2009; 165(1–3):148–155.

289 Ötker HM, Akmehmet-Balcïoglu I. Adsorption and degradation of enrofloxacin, a veterinary antibiotic on natural zeolite. J Hazard Mater. 2005; 122(3):251–258.

290 Braschi I, Blasioli S, Gigli L, Gessa CE, Alberti A, Martucci A. Removal of sulfonamide antibiotics from water: Evidence of adsorption into an organophilic zeolite Y by its structural modifications. J Hazard Mater. 2010; 178(1–3):218–225.

291 Krajisnik D, Dakovic A, Milojevic M, Malenovic A, Kragovic M, Bogdanovic DB, et al. Properties of diclofenac sodium sorption onto natural zeolite modified with cetylpyridinium chloride. Colloid Surf B-Biointerfaces. 2011; 83(1):165–172.

292 Martucci A, Pasti L, Marchetti N, Cavazzini A, Dondi F, Alberti A. Adsorption of pharmaceuticals from aqueous solutions on synthetic zeolites. Microporous Mesoporous Mat. 2012; 148(1):174–183.

293 Rakic V, Rajic N, Dakovic A, Auroux A. The adsorption of salicylic acid, acetylsalicylic acid and atenolol from aqueous solutions onto natural zeolites and clays: Clinoptilolite, bentonite and kaolin. Microporous Mesoporous Mat. 2013; 166:185–194.

294 Homem V, Alves A, Santos L. Amoxicillin removal from aqueous matrices by sorption with almond shell ashes. Intern J Environ Anal Chem. 2010; 90(14–15):1063–1084.

295 Dordio AV, Gonçalves P, Teixeira D, Candeias AJE, Castanheiro JE, Pinto AP, et al. Pharmaceuticals sorption behaviour in granulated cork for the selection of a support matrix for a constructed wetlands system. Intern J Environ Anal Chem. 2011; 91(7–8):615–631.

296 Villaescusa I, Fiol N, Poch J, Bianchi A, Bazzicalupi C. Mechanism of paracetamol removal by vegetable wastes: The contribution of π–π interactions, hydrogen bonding and hydrophobic effect. Desalination. 2011; 270(1–3):135–142.

18

Role of Bacterial Diversity on PPCPs Removal in Constructed Wetlands

María Hijosa-Valsero[1,4], Ricardo Sidrach-Cardona[2], Anna Pedescoll[1], Olga Sánchez[3] and Eloy Bécares[1]

[1] *Departamento de Biodiversidad y Gestión Ambiental, Facultad de Ciencias Biológicas y Ambientales, Universidad de León, León, Spain*
[2] *Instituto de Medio Ambiente, Universidad de León, León, Spain*
[3] *Departament de Genètica i Microbiologia, Edifici C, Universitat Autònoma de Barcelona, Cerdanyola del Vallès, Barcelona, Spain*
[4] *Present address: Instituto Tecnológico Agrario de Castilla y León (ITACyL), Centro de Biocombustibles y Bioproductos, Polígono Agroindustrial, León, Spain*

18.1   Introduction

Constructed wetlands (CWs) are low-cost wastewater treatment systems, which have been proved to be able to remove organic matter and emerging pollutants [1, 2]. However, due to the large surface area per inhabitant needed to reach the target water quality parameters [1], their construction is only feasible in small urban communities or as tertiary treatments dealing with a small fraction of conventional wastewater treatment plant (WWTP) effluents.

The mechanisms involved in pollutant removal in CWs can be divided into biological processes (microbiological degradation, biofilm adsorption, plant adsorption, plant uptake, and release of plant exudates) and physico-chemical processes (photodegradation, chemical degradation, and retention or adsorption by the gravel bed). Nevertheless, CWs are complex systems presenting several environments [3] and even microenvironments, where different physico-chemical conditions can reign, influencing the various above-mentioned removal processes [4–6]. These different environments inside the system depend on the specific design of the CW. Due to this complexity, removal mechanisms in CWs are not fully understood yet.

The role of microorganisms in pollutant removal in CWs is very important [1]. Currently, one of the most active research lines in the field of CWs is the characterization of bacterial communities [7–10], because identifying these organisms, and their associated metabolic pathways, could shed some light on degradation mechanisms in CWs and contribute to the optimization of CW design parameters and daily maintenance guidelines. Although the quantitative contribution of plants to organic matter removal is more limited than that of microorganisms, their presence favors directly or indirectly the elimination of some compounds [11, 12]. An important aspect frequently neglected in the studies focused on bacterial communities is the relation between pollutant removal and bacterial diversity in the reactors, and especially in CWs. Although the biodiversity–function debate has been frequently addressed in terrestrial ecosystems (see review on this aspect in [13]), only a few papers have studied the relationship between bacterial richness and pollutant removal in wastewater

Constructed Wetlands for Industrial Wastewater Treatment, First Edition. Edited by Alexandros I. Stefanakis.
© 2018 John Wiley & Sons Ltd. Published 2018 by John Wiley & Sons Ltd.

treatment [14]. To the best of our knowledge, only Calheiros et al. [15] and Dong and Reddy [16] have mentioned this aspect with regard to CWs.

In the present work, the microbial community of several small-scale CWs treating urban wastewater was characterized, with the objective of establishing possible relationships between microbial richness and pollutant removal, and between microbial richness and physico-chemical parameters inside the systems. These CWs had been monitored for several years for the removal of conventional quality parameters (chemical oxygen demand – COD, biological oxygen demand – BOD_5, total suspended solids – TSS, volatile suspended solids – VSS, total Kjeldahl nitrogen – TKN, ammonium nitrogen – NH_4–N, nitrate and orthophosphate) and pharmaceuticals and personal care products (PPCPs); and these previously obtained data were used for this new approach.

18.2 Mesocosm-Scale Experiences

18.2.1 Description of the Systems

Seven mesocosm-scale CWs were set up in the open air inside the facilities of the León WWTP, in the northwest of Spain (42°33′35.19″ N, 5°33′45.35″ W, 807 m a.s.l.). All CWs consisted of a fiberglass container (80 cm wide, 130 cm long, 50 cm high) with a surface area of approximately 1 m². The CWs differed in their design parameters, which are summarized in Figure 18.1. All the CWs presented horizontal flow. In May 2007, seedlings were collected in nearby wet areas and planted in wetlands CW1, CW2, CW3, CW5 and CW6 with a density of 50 plants/m². Wetlands CW1, CW2 and CW3 were planted with *Typha angustifolia*. Wetlands CW5 and CW6 were planted with *Phragmites australis*. Vegetation coverage was 100% in all these CWs. Wetlands CW4 and CW7 were left unplanted. The aerial part of the plants was harvested in October 2008 and October 2009. However, the living roots remained inside the beds and the plants grew again during the subsequent warm periods. Theoretical hydraulic retention time (HRT) values of tanks CW1, CW2, CW3, CW4, CW5, CW6 and CW7 were, respectively, 45.7, 69.0, 68.0, 61.5, 53.6, 42.2 and 34.6 h. Actual HRTs measured at the end of the experimental period (October 2010) were 37.4, 29.6, 37.8, 54, 37.4, 71.1 and 41.8 h, respectively [17].

León WWTP consists of a primary treatment (screening, sand removal, fat removal and primary clarifier) and a secondary treatment (plug-flow activated sludge with nitrification/denitrification and secondary clarifier). The WWTP was designed to treat the wastewater of 330,000 equivalent inhabitants with an inflow of 123,000 m³/d and a HRT of about 6 h. Urban wastewater coming from the primary clarifier of the WWTP was conducted to a homogenization tank of 0.5 m³ in order to feed all the CWs at a continuous flow rate of 50 L/d (input load 50 mm/d and a BOD_5 load 3 g/m²/d).

18.2.2 Sampling Strategy

The sampling was performed during the last operational period of the systems, specifically in summer 2010. The sampling strategy was different for every analyzed matrix:

(a) Aqueous influent and effluent grab samples were collected once a week while the CWs were functioning (n = 6, in July-September 2010) at the seven CWs, always on the same day and at the same time. Samples were collected in 1 L amber glass bottles, which were transported refrigerated

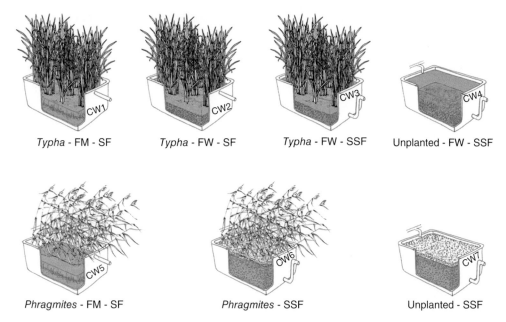

Figure 18.1 Schematic design characteristics of the CWs. Notes: FM: floating macrophytes and soilless, FW: free-water layer, SF: surface flow, SSF: subsurface flow. Systems CW1 and CW5 had a water layer of 30 cm and plant growth was supported by 20 cm long and 10 cm diameter garden-net cylinders (4 cm pore size) (see more details in Hijosa-Valsero et al. [4]). Systems CW2, CW3 and CW4 had a 25 cm layer of free-water (FW) over a 25 cm layer of siliceous gravel (d10 = 4 mm). Systems CW6 and CW7 consisted of a 45 cm siliceous gravel (d10 = 4 mm) layer, with an operational water depth of 40 cm.

(4°C) to the laboratory. These samples were used to analyze conventional pollutant and PPCP concentrations, and calculate their removal efficiencies.

(b) Physico-chemical parameters (temperature, pH, conductivity, dissolved oxygen and redox potential) were measured *in situ* at two different depths (5 cm below the water surface and 5 cm above the bottom of the tank) in each CW and the homogenization tank (n = 10, in July–September 2010).

(c) Interstitial water samples were taken at the end of the experimental season (in September 2010). For their collection, a 1.5 cm diameter × 1.5 m long iron bar was hammered vertically into four points at two different depths in the seven CWs, whenever it was necessary to cross the roots and/or gravel layer (Figure 18.2a), in order to create a channel to introduce a tube immediately after the bar had been removed. The tube was connected to a syringe, and 250 mL interstitial liquid were extracted and introduced in a glass bottle. Interstitial water samples were collected at two depths: 5 cm below the water surface and 5 cm above the tank bottom, and at four different positions from the loading point (Figure 18.2b). Then, those eight samples were merged to obtain an integrated sample, where the microbial community was analyzed.

(d) As for vegetal tissues and gravel samples, they were collected once the systems were dismantled (n = 1, in October 2010). However, for vegetal tissues, only an integrated sample of roots from the last longitudinal third of the CW was considered (the collected tissues corresponded to 15 cm of

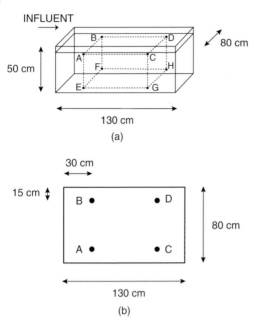

INFLUENT

Figure 18.2 Interstitial liquid sampling: (a) general diagram showing the sampling points (A, B, C, D, E, F, G and H) inside a CW. Points A, B, C and D were located 5 cm below the surface (regardless of gravel or water surface) and points E, F, G and H were 5 cm above the bottom of the tank. (b) Top view of a CW indicating the distances at which samples were taken.

the proximal part of the roots, i.e., the part which is closest to the stem, and included roots and rhizomes). Similarly, concerning substrate, an integrated sample of gravel from the last longitudinal third of the CW was taken from CW2, CW3, CW4, CW6 and CW7 (from the upper 15-cm of the gravel layer, regardless of the system configuration: FW-SSF or SSF). All these samples were used in the analysis of the microbial community. The fact of sampling gravel only from the upper part of the substrate layer does not imply a great error, since previous studies have shown that the majority of microbial biomass and bacterial activity is found in the first centimeters of the filter layer [18].

(e) For general biomass estimations (fresh weight and dry weight), all the macrophyte individuals were collected from the planted CWs (CW1, CW2, CW3, CW5 and CW6) when the systems were dismantled (in October 2010).

18.2.3 Analytical Methodology

Conventional quality parameters, like COD, BOD_5, TSS, VSS, TKN, NH_4-N, nitrate and orthophosphate were analyzed by using Standard Methods 5220 C, 5210 B, 2450 D, 2540 E, $4500-N_{org}$ B, $4500-NH_3$ C, $4500-NO_3^-$ D and 4500-P E, respectively [19]. PPCPs in wastewater samples were analyzed following a previously described SPE-GC/MS methodology [20, 21]. Temperature, pH, conductivity, dissolved oxygen and redox potential were measured *in situ* with probes (WTW, Weilheim, Germany).

The vegetal biomass (roots and stems) was collected and weighed in situ with a precision balance, representing the fresh weight. This biomass was then taken to the laboratory, where it was dried during 24 h at 65°C in an oven Model 374-A, P-Selecta (Abrera, Spain), obtaining then the dry weight (Table 18.1).

Table 18.1 Vegetal biomass measured at the end of the experiment (October 2010).

System	Leaves and stems		Roots	
	Fresh weight (kg/m²)	Dry weight (kg/m²)	Fresh weight (kg/m²)	Dry weight (kg/m²)
CW1 (*Typha*-FM-SF)	47.25	11.70	112.35	24.46
CW2 (*Typha*-FW-SF)	55.25	9.14	81.50	14.39
CW3 (*Typha*-FW-SSF)	54.77	14.08	74.21	10.10
CW4 (Unplanted-FW-SSF)	–	–	–	–
CW5 (*Phragmites*-FM-SF)	14.25	7.33	85.15	9.51
CW6 (*Phragmites*-SSF)	16.00	8.77	74.25	13.41
CW7 (Unplanted-SSF)	–	–	–	–

Microbial communities were characterized in gravel biofilm, plant roots and interstitial liquid. The denaturing gradient gel electrophoresis (DGGE) technique was applied. The procedure consists of the extraction of the microbial community from the matrix, followed by a DNA extraction and amplification by polymerase chain reaction (PCR) and, finally, a DGGE separation. This process is described in detail by Sidrach-Cardona [22]. All the bands appearing on the gel were counted, and every different band was considered an operational taxonomic unit (OTU). Since the number of OTUs in each studied environment can be roughly assimilated to the number of taxa, for the present study and as a simplification, microbial richness was considered as the number of OTUs.

18.3 Pollutant Concentrations and Removal Efficiencies in Mesocosms CWs

The global quality parameters of the influent wastewater (COD, BOD_5, TSS, VSS, $N-NH_4^+$, TKN and orthophosphate), as well as PPCP concentrations had been evaluated in previous studies [23, 24]. Table 18.2 summarizes the main characteristics of the influent wastewater, including physico-chemical parameters.

The removal efficiency for the above-mentioned pollutants in each CW is shown in Table 18.3. The variations in physico-chemical parameters in every treatment system were also recorded (Table 18.4). All these data were needed to elucidate the existence of possible relationships between removal performances in each wetland and the microbial communities.

18.4 Microbiological Characterization

The bands appearing in the DGGE analysis were counted separately for every environment (gravel, interstitial liquid and root surface) in each CW. The analysis was performed in duplicate and, in order to establish a richness value, for every pair of data, the highest value was chosen. The so calculated

Table 18.2 Average values and 0.95 confidence intervals for some pollutants and physico-chemical parameters in the urban wastewater used to feed the CWs in summer 2010. Data adapted from Hijosa-Valsero et al. [23]; Reyes-Contreras et al. [24].

PPCPs	Concentration
Ketoprofen (μg/L)	0.75 ± 0.11
Naproxen (μg/L)	1.74 ± 0.31
Ibuprofen (μg/L)	10.4 ± 2.94
Diclofenac (μg/L)	0.41 ± 0.05
Salicylic acid (μg/L)	15.9 ± 4.68
Caffeine (μg/L)	19.2 ± 1.43
Carbamazepine (μg/L)	0.99 ± 0.27
Methyl dihydrojasmonate (μg/L)	7.27 ± 1.41
Galaxolide (μg/L)	3.84 ± 1.07
Tonalide (μg/L)	0.25 ± 0.09
Organic matter and nutrients	
COD (mg/L)	131 ± 30
BOD_5 (mg/L)	74 ± 12
TSS (mg/L)	49 ± 6
VSS (mg/L)	44 ± 5
TKN (mg/L)	20 ± 2
NH_4-N (mg/L)	16 ± 2
NO_3^- (mg/L)	0.26 ± 0.09
Orthophosphate (mg/L)	2.29 ± 0.33
Physico-chemical parameters	
Temperature (°C)	19.4 ± 0.4
pH	7.0 ± 0.0
Conductivity (μS/cm)	587 ± 32
Dissolved oxygen (mg/L)	0.3 ± 0.1
Redox potential (mV)	-100 ± 21

microbial richness can be found in Table 18.5. Regarding the microbial community on the gravel, it was observed that CW4 (unplanted-FW-SSF) presented the highest richness, whereas CW2 (*Typha*-FW-SF) showed the lowest one. Surprisingly, the microbial richness of the interstitial liquid seemed to be inversely correlated with the gravel richness (i.e., when the gravel richness was high in a certain CW, its interstitial liquid richness was low; and vice versa). The microbial community in the interstitial liquid of CW1 and CW5 (both soilless systems) was not analyzed because these systems were working as hydroponic ones and no gravel was present. Concerning microbial richness on the roots, it was noticed that *Phragmites*-systems (CW5 and CW6) exhibited higher richness than *Typha*-systems (CW1, CW2 and CW3). Microbial richness was quite different depending on the studied environment (gravel, roots or liquid). In a horizontal surface flow (H-SF) CW used to treat

Table 18.3 Removal efficiencies (%) and 0.95 confidence intervals for the studied PPCPs and organic pollutants in the CWs (summer 2010). Data adapted from Hijosa-Valsero et al. [23]; Reyes-Contreras et al. [24].

	CW1	CW2	CW3	CW4	CW5	CW6	CW7
Ketoprofen	12 ± 42	n.r.	n.r.	79 ± 11	6 ± 69	n.r.	8 ± 49
Naproxen	47 ± 13	24 ± 50	92 ± 20	75 ± 13	71 ± 22	65 ± 3	55 ± 23
Ibuprofen	33 ± 23	n.r.	48 ± 20	74 ± 34	72 ± 20	32 ± 14	10 ± 38
Diclofenac	n.r.	10 ± 30	n.r.	1 ± 39	16 ± 54	n.r.	n.r.
Salicylic acid	95 ± 3	91 ± 3	95 ± 2	95 ± 1	91 ± 11	88 ± 18	89 ± 8
Caffeine	95 ± 2	84 ± 14	90 ± 15	99 ± 1	30 ± 128	96 ± 2	97 ± 3
Carbamazepine	53 ± 50	n.r.	n.r.	n.r.	n.r.	n.r.	n.r.
Methyl dihydrojasmonate	81 ± 7	83 ± 10	90 ± 6	87 ± 11	92 ± 5	85 ± 7	85 ± 11
Galaxolide	94 ± 3	66 ± 16	70 ± 14	67 ± 10	88 ± 6	82 ± 6	31 ± 43
Tonalide	69 ± 33	57 ± 39	55 ± 39	62 ± 13	76 ± 26	83 ± 5	n.r.
COD	61 ± 14	49 ± 17	66 ± 11	33 ± 20	67 ± 12	69 ± 12	55 ± 15
BOD_5	85 ± 6	84 ± 7	87 ± 6	78 ± 4	90 ± 3	87 ± 5	85 ± 7
TSS	84 ± 5	83 ± 4	85 ± 5	60 ± 12	85 ± 7	57 ± 17	85 ± 5
VSS	83 ± 5	83 ± 4	87 ± 3	59 ± 12	85 ± 5	77 ± 6	86 ± 4
TKN	10 ± 15	0.5 ± 9.2	6 ± 10	-22 ± 16	78 ± 12	88 ± 6	-1 ± 11
NH_4-N	-4 ± 16	-19 ± 15	-12 ± 19	-45 ± 30	84 ± 12	93 ± 6	-25 ± 19
NO_3^-	23 ± 32	24 ± 23	58 ± 18	59 ± 23	42 ± 27	-134 ± 187	44 ± 23
Orthophosphate	5 ± 26	-7 ± 25	-20 ± 20	-36 ± 22	29 ± 27	91 ± 6	-30 ± 16

n.r. = not removed.

a mixture of domestic wastewater, Li et al. [25] had already observed that the bacterial communities living in/on roots were considerably distinct from those occurring in the water.

In the present study, when comparing similar CWs, which only differed in the presence/absence of plants (like the pair CW3-CW4, or the pair CW6-CW7), there were no clear evidences to state whether planted CWs exhibited a higher microbial richness on gravel or in interstitial liquid than unplanted systems (Table 18.5). In general, there is no scientific consensus about this issue. For instance, whereas some authors have denied the influence of vegetal presence on CW microbial communities [18, 26, 27]; others have found a relationship between those two variables [28–30].

Regarding the connection between microbial communities and the type of CW, it has been observed that microbial biomass (i.e., the total amount of organisms) is within a similar range in horizontal SSF (H-SSF-CW), vertical SSF (V-SSF-CW) and SF-CWs [10]. However, the composition of microbial communities could be very different in every type of CW. In addition, the nature of the wastewater can also condition that community. Even nutrient concentrations in the influent could modify the microbial community [31, 32].

The comparison between vegetal species (CW1-CW5) revealed that the root community associated to *P. australis* was richer than that associated to *T. angustifolia* (Table 18.5). This is in agreement with previous works. In an H-SF-CW dealing with a mixture of domestic wastewater, Li et al. [25] observed that root-associated bacterial diversity and richness was higher in *P. australis* than in *T. angustifolia*;

Table 18.4 Average values and 0.95 confidence intervals for water physico-chemical parameters in every CW in summer 2010. All the values were measured in situ at two different depths: SUP, superior (5 cm below water surface), and INF, inferior (5 cm above the bottom of the tank). Data adapted from Hijosa-Valsero et al. [23].

Parameter	CW1	CW2	CW3	CW4	CW5	CW6	CW7
Temperature (°C) SUP	18.2 ± 0.6	17.9 ± 0.6	18.2 ± 0.7	17.8 ± 0.9	19.6 ± 0.8	19.6 ± 1.0	19.2 ± 1.0
Temperature (°C) INF	18.1 ± 0.7	19.8 ± 0.8	19.4 ± 0.9	18.4 ± 0.9	19.6 ± 0.8	19.8 ± 1.2	18.8 ± 0.9
pH SUP	7.1 ± 0.1	6.9 ± 0.1	6.9 ± 0.1	7.6 ± 0.0	6.9 ± 0.2	6.5 ± 0.0	7.2 ± 0.0
pH INF	6.9 ± 0.1	6.6 ± 0.1	6.4 ± 0.1	7.2 ± 0.0	6.7 ± 0.1	6.5 ± 0.0	7.1 ± 0.0
Conductivity (µS/cm) SUP	664 ± 50	650 ± 47	658 ± 30	631 ± 19	1271 ± 101	808 ± 123	644 ± 43
Conductivity (µS/cm) INF	662 ± 44	735 ± 84	683 ± 108	615 ± 17	1504 ± 175	1024 ± 177	653 ± 49
Dissolved oxygen (mg/L) SUP	0.2 ± 0.1	0.2 ± 0.1	0.2 ± 0.1	0.3 ± 0.1	0.3 ± 0.1	0.3 ± 0.1	0.3 ± 0.1
Dissolved oxygen (mg/L) INF	0.2 ± 0.1	0.2 ± 0.1	0.2 ± 0.1	0.1 ± 0.0	0.2 ± 0.1	0.2 ± 0.0	0.2 ± 0.0
Redox potential (mV) SUP	-101 ± 18	-71 ± 19	-75 ± 25	-15 ± 24	-58 ± 49	-20 ± 28	-80 ± 32
Redox potential (mV) INF	-126 ± 23	-201 ± 17	-134 ± 24	-132 ± 28	-107 ± 36	-102 ± 65	-155 ± 40

Table 18.5 Richness of the microbial community of every environment in CWs. Richness is expressed as number of bands in the DGGE analysis.

System	Gravel	Interstitial liquid	Roots
CW1 (*Typha*-FM-SF)	–	n.a.	26
CW2 (*Typha*-FW-SF)	28	51	28
CW3 (*Typha*-FW-SSF)	32	46	29
CW4 (Unplanted-FW-SSF)	38	40	–
CW5 (*Phragmites*-FM-SF)	–	n.a.	33
CW6 (*Phragmites*-SSF)	34	44	36
CW7 (Unplanted-SSF)	32	44	–

Note: n.a. Not analyzed.

in addition, communities also differed among plant species. Calheiros et al. [8] reported that distinct bacterial communities were found in two-stage H-SSF-CWs treating tannery wastewater planted with *P. australis* and *T. latifolia*, respectively. Arroyo et al. [28] compared mesocosm-scale CWs and concluded that the highest diversity and richness were found in those bacterial communities occupying mesocosms planted with *P. australis* in comparison to those planted with *T. latifolia*.

18.5 Link between Microbiological Richness and Pollutant Removal in CWs

It is generally agreed that microorganisms play a very important role in pollutant removal in CWs [1, 11, 33], both for macropollutants (like COD and BOD) and for micropollutants (like PPCPs). In the present work, a possible connection between microbial richness (number of different taxa inside a community) and the general removal performance in CWs was examined. Three different environments were considered separately for this study: gravel, interstitial liquid and roots. The richness of their respective communities, expressed as number of bands appearing in the gel after applying the DGGE technique, is shown in Table 18.5. In order to establish the possible connection between microbial richness and the removal efficiency of CWs, linear correlations were calculated.

18.5.1 Microbial Richness and Conventional Pollutant Removal

In this section, correlations between microbial richness in each CW environment and the overall conventional pollutant removal (percentage elimination of COD, BOD_5, TSS, VSS, $N-NH_4^+$, TKN and orthophosphate; see Table 18.3) were looked for. Only significant correlations ($p < 0.10$) will be shown and discussed.

18.5.1.1 Roots
In this case, positive correlations were found between microbial richness on the root biofilm and nutrient removal (NH_4^+-N, TKN and orthophosphate) (Figure 18.3). Previous researches had found positive correlations between bacterial richness/diversity and the removal of other common

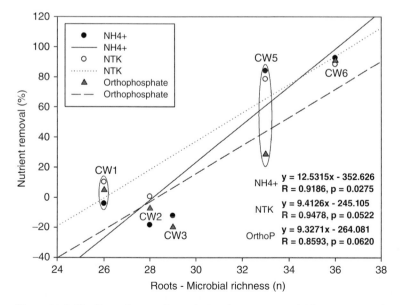

Figure 18.3 Significant ($p < 0.10$) correlations between microbial community richness in the studied environment (roots) and nutrient removal in CWs. Richness is expressed as number of bands in the DGGE analysis.

pollutants. For instance, Calheiros et al. [15] studied H-SSF-CWs treating tannery wastewater, and suggested that COD and BOD_5 were more easily eliminated in those systems with higher Shannon diversity indexes. On the contrary, Dong and Reddy [16] stated that there was no correlation between microbial richness or diversity values and nitrification processes in CWs treating swine wastewater. In the present work, it is interesting to notice that *Phragmites*-systems (CW5 and CW6) were the most efficient for nutrient removal (Table 18.3) and they were also the systems with the highest root microbial richness. These two facts contribute undoubtedly to the cited correlation. This agrees partly with the observations of Li et al. [25], who reported that many more species of bacteria involved in the total nitrogen cycle were observed in *P. australis* in comparison to *T. angustifolia*, but asserted that species involved in total phosphorus and organic matter removal were mainly found in *T. angustifolia*. However, it must be remembered that phosphorus elimination in CWs occurs mainly by sorption processes and not by biodegradation.

18.5.2 Microbial Richness and PPCP Removal

The existence of correlations between microbial richness in each CW environment and the overall PPCP removal (expressed as percentage; see Table 18.3) was evaluated. Only significant correlations ($p < 0.10$) are provided and discussed.

18.5.2.1 Gravel
Caffeine and naproxen were more efficiently removed when the microbial richness in the gravel environment was higher (Figure 18.4). Gravel microorganisms could be involved in the elimination of these substances.

18.5.2.2 Interstitial Liquid
On the contrary, a negative correlation appeared between microbial richness in the interstitial liquid and caffeine and naproxen removal (Figure 18.4). Interestingly, those wetlands with more gravel microbial richness (like CW4) are those with less richness in the interstitial liquid, and vice versa (Table 18.5). That explains the observation of lower caffeine and naproxen removal efficiencies in microbially rich interstitial environments.

18.5.2.3 Roots
There was a negative correlation between root microbial richness and carbamazepine removal (Figure 18.4). The highest richness was observed for the root communities of *P. australis* systems (CW5, CW6, see Table 18.5). Accordingly, carbamazepine is more efficiently removed by systems with lower root microbial richness, which are those planted with *T. angustifolia*. Therefore, *T. angustifolia* or its associated microorganisms (if somewhat less rich), contribute to carbamazepine removal. In any case, it cannot be said that root microbial richness is detrimental for PPCP removal. In fact, in a work studying a CW where *P. australis* and *T. angustifolia* were present, Li et al. [25] found that some bacterial species found exclusively on *T. angustifolia* roots had been typically reported to be involved in the metabolic or cometabolic pathways degrading chemical pollutants like biphenyl, polychlorinated biphenyls (PCBs), methanol, estradiol, halogenated ethenes and propenes, catechol, urea, tartrate isomers, etc.

It is remarkable that the removal of only three out of ten PPCPs analyzed is related to microbial richness. This fact could have several explanations: (a) Most PPCPs do not follow biological degradation routes, but chemical or physical degradation processes; this statement contradicts previously

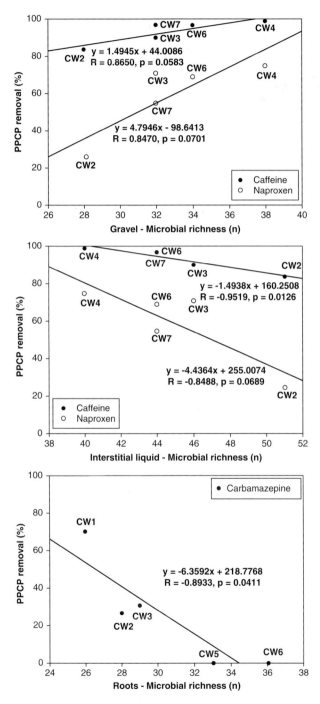

Figure 18.4 Significant (p < 0.10) correlations between microbial community richness in each studied environment and general PPCP removal in CWs. Richness is expressed as number of bands in the DGGE analysis.

published experimental results, a fact which weakens the validity of this hypothesis. It has been suggested that a biological phenomenon like the presence of plants in CWs contributes to a lesser or greater extent to the removal of ketoprofen, naproxen, ibuprofen, diclofenac, salicylic acid, galaxolide, tonalide and especially carbamazepine [5]. It has also been proposed that salicylic acid, caffeine, methyl dihydrojasmonate and galaxolide are probably removed by biodegradation or plant-mediated processes [24]. (b) The composition of microbial communities could be more important than its richness for the removal of certain anthropogenic organic compounds. This seems logical, because some bacterial species and strains are able to metabolize or co-metabolize persistent and dangerous chemical pollutants [34], whereas other species or strains cannot.

In any case, bacterial richness should be a determinant parameter when a compound needs various enzymes (from different bacteria) to be degraded. Therefore, those compounds which are known to suffer a biodegradation but do not show a correlation with microbial richness are perhaps being degraded by a single microbial taxon. Conversely, those PPCPs whose removal is correlated to microbial richness are maybe degraded by a group of different bacteria. The latter could be the case of caffeine and naproxen in gravel. However, the case of carbamazepine is a bit different, because its removal is probably related to the presence of *T. angustifolia*.

18.5.3 Effect of Physico-Chemical Parameters on Microbial Richness

Microbial communities are adapted to their environmental conditions; therefore it is expected that any alteration in physico-chemical parameters (temperature, insolation, pH, redox potential, dissolved oxygen, etc.) will cause a change in the composition of the microbial community. Moreover, physico-chemical parameters usually present spatial gradients inside CWs, thus creating micro-environments and allowing the existence of very different microbial communities in a relatively small system [3]. In this section, we tried to figure out whether microbial richness had any relationship with the measured physico-chemical parameters (temperature, pH, conductivity, dissolved oxygen and redox potential).

18.5.3.1 Gravel
The richness of the gravel biofilm community was only positively correlated with the concentration of dissolved oxygen in the upper part of the wetland (Figure 18.5). This means that the most oxygenated systems showed a greater gravel microbial richness.

18.5.3.2 Interstitial Liquid
On the other hand, as mentioned above (see Table 18.5), gravel richness and interstitial richness were inversely related. Because of that, a negative correlation was found between bacterial richness in the interstitial liquid and oxygen concentration in the upper part of the wetland (Figure 18.5).

18.5.3.3 Roots
The microbial community found in the root biofilm seemed to be more influenced by physico-chemical parameters. In fact, positive correlations were detected between richness and dissolved oxygen concentration, redox potential and temperature, and a negative correlation was recorded between richness and the pH value (Figure 18.5). Once more, this could be associated to the plant type, since *P. australis* systems (CW5 and CW6) presented the greatest root microbial richness (Table 18.5),

and also the highest water temperature, conductivity, dissolved oxygen concentration and redox potential, and the lowest pH (Table 18.4).

The influence of physico-chemical parameters in pollutant removal in CWs is evident. In small-scale V-SSF-CW planted with *P. australis*, higher redox potential values benefitted ammonia removal but limited total nitrogen removal [35]. Hijosa-Valsero et al. [5] studied several small-scale CWs and observed that temperature conditioned the removal of salicylic acid, caffeine, fragrances and nutrients; that the pH value was involved in the elimination of tonalide and nitrogen; and that dissolved oxygen concentration and redox potential affected the removal of some PPCPs, COD, BOD_5 and nutrients.

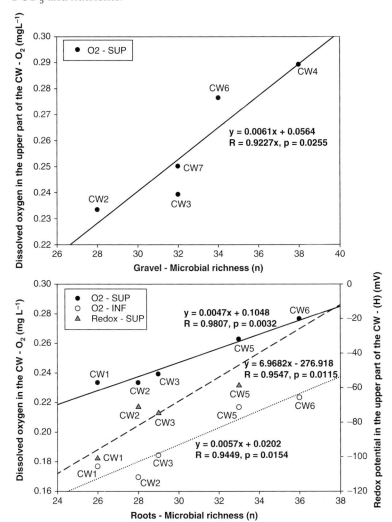

Figure 18.5 Significant ($p < 0.10$) correlations between microbial community richness in each studied environment and physico-chemical parameters inside CWs. Richness is expressed as number of bands in the DGGE analysis. Note: SUP, superior (5 cm below water surface), and INF, inferior (5 cm above the bottom of the tank).

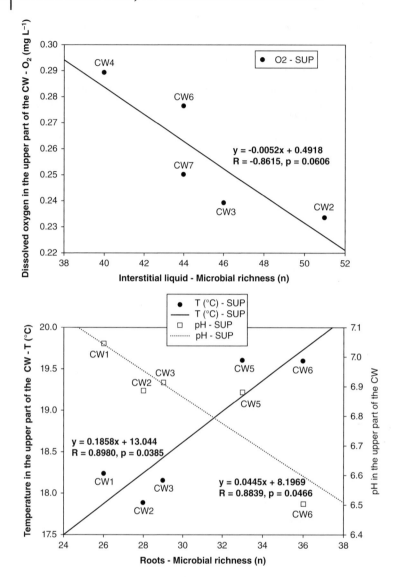

Figure 18.5 *(Continued)*

18.6 Mechanisms and Design Parameters Involved in PPCPs Removal

It is still difficult to describe the actual removal mechanisms for PPCPs because various removal pathways are simultaneously involved [36, 37], such as biodegradation, adsorption, plant uptake or photodegradation. For instance, the assessment of PPCP transformation products in pore water suggested some possible degradation routes in CWs. One of the main degradation routes for most compounds seems to be due to aerobic metabolism, since oxidized forms of certain compounds were

Table 18.6 Accepted removal mechanisms for the substances considered in this chapter.

Compound	Aerobic biodegradation	Anaerobic biodegradation	Adsorption	Photodegradation	Plant uptake	References
Analgesic anti-inflammatory drugs						
Ketoprofen				+		[4, 24, 54, 55]
Naproxen	+	+	+	+	+	[4, 24, 38, 53, 55–60]
Ibuprofen	+		+	+	+	[4, 20, 21, 24, 38, 53, 55–59, 61–63]
Diclofenac	+	+	+	+	+	[4, 24, 53, 55, 56, 58, 59, 61, 64–67]
Salicylic acid	+				+	[4, 21, 24, 38, 53]
Stimulant drugs						
Caffeine	+	+			+	[4, 21, 24, 53, 56–58, 65, 68–70]
Antiepileptic drugs						
Carbamazepine			+		+	[4, 20, 24, 55, 60, 62, 63, 71–76]
Fragrances						
Methyl dihydrojasmonate	+				+	[4, 38]
Galaxolide	+		+		+	[4, 21, 38, 65]
Tonalide	+		+		+	[4, 21, 38, 55, 56, 64, 72]

detected in the pore water [38, 39]. In fact, aerobic processes are known to be more efficient than anaerobic ones. Especially ibuprofen has been reported to be an easily biodegradable compound, whose removal increases with oxic conditions and plant presence [5, 40]. Table 18.6 shows the accepted removal mechanisms for the substances considered in this chapter. However, some removal mechanisms must predominate depending on wetland design [37]. Truu et al. [10] indicated that microbial mediated processes in CWs are dependent on hydraulic conditions, wastewater properties; including substrate and nutrient quality and availability, filter material, plants, and environmental factors.

Physico-chemical parameters influence both biotic and abiotic processes and determine the dominant microbiological populations present inside the CW and the metabolic pathways that PPCPs will take. In the systems considered here, high temperatures had a significant positive effect on the degradation of caffeine, naproxen, salicylic acid, methyl dihydrojasmonate, galaxolide and tonalide [4]. Microorganisms in CWs usually reach their optimal activity between 15–25°C [10]. Moreover, abiotic processes like adsorption are also temperature-dependent, since they are exothermic processes, hence favored by low temperatures. On the other hand, oxidized conditions favor biodegradation processes. At the experimental plant, high oxygen concentrations and redox potential significantly aided the degradation of diclofenac, ibuprofen, salicylic acid and carbamazepine; whereas low ones favored caffeine, methyl dihydrojasmonate, galaxolide and tonalide removal efficiencies [4]. The presence of oxygen or oxidized conditions enhances microbial processes depending on this electron acceptor, such as nitrification and aerobic respiration.

According to design characteristics, some PPCP removal was favored by plant presence. This fact could be related to the modification of redox values near the roots (usually creating aerobic environments). Plants are generally considered beneficial in CWs, by taking up and assimilating nutrients, by providing surface for biofilm, by O_2 pumping and release or by insulating against low temperatures [41,42]. Moreover, the activities of many microbiological enzymes depended on the presence or absence of plants in the substrate of CWs [43, 44].

The study of vegetal interactions with PPCPs also shed some light on CW functioning for the removal of organic pollutants [38]. Fragrances (methyl dihydrojasmonate, galaxolide and tonalide) and analgesics (naproxen, ibuprofen and salicylic acid) were detected and quantified in roots of both *Typha angustifolia* and *Phragmites australis* in the experimental CWs, proving that a portion of the studied PPCPs had been incorporated to vegetal tissues. Moreover, it is expectable that plants have taken up transformation products and generated their own either by endophytic bacteria [45] or plant metabolism [46]. Although *P. australis* proved more efficient than *T. angustifolia* in summer for the removal of ibuprofen, diclofenac, caffeine and methyl dihydrojasmonate (CW1 vs. CW5) (data not shown), there is no clear agreement on which species is most efficient at removing organic matter and nutrients from wastewater [47]. The ability of a plant to enhance pollutant removal depends not only on the typical species characteristics, but also on many other factors, such as microbial communities related to them, wastewater nature and climate conditions [4].

Some physico-chemical parameters (such as redox potential and insolation) in CWs are strongly influenced by the flow type. Some compounds are sensitive to photodegradation, such as ketoprofen [48, 49], naproxen [49–51], or diclofenac [51, 52]. The unplanted FW-SSF-CW (CW4) gave better results for the removal of naproxen and diclofenac than conventional SSF CW7. Hijosa-Valsero et al. [53] reported a removal efficiency of 85% for naproxen and 65% for diclofenac in a pond system during summer (a season with strong sunlight).

The great importance of the design configuration is noticeable for the removal of lipophilic fragrances (galaxolide and tonalide). These musks, mainly retained in CWs by adsorption processes [21, 54], should be more efficiently removed in conventional SSF-CWs (such as CW6 and CW7) due to the greater presence of gravel and hence larger adsorption surface. However, SF-CWs and hybrid systems remove galaxolide and tonalide more efficiently than horizontal SSF-CWs [21, 53, 54]. Microscopic algae and other suspended solids in the free-water layer of some kinds of CWs besides macrophyte leaves, stems and roots, could act as important adsorption surfaces.

18.7 Conclusions

The results obtained in the present work lead to the following conclusions:

- The vegetal species present in the CW determines the composition of the microbial community on the roots and the rhizosphere. Some specific microbial strains associated to a particular vegetal species could be responsible for the degradation of certain chemical pollutants.

- Nutrient removal in the studied CWs is directly related to microbial richness. In any case, the influence of other factors in the appearance of that relation cannot be discarded (such as CW configuration).
- In the case of PPCP pollution, the composition of the bacterial community (i.e., the identity of the taxa) could be more relevant than the bacterial richness (i.e., number of different taxa) as far as removal efficiency is concerned.
- The richness of the microbial community on the root biofilm is markedly influenced by physico-chemical parameters. However, this could be interrelated to the CW configuration and/or the plant species, since they can modify physico-chemical parameters inside the system.
- Differences in design, dimensions, vegetal species, wastewater nature, administered organic load, microbial diversity worldwide, physico-chemical and climatic parameters, etc. make it complicated to establish general patterns regarding CW microbial communities, as well as general organic matter degradation routes in which those microorganisms take part.

Acknowledgements

This study was funded by the Spanish Ministry of Science and Innovation (projects CTM2005-06457-C05-03 and CTM2008-06676-C05-03), by the Castilla y León Regional Government (projects LE009A07 and LE037A10-2) and by MAPFRE (project AG-180). The authors thank J.C. Sánchez for the maintenance of the systems. We thank C. Reyes-Contreras and Josep M. Bayona for PPCPs analyses (IDAEA-CSIC). We are grateful to L. Garrido and J. Mas for DGGE analyses (Universitat Autònoma de Barcelona). We thank Acciona Agua and Mancomunidad de Saneamiento de León y su Alfoz for their technical support.

References

1 Kadlec RH, Wallace S. Treatment Wetlands, 2nd edn. CRC Press, Taylor & Francis Group, Boca Raton, 2009.

2 Vymazal J. The use constructed wetlands with horizontal sub-surface flow for various types of wastewater. Ecol Eng. 2008; 35:1–17.

3 Imfeld G, Braeckevelt M, Kuschk P, Richnow HH. Monitoring and assessing processes of organic chemicals removal in constructed wetlands. Chemosphere. 2009; 74:349–362.

4 Hijosa-Valsero M, Matamoros V, Sidrach-Cardona R, Martín-Villacorta J, Bécares E, Bayona JM. Comprehensive assessment of the design configuration of constructed wetlands for the removal of pharmaceuticals and personal care products from urban wastewaters. Water Res. 2010; 44:3669–3678.

5 Hijosa-Valsero M, Sidrach-Cardona R, Martín-Villacorta J, Valsero-Blanco MC, Bayona JM, Bécares E. Statistical modelling of organic matter and emerging pollutants removal in constructed wetlands. Bioresour Technol. 2011; 102:4981–4988.

6 Wießner A, Kappelmeyer U, Kuschk P, Kästner M. Influence of the redox condition dynamics on the removal efficiency of a laboratory-scale constructed wetland. Water Res. 2005; 39:248–256.

7 Belila A, Snoussi M, Hassan A. Rapid qualitative characterization of bacterial community in eutrophicated wastewater stabilization plant by T-RFLP method based on 16S rRNA genes. World J Microbiol Biotechnol. 2012; 28:135–143.

8 Calheiros CSC, Duque AF, Moura A, Henriques IS, Correia A, Rangel AOSS, Castro PML. Changes in the bacterial community structure in two-stage constructed wetlands with different plants for industrial wastewater treatment. Bioresour Technol. 2009; 100: 3228–3235.

9 Ibekwe AM, Grieve CM, Lyon SR. Characterization of microbial communities and composition in constructed dairy wetland wastewater effluent. Appl Environ Microbiol. 2003; 69:5060–5069.

10 Truu M, Juhanson J, Truu J. Microbial biomass, activity and community composition in constructed wetlands. Sci Total Environ. 2009; 407:3958–3971.

11 Stottmeister U, Wießner A, Kuschk P, Kappelmeyer U, Kästner M, Bederski O, Müller RA, Moormann H. Effects of plants and microorganisms in constructed wetlands for wastewater treatment. Biotechnol Adv. 2003; 22:93–117.

12 Zhang D, Gersberg RM, Keat TS. Constructed wetlands in China. Ecol Eng. 2009; 35:1367–1378.

13 Loreau M, Naeem S, Inchausti P, Bengtsson J, Grime JP, Hector A, Hooper DU, Huston MA, Raffaelli D, Schmid B, Tilman D, Wardle DA. Biodiversity and ecosystem functioning: current knowledge and future challenges. Science. 2001; 294:804–808.

14 Briones A, Raskin L. Diversity and dynamics of microbial communities in engineered environments and their implications for process stability. Curr Opin Biotechnol. 2003; 14:270–276.

15 Calheiros CSC, Teixeira A, Pires C, Franco AR, Duque AF, Crispim LFC, Moura SC, Castro PML. Bacterial community dynamics in horizontal flow constructed wetlands with different plants for high salinity industrial wastewater polishing. Water Res. 2010; 44:5032–5038.

16 Dong X, Reddy GB. Ammonia-oxidizing bacterial community and nitrification rates in constructed wetlands treating swine wastewater. Ecol Eng. 2012; 40:189–197.

17 Pedescoll A, Sidrach-Cardona R, Sánchez JC, Carretero J, Garfí M, Bécares E. Design configurations affecting flow pattern and solids accumulation in horizontal free water and subsurface flow constructed wetlands. Water Res. 2013; 47:1448–1458.

18 Tietz A, Kirschner A, Langergraber G, Sleytr K, Haberl R. Characterisation of microbial biocoenosis in vertical subsurface flow constructed wetlands. Sci Total Environ. 2007; 380:163–172.

19 APHA-AWWA-WPCF. Standard Methods for the Examination of Water and Wastewater, 20th edn. American Public Health Association, Washington DC, 2001.

20 Matamoros V, García J, Bayona JM. Behavior of selected pharmaceuticals in subsurface flow constructed wetlands: a pilot-scale study. Environ Sci Technol. 2005; 39:5449–5454.

21 Matamoros V, Bayona JM. Elimination of pharmaceuticals and personal care products in subsurface flow constructed wetlands. Environ Sci Technol. 2006; 40:5811–5816.

22 Sidrach-Cardona R. Microbial communities in Constructed Wetlands and their relation to bacteria and pollutants removal. Doctoral dissertation. Departamento de Biodiversidad y Gestión Ambiental, Universidad de León, 2016.

23 Hijosa-Valsero M, Sidrach-Cardona R, Bécares E. Comparison of interannual removal variation of various constructed wetland types. Sci Total Environ. 2012; 430:174–183.

24 Reyes-Contreras C, Hijosa-Valsero M, Sidrach-Cardona R, Bayona JM, Bécares E. Temporal evolution in PPCP removal from urban wastewater by constructed wetlands of different configuration: A medium-term study. Chemosphere. 2012; 88:161–167.

25 Li YH, Zhu JN, Liu QF, Liu Y, Liu M, Liu L, Zhang Q. Comparison of the diversity of root-associated bacteria in *Phragmites australis* and *Typha angustifolia* L. in artificial wetlands. World J Microbiol Biotechnol. 2013; 24:1499–1508.

26 Gorra R, Coci M, Ambrosoli R, Laanbroek HJ. Effects of substratum on the diversity and stability of ammonia-oxidizing communities in a constructed wetland used for wastewater treatment. J Appl Microbiol. 2007; 103:1442–1452.

27 Iasur-Kruh L, Hadar Y, Milstein D, Gasith A, Minz D. Microbial population and activity in wetland microcosms constructed for improving treated municipal wastewater. Microb Ecol. 2010; 59:700–709.

28 Arroyo P, Ansola G, Sáenz de Miera LE. Effects of substrate, vegetation and flow on arsenic and zinc removal efficiency and microbial diversity in constructed wetlands. Ecol Eng. 2013; 51:95–103.

29 Calheiros CSC, Duque AF, Moura A, Henriques IS, Correia A, Rangel AOSS, Castro PML. Substrate effect on bacterial communities from constructed wetlands planted with *Typha latifolia* treating industrial wastewater. Ecol Eng. 2009; 35:744–753.

30 Faulwetter JL, Burr MD, Parker AE, Stein OR, Camper AK. Influence of season and plant species on the abundance and diversity of sulfate reducing bacteria and ammonia oxidizing bacteria in constructed wetland microcosms. Microb Ecol. 2013; 65:111–127.

31 Dong X, Reddy GB. Soil bacterial communities in constructed wetlands treated with swine wastewater using PCR-DGGE technique. Bioresour Technol. 2010; 101:1175–1182.

32 Gregory SP, Shields RJ, Fletcher DJ, Gatland P, Dyson PJ. Bacterial community responses to increasing ammonia concentrations in model recirculating vertical flow saline biofilters. Ecol Eng. 2010; 36:1485–1491.

33 Vymazal J. Removal of nutrients in various types of constructed wetlands. Sci Total Environ. 2007; 380:48–65.

34 Horvath RS. Microbial co-metabolism and the degradation of organic compounds in nature. Bacteriol Rev. 1972; 36(2):146–155.

35 Sun G, Zhu Y, Saeed T, Zhang G, Lu X. Nitrogen removal and microbial community profiles in six wetland columns receiving high ammonia load. Chem Eng J. 2012; 203:326–332.

36 Li Y, Zhu G, Ng WJ, Tan SK. A review on removing pharmaceutical contaminants from wastewater by constructed wetlands: Design, performance and mechanism. Sci Total Environ. 2014; 468–469:908–932.

37 Ávila C, García J. Chapter 6 – Pharmaceuticals and Personal Care Products (PPCPs) in the environment and their removal from wastewater through Constructed Wetlands. In: Persistent Organic Pollutants (POPs): Analytical Techniques, Environmental Fate and Biological Effects. Compr Anal Chem. 2015; 67:195–244.

38 Hijosa-Valsero M, Reyes-Contreras C, Domínguez C, Bécares E, Bayona JM. Behaviour of pharmaceuticals and personal care products in constructed wetland compartments: Influent, effluent, pore water, substrate and plant roots. Chemosphere. 2016; 145:508–517.

39 Gröning J, Held C, Garten C, Clauβnitzer U, Kaschabek SR, Schlömann M. Transformation of diclofenac by the indigenous microflora of river sediments and identification of a major intermediate. Chemosphere. 2007; 69:509–516.

40 Quintana JB, Weiss S, Reemtsma T. Pathways and metabolites of microbial degradation of selected acidic pharmaceutical and their occurrence in municipal wastewater treated by a membrane bioreactor. Water Res. 2005; 39:2654–2664.

41 Tanner CC. Plants as ecosystem engineers in subsurface-flow treatment wetlands. Water Sci Technol. 2001; 44:9–17.

42 Kyambadde J, Kansiime F, Gumaelius L, Dalhammar G. A comparative study of *Cyperus papyrus* and *Miscanthidium voilaceum*-based constructed wetlands for wastewater treatment in a tropical climate. Water Res. 2004; 38:475–485.

43 Zhang CB, Wang J, Liu WL, Zhu SX, Ge HL, Chang SX, Chang J, Ge Y. Effects of plant diversity on microbial biomass and community metabolic profiles in a full-scale constructed wetland. Ecol Eng. 2010; 36(1):62–68.

44 Zhang CB, Wang J, Liu WL, Zhu SX, Ge HL, Chang SX, Chang J, Ge Y. Effects of plant diversity on nutrient retention and enzyme activities in a full-scale constructed wetland. Bioresour Technol. 2010; 101(6):1686–1692.

45 Afzal M, Khan QM, Sessitsch A. Endophytic bacteria: prospects and applications for the phytoremediation of organic pollutants. Chemosphere. 2014; 117:232–242.

46 Schröder P, Daubner D, Maier H, Neustifter J, Debus R. Phytoremediation of organic xenobiotics – glutathione dependent detoxification in *Phragmites* plants from European treatment sites. Bioresour Technol. 2008; 99:7183–7191.

47 Brisson J, Chazarenc F. Maximizing pollutant removal in constructed wetlands: should we pay more attention to macrophyte species selection? Sci Total Environ. 2009; 407(13):3923–3930.

48 Lin A, Reinhard M. Photodegradation of common environmental pharmaceuticals and estrogens in river water. Environ Toxicol Chem. 2005; 24:1303–1309.

49 Pereira VJ, Linden KG, Weinberg HS. Evaluation of UV irradiation for photolytic and oxidative degradation of pharmaceutical compounds in water. Water Res. 2007; 41:4413–4423.

50 Valero M, Carrillo C. Effect of binary and ternary polyethilenenglycol and/or b-cyclodextrin complexes on the photochemical and photosensitizing properties of naproxen. J Photochem Photobiol. B 2004; 74:151–160.

51 Méndez-Arriaga F, Esplugas S, Jiménez J. Photocatalytic degradation of non-steroidal anti-inflammatory drugs with TiO_2 and simulated solar irradiation. Water Res. 2008; 42:585–594.

52 Andreozzi R, Marotta R, Paxéus N. Pharmaceuticals in STP effluents and their solar photodegradation in aquatic environment. Chemosphere. 2003; 50:1319–1330.

53 Hijosa-Valsero M, Matamoros V, Martín-Villacorta J, Bécares E, Bayona JM. Assessment of full-scale natural systems for the removal of PPCPs from wastewater in small communities. Water Res. 2010; 44:1429–1439.

54 Matamoros V, García J, Bayona JM. Organic micropollutant removal in a full-scale surface flow constructed wetland fed with secondary effluent. Water Res. 2008; 42:653–660.

55 Matamoros V, Salvadó V. Evaluation of the seasonal performance of a water reclamation pond-constructed wetland system for removing emerging contaminants. Chemosphere. 2012; 86:111–117.

56 Ávila C, Pedescoll A, Matamoros V, Bayona JM, García J. Capacity of a horizontal subsurface flow constructed wetland system for the removal of emerging pollutants: an injection experiment. Chemosphere. 2010; 81:1137–1142.

57 Matamoros V, Nguyen LX, Arias CA, Salvadó V, Brix H. Evaluation of aquatic plants for removing polar microcontaminants: a microcosm experiment. Chemosphere. 2012; 88:1257–1264.

58 Matamoros V, Arias CA, Nguyen LX, Salvadó V, Brix H. Occurrence and behavior of emerging contaminants in surface water and a restored wetland. Chemosphere. 2012; 88: 1083–1089.

59 Xu J, Wu L, Chang AC. Degradation and adsorption of selected pharmaceuticals and personal care products (PPCPs) in agricultural soils. Chemosphere. 2009; 77:1299–305.

60 Zhang DQ, Hua T, Gersberg RM, Zhu JF, Ng WJ, Tan SK. Carbamazepine and naproxen: fate in wetland mesocosms planted with *Scirpus validus*. Chemosphere. 2013; 91:14–21.

61 Calderón-Preciado D, Renault Q, Matamoros V, Cañameras N, Bayona JM. Uptake of organic emergent contaminants in spath and lettuce: an in vitro experiment. J Agric Food Chem. 2012; 60:2000–2007.

62 Dordio AV, Estêvão Candeias AJ, Pinto AP, Teixeira da Costa C, Carvalho AJP. Preliminary media screening for application in the removal of clofibric acid, carbamazepine and ibuprofen by SSF-constructed wetlands. Ecol Eng. 2009; 35:290–302.

63 Dordio A, Ferro R, Teixeira D, Palace AJ, Pinto AP, Dias CMB. Study on the use of *Typha* spp. for the phytotreatment of water contaminated with ibuprofen. Int J Environ Anal Chem. 2011; 91:654–667.

64 Ávila C, Reyes C, Bayona JM, García J. Emerging organic contaminant removal depending on primary treatment and operational strategy in horizontal subsurface flow constructed wetlands: influence of redox. Water Res. 2013; 47:315–325.

65 Hijosa-Valsero M, Matamoros V, Sidrach-Cardona R, Pedescoll A, Martín-Villacorta J, García J, Bayona JM, Bécares E. Influence of design, physico-chemical and environmental parameters on pharmaceuticals and fragrances removal by constructed wetlands. Water Sci Technol. 2011; 63:2527–34.

66 Park N, Vanderford BJ, Snyder SA, Sarp S, Kim SD, Cho J. Effective controls of micropollutants included in wastewater effluent using constructed wetlands under anoxic condition. Ecol Eng. 2009; 35:418–423.

67 Zhang DQ, Hua T, Gersberg RM, Zhu J, Ng WJ, Tan SK. Fate of diclofenac in wetland mesocosms planted with *Scirpus validus*. Ecol Eng. 2012; 49:59–64.

68 Dettenmaier EM, Doucette WJ, Bugbee B. Chemical hydrophobicity and uptake by plant roots. Environ Sci Technol. 2009; 43:324–329.

69 Dordio AV, Belo M, Teixeira DM, Carvalho AJP, Dias CMB, Picó Y, Pinto AP. Evaluation of carbamazepine uptake and metabolization by *Typha* spp., a plant with potential use in phytotreatment. Bioresour Technol. 2011; 102:7827–7834.

70 Zhang DQ, Gersberg RM, Hua T, Zhu JF, Goyal MK, Ng WJ, Tan SK. Fate of pharmaceutical compounds in hydroponic mesocosms planted with *Scirpus validus*. Environ Pollut. 2013; 181:98–106.

71 Dordio AV, Duarte C, Barreiros M, Carvalho AJP, Pinto AP, da Costa CT. Toxicity and removal efficiency of pharmaceutical metabolite clofibric acid by *Typha* spp. e potential use for phytoremediation? Bioresour Technol. 2009; 100:1156–1161.

72 Matamoros V, Arias C, Brix H, Bayona JM. Removal of pharmaceuticals and personal care products (PPCPs) from urban wastewater in a pilot vertical flow constructed wetland and a sand filter. Environ Sci Technol. 2007; 41:8171–8177.

73 Matamoros V, Arias C, Brix H, Bayona JM. Preliminary screening of small-scale domestic wastewater treatment systems for removal of pharmaceutical and personal care products. Water Res. 2009; 43:55–62.

74 Herklotz PA, Gurung P, Heuvel BV, Kinney CA. Uptake of human pharmaceuticals by plants grown under hydroponic conditions. Chemosphere. 2010; 78:1416–1421.

75 Shenker M, Harush D, Ben-Ari J, Chefetz B. Uptake of carbamazepine by cucumber plants: a case study related to irrigation with reclaimed wastewater. Chemosphere. 2011; 82:905–910.

76 Yu Y, Liu Y, Wu L. Sorption and degradation of pharmaceuticals and personal care products (PPCPs) in soils. Environ Sci Pollut Res. 2013; 20(6):4261–4267.

Part VII

Novel Industrial Applications

19

Dewatering of Industrial Sludge in Sludge Treatment Reed Bed Systems

S. Nielsen and E. Bruun

Orbicon A/S, Ringstedvej 20, Roskilde, Denmark

19.1   Introduction

Sewage sludge results from the treatment of wastewater originating from many sources including homes, industries, agro-industries and street runoff. Sewage sludges contain nutrients and organic matter and, due to these soil benefits, they are widely used as soil amendments. The beneficial effects of sludge application for agriculture and/or environmental purposes (forestry and land reclamation) are well known and documented (e.g., [1] and references therein). The use of organic wastes in agriculture is considered a way of maintaining or restoring the quality of soils, enlarging the slow cycling soil organic carbon pool. Sewage sludge, however, also contains contaminants including metals, pathogens, and organic pollutants [1, 2]. A wide variety of undesired substances, such as potentially trace elements and organic contaminants, can have adverse effects on the environment.

The production of sewage sludge has increased steadily, due to the growth of population connected to centralized wastewater treatment and the implementation of more strict environmental regulation [3]. For example, the yearly production of municipal sludge in the European Union (EU) – estimated to 5.5 million tons in 2005 – is expected to reach a production of 13 million tons of dry solid (t ds) by 2020 [4]. Sludge production from industry, food-industry and agriculture has simultaneously increased during the same period. By 2020, all waste in EU should be managed as a resource and landfilling of organic material should be virtually eliminated [4]. In order to achieve this for sewage sludge and to avoid landfilling or incineration, eco-sustainable technologies are necessary.

There are several technologies for sludge treatment (dewatering) in the market, ranging from traditional mechanical treatment by decanter and screw-press technology to the environmentally friendly, Sludge Treatment Reed Bed Systems (STRB). Sludge Treatment in Reed Bed Systems was developed in the late 1980s and has been providing a low cost and maintenance option for sludge dewatering and stabilization, at both small and larger treatment works [5, 6].

There are important differences in the environmental perspectives and costs involved in mechanical sludge dewatering followed by disposal on agricultural land compared to STRB. The capital cost of the STRB is typically higher than mechanical options, while the reverse is true for the system operating costs. A study by Nielsen [7], who compared the costs of mechanical treatment with STRB treatment of activated sludge (sludge production: 550 tons ds/yr) showed significant lifetime cost savings using a STRB solution when compared to a mechanical solution [8].

Constructed Wetlands for Industrial Wastewater Treatment, First Edition. Edited by Alexandros I. Stefanakis.
© 2018 John Wiley & Sons Ltd. Published 2018 by John Wiley & Sons Ltd.

In order to compare the two treatment methods, sizing, capital and operating cost and power use estimates were developed for comparison purposes (under Danish conditions). The study showed that a sludge strategy consisting of a new STRB would be approx. DKK 0.5–0.6 million cheaper per year than the option consisting of a new screw press or decanter, for sludge treatment and final disposal to agricultural land. Besides these economic benefits, and on the other hand to mechanical solutions, the STRB systems are also environmentally friendly, as the dewatering of sludge in STRB occurs without the use of chemicals and with a minimum of energy [8].

Sludge Treatment Reed Bed Systems are vertical flow constructed systems for the dewatering and stabilization of sludge. Sludge is loaded to the basins, which are typically vegetated with common reeds (*Phragmites australis* or *Typha*), and dewatered through passive drainage and evapotranspiration. Treatment of wastewater sludge in STRB is a widespread and common WWTP sludge treatment practice for both civil and industrial sludge in Europe. In Denmark the technology has been used for more than 28 years. The method for treatment and dewatering of sludge may influence the sludge quality considerably. For example STRB's are capable of degrading/reducing the concentrations of hazardous organic compounds, such as LAS, NPE, DEHP, and certain PAH's to a much higher degree than traditional mechanical treatment methods [9].

Many sludge types can be treated by the STRB technology including activated sludge or digested sludge, and waterworks (WW) sludge. Experience has shown that STRBs are capable of treating sludge with different qualities and with dry solid concentrations between 0.1% and up to 5% [10].

During the 1990s and onwards, stricter legislation has been brought into effect by the Danish Environmental Protection Agency (DEPA) to regulate the content of nutrients, heavy metals and hazardous organic compounds in sludge being spread on agricultural land [9]. In order to apply sludge on farmlands in Denmark the following criteria for the sludge quality have to be fulfilled (Table 19.1).

Table 19.1 The Danish and European Union (EU) legal limits for heavy metals and hazard organic compounds in sludge residue for agricultural use.

Limit values	Denmark		EU	
	BEK No. 1650 of 13/12/2006		86/278/EE EU Directive	ENV. E 3 (2000) Working document on sludge, 3rd draft
Heavy metals	mg/kg ds	mg/kg TP	mg/kg ds	mg/kg ds
Cadmium (Cd)	0.8	100	20–40	10
Copper (Cu)	1,000	–	1,000–1,750	1,000
Nickel (Ni)	30	2,500	300–400	300
Lead (Pb)	120	10,000	750–1,500	750
Zinc (Zi)	4,000	–	2,500–4,000	2,500
Mercury (Hg)	0.8	200	16–25	10
Chromium (Cr)	100	–	–	1,000
Organic CONTAMINANTS	**mg/kg ds**	**mg/kg TP**	**mg/kg ds**	**mg/kg ds**
LAS	1,300	–	–	2,600
PAH	3	–	–	6
NPE	10	–	–	50
DEHP	50	–	–	100

This chapter describes the experiences and results from STRBs and test systems treating industrial sludge with special focus on treatment of water works sludge. Industrial sludge may contain pollutants such as heavy metals, nutrients, hazardous organic compounds or fat in higher concentrations than domestic (household) sludge and is therefore often more difficult to treat and dewater or to dispose after treatment due to legal regulations.

19.2   Methodology

19.2.1   Description of an STRB

Sludge Treatment in Reed Bed Systems are vertical flow constructed systems for the dewatering and stabilization of sludge. Sludge treatment reed beds comprise a series of basins, which are vegetated typically with common reeds (*Phragmites australis*) in a filter [10]. The sludge residue remains on the surface of the filter in the basins and is mineralized through the natural biophysical interaction of plants, microbes and air, whilst water is removed from the sludge by both evapotranspiration and drainage through the sludge residue and filter (Figure 19.1) [10].

Vegetation and ventilation with air in STRB systems create favorable conditions for the transformation of degradable organic matter to a more stable humic form [11, 12]. As the organic matter mineralizes, the overall sludge volume reduces and the accumulating sludge residue becomes continually part of the filter in which the reeds grow. The height of the mineralized sludge residue layer increases with time with an accumulation rate of approximately 10–15 cm/yr, depending on the sludge quality. After typically 8 to 12 (up to 20) years of operation, the dewatered and mineralized sludge residue is excavated from the STRB basins and recycled as fertilizer or as soil conditioner.

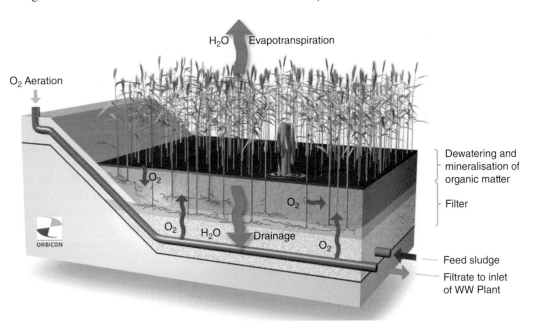

Figure 19.1 Cross-section of a STRB showing the sludge layer, filter and loading and drainage system (Copyright Orbicon, reproduced with permission).

Currently, STRBs are used primarily for treatment of sewage sludge from municipal wastewater treatment plants (WWTP), but the technology is also used for the treatment of water works (WW) sludge and sludge originating from the agro-industrial production. However, some types of sludge may not be suitable for dewatering in a STRB, e.g., if the sludge has a high concentration of fat and oil. Experience has shown that STRBs are capable of treating most types of sludge with dry solid concentrations between 0.1–5%.

19.2.2 Description of STRB Test-System

The dewatering and treatment efficiency of sludge in STRBs depends on the sludge quality, and the design and operation of the STRB. Before planning and establishing of new full-scale STRB plants, it is strongly recommended to test the sludge's dewatering properties in a test system with similar filter and setup as the intended full-scale system [7]. A test of the dewater ability of the sludge is especially important, if it is not 'normal' sludge from households, but industrial sludge with, for example, a high content of fat, oil, organic material or heavy metals.

In Denmark, Germany, Sweden, France and other countries in Europe the design and dimensioning of STRBs has been extremely variable during the last 20 years, even if they were treating the same sludge type. The number of basins in the different systems varied between 1–24 basins, basin areas between less than 100 m^2 to over 3,000 m^2 and the area load varied between 30 to over 100 kg ds/m^2/yr [7]. Consequently, the outcome of the sludge treatment has also varied. If the STRB has been properly designed and operated, high dry solid contents in the sludge residue has been obtained, while the opposite has been true at poorly designed and operated STRBs [9]. Some sludge types are difficult to dewater by both STRB and mechanical treatment technology.

A test system consists typically of 3–6 pilot-scale basins of 1–2 (up to 100) m^2 with filter, drain layer, drains, growth layer and reeds (*Phragmites*). During the trial period of typically 4–12 months (sometimes it is necessary with 2 years), each basin is loaded several times, followed by a resting period, where no sludge is added. The length of the loading and rest periods depends on the sludge characteristics, climate (cold climates need longer resting periods), and the age of the system and specific basin, the dry solid content and the thickness of the sludge residue.

Box 19.1 How to Operate an STRB

The operation of a STRB involves cycling through a number of reed beds basins. Under Danish climatic condition between 8–12 basins. Each basin could be loaded over a 1–2 week period, and followed by 5–10 weeks resting without any sludge loading. The length of the loading and rest periods depends on the sludge characteristics, climate, the age of the system, and the dry solid content in the feed sludge. The key to successful treatment is first a correct construction and dimensioning, but also monitoring these parameters and modifying the loading and resting program to ensure the operation. It is thus recommended to have an automated system for monitoring and operation, which continually updates and makes the necessary changes ensuring trouble free operations.

Although STRBs are simple in operation, their success is dependent on managing the loading cycles so that basins are not overloaded.

The sludge "capacity" of a basin for a single load cycle is determined by measuring how well the basin can be loaded whilst maintaining suitable drainage, which is gradually reduced during a loading cycle.

When the drainage efficiency reduces to a certain point, sludge loading is moved to another basin to allow the loaded basin to rest and recover before a new loading cycle begins [13].

The dewatering capacity (up to 40% ds under Danish climatic conditions) of an STRB system is comparable to energy intensive mechanical dewatering processes.

Advantages of a STRB compared to mechanical sludge treatment

- Beyond cost and energy savings, STRBs are documented to have a number of operational and environmental advantages over conventional mechanical dewatering processes.
- STRB do not require the addition of polymer coagulants, while achieving comparable or better dewatering capacity.
- STRB systems facilitate aerobic conversion of biodegradable organic matter to more stable forms, which also reduces the total sludge volume and formation of odorous substances.
- STRB also enhance the degradation of organic pollutants and pathogenic organisms in sludge and render sludge more suitable for land application as fertilizer [10, 14, 15].
- STRBs are reported to be more energy efficient [8, 16, 17] and even serve as carbon sinks due to carbon fixation by the vegetation cover [18]. Moreover, well-designed and well-managed STRBs have a lower climate impact than conventional treatment alternatives [19].

The main purpose of a test (phase 1) is to test whether the sludge would be suitable for further treatment in a STRB or not [7]. This will be answered by gathering information about:

- Sludge quality and characteristics.
- Dewatering efficiency of the sludge ($L/sec/m^2$).
- The sludge residue behavior (dry/crack up) in a trial bed.
- The growth of reeds. Is it possible to get the vegetation to grow in the sludge?

Later on in the test period in phase 2 the loading onto the reed beds will be much more intensive. Loading rates will vary in the test period so that the differences in load/rest ratio and area load can be more clearly defined. Different numbers of loading days and resting days will be tested. The main purpose in phase 2 is to test (ascertain the criteria) for the dimensioning and operation of a full-scale system [7]. For example:

- What loading (kg dry solid/m^2/yr) can the trial system treat?
- The number of basins that is necessary in order to get the desired resting period between loadings.
- How many days can we load in a loading period?
- How many days rest are necessary for drainage and drying (min and max)?
- Load and rest program in relation to sludge quality.
- Determine sludge residue and filtrate water quality.
- Summer and winter operation.
- Determine the sludge quality.
- Determine sludge residue and filtrate water quality.
- Finally, to give recommendations for up-stream changes on the works in order to reduce unwanted parameters in the feed sludge (e.g., heavy metal, fat and oil etc.)

19.3   Treatment of Industrial Sludge in STRB Systems

There is no sharp definition of "industrial sludge", but the term encompasses sludge originating from wastewater treatment at the industry's own wastewater treatment plant or sludge produced at a public WWTP with a large proportion of wastewater originating from industry and agro-industrial sludge. Treatment of wastewater from "industries", such as the food industry, aquaculture and waterworks, generates sludge that typically contains pollutants not present or in much higher concentrations than in normal sludge derived from domestic wastewater. These sludge types may be problematic and result, e.g., in reduction of the sludge dewatering, inhibition of reed growth or contamination of the sludge residue.

Examples of problematic compounds in sludge includes organic material such as fat and oil, heavy metals, nutrients such as phosphate and nitrate, and hazardous organic compounds – these are described in the following sections.

Of the different industrial sludge types, water works sludge typically have high content of iron or aluminum, but low content of organic material and fat and oil. Sludge generated from treatment of wastewater from the food industry (abattoir, dairies, aquaculture etc.) typically has a high proportion of organic material (carbohydrates, fat, oil, protein), heavy metals and nutrients (nitrogen and phosphorous), with high values of BOD (biological oxygen demand), COD (chemical oxygen demand) and suspended solids. Due to these constituents, the food industry sludges have a high potential to cause severe pollution problems to the environment.

19.3.1   Organic Material in Sludge

Insufficient dewatering of sludge in an STRB and in mechanical dewater equipment also, is often due to the quality of the sludge and sludge residual. The ratio between organic and inorganic material is an important parameter in the evaluation of sludge dewatering properties [9]. The organic content is determined typically by measuring the 'Loss on Ignition' (LOI).

The water retention capacity of sludge with a high proportion of organic material is many times larger than more inorganic sludge, as the content of organic material has major influence on the content of free water. The higher the organic material, the lower the content of free water and the higher the content of capillary water [20], which is more difficult to dewater [21].

When comparing the feed sludge's content of organic material (expressed as LOI) with the corresponding dry solid content in the sludge residue for a large number of STRB basins, there is a clear tendency that higher contents of organic material in the feed sludge results in lower dry solid contents in the sludge residue in the STRB (Figure 19.2).

19.3.2   Fats and Oil in Sludge

An important visual indication of a proper dewatering of the sludge is that the upper layer continually becomes cracked and broken during the resting phase. Sludge containing a high proportion of fats and oil often has a tendency not to crack and break open during the resting phase, reducing thereby the aeration from the surface considerably. If this cracking does not occur, it is due to ineffective dewatering of the sludge [21].

A high content of fats and oil in the feed sludge is especially important to address and reduce, since this, like organic material, has a pronounced effect on the dry solid content in the sludge residue in the STRB (Figure 19.3). Experience has shown that a fat content in the feed sludge above 5,000 mg/kg ds considerably reduces the dewatering and results in pronounced anaerobic conditions in the sludge residue, which becomes black and smelly [21].

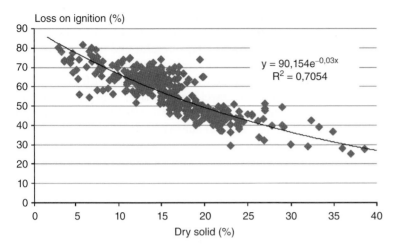

Figure 19.2 Correlation between organic content (loss on ignition) in the feed sludge and the dry solid in the sludge residue in the STRB [21]. Experience has shown that a LOI above 65% significantly reduces the dewatering of the sludge [21].

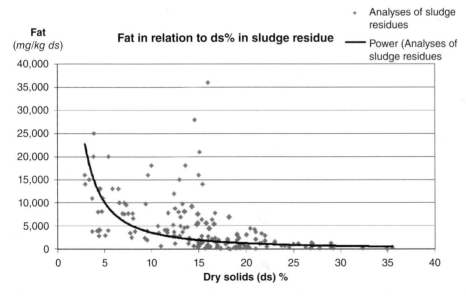

Figure 19.3 Correlation between content of fat in the feed sludge and the dry solid content in the sludge residue [21].

19.3.3 Heavy Metals in Sludge

Depending on the industry producing the wastewater, high contents of heavy metals may be present in sludge. The concentration of heavy metals may fluctuate over time, due to changes in the production at the industry producing the wastewater. Sludge, which normally has low levels of heavy metals below legal limits, may, e.g., if new industry is connected to the WWTP, rise to levels inhibitory for reeds or reach levels in the sludge residue, which exceed legal limits for agricultural use (Table 19.1).

Experience from Denmark has shown that the level of heavy metals in the sludge residue after 10 to 20 years of treatment in STRBs is in general low – well below the Danish and EU legal limit values [22]. Heavy metals are mainly bound to particles in the sludge residue and filter matrix, and the translocation of heavy metals out of the systems with the drainage is, therefore, limited [11, 15, 22, 23]. A study by Stefanakis and Tsihrintzis from 2012 [24] about heavy metals in an STRB showed that the major sink in the mass balance for heavy metals was the gravel drainage layer, while accumulation in the sludge layer and plant uptake was low. Less than 16% of the heavy metals left the bed through drainage. In Denmark, there have been cases with reeds dying out due to nickel pollution in the sludge residue at a WWTP with connected heavy industry.

19.3.4 Nutrients in Sludge

The sludge residue from an STRB is valuated as a fertilizer, since the phosphorus and nitrogen concentration is typically high due to anthropogenic origin of the sludge [25]. Nitrogen typically decreases during the treatment period and with the depth of the sludge residue in the basins, due to the microbial mediated processes of nitrification and denitrification [11]. In addition, some nitrogen leaves the system with the drainage water (filtrate water) mainly in the form of nitrate and is returned to the WWTP.

Total phosphorous concentration has been shown to increase for the whole sludge residue in an STRB compared to the feed sludge [15], due to the mineralization of organic material in the sludge residue. Part of the phosphorous interacts with iron and other constituents in the sludge and is, thus, bound in the sludge residue matrix.

19.3.5 Hazardous Organic Compounds in Sludge

Various hazardous organic compounds may be present in sludge, such as polyaromatic hydrocarbons (PAHs), di-2-ethylhexyl-phthalates (DEHPs), nonylphenol/nonylphenol ethoxylates (NPEs), linear alkyl benzene sulfonates (LASs), originating, respectively, from coal and tars, lubricating oil additives and detergents. Studies have shown that hazardous organic compounds are mineralized during the treatment in STRB [9, 26]. Unlike traditional mechanical dewatering, such as centrifuges, the long treatment time of the sludge residue in STRBs (typically 8–10 years or more) provides sufficient time for microbiological and non-biotic mineralization processes to effectively degrade and reduce most of the hazardous organic compounds. Even within a shorter timespan of 3–6 months, significant reductions have been shown [8, 22].

In a study of the mineralization of LAS and NPE in digested sludge treated in an STRB, a degradation of 98% of LAS and 93% of NPE was obtained under aerobic conditions ([9]; Danish Environmental Agency: Report no. 22, year 2000). The study showed that oxygen was the limiting factor in the degradation of the organic contaminants. Oxygen influx into the sludge improved the mineralization of LAS and NPE considerably, while mineralization under anaerobic conditions was very limited. In the same study, reductions of approximately 60 and 32% in an STRB were obtained for DEHP and PAH, respectively. The organic contaminants were not only mineralized in the upper layer of sludge residue, but in the whole depth. In storage experiments with anaerobically digested sludge (representing mechanical treatment) LAS, NPE, DEHP and PAH were only partly degraded in the top layer (0–20 cm) and below 20 cm, no degradation occurred.

19.4 Case Studies – Treatment of Industrial Sludge in Full-Scale and Test STRB Systems

The following sections present cases and results from treatment of sludge from water works and industry.

19.4.1 Case 1: Treatment of Industrial Sewage Sludge with High Contents of Fat

The industrial sewage sludge originates from WWTP treatment of wastewater with a large input from industry. Table 19.2 presents Danish STRBs systems treating sludge originating from treatment of wastewater with major input from industry: Tinglev STRB (abattoir), Kolding STRB (abattoir), Skagen test STRB (fishing industry), Skive STRB (abattoir) all Danish STRBs and in Kristianstad (Sweden) test STRB receiving sewage water from the food industry (dairies, abattoir representing (20–25%), chicken and others). These systems all receiving sludge with a high proportion of organic material. The sludge quality was characterized by a high loss of ignition (between 65 and 76% of dry solids), and with high contents of fat (15,000–30,000 mg/kg ds) and oil (2,300–7,000 mg/kg ds). Results of dewatering from these STRB systems are presented in Table 19.2 [21].

The high contents of fat and oil resulted in a dewatering profile with a maximum level approximately five to ten times lower (only 0.001–0.004 L/s/m^2) than observed for normal sewage sludge (Figure 19.4) in a well-functioning STRB system (Helsinge STRB in Denmark) typically with maximum levels of 0.015–0.020 L/s/m^2 (Figure 19.4). It has been shown that dewatering of sludge with LOI between 50 and 65% will have a maximal drainage in the order of 0.008–0.020 L/s/m^2 [21].

Sludge qualities with high contents of fat and oil often result in dewatering profiles, where the maximum level of the dewatering is very low and the dewatering does not decline to zero between the loads and after the last load, but has a long "tail" continuously remaining at a certain level until the next sludge load period of the specific basin. This means that the sludge residue in that basin does not have a resting phase in between loads, resulting in an anaerobic wet sludge residue.

Table 19.2 Dewatering efficiency in relation to the sludge – and sludge residue quality [21]. All examples below has feed sludge with high levels of fat.

STRB system	Major WW source	Dewatering L/sec/m^2	Dry solid (%) Sludge	Dry solid (%) Sludge residue	Loss on ignition (%) Sludge	Loss on ignition (%) Sludge residue	Fat in feed sludge mg/kg DS
Tinglev (F)	1	0.002–0.005	0.4–0.6	10–15	75	70	21,000
Kolding (F)	1	0.002–0.008	0.5–1.0	15–25	65	60	30,000
Skive (F)	1	0.001–0.004	0.8–1.2	2.9–7.1	76	–	15,000
Skagen (T)	2	0.001–0.003	0.5–3.0	5–14	75	80	16,000
Kristianstad (T)	3	0.003–0.020	0.5–2.5	8–14	70	75	29,000

Major wastewater source: (1) abattoir; (2) Fishing Industry; (3) Food-Industry (dairies, abattoir representing 20–25 %), chicken and others.
F: Full-scale; T: Test STRB.

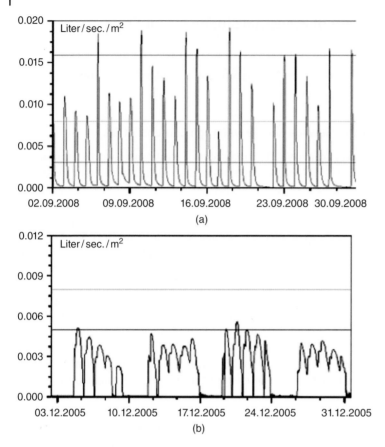

Figure 19.4 (A) Dewatering curve from Helsinge STRB and (B) Tinglev STRB. Helsinge STRB has a good dewatering of the sludge, while Tinglev STRB has a very low dewatering [21]. With permission of Obricon.

19.4.2 Case 2: Treatment of Industrial Sewage Sludge with High Contents of Heavy Metal (Nickel)

It is important to follow the feed sludge quality on a continuous basis, with regular sampling and analyses. If heavy metal concentrations, e.g., of nickel or another heavy metal exceeds the national legal limits (Table 19.1) for recycling sludge to agriculture in the feed sludge loaded to a STRB, the sludge residue cannot be applied on agricultural lands, but has to be deposited on a landfill or incinerated, which is much more expensive [22]. Potential inhibition of the reed growth or toxic effects is another reason for keeping a close eye on the sludge quality, here under the content of heavy metals in the feed sludge.

The Danish STRB (Stenlille) is an example of an STRB which has received sludge with high nickel levels (Figure 19.5). The vegetation (*Phragmites*) in the basins was visibly not thriving with thin yellow new shoots. The Danish national legal limits for sludge applied to agricultural lands is 30 mg/kg ds (Table 19.1).

Figure 19.5 Basin containing Nickel polluted sludge. Note the yellow color and poor reed coverage (06-06-2011) (Photo: Orbicon).

As can be seen in Figure 19.6, nickel in the feed sludge was on most occasions several times higher than the national limit (red line in Figure 19.6). Research has shown that nickel concentration in the reeds of more than 5 mg/kg ds may cause toxic effects on the plants [27]. The reeds at the specific STRB site had nickel concentrations of 9.7 mg/kg ds (Table 19.3).

In order to obtain good healthy vegetation and well-functioning STRB again, it is necessary to reduce the nickel concentration in the feed sludge either by the source or at the WWTP. In the specific

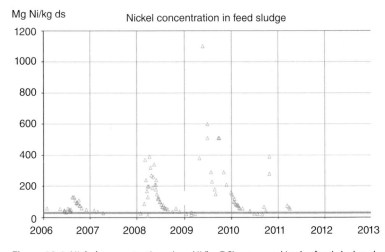

Figure 19.6 Nickel concentrations (mg Ni/kg DS) measured in the feed sludge during the period from 2006–2012. The Danish national legal limits for Nickel in sludge applied to agricultural lands is 30 mg Ni/kg DS (Gray line).

Table 19.3 The result of nickel analyses of reeds (*Phragmites australis*), and sludge residue sampled at different depth intervals.

Nickel concentration	mg Ni/kg ds
Reeds (*Phragmites*)	9.7
Sludge residue – top layer, (0–7 cm depth)	110
Sludge residue (7–14 cm depth)	150
Sludge residue (14–21 cm depth)	80

case, the industry, which caused the nickel pollution in the wastewater, closed for a few years after 2011 and the vegetation was re-established.

19.4.3 Case 3: Treatment of Water Works Sludge

Treatment and disposal of coagulated settled water works sludge presents a great difficulty for the water industry in Europe [28]. In general, dewatering characteristics are poor and the sludge is of limited beneficial use.

Water works (WW) sludge is generated during the purification and filtration processes of low solid waters from freshwater reservoirs, lakes or rivers or from groundwater reservoirs. For the purification process, coagulants, such as aluminium sulphate or polyaluminium chloride, and Iron based coagulants, such as ferric sulphate or ferric chloride, are used for removing impurities, after which the sludge is thickened and dewatered.

The sludge quality may considerably vary from one work to another [28], but all tend to be:

- Sticky
- Difficult to handle
- Often have an unpleasant odor (dependent on their source).

In general, most dewatered sludge goes to landfill at considerable costs [28]. Alternatively, sludge is dewatered onsite or it is transported to the nearest Waste Water Treatment Plan (WWTP) for dewatering. The typical method for dewatering of water works sludge is by mechanical treatment systems. However, in the last decade full-scale STRBs and test systems have been introduced for sludge dewatering of water works sludge.

In the following sections, results are presented from sludge treatment in STRB of Water Works Sludge from Hanningfield and Whitacre Water Work, briefly described below (Table 19.4).

Hanningfield Water Works (Northumbrian Water, England)
Currently the world's largest full-scale STRB system of 4.5 hectares and 16 basins for treatment of water works sludge is from Hanningfield Water Works. Hanningfield STRB was constructed full-scale in 2012, representing a treatment capacity of 1,275 tons dry solid per year and is situated approximately 3 km from the water works (Figure 19.7; Table 19.4).

The sludge quality from Hanningfield Water Works was originally tested in an STRB test system in the period 2008–2013 [21, 29] before the full-scale system was established (Figure 19.8). A trial was set up with six basins each of 20 m^2 at the Hanningfield Reservoir in Essex to examine the dewatering

Table 19.4 Treatment of water works sludge in STRB systems in England.

STRB	Period	No. of basins	Total basin area (m²)	Sludge type	Reference
Hanningfield WW STRB Test	2008–2013	6	120	Ferric sludge	[21, 29]
Hanningfield WW Full-scale STRB	2012–2014	16	42,500	Ferric sludge	[21, 29]
Lumley WW STRB Test	2010	3	3	Alum sludge	ARM/Orbicon
Whitacre WW STRB Test	2015–2016	3	3	Ferric sludge	ARM/Orbicon
Whitacre WW Full scale STRB	2016–2018	4	1,475	Ferric sludge	ARM/Orbicon

WW + Water Works.

processes of the sludge produced at the water work. The test system was built with a design comparable to a full-scale plant with reeds, ventilation, sludge input; reject water systems as well as filters and drains.

Whitacre Water Works (Severn Trent Water) and Test System

Whitacre Water Works (England) will by 2017 begin treating sludge by means of a full-scale STRB. Prior to the decision for a full-scale STRB, ferric sludge from Whitacre Water Works was tested in an STRB test system in a five-month period in 2015, with the purpose to clarify whether the sludge was suitable for treatment in a STRB. The Whitacre STRB pretest system was built with three basins each of 1 m² filter area with a design comparable to a full-scale plant with reeds, ventilation, sludge input,

Figure 19.7 Hanningfield STRB system treating sludge from Hanningfield Water Works situated 3 km from the STRB (Photo: Orbicon September, 2012).

Figure 19.8 Hanningfield test STRB (Photo: Orbicon 28.05.2011).

filtrate water systems as well as filters and drains. One container (IBC) was used for sludge holding before loading.

Whitacre Water Treatment Works generates sludge waste primarily from the de-sludging of the pulsator clarifiers producing primarily mineral ferric sludge and sludge from washing of the clarifiers. The existing treatment comprises coagulation with ferric sulphate.

19.4.3.1 Feed Sludge and Resulting Filtrate Quality
The feed sludge quality from WW are in general characterized by a low dry solid content and a high iron or aluminium content, due to the production process at the works using ferric- or aluminium sulphate/chloride for the coagulation process.

WW sludge is characterized by a much lower dry solid content (0.1–0.2% ds), five to ten times lower than typical activated sludge from WWTP. Moreover, the level of organic material and fats and oil is generally low in WW sludge (<1000 mg/kg ds) as the sludge is generated from a production process involving mainly surface water or groundwater. Based on experience, these parameters do not cause operational problems.

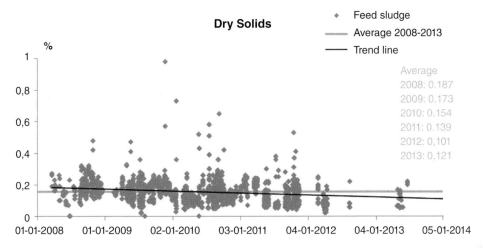

Figure 19.9 Total solid (%) in feed sludge at Hanningfield WW in the period from 2008–June 2013.

Dry solid content in feed sludge from Whitacre and Hanningfield WWs were in the range 0.1–0.5% ds (Figure 19.9) and for suspended solids in the range of 100–8,000 mg/L (Table 19.5). The data show that there is a large variation in the total solids loaded onto the basins (Figure 19.9). Therefore, it is very important to know the percentage of dry solids in each load in order to calculate the load (kg ds/m^2/yr) as precisely as possible. If an average "total solids" value is used to calculate the load, there is a risk that the load may be overestimated in some basins and underestimated in others.

The sludge from the two WWs was characterized by relatively low contents of organic material (as determined by LOI) between 20–40% of dry solids (Table 19.5). Sewage sludge usually has organic contents of approximately 50–70% determined by LOI.

The fats and oil content for Hanningfield WW was very low (on average 600 mg/kg ds, ranging from 10–2,400 mg/kg ds). The feed sludge had a high iron content around 250,000 mg/kg ds, ranging from 100,000 up to 400,000 mg/kg ds. The pH level is within the limits to allow good growing conditions for the reeds. There is nitrogen and phosphorus in the sludge, which makes it possible to reduce or even avoid fertilizer additions. After drainage the resulting filtrate water quality shows significant reductions in dry solids (70%), suspended solids (>95%), iron content (>95%) and total phosphorous (>95%) (Table 19.6).

19.4.3.2 Sedimentation and Capillary Suction Time

A sedimentation test of sludge is simple and quick first indication of the dewater ability of the feed sludge. Due to the usually low fat and oil, content in WW sludge, the sedimentation of suspended material proceeds rather fast, as shown in Figure 19.10 from Hanningfield STRB. The sludge settled to approximately 8 to 10 cm within 30 minutes. The filtrate water (above the settled sludge phase) is transparent without any coloring (Figure 19.10). As the sludge sedimentation characteristics may change over time due to upstream changes at the WW, it is recommended to measure the sedimentation repeatedly during a test period.

Another simple but good indication of the sludge dewater ability is the capillary suction time (CST). The CST is a measure of the dewatering properties of sludge. This method is essentially measuring

Table 19.5 Feed sludge quality parameters for Whitacre (Test) and Hanningfield (Test and Full-scale) WWs. Both works use ferric sulphate for the coagulation process.

Feed sludge from WW Sampling STRB site: Sampling period: Parameter	Unit	Whiteacre Test STRB Autumn 2015 Average (n = 2)	Hanningfield Test/Full-scale 2008–2013 Average (n > 25)	Interval range (both WW) (n > 25)
Dry solids	%	0.3	0.2	0.1–0.5
Suspended solids	mg/L	2,630	1,262	100–8,000
Loss on Ignition	%	40	23	10–40
pH	–	7.3	7.4	6.8 – 8.7
Fat and oil	mg/kg ds		604	10–2400
Iron (total) as Fe	mg/kg ds	258,530	232,800	100,000–400,000
Aluminum total	mg/kg ds	432	408	100–3,000
Total nitrogen	mg/kg ds		2,300	1,000–14,000
Total phosphorous	mg/kg ds		6964	1,500–11,000
Phosphate as P	mg/kg ds	7,241		2,000–8,000
Chloride	mg/kg ds	16,301	41,568	15,000–45,000
Calcium, total	mg/kg ds	33,555	98,143	32,000–290,000

Table 19.6 Filtrate water quality parameters after treatment in a STRB at Whitacre and Hanningfield WW.

Filtrate water from WW Sampling site: Sampling period: Parameter	Unit	Whiteacre Test STRB Autumn 2015 Average (n = 3)	Hanningfield Test/Full-scale 2008-2013 Average (n > 25)	Data range
Dry solids	%	0.06	0.05	0.001–0.06
Suspended solids	mg/L	93	0.01	0.001–0.05
pH		7.9	7.7	7.0-8.0
BOD$_5$	mg/L	5	2.4	1–36
COD total	mg/L	44	33	3–380
Iron (total) as Fe	mg/L	29	4.5	0–120
Total phosphorous as P	mg/L	0.1	0.2	0–4.6
Total nitrogen as N	mg/L		3.3	0–10
Chloride	mg/L	52	74	50–100

(a)   (b)

Figure 19.10 Sedimentation of Hanningfield WW sludge performed on sludge sampled 28.07.2009 (Photo: Orbicon).

how quickly the sludge wets a filter paper. A good dewatering ability corresponds to a CST range of 10–100 seconds (Figure 19.11).

The values of CST measured for WW sludge are in the higher end of what normally is observed for WWTP sewage sludge, which indicates that the sludge is somewhat more difficult to dewater. However, the higher CST values of the WW sludge did not affect the dewatering and the cracking of the sludge.

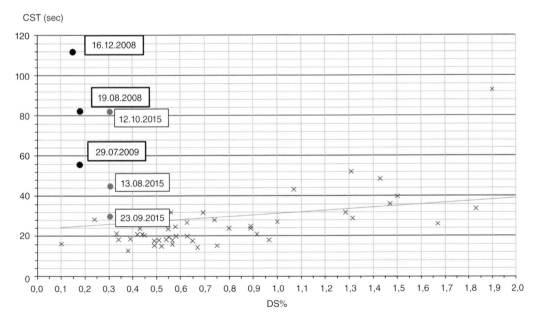

Figure 19.11 Capillary suction time (CST): measurement of surplus activated sludge and water works sludge. Red dots indicate Hanningfield WW and blue dots Whitacre WW.

The capillary suction time (CST) of Hanningfield and Whitacre WWs were both approximately 2–4 times (30–110 sec) the suction time in surplus activated sludge (10–40 sec). However, there are examples of digested sludge from WWTP with CST values of more than 1,000–2,000 sec. These types usually have a high content of fat and oil.

19.4.3.3 Sludge Volume Reduction and Sludge Residue Development

After a load of sludge, the dewatering phase results in the dry solid content of the sludge remaining on the filter surface as sludge residue, whereas the majority of its water content continues to flow vertically through the sludge residue. The water content is further reduced through evapotranspiration. The volume reduction is typically very high due to the low dry solid content in the sludge.

During the 5.5 months loading period in the Whitacre STRB test, there was a volume reduction of 97–98%. In the intensive loading periods with four days of loading, e.g., in basin 3 from 23rd–27th November, the sludge residue height increased fast, whereas the height was markedly reduced during the following resting period due to dewatering and evapotranspiration (Figure 19.12).

In the Hanningfield test STRB, the sludge volume reduction after 3–5 days of loading was of the same magnitude (>98%). In full-scale systems with much longer resting periods than in the test, the reduction is even more pronounced. The general experience from treatment of WW sludge in both test and full-scale systems is that the dry solid content (approx. 0.1–0.2%) in the feed sludge is concentrated 100–200 times to a dry solid content in the sludge residue of between >25% up to 40% ds [30]. In Hanningfield, dry solid contents of even above 50% ds in the sludge residue were obtained in some of the test basins.

The (visual) development of the sludge residue surface after loading and during the resting period is a very important indication of the dewater ability of the sludge. After a sludge load, the WW sludge typically cracks up quickly (Figures 19.13 and 19.14). A main reason for this is the low content of fat and oil and organic matter in WW sludge. For comparison, some types of sewage sludge – especially

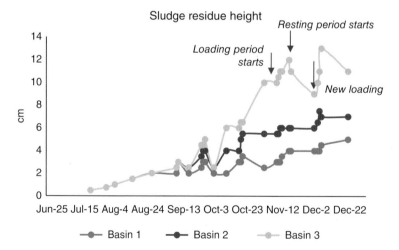

Figure 19.12 Whitacre STRB Test. Basin nos 1–3. Sludge residue height (cm) during the test period July–December 2015. Red arrows show the increase and decrease in sludge residue height during the load and subsequent resting period.

12.10.2015 (21 days since last load)   13.10.2015 (1 day since last load)

Figure 19.13 Examples of the sludge surface in Whitacre Test system. Pictures are taken 21 days after the last load (left) and 1 day after the load (right). Note the evident cracks in the sludge after only 1 days' rest (Photo: Orbicon, reproduced with permission).

Figure 19.14 Sludge residue in the Lagoon at Whitacre WW showing that the sludge residue is cracking up. (Photo: Orbicon, 13.08.2015, reproduced with permission.)

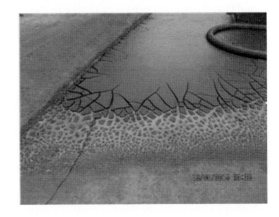

sludge originating from treatment of wastewater with a large proportion of dairy or abattoir wastewater (fat and oil) – will not crack up.

19.4.3.4   Filtrate Water Flow

The majority of the water in a sludge load continues to flow vertically through the sludge residue and filter to the drainage system. In general, the filtrate flow curve is one of the most important parameters to follow in the operation of an STRB (Figure 19.15). If the flow peak (L/sec/m^2) decreases over time after each resting period or if the flow curve tail does not decline to zero in the initial part of the rest period, this indicates that the STRB basin does not work as intended.

The filtrate water curves from Whitacre and Hanningfield test systems resembled filtrate water curves normally seen in full-scale systems for normal sewage sludge with an initial top, which then declines towards zero flow within few hours (Figure 19.15). The maximum flow rate was typically between 0.015–0.03 L/sec/m^2. The dewatering (L/sec/m^2) in the Hanningfield full-scale and test system has generally been good with no observed occurrences of surface clogging, slow dewatering or

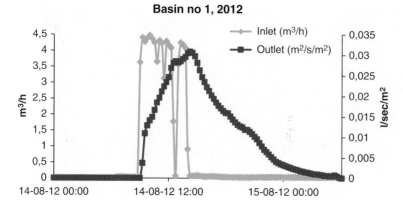

Figure 19.15 Hanningfield test STRB: four loads on the same day (blue curve, m^3/h) on test basin no. 1 and the resulting dewatering profile (red curve, $L/sec/m^2$), year 2012.

water ponding/no drainage through the sludge residue. The quality of the filtrate water strongly indicated that there is no bypass of sludge in the test basins.

Hanningfield: In some basins the maximum dewatering speed was over 0.025 $L/sec/m^2$, approximately two times higher than that seen from well-functioning STRB systems treating sludge from WWTP [30].

In summary the overall results from full-scale STRBs and test systems treating WW sludge has shown that WW sludge is treatable in an STRB system provided these systems are built and operated correctly and that the sludge quality is of the "normal" WW type.

19.5   Discussion and Conclusions

The sludge quality highly influences the dewater ability of sludge. STRBs are generally robust systems, which have been shown able to dewater many sludge types. Because mineralization takes place during the long treatment time in an STRB, the method has an advantage compared to traditional mechanical solutions. For example, STRBs reduce and degrade hazardous organic compounds effectively in contrast to mechanical solutions, where organic pollutants are not degraded [8].

19.5.1   Industrial Sludge

In general, higher concentrations of organic solids result in a lower dry solid percentage and more pronounced anaerobic conditions in the sludge residue. The level of fats and oil, together with the organic content and heavy metals, are main parameters to focus on and test for in industrial sludge. If the sludge quality contains high levels of fats and oil, it is important to address this upstream the STRB at the WWTP, e.g., by improving fat and grease removal and by reducing the content in the incoming wastewater to the WWTP.

Before investing in a full-scale STRB system, it is often advisable as a first step to establish STRB test systems to assess the sludge treatability in an STRB system. A thorough test in a test STRB provides valuable information about (1) if the sludge can be treated in an STRB; (2) the dimension criteria of a full-scale system, e.g., the areal load (kg ds/m^2/yr) and the number of basins.

Unlike organic compounds, heavy metals are not degraded in an STRB system, but are mainly bound in the sludge residue and filter. Heavy metal concentrations may be originated in wastewater from the industry, e.g., abattoirs constitute a larger proportion, than in normal wastewater. During the sludge treatment period in STRB, the concentrations of heavy metal may increase in the oldest layers because of the mineralization of organic solids (up to 20–25%) [22]. In some cases, i.e., in the oldest layer, it may exceed the legal national standards (Danish standards, BEK. No. 1650 of 13/12/2006) for soil application when expressed as "mg/kg ds" [15].

However, as shown by Nielsen and Bruun [22], the quality of the sludge residue in general, and even for sludge residues older than 20 years, complies with the EU and Danish regulations (BEK No. 1650 of 13/12/2006), when based on the contents of heavy metals in relation to phosphorus [22].

19.5.2 Water Works Sludge

Regarding WW sludge, the overall results from full-scale STRBs and test systems showed that WW sludge is treatable in an STRB system, provided these systems are built and operated correctly and that the sludge quality from the WW are of the "normal type" (low contents of organic material, and fat).

The sludge treatment results from Hanningfield and Whitacre WWs STRBs has shown that reeds were not inhibited by WWs sludge (e.g., high ferric) as the reeds grew well and in good health in both full-scale and test system. In summary:

- The dewatering was good. Shortly after the water has drained out of the system, the sludge residue surface cracks up very well. Even after 24–25 loading days in a row, the sludge cracks up after a few days' rest.
- The area load during the tests corresponded to between 30–45 kg ds/m^2/yr.
- The ferric sludge was dewatered to approximately 40% dry solids and desiccates in the trial beds.
- Filtrate water flow from the basins was good and had a curve with maximum flow of 0,02–0,03 L/m^2/sec shortly after loading, which then rapidly declined during following hours.
- Iron: There was no indication of iron in the filter layers and the concentration in the reject water was low during the 5 years of test operation. This indicates that the iron stays in the sludge residue and does not clog the drainpipes.
- Due to the high iron content caused by the production process at the works using ferric sulphate (Hanningfield) for the coagulation process, it is extra important to have aerobic conditions in the sludge residue in order to avoid iron precipitation as ochre in the filter and drain. Ochre might then clog the filter.
- Vegetation: The reeds in both tests (Hanningfield and Whitacre) showed in general to have been in good health during the whole test period and the feed sludge had no negative effect on the reed growth.

In summary, the sludge treatment results obtained from Hanningfield and Whitacre WWs in full-scale and test STRB systems showed that these were able to dewater the sludge satisfactorily.

Acknowledgements

The authors want to thank the technical staff at the Waste Water Treatment Plants and the Water Works for their technical assistance during operations and sampling. Data and results from the STRB test systems treating water works sludge have been obtained in collaboration between Essex and Suffolk Water, ARM Ltd and Orbicon A/S.

References

1 Mantovi P, Baldoni G, Toderi G. Reuse of liquid, dewatered, and composted sewage sludge on agricultural land: effects of long-term application on soil and crop. Water Res. 2005; 39(2):289–296.

2 Harrison EZ, Oakes SR, Hysell M, Hay A. Organic chemicals in sewage sludges. Sci Total Environ. 2006; 367(2):481–497.

3 Fytili D, Zabaniotou A. Utilization of sewage sludge in EU application of old and new methods – a review. Renew Sust Energ Rev. 2008; 12(1):116–140.

4 Salado R, Vencovsky D, Daly E, Zamparutti T, Palfrey R. Environmental, economic and social impacts of the use of sewage sludge on land. Part II: Report on Options and Impacts, 2010. Report by RPA, Milieu Ltd and WRc for the European Commission, DG Environment.

5 Nielsen S, Willoughby N. Sludge treatment and drying reed bed systems in Denmark. Water Environ J. 2005; 19(4):296–305.

6 Uggetti E, Ferrer I, Llorens E, Garcia J. Sludge treatment wetlands: a review on the state of the art. Bioresour Technol. 2010; 101:2905–2912.

7 Nielsen S. Assessment of sludge quality for treatment in sludge treatment reed bed system – trial systems. Water Sci Technol. 2016; (in press).

8 Nielsen S. Economic assessment of sludge handling and environmental impact of sludge treatment in a reed bed system. Water Sci Technol. 2015; 71(9):1286–1292.

9 Nielsen S. Mineralisation of hazardous organic compounds in a sludge reed bed and sludge storage. Water Sci Technol. 2005; 51(9):109–117.

10 Nielsen S. Sludge treatment and drying reed bed systems. Wastewater treatment in wetlands: theoretical and practical aspects. Water Sci Technol. 2007; 3–4:223–234.

11 Nielsen S, Peruzzi E, Macci C, Doni S, Masciandaro G. Stabilisation and mineralisation of sludge in reed bed systems after 10–20 years of operation. Water Sci Technol. 2014; 69(3):539–545.

12 Peruzzi E, Nielsen S, Macci C, Doni S, Iannelli R, Chiarugi M, Masciandaro G. Organic matter stabilization in reed bed systems: Danish and Italian examples. Water Sci Technol. 2013; 68(8):1888–1894.

13 Nielsen S. Sludge drying reed beds. Water Sci Technol. 2003; 48(5):101–110.

14 Nielsen S. Helsinge sludge reed bed system: reduction of pathogenic microorganisms. Water Sci Technol. 2007; 56(3):175–182.

15 Matamoros V, Nguyen LX, Arias CA, Nielsen S, Laugen MM, Brix H. Musk fragrances, DEHP and heavy metals in a 20 years old sludge treatment reed bed system. Water Res. 2012; 46(12):3889–3896.

16 Siracusa G, La Rosa AD. Design of a constructed wetland for wastewater treatment in a Sicilian town and environmental evaluation using the emergy analysis. Ecol Model. 2006; 197(3):490–497.

17 Zhou JB, Jiang MM, Chen B, Chen GQ. Emergy evaluations for constructed wetland and conventional wastewater treatments. Commun Nonlinear Sci Numer Simul. 2009; 14(4):1781–1789.

18 Dixon A, Simon M, Burkitt T. Assessing the environmental impact of two options for small-scale wastewater treatment: comparing a reedbed and an aerated biological filter using a life cycle approach. Ecol Eng. 2003; 20(4):297–308.

19 Olsson L, Larsen JD, Ye S, Brix H. Emissions of CO_2 and CH_4 from sludge treatment reed beds depend on system management and sludge loading. J Environ Manage. 2014; 141:51–60.

20 Kopp J, Dichtl N. Prediction of full-scale dewatering results by determining the water distribution of sewage sludges. Water Sci Technol. 2000; 42(9):141–149.

21 Nielsen, S. Sludge treatment reed bed facilities-organic load and operation problems. Water Sci Technol. 2011; 63(5):942–948.

22 Nielsen S, Bruun EW. Sludge quality after 10–20 years of treatment in reed bed systems. Environ Sci Pollut Res. 2015; 22(17):12885–12891.

23 Kołecka K, Nielsen S, Obarska-Pempkowiak H. The speciation of selected heavy metals of sewage sludge stabilized in reed basins. 11th International IWA Specialist Group Conference on Wetland Systems for Water Pollution Control 1–7, 2008. Indore, India.

24 Stefanakis AI, Tsihrintzis VA. Heavy metal fate in pilot-scale sludge drying reed beds under various design and operation conditions. J Hazard Mater. 2012; 213:393–405.

25 Kołecka K, Obarska-Pempkowiak J. Potential fertilizing properties of sewage sludge treated in the sludge treatment reed beds (STRB). Water Sci Technol. 2013; 68(6):1412–1418.

26 Federle TW, Itrich NR. Comprehensive approach for assessing the kinetics of primary and ultimate biodegradation of chemicals in activated sludge: application to linear alkylbenzene sulfonate. Environ Sci Technol. 1997; 31(4):1178–1184.

27 Allen SE.,. Chemical Analysis of Ecological Material, 2nd edn. Blackwell Scientific Publications, Oxford, 1989, 368 pp.

28 Nielsen S, Sellers TCP. Paper and presentation: Dewatering sludge originating in water treatment works in reed bed systems – 5 years of experience. 17th European Biosolids and Organic residuals Conference and Exibition, 2012. Leeds, UK.

29 Nielsen S. Sludge treatment in reed beds systems – development, design, experiences. Sustain Sanit Pract. 2012; 12:33–39.

30 Nielsen S, Cooper DJ. Dewatering sludge originating in water treatment works in reed bed systems. Water Sci Technol. 2011; 64(2):361–366.

20

Constructed Wetlands for Water Quality Improvement and Temperature Reduction at a Power-Generating Facility

Christopher H. Keller[1], Susan Flash[2] and John Hanlon[2]

[1] *Wetland Solutions, Inc., Gainesville, USA*
[2] *PurEnergy Operating Services, LLC, Syracuse, USA*

20.1   Introduction

PurEnergy Operating Services, LLC (PurEnergy), operates a 15 Megawatt (MW) biomass-fueled power-generating facility in the southeast United States. Operation of the facility produces an intermittent discharge of non-process, cooling tower and boiler blowdown, equipment wash water, and neutralized reverse osmosis brine. Event-based stormwater discharges from ash pile sedimentation ponds and drainage ditches blend with the industrial effluent and discharge to an unnamed tributary and from there to a major rural creek system.

Historically, the facility experienced periodic exceedances of effluent limitations for unionized ammonia, copper, pH, temperature, acute toxicity, and specific conductance. Facility staff and outside technical consultants conducted a treatment system alternatives evaluation to determine the preferred technical solution that could be implemented to improve final effluent water quality and achieve compliance with environmental regulations. The study recommended construction of a constructed treatment wetland system as the most cost-effective alternative to improve water quality for most of the parameters of interest. The proposed wetlands were not expected to reduce specific conductance, and dilution with cooler, pumped groundwater was initially recommended as the preferred method to achieve compliance with the applicable thermal water quality standard.

This chapter describes the basis of design for the constructed treatment wetland system and presents operational performance data for the period from startup in September 2013 through December 2015.

20.2   Basis of Design

Historical operational data were evaluated to provide an estimate of the likely inflow conditions for the proposed wetland. Effluent flow and water quality data were available for the period from January 2006 through December 2007 and again from January 2010 through June 2010. The facility was shut down during most of 2008 and 2009 for repairs.

Constructed Wetlands for Industrial Wastewater Treatment, First Edition. Edited by Alexandros I. Stefanakis.
© 2018 John Wiley & Sons Ltd. Published 2018 by John Wiley & Sons Ltd.

Table 20.1 Historical flow and effluent water quality data (January 2006–June 2010).

Parameter (units)	Standard	Mean	Range	Count	Number of exceedances
Flow (m^3/d)	NA	304	0–2,950	766	NA
Unionized ammonia (mg/L as NH$_3$)	0.02	0.01	0.001–0.04	101	6
Total ammonia (mg/L as N)	NA	0.19	0.027–0.51	101	NA
Total copper (mg/L)	0.03	0.034	0.0024–0.23	102	44
pH (standard units)	6.0–8.5	7.68	6.67–8.57	122	2
Specific conductance (μmhos/cm)	1,275	1,099	772–1,541	17	3

Table 20.1 shows summary statistics for pre-wetland flow and effluent water quality, along with applicable regulatory standards. Effluent flow averaged 304 m^3/d and ranged from 0–2950 m^3/d. Unionized ammonia averaged 0.01 mg/L (as NH$_3$) and ranged from 0.001–0.04 mg/L (as NH$_3$). The effluent unionized ammonia concentration exceeded the applicable regulatory standard of 0.02 mg/L (as NH$_3$) on six occasions. Total copper averaged 0.034 mg/L and ranged from 0.0024–0.23 mg/L, exceeding the standard of 0.03 mg/L 44 times during the period of study. Effluent pH averaged 7.68 standard units (s.u.) and ranged from 6.67–8.57 s.u., with only two exceedances of the maximum allowable pH of 8.5 s.u. Specific conductance averaged 1,099 μmhos/cm and ranged from 772–1,541 μmhos/cm with three exceedances of the 1,275 μmhos/cm standard. It should be noted that groundwater from the Floridan aquifer is the source of all process water used at the facility. The ambient groundwater has a specific conductance of about 400 μmhos/cm. The facility recycles water 3 to 4 times prior to discharge. When combined with the evaporative effects of the cooling tower and boiler, the recycling further concentrates the ions that contribute to specific conductance.

The discharge permit limited effluent water temperature to no more than 2.8°C (or 5°F) warmer than the water temperature measured at a background stream station. The regulatory standard is applied as the difference between instantaneous measurements rather than as the difference between daily (or longer) averages. Figure 20.1 compares the final effluent water temperature and the background stream temperature. The background stream exhibits temperature dynamics that follow the pattern of the average daily air temperature. By comparison, the effluent was much warmer than the background stream, with instantaneous measurement differences averaging 8.4°C and ranging from 0–20°C. However, the relationship between the background stream water temperature and air temperature indicated that, under the right environmental conditions, the effluent temperature could be substantially reduced through natural cooling mechanisms.

20.2.1 Design for Ammonia and Copper Reduction

The design method for ammonia and copper followed the first-order, area-based P-k-C* equation described by Kadlec and Wallace [1] and shown below.

$$\frac{C_2 - C^*}{C_1 - C^*} = \left(1 + \frac{k_{20}}{Pq}\right)^{-P}$$

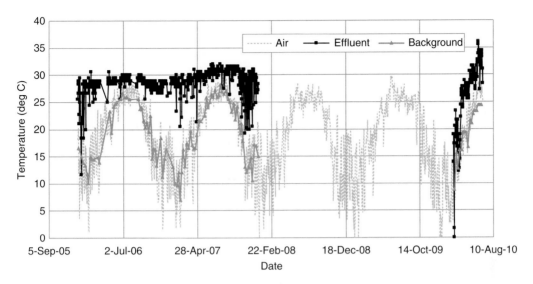

Figure 20.1 Effluent and background water temperature.

Where:

C_1 = average inlet concentration (mg/L)
C_2 = average outlet concentration (mg/L)
C^* = background concentration (mg/L)
k_{20} = first-order, area-based rate constant at 20°C (m/y)
q = average hydraulic loading rate (Q/A) (m/y)
P = apparent number of tanks-in-series (dimensionless)
Q = flow rate (m³/y)
A = wetland area (m²)

For certain parameters, where removal mechanisms are correlated with temperature or season, the value of k_{20} can be adjusted for the ambient water temperature using an Arrhenius equation:

$$k_T = k_{20}\theta^{T-20}$$

Where:

k_{20} = removal rate constant at 20°C (m/y)
θ = dimensionless constant
k_T = removal rate constant at temperature T (m/y)

Because treatment wetland area estimates generated with the *P-k-C** model represent long-term median performance, and daily, weekly, and monthly averages may be variable enough to cause excursions of permit limits, the target outflow concentrations were reduced using the excursion containment approach of Kadlec and Wallace [1]. Table 20.2 summarizes the parameters used in the *P-k-C** equation to estimate the required wetland area based on conservative ammonia and copper removal targets.

Table 20.2 Parameter selection for the *P-k-C** design method.

Design variable	Units	Ammonia	Copper
C_1	mg/L	0.20	0.035
C^*	mg/L	0.005	0.010
k_{20}	m/y	14.7	25
θ	Dimensionless	1.05	1.00
P	Dimensionless	2	2
T	°C	18.8	18.8
Initial target C_2	mg/L	0.08	0.025
90% Containment multiplier	Dimensionless	1.89	1.75
Final target C_2	mg/L	0.04	0.014

Sizing for ammonia removal established the final surface flow wetland area of 1.8 hectares (ha) at the design average inflow rate of about 380 m³/d. For this design flow and treatment area and assuming an average operating depth of about 0.25 m, the nominal hydraulic residence time was estimated to be 12 days.

20.2.2 Design for pH, Toxicity, and Specific Conductance

Natural freshwater wetlands exhibit pH values ranging from 6–7 s.u., and the organic compounds generated within both natural and constructed wetlands are a source of acidity that can buffer alkaline inflows [1]. Consequently, treatment wetland pH values are commonly circumneutral. Empirical data have shown that effluent toxicity is reduced or eliminated by the various processes that occur in treatment wetlands. At present, however, there are insufficient data and too much variability in test protocols to formulate a standardized wetland design equation for toxicity [1]. The design approach used for pH and toxicity was therefore presumptive, in that by sizing for copper and ammonia removal, the resulting residence time was assumed to be adequate to neutralize final effluent pH and reduce the most likely cause (unionized ammonia) of prior toxicity violations.

Specific conductance was not expected to be substantially changed with passage through the wetlands, except by dilution with rainfall. It would be possible to reduce specific conductance by decreasing the recycle rate or by diluting the final effluent with pumped groundwater, but this approach would increase total water withdrawals from the primary potable water aquifer in the region.

20.2.3 Design for Temperature Reduction

The design for temperature reduction was based on the energy budget approach described by Kadlec and Wallace [1]. Heat energy is added to water in the wetland compartment through several pathways. For most wetlands, the most dominant source of incoming heat energy is from net solar radiation. For this particular system however, considerable heat energy is also carried by the warm blowdown water that discharges from the cooling towers and boiler and is pumped to the wetland. Wetland heat energy outflows are usually dominated by evapotranspiration losses and convective nighttime cooling when air temperature is often lower than water temperature. Some heat energy also leaves with the water that discharges from the wetland to the receiving stream.

With the exception of the heat contained in the pumped inflow, the dominant energy inflows and outflows are driven by climatic variables. Net solar radiation is high during the day and effectively zero at night. Net solar radiation is also a function of latitude, time of year, and degree of cloud cover. The heat transferred by both evapotranspiration and convection is strongly influenced by wind speed and ambient air temperature.

As water flows through the wetland, the water temperature may change in response to the net effect of the various energy inflows and outflows. When the energy inflows and outflows are in equilibrium, the wetland water temperature reaches a thermal endpoint called the balance temperature. This means that no additional water temperature change is possible until there is a change in inflow or climatic conditions. If the relative humidity is 50%, then the balance temperature for wetland water is approximately equal to the average ambient air temperature. At higher humidity, the balance temperature is greater than the air temperature, and at lower humidity, the wetland water can actually be cooler than the air temperature. Kadlec and Wallace [1] report that the wetland balance temperature is typically reached within about 3 days of hydraulic residence time.

The design approach used to estimate the area necessary to achieve the desired level of ammonia and copper reduction resulted in a wetland with an approximate nominal hydraulic residence time of 12 days, implying that the balance temperature would likely be reached within the wetland and that the applied effluent could therefore be cooled as much as environmental conditions would allow. Typical monthly average balance temperatures were estimated using local meteorological data and the following equations:

$$P^{sat}(T_w) = RH * P^{sat}T_a + \frac{ET_o}{(1.96 + 2.60u)}$$

$$ln(P^{sat}) = 19.0971 - \frac{5349.93}{(T + 273.16)}$$

Where:

$P^{sat}(T_a)$ = saturation water vapor pressure at T_a (kPa)
$P^{sat}(T_w)$ = saturation water vapor pressure at T_w (kPa)
T_a = air temperature (°C)
T_w = water balance temperature (°C)
RH = relative humidity (fraction)
u = wind speed at 2 m elevation (m/s)
ET_o = potential evapotranspiration (mm/d)

Table 20.3 shows the estimated monthly balance temperature and the difference between the average air temperature and the balance temperature. These results show that the wetland would be expected to cool the effluent from 0.5 to 1.7°C below the average air temperature, except during the months of June through August. During the warmer, more humid summer months, the water temperature was estimated to be less than 1°C greater than the average air temperature. Daily average or instantaneous differences between the wetland water temperature and either the background stream temperature or air temperature would be subject to greater variability. Thus, it was not prudent to imply that the proposed wetland discharge would always meet the regulatory limit.

Table 20.3 Estimated monthly balance temperatures and differential between air and balance temperatures.

Month	T_a (°C)	ET_o (mm/d)	u (m/s)	RH	$P^{sat}(T_a)$ (kPa)	$P^{sat}(T_w)$ (kPa)	T_w (°C)	ΔT (°C)
January	10.2	1.4	1.98	0.74	1.24	1.11	8.52	1.68
February	11.0	1.8	1.96	0.73	1.31	1.22	9.95	1.05
March	15.2	2.5	1.98	0.71	1.72	1.58	13.9	1.30
April	18.3	3.5	1.86	0.70	2.10	1.98	17.4	0.90
May	23.0	4.1	1.74	0.74	2.81	2.72	22.5	0.50
June	25.8	4.4	1.43	0.80	3.32	3.42	26.3	−0.50
July	26.3	4.4	1.38	0.82	3.43	3.58	27.1	−0.80
August	26.2	3.8	1.44	0.84	3.40	3.54	26.8	−0.60
September	24.5	3.3	1.77	0.80	3.07	2.95	23.8	0.70
October	20.1	2.3	1.61	0.80	2.34	2.26	19.5	0.60
November	14.3	1.7	1.72	0.78	1.63	1.53	13.4	0.90
December	11.1	1.3	1.89	0.78	1.32	1.21	9.84	1.26

Data for T_a, ET_o, u, and RH from Florida Automated Weather Network Quincy station (September 2002–July 2010).

20.2.4 Process Flow and Final Design Criteria

Non-process wastewater is held in a covered, concrete basin where some radiant cooling can occur. Equipment to facilitate pH adjustment is available at the basin. Several times each day, depending on water levels, a portion of the basin volume is pumped to the constructed wetland for further treatment and discharge to the receiving waters. The pumped flow to the wetland splits at a manhole structure into two cells, each approximately 0.9 ha in size (140 m in length by 67 m in width). A perimeter embankment contains the water in the wetland and allows for the accumulation of approximately 0.3 m of direct rainfall above the treatment volume, while maintaining 0.3 m of freeboard. The embankments were designed with a minimum top width of 3.7 m to facilitate vehicle access and 4:1 (horizontal:vertical) side slopes for safety during maintenance activities. The wetland cells were designed with flat bottoms (no longitudinal slope) at a final elevation close to the existing seasonal high water table. The wetland cells were not lined. Table 20.4 summarizes general design criteria for the wetland cells. Figure 20.2 shows the design plan view and Figure 20.3 shows the cross section view for one of the two wetland cells.

20.3 Construction

Project construction commenced in October 2012. Major construction activities included the replacement of existing effluent pumps, construction of 840 m of 10-cm diameter P.V.C. inflow pipeline, installation of a wetland inflow splitter structure, clearing and grading of the two 0.94-ha wetland cells, installation of two outlet water level control structures with pressure transducer water level recorders, and construction of two outfall gravity pipelines to an existing conveyance ditch.

Table 20.4 Final design criteria.

Design parameter	Cell 1	Cell 2
Area (ha)	0.94	0.94
Length (m)	140	140
Width (m)	67	67
Aspect ratio	2	2
Nominal depth (m)	0.25	0.25
Design flow rate (m³/d)	190	190
Nominal hydraulic loading rate (cm/d)	2.0	2.0
Nominal hydraulic residence time (d)	12	12
Number of deep zones	3	3
Deep zone depth below cell bottom (m)	0.9	0.9

The south wetland cell was completed and placed in operation in September 2013 with all the flow routed through that cell until the north wetland cell was completed and brought online at the end of January 2014. Water depths were initially kept low (about 0.10 m) to allow the installed and naturally-recruited wetland plants to develop strong root systems and begin to expand in areal coverage. Because the completion of the north cell lagged the south cell and extended into the winter season, transplanted bare-root material was only used to establish vegetation in the south cell. Approximately 11,000 bare-root wetland propagules were installed in the south cell in July 2013, and consisted of softstem bulrush (*Schoenoplectus validus*), duck potato (*Sagittaria lancifolia*), and cattail (*Typha* spp.). The plant community in the north cell resulted solely from natural recruitment. Dominant vegetation in the north cell includes soft rush (*Juncus effusus*) and a variety of sedge species (*Carex* spp.).

The total construction cost was approximately $285,000 (U.S. dollars). The cost for surveying, engineering design, construction supervision, and permitting was approximately $50,000. Operational costs consist of electrical power for pumping to the wetland, water quality sampling and analysis, periodic embankment mowing, and routine report preparation for the environmental regulatory agency.

20.4 Operational Performance Summary

Operational performance data are summarized below for the period from September 2013 through December 2015.

20.4.1 Inflow and Outflow Rates and Wetland Water Depths

The average inflow rate was about 450 m³/d, which exceeded the design flow assumption of 380 m³/d. Wetland inflow rates were measured using a totalizing flow meter. Figure 20.4 shows the average daily discharge from wetland cells and water depths for the period from September 2013 through December 2015. Water surface elevations were measured using recording pressure transducers that reported

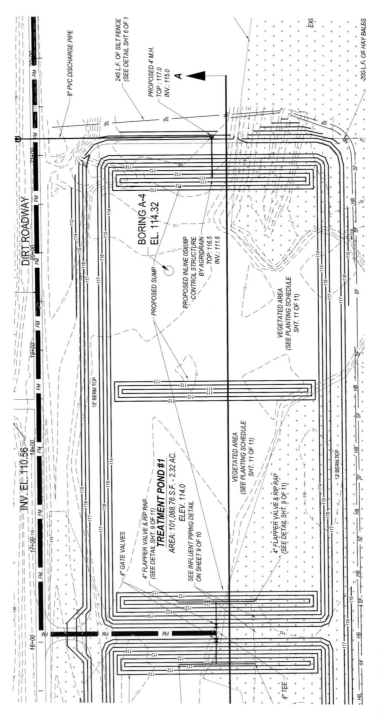

Figure 20.2 Partial site plan view.

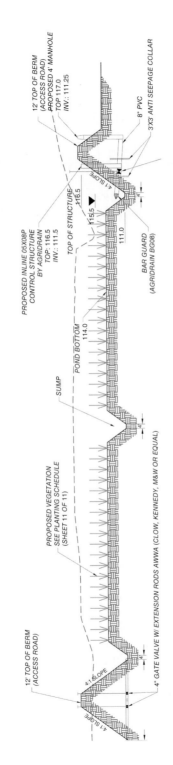

Figure 20.3 Partial cross-section view.

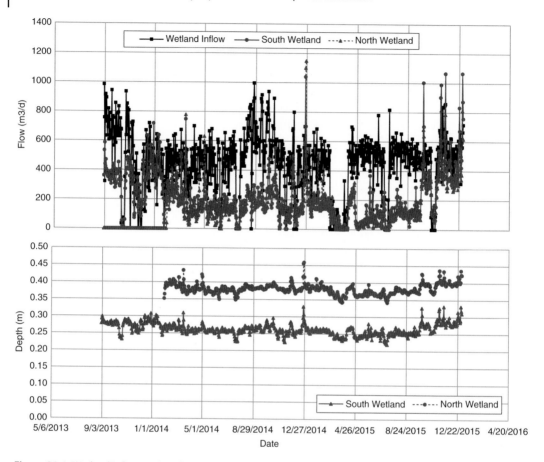

Figure 20.4 Wetland inflow and outflow rates and water depths (September 2013 through December 2015).

the depth of the water column above the sensor. Wetland water depths were estimated by subtracting the average wetland bottom elevation from the daily average water surface elevation. Wetland outflow rates were then estimated from the water surface elevations and control weir elevations using a rectangular weir equation that was calibrated by Chun and Cooke [2] for the specific type of structures that were installed.

The discharge from the south and north cells averaged 218 and 173 m^3/d, respectively. The difference between annual rainfall and evapotranspiration at this location led to a net gain in hydrologic inflows; however the net rainfall contribution was only about 5% of the pumped inflow volume. Discrepancies between pumped inflow volumes and calculated wetland outflow volumes were attributed to instrument error and minor seepage losses to the groundwater table. Water depths averaged 0.26 m in the south cell and 0.38 m in the north cell. Though there were several brief interruptions in flow delivery to the wetland cells, neither dried out during the period of record.

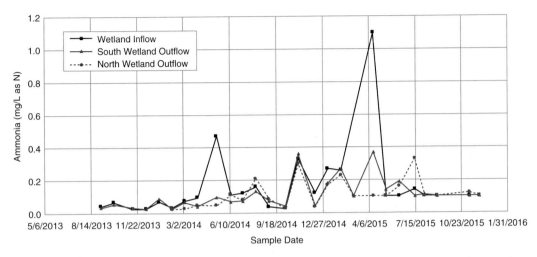

Figure 20.5 Wetland inflow and outflow ammonia concentrations (September 2013 through December 2015).

20.4.2   Ammonia

Figure 20.5 shows wetland total ammonia data from grab samples collected at the effluent pump station (wetland inflow) and outlet water control structures from the two wetland cells. Wetland inflow total ammonia concentrations averaged 0.16 mg/L (as N) and ranged from 0.026–1.1 mg/L. Wetland outflow concentrations ranged from 0.026–0.37 mg/L and averaged 0.11 mg/L from the south cell and 0.12 mg/L from the north cell. The removal efficiency for ammonia was low (28%), as a direct result of the low inflow concentration. Unionized ammonia, which increases with increasing pH and water temperature and is toxic to aquatic fauna, averaged less than 0.001 mg/L (as N) at the wetland outflow stations and was consistently well-below the regulatory limit (0.02 mg/L as NH_3 or 0.016 mg/L as N) due to the combined effects of total ammonia removal, pH neutralization, and effluent cooling as water flowed through the treatment system.

20.4.3   Copper

Figure 20.6 shows wetland inflow and outflow total copper concentration data. Wetland inflow total copper concentrations averaged 0.026 mg/L and ranged from 0.0079–0.13 mg/L. Wetland outflow concentrations ranged from 0.0018–0.0098 mg/L and averaged 0.0035 mg/L from the south cell and 0.0031 mg/L from the north cell. The estimated concentration reduction efficiency was 88%.

20.4.4   pH

Figure 20.7 shows wetland inflow and outflow pH measurements that were collected in conjunction with grab samples for analytical water quality parameters. Wetland inflow pH averaged 8.06 standard units (s.u.) and ranged from 6.74–8.56 s.u. Outflows from the south wetland cell averaged 7.20 s.u.

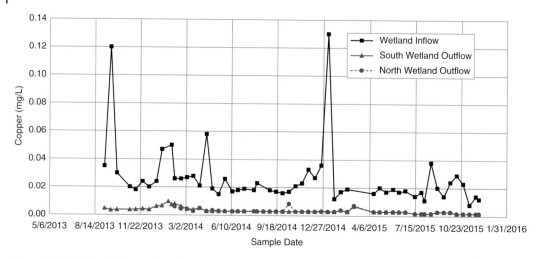

Figure 20.6 Wetland inflow and outflow copper concentrations (September 2013 through December 2015).

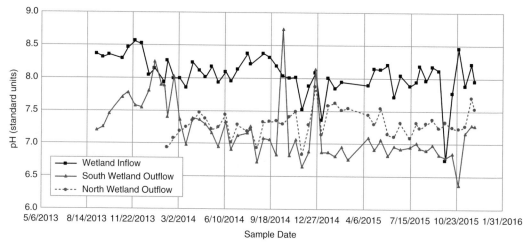

Figure 20.7 Wetland inflow and outflow pH measurements (September 2013 through December 2015).

and ranged from 6.37–8.75 s.u. Outflows from the north wetland cell averaged 7.30 s.u. and ranged from 6.83–7.87 s.u.

20.4.5 Temperature

Effluent temperature was consistently reduced with passage through the wetland cells. Figure 20.8 shows wetland inflow and outflow water temperatures based on grab sample data collected at the time that routine sampling was conducted for the other analytical parameters described above. Figure 20.9 shows the differential between average monthly wetland inflow and outflow temperatures. Monthly average effluent cooling averaged 6.01°C and ranged from <1°C to 11.6°C.

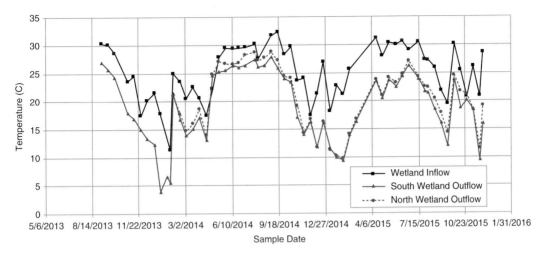

Figure 20.8 Wetland inflow and outflow temperature (September 2013 through December 2015).

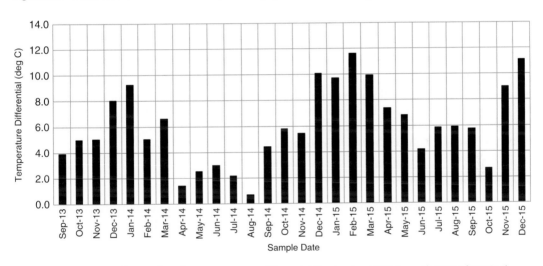

Figure 20.9 Wetland inflow and outflow temperature differential (September 2013 through December 2015).

As part of the renewal process for the facility's industrial wastewater discharge permit, a detailed temperature study was conducted to document daily temperature variability at the inflow to the wetland, at the wetland outlet, and in nearby reference stream systems. Figure 20.10 compares the daily wetland outflow temperature to the balance temperature estimated using the procedure described above. This analysis showed that the wetland effluent was cooled nearly to the extent possible based on local meteorological conditions. Data collected further downstream in the receiving system showed that additional cooling occurred from shading by canopy vegetation and seepage inputs of cooler groundwater. By the time that the effluent-dominated receiving stream merged with a larger creek system, there was no measurable temperature difference between the water bodies.

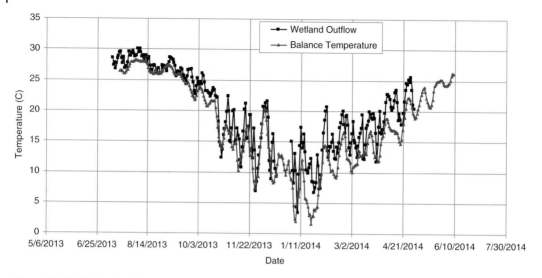

Figure 20.10 Wetland outflow and estimated balance temperatures (July 2013 through April 2014).

20.4.6 Whole Effluent Toxicity

Whole effluent toxicity (WET) monitoring is routinely required for surface water discharges. In this case, the analyses consisted of 96-hour acute static renewal multi-concentration tests using the water flea (*Ceriodaphnia dubia*) and bannerfin shiner (*Cyprinella leedsi*). Following completion of construction and routing of all industrial effluent through the wetland system, the facility passed each toxicity test (n = 8) conducted between September 2013 and December 2015.

20.4.7 Specific Conductance

As was expected during the design process, specific conductance exhibited minimal change with passage through the wetland system. Inflow specific conductance averaged 1,420 μmhos/cm and wetland outflows averaged 1,343 μmhos/cm with the slight dilution attributed to rainfall effects (Figure 20.11).

20.5 Discussion

Implementation of the treatment wetland project successfully and cost-effectively resolved most of the facility's water quality compliance challenges that were in effect at the time that the project was designed and constructed. As described above, the water quality parameters of interest included unionized ammonia, copper, pH, whole effluent toxicity, specific conductance, and temperature.

The wetland performance data demonstrated that water quality compliance standards were met for unionized ammonia, copper, pH, and toxicity. The risk of exceeding the unionized ammonia standard has been greatly reduced due to bulk removal of ammonia nitrogen (by driving ammonia through the nitrification and denitrification pathways) and the complementary effects of the wetland lowering pH and reducing water temperature. There was a single exceedance for pH since the wetland was placed

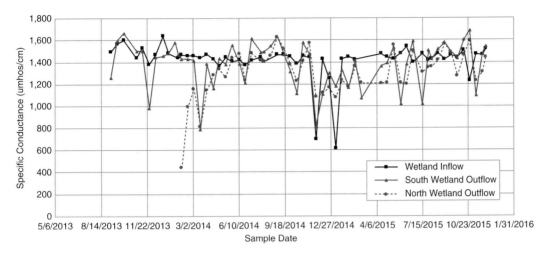

Figure 20.11 Wetland inflow and outflow specific conductance (September 2013 through December 2015).

in operation in September 2013. However, wetland outflow pH was in compliance, but filamentous algal growth in the immediate vicinity of the final discharge monitoring station, several hundred feet downstream, raised the average daily pH due to the consumption of dissolved carbon dioxide (a weak acid) associated with photosynthesis.

Specific conductance was minimally reduced with passage through the wetland and remains a parameter of interest for the facility and the regulatory agency. A previously-recommended approach to lowering effluent specific conductance was to blend the final effluent with pumped groundwater. While this approach could certainly be used to meet the effluent limitation, the management team did not consider it to be an environmentally-preferred solution, particularly because a separate regulatory permit required the implementation of a water conservation plan to reduce total groundwater consumption and maximize efficient use of water within the facility. The facility is instead studying the biology of the receiving stream with the expectation that the results will demonstrate that the macroinvertebrate population is consistent with that of a healthy aquatic system, despite the slightly elevated specific conductance. The surface water discharge regulations allow the adoption of an alternative standard, when such a demonstration of aquatic ecosystem health is made.

The wetland cooling data and the ancillary receiving stream studies showed that the effluent temperature was greatly reduced compared to historical conditions. The regulatory agency subsequently agreed with these findings and removed effluent water temperature as a regulated parameter when the discharge permit was renewed in October 2015. This change eliminated the need to operate and maintain continuous temperature logging equipment at the point of discharge and at a remote background monitoring station, saving staff time and associated expenses. Similar to the prior strategy to use groundwater to dilute specific conductance, the consumption of cooler groundwater would no longer have to be considered as a means to reduce effluent temperature.

The constructed wetland project has achieved all of the water quality improvement objectives for which it was designed for lower cost than the conventional treatment alternatives that were considered. The wetland requires limited and straight-forward interaction for operations and maintenance, allowing the facility staff to focus their time and efforts on their primary mission of power generation, rather than industrial wastewater treatment.

References

1 Kadlec, RH, Wallace, SD. Treatment Wetlands, 2nd edn. Boca Raton: CRC Press; 2009.
2 Chun JA, Cooke RA. Calibrating Agridrain Water Level Control Structures Using Generalized Weir and Orifice Equations. Appl Eng Agric. 2008. 24(5):595–602.

21

Recycling of Carwash Effluents Treated with Subsurface Flow Constructed Wetlands

A. Torrens[1], M. Folch[1], M. Salgot[1] and M. Aulinas[2]

[1] Soil Science Laboratory, Faculty of Pharmacy, University of Barcelona, Barcelona, Spain
[2] Life MinAqua PM, Grup Fundació Ramon Noguera, Girona, Spain

21.1 Introduction

In some Mediterranean regions such as Catalonia, urban development is putting existing water resources at risk. In these zones, any attempt to reclaim, recycle and reuse water is considered a "win–win" strategy by both enhancing water supplies and reducing pollution. Such win–win strategies can be implemented in many industrial sites and for activities that involve tap water consumption [1].

A sector contributing to high water consumption is commercial car washing. Currently, vehicle washing facilities are widely spread throughout all urban areas in developed countries. Despite their significant negative environmental impacts that result from the high consumption of resources (water and energy) and generation of waste, there are very few facilities that are committed to using innovative solutions to address this problem. Efforts to improve water use efficiency should be placed in conjunction with maintaining or improving water quality. Therefore, water consumption and pollution should be reduced by better management and technical improvements in the treatment and recycling of wastewater. The car wash industry poses additional environmental threats due to the use of detergents with surfactants and additives.

The car wash industry appears today to be more conscious of the need for wastewater treatment and water reclamation. Environmental legislation and guidelines concerning this specific issue have been released worldwide. In Queensland, Australia, it is forbidden to use more than 70 L of fresh water per vehicle, and some European countries restrict water consumption to 60–70 L per car and/or impose reclamation percentages (70–80%). In the Netherlands and Scandinavian countries, 60–70 L/car is the maximum amount of fresh water consumption allowed. The recycling of 80% of car wash effluent is compulsory in Germany and Austria [2].

The majority of car wash facilities in Europe and Spain do not recycle their wastewater; they treat the wastewater in order to meet the established thresholds to discharge to the sewer system or the receiving media. Currently, car wash facilities in the city of Girona use tap water, and each car wash uses approximately 320 L of water [3]. The total consumption of water in the car wash industry mainly depends on the type of vehicle cleaning system. Table 21.1 shows water consumption according to

Constructed Wetlands for Industrial Wastewater Treatment, First Edition. Edited by Alexandros I. Stefanakis.
© 2018 John Wiley & Sons Ltd. Published 2018 by John Wiley & Sons Ltd.

Table 21.1 Water consumption average according to the type of car wash.

Vehicle cleaning system	Water consumption (L/washing)	Water consumption (L/washing) Life MinAqua (water audits)
Conveyor car	100–350	252–295
Large vehicles	200–650	66
Self-service box	70–80	440
Hand-wash	50–500	–

type of car wash, expressed in liters per washing. The figures are based on data from the literature [4] and the experience of the Life MinAqua project (2012–2016), as it is shown in the last column.

The criteria for vehicle wash reclamation systems must include public acceptance, aesthetic quality of reclaimed water, microbiological risk and chemical issues. Reports by the International Carwash Association indicate that the water quality of vehicle washes should be sufficiently high such that the vehicles and wash equipment are not damaged (chemical risks include corrosion, scaling and spot formation), the microbial risk to operators and users must be minimal, and the aesthetic conditions must be acceptable [5]. Therefore, controlling the microbiological risk of reclaimed water is an important issue in the car wash industry. In addition to bacterial indicators, Legionella content is an important parameter that must be controlled if the installation has any equipment that can produce aerosols.

The car wash process consists of the following steps: (1) application of a degreasing agent all over the surface of the vehicles; (2) addition of acid and alkaline cleansers; and (3) a coating [6]. Effluents from car wash facilities contain a number of pollutants such as sand, dust, detergents/surfactants, organic matter, oil, fat, oil/water emulsions, carbon, asphalt and salts [7]. This effluent also presents high levels of turbidity, organic matter, phosphorous and nitrogen compounds, plasticizers, brake dust from rubber linings and various heavy metals [4, 5]. Few studies have characterized car wash effluents, and even fewer from a microbiological point of view. However, the study conducted by Zaneti et al. [5] demonstrates that car wash effluent includes high concentrations of bacterial indicators (fecal coliforms).

Conventional treatment methods such as series of settler tanks and hydrocarbon separators are often used when wastewater needs to be discharged into the environment or to the sewerage. If wastewater reclamation is envisaged, the effluent must be treated to meet an acceptable level of water quality such that it can be recycled; thus, higher quality effluents are needed. In the car wash industry, a typical approach used for reclamation systems entails usually physical–chemical treatment, i.e., flocculation–sedimentation and direct filtration. Some suppliers have developed recycling equipment based on flocculation–coagulation processes and compact filtration systems. According to Brown [8], car wash wastewater reclamation requires the separation of sand, gravel, oils and fats prior to reuse. Additional treatment processes can be employed to strengthen the quality of the reclaimed water, such that it can be used in different washing stages (pre-soak, wash, rocker panel/undercarriage, first rinse, and final rinse). Some processes and technologies that have been proposed and tested include reverse osmosis, nanofiltration, ultrafiltration, flocculation–sedimentation and flocculation–flotation [2]. Filtration treatments with activated carbon, ozone and ultrafiltration are being studied along with electrochemical methods such as anodic oxidation with diamond and lead dioxide anodes [9, 10].

As far as the authors know, natural technologies have never been applied to treat these effluents, even though Constructed Wetlands (CWs) are used to treat effluents with similar characteristics (e.g., urban runoff) or those even less biodegradable (e.g., petroleum and oil industry). Therefore, in this context it is proposed the use of natural treatments to reclaim wastewater from car wash facilities.

The general objective of the pilot project study presented in this chapter is to demonstrate that natural treatments can effectively reclaim wastewater from car wash effluents facilities. Specifically, the purpose of the study is to evaluate the viability of different Subsurface Flow Constructed Wetlands (SSFCWs) configurations to treat the effluent from car wash facilities for internal recycling and reduce tap water consumption.

21.2 Case Study: Description

Montfullà's car wash facility, from the special employment center owned by the Grup Fundació Ramon Noguera, is located at Montfullà's industrial park in Bescanó, province of Girona (Catalonia, Spain). This facility was opened in 2011 and has a conveyor or pull-along car wash, a gantry car wash for commercial and large vehicles (buses, trucks, etc.) and a self-service car wash facility. The area's average yearly rainfall is between 700 and 900 mm, and average monthly temperatures range from 1 –24°C. Three pilot plants were purposely designed and constructed for the project: one Vertical Flow Constructed Wetland (VFCW), one Horizontal Flow Constructed Wetland (HFCW) and one Infiltration–Percolation (IP) filter with subsurface drip irrigation equipped with disk filters before the IP. Figure 21.1 presents the layout of the whole pilot installation. As mentioned before, the aim of this chapter is to focus only on the results obtained with the Subsurface Flow Constructed Wetlands configurations used.

21.2.1 Pilot Vertical Flow Constructed Wetland

The VFCW is made up of a movable container filled with filtering and draining material. This pilot treats wastewater directly from the settling tank (ST) of the facility. Wastewater is discontinuously applied onto the surface through a distribution system (Figure 21.2). The water percolates through the VFCW, and is collected by a draining system connected to the sewer. Figure 21.3 shows a view of this equipment and Table 21.2 summarizes the general characteristics of this pilot.

21.2.2 Pilot Horizontal Flow Constructed Wetland

The HFCW pilot system initially consisted of an elevated storage tank receiving wastewater from the ST by means of the main pump of the system. This 1,000 L tank used gravity to send water into the HFCW. The feed flow was manually regulated by opening or closing a valve. However, it was not useful to adequately regulate the flow because solids clogged it. These solid deposits caused a progressive decrease in flow and made it impossible to control and regulate it. A peristaltic pump was installed to feed the HFCW and in order to better regulate the inflow.

The HFCW container is divided into compartments (Figure 21.4). Water is applied under the surface in the inlet area. The water flows through the HFCW and is collected on the opposite side. The outlet system has and adjustable pipe to regulate the level inside the HFCW (Figure 21.4). From the HFCW, the treated water can be discharged into the sewer system by gravity or be pumped to the

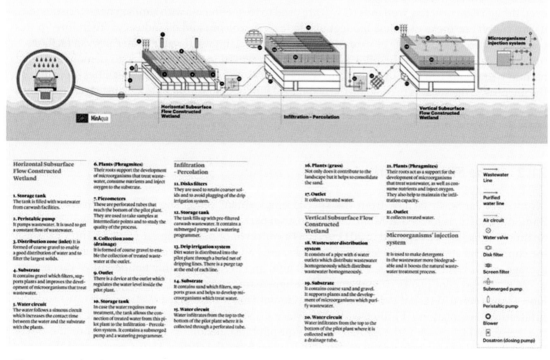

Horizontal Subsurface Flow Constructed Wetland

1. Storage tank
The tank is filled with wastewater from carwash facilities.

2. Peristaltic pump
It pumps wastewater. It is used to get a constant flow of wastewater.

3. Distribution zone (inlet) It is formed of coarse gravel to enable a good distribution of water and to filter the largest solids.

4. Substrate
It contains gravel which filters, supports plants and improves the development of microorganisms that treat wastewater.

5. Water circuit
The water follows a sinuous circuit which increases the contact time between the water and the substrate with the plants.

6. Plants (Phragmites)
Their roots support the development of microorganisms that treat wastewater, consume nutrients and inject oxygen to the substrate.

7. Piezometers
These are perforated tubes that reach the bottom of the pilot plant. They are used to take samples at intermediate points and to study the quality of the process.

8. Collection zone (drainage)
It is formed of coarse gravel to enable the collection of treated wastewater at the outlet.

9. Outlet
There is a device at the outlet which regulates the water level inside the pilot plant.

10. Storage tank
In case the water requires more treatment, the tank allows the connection of treated water from this pilot plant to the infiltration - Percolation system. It contains a submerged pump and a watering programmer.

Infiltration - Percolation

11. Disks filters
They are used to retain coarser solids and to avoid plugging of the drip irrigation system.

12. Storage tank
The tank fills up with pre-filtered carwash wastewater. It contains a submerged pump and a watering programmer.

13. Drip irrigation system
Dirt water is distributed into the pilot plant through a buried net of dripping lines. There is a purge tap at the end of each line.

14. Substrate
It contains sand which filters, supports grass and helps to develop microorganisms which treat water.

15. Water circuit
Water infiltrates from the top to the bottom of the pilot plant where it is collected through a perforated tube.

16. Plants (grass)
Not only does it contribute to the landscape but it helps to consolidate the sand.

17. Outlet
It collects treated water.

Vertical Subsurface Flow Constructed Wetland

18. Wastewater distribution system
It consists of a pipe with 6 water outlets which distribute wastewater homogeneously which distribute wastewater homogeneously.

19. Substrate
It contains coarse sand and gravel. It supports plants and the development of microorganisms which purify wastewater.

20. Water circuit
Water infiltrates from the top to the bottom of the pilot plant where it is collected with a drainage tube.

21. Plants (Phragmites)
Their roots act as a support for the development of microorganisms that treat wastewater, as well as consume nutrients and inject oxygen. They also help to maintain the infiltration capacity.

22. Outlet
It collects treated water.

Microorganisms' injection system
It is used to make detergents in the wastewater more biodegradable and it boosts the natural wastewater treatment process.

— Wastewater Line

— Purified water line

— Air circuit

⊙ Water valve

▢ Disk filter

▨ Screen filter

⬒ Submerged pump

◯ Peristaltic pump

● Blower

▣ Dosatron (dosing pump)

Figure 21.1 Pilot plant layout (modified from [11]).

Figure 21.2 Pilot VFCW distribution system.

IP system. Figure 21.5 shows a view of the HFCW pilot. Table 21.3 shows a summary of the general features of the HFCW pilot.

21.2.3 Operation and Monitoring

The two pilot CWs treated wastewater from a setting tank and operated in parallel during the entire monitoring period (2 years). The HFCW was fed continuously whereas the VFCW was intermittently

Figure 21.3 View of the pilot VFCW.

Table 21.2 Summary of the pilot VFCW characteristics.

Operation	Discontinuous by batches
Feeding mode	Instant flow at the application point: 4.8 m³/h
Distribution system	Overground pipeline with 6 outlets
Container	Built in steel
Container size	The container has a total surface of 10.58 m² – Total length: 4.6 m – Total width: 2.3 m – Height: 1.3 m
Filtering material	Two layers of filtering material: – Top layer of calibrated fine sand ($d_{10} = 0.23$, CU = 3.2, fines content < 3%) 0.40 m height – Bottom layer of fine gravel (2–8 mm) 0.50 m height
Draining material	Transition layer 0.1 m (7–12 mm gravel) Draining layer 0.2 m (25–40 mm gravel)
Vegetation	*Phragmites australis*
Outlet structure	PVC Pipeline

fed. The HFCW was fed with a peristaltic pump; three different hydraulic loads (HL) were applied: 1.4 cm/day, 7.5 cm/day and 14 cm/day. For the VFCW pilot, three HL were tested: 4.5 cm/day, 19 cm/day and 36 cm/day and for each HL, two dosing modes were tested for a different number of applications (4 or 8 batches per day).

Water quality was monitored by systematically taking samples from the pilots (inlets, outlets). The sampling frequency was weekly, biweekly, monthly or bimonthly depending on the parameter and the sampling points. The parameters were analyzed by using a multi-parametric probe: pH, electrical conductivity (EC), water temperature, redox, turbidity and dissolved oxygen (DO) or in laboratory (LABAQUA S.A.): COD, dCOD, DBO$_5$, SS, VSS, TKN, N-NH$_4^+$, N-NO$_3^-$, P-PO$_4^{3-}$, sulfates (SO$_4^{2-}$),

Figure 21.4 Pilot HFCW container and outlet device.

Figure 21.5 View of the pilot HFCW.

chlorides (Cl⁻), calcium (Ca²⁺), magnesium (Mg²⁺), alkalinity, total surfactants (anionic, cationic, non-ionic), oil, fats, hydrocarbons, *E. coli* and *Legionella* spp. Additionally one annual sample was taken at the pilots' inlets for helminths eggs (*Ancylostoma*, *Trichuris* and *Ascaris*).

21.3 Case Study: Results and Discussion

21.3.1 Influent Characterization

During the study period, the pump that feeds the pilot system was placed between 30 and 120 cm from the bottom of the ST (between 80–120 cm from the base of the basin the first year, and nearer

Table 21.3 Summary of the pilot HFCW characteristics.

Operation	Continuous
Feeding mode	Peristaltic pump
Container	Built in steel; interior compartments of 2×0.6 m
Size	The pilot has a total surface of 10.58 m^2
	– Total pilot length: 4.6 m
	– Total pilot width: 2.3 m
	– Height: 0.6 m
Filtering material	Inlet and outlet areas (25–40 mm gravel; 1 m length of this gravel is placed in the inlet zone and 0.5 m length in the outlet zone)
	Filtering zone (12–18 mm gravel)
Vegetation	*Phragmites australis*
Outlet device	Adjustable level pipe

the bottom at 30–40 cm from the base the second year). Therefore, the wastewater that was pumped was partially settled in relation to the effluent coming directly from the car wash facility. Table 21.4 shows pilot influent wastewater quality (average value, maximum and minimum values) for the two years of monitoring.

As Table 21.4 shows, the car wash effluent showed great differences throughout the year of the study. All parameters, especially SS, showed significant variations, due to variable dirt levels contributed by each vehicle and type of vehicle (cars, trucks), and, more importantly, to the height the pump was placed in the ST and the time of sampling. The average percentage of VSS in reference to SS in the influent to the pilots was about 41%, which implies a significant amount of mineral solids (mineral solids adhered to car wheels and tires). When the pump was placed near the bottom of the ST (year 2), the average SS was much higher. Four additional wastewater characterization campaigns were made in samples taken directly before the ST. These samples presented about 10% of VSS in reference to SS (SS values about 977 mg/L and VSS of 112 mg/L). Therefore, non-organic SS were about 90% (basically sand and fines) and settling of these particles was relatively quick.

The BOD$_5$/COD ratio was close to 0.3, which is a medium biodegradability index, lower than urban wastewater (usually above 0.4) [12]. These values are different from urban wastewater or other types of industrial wastewater (e.g., agrofood industry), which have a higher percentage of organic material and are more biodegradable. This type of wastewater will, therefore, be more difficult to degrade biologically than urban or industrial wastewater with higher BOD$_5$ contents. In addition to biodegradability, other mechanisms such as filtration and sedimentation will be important to treat this water. Long-term studies would be necessary to analyze the accumulation of these inorganic particles at the filtering matrix. Table 21.5 shows usual car wash and urban wastewater values for certain parameters [5].

Analytical results from the literature [2, 5, 13, 14] are very similar to those obtained in the water sampling characterization at the inlet of the ST of this project [15]. However, if data from the water arriving to the pilots is compared to literature data, the values are lower in the pilot's influent, especially for organic matter and solids. This is probably because in the mentioned studies, samples were taken before any pretreatment, without settling. It must be also pointed out that the values included in

Table 21.4 Influent wastewater characterization.

Parameters	Units	n	Average (whole period)	Average (year 1)	Average (year 2)	Max	Min
Temperature	°C	67	18.7	18.3	19.1	27.8	9.5
pH		67	7.9	8.0	7.9	9.3	6.7
Redox	mV	67	86.2	88.4	84.0	225	-47
EC	µS/cm	67	548	503	593	1259	179
DO	%	67	10.2	14.5	5.9	65.2	0.0
DO	mg/L	67	0.9	1.2	0.6	5.9	0.0
Turbidity	FNU	67	114	85.0	144	265	33.8
COD	mg/L	66	71.6	48.8	94.4	438	bdl
$_d$COD	mg/L	66	32.4	22.3	42.5	190	bdl
$_p$COD	mg/L	66	41.3	30.8	51.9	346	5.0
BOD_5	mg/L	66	18.8	14.0	23.6	70	bdl
SS	mg/L	66	63.8	41.0	86.6	421	bdl
VSS	mg/L	31	25.6	15.4	35.9	210	bdl
TKN	mg/L	31	3.8	4.2	3.5	34.2	bdl
$N-NO_3^-$	mg/L	31	2.1	2.5	1.8	14.8	bdl
$N-NH_4^+$	mg/L	31	0.6	0.3	1.0	3.3	bdl
$P-PO_4^{3-}$	mg/L	31	0.7	0.4	1.0	6.5	bdl
$S-SO_4^{2-}$	mg/L	15	48.7	51.4	46.0	157	31.8
Cl^-	mg/L	15	51.5	58.0	45.1	250	20.1
Ca^{2+}	mg/L	15	59.6	54.3	64.9	79.2	50.6
Mg^{2+}	mg/L	15	9.8	8.8	10.8	12.5	8.5
Alkalinity	mg/L $CaCO_3$	32	174	160	187	239	58.1
Anionic surfactants	mg/L	32	bdl	bdl	bdl	0.9	bdl
Cationic surfactants	mg/L	32	bdl	bdl	bdl	0.4	bdl
Non-ionic surfactants	mg/L	32	0.4	bdl	0.4	1.4	bdl
Hydrocarbons, oil and fats	mg/L	6	0.3	0.2	0.4	0.6	bdl
E. coli	CFU/100 mL	64	2382	1262	3503	59000	0
Legionella spp.	CFU/L	8	0	0.0	0.0	0	0
Nematode eggs	Eggs/10L	2	0	0.0	0.0	0	0

Max = maximum, Min = minimum, SD = standard deviation, bdl = below detection limit.

the literature [2] vary greatly, which shows the importance of obtaining and analyzing a large number of samples to be able to draw reliable conclusions.

Regarding nutrients, concentration in nitrogen and phosphate forms in the influent was low and similar to that described in the literature. With respect to the concentration of the three types of surfactants analyzed, it was lower than expected (only non-ionic surfactants and in low concentrations were found: maximum 1.4 mg/L), which may be due to its rapid biodegradability, high dilution

Table 21.5 Urban and car wash wastewater characteristics (adapted from [5]).

	COD (mg/L)	Surfactants (mg/L)	Total phosphorus (mg/L)	Total nitrogen (mg/L)
Urban wastewater	430	4	7	40
Car wash wastewater	191	21	1	9
Bus wash wastewater	307	6.3	8.5	5
Truck wash wastewater	600	21	8.5	30

and an optimized detergent dosing. Other wastewater characterizations were additionally carried out in parallel in several car washing facilities in the same project [16], which presented surfactant concentrations slightly larger: 2.6 mg/L of non-ionic surfactants (these values are from a car wash facility in Girona managed by the same company as the Montfullà facility). The average of 15 sampling campaigns conducted in another car wash facility in the Basque Country (within the same project) gave a concentration of 0.25 mg/L for non-ionic surfactants and bdl for anionic surfactants.

The hydrocarbons, oils and fats contents were low (average 0.3 mg/L and maximum 0.6 mg/L). Again, these results are different from those of the literature (oils and fats values close to 5 mg/L), and from the data obtained from initial characterization at the inlet of the ST (hydrocarbons average of 2.1 mg/L and oils and fats average of 13.9 mg/L). These results can be explained because these components are less dense and float, so they move to the second ST instead of being applied to the tertiary treatment systems. Therefore, it can be estimated that a significant part of these components is not injected by the pump into the pilots. In fact, the initial water characterization results [15] showed values before the ST of 1–4 mg/L of hydrocarbons and 12–15 mg/L of fats, and at the outlet of the third ST (before the hydrocarbon separator) values of 12–14 mg/L of fats and 1–2 mg/L of hydrocarbons. Therefore, only about 10% of the fats and oils from the initial effluent were sent to the pilots.

In terms of microbiological parameters, pH, conductivity and most soluble substances (chlorides, sulfates, calcium and magnesium) results are in accordance with literature data [2, 5, 13].

21.3.2 Effluent Quality for Recycling

The basic purpose of the CW prototypes is to generate treated wastewater of sufficient quality to be recycled at car wash facilities (in the most appropriate wash stages). The main purpose is to recycle the wastewater treated by the pilots for the same purpose in the same facility to save water. However, there is no mandatory regulation for recycling in these facilities in Spain. There are some recommendations on quality from companies (for internal use) that water should have tap or recycled water to be used in car washing equipment. Since there is no legal regulation setting the specific limit for most of the recycling parameters, this study has taken as one of the quality targets the values included in Royal Decree 1620/2007 "establishing the legal regime for reusing treated water." In fact, Annex IA of this RD shows the quality criteria for reusing water according to different uses. This annex indicates the quality required for 1. Urban use/Quality 1.2/Services/ d) Industrial vehicle washing. Table 21.6 shows the quality criteria for urban wastewater reuse, quality 1.2 Services (including industrial vehicle washing). These values allow comparison of the water quality obtained in the study with the decree values and ensure that the water to be recycled does not pose a health risk.

Table 21.6 Quality required according to RD 1620/2007 for urban use quality 1.2. Services.

Water use	Maximum acceptable value				
	Intestinal nematodes	*E. coli*	SS	Turbidity	Other criteria
1.2. Quality Services					
a) Urban green areas watering b) Street washing c) Fire-fighting systems d) Industrial vehicle washing	1 egg/10L	200 CFU/100mL	20 mg/L	10 FNU	*Legionella* spp. 100 CFU/L (if there is risk of aerosolization)

21.3.3 Performance of the Constructed Wetland Pilots

21.3.3.1 Horizontal Flow Constructed Wetland

Table 21.7 shows the overall water quality results for the HFCW effluent (for all applied HLs).

Regarding the organic matter parameters, very high performances were also obtained (especially in terms of BOD_5), offering an effluent with very low concentrations for these parameters. The nutrients were eliminated or transformed in a distinct way depending on the parameter: the few nitrates entering the HFCW were almost entirely eliminated by plant absorption mechanisms and/or via denitrification and volatilization [17]. Phosphates were completely removed. In fact, concentrations of these nutrients were very low in the inlet; therefore, they were almost completely absorbed by plants. This can be observed from the uneven plant growth (see Figure 21.6).

As seen in Figure 21.6, the plants in the area closest to the inlet have considerably greater growth than those in the outlet area, with a gradual descent. This may be explained by the fact that the few nutrients found in the inlet wastewater were absorbed by the plants at the beginning of the wetland, therefore are gradually depleted as the water circuit progresses. TKN and ammonia removal were almost 100%, but it is important to once again point out the low inlet concentrations. The soluble ions, such as the calcium, magnesium, sulfate and alkalinity did not vary substantially. The EC was slightly

Figure 21.6 *Phragmites australis* in the pilot HFCW.

Table 21.7 HFCW effluent quality.

Parameters	Units	n	Average	Max	Min	dl	%<dl
Temperature	°C	67	17.6	28.1	6.1	−5	0
pH		67	7.2	7.9	6.4	0	0
Redox	mV	67	−7.8	170	−90.4	±2,000	0
EC	µS/cm	67	564	1,035	306	0	0
DO	%	67	12.8	85.9	0.3	−	0
DO	mg/L	67	1.1	6.7	0	0	0
Turbidity	FNU	67	2.9	13.7	0.1	0	1.3
COD	mg/L	66	16.9	58	bdl	10	31.7
$_d$COD	mg/L	66	10	40	bdl	10	68.3
$_p$COD	mg/L	66	7.75	34	0	−	−
BOD_5	mg/L	66	5.1	15	bdl	5	52.1
SS	mg/L	66	3.9	29	bdl	3	67.3
VSS	mg/L	31	bdl	9	bdl	3	90.2
TKN	mg/L	31	bdl	8.1	bdl	3	87.1
$N-NO_3^-$	mg/L	31	1.1	7.7	bdl	0.5	56.2
$N-NH_4^+$	mg/L	31	bdl	0.3	bdl	0.1	40.6
$P-PO_4^{3-}$	mg/L	31	bdl	0.7	bdl	0.1	80.4
$S-SO_4^{2-}$	mg/L	15	40.3	73.8	28.4	5	0
Cl^-	mg/L	15	44.2	82.6	22.2	10	0
Ca^{2+}	mg/L	15	57.9	71	42.6	2	0
Mg^{2+}	mg/L	15	10	11.8	7.8	2	0
Alkalinity	mg/L $CaCO_3$	32	193	321	83.6	5	0
Anionic surfactants	mg/L	32	bdl	0	bdl	0.1	100
Cationic surfactants	mg/L	32	bdl	0	bdl	0.2	100
Non-ionic surfactants	mg/L	32	bdl	0	bdl	0.5	100
Hydrocarbons oil and fats	mg/L	6	bdl	0.1	bdl	0.05	66.7
E. coli	CFU/100 mL	64	131	3000	0	−	−
Legionella spp.	CFU/L	8	0	0	0	−	−

Max = maximum; Min = minimum; SD = standard deviation; dl = detection limit, %< dl = percentage below detection limit.

higher in the HFCW outlet especially in summer months when evapotranspiration was higher. A hydraulic monitoring of the inlet and outlet flow during the same day was carried out in order to verify any water loss caused by evapotranspiration. A certain amount of water was lost between the inlet and the outlet: from 8% to 21.8%. These losses may be due to evapotranspiration of water in the HFCWs due to the long HRTs. With data from the two years of study, it may be concluded that the yearly average flow loss has been approximately 15%. These water losses are to be expected based on the climate of the site and the plant growth [18].

The anionic and cationic surfactants were bdl in the inlet and outlet of the HFCWs. For the non-ionic detergents, they were always found to be bdl in HFCWs outlet. Removal of non-ionic surfactants from municipal wastewater using HFCW was studied by Sima and Holcova [19]. Non-ionic surfactants were removed with a high efficiency reaching almost 100% elimination. The study found that non-ionic surfactants were degraded both under aerobic and anaerobic conditions. However, because of the low concentrations of surfactants in our study, it was not possible to go in depth in the study of degradation/removal of surfactants in HFCWs. The low concentration of detergents in the HFCW inlet was due to the low doses applied and the high biodegradability of the detergents employed in the facility.

Hydrocarbons, oils and fats were almost entirely eliminated in the HFCW, with only one sample having a value over the detection limit, 0.1 mg/L. However, it must be noted that the influent concentrations were very low and that almost all oils and fats applied accumulated in the HFCW inlet area and clogged it. Polyaromatic hydrocarbons can be eliminated in CWs by various mechanisms [20, 21] not detailed here. Regarding values of dissolved oxygen and redox potential in the HFCW effluent, it was observed that the concentration of oxygen decreased gradually in the HFCW and that the redox potential had slightly negative values at the outlet (−7), indicating that the HFCW also presented anoxic/anaerobic areas.

Finally, regarding microbiological parameters, the average elimination of *E. coli* was 1.4 Ulog, a value that is quite characteristic of these systems (between 1–2 Ulog) [22–24].

In Table 21.8, the HFCW pilot effluent values are compared with those established by RD 1620/2007. The RD 1620/2007 is very strict with respect to self-management plans of the facilities. These self-management plans include a determined number of samples per time unit for the parameters indicated for each use. In this study there has not been strict compliance with this number of samples; therefore the calculations were made taking into account the number of analyses in our study. In Annex I.C of RD 1620/2007, it is established that 90% of the samples may not exceed the established values. Furthermore, it is indicated that the results' deviation may not exceed a certain maximum established thresholds (Maximum Allowable Value): 1 Ulog in the case of *E. coli* and *Legionella* spp.; 50% for SS and 100% for turbidity and intestinal nematodes.

In the HFCW effluent, for turbidity, only 4% of the samples presented higher values than those recommended, with these values being lower than 20 FNU, thus complying with the criteria for this parameter. For *E. coli,* 12% of the samples had values which exceeded 200 CFU/100mL, with the maximum deviation allowed being 10% of the samples. These results indicate that for greater safety, a final chlorination or other disinfection step is recommended after the HFCW. A final chlorination would allow residual chlorine which will act as a disinfectant in the pipes; thereby preventing *E. coli* recontamination or growths. It would also serve to help controlling *Legionella* spp.

Table 21.8 Comparison of HFCW pilot effluent quality with the limits established in RD 1620/2007.

	Nematodes	*E. coli*	SS	Turbidity	*Legionella* spp.
	Eggs/L	CFU/100mL	mg/L	FNU	CFU/L
RD Value	1	200	20	10	100
% > RD	0	12	0	4	0

After one year of operation, deposits were formed in different parts of the pilot system (valves, and HFCW inlet area). These deposits accumulated slowly in the HFCW inlet zone (Figure 21.7), clogging the system after one year of operation (water flooding in the inlet). After sampling, the deposits were removed, and the inlet gravel was cleaned.

These "biosolid" accumulations (black color) were characterized (Tables 21.9 and 21.10). The results of the accumulated material characterization show VS percentages lower than the percentages found in sludge, and algae or swine slurry deposits [25]. As an example, the percentage of VS in swine slurry or algae deposits ranged from 40–60. The percentage of VS of the deposits in the car washing pilots was only 17. The percentage of total carbon is also low (9.9% TOC). Therefore, the percentage of organic matter was not high. DM (dry matter) values were low (17.1%) representing a material with high water content. The percentage of oils and fats is of 0.081%. As discussed before, most oils and fats were not pumped to the pilots (floated and passed on to the second ST). The high concentration of metals (particularly copper and zinc) is also remarkable. Metal monitoring will be carried out in

Figure 21.7 Material settling in the inlet (HFCW).

Table 21.9 Deposit characterization.

Parameter	Value
DM	17.1%
TOC	9.9%
VS	17.9% DM
Oils and fats	0.08% DM
pH	5.9

Table 21.10 Deposit characterization (metals).

Metal	Value (mg/Kg DM)
Antimony	41
Arsenic	<10
Cadmium	<2.0
Copper	3,623
Tin	68
Mercury	<0.20
Nickel	57
Lead	114
Selenium	<10
Thallium	<4
Tellurium	4
Zinc	1,642

the second stage of the project. There are numerous experiences with CWs (e.g., [26]) demonstrating their ability to eliminate them by using macrophytes (phytoremediation).

The HFCW clogging occurred with low organic and SS surface loading rates values. SS surface loading rates were on average of 2.9 g $SS/m^2/d$ with a maximum 8.3 g $SS/m^2/d$. COD loading rates were on average of 3.5 g $COD/m^2/d$ with a maximum 10.1 g $COD/m^2/d$. HFCWs having a probability of clogging is well known; however, these technology usually receive much higher SS and organic loadings than the studied pilot without clogging problems. The fact that the inlet distribution of the modular HFCW has only one pipe makes this system more sensitive to clogging in the distribution area. However, similar HFCWs (with the same inlet design) treating domestic wastewater did not have any clogging problems with higher organic and SS surface loading rates [27]. This means that fat and oil content (even at low concentrations) quickly clog HFCWs inlet areas. As a result, it is recommended that fats and oils be removed before application to HFCWs.

21.3.3.2 Vertical Flow Constructed Wetland

Table 21.11 shows the results of the outlet water quality in the VFCW for all the applied HLs.

The VFCW performed very efficiently, offering a very good effluent quality in terms of physico-chemical and microbiological parameters. The efficiency was high, particularly with regards to the COD and BOD_5. In terms of SS and turbidity, the outlet qualities were good, although variable, and the average turbidity was 10 FNU. The most important SS elimination mechanism in VFCW is filtration [28]. Table 21.12 shows the percentage removal for the SS, turbidity and the organic matter parameters for the first and second year of operation.

The SS removal was about 80% (71% the first year of operation and 96% the second year). Regarding organic matter parameters, the results (% removals) were also quite high, between 75 and 100.

The VFCW was designed to create a "biosolid" surface layer, as the French VFCW model. This layer of organic solids can reduce the infiltration velocities and increase filtration, leading to a larger retention of particulate substances. In the case of this pilot plant, in the VFCW, the water received

Table 21.11 VFCW effluent quality.

Parameters	Units	n	Average	Max	Min	dl	%<dl
Temperature	°C	75	18.7	28.3	8.4	−5	0
pH		75	7.35	8	6.8	0	0
Redox	mV	75	115.5	236	−57	±2,000	0
EC	µS/cm	75	604	1,051	−10.4	0	0
DO	%	75	43.3	784	8.7	–	0
DO	mg/L	75	3.9	78.7	0.7	0	0
Turbidity	FNU	75	10.4	21.6	0	0	0
COD	mg/L	71	11.8	68	bdl	10	48.4
$_d$COD	mg/L	71	bdl	32	bdl	10	81.7
$_p$COD	mg/L	71	4.3	58	0	–	–
BOD_5	mg/L	71	bdl	11	bdl	5	81.7
SS	mg/L	71	7.5	59	bdl	3	46.5
VSS	mg/L	35	bdl	6	bdl	3	89
TKN	mg/L	35	bdl	16.2	bdl	3	84.8
$N\text{-}NO_3^-$	mg/L	34	5.1	31.2	bdl	0.5	11.8
$N\text{-}NH_4^+$	mg/L	34	bdl	0.8	bdl	0.1	48.8
$P\text{-}PO_4^{3-}$	mg/L	33	bdl	1.6	bdl	0.1	78.3
$S\text{-}SO_4^{2-}$	mg/L	17	45.3	100.7	27	5	0
Cl^-	mg/L	17	45.9	103.9	21.4	10	0
Ca^{2+}	mg/L	17	63.2	95.5	49.5	2	0
Mg^{2+}	mg/L	17	9.75	16.3	7.6	2	0
Alkalinity	mg/L $CaCO_3$	34	180.1	291.1	128.1	5	0
Anionic surfactants	mg/L	34	bdl	0	bdl	0.1	100
Cationic surfactants	mg/L	34	bdl	0	bdl	0.2	100
Non-ionic surfactants	mg/L	34	bdl	0	bdl	0.5	100
Hydrocarbons, oil and fats	mg/L	10	bdl	0	bdl	0.1	100
E. coli	CFU/100 mL	71	19	600	0	–	–
Legionella spp.	CFU/L	8	120	1200	0	–	–

Max = maximum; Min = minimum; SD = standard deviation; dl = detection limit; %< bdl = percentage below detection limit.

from the car washing facility had few organic solids, thus the layer forming on the surface was quite thin in the first year of operation. This leads to a faster infiltration and lower filtration. Across the same batch, the VFCW effluent turbidity was observed to vary depending on the specific time of sampling after application during the first year (Table 21.13). Part of the water exits after a few minutes that presents higher turbidity and SS concentration.

Table 21.12 VFCW % removal for the first and second year of operation (SS, VSS, Turbidity and organic matter parameters).

Parameter	% Removal (average)	
	Year 1	Year 2
Turbidity	80.9	96.1
SS	71.5	96.1
VSS	86.3	100
COD	78.5	86.0
$_d$COD	75.7	100
$_p$COD	76.3	93.1
BOD_5	80.3	100

Table 21.13 Turbidity changes based on time of sampling (HL = 36 cm/day, 4 batches/day, first year of operation).

Sampling time	Turbidity (FNU)
Prior to batch (small outflow)	2.6
Immediately following batch (very high outflow)	24.9
30 minutes after batch (average outflow)	7.3

During the two years of operation, there were no signs of filter clogging, even with HLs of 36 cm/day. Nitrates were higher in the outlet of the VFCW as compared to the inlet, due to the fact that the system is aerobic and oxidizes the ammonia into nitrates [29]. The low TKN, ammonia and phosphate loads entering the VFCW were completely eliminated. Unlike the HFCW, the plants developed similarly across the VFCW filter (given that the distribution of water on the surface is similar across the points of the filter). Therefore, the plants may absorb nutrients in a similar way across the entire bed (see Figure 21.8).

The soluble ions, such as the calcium, magnesium, sulfate cations or alkalinity did not vary considerably during the treatment. Salinity also showed no significant variations. The anionic and cationic surfactants were bdl in the VFCW influents and effluents. As for the non-ionic surfactants, elimination reached 100% (it should be noted that on a very few occasions the influent values were greater than the detection limit values). Non-ionic surfactants can be degraded both under aerobic and anaerobic conditions in CWs [19]. Therefore, these surfactants could be removed and/or transformed in VFCWs. Because of the low concentrations of surfactants on VFCW influent and effluent, it was not possible to go deeper into the study of degradation/removal of surfactants in VFCWs.

Hydrocarbons, oils and fats were removed in the pilot. Once again, it should be noted that the inlet concentrations are very low. Polycyclic aromatic hydrocarbons can be eliminated in VFCWs

Figure 21.8 *Phragmites* in the pilot VFCW.

[20]. However, it was not possible to go into detail in the study of the removal of these compounds in the pilots due to their low concentrations. Unlike the HFCW where the formation of oily looking deposits was observed in the inlet area, no major accumulations in the VFCW filter which may lead to clogging have been observed.

When observing the outlet values and results of dissolved oxygen elimination and redox potential, it was found that the concentration of oxygen increased considerably and the redox potential had slightly greater positive values than in the inlet. These results indicate that the VFCW is an aerobic system that oxygenates the influent.

The "limit" for the performance of the HFCW as well as of the VFCW is the clogging. The media clogging occurs when the conductivity of the filter material is reduced. The increase in biomass and the development of biofilm and microorganisms leads to a strong reduction of the presence of oxygen in the lower layer and a resulting decrease in efficiency yields for all oxidizing processes (nitrification, carbon oxidation, elimination of pathogens). The German guidelines [30] specify that the VFCW must be designed with the following values for their proper operation: maximum organic load of 20 g $COD/m^2/d$ with a maximum concentration of 100 mg/L of SS, loading of SS = 5 $g/m^2/d$, CH <8 cm/day in winter and 12 cm/day in summer. However, the French model supports higher values. In the first stage of the French VFCWs, the design and the special operating conditions allow an organic load of 100 g $COD/m^2/d$ [31].

There was no clogging in the VFCWs. As indicated before, part of influent solids were retained on the filter surface and created a thin layer, as is common in these systems. Moreover, distribution of the VFCW water makes it less likely to clog, as the water is distributed evenly all over the surface of the filter. It should be noted that the high efficiency of the VFCW treating car wash effluent and the oxygenation capacity even with the high loads of the second year of operation (the average loads applied the first year were around 15 g $COD/m^2/d$ and during the second year of about 35 g $COD/m^2/d$). This indicates that the loads of 35 g $COD/m^2/d$ are not limiting for this VFCW design with this type of influent [16, 32].

For the microbiological parameters, there was a high *E. coli* elimination, with an average of 2.3 Ulog. These values are slightly greater than those normally observed in the VFCW which tend to range between 1 and 2 Ulog [33]. However, in the studied VFCW the filtration media was deeper (90 cm) and the granulometry of the top sand layer was finer. Removals greater than 2 Ulogs for *E. coli* or fecal coliforms has been observed in infiltration-percolation systems with 150 cm of similar sand granulometry [34, 35]. *Legionella* spp. was always absent in the inlet, and presented one positive in the outlet of the VFCW, with only one value of 1200 CFU/L. When comparing with the RD 1620/2007, this value represents 3 Ulog, and therefore it is within the range of the maximum acceptable threshold as indicated in Annex 1C of the RD. Despite this, the feeding system was disinfected with chlorine and the sampling was repeated after 15 days, with negative results at the outlet of the VFCW and at several points of the car wash facility. Table 21.14 compares the effluent values of the VFCW pilot with those from the RD 1620/2007.

For turbidity, 32% of the samples had higher values than indicated in the RD, mainly in the first year of operation. Although the values are very close to 10 (the average is 15), they do not comply with the regulations with regard to this parameter. The SS complies with the regulations (less than 10% of the values exceed the limits). The follow-up of the VFCW performances for a longer time period would be necessary. The organic layer would increase with time, thereby increasing filtration and providing higher percentage removal of suspended solids and turbidity. For *E. coli*, the results are almost always lower than the value of 200 CFU/100mL established by RD 1620/2007. As for *Legionella* spp., the facility also fails to comply with the thresholds; in one of the five samples there was a value of 1,200 CFU/L. These results indicate that it would be useful to add a final disinfection treatment (such as chlorination) as indicated before. This chlorination would also provide residual chlorine which would serve to disinfect the pipes thereby preventing pathogen recontamination.

21.3.3.3 Comparison of Performances

Table 21.15 shows the average removal percentages (physico-chemical parameters) for the two SSFCW pilots.

The two SSFCW pilots presented very high removal percentage for turbidity and for all of the organic matter and SS parameters. Even when considering that the HLs applied to the VFCW are greater than those of the HFCW, the VFCW performance is slightly greater than that of the HFCW. This may be due to the fact that the oxidation of the organic matter occurs more efficiently in the VFCW, through the application of batches at high flow rates. The vertical flow and operation method favors filter oxidation. The aerobic functioning of the system via batches permits an efficient oxidation of the dissolved organic matter.

HFCW presented average SS and turbidity percentage removal slightly higher than the VFCW. However, if we compare the percentage removal of the second year of operation, the reductions are similar for both pilots (97.9% and 96.1% for turbidity at HFCW and VFCW, respectively; 100% and 96.1% of SS removal at HFCW and VFCW, respectively). The VFCW was designed to form a layer of coarse solids on the surface. Again, it is indicated that this layer of organic solids reduces infiltration

Table 21.14 Comparison of the VFCW outlet quality with that of RD 1620/2007.

	Nematodes	*E. coli*	SS	Turbidity	*Legionella* spp.
	Eggs/L	CFU/100mL	mg/L	FNU	CFU/L
RD Value	1	200	20	10	100
% > RD	0	3	7	32	10

Table 21.15 Comparison of performance of the pilots for physicochemical parameters (% removal).

Parameter	HFCW	VFCW
COD	73.3	78.5
$_d$COD	72.2	75.7
$_p$COD	70.8	76.3
BOD$_5$	81.3	80.3
Turbidity	96.7	80.9
SS	88.5	71.5
VSS	88.8	86.3
TKN	100[a]	100[a]
N-NO$_3^-$	80.1	−56.9[b]
N-NH$_4^+$	100	100
P-PO$_4^{3-}$	89.3	87.5
Anionic surfactants	[c]	[c]
Cationic surfactants	[c]	[c]
Non-ionic surfactants	100[a]	100[a]
Hydrocarbons, oils and fats	75.2	100[a]

[a] The average outlet values were bdl.
[b] nitrification
[c] the average inlet and outlet values were bdl.

rates and increases filtration, which results in a large retention of particulate matter. In the second year of operation this layer of almost 1 cm was formed, allowing a better and slower filtration, and more homogeneous yields. Additionally, the colonization of *Phragmites* roots throughout the filter in the second year and a higher biomass inside the filter (due to the application of higher organic loads) has also favored greater efficiency.

The VFCW nitrified the effluents, while the HFCW, on the other hand, presented anaerobic conditions and therefore did not produce nitrates. The three pilots eliminated TKN and ammonia almost entirely (values quite always bdl). The same occurred with phosphates. Regarding calcium, magnesium and sulfate ions, there were no major variations in any of the three pilots. No further variations were found for alkalinity, with a slight increase being found in the effluent from all three pilot plants.

The (non-ionic) surfactants from the influent of the pilot plants were completely eliminated (100%) by the two prototypes, but the concentrations of these surfactants in the influent were quite low. The same results were found for oils, fats and hydrocarbons: they were eliminated very efficiently in the two pilot plants; however, their influent concentrations were quite low.

Regarding the microbiological parameters (Table 21.16), the VFCW showed the greatest removal.

The VFCW presented an average reduction of 2.3 Ulog for *E. coli*. However, for HFCW it was 1.2 Ulog. HFCW presented lower Ulog removals even with lower HLs. The removal mechanisms for *E. coli* seem to be more effective in aerobic systems with finer particle media (sand) and high depths of the filtering media such as in the VFCW.

Table 21.16 Comparison of performances of the pilot plants for the microbiological parameters (Ulog).

Parameter	HFCW	VFCW
E. coli	1.2	2.3
Legionella spp.	Absent	Absent in the inlet and 1 positive in the outlet

21.4 Design and Operation Recommendations

21.4.1 Horizontal Flow Constructed Wetland

The HFCW (with the design and operation studied) was not proved to be the better of the two CW technologies for the treatment of this type of water, for the following reasons:

1) Greater sensitivity to clogging especially the inlet area.
2) Lower elimination of *E. coli*, and very sensitive removal with high hydraulic loads.
3) The feeding of the HFCW must be continuous for better operation. If they cannot be fed by gravity, in car wash facilities producing low wastewater volumes, there is a difficulty in operation and maintenance of the pumps which affects wastewater with a high concentration of solids.
4) The maximum tested HL that reaches the quality limits for recycling in the car washing facility was approximately 10 cm/d (much lower than the HL for the VFCW: 36 cm/d).
5) Irregular growth of *Phragmites australis* (high growth near the entrance zone and very low in the areas near the outlet) is due to the low concentration of nutrients in this type of wastewater, and the way of feeding the HFCW. Nutrients are used by plants in the first stages of the wetland.
6) Higher evapotranspiration (approximately 20% of the water volume is lost) results in less availability of treated water for recycling and higher EC.

However, the HFCW is simpler to operate and presents fewer operation costs. If a HFCW must be implemented for treating carwash effluent, it is necessary to:

- Remove oil and fats.
- Remove inorganic solids.
- Design a different inlet area to that of the studied pilot (much larger) with a coarser water distribution system per channel throughout this section. Thus, the water circuit design of the pilot would have to be designed in a completely different way.

Moreover, a final disinfection step (e.g., chlorination) is required to ensure free chlorine (0.5–2 mg chlorine/L) at the water recycled storage tank before being used in the carwash machinery.

From the study of water consumption per vehicle of the Montfullà car washing facility and the performance of the pilots, the area (m²) of HFCWs needed to reach the quality values for recycling has been determined. The consumption data considered have been: general vehicles (cars + industrial): 350 L/unit wash; cars: 250 L/car; industrial vehicles (trucks, buses): 450 L/vehicle. The HFCW area needed is [36]:

- General vehicles ≈ 2.5 m^2/vehicle
- Cars ≈ 1.8 m^2/car
- Industrial vehicles ≈ 3.2 m^2/vehicle.

21.4.2 Vertical Flow Constructed Wetland

This technology with the design and operation studied, has proved to be the best option for the treatment of car wash effluent. The only limiting factor is the variability in turbidity values, during the first year of operation. This variability decreases in a "natural" way with the operation time. The variability of turbidity quality could also be reduced by further fractionating the daily application of water. Less variability was observed with eight batches per day compared to four batches. Thus, a minimum of eight batches per day is recommended (with a maximum of 15 so as not to saturate the medium and thus entering into anaerobic conditions). Another option to obtain a more constant quality would be to operate the VFCW more closely than is done with an infiltration percolation system, with the application of a lower instantaneous flow for a longer time. Even so, this possibility should be studied at pilot level, with an adequate water distribution designed to ensure a good surface distribution; aerobic conditions would also have to be ensured.

Although no clogging problems have been described for this technology, it is advised that the recommendations for HFCW be followed:

- Implement an efficient pretreatment for the removal of oils and grease and inorganic solids, which would substantially extend the life of the filter and avoid possible problems of clogging.
- A final disinfection step for the recycled water before being used in the car wash machinery.

As with HFCW, the area (m^2) of VFCW needed to reach quality values for recycling was determined [36]:

- General vehicles ≈ 1 m^2/vehicle
- Cars ≈ 0.7 m^2/car
- Industrial vehicles ≈ 1.3 m^2/vehicle.

21.5 Conclusions

The main outcome of the study is the viability of the application of subsurface flow constructed wetlands to treat the effluents from car wash facilities once their design and operation have been optimized. The results show that car wash effluents have high concentrations of inorganic suspended solids, very variable concentrations of *E. coli* and organic matter, low concentrations of nutrients, and the presence of hydrocarbons, fats and oils. The studied car wash facility effluent contains non-ionic surfactants, but at lower concentrations than expected due to the high biodegradability of the used detergents and the low dosing of products.

The two types of constructed wetland performed very efficiently with respect to turbidity, organic matter and suspended solids. Nutrient concentrations were quite low in the influent, thus *Phragmites australis* growth in the CW pilot plants was much slower than for other types of wastewater (especially in the HFCWs in the areas farthest from the residual water inlet). The VFCW nitrified the effluents. The HFCW presented more anaerobic conditions. The nutrient concentrations were very low, usually bdl. No major variations were found in any of the pilot plants for the parameters calcium, magnesium, sulfates and alkalinity.

Surfactants present in the inlet of the pilot plants (non-ionic detergents) were fully removed in the pilots, but it should be noted that the concentrations of these detergents were quite low in the influent. The same occurred with the oils, fats and hydrocarbons: they were efficiently eliminated by the two pilot plants, but their influent concentrations were very low, since the majority of these components pass through a second decanter and did not reach the pilot plants.

In general, *E. coli* was removed to acceptable limits for recycling, with concentrations lower than the limits established in the legislation for reclaimed water reuse in Spain. The removal of *E. coli* was much greater in the VFCW than in the HFCW. *Legionella* spp. was always absent in the inlet and outlet of the pilot plants except once in the effluent of VFCW. A final disinfection (i.e., chlorination) in the storage tank is recommended to minimize the microbiological risk, and in order to ensure the presence of (residual) disinfectant in the recycling pipes.

The oil and fats content of the car wash effluents as well as the inorganic suspended solids made pretreatment necessary in order to avoid media clogging. This is especially important for the performance of the HFCW. The VFCW worked without any clogging problems throughout the study for all of the applied loads, even without resting periods.

The study revealed the capability of SSFCWs, mainly the VFCW, for the treatment of car wash wastewater previously settled and for which oils and fats were previously eliminated. They may produce a very high quality effluent that may be recycled within the system (in processes requiring less exigent water quality and that have greater water consumption: prewash with hand-held lances and first wash step with brush arches). SSFCWs, mainly VFCW, have shown resilience to load and hydraulic fluctuations, to chemical pollutants and to variable environmental conditions; they are simple to operate and are maintained with minimum energy requirements and with an added aesthetical value.

More detailed studies should be carried on the effect of certain potentially corrosive pollutants (metals, salts) on the machinery used in the facilities.

References

1 Al-Odwani A, Ahmed M, Bou-Hamad S. Car wash water reclamation in Kuwait. Desalination. 2007; 206(1–3):17–28.

2 Zaneti R, Etchepare R, Rubio J. Car wash wastewater reclamation. Full-scale application and upcoming features. Resour Conserv Recycl. 2011; 55:953–959.

3 L.E.Q.U.I.A. (Laboratori d'Enginyeria Química i Ambiental). Conveni de col.laboració científica entre la Universitat de Girona i la Fundació Ramon Noguera. Informe final. Internal report. Universitat de Girona, Spain, 2008.

4 Janik A, Kupiec H. Trends in Modern Car Washing. Pol J Env Stud. 2007; 16(6):927–931.

5 Zaneti R, Etchepare R, Rubio J. More environmentally friendly vehicle washes: water reclamation. J Clean Prod. 2012; 37:115–124.

6 Paxéus N. Vehicle Washing as a Source of organic Pollutants in Municipal Wastewater. Water Sci Technol. 1996; 33(6):1–8.

7 Hamada T, Miyazaki Y. Reuse of carwash water with a cellulose acetate ultrafiltration membrane aided by flocculation and activated carbon treatments. Desalination. 2004; 169(3):257–267.

8 Brown C. Water effluent and solid waste characteristics in the professional car wash industry. Report for the International Carwash Association (ICA), Chicago, USA, 2002.

9 Panizza M, Cerisola G. Applicability of electrochemical methods to carwash wastewaters for reuse. Part 1: Anodic oxidation with diamond and lead dioxide anodes. J Electroanal Chem. 2010; 638(1):28–32.

10 Kiran SA, Arthanareeswaran G, Thuyavan YL, Ismail AF. Influence of bentonite in polymer membranes for effective treatment of car wash effluent to protect the ecosystem. Ecotox Environ Safe. 2015; 121:186–192.

11 MinAqua. Good practices guide for car wash installations. Grup Fundació Ramon Noguera, 2016. Available at http://minaqua.org/wp8c/wp-content/uploads/2016/12/Good-practices-guide-for-car-wash-installations-ENG-bxa.pdf.

12 Tchobanoglous F, Burton L, Stensel HD. Wastewater Engineering: Treatment and Reuse. 4th ed. Metcalf & Eddy Inc., McGraw-Hill: New York, USA, 2003.

13 Bhatti Z, Mahmood Q, Raja I, Malik A, Khan M, Wu D. Chemical oxidation of carwash industry wastewater as an effort to decrease water pollution. Phys Chem Earth. 2011; 36(9–11):465–469.

14 Zaneti R, Etchepare R, Rubio J. Car wash wastewater treatment and water reuse – a case study. Water Sci Technol. 2013; 67(1):82–88.

15 MinAqua. Informe de resultados de la caracterización inicial de las aguas residuales. Grup Fundació Ramon Noguera, 2014. www.minaqua.org/wp5/wpcontent/ uploads/2012/12/A6_Informe entregable_vf.pdf.

16 MinAqua. Informe de resultados del monitoreo de la eficiencia de los pilotos I (Pilotos Montfullà, primer año de funcionamiento). Grup Fundació Ramon Noguera, 2015.

17 Vymazal J, Kröpfelová L. Removal of organics in constructed wetlands with horizontal sub-surface flow: A review of the field experience. Sci Total Environ. 2009; 407(13):3911–3922.

18 Milani M, Toscano A. Evapotranspiration from pilot-scale constructed wetlands planted with Phragmites australis in a Mediterranean environment. J Environ Sci Health. 2013; 48(5):568–580.

19 Sima J, Holcova V. Removal of Nonionic Surfactants from Wastewater Using a Constructed Wetland. Chem Biodivers. 2011; 8(10):1819–1832.

20 Vymazal J. Constructed wetlands for treatment of industrial wastewaters: A review. Ecol Eng. 2014; 73:724–751.

21 Xu L, Liuanbin R, Zhao Y, Doherty L, Hu L, Hao X. A review of incorporation of constructed wetland with other treatment processes. Chem Eng J. 2015; 279(1):220–230.

22 Huertas E. Regeneració i reutilització d'aigües residuals. Tecnologia, control i risc. PhD thesis, Universitat de Barcelona, Spain, 2009.

23 Torrens A, Folch M, Sasa J, Lucero M, Huertas E, Molle P, Boutin C, Salgot M. Removal of bacterial and viral indicators in horizontal and vertical subsurface flow constructed wetlands 12th IWA International Conference on Wetland Systems for Water Pollution Control, 3–8 October 2010, San Servolo, Venice, Italy.

24 Sasa J. Influencia de los periodos de llenado y vaciado en la eliminación de nutrients y desinfección en zonas humedas construidas de flujo subsuperficial horizontal. PhD Thesis, Universitat de Barcelona, Spain, 2014.

25 Torrens A. Subsurface flow constructed wetlands for the treatment of wastewater from different sources. Design and Operation. PhD thesis, Universitat de Barcelona, Spain, 2015. Available online at: http://www.tesisenred.net/bitstream/handle/10803/380738/ATA_THESIS.pdf?sequence=1&isAllowed=y

26 Gillespie WB, Bradley HW, Rodgers JH, Cano M, Dorn PB. Transfers and transformations of zinc in constructed wetlands: mitigation of a refinery effluent. Ecol Eng. 2000; 14(3):279–292.

27 Torrens A, Folch M, Bayona C, Salgot M. Upgrading quality of wastewater by means of decentralized natural treatment systems Sustainable management of environmental issues related to water stress in Mediterranean islands. Final Conference and Stakeholders event. Mediwat Project. Palermo, Italy, 2013, pp. 136–144.

28 Torrens A, Molle P, Boutin C, Salgot M. Impact of design and operation variables on the performance of vertical-flow constructed wetlands and intermittent sand filters treating pond effluent. Water Res. 2009; 43:1851–1858.

29 Molle P, Liénard A, Grasmick A, Iwema A. Effect of reeds and feeding operations on hydraulic behaviour of vertical flow constructed wetlands under hydraulic overloads. Water Res. 2006; 40(3):606–612.

30 ATV-A 262. Arbeitsblatt ATV-A 262. Grundsätze für Bemessung, Bau und Betrieb von Pflanzenbeeten für kommunales Abwasser bei Ausbaugrössen bis 1000 Einwohnerwerte. Gesellschaft zur Forderung der Abwassertechnik d.V., Germany, 1998.

31 Bresciani R, Masi F. Manuale pratico di fitodepurazione. Terra Nuova eds. Città di Castello, Italy, 2013.

32 MinAqua. Informe de resultados del monitoreo de la eficiencia de los pilotos II (Pilotos Montfullà, segundo año de funcionamiento), 2016.

33 Torrens A, Molle P, Boutin C, Salgot M. Removal of bacterial and viral indicators in vertical flow constructed wetlands and intermittent sand filters. Desalination. 2009; 246:169–178.

34 Folch M. Tratamiento terciario de aguas residuales por infiltración-percolación. PhD thesis, Universitat de Barcelona, Spain, 1999.

35 Brissaud F, Salgot M, Folch M, Auset M, Huertas E, Torrens A. Wastewater infiltration percolation for water reuse and receiving body protection: thirteen years' experience in Spain. Water Sci Technol. 2007; 55(7):227–223.

36 MinAqua. Descripción de los procedimientos de escalado. Demonstration project for water in car wash premises using innovative detergents and soft treatment systems, 2016. http://www.minaqua.org/wp14/wp-content/uploads/2012/12/B8.3_Descripci%C3%B3n-procedimientos-escalado_vf.pdf.

22

Constructed Wetland-Microbial Fuel Cell: An Emerging Integrated Technology for Potential Industrial Wastewater Treatment and Bio-Electricity Generation

Asheesh K. Yadav[1], Pratiksha Srivastava[3], Naresh Kumar[2], Rouzbeh Abbassi[3] and Barada Kanta Mishra[1]

[1] Department of Environmental and Sustainability, CSIR-Institute Minerals and Materials Technology, Bhubaneswar, India
[2] Department of Geological and Environmental Sciences, Stanford University, Stanford, USA
[3] Australian Maritime College (AMC), University of Tasmania, Launceston, Australia

22.1   Introduction

Constructed wetlands (CWs) or treatment wetlands are man-made engineered systems that are designed and employed to improve the water quality. CWs run with relative low external energy requirements and are easy to operate and maintain. As a result, CWs have been set up all over the world as an alternative to conventional wastewater treatment systems [1, 2]. In general, a CW comprises a basin filled with filter media and planted with water loving plants. The use of CW for wastewater treatment was first tested in Germany in the 1950s [3], but full-scale CWs were only built during the late 1960s [4]. Initially CWs were used to treat municipal wastewaters but later also used to treat industrial and agricultural wastewaters, landfill leachate, and storm water runoff [2]. Among various type of CWs, horizontal subsurface flow constructed wetland (HSSF-CW) is the most widely used due to its design simplicity. HSSF-CWs are mainly anaerobic (deeper or inner portion of the bed) in nature [5], but have some spatial redox variations [6]. The upper part of the wetland remains under aerobic conditions because of its close contact with the atmosphere, providing redox gradients of about 0.5 V [6–9].

CWs are generally slow in treatment, e.g., due to their anaerobic conditions in HSSF-CWs, where inferior or limited electron acceptors are available. Thus, CW application needs a large land area compared to other intensive technologies such as activated sludge process. This is one of the biggest challenges in extensive application of CW technology. There have been attempts to accelerate the treatment performance of CWs, which can be categorized into (i) operation strategies such as effluent re-circulation, artificial aeration, tidal operation, drop aeration, flow direction reciprocation, earthworm integration, bioaugmentation; (ii) configuration innovations such as circular-flow towery hybrid, baffled subsurface-flow; and (iii) electron donor supplementation [10]. Recently, a new approach by integrating microbial fuel cells (MFCs) into constructed wetlands has been commenced to enhance the treatment efficiency of CWs [11, 12]. The basic idea behind this approach is to provide extra electron acceptor in the form of conductive material (anode electrode) in anaerobic zone of the CWs to promote the anaerobic reaction. This approach is gaining a lot of interest lately, as it has dual advantage like performance intensification of CWs and electricity generation.

The concept of MFC is not new; Michael C. Potter had demonstrated the concept of MFC in 1911 by demonstrating electricity generation using *Escherichia coli* culture [13]. In Potter's experiment, reported electricity was not very high, therefor, not much attention was given to this concept. In 1931, Cohen [14] revived the concept with more voltage output using stacked MFCs. There was not much progress on this subject since then. However, recent energy crises and increasing awareness of global warming have given a new hope for this technology.

An MFC can be distinct as a two chambers device, i.e., one anaerobic chamber and another one aerobic chamber. The anaerobic chamber is also known as the anode chamber since the anode electrode is placed here, while the aerobic chamber is known as the cathode chamber due to the presence of the cathode electrode (Figure 22.1). In the anaerobic chamber (anode chamber), microbes work as bio-catalysts to oxidize substrates and produce electrons and protons. On the other hand, in the aerobic chamber (cathode chamber), reduction reactions occur. The anaerobic chamber connects with the aerobic chamber through an electrical wire circuit consisting of resistance or load. This circuit allows the flow of electrons. Both chambers also connect through a proton exchange membrane (PEM), which only allows passing of the protons generated during oxidation of substrate.

In an MFC, microorganisms donate electrons to the anode electrode which pass them to the cathode through a circuit where it reacts with oxidant (such as oxygen) and protons. Consequently, due to the flow of electrons through the electrical circuit, electricity is generated. In a nutshell, the anode works as electron acceptor and the cathode works as an electron donor in MFCs. The said electrode conditions make MFC a unique device which involves both oxidation and reduction along with electric current generation. The working principle of an MFC can be explained by the following anode

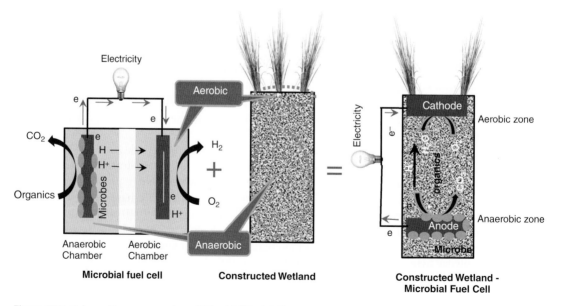

Figure 22.1 Schematic representation of Microbial Fuel Cell, Constructed Wetlands and Integrated Constructed Wetland-Microbial Fuel Cell.

and cathode reactions [for the case of acetate as the organic substrate (reducer) and oxygen as the oxidant]:

$$Anode: \ CH_3COOH + 2H_2O \rightarrow 2CO_2 + 8H^+ + 8e^- \qquad (22.1)$$

$$Cathode: \ O_2 + 4H^+ + 4e^- \rightarrow 2H_2O \qquad (22.2)$$

Since both the reactions occur at two separate places, electricity generation is possible. In the anodic chamber, anaerobic microbes (known as exoelectrogen) oxidize the acetate and donate the electrons to the anode. The electrons flow from the anode to the cathode through an external circuit. Charge neutrality of the MFC is maintained because protons drift from the anode to the cathode in the aqueous solution through PEM.

MFCs have the special advantage of using low-grade substrates such as wastewater for energy generation [15]. Domestic, industrial, and agricultural wastewater can also be a potential source of substrate for MFCs [15, 16].

Similar to MFC, the CW also involves aerobic and anaerobic zone dependent processes during wastewater treatment. The presence of both aerobic and anaerobic conditions in CW is suitable for developing *in-situ* MFC in the CW. Thus, integration of MFC into CW can provide additional electron acceptors to the anaerobic zone of the CW.

22.2 The Fundamentals of MFC and Microbial Electron Transfer to Electrode

MFCs are a kind of bio-electrochemical system (BES) that generates electricity by means of electrochemically active microorganisms. All types of BES depend on the action of electroactive bacteria (EAB). EAB are also known as exoelectrogens, electrogens, electricegens, exoelectrogenic, or anode respiring bacteria. EAB are the microorganisms, which are capable of donating the electrons to the solid external electrode, either *via* direct or indirect mechanisms [17, 18]. Various EAB sources have been reported including marine or fresh water sediments [19, 20], aerobic/anaerobic wastewater treatment sludge, wastewater [21–24], as well as various manures [25, 26].

Some of the EABs are capable of changing their metabolic preferences of soluble electron donors such as glucose, acetate, hydrogen or acceptors such as nitrate, oxygen, fumarate with a solid electron donor or acceptor via direct electron transfer (DIET/DET) mechanism [27]. This type of electron transfer is likely either by outer-membrane redox proteins and cytochrome cascades, or by conductive nanowires [28, 29].

In some cases, bacteria are not in direct contact with the electron acceptors. Such conditions force the bacteria to transfer their electron *via* Indirect Electron Transfer (IET) mechanism. In such a condition, bacterial metabolism produces an electron shuttle or mediator for extracellular electron transfer [30]. In IET, EAB employ mediator molecules generated by bacteria themselves or by an artificially provided one [31]. Figure 22.2 exemplifies the direct and indirect electron transfer to electrodes in MFC.

Until now, *Shewanella* [32] and *Geobacter* [33] genera have been well recognised as EAB. Among these, *Geobacter* species have shown their potential to direct electron transfer and indirect electron transfer through conductive pili or nanowires [34]. Since the electrons move through conductive electrical wire to the cathode in the MFC, the the MFC is utilized as an alternative technology to harvest energy directly from wastewater in the form of electricity [35–37].

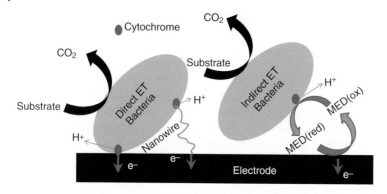

Figure 22.2 Schematic illustration of direct and indirect electron transfer to electrode by EAB [adapted from (18)].

In an MFC, EAB oxidizes the organic and inorganic substrates and produces electrons and protons. The generated electrons are transferred to the anode by the above described mechanism from where they flow via a conductive material or electrical wire and a resistor to meet a higher redox electron acceptor such as oxygen, at the cathode. The protons also travel towards the cathode [38, 39].

In order to make use of the anode as the final electron acceptor by EABs, no acceptor with higher redox potential should be present in their surrounding area. There must be sufficient redox gradient between anode and cathode to drive the electron through the electrical circuit connecting these electrodes [38, 39].

In some MFC approaches, a proton exchange membrane (PEM) between the anode zone and the cathode zone is provided which stops oxygen diffusion from the cathode zone to the anode zone and only allows the cation, e.g., protons to transfer between the anode and the cathode zones. In other types of approaches, the natural redox gradient of the natural or semi-natural environments is utilized. The sediment or benthic microbial fuel cells (sMFC) are the examples of the later type of MFCs. In nature, there are several environments such as rice paddy fields, natural wetlands, ponds, lakes, rivers, and marine environments where natural redox gradients exist and can be exploited.

The sediment or benthic microbial fuel cells exploit such naturally occurring redox gradients by burying an anode in water saturated soil/sediment (anaerobic zone) and connecting the same to a cathode placed in the overlying aerobic zone [40, 41] or at the air–soil–water interface [42]. The EAB existing in the vicinity of the anode zone, oxidizes the organic or inorganic matter present in the surrounding environment and produces electricity in the process. In recently developed constructed wetlands, microbial fuel cells systems can also be classified as sediment or benthic microbial fuel cells [12, 43–45].

22.3 State of the Art of CW-MFCs

22.3.1 Design and Operation of CW-MFCs

The CWs and MFC coupling have been studied in past years, but their implementation as a suitable wastewater treatment alternative is still in the developmental stage [45]. Most of the studies with CW-MFCs focus on the electrical current generated by optimizing the high redox gradient using synthetic wastewater, swine slurry wastewater, or municipal wastewater [45]. Industrial wastewater treatment is not being studied rigrously in CW-MFCs. Doherty et al. [45] have reviewed the various CW-MFCs systems in detail on various operational and design aspects. In reported studies, flow direction of the wastewater is the most frequently studied area. Most of the studies deal with CW-MFCs where an anode is buried in the lower anaerobic part and a cathode is placed in the upper aerobic surface near the vicinity of the plant rhizosphere. Such an arrangement provides the maximum redox gradient in the system. To Create a sharp redox gradient between anode and cathode area, an artificial arrangement such as a glass-wool separator was used [12, 44].

Typical MFC uses a PEM between the anaerobic (anode) and aerobic (cathode) zones for making a greater cell force. Such an arrangement would incur additional cost and is associated with other operational difficulties. Apart from MFC structural design, MFC performance is dependent on biological, chemical, and electrical factors. Various parameters, which affect the MFC performance are listed by Rabaey and Verstraete [46] and Min and Logan [38] such as: (a) substrate conversion rate; (b) over potentials at the anode; (c) over potentials at the cathode; (d) proton exchange membrane related factors; and (e) internal resistance of the MFCs. Nevertheless, Logan [47] has reported other influential operational variables on the performance of MFCs such as the concentration of chemical oxygen demand (COD) in the anodic chamber, pH or temperature, surface area of the electrodes, electrode materials and relative distance between electrodes.

In recent studies, where wastewaters were fed in upflow mode in CW-MFC, no separator was used, as wastewater firstly comes in contact with the anaerobic zone and then in the aerobic zone where the anode and cathode are placed, respectively [43, 48, 49]. However, the upflow regime without separator in CW-MFCs needs sufficient distance between anode and cathode, which results in higher ohmic resistance and less electrical current generation [50–54]. Doherty et al. [45] used glass-wool as a separator and kept anode and cathode close to each other in their CW-MFCs; the wastewater was fed simultaneously upflow into the anode and downflow into the cathode. With the above operation and design strategies, they were able to enhance the maximum power density by 70%, but the system was prone to low performance under higher organic loads due to limited oxygen availability for the cathode reaction. Furthermore, a horizontal flow constructed wetland-microbial fuel cell using a bentonite layer to separate the lower anaerobic anode compartment and the upper aerobic cathode compartment was designed and operated by recirculating the wastewater flow in this system from the bottom (anode compartment) to the top (cathode compartment) [55]. Villasenor et al. [55] reported that power generation was lower at higher organic loads due to fewer amounts of organics being removed in the anode zone; it reached the cathode zone where it may have taken up the oxygen and obstructed the cathode reaction.

Most of the studies with CW-MFCs have used graphite or carbon as electrode materials. Srivastava et al. [56] have used Pt catalyst based carbon cloths at the cathode. These electrode materials were mostly used in granular form. In CWs, the granular form of these materials may replace the normal filter materials of CW, i.e., stone gravel or sand particles. Additionally, the use of graphite or charcoal can provide additional removal due to their adsorption capacity for pollutants such as organics and other ions [57].

22.3.2 Performance Evaluation of Various CW-MFCs

A comparative performance of various CW–MFCs based studies was compiled and is presented in Table 22.1. The first integrated CW–MFC technology was developed by placing the electrodes in

Table 22.1 Reported performance of CW-MFCs.

Wastewater flow	Liquid volume (L)	Electrode material	Initial COD (mg/L) and (% removal)	Max. power	Reference
Vertical flow	5.4	Anode–Graphite plate Cathode–Graphite plate	1,500 (74.9)	15.7 mW/m^2	[12]
Vertical upflow	3.7	Anode–Graphite plates Cathode–Graphite plates	1,058 (76.5)	9.4 mW/m^2	[44]
Vertical flow	12.4	Anode–Granular Activated Carbon Cathode–Granular Activated Carbon	180 (86)	0.302 W/m^3	[48]
Horizontal subsurface flow	96	Anode–Graphite plates Cathode–Graphite plates	250 (80–100)	0.15 mW/m^2	[55]
Vertical flow	12.4	Anode–Granular Activated Carbon Cathode–Granular Activated Carbon	193–205 (94.8)	12.42 mW/m^2	[49]
Vertical flow	–	Anode–Granular Activated Carbon Cathode–Granular Activated Carbon	300 (72.5)	0.852 W/m^3	[59]
Vertical upflow	8.1	Anode–Granular Graphite Cathode–Granular Graphite	411–854 (64)	0.268 W/m^3	[60]
Vertical upflow	–	Anode–Carbon Felt Cathode–Carbon Felt	314.8 (100)	6.12 mW/m^2	[62]
Simultaneous upflow–downflow	8.1	Anode–Graphite Granules Cathode–Graphite Granules	583 (64)	0.276 W/m^3	[61]
Vertical flow	1.8	Anode–Granular Activated Carbon Cathode–Granular Activated Carbon	770–887 (90.9)	43.63 mW/m^3	[56]
Vertical flow	1.8	Anode–Granular Graphite Cathode–Granular Graphite	770–887 (80.9)	0.10 mW/m^3	[56]

(Continued)

Table 22.1 (Continued)

Wastewater flow	Liquid volume (L)	Electrode material	Initial COD (mg/L) and (% removal)	Max. power	Reference
Vertical flow	1.8	Anode–Granular Graphite Cathode–Pt coated Carbon Cloth	770–887 (84)	320.8 mW/m^3	[56]
Vertical flow	1.87	Anode–Granular Activated Charcoal Cathode–Pt coated Carbon cloth	770–887 (91.4)	92.48 mW/m^3	[56]
Horizontal subsurface flow	–	Anode–Cylindrical Graphite rod Cathode–Cylindrical Graphite rod	323 (61)	36 mW/m^2	[63]
Vertical up flow	–	Anode-activated carbon Cathode-activated carbon	624 (99)	93 mW/m^3	[64]
Vertical flow	1.5	Anode–Powder activated carbon Cathode–Granular graphite	500 (80)	87.79 mW/m^2	[65]
Horizontal subsurface flow	–	Anode–Cylindrical graphite rod Cathode–Cylindrical graphite rod	323 (60.6)	131 mWh/m^2/d	[66]

the aerobic and anaerobic zones of the CW [11]. A miniature floating macrophyte based ecosystem integrated with fuel cells was designed by Venkata Mohan et al. [58] and reported the current generation of 224.9 mA/m^2, 86.7% COD removal and 72.3% volatile fatty acids (VFA) removal. Yadav et al. [12] developed another vertical flow CW-MFC and used the same for dye containing wastewater treatment, achieving a power density of 15.7 mW/m^2 and COD removal of 74.9%.

Liu et al. [49] developed a constructed wetland coupled with MFC on the principles of photosynthetic MFC by utilizing root exudates of *Ipomoea aquatica* as part of fuel. They achieved the maximum power density of 12.42 mW/m^2 produced from the CW-MFC planted with *I. aquatica*, which was 142% higher than the 5.13 mW/m^2 obtained from the CW-MFC without plants.

A power density of 9.4 mW/m^2 and 76.5% COD removal was reported by treating real wastewater in continuous upflow CW-MFC [44].

Fang et al. [48] developed another CW-MFC and used the same for treating dye containing wastewater of 180 mg/L initial COD concentration and producing power density of 0.302 W/m^3 and 86% COD. Fang et al. [59] reported power density of 0.852 W/m^3 and 72.5% COD removal in an MFC coupled constructed wetland system using azo dye containing wastewater treatment. Villasenor et al.

[55] developed a horizontal subsurface flow CW-MFC and achieved a maximum power density of $0.15 \, mW/m^2$ and 80–100% of COD removal during the synthetic wastewater treatment.

Alum sludge-based CW-MFCs were designed and tested for swine slurry wastewater treatment with electricity generation [60]. A maximum power density of $0.268 \, W/m^3$ and COD removal of 64% were achieved in this study during the treatment of the slurry wastewater. Doherty et al. [61] explored the effects of electrode spacing and flow direction of wastewater in CW-MFCs for simultaneously power generation and wastewater treatment. A maximum power of $0.276 \, W/m^3$ was achieved in simultaneous upflow of wastewater into the anode and downflow of wastewater into the cathode. Oon et al. 62] has designed another upflow constructed wetland–microbial fuel cell (UFCW-MFC) for simultaneous wastewater treatment and electricity generation. The study has reported the maximum power density of $6.12 \, mW/m^2$, with removal efficiencies of 100%, 40% and 91% for COD, NO_3^- and NH_4^+, respectively.

22.4 Potential Industrial Wastewater Treatment in CW-MFCs

The composition of municipal and industrial wastewater is not identical in most cases. Industrial wastewater contains a much higher concentration of organics, solids, and nutrients than municipal wastewater [67] and therefore, the use of constructed wetlands for industrial wastewater treatment requires some kind of pretreatment in such cases. In general, the BOD/COD ratio is used as a parameter to determine the biological degradability of wastewater. If this ratio is greater than 0.5, the wastewater, such as wastewaters from breweries, food industry, abattoirs, dairies etc., is far more biodegradable [67]. In general, the BOD/COD ratio for these wastewaters ranges between 0.6 and 0.7, but could be as high as 0.8.

Both the CW and MFC technology are based on biological treatment. Thus, these technologies need a wastewater which must show a good biodegradability. To date, no industrial wastewater treatment is done using CW-MFC technology but both technologies are applied to industrial wastewater treatment individually. Thus, there is a strong possibility of industrial wastewater treatment in CW-MFC technology with improved performance as MFC components can catalyze the anaerobic reactions of CW technology [56].

One of the main reasons for MFC incorporation in conventional CWs is because of the flow of electrons from the anaerobic zone to aerobic zones. Such flow of electrons can enhance the contaminant removal from the wastewater. For example, during oxidation of organics in the anaerobic zone of CW-MFC, the anode can be used as an electron acceptor and these electrons can then be transported to the cathode to complete the reduction reactions, such as oxygen reduction. This type of synergy between organic oxidation and flow of electrons (electricity production) via the MFC arrangement can enhance the treatment in the CW-MFCs compared to traditional CWs. Most of the reported studies with CW-MFCs have given attention to the anode part and, until now, the cathode part of the CW-MFC has not being focused on from the point of view of wastewater treatment.

The cathode part performs the reduction reaction by offering the donation of the electron. It can help in improving the de-nitrification reaction or producing hydrogen peroxide (H_2O_2) during the oxygen reduction reaction [68, 69]. It is important to mention that H_2O_2 is a strong oxidizing substance and is extensively used in advanced oxidation. The generation of H_2O_2 may also delay the clogging of CW by oxidizing some of the clogging material. As a result of H_2O_2 production at the cathode, there can be an enhancement in wastewater treatment efficiency of CW. There are also other benefits of cathode in MFCs. For example, it can be used for heavy metal removal through reductive precipitation [70]. This reaction can provide an opportunity to precipitate the heavy metals from wastewaters. In brief, the anode and cathode of CW-MFC technology can help in catalyzing pollutant removal processes to improve the wastewater treatment compared to conventional CWs.

In the reported literature, COD was typically used to prove the capability of MFC incorporation into CW for wastewater treatment. Srivastava et al. [56] found that MFC integration contributed 27–49% enhanced COD removal in CW-MFC system compared to traditional CW. Fang et al. [48] has shown 12.6% improvement in dye removal in CW-MFC system compared to traditional CW. Similarly, Doherty et al. [45] found that 33% of the total COD was removed in the anode zone, which only resided in 13.6% of the liquid volume of CW-MFC system in their study.

The current section of the chapter describes various industrial wastewaters treatment, where either MFC or CW was applied individually. It further gives prospective view of their treatment in CW-MFC. There are some common industrial wastewaters including brewery, winery, coking, dairy, distillery, petroleum refinery, tannery, slaughter houses, starch production, and pulp and paper studied in both MFC and CW technology for their treatment (Table 22.2).

Table 22.2 An overview of treatment performance of MFCs and CWs at individual level for various industrial wastewaters.

Industry wastewater	MFC Technology			Constructed Wetlands technology		
	Electricity generation	% Removal (initial concentration)	Reference	Type of CW	% Removal (initial concentration)	Reference
Brewery wastewater	669 mW/m^2	20.7% (1,501 mg COD/L)	[71]	HF	25% (150 mg COD/L)	[81]
Winery wastewater	31.7 Wh/m^3	65% (2,200 mg COD/L)	[72]	HF	49–79% (–)	[82]
Coking wastewater	538 mW/m^2	50% (3,200 mg COD/L)	[73]	HF	35-52% (140 mg COD/L)	[83]
Dairy wastewater	621.1 mW/m^2	90.5% (53.2 kg COD/m^3/d)	[74]	HF	69% BOD (25.3 mg BOD/ L)	[84]
Distillery wastewater	202 mW/m^2	71.8% (3,200 mg COD/L)	[75]	HF	64%	[85]
Petroleum refinery wastewater	330.4 mW/cm^3	64% (250 mg COD /L)	[76]	VF	65% (232 mg COD/L)	[86]
Tannery wastewaters	7 mW/m^2	88% (1,100 mg COD/L)	[77]	HF	63.4% (2,104 mg COD/L)	[87]
Slaughterhouse wastewater	578 mW/m^2	93% (4,850 mg COD/L)	[78]	HF	98% (349 mg COD/L)	[88]
Starch production wastewater	239.4 mW/m^2	98% (4,852 mg COD/L)	[79]	VF-HF	94% (54,065 mg COD/L)	[89]
Pulp and paper mill wastewater	40.2 mW/m^2	55% (4,100 mg COD/L)	[80]	HF	84% (1,084 mg COD/L)	[90]

The wastewaters from slaughter houses, chemical, brewery, food processing, and other industries are the most frequently studied wastewater in MFCs and other BESs [91–93]. The widespread footprints of these industries globally and the necessity of the treatment of these wastewater, which contains high organic matter may possibly be the reason for the higher number of studies to generate bio-electrochemical energy in MFCs using these wastewaters. High biodegradability and abundance of availability of wastewaters of beverage and food processing industries, such as brewery, winery, dairy, vegetable, meat and others, make them suitable for treatment in MFCs [94, 95]. Additionally, these industrial wastewaters generally do not contain any microbial growth inhibiting agents [95]. Most of the MFC studies are performed using industrial wastewaters containing COD concentrations ranging from 3,000 to 5,000 mg/L [96–98].

On the other hand, wastewaters from the above industries are also being treated in CWs due to their high biodegradability and abundant availability. These similarities can also be the basis for treating such wastewaters in CW-MFCs. Until now, electricity generation in MFC has not achieved a usable scale; therefore, the loss of electricity generation during such wastewater treatment will not be an issue of serious concern, if such wastewaters are treated in CW-MFCs. In fact, there are no reports so far, which indicate the adverse effects of MFC integration into CW on treatment performance. Rather, reports indicate the improvement in treatment performance of CW after incorporating the MFC [45, 48, 56]. Thus, CW-MFC can be encouraged for some industrial wastewater treatment, as shown in Table 22.2.

22.5 Challenges in Generating Bio-Electricity in CW-MFCs During Industrial Wastewater Treatment

Microbial fuel cell technology appears to be a promising biotechnological process, but there are several operational, scale-up and cost-related challenges, which limit its real world applications. To design practically viable MFCs, a scaled-up system is needed. It requires not only escalating the MFC size and treatment efficiency to a real world application level, but also to attain the energy generation to the level of real world application [47]. Although theoretically neutral or positive energy balance has been demonstrated, there is no real operation of energetically self-sustained MFCs for wastewater treatment [96]. Thus, the leading challenge of MFC technology for wastewater treatment application is to scale up the reactor size and energy output.

So far, during wastewater treatment with MFC technology, maximum current densities ranging from 10 to 25 A/m^2 for milliliter-scale systems and 6 A/m^2 or lower for liter- or higher-scale systems have been achieved [99]. Clauwaert et al. [100] described that for an economically viable application of MFCs, power density of about 1 kW/m^3 or equivalent current density of 5,000 A/m^3 of total anolyte volume or 50 A/m^2 of projected anode surface area would be required. This section of the chapter describes some of the wastewater treatment challenges related to MFC technology.

The explicit limitations related to MFC scaling up include: high internal resistance, high material cost, pH buffering, and low efficiency of mixed culture biofilm on an electrode [47, 101]. For removing these limitations from the MFC technology, a joint effort in reactor engineering, material development and biological manipulation is needed. There is already knowledge available with previous experience of scaling up efforts which should be utilized for potential reactor design [102].

In such efforts, the biggest lesson is to keep the distance of the electrode as minimum as possible and to ensure an effective transfer of electrons to minimize internal resistance. To begin with, a close distance between the anode and the cathode electrodes, and an efficient transfer of ions should be ensured to decrease internal resistance in any upcoming larger MFCs [102]. This is imperative, because a wider distance between two electrodes and a pH gradient are generally the main reasons mentioned in previous studies for high internal resistance in the large-scale systems [101]. However, reducing the distance between two electrodes can also enhance the possibility of oxygen diffusion into the anode zone. In this direction, a separator can help in enhancing the performance. Other issues such as a high surface area electrode can provide better opportunities to EABs to donate electrons to an electrode, which will lead to more wastewater treatment and electricity generation [103].

22.6 Future Directions

There is no doubt that CW-MFC has vast potential for wastewater treatment and electricity generation. To date, most of the studies have been conducted with synthetic wastewater and the main focus of these studies has been bio-electricity generation and optimization of the operational conditions. There is a strong possibility of industrial wastewater treatment in CW-MFC technology with improved treatment performance, thus, this aspect should also be explored further. The effect of MFC implantation into CW is also poorly studied so far. However, in CW-MFC, conductive material can work as electron donor and electron acceptor. Therefore, there is the possibility that conductive materials can manipulate the microbiology and associated processes of the CW. This effect should be studied in further detail.

There are a few studies [56], which indicate that integration of the MFC into CW improves the anaerobic oxidation of organics in CW. The results obtained in these studies can have significant positive impacts, particularly towards faster and compact CW development with less aerial footprints. It can also make a positive impact in accelerating the anaerobic reaction particularly in cold climates. There are also some studies indicating that methane production can be hampered by incorporating a plant-MFC into a rice microcosm [104]. The anode proposed as electron acceptor in plant-MFCs and CW-MFCs, provides a more favorable electron acceptor and limits the growth of methanogens due to competition with EAB [48, 105]. Therefore, production of greenhouse gas must also be explored in future CW-MFC studies. The anode may also help in oxidation of recalcitrant pollutants, thus, it must also be addressed in further investigations on CW-MFCs. The cathode portion of the MFC may work as an electron donor, which can provide the possibility of higher and more efficient reduction reaction at low cost. Therefore, this aspect should also be explored for de-nitrification, heavy metal removal and recovery and hydrogen peroxide production, etc.

The CW-MFC is an upcoming promising technology for wastewater treatment and bio-electricity generation, though at present the power level is not sufficient for any significant real world applications. To enhance the level of electricity generation, new operational approaches must be tested. At present, the level of current and voltage production in CW-MFC can be useful in biosensor development for assessing the health of the CW-MFC system through digital monitoring. Figure 22.3 summarizes some of the future directions of CW-MFC technology.

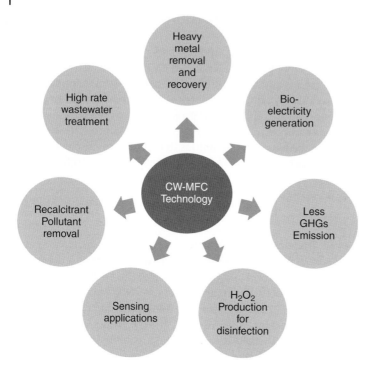

Figure 22.3 CW-MFCs and future direction.

Acknowledgements

AKY greatly acknowledges SERB, Government of India for Fast Track Project Grant (SR/FTP/ETA-0058/2011) and Funding from CSIR, in the form of a XII five-year network project on 'Sustainable Waste Management Technologies for Chemical and Allied Industries' (SETCA, CSC-0113).

References

1 Garcia J, Rousseau DPL, Morato J, Lesage E, Matamoros V, Bayona JM. Contaminant removal processes in subsurface-flow constructed wetlands: a review. Crit Rev Environ Sci Technol. 2014; 40:561–661.

2 Vymazal J. Constructed wetlands for wastewater treatment: five decades of experience. Environ Sci Technol. 2011; 45:61–69.

3 Seidel K.. Zur Problematik der Keim- und Pflanzengewasser. Verh Internat Verein Limnol. 1961; 14:1035–1039.

4 De Jong J. The purification of wastewater with the aid of rush or reed ponds. In: Tourbier J, Pierson RW, eds. Biological Control of Water Pollution. Pennsylvania University Press, Philadelphia, 1976; pp. 133–139.

5 Baptista JDC, Donnelly T, Rayne D, Davenport RJ. Microbial mechanisms of carbon removal in subsurface flow wetlands. Water Sci Technol. 2003; 48(5):127–34.

6 Garcı́ a J, Ojeda E, Sales E, Chico F, Pı́ riz T, Aguirre P, Mujeriego R. Spatial variations of temperature, redox potential, and contaminants in horizontal 614 flow reed beds. Ecol Eng. 2003; 21(2–3):129–142.

7 Dušek J, Picek T, Čížková H. Redox potential dynamics in a horizontal subsurface flow constructed wetland for wastewater treatment: Diel, seasonal and spatial fluctuations. Ecol Eng. 2008; 34(3):223–232.

8 Pedescoll A, Sidrach-Cardona R, Sánchez JC, Bécares E. Evapotranspiration affecting redox conditions in horizontal constructed wetlands under Mediterranean climate: Influence of plant species. Ecol Eng. 2013; 672(58):335–343.

9 Bezbaruah AN, Zhang TC. pH, redox, and oxygen microprofiles in rhizosphere of bulrush (*Scirpus validus*) in a constructed wetland treating municipal wastewater. Biotechnol Bioeng. 2004; 88:60–70.

10 Wu S, Kuschk P, Brix H, Vymazal J, Dong RA. Development of constructed wetlands in performance intensifications for wastewater treatment: a nitrogen and organic matter targeted review. Water Res. 2014; 57:40–55.

11 Yadav AK. Design and development of novel constructed wetland cum microbial fuel cell for electricity production and wastewater treatment. 12th IWA International Conference on Wetland Systems for Water Pollution Control 2010, Venice, Italy, 4–9 October.

12 Yadav AK, Dash P, Mohanty A, Abbassi R, Mishra BK. Performance assessment of innovative constructed wetland-microbial fuel cell for electricity production and dye removal. Ecol Eng. 2012; 47:126–131.

13 Potter MC. Electrical effects accompanying the decomposition of organic compounds. Proc R Soc Lond B Biol Sci. 1911; 84:160–276.

14 Cohen B. The bacterial culture as an electrical half-cell. J Bacteriol. 1931; 21:18–19.

15 Pant D, Van Bogaert G, Diels L. A review of the substrates used in microbial fuel cells (MFCs) for sustainable energy production. Bioresour Technol. 2010; 101:1533–1543.

16 Rismani-Yazdi H, Carver SM, Christy AD. Cathodic limitations in microbial fuel cells: an overview. J Power Sources .2008; 180:683–694.

17 Borole AP, Reguera G, Ringeisen B, Wang ZW, Feng Y, Kim BH. Electroactive biofilms: Current status and future research needs. Energy Environ Sci. 2011;4(12):4813.

18 Schrôder U, Harnisch F, Angenent LT. Microbial electrochemistry and technology: terminology and classification. Energy Environ. Sci. 2015; 8(2):513–519.

19 Lovley DR. Extracellular electron transfer: Wires, capacitors, iron lungs, and more. Geobiology. 2008; 6(3):225–231.

20 Risgaard-Petersen N, Damgaard LR, Revil A, Nielsen LP. Mapping electron sources and sinks in a marine biogeobattery. J Geophys Res G: Biogeosci. 2014; 119(8):1475–1486.

21 Villano M, Aulenta F, Beccari M, Majone M. Start-up and performance of an activated sludge bioanode in microbial electrolysis cells. Chem Eng Trans. 2012; 27:109–114.

22 Gao C, Wang A, Wu WM, Yin Y, Zhao YG. Enrichment of anodic biofilm inoculated with anaerobic or aerobic sludge in single chambered air-cathode microbial fuel cells. Bioresour Technol. 2014; 167:124–32.

23 Escapa A, San-Martín MI, Morán A. Potential use of microbial electrolysis cells in domestic wastewater treatment plants for energy recovery. Frontiers Energy Res. 2014; 2:1–10.

24 Velvizhi G, Venkata Mohan S. Bioelectrogenic role of anoxic microbial anode in the treatment of chemical wastewater: Microbial dynamics with bioelectro-characterization. Water Res. 2015; 70:52–63.

25 Min B, Kim J, Oh S, Regan JM, Logan BE. Electricity generation from swine wastewater using microbial fuel cells. Water Res. 2005; 39(20):4961–4968.

26 Vilajeliu-Pons A, Puig S, Pous N, Salcedo-Dávila I, Baðeras L, Balaguer MD, Colprim J. Microbiome characterization of MFCs used for the treatment of swine manure. J Hazard Mater. 2015; 288:60–68.

27 Erable B, Duţeanu N, Ghangrekar M, Dumas C, Scott K. Application of electroactive biofilms. Biofouling. 2010; 26(1):57–71.

28 Bonanni PS, Schrott GD, Robuschi L, Busalmen JP. Charge accumulation and electron transfer kinetics in *Geobacter sulfurreducens* biofilms. Energy Environ Sci, 2012; 5(3):6188.

29 Busalmen JP, Esteve-Nuðez A, Feliu JM. Whole cell electrochemistry of electricity-producing microorganisms evidence an adaptation for optimal exocellular electron transport. Environ Sci Technol. 2008; 42(7):2445–2450.

30 Arends JBA., Verstraete W. 100 years of microbial electricity production: three concepts for the future. Microb Biotechnol. 2012; 5(3):333–346.

31 Mao L, Verwoerd WS. Selection of organisms for systems biology study of microbial electricity generation: a review. Int J Energy Environ Eng. 2013; 4:17.

32 Ringeisen BR, Henderson E, Wu PK, Pietron J, Ray R, Little B, Biffinger JC, Jones-Meehan JM. High power density from a miniature microbial fuel cell using *Shewanella oneidensis*. Environ Sci Technol. 2006; 40:2629–2634.

33 Kiely PD, Regan JM, Logan BE. The electric picnic: synergistic requirements for exoelectrogenic microbial communities. Curr Opin Biotech. 2011; 22:378–385.

34 Reguera G, Nevin KP, Nicoll JS, Covalla SF, Woodard TL, Lovley DR. Biofilm and nanowire production leads to increased current in *Geobacter sulfurreducens* fuel cells. Appl Environ Microbiol. 2006; 72:7345–7348.

35 Du Z, Li H, Gu T. A state of the art review on microbial fuel cells: A promising technology for wastewater treatment and bioenergy. Biotechnol Adv. 2007; 25:464–482.

36 Lefebvre O, Uzabiaga A, Chang IS, Kim BH, Ng HY. Microbial fuel cells for energy self-sufficient domestic wastewater treatment-a review and discussion from energetic consideration. Appl Microbiol Biotech. 2011; 89:259–270.

37 Min B, Logan BE. Continuous electricity generation from domestic wastewater and organic substrates in a flat plate microbial fuel cell. Environ Sci Technol. 2004; 38:5809–5814.

38 Logan BE, Hamelers B, Rozendal R, Schröder U, Keller J, Freguia S, Aelterman P, Verstraete W, Rabaey K. Microbial fuel cells: methodology and technology. Environ Sci Technol. 2006; 40:5181–5192.

39 Rabaey K, Rodríguez J, Blackall LL, Keller J, Gross P, Batstone D, Verstraete W, Nealson KH. Microbial ecology meets electrochemistry: electricity-driven and driving communities. ISME J. 2007; 1:9–18.

40 Reimers CE, Tender LM, Fertig S, Wang W. Harvesting energy from the marine sediment-water interface. Environ Sci Technol. 2001; 35:192–195.

41 Tender LM, Reimers CE, Stecher HA, Holmes DE, Bond DR, Lowy DA, Pilobello K, Fertig SJ, Lovley DR. Harnessing microbially generated power on the seafloor. Nat Biotechnol. 2002; 20:821–825.

42 Huang DY, Zhou SG, Chen Q, Zhao B, Yuan Y, Zhuang L. Enhanced anaerobic degradation of organic pollutants in a soil microbial fuel cell. Chem Eng J. 2011; 172(2):647e–653.

43 Corbella C, Garfí M, Puigagut J. Vertical redox profiles in treatment wetlands 586 as function of hydraulic regime and macrophytes presence: Surveying the optimal 587 scenario for microbial fuel cell implementation. Sci Total Environ. 2014; 470–471:754–758.

44 Zhao Y, Collum S, Phelan M, Goodbody T, Doherty L, Hu Y. Preliminary investigation of constructed wetland incorporating microbial fuel cell: Batch and continuous flow trials. Chem Eng J. 2013; 229:364–370.

45 Doherty L, Zhao Y, Zhao X, Hu Y, Hao X, Xu L, Liu R. A review of a recently emerged technology: Constructed wetland – Microbial fuel cells. Water Res. 2015; 85:38–45.

46 Rabaey K, Verstraete W. Microbial fuel cells: novel biotechnology for energy generation. Trends Biotechnol. 2005; 23:291–298.

47 Logan BE. Scaling up microbial fuel cells and other bioelectrochemical systems. Appl Microbiol Biotechnol. 2010; 85:1665–1671.

48 Fang Z, Song H, Cang N, Li X. Performance of microbial fuel cell coupled constructed wetland system for decolorization of azo dye and bioelectricity generation. Bioresour Technol. 2013; 144:165–171.

49 Liu S, Song H, Li X, Yang F. Power generation enhancement by utilizing plant photosynthate in microbial fuel cell coupled constructed wetland system. Int J Photoenergy. 2013; 172010.

50 Cheng S, Liu H, Logan BE. Increased power generation in a continuous flow MFC with advective flow through the porous anode and reduced electrode spacing. Environ Sci Technol. 2006; 40:2426–2432.

51 Liu H, Cheng S, Huang L, Logan B. Scale-up of membrane-free single chamber microbial fuel cells. J Power Sources. 2008; 179:274–279.

52 Fan Y, Han SK, Liu H. Improved performance of CEA microbial fuel cell with increased reactor size. Energy Environ Sci. 2012; 5:8273–8280.

53 Ahn Y, Logan BE. A multi-electrode continuous flow microbial fuel cell with separator electrode assembly design. Appl Microbiol Biotechnol. 2012; 93:2241–2248.

54 Srikanth S, Venkata Mohan S. Influence of terminal electron acceptor availability to the anodic oxidation on the electrogenic activity of microbial fuel cell (MFC). Bioresour Technol. 2012; 123:480–487.

55 Villasenor J, Capilla P, Rodrigo MA, Canizares P, Fernandez FJ. Operation of a horizontal subsurface flow constructed wetland-microbial fuel cell treating wastewater under different organic loading rates. Water Res. 2013; 47:6731–6738.

56 Srivastava P, Yadav AK, Barada Kanta Mishra BK. The effects of microbial fuel cell integration into constructed wetland on the performance of constructed wetland. Bioresour Technol. 2015; 195:223–230.

57 Dordio AV, Carvalho A. Organic xenobiotics removal in constructed wetlands, with emphasis on the importance of the support matrix. J Hazard Mater. 2013; 252:272–292.

58 Venkata Mohan S, Mohanakrishna G, Chiranjeevi P. Sustainable power generation from floating macrophytes based ecological microenvironment through embedded fuel cells along with simultaneous wastewater treatment. Bioresour Technol. 2011; 102:7036–7042.

59 Fang Z, Song H, Cang N, Li X. Electricity production from azo dye wastewater using a microbial fuel cell coupled constructed wetland operating under different operating conditions. Biosens Bioelectron. 2015; 68:135–141.

60 Doherty L, Zhao X, Zhao Y, Wang W. The effects of electrode spacing and flow direction on the performance of microbial fuel cell-constructed wetland. Ecol Eng. 2015; 79:8–14.

61 Doherty L, Zhao Y, Zhao X, Wang W. Nutrient and organics removal from swine slurry with simultaneous electricity generation in an alum sludge-based constructed wetland incorporating microbial fuel cell technology. Chem Eng J. 2015; 266:74–81.

62 Oon Y, Ong S, Ho L, Wong Y, Oon Y, Lehl HK, Thung W. Hybrid system up-flow constructed wetland integrated with microbial fuel cell for simultaneous wastewater treatment and electricity generation. Bioresour Technol. 2015; 186:270–275.

63 Corbella C, Guivernau M, Viñas M, Puigagut J. Operational, design and microbial aspects related to power production with microbial fuel cells implemented in constructed wetlands. Water Res. 2015; 84:232–242.

64 Oon Y, Ong S, Ho L, Wong Y, Dahalan FA, Oon Y, Lehl HK, Thung W. Synergistic effect of up-flow constructed wetland and microbial fuel cell for simultaneous wastewater treatment and energy recovery. Bioresour Technol. 2016; 203:190–197.

65 Xu L, Zhao Y, Liam Doherty L, Hu Y, Hao X. Promoting the bio-cathode formation of a constructed wetland-microbial fuel cell by using powder activated carbon modified alum sludge in anode chamber. Sci Report. 2016; 6:26514.

66 Corbella C, Garfí M, Jaume P. Long-term assessment of best cathode position to maximise microbial fuel cell performance in horizontal subsurface flow constructed wetlands. Sci Total Environ. 2016; 563–564:448–455.

67 Vymazal J. Constructed wetlands for treatment of industrial wastewaters: A review. Ecol Eng. 2014; 73:724–751.

68 Srivastava P. Microbial electrolysis: Possibility of denitrification in constructed wetlands. 15th IWA International Conference on Wetland Systems for Water Pollution Control, 4-9 September 2016, Gdańsk, Poland, pp. 1010–1015.

69 Yan H, Saito T, Regan JM. Nitrogen removal in a single-chamber microbial fuel cell with nitrifying biofilm enriched at the air cathode. Water Res. 2012; 46(7):2215–2224.

70 Wang H, Ren ZJ. Bioelectrochemical metal recovery from wastewater: a review. Water Res. 2014, 66:219–232.

71 Wen Q, Wu Y, Zhao L, Sun Q. Production of electricity from the treatment of continuous brewery wastewater using a microbial fuel cell. Fuel 2010; 89:1381–1885.

72 Cusick RD, Kiely PD, Logan B.E. A monetary comparison of energy recovered from microbial fuel cells and microbial electrolysis cells fed winery or domestic wastewater. Int J Hydrogen Energy. 2010; 35:8855–8861.

73 Huang L, Yang X, Quan X, Chen J, Yang F. A microbial fuel cell–electrooxidation system for coking wastewater treatment and bioelectricity generation. J Chem Technol Biotechnol. 2010; 85:621–627.

74 Mansoorian HJ, Mahvi AH, Jafari AJ, Khanjani J. Evaluation of dairy industry wastewater treatment and simultaneous bioelectricity generation in a catalyst-less and mediator-less membrane microbial fuel cell. J Saudi Chem Soc. 2014; 20(1):88–100.

75 Samsudeen N, Radhakrishnan TK, Matheswaran M. Bioelectricity production from microbial fuel cell using mixed bacterial culture isolated from distillery wastewater. Bioresour Technol. 2015; 195:242–247.

76 Guo X, Zhan Y, Chen C, Cai B, Wang Y, Guo S. Influence of packing material characteristics on the performance of microbial fuel cells using petroleum refinery wastewater as fuel. Renew Energy. 2016; 87:437–444

77 Sawasdee V, Pisupaisal N. Simultaneous pollution treatment and electricity generation of tannery wastewater in air-cathode single chamber MFC. Int J Hydrog Energy. 2016; 41(35):15632–15637.

78 Katuri KP, Enright AM, O'Flaherty V, Leech D. Microbial analysis of anodic biofilm in a microbial fuel cell using slaughterhouse wastewater. Bioelectrochemistry. 2011; 87:164–171.

79 Lu N, Zhou S, Zhuang L, Zhang J, Ni J. Electricity generation from starch processing wastewater using microbial fuel cell technology. Biochem Eng J. 2009; 43:246–251.

80 Krishna KV, Sarkar O, Venkata Mohan S. Bioelectrochemical treatment of paper and pulp wastewater in comparison with anaerobic process: Integrating chemical coagulation with simultaneous power production. Bioresour Technol. 2014; 174:142–151.

81 Jones CLW, Britz P, Davies MTT, Scheepers R, Cilliers A, Crous L, Laubscher R. The wealth in brewery effluent – water and nutrient recovery using alternative technologies. 15th International Water Technology Conference (IWTC 2011), Alexandria, Egypt, 2011.

82 Grismer ME, Carr MA, Shepherd HL. Evaluation of constructed wetlands treatment performance for winery wastewater. Water Environ Res. 2003; 75(5):412–421.

83 Jardinier N, Blake G, Mauchamp A, Merlin G. Design and performance of experimental constructed wetlands treating coke plant effluents. Water Sci Technol. 2001; 44(11–12):485–491.

84 Idris SM, Jones PL, Salzman SA. Evaluation of the giant reed (*Arundo donax*) in horizontal subsurface flow wetlands for the treatment of dairy processing factory wastewater. Environ Sci Pollut. 2012; 19:3525.

85 Billore SK, Singh N, Ram HK, Sharma JK, Singh VP, Nelson RM, Das P. Treatment of a molasses based distillery effluent in a constructed wetland in central India. Water Sci Technol. 2001; 44(11/12):441–448.

86 Mustapha HI, Bruggen JJAV, Lens PNL. Vertical subsurface flow constructed wetlands for polishing secondary Kaduna refinery wastewater in Nigeria. Ecol Eng. 2015; 84:588–595.

87 Dotro G, Castro S, Tujchneider O, Piovano N, Paris M, Faggi A, Palazolo P, Larsen D, Fitch M. Performance of pilot-scale constructed wetlands for secondary treatment of chromium-bearing tannery wastewaters. J Hazard Mater. 2012; 239e240:142–151.

88 Lavigne RL, Jankiewicz J. Artificial wetland treatment technology and its use in the Amazon river forests of Ecuador. 7th International Conference Wetland Systems for Water Pollution Control, University of Florida, Gainesville, 2000, pp. 813–820.

89 Kato K, Inoue T, Ietsugu H, Koba T, Sasaki H, Miyaji N, Yokota T, Sharma PK, Kitagawa K, Nagasawa T. Design and performance of hybrid reed bed systems for treating high content wastewater in the cold climate. In: Masi F, Nivala J, eds. 12th International Conference Wetland Systems for Water Pollution Control, IWA, IRIDRA Srl and Pan Srl Padova, Italy, 2010, pp. 511–517.

90 Choudhary AK, Kumar S, Sharma C, Kumar V. Green technology for the removal of chloro-organics from pulp and paper mill wastewater. Water Environ Federation. 2015; 87(7):660–669.

91 Katuri KP, Enright AM, O'Flaherty V, Leech D. Microbial analysis of anodic biofilm in a app microbial fuel cell using slaughterhouse wastewater. Bioelectrochem. 2012; 87:164–171.

92 Li XM, Cheng KY, Selvam A, Wong JWC. Bioelectricity production from acidic food waste leachate using microbial fuel cells: effect of microbial inocula. Process Biochem. 2013; 48(2):283–288.

93 Li WW, Sheng GP, Yu HQ. Electricity generation from food industry wastewater using microbial fuel cell technology. In: Food industry wastes: assessment and recuperation of commodities. 2013;249–261.

94 Digman B, Kim DS. Review: alternative energy from food processing wastes. Environ Prog. 2008; 27(4):524–537.

95 Guo J, Yang C, Peng L. Preparation and characteristics of bacterial polymer using pre-treated sludge from swine wastewater treatment plant. Bioresour Technol. 2014; 152:490–498.

96 Zhang F, Ge Z, Grimaud J, Hurst J, He Z. Long-term performance of liter scale microbial fuel cells treating primary effluent installed in a municipal wastewater treatment facility. Environ Sci Technol. 2013; 47(9):4941–4948.

97 Zhang X, Zhu F, Chen L, Zhao Q, Tao G. Removal of ammonia nitrogen from wastewater using an aerobic cathode microbial fuel cell. Bioresour Technol 2013; 146:161–168.

98 Zhuang L, Yuan Y, Wang Y, Zhou S. Long-term evaluation of a 10-liter serpentine-type microbial fuel cell stack treating brewery wastewater. Bioresour Technol. 2012; 123:406–412.

99 Rabaey K, Butzer S, Brown S, Keller J, Rozendal RA. High current generation coupled to caustic production using a lamellar bioelectrochemical system. Environ Sci Technol. 2010; 44(11):4315–4321.

100 Clauwaert P, Aelterman P, De Schamphelaire L, Carballa M, Rabaey K, Verstraete W. Minimizing losses in bio-electrochemical systems: the road to applications. Appl Microbiol Biotechnol. 2008; 79(6):901–913.

101 Fornero JJ, Rosenbaum M, Angenent LT. Electric power generation from municipal, food, and animal wastewaters using microbial fuel cells. Electroanalysis. 2010; 22:832–843.

102 Li WW, Yu H, Zhen Q, He Z. Towards sustainable wastewater treatment by using microbial fuel cells-centered technologies. Energy Environ Sci. 2014; 7:911–924.

103 Logan BE, Rabaey K. Conversion of wastes into bioelectricity and chemicals by using microbial electrochemical technologies. Science. 2012; 337:686–690.

104 Arends JB, Speeckaert J, Blondeel E, De Vrieze J, Boeckx P, Verstraete W, Rabaey K, Boon N. Greenhouse gas emissions from rice microcosms amended with a plant microbial fuel cell. Appl Microbiol Biotechnol. 2014; 98:3205–3217.

105 Timmers RA, Rothballer M, Strik DP, Engel M, Schulz S, Schloter M, Hartmann A, Hamelers B, Buisman C. Microbial community structure elucidates performance of *Glyceria maxima* plant microbial fuel cell. Appl Microbiol Biotechnol. 2012; 94:537–548.

23

Constructed Wetlands for Stormwater Treatment from Specific (Dutch) Industrial Surfaces

Floris Boogaard[1,2], Johan Blom[3] and Joost van den Bulk[3]

[1] *Hanze University of Applied Sciences (Hanze UAS), Zernikeplein, Groningen, The Netherlands*
[2] *Department of Water Management, Faculty of Civil Engineering and Geosciences, Delft University of Technology, Delft, The Netherlands*
[3] *Tauw bv, Handelskade, The Netherlands*

23.1 Introduction

Constructed wetlands are one type of Sustainable Urban Drainage System (SUDS) that have been used for decades. They provide stormwater conveyance and improve stormwater quality. European regulations for water quality dictate lower concentrations for an array of dissolved pollutants. The increase in the ambitions of the removal efficiency for these systems on industrial areas requires a better understanding of the characteristics of stormwater and the functioning of constructed wetlands as SUDS.

For a detailed view on the achievements of constructed wetlands for stormwater on industrial sites, knowledge on stormwater quality and characteristics is essential as described in the next paragraphs:

- Stormwater quality
- Industrial stormwater quality
- Fraction of pollutants attached to particles
- Research on suspended solids
- Particle size distribution.

The removal efficiency of constructed wetlands is considered in paragraph 23.2.4. Special attention is given to the Dutch situation as an example, since recent monitoring on the characteristics have led to an abundance of data on the quality and characteristics of stormwater and new insights on the treatability of stormwater.

23.2 Stormwater Characteristics

23.2.1 Stormwater Quality in Urban Areas

The stormwater quality of industrial areas is highly dependent on the activities at the industrial site and measures taken to prevent emissions. Not many measurements are published since companies

Table 23.1 (Inter-)national stormwater quality data from residential areas [2].

Substance	Unit	Dutch[a] Mean	USA NSQD[b] Median	Europe/Germany ATV Database[c] Mean	Worldwide[d] Mean
TSS	mg/L	36	48	141	150
BOD	mg/L	7.1	9	13	
COD	mg/L	37	55	81	
TKN	mg N/L	2.8	1.4	2.4	2.1
TP	mg P/L	0.6	0.3	0.42	0.35
PB	µg/L	32.2	12	118	140
Zn	µg/L	52.2	73	275	250
Cu	µg/L	8	12	48	50

[a] [3] updated Dutch STOWA database (first version 3.1.2013, updated 2018) based on data monitoring projects in the Netherlands, residential and commercial areas, with n ranging from 26 (SS) to 684 (Zn);
[b] [4] NSQD monitoring data collected over nearly a ten-year period from more than 200 municipalities throughout the USA. The total number of individual events included in the database is 3.770 with most in the residential category (1.069 events);
[c] [5] ATV database, partly based on the US EPA nationwide urban runoff programme (NURP), with n ranging from 17 (TKN) to 178 (SS);
[d] [6] Typical pollutant concentrations based on review of worldwide [7] and Melbourne [8] data.

are cautious about generating bad publicity. Basic knowledge of stormwater quality from surfaces in commercial and residential areas can give a quick insight regarding substances and concentrations that can be found in stormwater. In paragraph 23.2.2, special attention to concentrations from industrial areas is given, but must be regarded as a rough indication since the stormwater quality is highly dependent on the specific at the industrial site.

The quality and characteristics of stormwater can strongly differ per country, location and even between and during stormwater events [1]. International data of stormwater quality from USA, Australia, and Europe is given in Table 23.1.

To give an indication of the content of databases, a short description of the Dutch stormwater quality database is given, one of the largest databases in the world. For the Dutch situation, research monitoring data was collected over a 15-year period (the earliest measurement in the database is from 1999) from more than 60 municipalities and over 200 locations throughout the country. The total number of individual events included in the database now is 8,300. The national database of all collected stormwater monitoring data allows for a scientific analysis of the data and information and recommendations for improving the quality monitoring. Each data set has gone through a quality assurance/quality control review based on reasonableness of data, extreme values, relationships among parameters, sampling methods and a review of the analytical methods [3].

Most data on the characterization of stormwater quality (contaminants concentration, particle size distribution of suspended sediment, fraction bound to suspended solids) was found by sampling stormwater during rainfall. Most of the samples were analyzed in certified laboratories according to standard methods and standard quality control/assurance procedures.

Preferably data from well-described stormwater research sites were used for the database (peer reviewed journals). In addition the following information was entered: Aim of the research, site descriptions (state, municipality, land use components), and sampling information (date, season, sampling method, sample type) with links to the original research reports and articles.

23.2.2   Industrial Stormwater Quality

Figure 23.1 shows stormwater concentrations from several industrial sites in the Netherlands. In this research, categories were separately analyzed according to national guidelines on environmental management. The categories are determined by emission of particles, odor, noise and possible risk for the surrounding area. Category 1 includes office buildings, Category 2 are areas with car showrooms, parking areas, garages. Category 3 includes industries dealing with construction and demolition waste or galvanizing. Categories 4, 5 and 6 are the most dangerous Categories with steel factories or oil refineries. Of course the data show a lot of variability within the different industries, but it was concluded that the categories could be a first indication of the stormwater quality from industrial sites. On most parameters, categories 1 and 2 are cleaner than the Category 3 and Categories 4, 5 and 6. Average concentrations of heavy metals such as nickel (Ni) and copper (Cu), and nutrients (Nkj) exceed the MAC (Maximum Acceptable Level) concentration and therefore treatment of the stormwater is advised.

23.2.3   Fraction of Pollutants Attached to Particles

Treatability of stormwater runoff by sedimentation depends on the degree of pollutants bound to particles. Distribution between dissolved and particle-bound pollution load can be determined by comparing the total concentration in samples with the filtered sample (0.45 μm).

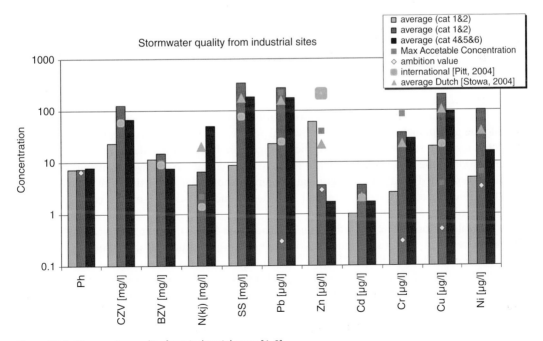

Figure 23.1 Stormwater quality from industrial areas [4, 9].

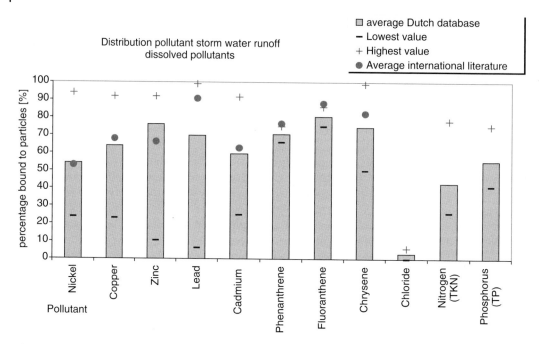

Figure 23.2 Distribution of pollutants in stormwater (90 samples from 25 locations) [3].

Figure 23.2 shows the average values of pollutants bound to suspended solids in stormwater from roofs and roads in residential areas. The plus and minus give the range of the data values, which indicate a large variability in the ability of pollutants to bind to suspended solids. The dot gives the typical average value found throughout the world, which was taken from comparable international studies [10].

From Figure 23.2, the pollutant behavior can be derived. Nutrients are less bound to particles than most of the heavy metals and PAH and, therefore, harder to retain than other contaminants. Within a certain pollutant group, such as metals, the individual pollutants have their own specific behavior. The average Dutch research results are similar to the average from international data [3]. Figure 23.2 also gives an indication of the maximum removal efficiency rate that can be achieved by using settlement devices. To get a detailed insight of the removal efficiency, knowledge of particle size distribution of suspended sediment in stormwater is required, in order to find out which particles can be captured by settlement facilities.

As heavy metals are bound in the order of 65% (lead up to 90%), a higher removal rate with settlement basins (removal only of suspended solids and not solved pollutants) should not be expected, but is rarely determined in the field. When an 80% removal rate is needed to achieve the WFD goal for copper, then it is unlikely that this quality standard will be achieved with sedimentation basins only since copper is bound for 65% (average) to suspended solids. Therefore, additional purification systems to filtration are needed, such as adsorption or phytoremediation of pollutants in sustainable urban drainage systems such as constructed wetlands. Information on the particle size distribution is also an important part for a detailed insight into treatability of stormwater.

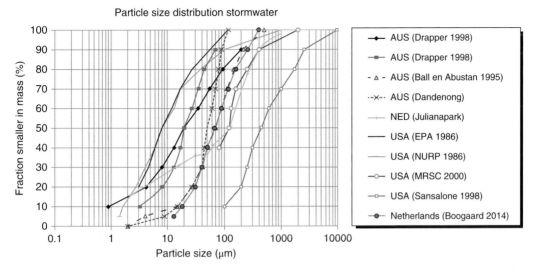

Figure 23.3 Particle size distributions observed around the world in stormwater.

23.2.3.1 Particle Size Distribution

To obtain detailed information on the achievements of settlement and filtering of sustainable urban drainage systems, an examination of particle size distribution is advised. Measurements at several locations around the world were taken to determine the particle size distribution. The results are given in Figure 23.3. The particle size distribution highly varies with each different stormwater drainage location. To get a general idea; roughly half of the mass consists of particles smaller than 90 μm. These fine particles will hardly be removed by settlement facilities [11] and need filter or adsorption systems such as constructed wetlands.

23.2.4 Removal Efficiency

The removal efficiency of constructed wetlands for stormwater derived from existing monitoring results differ from study to study, but are mostly within the ranges of international literature. Not many monitoring results of wetlands at specific industrial areas are available, but the removal efficiency is expected to be within this range as from residential and commercial surfaces.

If we compare Figure 23.4 with Figure 23.2 (distribution of pollutants in Dutch stormwater), we can see a correlation between the removal efficiency and the amount of bound particles. Nutrients are less bound to particles and have lower removal efficiency than heavy metals. The removal efficiency of constructed wetlands derived from existing monitoring results differ from study to study, but are mostly within the ranges of international literature.

23.3 Best Management Practices of (Dutch) Wetlands at Industrial Sites

Most stormwater from industrial sites in the world will be transferred to the wastewater system treatment plant and finally directly to surface water. Wetlands can be implemented after the WWTP

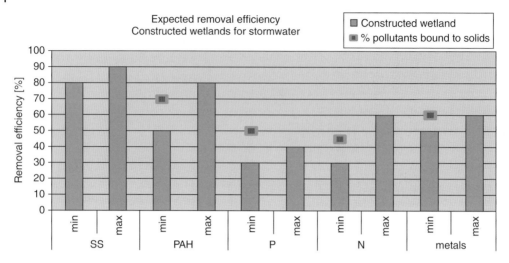

Figure 23.4 Expected removal efficiency on wetlands for stormwater [12].

mostly referred to as 'waterharmonicas'. A waterharmonica is a (natural) constructed wetland as well an ecological engineering solution for upgrading well-treated wastewater with relative low carbon loads. It is a special combination through a customized selection of constructed natural processes for: biological filtration by Daphnia, phototrophic processes in algae mats on reed stems, oxygenation during day time by water plants, introducing food chains, ecotoxicological aspects, natural and recreational values, water buffering, nutrient removal, etc. [13]. The waterharmonica can be found all over the world (Figure 23.5) with a high density in the Netherlands (total land area is only 41,543 km^2).

The Netherlands counts over 15 full scale waterharmonica applications for ecological upgrading of 1,000–40,000 m^3/d treated wastewater, with five more currently under design (Figure 23.6). The Netherlands is mostly a "man-made country", so not many natural wetlands are present. Therefore it has a high density of constructed wetlands without the relation to WWTP, i.e., wetlands that are used to purify water from different origins and surfaces that are expanded over the country (Figure 23.6).

In this paragraph, some unique constructed wetlands for stormwater purification at industrial sites are presented. They can be regarded as Best Management Practices (BMPs), since they stand out for being the biggest wetland, showing implementation of high removal permeable levees, or situated in the most polluted area in the Netherlands, in order to ensure the water quality of a closed water system of a park that is mainly used for recreation:

1) Constructed wetland Amsterdam westergasfabriekterrein.
2) Constructed wetland Oostzaan: multifunctional high removal efficient.
3) Constructed wetland Hoogeveen, oude Diep.

Detailed information with videos, photos and research documents from these cases are available on the tool www.climatescan.nl. This tool is used for international knowledge exchange: available for all, and everybody is encouraged to add functioning SUDs as constructed wetlands to this public database [14].

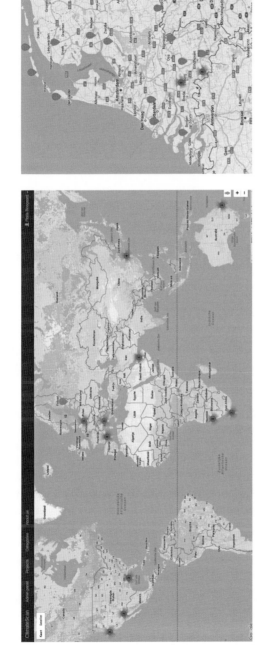

Figure 23.5 Mapped constructed wetlands "waterharmonicas" around the world (left) and the Netherlands (right) (source: https://www.climatescan.nl/#filter-23).

Figure 23.6 Mapped constructed wetlands in The Netherlands for waste and stormwater (source: https://www .climatescan.nl/#filter-9) [14].

23.3.1   Amsterdam Westergasfabriekterrein

In this paragraph, the horizontal wetland in Amsterdam in Westergasfabriekpark, namely, a closed water system in a park is discussed, this being one of the most polluted industrial sites of the Netherlands.

At the end of the 19th century, the Imperial Continental Gas Association (ICGA) built a gas factory complex in Amsterdam. By the time the factory shut down, the site was heavily polluted, making it difficult to find a new purpose for the area. The redevelopment of the site demanded an integral approach and much care was given to the (polluted) water system. The water system in this transformation from a "polluted no-go area" to a multifunctional park needed to be self-sufficient, in order to prevent spreading of polluted (ground-)water to the areas around it. In order to accomplish this goal, the water is stored and purified in a water system with constructed wetland. A small area is used as a bathing area, which is filled with drinking water [11]. The constructed wetland area is one of the key elements of the water system to purify the water in the park that is intensively used for recreation. The Westergasfabriek is regarded as a model for redevelopment, far beyond the Netherlands' borders (Figure 23.7).

23.3.2   Constructed Wetland Oostzaan: Multifunctional High Removal Efficiency

In the industrial site Brombach in Oostzaan, a horizontal wetland is implemented. Unique is the end filtration step with Lava and Olivine (Mg,Fe_2SiO_4) before the water is discharged to the surface water. Doubt about the effluent quality from this industrial site was the main reason for the implementation

Figure 23.7 Example of horizontal wetland in Amsterdam in Westergasfabriekpark for closed system [11].

Table 23.2 Characteristics of the constructed wetland in Oostzaan.

Functions of wetland and storage at industrial area Oostzaan	Stormwater purification, water storage, recreation and ecology
Connected industrial surface to wetland area	68,700 m²
Maximum storage capacity	3,400 m³
Length average	180 m
Lowest height bottom wetland	NAP −1,2
Surface stormwater storage and wetland	1,224 m²
Storage capacity wadi	634 m³
Water level in surrounding surface water system(polder)	Between NAP −1,43 until −1,46 m.
Dimensions of permeable treatment levee	2.5 × 5 × 1.2 m
Material of permeable treatment levee	Lava and Olivine (($Mg,Fe)_2SiO_4$)
Involved stakeholders	Municipality Oostzaan, Water authority Hoogheemraadschap Hollands Noorderkwartier, consulting agency Tauw

of this high removal efficiency wetland (Table 23.2). The wetland is created as a multifunctional wetland to increase the value of the district for ecology and recreation, next to water storage and water quality improvement, and risk management of possible calamities at industrial sites (Figure 23.8).

Figure 23.8 Example of horizontal wetland with extra purification with Lava and Olivine (more photos and videos on: https://www.climatescan.nl/projects/101/detail [14]).

23.3.3 Constructed Wetland Hoogeveen, Oude Diep

Stormwater from an industrial area is collected and treated in the horizontal flow constructed wetland. The stormwater is treated alongside surface water from an agricultural area. The total wetland area is 60,000 m^2 (Figure 23.9).

The residence time of the water in the filter is not exactly known, but is estimated to be around 20 days. The main goal of the filter is the treatment of storm water and sewerage overflows to reduce the pollution of the surface water "Het Oude Diep". From 2011 to 2016 the filter was monitored. Every month samples were taken and analyzed. The results are shown in Table 23.3. Note that horizontal flow constructed wetlands can show negative removal efficiencies due to: measurement method (grab samples), emissions from top soil layer or vegetation, feces from birds or other wildlife etc.

The resulting removal efficiencies of compartments 1 and 2 are shown in Figure 23.10.

The results show that the Hoogeveen filter mainly removes metals (aluminum, cobalt, copper, iron, nickel, lead, vanadium and zinc). The removal of nitrogen and phosphorus are low to negligible, but it has to be kept in mind that the sewerage overflows are not monitored effectively. The results are in general in line with international research results on removal efficiencies. As shown in Figure 23.4, nutrients are less bound to particles and in most cases have a lower removal efficiency than heavy metals. Note that the removal efficiency of constructed wetlands derived from existing monitoring results differ from study to study, but are mostly within the ranges of internationa lliterature.

23.3.4 Cost

The average monetary cost of implementation of vertical flow drainage wetlands are usually in the order of 50–100 €/m^2 field area. A large part of the costs is used for the filling of the filter (1 m^3/m^2) and earth movement. Surface flow wetlands usually require much lower costs of implementation. Usually the costs are 10–20 €/m^2 field area. The costs of the reeds are an important variable: 15 €/field area. The construction costs of several wetlands are described in Figure 23.11.

The average cost of implementation of the vertical flow wetlands was in the order of 55 €/field area, in contrast to the cost of the surface flow wetlands (15 €/field area).

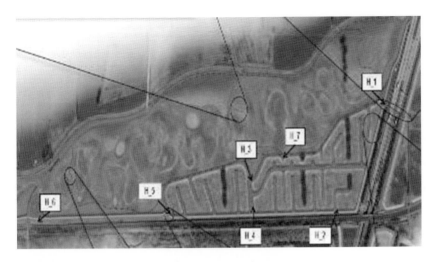

Figure 23.9 Example of one of the largest horizontal flow wetlands in Hoogeveen; top view of the wetland (left), information of this wetland (right) (http://www.climatescan.nl/page?details=91).

23.3.5 Choosing Best Location(s) of Wetlands on Industrial Areas

Since the stormwater of industrial areas can be polluted, constructed wetlands or other purification methods are implemented as end-of-the-pipe measures at drainage outlets. Surface flow of stormwater on the surfaces of the industrial sites should also be taken into account: at times of intensive rainfall (which will become more frequent due to climate change) or calamity situations during firefighting. These polluted surface stormwater flows can potentially be an environmental disaster to soil, groundwater and or surface water. In order to map these storm water flows, a Digital Elevation Model (DEM) or quick scan can be performed to select the right location(s) for a purification method such as constructed wetlands (Figure 23.12).

Table 23.3 Monitoring results of influent and effluent Hoogeveen 2011–2016.

Parameter	Unit	Influent compartment 1	Influent compartment 2	Effluent compartment 1 (influent compartment 3)	Effluent compartment 2 (influent compartment 3)	Effluent compartment 3	Effluent standard used for the design (based on MKN / MTR in 2011)
BOD	mg/l	1.8	3.6	2.2	4.7	2.0	
Phosphorus							
Phosphate-P	mg/L	0.05	0.13	0.06	0.11	0.07	
Total-P	mg/L	0.11	0.27	0.12	0.34	0.16	0.15
Nitrogen							
Ammonium	mg/L	0.34	0.77	0.34	1.18	0.40	
N-Kjeldahl	mg/L	1.2	2.3	1.4	3.0	1.4	
Nitrite	mg/L	0.02	0.03	0.02	0.02	0.02	
Nitrate	mg/L	0.25	0.38	0.54	0.21	0.32	
Total-N	mg/L	1.4	2.7	1.9	3.2	1.7	2.2
Metals							
Aluminium	μg/L	98	173	79	59	71	
Barium	μg/L	34	35	33	32	33	
Calcium	mg/L	50	34	47	41	47	
Cobalt	μg/L	0.3	0.3	0.3	0.2	0.2	
Chromium	μg/L	1.4	1.7	1.3	1.6	1.4	3.4
Copper	μg/L	2.7	4.1	2.0	1.6	1.9	3.8
Iron	mg/L	4.5	4.7	4.3	4.9	3.8	
Magnesium	mg/L	5.5	4.4	4.3	4.6	5.3	
Manganese	μg/L	154	180	165	196	207	
Nickel	μg/L	1.4	1.7	1.3	1.2	1.3	20
Lead	μg/L	1.1	1.9	0.7	0.5	0.5	7.2
Strontium	μg/L	156	121	145	139	145	
Vanadium	μg/L	1.4	1.9	1.2	1.0	1.1	
Zinc	μg/L	13	21	12	10	10	7.8

Most quick scans such as CLOUDS (Calamity Levels Of Urban Drainage systems) are based on only the following readily available data [15]:

- Accurate DEM (Digital Elevation Model) with four points per square meter and a vertical accuracy of several centimeters; this provides an insight in the surface elevation.
- GIS-map with houses, streets and waterways.

The resulting maps show the expected water depths for cloudbursts and the main stream lines of the above groundwater flow during intensive rainfalls or firefighting.

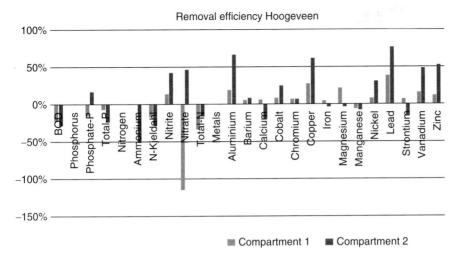

Figure 23.10 Removal efficiency of the Constructed wetland at Hoogeveen.

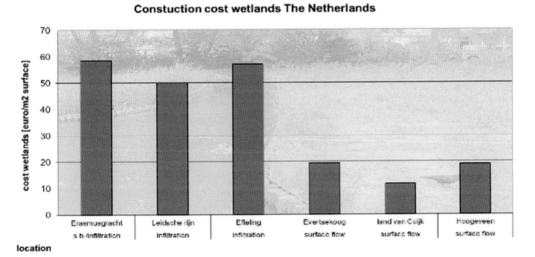

Figure 23.11 Monitored construction costs of some wetlands in the Netherlands [1].

23.4 Innovation in Monitoring Wetlands

Most efficiency studies on wetlands look at inputs and outputs of the water quantity and quality as the main parameter for determining the removal efficiency or hydraulic capacity of wetlands. Two short examples are given in this paragraph as innovating monitoring methods on wetlands.

23.4.1 Innovative Determination of Long-Term Hydraulic Capacity of Wetlands

The hydraulic capacity and clogging of wetlands in the long-term are hardly monitored. A new innovating monitoring method to determine the hydraulic capacity of wetlands is full-scale testing.

Figure 23.12 Top: fire at Moerdijk (the Netherlands) leading to enviromental disaster. Bottom: surface model at industrial site to calculate the outlfow of water during intensive rainfall or emission during calamities to locate the best locations for constructed wetlands.

In such a test, a small constructed wetland or bioswales is filled with a certain amount of water, where the detention time is measured. A review of horizontal wetlands and wet bioswales that have been functioning in the Netherlands for more than 10 years showed that the hydraulic discharge and infiltration capacity can still be sufficient without maintenance (Figure 23.13). Most of the small horizontal wetlands and wet bioswales will treat the water within 1 day, emptying the storage capacity for a new stormwater event. In some cases, the construction is altered to induce the detention time when higher removal efficiency was needed.

Figure 23.13 Full-scale testing of horizontal constructed wetland in Purmerend [2]. Reproduced with permission.

23.4.2 Innovating Monitoring of Removal Efficiency and Eco-Scan

When the removal efficiency needs to be improved, more knowledge-based systems are needed. In order to gain a better understanding of spatial issues in constructed wetlands (e.g., thickness of deposits and water quality parameters), an innovative monitoring tool can be applied to determine the water quality and, in addition, the ecological state of the wetlands. The tool is a semi-autonomous underwater drone (Figure 23.14). The drone is equipped with sensors for pressure (depth), temperature, conductivity, nitrate, ammonium, dissolved oxygen and turbidity. In addition to the data from the sensors, the drone can also collect video images, which are used for eco-scans. The ecoscans show detailed information on the development of organisms and vegetation in time from which positive and negative developments may be derived in order to optimize the constructed wetland. The 3D detailed information of water quality, instead of a single grab sample, shows an in-depth investigation of the wetland and possible improvement areas in the wetland.

The underwater drone proved to be a cost-effective tool and gave a quick insight into the spatial variation of selected performance parameters. As a side effect, the drone provides video footage of the underwater ecology and biodiversity. These drones can be navigated to areas within the constructed wetland that are usually omitted in monitoring, thus extending the knowledge on the wetland.

23.5 Conclusions and Recommendations

23.5.1 Conclusions

Regarding the characteristics of stormwater quality and aims for removal efficiency for achieving water quality goals in Europe, purification methods based on only settling as the primarily treatment process, will not be able to achieve the required removal efficiencies. An additional stormwater treatment step with filtration or adsorption will be necessary such as that provided by constructed wetlands. The removal efficiency of constructed wetlands derived from existing monitoring results

Attached Equipment

(1) In-situ TROLL 9500 Sensors:
- Nitrate and Ammonium ISE
- Rugged Dissolved Oxygen

(2) CTD Driver:
- Temperature
- Pressure
- Conductivity

(3) Diving light

(4) HD Video Camera (GoPro 3+)

Figure 23.14 Aqua drones used for water quality and ecoscans as applied in Hoogeveen wetland [16]. Reproduced with permission.

differs from study to study, but is mostly within the ranges of the international literature. Nearly all the results come from input–output studies.

The average monetary cost of implementation of vertical drainage wetlands were in the order of 55 €/field area in contrast to the cost of the horizontal wetlands (15 €/field area).

The hydraulic capacity and removal efficiency are in most cases not monitored because of the lack of budget. Cost-effective innovative monitoring can be acquired. Two examples are regarded in this chapter. Aquatic drones can be a cost-effective solution providing spatial variation of quality and video footage of biodiversity. The hydraulic capacity of small-scale constructed wetlands and

Figure 23.15 Unaccessible wetland in UK with warning signs (left). Augmented reality as an interactive communication tools to show the pros and cons of constructed wetlands and get more engagement from stakeholders (right) [14]. Reproduced with permission.

bioswales can be monitored by full-scale testing. In the research it was found that there was a wide variation of the infiltration capacities, mostly leading to a detention time of less than 2 days.

23.5.2 Recommendations

- In most cases the constructed wetlands were shown or perceived to be effective. Long-term performance, however, remains an issue. New guidelines should be set up for the design, implementation and maintenance for Dutch wetlands.
- DEM models are advised that show the main stream lines of the above groundwater flow during intensive rainfalls or firefighting in order to choose wisely the location(s) of new constructed wetlands.
- Good examples of BMPs are needed for international knowledge exchange (see BMPs of constructed wetlands on industrial areas in this chapter).
- Use of interactive communication tools is recommended to show the benefits (and not highlighting only the risks) of constructed wetlands and to obtain more engagement from stakeholders (Figure 23.15).
- Data and videos from constructed wetlands are available in an online tool (www.climatescan.nl). This tool is available for all, and everybody is encouraged to add functioning SUDs to this public database.

References

1 Boogaard FC, Lemmen G. Facts about the quality of stormwater. In Dutch: De feiten over de kwaliteit van afstromend regenwater), STOWA, 2007.
2 Boogaard FC. Stormwater characteristics and new testing methods for certain sustainable urban drainage systems in The Netherlands. Delft, 2015.

3 Boogaard F, van de Ven F, Langeveld J, van de Giesen N. Selection of SUDS based on storm water quality characteristics. Challenges. 2014; 5(1):112–122.

4 Pitt R. National Stormwater Quality Database (NSQD). Department of Civil and Environmental Engineering, University of Alabama: Tuscaloosa, AL, USA, 2004.

5 Fuchs S, Brombach H, Wei BG. New Database on urban runoff pollution. 5th International Conference Sustainable Techniques and Strategies in Urban Water Management, Lyon, France, 6–10 June 2004, pp. 145–152.

6 Bratieres K, Fletcher TD, Deletic A, Zinger Y. Nutrient and sediment removal by storm water biofilters: A large-scale design optimisation study. Water Res. 2008; 42:3930–3940.

7 Duncan H. Urban stormwater quality: a statistical overview. CRC for Catchment Hydrology; 1999.

8 Stone M, Marsalek J. Trace metal composition and speciation in street sediment. Water Air Soil Pollut. 1996; 87:149–168.

9 Boogaard F, van der Hulst W. Characteristics of stormwater on industrial areas. In Dutch: omgaan met hemelwater bij bedrijfs-en bedrijventerreinen, rapportnummer 2004-23, ISBN 90.5773.257.2, Stowa, Utrecht, 2004.

10 Walker D, Passfield F, Phillips S, Botting J, Pitrans H. Stormwater sediment properties and land use in Tea Tree Gully, South Australia. 17th AWWA Federal Convention, Melbourne, 1997.

11 Boogaard FC. Watermangement applied to westergasfabriekpark. In Dutch: Waterbeheer 21e eeuw toegepast op stadspark, Neerslag 2004/1.

12 Wilson S, Bray R, Cooper P. Sustainable drainage systems, hydraulic, structural and water quality advice; London, 2004, CIRIA C609.2004 RP663 ISBN 0-86017-609-6.

13 Kampf R, van den Boomen R. Natural constructed wetlands between well-treated waste water and usable surface water, waterharmonicas in the Netherlands (996-2012). ISBN 978.90.5773.599.8 rapport 08, 2013.

14 Boogaard F, Tipping J, Muthanna T, Duffy A, Bendall B, Kluck J. Available at https://www.climatescan.nl/uploads/projects/2111/files/231/icud2017%20websbased%20climatescan%20Boogaard%20Bendal%20Muthanna%20Duffy%20Kluck%20Tipping.pdf. Web-based international knowledge exchange tool on urban resilience and climate proofing cities: climatescan. 14th IWA/IAHR international conference on urban drainage (ICUD), (10-15) September 2017 Prague.

15 Kluck J, Boogaard FC, Goedbloed D, Claassen M. Storm water flooding Amsterdam, from a quick scan analyses to an action plan. International water week, Amsterdam, 2015.

16 de Lima RLP, Boogaard FC, de Graaf RE. Innovative dynamic water quality and ecology monitoring to assess about floating urbanization environmental impacts and opportunities. International water week, Amsterdam, 2015.

Part VIII

Managerial and Construction Aspects

24

A Novel Response of Industry to Wastewater Treatment with Constructed Wetlands: A Managerial View through System Dynamic Techniques

Ioannis E. Nikolaou[1] and Alexandros I. Stefanakis[2,3,4]

[1] Business Economic and Environmental Technology Lab, Department of Environmental Engineering, Democritus University of Thrace, Xanthi, Greece
[2] Bauer Resources GmbH, BAUER-Strasse 1, Schrobenhausen, Germany
[3] Department of Engineering, German University of Technology in Oman, Athaibah, Oman
[4] Bauer Nimr LLC, Muscat, Oman

24.1   Introduction

The concept of Corporate Social Responsibility (CSR) has lately gained a great momentum from a range of international institutions (e.g., the UN and the EU). They have launched many institutional documents to assist industries in introducing CSR concerns into their strategic management (e.g., UN Global Compact, the EC's Directives, [1, 2] and GRI-G4 [3]). Such documents outline the general principles for industries to advance their processes, and to civilize the human rights of employees within their boundaries. Another significant influence of institutional documents is on environmental and economic issues in the context of sustainable development.

A variety of definitions describe the CSR content with emphasis on ethical and social aspects (e.g., local community's rights, accountability issues, and anticorruption matters). Dahlsrud [4] identifies 37 definitions of CSR, which put more emphasis on the stakeholder aspect. The most important question is why industries design CSR strategies. Some recognized explanations refer to: (a) legislative and regulatory regime (compliance-driven); (b) profit-seeking actions (profit-driven); (c) environmental sensitivities (environmentally-driven); (d) achieving sustainability goals to simultaneously address environmental, economic and social issues (sustainability-driven); and (e) an overall approach (holistically driven) [5].

Regardless of the extensive concerns that are included in the recent institutional document/guidelines of CSR, some of them have offered certain instructions for wastewater treatment issues as one basic element of environmental protection. The importance of wastewater treatment varies among the industry sectors. Current literature firstly examines the framework in which the concepts of CSR and wastewater treatment could be combined, and secondly looks for the ways in which current CSR guides have addressed wastewater treatment [6, 7].

Furthermore, two general facets are significant to the relative debate; the strategic-regulatory view and the ethical view of industry to adopt CSR and wastewater projects. The former focuses on the view that an industry invests in wastewater treatment practices on a strategic/mandatory basis in order to eliminate potential risks. These might be associated with potential work and social problems and so the regulatory regime is addressed (e.g., health and safety of staff and hygiene risks of the local communities) [8, 9]. The latter explains the implementation of wastewater practices from industries, as a result of ethical motivations to contribute to environmental protection based mainly on the intrinsic value of the natural environment remaining sustainable for future generations, and not merely as an essential tool to improve the financial position of the industry [10].

This chapter proposes a system dynamic model, which relies on the CSR idea, the balanced scorecard context and Constructed Wetlands (CWs) systems, to assist industries and scholars in improving their understanding regarding wastewater, CWs and industries' financial performance. The idea originates from the first facet, which recognizes wastewater treatment strategy as an essential tool for advancing the operation and improving corporate financial performance. The theoretical background of the model proposed has emerged from the Business Case Theory. In particular, the model proposed stems from a portfolio of concepts such as CWs, Carroll's CSR pyramid and the Elkington triple-bottom-line approach, system dynamic theory and the balanced scorecard. These concepts are combined in the model proposed through the system dynamic Stella software. Finally, some scenarios are conducted to test the model.

The rest of the chapter involves four sections. The first section includes the theoretical underpinning of current CSR and CWs literature. The second section describes the frame, which is designed in order to develop a model to assist industries in incorporating CWs into their strategic management. The third section includes some examples, which are based on certain scenarios and, finally, the fourth section describes a typology of decision making. The final section describes the conclusions and discussion.

24.2 Theoretical Underpinning

This section aims to join various parts of the literature of environmental engineering and management into the context of industry. The main goal is to transfer wastewater practices into the strategic management of industries and translate these practices into the financial language of industries. The dichotomy between engineering and management science has often been detected and many attempts have been made to offer a common place to combine such principles [14]. Thus, some crucial, essential and useful bodies of literature are discussed. Firstly, the basis of this chapter and the theoretical background is the description of CWs, where it is proposed as the central practice for industries achieving the wastewater goals through CSR strategies, either as a result of regulatory requirements or as an outcome of strategic options to identify new entrepreneurial opportunities or gain competitive advantage. Secondly, an analysis regarding economic and environmental benefits of adopting CWs will be made. Actually, a limited number of studies have focused on the economic benefits and costs of CWs, mainly in relation to conventional treatment techniques of wastewater. Thirdly, some industrial examples regarding wastewater treatment through CWs are examined, such as the mining industry and the metal industry. Finally, an analysis regarding the CSR concept and wastewater treatment has been made. In this point, an effort has been made to identify in the literature certain practices and indicators, which are utilized by the industry to evaluate wastewater effluents.

24.2.1 Constructed Wetlands – A Short Review

In general, CWs are a human construction, aiming at mimicking the processes of natural wetlands, which are nutrient reservoirs and control areas for organic and inorganic pollutants [12]. There are several physical, biological and chemical processes taking place in CWs during wastewater treatment. The performance of wastewater treatment with CWs can also be influenced by climatic conditions (e.g., temperature variations) [12–14].

A classical classification of CWs, in relation to water flow patterns, is made in two general categories; surface and sub-surface wastewater flow [12, 13, 15]. According to the vegetation type, CWs can be classified into three general categories: (i) free-floating macrophyte-based systems; (ii) emergent macrophyte-based systems; and (iii) submerged macrophyte-based systems [12, 15, 16]. The first category mainly employs free-floating aquatic macrophytes, which aim at removing nutrients and improving the performance of stabilization ponds [17]. The second category includes emergent macrophyte-based systems, which could be developed through three potential situations: free water surface flow, subsurface horizontal flow, and vertical flow. The third category includes a limited number of examples regarding submerged macrophytes for wastewater treatment. The majority propose this type of macrophyte as a suitable tool for wastewater treatment after primary and secondary treatment.

Furthermore, another main classification is between vertical and horizontal flow CWs [12]. The typical construction of horizontal flow CWs includes a large sand/gravel-filled tank with vegetation, which is mainly utilized for secondary or tertiary wastewater treatment. The construction of these systems have been described as *"beneath the soil surface and from the inlet to the outlet of a gravel bed planted with wet-land vegetation"* [18–20]. The vertical flow CWs consist of essentially three components, such as wastewater pretreatment, the CWs beds and an effluent ditch [12, 14]. It is worth noting that operational characteristics and design structure could play a critical role in the effectiveness and performance of vertical flow CWs. Stefanakis and Tsihrintzis [14] also place emphasis on hydraulic load, the plant species and wastewater source as critical factors which influence the performance of CWs.

24.2.2 Constructed Wetlands: An Economic–Environmental Approach

Today, CWs are considered a low cost and environmentally friendly solution for treating wastewater loads of municipalities and industries. Actually, the majority of current studies have focused on examining the prospects of municipalities to utilize such methods and comparing the financial costs of these systems with conventional methods of treating wastewater [21, 22]. The conventional methods for municipal wastewater treatment (e.g., activated sludge method, anaerobic digestion) seem to require high financial investments in order to acquire proper infrastructures [12, 13, 23]. At the municipal level, a number of environmental and economic benefits have been proposed, such as improvement of water quality, better vegetation productivity and energy savings [12, 13, 24].

In general, a number of different categories of costs are necessary to be estimated in order to examine the viability of CWs as a solution for wastewater treatment and to be preferable in relation to conventional wastewater methods. Some cost categories of CWs are translated in monetary terms and energy terms [12, 23]. The former analysis takes into account capital costs (e.g., land acquisition, equipment, inflation, and engineering consistency) and operational and maintenance costs (e.g., maintenance, electricity, labor and sludge disposal). The latter analysis includes embodied energy for capital costs (e.g., the energy intensity of real estate, stone and clay products, holding tank, installation, sludge thickener, and sludge drying bed) and embodied energy for annual operational and maintenance cost (e.g., the energy intensity per labor, chemicals and electricity). In this logic, it has been considered that CWs are the most sustainable solution for wastewater recycling because of the low cost, use of solar radiation, and low maintenance needs [12, 13, 25].

Many studies presented the benefits from the development of CWs, such as decreasing of industries' pressures on various aspects of the natural environment. For example, a range of success stories in the Netherlands has been examined and the high improvement in water quality identified, as a result of decreasing nutrient budgets, COD and BOD, bacterial pollution and contributing of plant uptake and soil bacterial to the total nutrient removal [26]. From an ecological point of view, it is argued that the operation of CWs facilitates water reuse and contributes to food chain sustainability of phyto- and zooplankton [27]. In particular, they outline the transformation of nutrients in wastewater to plant biomass. Additional benefit of CWs is the creation of a new habitat for ecosystems [12].

24.2.3 Constructed Wetlands: An Industrial Viewpoint

The CWs have lately been recognized as an effective environmental technology for various industries to treat their wastewaters [12, 28, 29]. This book already provides different industrial sectors with CWs applications. Some good examples of CWs have been implemented in the mining industry, which focus mainly on restricting acid mine drainage, one of the most significant industrial pollutants [30]. The findings show that CWs play a critical role in the efforts of the mining industry to remove heavy metals and stop the spreading of metal contamination in the water surface and subsurface water bodies.

Analogous suggestions for adopting CWs to treat wastewater of industrial sites have been made by many scholars [31–34]. Khan et al. [28] implemented pilot CWs systems in an industrial estate in Pakistan, where many hazardous materials are detected in waters, such as effluents from textiles, ghee, cooking oil, marble, steel, soap and plastics industries. The findings showed that Pb and Ni concentrations have been reduced through CWs. Vrhovšek et al. [31] have developed three CWs pilot systems in Slovenia to purify wastewater from the food processing industry. The findings showed that COD and BOD have been efficiently reduced. Groundwater contaminated with petroleum derivatives and hydrocarbons has also been effectively remediated using wetland systems in a large industrial site in Germany [33].

Finally, some good examples have been identified in the agricultural sector, which faces various pollution problems regarding water resources such as agricultural wastewater and agricultural cropland runoff [34–37]. These agricultural effluents play a critical role in soil contamination and the quality of the local water reservoir. The problem with agricultural pollution is its non-point source character and, of course, the difficulty to commit liable agents to undertake suitable practices to restore the environmental damage. An appropriate policy, for the agricultural sector to address these problems, is the construction of CWs at various agricultural points.

24.2.4 CWs Through a CSR Glance

Numerous industrial associations, shareholders, investors and managers have lately highlighted that water issues are very important for industries' day-to-day operations. Various interesting justifications have been given to enlighten their propensity to invest those industries, which take care of their responsibilities in wastewater management and, in general, in the quality of water resources. The two prevailing explanations rely on: (a) the voluntary trend of agents to adopt wastewater treatment systems and water reuse, recycling and reduction strategies (motivation-based strategy); and (b) the compulsory adoption of wastewater systems to avoid capital costs from penalties due to non-compliance with regulations (regulatory-based strategy).

The former strategy implies industries being proactive, whereby they voluntarily invest in wastewater treatment systems in order to exploit new opportunities and reap benefits. The potential benefits might be classified in three general groups, such as risk reduction (e.g., water risks, financial risks), operational costs decreasing (e.g., less costs for fresh water) and gaining benefits (e.g., reputation). The minimization of industrial risks may pertain either to the decrease of financial costs for compliance with regulatory requirements and the avoidance of paying possible penalties due to environmental legislation violations, as well as the elimination of the probability of facing a severe environmental regulatory regime in the future. Similarly, industries could face risks, such as water scarcity (now or in the future) for their production processes or some discontinuities in their supply chain, where key raw materials for industry's operations are derived from countries affected by extreme weather events (e.g., droughts or floods). Thus, the voluntary strategies of industries aim to address such problems by treating wastewater through the classical triple goal: recycling, reusing and reduction of water resources.

Another significant goal of industries is the improvement of their reputation and identification of a niche market with environmentally sensitive consumers. A representative example is the mining industry, which voluntarily implements wastewater practices to obtain the *"license to operate"* (increases reputation) from local communities. This is a classical problem which the mining industry faces, since people of the local community strongly criticize its responsibility to the natural environment, even in the case where a complete obeyance of the mining industry to the regulatory requirements takes place [38]. The adoption of such strategies assists the mining industry in gaining the confidence of local communities and avoiding potential barriers on its operation.

A reactive nature of the second strategy implies that industries implement wastewater treatment systems in order to attain the requirements of the law. The economic community considers this drift as long-term, since industries plan strategies for protecting the natural environment on a compulsory basis and miss the potential benefits from a voluntary environmental management strategy like the first mover advantage and win–win benefits [39]. This trend is taken into consideration largely in engineering and policy making scientific fields, which face industries as good citizens who are liable to obey with law. Nevertheless, in the real world, industries face regulations and legislations as probable risks (regulatory risks), which should be addressed by abstracting financial resources from long-run organized investment plans and by losing a part of the competitiveness regarding their competitors.

It is worth noting that, the proactive strategy gained greater attention in the relative literature, given that industries have lately had the opportunity to formally certify environmentally friendly strategies and, of course, to offer a clear and reliable signal to consumers who are willing to buy from (and reward such) better environmentally friendly industries, such as China Water Conservation Certification, WaterSense, and the Water Efficiency Product Labeling Scheme. Despite these separate water labels, wastewater and water management strategies are also identified as parts of general CSR strategies of industries. Namely, industries adopt wastewater systems to treat their wastewater effluents as a general industry strategy to contribute to sustainable development. A typical policy has been launched by the EU [1], which proposes a triple-bottom-line approach to encourage and facilitate industries to make clear contributions to sustainable development by *incorporating economic, environmental (including wastewater treatment strategies)* and social concerns into their strategic management. Moreover, many international organizations (e.g., UN, GRI) have proposed CSR guidelines and CSR labels to make formal and reliable the voluntary efforts of industries to address environmental problems. Current literature provides evidence and success stories regarding positive influences of such strategies on reputation and market share of industries [40]. The handling of wastewater effluents should also be a significant task for industries, since under certain conditions many new entrepreneurial opportunities seem to arise such as water recycling, water reuse, selling recycling water in the agricultural sector and new products developing from sludge.

24.3 Methodology

This section analyzes some fundamental steps of the methodology. Firstly, the structure of the methodology proposed is shown by diagrammatic illustration. Secondly, a mixture of CSR strategy and CWs technology is made by using Carroll's CSR pyramid, Elkington triple-bottom-line approach and literature of wastewater/CWs. Thirdly, a connection of the CSR-CWs agenda and balanced scorecard idea has been made. Fourth, a system dynamic decision making model is designed by using balanced scorecard CSR-CWs agenda. Finally, three basic scenarios have been tested as exemplars for describing certain types of decision makers in such topics.

24.3.1 Research Structure

This section provides a concise explanation of the fundamental steps of the methodology for developing a dynamic model to assist decision making of industries in designing CWs strategies within a CSR context. Figure 24.1 includes five key steps of the methodology. The first step (Figure 24.1: S_1) includes a short review of CSR and CWs literature. The aim of this step is to make

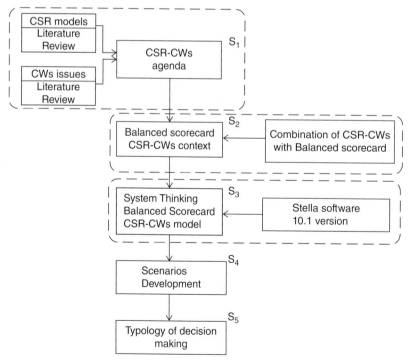

Figure 24.1 The research structure.

a combination of these literature bodies in a novel CSR-OHS agenda. Exclusively, Carrols' pyramid, the triple-bottom-line approach [41, 42] and CWs issues will be discussed [12, 14]. It also provides a nexus between these types of literature by outlining a new research agenda including critical parameters regarding CSR-CWs issues.

The second step (Figure 24.1: S_2) aims to bring together the key parameters of the CSR-CWs agenda with a balanced scorecard idea, in order to assist industries in improving the decision making regarding wastewater issues within a CSR context. The balanced scorecard is considered a suitable tool to link qualitative and quantitative information as well as financial and non-financial information [43].

The third step shifts from a static to a dynamic version of the balanced scorecard. In particular, a system dynamic model is developed, which is based on CSR-CWs balanced scorecard. The fourth step provides four different scenarios to examine what will happen *("what if")* if some critical parameters of the CSR-CWs agenda change [44]. Finally, the final step tries to develop some types of decision making.

24.3.2 The CSR-CWs Agenda

An extensive literature has focused on the content of CSR. A variety of different CSR terms are utilized, such as corporate sustainability, ethical firm, triple-bottom-line, corporate citizenship, sustainable entrepreneurship, and stakeholder firm with a different focus on social and environmental aspects [4, 45]. The CSR is considered a luminous and suitable theoretical construction for strengthening the bonds between industries and societies [46]. Several theoretical explanations have been suggested to enlighten CSR: economic-based theories, political-based theories, social-based theories and ethical-based theories [47]. The most popular terms that have been recently utilized are CSR and corporate sustainability (CS). These terms are sometimes considered identical and sometimes divergent [48]. The CSR emphasizes mainly social and ethical aspects, while the CS focuses primarily on protection of natural resources. Finally, we could say that these terms are met within the triple-bottom-line approach [49].

This methodology hypothesizes that industries incorporate CW issues into a CSR agenda by using a combination of triple-bottom-line approach and Carroll's CSR pyramid. This approach is based on the Business Case Approach, where CSR is faced as a strategic tool which assists industries in designing and taking advanced decisions for CSR practices and corporate financial positions [50]. Figure 24.2 shows the reasonable and sequential steps for defining a new CSR-CWs agenda. In particular, the top square indicates an effort to combine Elkington's triple-bottom-line approach and Carroll's CSR pyramid model. The Elkingtons's CSR model encompasses issues in relation to the three pillars of sustainable development such as economic, environment and social. The proposed CSR models aim to facilitate a primary assessment of a list of CSR tools and guidelines for emerging CWs and wastewater issues within the triple-bottom-line approach and CSR pyramid. The combination and analysis of such CSR tools might provide a new form of CSR-CWs agenda.

24.3.3 CSR-CWs Balanced Scorecard

This section provides a modified balanced scorecard by incorporating and allocating in its four classical dimension certain issues of the CSR-CWs agenda. The selection of a balanced scorecard is derived from both the suitability and popularity of this tool for corporate strategic management and its ability to bring together financial and non-financial information [43], which is necessary in order to manage

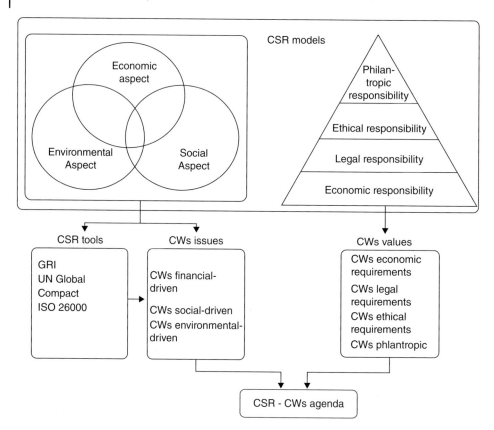

Figure 24.2 CSR-CWs logic.

CSR-CWs issues. A traditional balanced scorecard tool has four dimensions, which assist industries in managing, designing, measuring and changing their performance.

The first dimension is the financial which embraces information regarding revenue growth, costs, financial capital and asset utilization. In the case of the CSR-CWs agenda, this dimension includes mainly costs and revenues which are emerged from CSR-CWs agenda implementation. The second dimension deals with customers issues from several different facets such as customer satisfaction, customer retention, new customer acquisition, customer profitability [43]. This could be translated under CSR-CWs as customer retention thanks to the adoption CSR-CWs topics, to attract sensitive customers for issues of CSR-CWs, and higher profitability of customers who demand CSR-CWs topics. However, Kaplan and Norton [43] support the fact that the customers' value might be associated with three factors, product/service attributes (e.g., quality and price), customer relationship (e.g., brand equity) and image and reputation (e.g., trust). Consequently, a critical question is how do these factors affect and are affected by CSR-CWs topics? In the case of CSR, Tsalis et al. [51] identify that the customer dimension should be expanded by incorporating several groups of the industries' stakeholders, since the CSR literature recommends shifting from one group of stakeholders (e.g., shareholders) to many groups of stakeholders (e.g., customers, employees, local community) [52, 53]. The model proposed considers for this dimension two main groups: customers and local communities.

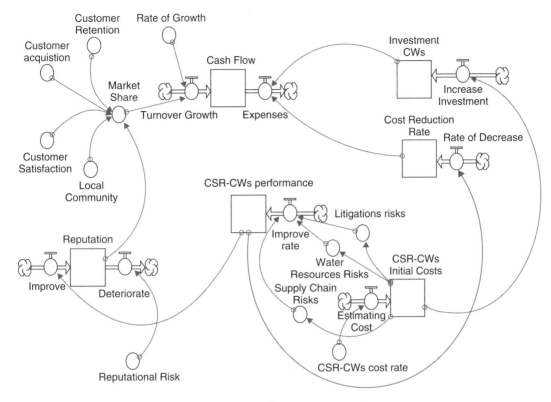

Figure 24.3 The CSR-CWs model (designed in Stella Software Version 10.0.6).

The third dimension pertains to the internal business process, which focuses on the holistic value chain consisting of three sequential parts: innovation cycle, operation cycle, and service cycle. This perspective might be translated in the language of the CSR-CWs agenda as incorporate topics regarding CSR and CWs into innovation, operation and service cycle. The final dimension is learning and growth, where industries try to find ways to cover gaps regarding the outdated capabilities of employees, improve information channels among different external stakeholder groups (e.g., local communities) and provide equal opportunities for employees.

24.3.4 CSR-CWs Balanced Scorecard System Dynamic Model

Many of the current strategic management tools offer a dynamic depiction of the state by using feedback loops to examine the performance and efficiency of strategy implementation. It is supported that the balanced scorecard is an efficient tool for designing and achieving short, meso- and long-run strategic management goals of industrial units. Despite the promises of the initiators of the balanced scorecard for its potential to link lagging and leading indicators [43], the long-run decision of making feedback loops between lagging and leading indicators will be better achieved by dynamic advance of indicators through a system dynamic software [54, 55]. In this logic, some indicative academic works in the field of corporate sustainability and CSR have been launched by many scholars [51, 56, 57].

Figure 24.3 illustrates a system dynamic balanced scorecard for the CSR-CWs agenda. It is hypothesized that the core *Financial Dimension* includes two fundamental variables: the *Cash Flow* of industry and the *CSR-CWs Initial Cost*, which is related to initial construction costs of the CWs strategy of the industry, in order to treat wastewater loads (e.g., land acquisition, design and implementation). In this variable, many other costs are also included (through inflow procedures), such as maintenance and operational costs which increase or decrease the final accumulated costs in this stock during the time horizon of operation. Additionally, a progress in CSR-CW performance might imply a positive association with *Market Share* and *Turnover Growth* of industry. There is evidence that industries implementing a high-quality CSR agenda (including excellent wastewater treatment) seem to draw reputational advantages and attract new consumers by strengthening the loyalty and trust of consumers. Additionally, a well-designed CSR-CWs strategy might lead industries to reduce operational costs in the long-run period. Despite the initial costs of construction and annual costs of maintenance and operation, wastewater treatment with CWs might eliminate water use per annum and wastewater potential penalties. The annual costs could be also decreasing through intellectual capital development and productivity of employees.

The *Stakeholder Dimension* includes *Customer Satisfaction, Customer Acquisition, Customer Retention, Market Share and Local Community*. This implies that minimization of reputational risks (e.g., free wastewater discharge) may positively influence customers' preference for the product from the industry with better CSR-CWs agenda and wastewater performance. A positive relationship could be identified between local community perception for industry and CSR-CWs performance. This means that good wastewater treatment by the industry might provide a clear signal to the local community, which legitimizes the operation of industry (social "license to operate"). It also includes variables in relation to CWs, which affect industry's risks (e.g., Litigation Risk, Water Resource Risks, Supply Chain Risks). The improvement of the overall picture of industries might play a role in the decision of investors who ask for sustainable investments. These potential prospects of stakeholders and industries can explain the positive or negative feedbacks of the model presented in Figure 24.3.

The *Internal Business Process* is consisted by CSR-CWs performance. This implies that the improvement of industries' performance is subsequent to an entire accomplishment of the CSR-CWs goals. The *Organization Learning and Growth* dimension is translated as CSR-CWs Initial Costs and *Investment CWs*. This includes the budget expenditures of industry for CWs investment in training programs to make employees able to face potential wastewater risks. This implies also the creation of new and strengthening the capabilities and resources of industry, which is a result of internal efforts of the industry to face wastewater issues.

24.3.5 Some Certain Scenario Developments

The latest risky and uncertain financial environment, which has come about, has brought scenario planning to the forefront. Many definitions and tools have been suggested in the present literature for scenario planning, such as probabilistic modified trends, thinking analysis, learning, institutive logic, and the La Prospective methodologies [58]. There are two general categories of scenarios; (a) exploratory scenarios, which take into consideration historical data to forecast the prospects of a topic; and (b) normative scenarios, which follow intuitive logic and alternative visions to describe the potential future events [59]. According to Postma and Liebl [60], scenario methodologies follow two essential principles: (a) consistency and (b) causality.

The former (normative) scenario technique is more suitable for the case of the proposed model given that the majority of variables and feedback loops are primarily designed through data arisen from the CSR and CWs literature (mainly from historical and secondary data). This model is based on two criteria to design scenarios. The first criterion pertains to the proactive or reactive driving force of industries to implement CW practices through CSR strategy. A proactive move of industries to use CWs for wastewater treatment might be the result of their existing proclivity to voluntarily adopt CSR strategies in order to moderate potential new legislative requirements for environmental protection or to improve their reputation and profitability. A reactive character of industries might be an inactive response towards legal requirements by incorporating wastewater issues into an industry operation without an organized plan to reap financial and reputational benefits.

The comparison of such criteria offers four significant potential scenarios. The key variables of the first scenario are the *Investment CWs, Cost Reduction Rate* and *Reputation* (Proactive character of Industry – The Business Case Approach). The second scenario is based on the variables *CSR-CWs Cost* and *Reputational Risk* (Proactive character of Industry – The Ethical Case Approach). The third scenario affects the variables of *CSR-OHS Cost* (Reactive character of Industry – The Business Case Approach). The last scenario implies deals with variables of *CSR-CWs Cost* and *Reputational Risk* changes (Reactive character of industry – The Ethical Case Approach).

24.4 Test of Scenarios and a Typology Construction for Decision Making

Two tasks have been carried out in this section. The first of these is to demonstrate the findings of the four scenarios as emerging from system dynamic Stella software. The latter task is to develop a typology for the decision making of industries' managers.

24.4.1 Scenario Analysis

24.4.1.1 The Proactive Industry – The Business Case Approach

The first scenario examines the tendencies of some significant variables (*Reputation, CSR-CWs Performance* and *Cash Flow*) in the case where some key variables have gradually changed (*Investment CWs, Cost Reduction Rate* and Improvement in *Reputation*). Empirical findings indicate that industries voluntarily employ CSR-CWs issues in order to advance their status and reduce future *Cash Flow* expenditures. Figure 24.4 depicts that the industry raises *Cash Flow* and improves *CSR-CWs Performance* and *Reputation*.

The findings are justified by the rationality that industries shift voluntarily to implement an innovative CSR-CWs agenda mainly to go beyond the regulatory requirements and to exploit new financial opportunities. Many scholars support that under certain conditions voluntary strategies will encourage industries to identify win–win solutions [39]. Namely, industries could recognize financial (increase revenues) and environmental benefits (decrease BOD, COD). Indeed, the discussion of this relationship is widespread in the corporate environmental management and CSR literature [61].

24.4.1.2 Proactive Industry – The Ethical Case Approach

The ethical case approach of industry refers to their propensity to voluntarily adopt CSR-CW strategies with the aim of solving societal matters rather than to seek for financial benefits. The relationship

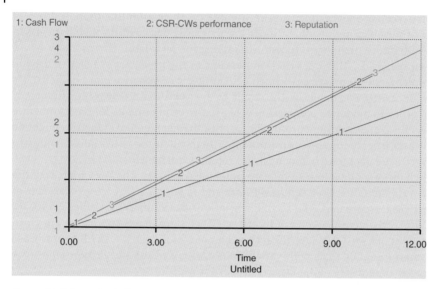

Figure 24.4 Proactive Industry – Business case trends.

between ethical and social responsibility matters seems dominated by SMEs and mainly small family industries, where owners are closer to local communities. These cases outline new social norms and cultural values between industry owners and society agents (non-economic values) [62]. Thus, local industries will invest in the CSR-CWs agenda by increasing the *CSR-CWs Cost* and reducing the perceived *Reputational Risk* of societies (Figure 24.5). This also implies a constant *Reputation*, improved *CSR-CWs Performance* and enhanced *Cash Flow*. Although the hypothesis for constant *Reputation* seems to be hardly conservative, nevertheless it is based on the rationality that industries are not

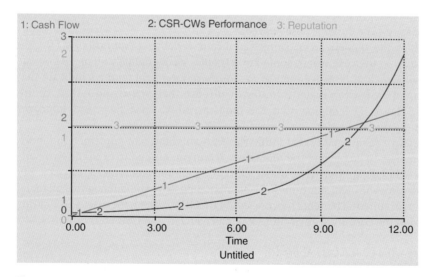

Figure 24.5 Proactive Industry – Ethical approach.

primarily interested in translating their good image into money. This implies that these industries do not invest in communication channels with consumers since the ethical element prevails over the financial element.

24.4.1.3 Reactive Industry – The Business Case Approach

A high set of industries implement CSR strategies in order to comply with legal requirements. This is considered as a passive reaction of industries only satisfying the minimum requirements of laws regarding social and environmental topics. Many scholars have lately supported the compulsory nature of CSR strategies [63]. This scenario assumes that only the cost of CWs is changed. This implies an increase in *Cash Flow, CSR-CWs Performance* and a decrease in *Reputation* (Figure 24.6). The fall of industry *Reputation* might be due to the lack of necessary promotion of *CSR-CWs Performance*, since industry adopts these strategies only as regulatory requirements and not as a beneficial strategy which aims to offer financial profits.

24.4.1.4 Reactive Industry – The Ethical Case Approach

The logical basis of this scenario is that industry adopts reactive CSR-CWs strategies under an ethical setting. Specifically, the differentiation of this scenario from the previous one is that industries always comply with the law and respond to each requirement. Actually, industries confront many dilemmas in their everyday operations. In the Business Case Approach, industries implement strategies only in response to the requirements of laws, while the second option, the Ethical Case Approach, is the capability of choice, whereby industries may avoid (temporarily or permanently) all or part of the legal requirements [64]. This scenario shows an increase in *CSR-CWs Cost* (to satisfy every requirement of the law) and zero *Reputational Risk* given that industries encompass the entire aspects of the law and unexpected or extreme events may be unlikely to occur. This scenario has been differentiated from the previous only in the trends of *Reputation*. The *Reputation* has a steady route as consumers recognize the efforts of industries to be honest by complying with the legal requirements in contrast

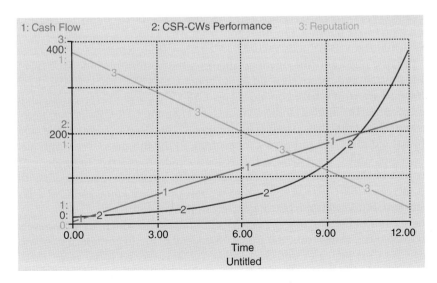

Figure 24.6 Reactive industry – The Business Case approach.

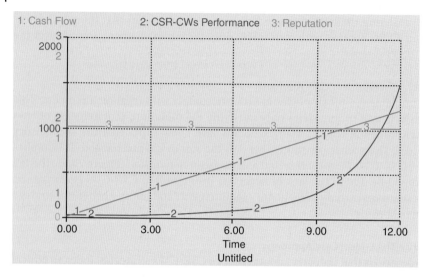

Figure 24.7 Reactive Industry – The Ethical Case approach.

to the previous scenario, where society is aware of the unethical operation of industries (Figure 24.7). However, of course, consumers do not reward industries only because of their compliance with the law. Additionally, it is obvious that the *CSR-CWs Performance* and *Cash Flow* will increase.

24.4.2 A Typology of Industry Decision Making in CSR-CWs Agenda

The scenario analysis indicates some basic types of decision making and strategy of industries in their effort to both responses to the needs of local societies for a clean natural environmental and a modern economy for higher competitiveness. Figure 24.8 classifies four types of industry decision making-strategy. The horizontal axis outlines the direction from The Ethical Case Approach to Business Case Approach, while the vertical axis records a direction from Reactive to Proactive Approach of industries. The first quadrant (the opening square in the bottom) includes the types of industries who are both ethically and reactively – driven to adopt CSR-CWs issues. Under these circumstances, the hypothetical scenario is based on increasing *CSR-CWs Costs* and the elimination of *Reputational Risk* (tend to zero). The findings indicate that *Cash Flow* and *CSR-CWs Performance* have an increasing trend, while *Reputation* has a straight line progress. The second quadrant (the next square in the horizontal axis) indicates that industries concentrate to the Business Case Approach and reactively – driven to adopt CSR-CWs issues. In this scenario, industries strive to implement CSR-CWs issues with the aim of aligning an industry's operation with legislative requirements. To attain this aim, industries increase *CSR-CWs Costs*. The results show that *Cash Flows* and *CSR-CWs Performance* could be increased, but *Reputation* may negatively slope.

The third quadrant (the second in the horizontal axis) refers to industries which are proactively and ethically – driven to implement issues from the CSR-CWs agenda. The hypothetical scenario conjectures that the *CSR-CWs Cost* increases and *Reputational Risk* is tended to be zero. These strategies of industries might lead to a rise in *Cash Flow* and *CSR-CWs Performance* and a steady trend of *Reputation*. The final quadrant (diatonically to the top) indicates industries which proactively

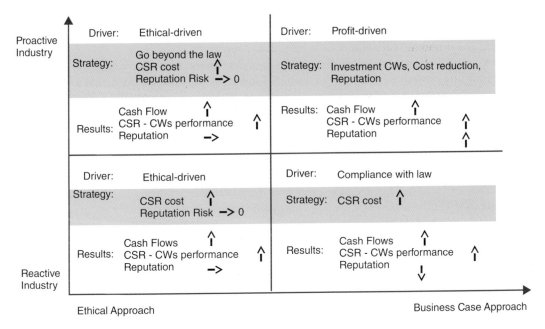

Figure 24.8 Various types of industries decision making.

invest in CSR-CWs issues so as to exploit financial benefits. It is hypothesized that *Investment CWs*, *Cost Reduction Rate* and *Reputation* will increase. The results show that key parameters (*Cash Flow*, *CSR-CWs Performance* and *Reputation*) will progress positively.

24.5   Conclusion and Discussion

This chapter develops a multi-step decision making model to assist industries in introducing Corporate Social Responsibility (CSR)–Constructed Wetlands (CWs) topics into their strategic management to face wastewater problems. This model makes five contributions to the current literature. The first contribution is related to a novel frame developed to make the option of wastewater issues within CSR context straightforward and unbiased. This frame offers guarantees to the industries for selecting wastewater issues both as a consequence of regulation and self-regulation. The Carroll's pyramid and Elkington's triple-bottom-line approach are two vital CSR models, which structure the starting point of selection the critical wastewater topics.

The second contribution concerns the transformation of a strategic management tool from industry's wide-ranging needs for facing wastewater problems. Thus, a strategic balanced scorecard is developed for CSR-CWs issues. This transfers the issues regarding CWs into four dimensions of the balanced scorecard: Financial, Costumers (Stakeholder dimension for the purpose of this chapter), Learning and Growth and Internal Processes. This contributes to the strategic management in the CSR-CWs agenda and facilitates managers to design appropriate and sufficient strategic plans to cope with potential risks [65].

The third contribution focuses on developing a system dynamic model, which is based on a balanced scorecard outline. The dynamic relationships of CWs – wastewater variables and general management and economic variables of industries have been made.

The fourth contribution pertains to the design and development of four normative scenarios regarding the future prospects of some key variables (*Cash Flow, CSR-CWs Performance* and *Reputation*) under certain strategic conditions (e.g., increasing *CWs Investment* and *CSR-CWs Costs*). This assists scholars and practitioners to understand many different strategies when proposing or implementing environmental strategies. The findings explain that the proactive adoption of CSR-CWs by industries might present many benefits such as increasing *Cash Flow, CSR-CWs Performance* and *Reputation*. The reactive response of industries to CSR-CWs issues seems to provide benefits on *CSR-CWs Performance* and *Cash Flow*; nevertheless *Reputation* remains constant. Some interesting differentiations are identified between the cases where industries are motivated by ethical criteria or by economic criteria (The Business Case Approach). The latter case seems to offer more benefits for industries.

The last contribution relies on the fact that corporate environmental management issues are associated with environmental engineering topics. An attempt has been made to connect wastewater issues with management of industry to improve clear economic parameters such as *Reputation, Cash Flow* and *CSR-CWs Performance*.

Obviously, like any other study, some limitations exist here, such as some feedback loops (e.g., *Reputation*) intuitively designed. Namely, the percentage influences of cost implementation of CWs and *CWs Performance* (e.g., reduction of BOD, COD) with *Reputation* and *Customers* are made through literature review, experience and mental perception of modelers and not through accurate relationships. It is worth noting that this is not an important weakness for management science and the system theory field since the majority of variables are qualitative and the interest focus is on identifying certain trends and not accurate numbers. From the engineering science viewpoint, accurate figures are necessary and this could be a good field for future research, where many studies could be conducted on a specific industry to identify accurate quantitative relationships in addition to environmental engineering (e.g., BOD, COD) with economics (e.g., Costs) and management variables (e.g., reputation, customer attractiveness). Another limitation is that the majority of data used have been collected from the current literature and similar examples, either ecological engineering literature or corporate environmental management literature. This could be a good field for future research, where some studies have been carried out to draw primary data from a specific industry.

References

1 EC. Green Paper: Promoting a European Framework for Corporate Social Responsibility. Commission of the European Communities, Brussels; 2001.

2 EC. Communication from the Commission to the European Parliament, the Council, the European Economic and Social Committee and the Committee of the Regions. Brussels; 2011.

3 Global Reporting Initiative. GRI Sustainability Reporting Guidelines (G4) and the European Directive on non-financial information disclosure. GRI; 2014

4 Dahlsrud A. How corporate social responsibility is defined: an analysis of 37 definitions. Corp Soc Resp Env Ma. 2008; 15:1–13.

5 Van Marrewijk M. Concepts and definitions of CSR and corporate sustainability: between agency and communion. J Bus Ethics. 2003; 44:95–105.

6 Wilburn K, Wilburn R. Using global reporting initiative indicators for CSR programs. J Global Sustain. 2013; 4(1):62–75.

7 Cook N, Sarver E, Krometis LA. Putting corporate social responsibility to work in mining communities: exploring community needs for central wastewater treatment. Resources. 2015; 4(2):185–202.

8 Larsson DGJ, Fick J. Transparency throughout the production chain – a way to reduce pollution from the manufacturing of pharmaceuticals? Regul Toxicol Pharm 2009; 53:161–163.

9 Mosse KPM, Patti AF, Christen EW, Cavagnaro TR. Review: winery wastewater quality and treatment option in Australia. Austr J Grape Wine R. 2011; 17(2):111–122.

10 Siltaoja ME. Value priorities as combining core factors between CSR and reputation – a qualitative study. J Bus Ethics. 2006; 68(1): 91–111.

11 Evangelinos K, Allan S, Jones K, Nikolaou EI. Environmental management practices and engineering science: a review and typology for future research. Integr Environ Asses. 2014; 10(2):153–163.

12 Stefanakis AI, Akratos CS, Tsihrintzis VA. Vertical Flow Constructed Wetlands: Eco-engineering Systems for Wastewater and Sludge Treatment. Amsterdam: Elsevier Science; 2014.

13 Stefanakis AI. Constructed Wetlands: description and benefits of an eco-tech water treatment system. In: McKeown E, Bugyi G, eds. Impact of Water Pollution on Human Health and Environmental Sustainability. Hershey – Pennsylvania, USA: Information Science Reference (IGI Global); 2016, pp. 281–303.

14 Stefanakis AI, Tsihrintzis VA. Effects of loading, resting period, temperature, porous media, vegetation and aeration on performance of pilot-scale vertical flow constructed wetlands. Chem Eng J. 2012; 181–182:416–430.

15 Vymazal J. Removal of nutrients in various types of constructed wetlands. Sci Total Environ. 2007; 380(1–3):48–65.

16 Brix H. Use of constructed wetlands in water pollution control: historical development, present status, and future perspectives. Water Sci Technol. 1994; 30(8):209–223.

17 Reddy KR, De Busk TA. State-of-the-art utilization of aquatic plants in water pollution control. Water Sci Technol. 1987; 19(10):61–79.

18 Rizzo A, Langergraber G, Galvão A, Boano F, Raelli R., Ridolfi L. Modelling the response of laboratory horizontal flow constructed wetlands to unsteady organic loads with HYDRUS-CWMI. Ecol Eng. 2014; 68:209–213.

19 Stefanakis AI, Tsihrintzis VA. Effect of outlet water level raising and effluent recirculation on removal efficiency of pilot-scale Horizontal Subsurface Flow Constructed Wetlands. Desalination. 2009; 248(1–3): 961–976.

20 Stefanakis AI, Akratos CS, Tsihrintzis VA. Effect of wastewater step-feeding on removal efficiency of pilot-scale Horizontal Subsurface Flow Constructed Wetlands. Ecol Eng. 2011; 37(3):431–443.

21 Greenway M. Nutrient content of wetland plants in constructed wetlands receiving municipal effluents in tropical Australia. Water Sci Technol. 1997; 35:135–142.

22 Merlin G, Pajean JL, Lissolo T. Performances of constructed wetlands for municipal wastewater treatment in rural mountainous areas. Hydrobiologia 2002; 466(1):87–98.

23 Ko JY, Daya JW, Lanea RR, Day JN. A comparative evaluation of money-based and energy-based cost–benefit analyses of tertiary municipal wastewater treatment using forested wetlands vs. sand filtration in Louisiana. Ecol Econ. 2004; 49:331–347.

24 Day JW, Arancibia AY, Mitsch WJ, Lara-Dominguez AL, Day JN, Ko JY, Lane RR, Lindsey J, Lomeli DZ. Using ecotechnology to address water quality and wetland habitat loss problems in the Mississippi basin: a hierarchical approach. Biotechnol Adv. 2003; 22:135–159.

25 Welker AG, Dougherty JM, McHenry JL, Van Loon WA. Treatment variability for wetland wastewater treatment design in cold climates. Ecol Eng. 2002; 19:1–11.

26 Verhoeven JTA, Meuleman AFM. Wetlands for wastewater treatment: opportunities and limitations, Ecol Eng. 1999; 12:5–12.

27 Rousseau DPL, Lesage E, Story A, Vanrollghem PA, De Pauw N. Constructed wetlands for water reclamation. Desalination. 2008; 218:181–189.

28 Khan S, Ahmad I, Shah MT, Rehman S, Khaliq A. Use of constructed wetlands for the removal of heavy metals from industrial wastewater. J Environ Manage. 2009; 90:3451–3457.

29 Shubiao W, Wallace S, Brix H, Kuschk P, Kipkemoi Kirui W, Masi F, Dong R. Treatment of industrial effluents in constructed wetlands: Challenges, operational strategies and overall performance. Environ Pollut. 2015; 201:107–120.

30 Sheoran AS, Sheoran V. Heavy metal removal mechanism of acid mine drainage in wetland: a critical review. Mineral Eng 2006; 19:105–116.

31 Vrhovšek D, Kukanja V, Bulc T. Constructed Wetland (CW) for industrial wastewater treatment. Water Resour. 1996; 30(10):2287–2292.

32 Chen TY, Kao CM, Yen TY, Chien HY, Chao AC. Application of a constructed wetlands for industrial wastewater treatment: a pilot-scale study. Chemosphere. 2006; 64:497–502.

33 Stefanakis AI, Seeger E, Dorer C, Sinke A, Thullner M. Performance of pilot-scale horizontal subsurface flow constructed wetlands treating groundwater contaminated with phenols and petroleum derivatives. Ecol Eng. 2016; 95:514–526.

34 Tatoulis T, Stefanakis AI, Frontistis Z, Akratos CS, Tekerlekopoulou AG, Mantzavinos D, Vayenas DV. Treatment of table olive washing water using trickling filters, constructed wetlands and electrooxidation. Environ Sci Pollut R. 2016; 1-8. DOI 10.1007/s11356-016-7058-6.

35 Vymazal J, Březinová T. The use of constructed wetlands for removal of pesticides from agricultural runoff and drainage: A review. Environ Int 2015; 75:11–20.

36 Rozema ER, Van der Zaag AC, Wood JD, Drizo A, Zheng Y, Madani A, Gordon RJ. Constructed Wetlands for Agricultural Wastewater Treatment in Northeastern North America: A Review. Water. 2016; 8(5): 173.

37 Sultana MY, Akratos CS, Vayenas DV, Pavlou S. Constructed wetlands in the treatment of agro-industrial wastewater: A review. Hemijska industrija 2015; 69(2): 127–142.

38 Moffat K, Zhang A. The paths to social license to operate: An integrative model explaining community acceptance of mining. Resour Policy 2014; 39:61–70.

39 Porter EM, Van der Linde C. Green and competitive: ending the stalemate. Harvard Business Rev. 1995; 121–134.

40 Lii YS, Lee M. Doing right leads to doing well: when the type of CSR and reputation interact to affect consumer evaluations of the firm. J Bus Ethics. 2012; 105(1):69–81.

41 Pedersen ER. Modelling CSR: how managers understand the responsibilities of business towards society. J Bus Ethics. 2010; 91:155–166.

42 Carroll AB, Shabana KM. The business case for corporate social responsibility: a review of concepts, research and practice. Int J Man Rev. 2009; 86–110.

43 Kaplan RS, Norton DP. Linking the balanced scorecard to strategy. Calif Manage Rev. 1996; 39(1): 53–79.

44 Borjeson L, Hojer M, Dreborg KH, Ekvall T, Finnveden G. Scenario types and techniques: towards a user's guide. Futures. 2006; 38:723–739.

45 Van Marrewijk M. Concepts and Definitions of CSR and Corporate Sustainability: Between Agency and Communion. J Bus Ethics. 2003; 44(2):95–105.

46 Okoye A. Theorising corporate social responsibility as an essentially contested concept: is a definition necessary? J Bus Ethics. 2009; 89:613–627.

47 Garriga EM, Mele D. Corporate Social Responsibility Theories: Mapping the Territory. J Bus Ethics. 2004; 53:51–71.

48 Hahn T, Scheermesser M. Approaches to Corporate Sustainability among German Companies. Corp Soc Resp Env Ma. 2006; 13:150–165.

49 Elkington J. Partnerships from cannibals with forks: The triple bottom line of 21st-century business. Environ Qual Manage. 1998; 8(1):37–51.

50 Weber M. The business case for corporate social responsibility: A company-level measurement approach for CSR. Eur Manag J. 2008; 26:247–261.

51 Tsalis AT, Nikolaou EI, Grigoroudis E, Tsagarakis PK. A dynamic sustainability balanced scorecard methodology as a navigator for exploring the dynamics and complexity of corporate sustainability strategy. Civ Eng Environ Syst. 2015; 32(4):281–300.

52 Freeman RE. The politics of stakeholder theory: some future directions. Bus Ethics Q. 1994; 4(4):409–421.

53 Munilla LS, Miles MP. The corporate social responsibility continuum as a component of stakeholder theory. Bus Soc Rev. 2005; 110(4):371–387.

54 Nielsen S, Nielsen EH. System dynamics modelling for a balanced scorecard. Manage Res News. 2008; 31(3):169–188.

55 Barnabè F. A "system dynamics-based Balanced Scorecard" to support strategic decision making. Int J Prod Per Man. 2011; 60(5):446–473.

56 Liu Q, Li X, Hassall M. Evolutionary game analysis and stability control scenarios of coal mine safety inspection system in China based on system dynamics. Safety Sci. 2015; 80:13–22.

57 Nikolaou I, Evangelinos K, Leal FW. A system dynamic approach for exploring the effects of climate change risks on firms' economic performance. J Clean Prod. 2015; 103(15):499–506.

58 Bradfield R, Wright G, Burt G, Cairns G, Van Der Heijden G. The origins and evolution of scenario techniques in long range business planning. Futures. 2015; 37:795–812.

59 Godet M. The art of scenarios and strategic planning: tools and pitfalls. Technological Forecasting and Social Change. 2000; 65:3–22.

60 Postma TJBM, Liebl F. How to improve scenario analysis as a strategic management tool? Technol Forecast Soc. 2005; 72:161–173.

61 Buysse K, Verbeke A. Proactive environmental strategies: a stakeholder management perspective. Strategic Manage J. 2003; 24(5):453–470.

62 Fassin Y, Van A, Buelnes M. Small-business owner-managers' perceptions of business ethics and CSR-related concepts. J Bus Ethics. 2011; 98(3):425–453.

63 Vandekerckhove WMS, Commers R. Beyond voluntary/ mandatory juxtaposition. Towards a European framework on CSR as network governance. Soc Respons J. 2005; 1(1/2): 98–103.

64 Khanna M, Anton WRQ. Corporate Environmental Management: Regulatory and Market-Based Incentives. Land Econ. 2002; 78(4):539–558.

65 Gallagher M. Business Continuity Management How to protect your company from danger. Edinburgh: Pearson Education Limited; 2003.

25

A Construction Manager's Perception of a Successful Industrial Constructed Wetland Project

Emmanuel Aboagye-Nimo[1], Justus Harding[2] and Alexandros I. Stefanakis[2,3,4]

[1] School of Environment and Technology, University of Brighton, Brighton, UK
[2] Bauer Resources GmbH, Schrobenhausen, Germany
[3] Department of Engineering, German University of Technology in Oman, Athaibah, Oman
[4] Bauer Nimr LLC, Muscat, Oman

25.1 Key Performance Indicators for Construction Projects

Construction projects, whether large or small, have some form of measurements to identify how successful they have been. One very popular concept used to assess such performances in construction is the adaptation of "Key Performance Indicators" (KPIs). By definition, KPIs are compilations of data measures used to assess the performance of a construction operation [1]. In other words, they are methods that management use to evaluate and assess the performance of the given task or activity. These evaluations are very useful when comparing actual outcome with estimated performance in many aspects including effectiveness, efficiency and quality of workmanship and final product [1, 2].

Performance indicators in construction have often been reviewed through academic works and industry-related publications. All these bodies of work have often suggested similar areas of focus, such as clients' needs, public interests, cost management and safety considerations. Table 25.1 presents excerpts of findings from performance indicators found over the years.

Performance indicators in construction projects have remained consistent over the years. The key recurring areas revolve around stakeholder interests (e.g., clients and public), productivity (time and cost) and safety considerations. By and large, these performance indicators and measurements are integral to projects in a similar manner to business management. Setting out clear targets from the onset of the project helps in the achievement of success. In other words, if the targets are not set, groups and individuals involved in the projects will not be able to assess their accomplishments.

The Building Research Establishment (BRE), a former UK government establishment uses benchmarking as a means of measuring project performance and subsequently achieving continuous improvement. Benchmarking provides a "yardstick" by which to judge your performance [8]. They recommend using bespoke KPIs for organisations that are new to the industry (i.e., organisations without prior project records) in order to be able to assess their progress and inevitably, project success or otherwise. They offer a KPI engine that provides comprehensive support for collecting, reporting and analyzing data. This engine allows organizations to benchmark their projects against a range of data sets including the following:

Table 25.1 Performance indicators for industry measures [3].

Latham, 1994 [4]	Egan, 1998 [5]	Construction Productivity Network, 1998 [6]	Construction Industry Board, 1998 [7]
Client satisfaction	Construction cost	People	Capital cost
Public interest	Construction time	Processes	Construction time
Productivity	Defects	Partners	Time predictability
Project performance	Client satisfaction (product)	Products	Cost predictability
Quality	Client satisfaction (service)		Defects
Research and development	Profitability		Safety
Training and recruitment	Productivity		Productivity
Financial	Safety		Turnover and profitability
	Cost predictability (construction)		Client satisfaction
	Time predictability (construction)		
	Cost predictability (design)		
	Time predictability (design		

- Client satisfaction
- Defects
- Construction time and cost
- Productivity
- Profitability
- Health and safety
- Employee satisfaction
- Staff turnover
- Sickness absence
- Working hours
- Qualifications and skills
- Impact on environment
- Whole life performance
- Waste
- Commercial vehicle movements.

The output of the analysis is presented in tables, graphs and action plans. For organizations and construction teams without a firm grasp on performance indicators and measurement techniques; this serves as a very useful tool that can be used until a firm organizational strategy is put in place. Organizational strategies may be fully developed as the organizations gain experience in the industry and are subsequently able to create their own KPIs and benchmarking techniques.

25.2 Function and Values of Constructed Wetlands

Constructed wetlands are wastewater treatment systems, and are constructed in such a way to improve the water quality including domestic wastewater, agricultural wastewater, mine drainage, petroleum refinery wastes and pretreated industrial wastewater [9]. Constructed wetlands can be effective and provide a number of functions and values, though not all wetlands provide the same values, but all wetlands contribute to [9–13]:

- improvement of water quality;
- support of wildlife habitat;
- cost-effectiveness as for lower energy consumption;
- heat storage and release;
- trapping of sediments and other substances;
- ease of use, maintenance and operation;
- cycling of nutrients and other materials;
- education and research.

As discussed in the previous section, one of the crucial performance indicators of every construction is the client's needs, time and cost considerations and finally safety practices. Thus, in order to achieve any of the functional requirements in the above list, stakeholders expect the project to be economical, safe and produced on time.

Beyond the functions discussed above, other interrelated factors must come together in order for a Constructed Wetland project to achieve its expected performance level. These factors depend on clients, legislation and society. The factors range from aesthetic value to users' health and safety. Design and service life, project cost, quality of finished project and environmental impact must all be considered if the project is to achieve its desired performance level [14].

Performance gaps can be identified at several points in the projects. At the inception of projects, clients and even designers may have over-ambitious ideas that cannot be realized. However, experience and feedback processes can be used to identify such issues. Some of the challenges include those in the design phase, when performance targets are set (taking into consideration legal effluent standards etc.). The very important construction phase presents the most difficulty, as shortfalls often lead to compromise of the ambitious ideas conceived from the inception of the project. During the handover stage, workers tend to rush as they are pressured to commission the project and handover to the end-user. This rush can often affect end-user training and poor quality in final finishes. The final phase to identify performance gaps occurs during the system operation. At this stage, operators of the facility may need support to ensure the Constructed Wetland is set up for efficient long-term operation.

25.2.1 Constructed Wetland Components

Constructed wetland design is based on the integration of vegetation, substrate media, and hydraulic characteristics that are combined to remove the various constituents from wastewater. All these components come together to fulfil the requirements of clients. Successful performance of a natural treatment system not only relies on a good design, but also proper construction and sound operation. Emphasis should be given to the enhancement of engineering efficiency, economy of scale, supply chain management and, more importantly, to the in-country value supporting nationalization processes. Under these sections, there are difference components:

Vegetation Both emergent/submergent plants (higher plants/reeds) and non-vascular plants (algae) are important in constructed wetlands. The root system of the plants is a critical parameter in the wetland's operation and efficiency, since it supplies oxygen and supports the microbial community that degrades the various wastewater pollutants, while maintaining the hydraulic conductivity of the bed [9].

Substrates   Substrates used to construct wetlands include soil, sand, gravel, rock, and organic materials such as compost. Sediments and litter then accumulate in the wetland because of the low water velocities and high productivity typical of wetlands. The substrates, sediments, and litter are important for several reasons [9]:

- They support many of the living organisms in wetlands.
- They support the growth of the planted macrophytes.
- Substrate permeability affects the movement of water through the wetland.
- They stabilize the wetland bed (interaction effects with developed plant roots).
- Substrates provide storage for many contaminants.
- They provide filtration effects.
- Accumulation of litter increases the amount of organic matter in the wetland, which provides sites for material exchange and microbial attachment, and is a source of carbon, which is required for some pollutant removal processes in wetland systems.

Hydraulic characteristics   Wastewater treatment by constructed wetlands depends on the retention time within the system. The actual retention time within a constructed wetland is usually unknown and may differ from the theoretically calculated, given that it is a function of the wastewater flow path and an extent of wastewater interactions with the wetland porous media and vegetation. Moreover, proper design of the wetland system (e.g., selection of appropriate media porosity, hydraulic load etc.) is crucial for the effective performance of the system.

25.3   Clear Deliverables of Project

In addition to being able to deliver a project that addresses client needs, regulations and societal necessities, every construction project manager or leader needs to understand the practicalities of the various milestones they set in their projects. Deliverables are the products, services, and results that a project produces. Thus, they are the cornerstone to project success. These help project teams identify the feasibility and practicalities of the projects. The key to the success of each phase in the process is the production of project deliverables [15]. These include reports and documentation associated with each phase, e.g., engineering reports, proposals, design drawings and design documents. They act as the agent, which ensures the enactment of each phase as planned, concluding with the presentation of the deliverables at each end of phase review.

Deliverables are active working documents, which are subject to change throughout the majority of construction process. They can be in one of the following states [15]:

- **Initial:** preliminary information is presented;
- **Updated:** current information is updated;
- **Revised:** major changes/decisions will significantly alter the content and context of the deliverable; and

- **Finalized:** the information presented is agreed and is unlikely to change throughout the duration of the project.

Since deliverables are active documents, it is important that all involved parties in the project have "real time" access to any changes in order not to create any ambiguities. Scope of work forms an integral part of a constructed wetland and includes, but is not limited to, the following:

- Characterization of wastewater through sampling and analyses program.
- Identification of pretreatment requirements: a comprehensive review of existing water separation facilities for wastewater, including assessment of treatment effectiveness.
- Specifying all required facilities to enable treated water re-use in irrigation or other requirements.
- Preparation of construction documents with a review process, including:
  - Construction grade survey
  - Geotechnical site evaluation
  - Permitting and approvals
  - Final site selection
  - Preparation of specifications
  - Integration engineering design including controls specification, permit review and preparation
  - Cost estimates that include life cycle analysis.
- Development of Operational Manual: key element of final design.
- Final design manual.
- Integrate appropriate elements into design to allow fully-functional treatment facility to be used at demonstration, safe for visitors and environmentally friendly.
- Specify sampling and monitoring requirements.
- Specify protocols.

25.3.1 Health and Safety Considerations in Construction Projects

Most countries with developed infrastructural planning methods have comprehensive safety frameworks. The Health and Safety Executive (HSE) offers guidelines that broadly prescribe the general duties for employers, employees and the self-employed through the CDM regulations 2015. Due to the common practices amongst Commonwealth nations, many countries around the world have adopted aspects of these regulations. Furthermore, the CDM regulations are built upon EU directives of workplace safety.

Historically, the fundamental principles on which CDM Regulations are based are as follows [16, 17]:

- Safety must be systematically considered throughout the course of the project.
- Every member who contributes to the health and safety of a project needs to be included.
- Proper planning and coordination need to be undertaken from the commencement of the project.
- Individuals in charge of the provision of health and safety need to be competent.
- Communication and sharing of information between all parties must be undertaken.
- A formal record of safety information for future use must be made.

As stated, under these regulations, not only is the contractor responsible for the health and safety of workers; all stakeholders, including the client have a duty to ensure works and activities are carried out under safe conditions. In contrast, contractors were also left with the sole responsibility of handling health and safety matters under previous regulations. Table 25.2 presents the various stakeholders and their roles under CDM Regulations 2015.

25.3.2 Hazard Identification and Risk Screening

The hazard identification and risk screening process is a key to the effective implementation of HSE Management Systems (HSEMS). The processes described ensure that all hazards and potential effects for the construction of the wetland are fully evaluated. The key processes are shown in Table 25.3.

25.3.3 Securing the Project

Securing construction materials and components are of high priority during the construction phase. This is because of the financial worth of these materials and components. As a result, site storage and handling of equipment and materials are considered as high risk aspects of projects [19]. Estimates from the United States alone indicate that between 1 and 4 billion dollars' worth of materials, tools, as well as large and small equipment are stolen every year from construction sites [20, 21]. With such high figures, it is therefore understandable why securing projects against theft and vandalism is not taken lightly.

Theft This is when people steal from construction sites. Many construction sites (large, medium or small) experience high levels of thefts annually. The compilation of financial losses incurred by construction firms is so significant that many contractors have to pay high premiums for their risk insurance coverage. Workers have to implement proactive measures to help prevent theft of equipment, materials and building components. The use of lockboxes, security fencing, warning signs, removal of unused equipment and use of night security forces are all measures that are used. With regard to theft of machinery and equipment, workers park equipment and machinery in well-lit areas and in a specific formation in order to prevent thieves from driving them off site. Finally, some companies modify the ignition or fuel lines of their machinery so that they cannot be stolen.

Vandalism This is considered to be a nuisance crime. However, this is does not cause as much financial loss as theft, but is still a cut-back on profits. The types of vandalism found on construction sites include the following:

- broken glass;
- destruction of in-place materials;
- damage to construction equipment; and
- vehicle damage vandalism.

Small to medium-sized companies have been found to experience higher losses from incidents of vandalism. This may be as a result of the inability to provide high level surveillance and security on their projects.

25.4 Critical Points in Constructing Wetlands

There has been much research on constructed wetlands, but the optimal design of constructed wetlands for various applications has not yet been determined [9]. This means that there is not a widely

Table 25.2 Summary of duties under CDM Regulations 2015 (adapted from [18], with permission).

CDM Duty holders[a] – Who are they?	Main duties
Commercial clients – Organizations or individuals for whom a construction project is carried out that is done as part of a business.	Make suitable arrangements for managing a project, including making sure: • other duty holders are appointed as appropriate • sufficient time and resources are allocated. Make sure: • relevant information is prepared and provided to other duty holders • the principal designer and principal contractor carry out their duties • welfare facilities are provided.
Domestic clients – People who have construction work carried out on their own home (or the home of a family member) that is not done as part of a business[b]	Though in scope of CDM 2015, their client duties are normally transferred to: • the contractor for single contractor projects • the principal contractor for projects with more than one contractor. However, the domestic client can instead choose to have a written agreement with the principal designer to carry out the client duties.
Designers – Organizations or individuals who as part of a business, prepare or modify designs for a building, product or system relating to construction work.	When preparing or modifying designs, eliminate, reduce or control foreseeable risks that may arise during: • construction • the maintenance and use of a building once it is built. Provide information to other members of the project team to help them fulfil their duties.
Principal designers – Designers appointed by the client in projects involving more than one contractor. They can be an organization or an individual with sufficient knowledge, experience and ability to carry out the role.	Plan, manage, monitor and coordinate health and safety in the pre-construction phase of a project. This includes: • identifying, eliminating or controlling foreseeable risks • ensuring designers carry out their duties. Prepare and provide relevant information to other duty holders. Liaise with the principal contractor to help in the planning, management, monitoring and coordination of the construction phase.
Principal contractors – Contractors appointed by the client to coordinate the construction phase of a project where it involves more than one contractor.	Plan, manage, monitor and coordinate health and safety in the construction phase of a project. This includes: • liaising with the client and principal designer • preparing the construction phase plan • organising cooperation between contractors and coordinating their work. Make sure: • suitable site inductions are provided • reasonable steps are taken to prevent unauthorised access • workers are consulted and engaged in securing their health and safety • welfare facilities are provided.
Contractors – Those who carry out the actual construction work, contractors can be an individual or a company.	Plan, manage and monitor construction work under their control so it is carried out without risks to health and safety. For projects involving more than one contractor, coordinate their activities with others in the project team – in particular, comply with directions given to them by the principal designer or principal contractor. For single contractor projects, prepare a construction phase plan.

(Continued)

Table 25.2 (Continued)

CDM Duty holders[a] – Who are they?	Main duties
Workers – Those working for or under the control of contractors on a construction site.	Workers must: • be consulted about matters which affect their health, safety and welfare • take care of their own health and safety, and of others who might be affected by their actions • report anything they see which is likely to endanger either their own or others' health and safety • cooperate with their employer, fellow workers, contractors and other duty-holders.

[a] Organizations or individuals can carry out the role of more than one duty holder, provided they have the skills, knowledge, experience and (if an organization) the organizational capability necessary to carry out those roles in a way that secures health and safety.

[b] CDM 2015 applies if the work is carried out by someone else on the domestic client's behalf. If the householder carries out the work themselves, it is classed as DIY and CDM 2015 does not apply.

Table 25.3 Hazard identification and risk screening processes considered during the a constructed wetland project.

	Integrated Hazard Identification and Risk Screening
1	Carry out a HAZID exercise with project personnel to identify the hazards associated at each site/facility and quantify the risk into High, Medium, and Low Categories.
2	Carry out and ENVID exercise with project personnel to identify the hazards associated at each site/facility and quantify the risk into High, Medium, and Low Categories.
3	Carry out an Occupational Health Risk Assessment (OHRA) exercise with project personnel to identify the hazards associated at each site/facility and quantify the risk into High, Medium, and Low Categories.
4	Carry out Risk Screening
5	Populate Hazard and Effects Register
6	Develop Bowtie Diagrams

accepted or applied design. Each wetland design is based on the specific wastewater characteristics and origin, as well as client needs and targets. As mentioned earlier, this chapter will not focus on the operational aspects of the wetlands, but the design and construction phases. Since wetland designs attempt to mimic natural wetlands in overall structure, many of the considerations made while designing them are relatively different from traditional construction projects. Some of the main design considerations for successfully constructing wetlands include [22]:

• Keep the design simple. Complex technological approaches often invite failure.
• Design for minimal maintenance.
• Design the system to use natural energies, such as gravity flow.
• Design for the extremes of weather and climate, not the average. Storms, floods, and droughts are to be expected and planned for, not feared.
• Design the wetland with the landscape, not against it. Integrate the design with the natural topography of the site.

- Avoid over-engineering the design with rectangular basins, rigid structures and channels, and regular morphology.
- Give the system time. Wetlands do not necessarily become functional overnight and several years may elapse before performance reaches optimal levels. Strategies that try to short-circuit the process of system development or to over-manage often fail.
- Design the system for function, not form. For instance, if initial plantings fail, but the overall function of the wetland, based on initial objectives, is intact, then the system has not failed.

The key success points for constructing wetlands and traditional construction projects have some similarities in some areas, but could not be more different in others. In some construction projects, designers tend to use simple designs that can be produced by the construction teams. However, in many other cases, designers showcase their creativity and produce outstanding projects that require very few and capable teams to bring to reality. Furthermore, clients' briefs often produce challenges because of what they want, how they want it and when they want it. Making use of natural energies is commended in construction, e.g., making use of sunlight and rainwater harvesting techniques. These fall within the broad categorization of sustainable designs. However the design may not always embrace nature as planned and, hence, may have to be altered accordingly. With regard to extreme weathers, any infrastructure is designed using the knowledge of existing structures in the area in addition to the environmental contexts. For example, a construction near the coastline will take salt spray from the sea into consideration and, thus, not use components and materials that are highly susceptible to rust. Although over-engineering is regarded as a waste of resources, designs that do not conform to the norm are commended.

The final two points are critical in both designs. In traditional construction, time is often limited and workers are forced to finish in very tight timeframes. However, constructed wetlands are designed to offer allowance for time. Some designs can be given years to deliver their purpose. Finally, constructed wetlands are designed for function. However, many construction designs are based on ostentatious ideas. The function of every aspect of the construction should deliver as expected, unlike constructed wetlands where the bigger picture is considered, i.e., "If initial plantings fail, but the overall function of the wetland, based on initial objectives is intact, then the system has not failed".

25.5   Summary

This chapter has briefly discussed the areas that are considered to be critical to construction success of a constructed wetland project. The key performance indicators of construction projects, i.e., setting up benchmarks and other means of measuring if the project has been able to deliver all the proposed objectives in the project were discussed. In addition, the functions of constructed wetlands and their various components were presented, highlighting how important it is for every facet of the constructed wetland to carry out a specific role and possess unique qualities. While constructed wetlands should perform a certain function, the design and construction phases of projects need constant review and these must be recorded in project documents and portfolios, namely project deliverables.

One area that both traditional construction and constructed wetlands consciously try to adopt is natural resources. Both areas prioritize environmental impacts of the projects although by sheer size of projects, the constructed wetland is less harmful to the environment. Securing project sites is also of much concern to construction teams. This is not intensively highlighted in constructed wetlands projects probably because of the location of the projects, e.g., usually away from human settlements to blend in with nature. The health and safety aspects considered in constructed wetlands is mainly about public health as an aftermath of the wetlands treating the municipal or industrial wastewater. Health, Safety and Environmental Services form an integral part of the design, construction and operation of a constructed wetland. The actual practices of the teams that build constructed wetlands can be explored in great detail for empirical evidence in future.

References

1 Cox RF, Issa RR, Ahrens D. Management's perception of key performance indicators for construction. J Constr Eng Manag. 2003; 129(2):142–151.
2 Chan AP, Chan AP. Key performance indicators for measuring construction success. Benchmarking: An International Journal 2004; 11(2):203–221.
3 Takim, R and Akintoye, A. Performance indicators for successful construction project performance. In: Greenwood D, ed. 18th Annual ARCOM Conference, University of Northumbria. Association of Researchers in Construction Management, Vol. 2, 2–4 September 2002, pp. 545–555.
4 Latham M. Constructing the team: a joint review of procurement and contractual arrangements in the UK construction industry final report. London: HMSO, 1994.
5 Egan J. Rethinking construction: report of the construction task force on the scope for improving the quality and efficiency of UK construction. Department of the Environment, Transport and the Region, London, 1998.
6 Construction Productivity Network. Conference report on performance measurement in construction. Report CPN816L. London: Construction Industry Research and Information Association, 1998.
7 Construction Industry Board. Key performance indicators for the construction industry. London: Construction Industry Board, 1998.
8 BRE. Key performance indicators (KPI's) for the construction industry. 2016. Available: https://www.bre.co.uk/page.jsp?id=1478 [accessed 01-10-2016].
9 Stefanakis AI, Akratos CS, Tsihrintzis VA. Vertical Flow Constructed Wetlands: Eco-engineering Systems for Wastewater and Sludge Treatment. Amsterdam: Elsevier Science; 2014.
10 MEA (Millennium Ecosystem Assessment). Ecosystems and Human Well-being: Wetlands and Water Synthesis. MEA, World Resources Institute, Washington D.C, 2005. (www.millenniumassessment.org).
11 De Groot R, Stuip M, Finlayson M, Davidson N. Valuing Wetlands: Guidance for valuing the benefits derived from wetland ecosystem services. Ramsar Technical Report No. 3, CBD Technical Series No. 27. Ramsar Convention Secretariat, Gland, Switzerland, 2006.
12 Ghermandi A, Van den Bergh JCJM, Brander LM, De Groot HLF, Nunes P. Values of natural and human-made wetlands: A meta-analysis. Water Resour Res. 2010; 46(W12516):12.
13 Schuyt K, Brander L. The economic values of the world's wetlands. WWF International. Prepared with support from the Swiss Agency for the Environment, Forests and Landscape (SAEFL), Gland/Amsterdam, 2004.
14 Emmitt S, Gorse CA. Barry's introduction to construction of buildings. John Wiley & Sons, 2013.

15 Kagioglou M, Cooper R, Aouad G, Sexton M. Rethinking construction: the generic design and construction process protocol. Eng Constr Architect Manage. 2000; 7(2):141–153.

16 Baxendale T, Jones O. Construction design and management safety regulations in practice – progress on implementation. Int J Proj Manag. 2000; 18(1):33–40.

17 Joyce R. CDM Regulations 2007 explained. London: Thomas Telford; 2007.

18 HSE. Managing health and safety in construction. 2016. Available at: http://books.hse.gov.uk/hse/public/saleproduct.jsf?catalogueCode=9780717666263 [accessed 01-10-2016].

19 Clarke RV, Goldstein H. Reducing theft at construction sites: Lessons from a problem-oriented project. Crime Prev Studies. 2002; 13:89-130.

20 Barrios J. Building frenzy spurs rise in home site theft. Arizona Daily Star, 2005, p. A1.

21 Berg R, Hinze J. Theft and vandalism on construction sites. J Constr Eng M. 2005; 7(131):823–833.

22 Mitsch WJ. Landscape design and the role of created, restored, and natural riparian wetlands in controlling nonpoint source pollution. Ecol Eng. 1992; 1(1–2):27–47.

Index

Constructed Wetlands for Industrial Wastewater Treatment, First Edition. Edited by Alexandros I. Stefanakis.
© 2018 John Wiley & Sons Ltd. Published 2018 by John Wiley & Sons Ltd.